Asymptotics and Special Functions

AKP CLASSICS

Asymptotics and Special Functions

Frank W. J. Olver

University of Maryland
College Park, Maryland
and
National Institute of Standards and Technology
Gaithersburg, Maryland

A K Peters
Wellesley, Massachusetts

Editorial, Sales, and Customer Service Office

A K Peters, Ltd.
289 Linden Street
Wellesley, MA 02181

Library of Congress Cataloging-in-Publication Data

Olver, Frank W. J., 1924-
Asymptotics and special functions / Frank W. J. Olver.
p. cm. -- (AKP classics)
Originally published: New York: Academic Press, 1974.
Includes bibliographical references and index.
ISBN: 1-56881-069-5
1. Functions, Special. 2. Asymptotic expansions. 3. Differential equations--Numerical solutions. I. Title. II. Series.
QA351.048 1997 97-377
515'.5--dc21 CIP

Printed in the United States of America
01 00 99 98 97 10 9 8 7 6 5 4 3 2 1

To the memory of
my daughter Linda (1953–1965)

CONTENTS

3 Integrals of a Real Variable

4 Contour Integrals

5 Differential Equations with Regular Singularities; Hypergeometric and Legendre Functions

6 The Liouville–Green Approximation

PREFACE TO A K PETERS EDITION

Academic Press ceased publication of *Asymptotics and Special Functions* in 1995. I am most grateful for the care and consideration exercised by this company extending over twenty-two years of publication. I am equally grateful to the publishing company of A K Peters for agreeing to reprint and publish the whole volume. I have taken this opportunity to correct several errors, mostly minor, and I am indebted to T. M. Dunster, J. M. Melenk, and R. B. Paris for drawing my attention to some of these mistakes.

There have been significant advances in the theory of asymptotic analysis and special functions during the past twenty-three years. It is my hope, however, that *Asymptotics and Special Functions* will continue to provide a useful broad introduction as well as a work of reference. For readers who need to pursue aspects of these topics more deeply, several excellent books have appeared in the meantime, including *Special Functions of Applied Mathematics* by B. C. Carlson, Academic Press, New York, 1977; *Computation with Recurrence Relations* by J. Wimp, Pitman, London, 1984; *Special Functions of Mathematical Physics* by A. F. Nikiforov and V. B. Uvarov, Birkhäuser, Basel, 1988; *Asymptotic Approximations of Integrals* by R. Wong, Academic Press, New York, 1989; *Basic Hypergeometric Series* by G. Gasper and M. Rahman, Cambridge University Press, London, 1990; and *Special Functions: An Introduction to the Classical Functions of Mathematical Physics* by N. M. Temme, John Wiley & Sons, New York, 1996. At the present time particularly active areas of research include exponential and hyper asymptotics and algorithms for the automatic computation of the special functions. Short introductory surveys to these topics will be found in the articles "Stokes phenomenon demystified" by R. B. Paris and A. D. Wood, *IMA Bulletin*, v. 31 (1995), pp. 21–28, and "Numerical evaluation of special functions" by D.W. Lozier and F. W. J. Olver, *AMS Proceedings of Symposia in Applied Mathematics*, v. 48 (1994), pp. 79–125. Research papers will be found in recent issues of journals devoted to applied mathematics and theoretical physics, especially *Proceedings of the Royal Society of London, Series A*, and *Methods and Applications of Analysis*.

Frank W. J. Olver
College Park, Maryland
October, 1996

PREFACE

Classical analysis is the backbone of many branches of applied mathematics. The purpose of this book is to provide a comprehensive introduction to the two topics in classical analysis mentioned in the title. It is addressed to graduate mathematicians, physicists, and engineers, and is intended both as a basis for instructional courses and as a reference tool in research work. It is based, in part, on courses taught at the University of Maryland.

My original plan was to concentrate on asymptotics, quoting properties of special functions as needed. This approach is satisfactory as long as these functions are being used as illustrative examples. But the solution of more difficult problems in asymptotics, especially ones involving uniformity, necessitate the use of special functions as approximants. As the writing progressed it became clear that it would be unrealistic to assume that students are sufficiently familiar with needed properties. Accordingly, the scope of the book was enlarged by interweaving asymptotic theory with a systematic development of most of the important special functions. This interweaving is in harmony with historical development and leads to a deeper understanding not only of asymptotics, but also of the special functions. Why, for instance, should there be four standard solutions of Bessel's differential equation when any solution can be expressed as a linear combination of an independent pair? A satisfactory answer to this question cannot be given without some knowledge of the asymptotic theory of linear differential equations.

A second feature distinguishing the present work from existing monographs on asymptotics is the inclusion of error bounds, or methods for obtaining such bounds, for most of the approximations and expansions. Realistic bounds are of obvious importance in computational applications. They also provide theoretical insight into the nature and reliability of an asymptotic approximation, especially when more than one variable is involved, and thereby often avoid the need for the somewhat unsatisfactory concept of generalized asymptotic expansions. Systematic methods of error analysis have evolved only during the past decade or so, and many results in this book have not been published previously.

The contents of the various chapters are as follows. Chapter 1 introduces the basic concepts and definitions of asymptotics. Asymptotic theories of definite integrals containing a parameter are developed in Chapters 3, 4, and 9; those of ordinary linear differential equations in Chapters 6, 7, 10, 11, 12, and 13; those of sums and sequences in Chapter 8. Special functions are introduced in Chapter 2 and developed in most of the succeeding chapters, especially Chapters 4, 5, 7, 8, 10, 11, and 12. Chapter 5 also introduces the analytic theory of ordinary differential equations. Finally, Chapter 14 is a brief treatment of methods of estimating (as opposed to bounding) errors in asymptotic approximations and expansions.

An introductory one-semester course can be based on Chapters 1, 2, and 3, and the first parts of Chapters 4, 5, 6, and 7.† Only part of the remainder of the book can be covered in a second semester, and the selection of topics by the instructor depends on the relative emphasis to be given to special functions and asymptotics. Prerequisites are a good grounding in advanced calculus and complex-variable theory. Previous knowledge of ordinary differential equations is helpful, but not essential. A course in real-variable theory is not needed; all integrals that appear are Riemannian. Asterisks (*) are attached to certain sections and subsections to indicate advanced material that can be bypassed without loss of continuity. Worked examples are included in almost all chapters, and there are over 500 exercises of considerably varying difficulty. Some of these exercises are illustrative applications; others give extensions of the general theory or properties of special functions which are important but straightforward to derive. On reaching the end of a section the student is strongly advised to read through the exercises, whether or not any are attempted. Again, a warning asterisk (*) is attached to exercises whose solution is judged to be unusually difficult or time-consuming.

All chapters end with a brief section entitled *Historical Notes and Additional References*. Here sources of the chapter material are indicated and mention is made of places where the topics may be pursued further. Titles of references are collected in a single list at the end of the book. I am especially indebted to the excellent books of de Bruijn, Copson, Erdélyi, Jeffreys, Watson, and Whittaker and Watson, and also to the vast compendia on special functions published by the Bateman Manuscript Project and the National Bureau of Standards.

Valuable criticisms of early drafts of the material were received from G. F. Miller (National Physical Laboratory) and F. Stenger (University of Utah), who read the entire manuscript, and from R. B. Dingle (University of St. Andrews), W. H. Reid (University of Chicago), and F. Ursell (University of Manchester), who read certain chapters. R. A. Askey (University of Wisconsin) read the final draft, and his helpful comments included several additional references. It is a pleasure to acknowledge this assistance, and also that of Mrs. Linda Lau, who typed later drafts and assisted with the proof reading and indexes, and the staff of Academic Press, who were unfailing in their skill and courtesy. Above all, I appreciate the untiring efforts of my wife Grace, who carried out all numerical calculations, typed the original draft, and assisted with the proof reading.

† For this reason, the first seven chapters have been published by Academic Press as a separate volume, for classroom use, entitled *Introduction to Asymptotics and Special Functions*.

1

INTRODUCTION TO ASYMPTOTIC ANALYSIS

1 Origin of Asymptotic Expansions

1.1 Consider the integral

$$F(x) = \int_0^\infty e^{-xt} \cos t \, dt \tag{1.01}$$

for positive real values of the parameter x. Let us attempt its evaluation by expanding $\cos t$ in powers of t and integrating the resulting series term by term. We obtain

$$F(x) = \int_0^\infty e^{-xt}\left(1 - \frac{t^2}{2!} + \frac{t^4}{4!} - \cdots\right) dt \tag{1.02}$$

$$= \frac{1}{x} - \frac{1}{x^3} + \frac{1}{x^5} - \cdots. \tag{1.03}$$

Provided that $x > 1$ the last series converges to the sum

$$F(x) = \frac{x}{x^2+1}.$$

That the attempt proved to be successful can be confirmed by deriving the last result directly from (1.01) by means of two integrations by parts; the restriction $x > 1$ is then seen to be replaceable by $x > 0$.

Now let us follow the same procedure with the integral

$$G(x) = \int_0^\infty \frac{e^{-xt}}{1+t} dt. \tag{1.04}$$

We obtain

$$G(x) = \int_0^\infty e^{-xt}(1 - t + t^2 - \cdots) \, dt$$

$$= \frac{1}{x} - \frac{1!}{x^2} + \frac{2!}{x^3} - \frac{3!}{x^4} + \cdots. \tag{1.05}$$

This series diverges for all finite values of x, and therefore appears to be meaningless.

Why did the procedure succeed in the first case but not in the second? The answer is not hard to find. The expansion of $\cos t$ converges for all values of t; indeed it converges uniformly throughout any bounded t interval. Application of a standard theorem concerning integration of an infinite series over an infinite interval† confirms that the step from (1.02) to (1.03) is completely justified when $x > 1$. In the second example, however, the expansion of $(1+t)^{-1}$ diverges when $t \geqslant 1$. The failure of the representation (1.05) may be regarded as the penalty for integrating a series over an interval in which it is not uniformly convergent.

1.2 If our approach to mathematical analysis were one of unyielding purity, then we might be content to leave these examples at this stage. Suppose, however, we adopt a heuristic approach and try to sum the series (1.05) numerically for a particular value of x, say $x = 10$. The first four terms are given by

$$0.1000 - 0.0100 + 0.0020 - 0.0006, \tag{1.06}$$

exactly, and the sum of the series up to this point is 0.0914. Somewhat surprisingly this is very close to the correct value $G(10) = 0.09156\ldots$.‡

To investigate this unexpected success we consider the difference $\varepsilon_n(x)$ between $G(x)$ and the nth partial sum of (1.05), given by

$$\varepsilon_n(x) = G(x) - g_n(x),$$

where

$$g_n(x) = \frac{1}{x} - \frac{1!}{x^2} + \frac{2!}{x^3} - \cdots + (-)^{n-1}\frac{(n-1)!}{x^n}.$$

Here n is arbitrary, and $\varepsilon_n(x)$ is called the *remainder term*, *error term*, or *truncation error* of the partial series, or, more precisely, the nth such term or error. Since

$$\frac{1}{1+t} = 1 - t + t^2 - \cdots + (-)^{n-1}t^{n-1} + \frac{(-)^n t^n}{1+t},$$

substitution in (1.04) yields

$$\varepsilon_n(x) = (-)^n \int_0^\infty \frac{t^n e^{-xt}}{1+t}\,dt. \tag{1.07}$$

Clearly,

$$|\varepsilon_n(x)| < \int_0^\infty t^n e^{-xt}\,dt = \frac{n!}{x^{n+1}}. \tag{1.08}$$

In other words, the partial sums of (1.05) approximate the function $G(x)$ with an error that is numerically smaller than the first neglected term of the series. It is also

† Bromwich (1926, §§175–6). This theorem is quoted fully later (Chapter 2, Theorem 8.1).

‡ Obtainable by numerical quadrature of (1.04) or by use of tables of the exponential integral; compare Chapter 2, §3.1.

clear from (1.07) that the error has the same sign as this term. Since the next term in (1.06) is 0.00024, this fully explains the closeness of the value 0.0914 of $g_4(10)$ to that of $G(10)$.

1.3 Thus the expansion (1.05) has a hidden meaning: it may be regarded as *constituting a sequence of approximations* $\{g_n(x)\}$ *to the value of* $G(x)$. In this way it resembles a convergent expansion, for example (1.03). For in practice we cannot compute an infinite number of terms in a convergent series; we stop the summation when we judge that the contribution from the tail is negligibly small compared to the accuracy required. There are, however, two important differences. First, $\varepsilon_n(x)$ cannot be expressed as the sum of the tail. Secondly, by definition the partial sum of a convergent series becomes arbitrarily close to the actual sum as the number of terms increases indefinitely. With (1.05) this is not the case: for a given value of x, successive terms $(-)^s s!/x^{s+1}$ diminish steadily in size as long as s does not exceed $[x]$, the integer part of x. Thereafter they increase without limit. Correspondingly, the partial sums $g_n(x)$ at first approach the value of $G(x)$, but when n passes $[x]$ errors begin to increase and eventually oscillate wildly.†

The essential difference, then, is that whereas the sum of a convergent series can be computed to arbitrarily high accuracy with the expenditure of sufficient labor, the accuracy in the value of $G(x)$ computed from the partial sums $g_n(x)$ of (1.05) is restricted. For a prescribed value of x, the best we can do is to represent $G(x)$ by $g_{[x]}(x)$. The absolute error of this representation is bounded by $[x]!/x^{[x]+1}$, and the relative error by about $[x]!/x^{[x]}$.

Although the accuracy is restricted, it can be extremely high. For example, when $x = 10$, $[x]!/x^{[x]} \fallingdotseq 0.36 \times 10^{-3}$.‡ Therefore when $x \geqslant 10$, the value of $G(x)$ can be found from (1.05) to at least three significant figures, which is adequate for some purposes. For $x \geqslant 100$, this becomes 42 significant figures; there are few calculations in the physical sciences that need accuracy remotely approaching this.

So far, we have considered the behavior of the sequence $\{g_n(x)\}$ for fixed x and varying n. If, instead, n is fixed, then from (1.08) we expect $g_n(x)$ to give a better approximation to $G(x)$ than any other partial sum when x lies in the interval $n < x < n+1$.§ Thus, no single approximation is "best" in an overall sense; each has an interval of special merit.

1.4 The expansion (1.05) is typical of a large class of divergent series obtained from integral representations, differential equations, and elsewhere when rules governing the applicability of analytical transformations are violated. Nevertheless, such expansions were freely used in numerical and analytical calculations in the eighteenth century by many mathematicians, particularly Euler. In contrast to the foregoing analysis for the function $G(x)$ little was known about the errors in approximating functions in this way, and sometimes grave inaccuracies resulted.

† For this reason, series of this kind used to be called *semiconvergent* or *convergently beginning*.

‡ Here and elsewhere the sign $\fallingdotseq$ denotes approximate equality.

§ Since (1.08) gives a bound and not the *actual* value of $|\varepsilon_n(x)|$, the interval in which $g_n(x)$ gives the best approximation may differ slightly from $n < x < n+1$.

Early in the nineteenth century Abel, Cauchy, and others undertook the task of placing mathematical analysis on firmer foundations. One result was the introduction of a complete ban on the use of divergent series, although it appears that this step was taken somewhat reluctantly.

No way of rehabilitating the use of divergent series was forthcoming during the next half century. Two requirements for a satisfactory general theory were, first, that it apply to most of the known series; secondly, that it permit elementary operations, including addition, multiplication, division, substitution, integration, differentiation, and reversion. Neither requirement would be met if, for example, we confined ourselves to series expansions whose remainder terms are bounded in magnitude by the first neglected term.

Both requirements were satisfied eventually by Poincaré in 1886 by defining what he called *asymptotic expansions*. This definition is given in §7.1 below. As we shall see, Poincaré's theory embraces a wide class of useful divergent series, and the elementary operations can all be carried out (with some slight restrictions in the case of differentiation).

2 The Symbols ~, o, and O

2.1 In order to describe the behavior, as $x \to \infty$, of a wanted function $f(x)$ in terms of a known function $\phi(x)$, we shall often use the following notations, due to Bachmann and Landau.† At first, we suppose x to be a real variable. At infinity $\phi(x)$ may vanish, tend to infinity, or have other behavior—no restrictions are made.

(i) If $f(x)/\phi(x)$ tends to unity, we write

$$f(x) \sim \phi(x) \qquad (x \to \infty),$$

or, briefly, when there is no ambiguity, $f \sim \phi$. In words, *f is asymptotic to ϕ, or ϕ is an asymptotic approximation to f.*

(ii) If $f(x)/\phi(x) \to 0$, we write

$$f(x) = o\{\phi(x)\} \qquad (x \to \infty),$$

or, briefly, $f = o(\phi)$; in words, f is of order less than ϕ.‡

(iii) If $|f(x)/\phi(x)|$ is bounded, we write

$$f(x) = O\{\phi(x)\} \qquad (x \to \infty),$$

or $f = O(\phi)$; again, in words, *f is of order not exceeding ϕ.*

Special cases of these definitions are $f = o(1)$ $(x \to \infty)$, meaning simply that f vanishes as $x \to \infty$, and $f = O(1)$ $(x \to \infty)$, meaning that $|f|$ is bounded as $x \to \infty$.

† Landau (1927, Vol. 2, pp. 3–5).

‡ In cases in which $\phi(x)$ is not real and positive, some writers use modulus signs in the definition, thus $f(x) = o(|\phi(x)|)$. Similarly in Definition (iii) which follows.

As simple examples

$$(x+1)^2 \sim x^2, \qquad \frac{1}{x^2} = o\left(\frac{1}{x}\right), \qquad \sinh x = O(e^x).$$

2.2 Comparing (i), (ii), and (iii), we note that (i) and (ii) are mutually exclusive. Also, each is a particular case of (iii), and when applicable each is more informative than (iii).

Next, the symbol O is sometimes associated with an interval $[a, \infty)$† instead of the limit point ∞. Thus

$$f(x) = O\{\phi(x)\} \qquad \text{when} \quad x \in [a, \infty) \tag{2.01}$$

simply means that $|f(x)/\phi(x)|$ is bounded throughout $a \leqslant x < \infty$. Neither the symbol $\sim$ nor o can be used in this way, however.

The statement (2.01) is of existential type: it asserts that there is a number K such that

$$|f(x)| \leqslant K|\phi(x)| \qquad (x \geqslant a), \tag{2.02}$$

without giving information concerning the actual size of K. Of course, if (2.02) holds for a certain value of K, then it also holds for every larger value; thus there is an infinite set of possible K's. The least member of this set is the supremum (least upper bound) of $|f(x)/\phi(x)|$ in the interval $[a, \infty)$; we call it the *implied constant* of the O term for this interval.

2.3 The notations $o(\phi)$ and $O(\phi)$ can also be used to denote the *classes* of functions f with the properties (ii) and (iii), respectively, or *unspecified* functions with these properties. The latter use is generic, that is, $o(\phi)$ does not necessarily denote the same function f at each occurrence. Similarly for $O(\phi)$. For example,

$$o(\phi) + o(\phi) = o(\phi), \qquad o(\phi) = O(\phi).$$

It should be noted that many relations of this kind, including the second example, are not reversible: $O(\phi) = o(\phi)$ is false. Relations involving $\sim$ are always reversible, however.

An instructive relation is supplied by

$$e^{ix}\{1+o(1)\} + e^{-ix}\{1+o(1)\} = 2\cos x + o(1). \tag{2.03}$$

This is easily verified by expressing $e^{\pm ix}$ in the form $\cos x \pm i \sin x$ and recalling that the trigonometric functions are bounded. The important point to notice is that the right-hand side of (2.03) cannot be rewritten in the form $2\{1+o(1)\}\cos x$, for this would imply that the left-hand side is *exactly* zero when x is an odd multiple of $\frac{1}{2}\pi$. In general this is false because the functions represented by the $o(1)$ terms differ.

† Throughout this book we adhere to the standard notation (a, b) for an open interval $a < x < b$; $[a, b]$ for the corresponding closed interval $a \leqslant x \leqslant b$; $(a, b]$ and $[a, b)$ for the partly closed intervals $a < x \leqslant b$ and $a \leqslant x < b$, respectively.

Ex. 2.1† If ν has any fixed value, real or complex, prove that $x^\nu = o(e^x)$ and $e^{-x} = o(x^\nu)$. Prove also that‡ $\ln x = o(x^\nu)$, provided that $\operatorname{Re} \nu > 0$.

Ex. 2.2 Show that

$$x + o(x) = O(x), \qquad \{O(x)\}^2 = O(x^2) = o(x^3).$$

Ex. 2.3 Show that

$$\cos\{O(x^{-1})\} = O(1), \qquad \sin\{O(x^{-1})\} = O(x^{-1}),$$

and

$$\cos\{x+\alpha+o(1)\} = \cos(x+\alpha) + o(1),$$

where α is a real constant.

Ex. 2.4 Is it true that

$$\{1+o(1)\}\cosh x - \{1+o(1)\}\sinh x = \{1+o(1)\}e^{-x}?$$

Ex. 2.5 Show that

$$O(\phi)O(\psi) = O(\phi\psi), \qquad O(\phi)o(\psi) = o(\phi\psi), \qquad O(\phi) + O(\psi) = O(|\phi|+|\psi|).$$

Ex. 2.6 What are the implied constants in the relations

$$(x+1)^2 = O(x^2), \qquad (x^2-\tfrac{1}{2})^{1/2} = O(x), \qquad x^2 = O(e^x),$$

for the interval $[1, \infty)$?

Ex. 2.7 Prove that if $f \sim \phi$, then $f = \{1+o(1)\}\phi$. Show that the converse holds provided that infinity is not a limit point of zeros of ϕ.

Ex. 2.8 Let $\phi(x)$ be a positive nonincreasing function of x, and $f(x) \sim \phi(x)$ as $x \to \infty$. By means of the preceding exercise show that

$$\sup_{t \in (x,\infty)} f(t) \sim \phi(x) \qquad (x \to \infty).$$

3 The Symbols ~, o, and O (continued)

3.1 The definitions of §2.1 may be extended in a number of obvious ways. To begin with, there is no need for the asymptotic variable x to be continuous; it can pass to infinity through any set of values. Thus

$$\sin\left(\pi n + \frac{1}{n}\right) = O\left(\frac{1}{n}\right) \qquad (n \to \infty),$$

provided that n is an integer.

Next, we are not obliged to concern ourselves with the behavior of the ratio $f(x)/\phi(x)$ solely as $x \to \infty$; the definitions (i), (ii), and (iii) of §2.1 also apply when x tends to any finite point, c, say. For example, if $c \neq 0$, then as $x \to c$

$$\frac{x^2-c^2}{x^2} \sim \frac{2(x-c)}{c} = O(x-c) = o(1).$$

† In Exercises 2.1–2.5 it is assumed that large positive values of the independent variable x are being considered.

‡ $\ln x \equiv \log_e x$.

We refer to c as the *distinguished point* of the asymptotic or order relation.

3.2 The next extension is to complex variables. Let $\mathbf{S}$ be a given infinite sector $\alpha \leqslant \operatorname{ph} z \leqslant \beta$, $\operatorname{ph} z$ denoting the *phase* or *argument* of z. Suppose that for a certain value of R there exists a number K, *independent of* $\operatorname{ph} z$, such that

$$|f(z)| \leqslant K|\phi(z)| \qquad (z \in \mathbf{S}(R)), \tag{3.01}$$

where $\mathbf{S}(R)$ denotes the intersection of $\mathbf{S}$ with the annulus $|z| \geqslant R$. Then we say that $f(z) = O\{\phi(z)\}$ as $z \to \infty$ in $\mathbf{S}$, or, equivalently, $f(z) = O\{\phi(z)\}$ in $\mathbf{S}(R)$. Thus the *symbol O automatically implies uniformity with respect to* $\operatorname{ph} z$.† *Similarly for the symbols ~ and o.*

For future reference, the point set $\mathbf{S}(R)$ just defined will be called an *infinite annular sector* or, simply, *annular sector*. The vertex and angle of $\mathbf{S}$ will also be said to be the *vertex* and *angle* of $\mathbf{S}(R)$.

The least number K fulfilling (3.01) is called the *implied constant* for $\mathbf{S}(R)$. Actually there is no essential reason for considering annular sectors, the definitions apply equally well to any *region* (that is, point set in the complex plane) having infinity or some other distinguished point as a limit point; compare Exercise 3.2 below.

3.3 An important example is provided by the tail of a convergent power series:

Theorem 3.1 *Let* $\sum_{s=0}^{\infty} a_s z^s$ *converge when* $|z| < r$. *Then for fixed n,*

$$\sum_{s=n}^{\infty} a_s z^s = O(z^n)$$

in any disk $|z| \leqslant \rho$ *such that* $\rho < r$.

To prove this result, let ρ' be any number in the interval (ρ, r). Then $a_s \rho'^s \to 0$ as $s \to \infty$; hence there exists a constant A such that

$$|a_s| \rho'^s \leqslant A \qquad (s = 0, 1, 2, \ldots).$$

Accordingly,

$$\left|\sum_{s=n}^{\infty} a_s z^s\right| \leqslant \sum_{s=n}^{\infty} A \frac{|z|^s}{\rho'^s} = \frac{A\rho'^{(1-n)}|z|^n}{\rho' - |z|} \leqslant \frac{A\rho'^{(1-n)}}{\rho' - \rho} |z|^n.$$

This establishes the theorem.

A typical illustration is supplied by

$$\ln\{1 + O(z)\} = O(z) \qquad (z \to 0).$$

3.4 An asymptotic or order relation may possess uniform properties with respect to other variables or parameters. For example, if u is a parameter in the interval $[0, a]$, where a is a positive constant, then

$$e^{(z-u)^2} = O(e^{z^2})$$

† Not all writers use O and the other two symbols in this way.

as $z \to \infty$ in the right half-plane, uniformly with respect to u (and ph z). Such regions of validity are often interdependent: $u \in [-a, 0]$ and the left half of the z plane would be another admissible combination in this example.

Ex. 3.1 If δ denotes a positive constant, show that $\cosh z \sim \frac{1}{2}e^z$ as $z \to \infty$ in the sector $|\mathrm{ph}\, z| \leqslant \frac{1}{2}\pi - \delta$, but not in the sector $|\mathrm{ph}\, z| < \frac{1}{2}\pi$.

Ex. 3.2 Show that $e^{-\sinh z} = o(1)$ as $z \to \infty$ in the half-strip $\mathrm{Re}\, z \geqslant 0$, $|\mathrm{Im}\, z| \leqslant \frac{1}{2}\pi - \delta < \frac{1}{2}\pi$.

Ex. 3.3 If p is fixed and positive, calculate the implied constant in the relation $e^{-z} = O(z^{-p})$ for the sector $|\mathrm{ph}\, z| \leqslant \frac{1}{2}\pi - \delta < \frac{1}{2}\pi$, and show that it tends to infinity as $\delta \to 0$.

Ex. 3.4 Assume that $\phi(x) > 0$, p is a real constant, and $f(x) \sim \phi(x)$ as $x \to \infty$. With the aid of Theorem 3.1 show that $\{f(x)\}^p \sim \{\phi(x)\}^p$ and $\ln\{f(x)\} \sim \ln\{\phi(x)\}$, provided that in the second case $\phi(x)$ is bounded away from unity.

Show also that $e^{f(x)} \sim e^{\phi(x)}$ may be false.

Ex. 3.5 Let x range over the interval $[0, \delta]$, where δ is a positive constant, and $f(u, x)$ be a positive real function such that $f(u, x) = O(u)$ as $u \to 0$, uniformly with respect to x. Show that

$$\{x + f(u, x)\}^{1/2} = x^{1/2} + O(u^{1/2})$$

as $u \to 0$, uniformly with respect to x.

4 Integration and Differentiation of Asymptotic and Order Relations

4.1 As a rule, asymptotic and order relations may be *integrated*, subject to obvious restrictions on the convergence of the integrals involved. Suppose, for example, that $f(x)$ is an integrable function of the real variable x such that $f(x) \sim x^\nu$ as $x \to \infty$, where ν is a real or complex constant. Let a be any finite real number. Then as $x \to \infty$, we have

$$\int_x^\infty f(t)\, dt \sim -\frac{x^{\nu+1}}{\nu+1} \qquad (\mathrm{Re}\, \nu < -1), \tag{4.01}$$

and

$$\int_a^x f(t)\, dt \sim \begin{cases} \text{a constant} & (\mathrm{Re}\, \nu < -1), \\ \ln x & (\nu = -1), \\ x^{\nu+1}/(\nu+1) & (\mathrm{Re}\, \nu > -1). \end{cases} \tag{4.02}$$

To prove, for example, the third of (4.02), we have $f(x) = x^\nu\{1 + \eta(x)\}$, where $|\eta(x)| < \varepsilon$ when $x > X > 0$, X being assignable for any given positive number ε. Hence if $x > X$, then

$$\int_a^x f(t)\, dt = \int_a^X f(t)\, dt + \frac{1}{\nu+1}(x^{\nu+1} - X^{\nu+1}) + \int_X^x t^\nu \eta(t)\, dt,$$

and so

$$\frac{\nu+1}{x^{\nu+1}} \int_a^x f(t)\, dt - 1 = \frac{\nu+1}{x^{\nu+1}} \int_a^X f(t)\, dt - \frac{X^{\nu+1}}{x^{\nu+1}} + \frac{\nu+1}{x^{\nu+1}} \int_X^x t^\nu \eta(t)\, dt.$$

The first two terms on the right-hand side of the last equation vanish as $x \to \infty$, and the third term is bounded by $|\nu+1|\varepsilon/(1+\operatorname{Re}\nu)$. The stated result now follows.

The results (4.01) and (4.02) may be extended in a straightforward way to complex integrals.

4.2 *Differentiation* of asymptotic or order relations is not always permissible. For example, if $f(x) = x+\cos x$, then $f(x) \sim x$ as $x \to \infty$, but it is not true that $f'(x) \to 1$. To assure the legitimacy of differentiation further conditions are needed. For real variables, these conditions can be expressed in terms of the monotonicity of the derivative:

Theorem 4.1[†] *Let $f(x)$ be continuously differentiable and $f(x) \sim x^p$ as $x \to \infty$, where p ($\geqslant 1$) is a constant. Then $f'(x) \sim px^{p-1}$, provided that $f'(x)$ is nondecreasing for all sufficiently large x.*

To prove this result, we have $f(x) = x^p\{1+\eta(x)\}$, where $|\eta(x)| \leqslant \varepsilon$ when $x > X$, assignable and positive, ε being an arbitrary number in the interval $(0, 1)$. If $h > 0$, then

$$hf'(x) \leqslant \int_x^{x+h} f'(t)\,dt = f(x+h) - f(x)$$

$$= \int_x^{x+h} pt^{p-1}\,dt + (x+h)^p\eta(x+h) - x^p\eta(x)$$

$$\leqslant hp(x+h)^{p-1} + 2\varepsilon(x+h)^p.$$

Set $h = \varepsilon^{1/2}x$. Then we have

$$f'(x) \leqslant px^{p-1}\{(1+\varepsilon^{1/2})^{p-1} + 2p^{-1}\varepsilon^{1/2}(1+\varepsilon^{1/2})^p\} \qquad (x > X).$$

Similarly,

$$f'(x) \geqslant px^{p-1}\{(1-\varepsilon^{1/2})^{p-1} - 2p^{-1}\varepsilon^{1/2}\} \qquad (x > X/(1-\varepsilon^{1/2})).$$

The theorem now follows.

Another result of this type is stated in Exercise 4.4 below. It should be appreciated, however, that monotonicity conditions on $f'(x)$ are often difficult to verify in practice because $f'(x)$ is the function whose properties are being sought.

4.3 In the complex plane, differentiation of asymptotic or order relations is generally permissible in subregions of the original region of validity. An important case is the following:

Theorem 4.2[‡] *Let $f(z)$ be holomorphic[§] in a region containing a closed annular sector* **S**, *and*

$$f(z) = O(z^p) \qquad (or\ \ f(z) = o(z^p)) \tag{4.03}$$

† de Bruijn (1961, §7.3).
‡ Ritt (1918).
§ That is, analytic and free from singularity.

as $z \to \infty$ *in* **S**, *where p is any fixed real number. Then*

$$f^{(m)}(z) = O(z^{p-m}) \qquad (\text{or} \;\; f^{(m)}(z) = o(z^{p-m})) \tag{4.04}$$

as $z \to \infty$ *in any closed annular sector* **C** *properly interior to* **S** *and having the same vertex.*

The proof depends on Cauchy's integral formula for the mth derivative of an analytic function, given by

$$f^{(m)}(z) = \frac{m!}{2\pi i} \int_{\mathscr{C}} \frac{f(t)\,dt}{(t-z)^{m+1}}, \tag{4.05}$$

in which the path $\mathscr{C}$ is chosen to be a circle enclosing $t = z$. The essential reason z is restricted to an interior region in the final result is to permit inclusion of $\mathscr{C}$ in **S**.

Since $|z - \text{constant}|^p \sim |z|^p$, the vertex of **S** may be taken to be the origin without loss of generality. Let **S** be defined by $\alpha \leqslant \operatorname{ph} z \leqslant \beta$, $|z| \geqslant R$, and consider the annular sector **S**′ defined by

$$\alpha + \delta \leqslant \operatorname{ph} z \leqslant \beta - \delta, \qquad |z| \geqslant R',$$

where δ is a positive acute angle and $R' = R/(1-\sin\delta)$; see Fig. 4.1. By taking δ small enough we can ensure that **S**′ contains **C**. In (4.05) take $\mathscr{C}$ to be $|t-z| = |z|\sin\delta$. Then

$$|z|(1-\sin\delta) \leqslant |t| \leqslant |z|(1+\sin\delta).$$

Hence $t \in$ **S** whenever $z \in$ **S**′. Moreover, if K is the implied constant of (4.03) for **S**, then

$$|f^{(m)}(z)| \leqslant \frac{m!}{(|z|\sin\delta)^m} K|z|^p(1 \pm \sin\delta)^p,$$

the upper or lower sign being taken according as $p \geqslant 0$ or $p < 0$. In either event $f^{(m)}(z)$ is $O(z^{p-m})$, as required. The proof in the case when the symbol O in (4.03) and (4.04) is replaced by o is similar.

We have shown, incidentally, that the implied constant of (4.04) in **S**′ does not

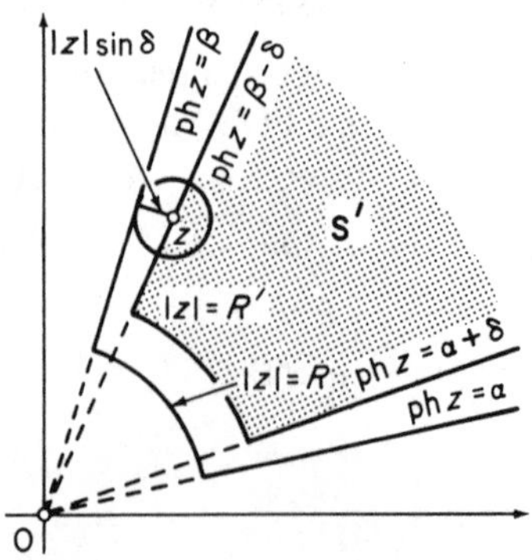

Fig. 4.1 Annular sectors **S**, **S**′.

exceed $m!(\csc\delta)^m(1\pm\sin\delta)^p K$, but because this bound tends to infinity as $\delta\to 0$, we cannot infer that (4.04) is valid in **S**.

Ex. 4.1 Show that if $f(x)$ is continuous and $f(x)=o\{\phi(x)\}$ as $x\to\infty$, where $\phi(x)$ is a positive nondecreasing function of x, then $\int_a^x f(t)\,dt = o\{x\phi(x)\}$.

Ex. 4.2 It may be expected that in the case $\operatorname{Re}\nu=-1$, $\operatorname{Im}\nu\neq 0$, the result corresponding to (4.02) would be $\int_a^x f(t)\,dt=O(1)$. Show that this is false by means of the example $f(x)=x^{i\mu-1}+(x\ln x)^{-1}$, where μ is real.

Ex. 4.3 If u and x lie in $[1,\infty)$, show that

$$\int_x^\infty \frac{dt}{t(t^2+t+u^2)^{1/2}} = \frac{1}{x}+O\left(\frac{1}{x^2}\right)+O\left(\frac{u^2}{x^3}\right).$$

Ex. 4.4 Suppose that $f(x)=x^2+O(x)$ as $x\to\infty$, and $f'(x)$ is continuous and nondecreasing for all sufficiently large x. Show that $f'(x)=2x+O(x^{1/2})$. [de Bruijn, 1961.]

Ex. 4.5 In place of (4.03) assume that $f(z)\sim z^\nu$, where ν is a nonzero real or complex constant. Deduce from Theorem 4.2 that $f'(z)\sim \nu z^{\nu-1}$ as $z\to\infty$ in **C**.

Ex. 4.6 Let **T** and **T′** denote the half-strips

$$\mathbf{T}\ :\quad \alpha\leqslant \operatorname{Im}z\leqslant\beta,\qquad \operatorname{Re}z\geqslant\rho,$$

$$\mathbf{T}'\ :\quad \alpha+\delta\leqslant \operatorname{Im}z\leqslant\beta-\delta,\qquad \operatorname{Re}z\geqslant\rho,$$

where $0<\delta<\frac{1}{2}(\beta-\alpha)$. Suppose that $f(z)$ is holomorphic within **T**, and $f(z)=O(e^z)$ as $z\to\infty$ in **T**. Show that $f'(z)=O(e^z)$ as $z\to\infty$ in **T′**.

Ex. 4.7 Show that the result of Exercise 4.6 remains valid if both terms $O(e^z)$ are replaced by $O(z^p)$, where p is a real constant.

Show further that $f'(z)=O(z^{p-1})$ is false by means of the example z^pe^{iz}.

5 Asymptotic Solution of Transcendental Equations: Real Variables

5.1 Consider the equation

$$x+\tanh x = u,$$

in which u is a real parameter. The left-hand side is a strictly increasing function of x. Hence by graphical considerations there is exactly one real root $x(u)$, say, for each value of u. What is the asymptotic behavior of $x(u)$ for large positive u?

When x is large, the left-hand side is dominated by the first term. Accordingly, we transfer the term $\tanh x$ to the right and treat it as a "correction":

$$x = u-\tanh x.$$

Since $|\tanh x|<1$, it follows that

$$x(u)\sim u\qquad (u\to\infty). \tag{5.01}$$

This is the first approximation to the root. An immediate improvement is obtained by recalling that $\tanh x = 1+o(1)$ as $x\to\infty$; thus

$$x = u-1+o(1)\qquad (u\to\infty). \tag{5.02}$$

To derive higher approximations we expand $\tanh x$ in a form appropriate for large x, given by

$$\tanh x = 1 - 2e^{-2x} + 2e^{-4x} - 2e^{-6x} + \cdots \qquad (x > 0),$$

and repeatedly substitute for x in terms of u. From (5.02) it is seen that $e^{-2x} = O(e^{-2u})$.† Hence with the aid of Theorem 3.1 we obtain

$$x = u - 1 + O(e^{-2x}) = u - 1 + O(e^{-2u}).$$

The next step is given by

$$\begin{aligned} x &= u - 1 + 2\exp\{-2u+2+O(e^{-2u})\} + O(e^{-4x}) \\ &= u - 1 + 2e^{-2u+2} + O(e^{-4u}). \end{aligned} \tag{5.03}$$

Continuation of the process produces a sequence of approximations with errors of steadily diminishing asymptotic order. Whether the sequence converges as the number of steps tends to infinity is not discernible from the analysis, but the numerical potential of the process can be perceived by taking, for example, $u = 5$ and ignoring the error term $O(e^{-4u})$ in (5.03). We find that $x = 4.0006709\ldots$, compared with the correct value $4.0006698\ldots$, obtained by standard numerical methods.‡

5.2 A second example amenable to the same approach is the determination of the large positive roots of the equation

$$x \tan x = 1.$$

Inversion produces

$$x = n\pi + \tan^{-1}(1/x),$$

where n is an integer and the inverse tangent has its principal value. Since the latter is in the interval $(-\frac{1}{2}\pi, \frac{1}{2}\pi)$, we derive $x \sim n\pi$ as $n \to \infty$.

Next, when $x > 1$,

$$\tan^{-1}\frac{1}{x} = \frac{1}{x} - \frac{1}{3x^3} + \frac{1}{5x^5} - \frac{1}{7x^7} + \cdots.$$

Hence $x = n\pi + O(x^{-1}) = n\pi + O(n^{-1})$. The next two substitutions produce

$$x = n\pi + \frac{1}{n\pi} + O\left(\frac{1}{n^3}\right), \qquad x = n\pi + \frac{1}{n\pi} - \frac{4}{3(n\pi)^3} + O\left(\frac{1}{n^5}\right).$$

And so on.

5.3 A third example is provided by the equation

$$x^2 - \ln x = u, \tag{5.04}$$

in which u is again a large positive parameter. This differs from the preceding

† It should be observed that this relation cannot be deduced directly from (5.01).

‡ Error bounds for (5.03) are stated in Exercise 5.3 below.

examples in that the "correction term" $\ln x$ is unbounded as $x \to \infty$. To assist with (5.04) and similar equations we establish the following simple general result:

Theorem 5.1 *Let $f(\xi)$ be continuous and strictly increasing in an interval $a < \xi < \infty$, and*

$$f(\xi) \sim \xi \qquad (\xi \to \infty). \tag{5.05}$$

Denote by $\xi(u)$ the root of the equation

$$f(\xi) = u \tag{5.06}$$

which lies in (a, ∞) when $u > f(a)$. Then

$$\xi(u) \sim u \qquad (u \to \infty). \tag{5.07}$$

Graphical considerations show that $\xi(u)$ is unique, increasing, and unbounded as $u \to \infty$. From (5.05) and (5.06) we have $u = \{1+o(1)\}\xi$ as $\xi \to \infty$, and therefore, also, as $u \to \infty$. Division by the factor $1+o(1)$ then gives $\xi = \{1+o(1)\}u$, which is equivalent to (5.07).

5.4 We return to the example (5.04). Here $\xi = x^2$ and $f(\xi) = \xi - \frac{1}{2}\ln\xi$. Therefore $f(\xi)$ is strictly increasing when $\xi > \frac{1}{2}$, and the theorem informs us that $\xi \sim u$ as $u \to \infty$; equivalently,

$$x = u^{1/2}\{1+o(1)\} \qquad (u \to \infty).$$

Substituting this approximation into the right-hand side of

$$x^2 = u + \ln x, \tag{5.08}$$

and recalling that $\ln\{1+o(1)\}$ is $o(1)$, we see that

$$x^2 = u + \tfrac{1}{2}\ln u + o(1),$$

and hence (Theorem 3.1)

$$x = u^{1/2}\left\{1 + \frac{\ln u}{4u} + o\left(\frac{1}{u}\right)\right\}.$$

As in §§5.1 and 5.2, the resubstitutions can be continued indefinitely.

Ex. 5.1 Prove that the root of the equation $x \tan x = u$ which lies in the interval $(0, \frac{1}{2}\pi)$ is given by

$$x = \tfrac{1}{2}\pi(1-u^{-1}+u^{-2}) - (\tfrac{1}{2}\pi - \tfrac{1}{24}\pi^3)u^{-3} + O(u^{-4}) \qquad (u\to\infty).$$

Ex. 5.2 Show that the large positive roots of the equation $\tan x = x$ are given by

$$x = \mu - \mu^{-1} - \tfrac{2}{3}\mu^{-3} + O(\mu^{-5}) \qquad (\mu\to\infty),$$

where $\mu = (n+\frac{1}{2})\pi$, n being a positive integer.

Ex. 5.3 For the example of §5.1, show that when $u > 0$

$$x = u - 1 + 2\vartheta_1 e^{-2u+2},$$

and hence that

$$x = u - 1 + 2e^{-2u+2} - 10\vartheta_2 e^{-4u+4},$$

where ϑ_1 and ϑ_2 are certain numbers in the interval $(0, 1)$.

Ex. 5.4 Let $M(x)\cos\theta(x) = \cos x + o(1)$ and $M(x)\sin\theta(x) = \sin x + o(1)$, as $x \to \infty$, where $M(x)$ is positive and $\theta(x)$ is real and continuous. Prove that

$$M(x) = 1 + o(1), \qquad \theta(x) = x + 2m\pi + o(1),$$

where m is an integer.

Ex. 5.5 Prove that for large positive u the real roots of the equation $xe^{1/x} = e^u$ are given by

$$x = \frac{1}{u} - \frac{\ln u}{u^2} + \frac{(\ln u)^2}{u^3} + O\left(\frac{\ln u}{u^3}\right), \qquad x = e^u - 1 - \tfrac{1}{2}e^{-u} + O(e^{-2u}).$$

Ex. 5.6 (*Error bound for Theorem* 5.1) Let ξ be positive and $f(\xi)$ a strictly increasing continuous function such that

$$|f(\xi) - \xi| < k\xi^{-p},$$

k and p being positive constants. Show that if $u > 0$ and a number δ can be found such that $\delta \in (0, 1)$ and $\delta(1-\delta)^p \geqslant ku^{-p-1}$, then the positive root of the equation $f(\xi) = u$ lies in the interval $(u - u\delta, u + u\delta)$. Deduce that if l is an arbitrary number exceeding k, then the root satisfies

$$|\xi - u| < kl^p\{(l-k)u\}^{-p},$$

provided that $u > l(l-k)^{-p/(p+1)}$.

Ex. 5.7 Show that for large u the positive root of the equation $x \ln x = u$ is given by

$$x(u) \sim u/\ln u.$$

Show also that when $u > e$

$$\frac{u}{\ln u} < x(u) \leqslant \left(1 + \frac{1}{e}\right)\frac{u}{\ln u}.$$

6 Asymptotic Solution of Transcendental Equations: Complex Variables

6.1 Suppose now that $f(z)$ is an analytic function of the complex variable z which is holomorphic in a region containing a closed annular sector $\mathbf{S}$ with vertex at the origin and angle less than 2π. Assume that

$$f(z) \sim z \qquad (z \to \infty \text{ in } \mathbf{S}). \tag{6.01}$$

Then the relation

$$u = f(z) \tag{6.02}$$

maps $\mathbf{S}$ onto an unbounded region $\mathbf{U}$, say. The essential difficulty in establishing a result analogous to Theorem 5.1 is to restrict z and u in such a way that these variables have a one-to-one relationship.

Theorem 6.1 *Let* $\mathbf{S}_1$ *and* $\mathbf{S}_2$ *be closed annular sectors with vertices at the origin,* $\mathbf{S}_1$ *being properly interior to the given annular sector* $\mathbf{S}$, *and* $\mathbf{S}_2$ *being properly interior to* $\mathbf{S}_1$.

(i) *If the boundary arcs of* $\mathbf{S}_1$ *and* $\mathbf{S}_2$ *are of sufficiently large radius, then equation* (6.02) *has exactly one root* $z(u)$ *in* $\mathbf{S}_1$ *for each* $u \in \mathbf{S}_2$.

(ii) $z(u) \sim u$ *as* $u \to \infty$ *in* $\mathbf{S}_2$.

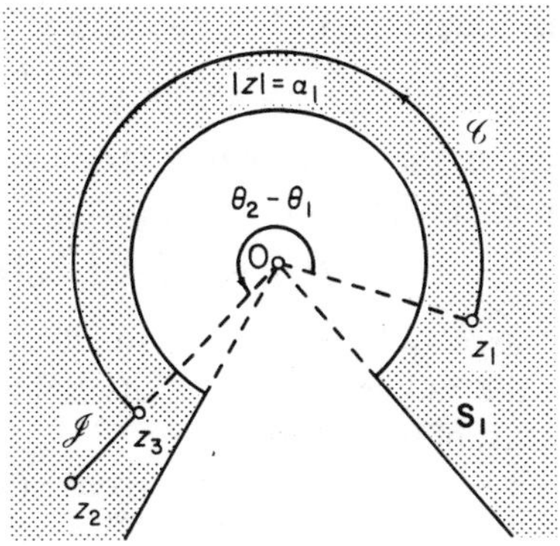

Fig. 6.1 t plane.

To establish this result write

$$f(z) = z + \xi(z).$$

From (6.01) and Ritt's theorem (§4.3) it follows that $\xi'(z) = o(1)$ as $z \to \infty$ in $\mathbf{S}_1$. Let z_1 and z_2 be any two distinct points of $\mathbf{S}_1$, labeled in such a way that $|z_1| \leqslant |z_2|$. Then

$$f(z_2) - f(z_1) = (1+\vartheta)(z_2 - z_1), \tag{6.03}$$

where $\vartheta = \{\xi(z_2) - \xi(z_1)\}/(z_2 - z_1)$.

6.2 The first step is to prove that when the radius a_1, say, of the boundary arc of $\mathbf{S}_1$ is sufficiently large, $|\vartheta| < 1$ for all z_1 and z_2 in $\mathbf{S}_1$. Clearly

$$|\vartheta| = \left| \frac{1}{z_2 - z_1} \int_{z_1}^{z_2} \xi'(t)\, dt \right| \leqslant \frac{l(z_1, z_2)}{|z_2 - z_1|} \delta,$$

where δ is the maximum value (necessarily finite) of $|\xi'(z)|$ in $\mathbf{S}_1$, and $l(z_1, z_2)$ is the length of the path of integration.

Figure 6.1 shows that in certain cases we cannot integrate along the join† of z_1 and z_2 and keep within $\mathbf{S}_1$; in this diagram $\theta_1 = \operatorname{ph} z_1$ and $\theta_2 = \operatorname{ph} z_2$. However, as integration path we can always take the circular arc $\mathscr{C}$ centered at $t = 0$ and extending from z_1 to $z_3 \equiv |z_1| e^{i\theta_2}$, together with the join $\mathscr{J}$, say, of z_3 and z_2. Since the angle $z_1 z_3 O$ is less than $\frac{1}{2}\pi$ in all cases, both $|z_3 - z_1|$ and $|z_3 - z_2|$ are bounded by $|z_2 - z_1|$. Denoting the angle of $\mathbf{S}_1$ by σ, we have

$$\frac{l(z_1, z_2)}{|z_2 - z_1|} \leqslant \frac{\text{length of } \mathscr{J}}{|z_3 - z_2|} + \frac{\text{length of } \mathscr{C}}{|z_3 - z_1|} = 1 + \frac{|\theta_2 - \theta_1|}{2 \sin|\frac{1}{2}\theta_2 - \frac{1}{2}\theta_1|} < k,$$

where $k = 1 + \frac{1}{2}\sigma \csc(\frac{1}{2}\sigma)$ and is finite, since $\sigma < 2\pi$.

Thus $|\vartheta| \leqslant k\delta$. As $a_1 \to \infty$, we have $\delta \to 0$. Hence $|\vartheta| < 1$ for sufficiently large a_1, as required.

6.3 Reference to (6.03) shows that $f(z_1) \neq f(z_2)$. Accordingly, equation (6.02) maps $\mathbf{S}_1$ conformally on a certain u domain $\mathbf{U}_1$, say.

† *Join* means the straight-line connection.

Consider the boundaries of $\mathbf{U}_1$. For large $|z|$, we have

$$\operatorname{ph}\{f(z)\} = \operatorname{ph} z + \operatorname{ph}\{1+z^{-1}\xi(z)\} = \operatorname{ph} z + o(1).$$

Hence in a neighborhood of infinity $\mathbf{U}_1$ contains $\mathbf{S}_2$. The other boundary of $\mathbf{U}_1$ corresponds to the arc $|z| = a_1$. On this arc

$$|f(z)| = a_1|1+z^{-1}\xi(z)| \leqslant 2a_1,$$

for sufficiently large a_1. Therefore the annular sector $\mathbf{S}_2$ is entirely contained in $\mathbf{U}_1$, provided that the radius a_2, say, of its boundary arc is sufficiently large. This establishes Part (i) of the theorem.

To prove Part (ii), we observe that given ε (>0), a_1 can be chosen so that

$$|z^{-1}\xi(z)| < \varepsilon(1+\varepsilon)^{-1}$$

when $z \in \mathbf{S}_1$. Then

$$\left|\frac{z(u)}{u} - 1\right| = \left|\frac{z^{-1}\xi(z)}{1+z^{-1}\xi(z)}\right| < \frac{\varepsilon(1+\varepsilon)^{-1}}{1-\varepsilon(1+\varepsilon)^{-1}} = \varepsilon.$$

The condition $z(u) \in \mathbf{S}_1$ can be satisfied for all $u \in \mathbf{S}_2$, again by making a_2 large enough. The proof of Theorem 6.1 is now complete.

Ex. 6.1 Show that if m is an integer or zero, then in the sector $(m-\frac{1}{2})\pi \leqslant \operatorname{ph} z \leqslant (m+\frac{1}{2})\pi$ the large zeros of the function $z\tan z - \ln z$ are given by

$$z = n\pi e^{m\pi i}\left[1 + \frac{\ln(n\pi)+m\pi i}{(n\pi)^2} + O\left\{\frac{(\ln n)^3}{n^4}\right\}\right],$$

where n is a large positive integer.

7 Definition and Fundamental Properties of Asymptotic Expansions

7.1 Let $f(z)$ be a function of the real or complex variable z, $\sum a_s z^{-s}$ a formal power series (convergent or divergent), and $R_n(z)$ the difference between $f(z)$ and the nth partial sum of the series; thus

$$f(z) = a_0 + \frac{a_1}{z} + \frac{a_2}{z^2} + \cdots + \frac{a_{n-1}}{z^{n-1}} + R_n(z). \tag{7.01}$$

Suppose that for each fixed value of n

$$R_n(z) = O(z^{-n}) \tag{7.02}$$

as $z \to \infty$ in a certain unbounded region $\mathbf{R}$. Then, following Poincaré (1886), we say that the series $\sum a_s z^{-s}$ is *an asymptotic expansion of* $f(z)$, and write†

$$f(z) \sim a_0 + \frac{a_1}{z} + \frac{a_2}{z^2} + \cdots \qquad (z \to \infty \text{ in } \mathbf{R}). \tag{7.03}$$

† The sense in which the symbol $\sim$ is now being used differs from that of §§2 and 3. To avoid all possible confusion some writers use $\approx$ for asymptotic expansions and confine $\sim$ to asymptotic approximations.

We call z the *asymptotic variable*, and the implied constant of the O term (7.02) for $\mathbf{R}$ the nth *implied constant* of the asymptotic expansion for $\mathbf{R}$.

If the relation (7.02) holds only when $n \leqslant N$, say, or, more generally, if $R_n(z) = o(1/z^{n-1})$ for $n \leqslant N$, then we say that (7.03) is an *asymptotic expansion to N terms*. We shall assume, however, that such a restriction applies only when specifically stated.

From Theorem 3.1 (with z replaced by $1/z$) it is seen that if the series $\sum a_s z^{-s}$ converges for all sufficiently large $|z|$, then it is the asymptotic expansion of its sum, defined in the usual way, without restriction on $\operatorname{ph} z$. Naturally, however, greater interest attaches to asymptotic expansions that diverge. An example has been provided by (1.05); this is a consequence of (1.08).

7.2 Theorem 7.1 *A necessary and sufficient condition that $f(z)$ possesses an asymptotic expansion of the form* (7.03) *is that for each nonnegative integer n*

$$z^n\left\{f(z) - \sum_{s=0}^{n-1} \frac{a_s}{z^s}\right\} \to a_n \tag{7.04}$$

as $z \to \infty$ in $\mathbf{R}$, uniformly with respect to $\operatorname{ph} z$.

Clearly (7.04) implies (7.02); this is the sufficiency condition. To verify necessity we have from (7.01) and (7.02)

$$z^n R_n(z) = z^n\left\{\frac{a_n}{z^n} + R_{n+1}(z)\right\} \to a_n \qquad (z \to \infty),$$

which is equivalent to (7.04).

Immediate corollaries of Theorem 7.1 are:

(i) (Uniqueness property) *For a given function $f(z)$ and region $\mathbf{R}$, there is at most one expansion of the form* (7.03).

(ii) *The* nth *implied constant of* (7.03) *for the region $\mathbf{R}$ cannot be less than $|a_n|$.*

7.3 *The converse of Corollary* (i) *of* §7.2 *is false.* Consider the asymptotic expansion of e^{-z} in the sector $|\operatorname{ph} z| \leqslant \frac{1}{2}\pi - \delta < \frac{1}{2}\pi$. Since, for each n, $z^n e^{-z} \to 0$ as $z \to \infty$ in this region, the relation (7.04) yields $a_n = 0$, $n = 0, 1, \ldots$. Thus

$$e^{-z} \sim 0 + \frac{0}{z} + \frac{0}{z^2} + \cdots \qquad (|\operatorname{ph} z| \leqslant \tfrac{1}{2}\pi - \delta). \tag{7.05}$$

Now let $a_0, a_1, a_2, \ldots$ denote any given sequence of constants. If one function $f(z)$ exists such that†

$$f(z) \sim a_0 + \frac{a_1}{z} + \frac{a_2}{z^2} + \cdots \qquad (z \to \infty \text{ in } |\operatorname{ph} z| \leqslant \tfrac{1}{2}\pi - \delta),$$

then there is an infinity of such functions, again because the relation $z^n e^{-z} \to 0$ shows that an arbitrary constant multiple of e^{-z} can be added to $f(z)$ without affecting the coefficients in the expansion.

† We shall see later (§9) that this condition is always satisfied.

The lack of uniqueness of the function represented by an asymptotic expansion contrasts with the sum of a convergent series. We used the sector $|\operatorname{ph} z| \leqslant \frac{1}{2}\pi - \delta$ for illustration; other sectors (of finite angle) can be treated by using $\exp(-z^p)$ in place of e^{-z}, p being a suitably chosen positive constant.

7.4 It may happen that although a function $f(z)$ has no asymptotic expansion of the form (7.03) in a given region, the quotient $f(z)/\phi(z)$, where $\phi(z)$ is a given function, possesses such an expansion. In this case we write

$$f(z) \sim \phi(z) \sum_{s=0}^{\infty} \frac{a_s}{z^s}.$$

Except when $a_0 = 0$, the leading term $a_0 \phi(z)$ provides an asymptotic approximation to $f(z)$ in the sense of §§2 and 3:

$$f(z) \sim a_0 \phi(z).$$

In a similar way, if the difference $f(z) - \phi(z)$ has an asymptotic expansion $\sum a_s z^{-s}$, then we write

$$f(z) \sim \phi(z) + \sum_{s=0}^{\infty} \frac{a_s}{z^s}.$$

Examples of this form of representation are supplied by analytic functions $f(z)$ having a pole at the point at infinity; if the order of the pole is n, then $\phi(z)$ is a polynomial in z of degree n.

7.5 In the situation mentioned in the last sentence the asymptotic expansions converge for sufficiently large $|z|$. This result is not as special as it might appear, however:

Theorem 7.2 *In a deleted neighborhood of infinity*† *let $f(z)$ be a single-valued holomorphic function, and*

$$f(z) \sim \sum_{s=n}^{\infty} \frac{a_s}{z^s} \tag{7.06}$$

as $z \to \infty$ for all $\operatorname{ph} z$, *n being a fixed integer (positive, zero, or negative). Then this expansion converges throughout the neighborhood, and $f(z)$ is its sum.*

To prove this result, let $|z| > R$ be the given neighborhood, and

$$f(z) = \sum_{s=-\infty}^{\infty} \frac{b_s}{z^s}$$

the corresponding Laurent expansion. This series converges when $|z| > R$, and

$$b_s = \frac{1}{2\pi i} \int_{|z|=\rho} f(z) z^{s-1}\, dz, \tag{7.07}$$

† That is, a neighborhood of infinity less the point at infinity.

for any value of ρ exceeding R. From (7.06) we have $f(z) = O(z^{-n})$ as $z \to \infty$. Letting $\rho \to \infty$ in (7.07) we deduce that b_s vanishes when $s < n$. Thus

$$f(z) = \sum_{s=n}^{\infty} \frac{b_s}{z^s}.$$

This convergent expansion is also an asymptotic expansion (Theorem 3.1), and since the asymptotic expansion of $f(z)$ is unique it follows that $a_s = b_s$. This completes the proof.

7.6 The final result in this section is immediately derivable from Theorem 7.2:

Theorem 7.3 *Let $f(z)$ be single valued and holomorphic in a deleted neighborhood of infinity. Assume that* (7.06) *holds in a closed sector* **S**, *and also that this expansion diverges for all finite z. Then the angle of* **S** *is less than* 2π *and $f(z)$ has an essential singularity at infinity.*

It needs to be emphasized that Theorems 7.2 and 7.3 apply only to functions that are *single valued.* If $f(z)$ has a branch point at infinity, then it *can* possess a divergent asymptotic expansion in a phase range exceeding 2π.

Ex. 7.1 Show that the definition of an asymptotic expansion is unaffected if we substitute

$$R_n(z) = o(1/z^{n-p}) \qquad (n = N, N+1, \ldots)$$

for (7.02), p being any fixed positive number and N any fixed nonnegative integer.

Ex. 7.2 Show that none of the functions $z^{-1/2}$, $\sin z$, and $\ln z$ possesses an asymptotic expansion of the form (7.03).

Ex. 7.3 Construct an example of a single-valued function that has an essential singularity at infinity and a convergent asymptotic expansion in the sector $|\mathrm{ph}\, z| \leq \frac{1}{2}\pi - \delta < \frac{1}{2}\pi$.

8 Operations with Asymptotic Expansions

8.1 (i) *Asymptotic expansions can be combined linearly.* Suppose that

$$f(z) \sim \sum_{s=0}^{\infty} f_s z^{-s}, \qquad g(z) \sim \sum_{s=0}^{\infty} g_s z^{-s},$$

as $z \to \infty$ in regions **F** and **G**, respectively. Then if λ and μ are any constants

$$\lambda f(z) + \mu g(z) \sim \sum_{s=0}^{\infty} (\lambda f_s + \mu g_s) z^{-s} \qquad (z \to \infty \text{ in } \mathbf{F} \cap \mathbf{G}).$$

This follows immediately from the definition.

(ii) *Asymptotic expansions can be multiplied.* That is,

$$f(z) g(z) \sim \sum_{s=0}^{\infty} h_s z^{-s} \qquad (z \to \infty \text{ in } \mathbf{F} \cap \mathbf{G}),$$

where

$$h_s = f_0 g_s + f_1 g_{s-1} + f_2 g_{s-2} + \cdots + f_s g_0.$$

For if $F_n(z)$, $G_n(z)$, and $H_n(z)$ denote the remainder terms associated with the nth partial sums of the expansions of $f(z)$, $g(z)$, and $f(z)g(z)$, respectively, then

$$H_n(z) = \sum_{s=0}^{n-1} \frac{f_s}{z^s} G_{n-s}(z) + g(z)F_n(z) = O\left(\frac{1}{z^n}\right).$$

(iii) *Asymptotic expansions can be divided.* For if $f_0 \neq 0$ and $|z|$ is sufficiently large, then

$$\frac{1}{f(z)} = \frac{1}{f_0 + F_1(z)} = \sum_{s=0}^{n-1} \frac{(-)^s}{f_0^{s+1}} \left\{ \frac{f_1}{z} + \cdots + \frac{f_{n-1}}{z^{n-1}} + F_n(z) \right\}^s + \frac{(-)^n \{F_1(z)\}^n}{f_0^n \{f_0 + F_1(z)\}}.$$

Since $F_1(z) = O(z^{-1})$ and $F_n(z) = O(z^{-n})$, it follows that

$$\frac{1}{f(z)} = \sum_{s=0}^{n-1} \frac{k_s}{z^s} + O\left(\frac{1}{z^n}\right) \qquad (z \to \infty \text{ in } \mathbf{F}),$$

where $f_0^{s+1} k_s$ is a polynomial in $f_0, f_1, \ldots, f_s$. Since n is arbitrary, this means that the asymptotic expansion of $1/f(z)$ certainly exists.

The coefficients k_s can be found by this process, but as in the case of convergent power series they are more conveniently calculated from the recurrence relation

$$f_0 k_s = -(f_1 k_{s-1} + f_2 k_{s-2} + \cdots + f_s k_0) \qquad (s = 1, 2, \ldots)$$

obtained by use of the identity $f(z)\{1/f(z)\} = 1$. The first four are given by

$$k_0 = 1/f_0, \qquad k_1 = -f_1/f_0^2,$$

$$k_2 = (f_1^2 - f_0 f_2)/f_0^3, \qquad k_3 = (-f_1^3 + 2f_0 f_1 f_2 - f_0^2 f_3)/f_0^4.$$

The necessary modifications when $f_0 = 0$ are straightforward.

8.2 (iv) *Asymptotic expansions can be integrated.* Suppose that for all sufficiently large values of the positive real variable x, $f(x)$ is a continuous real or complex function with an asymptotic expansion of the form

$$f(x) \sim f_0 + \frac{f_1}{x} + \frac{f_2}{x^2} + \cdots.$$

Unless $f_0 = f_1 = 0$ we cannot integrate $f(t)$ directly over the interval $x \leqslant t < \infty$ because of divergence. However, $f(t) - f_0 - f_1 t^{-1}$ is $O(t^{-2})$ for large t and therefore integrable. Integrating the remainder terms in accordance with §4.1, we see that

$$\int_x^\infty \left\{ f(t) - f_0 - \frac{f_1}{t} \right\} dt \sim \frac{f_2}{x} + \frac{f_3}{2x^2} + \frac{f_4}{3x^3} + \cdots \qquad (x \to \infty).$$

Next, if a is an arbitrary positive reference point then

$$\int_a^x f(t)\,dt = \left(\int_a^\infty - \int_x^\infty \right) \left\{ f(t) - f_0 - \frac{f_1}{t} \right\} dt + f_0(x-a) + f_1 \ln\left(\frac{x}{a}\right)$$

$$\sim A + f_0 x + f_1 \ln x - \frac{f_2}{x} - \frac{f_3}{2x^2} - \frac{f_4}{3x^3} - \cdots$$

as $x \to \infty$, where

$$A = \int_a^\infty \left\{ f(t) - f_0 - \frac{f_1}{t} \right\} dt - f_0 a - f_1 \ln a.$$

These results can be extended to analytic functions of a complex variable that are holomorphic in, for example, an annular sector. The branch of the logarithm used must be continuous.

8.3 (v) *Differentiation of an asymptotic expansion may be invalid.* For example,† if $f(x) = e^{-x} \sin(e^x)$ and x is real and positive, then

$$f(x) \sim 0 + \frac{0}{x} + \frac{0}{x^2} + \cdots \qquad (x \to \infty).$$

But $f'(x) \equiv \cos(e^x) - e^{-x} \sin(e^x)$ oscillates as $x \to \infty$, and therefore, by Theorem 7.1, has no asymptotic expansion of the form (7.03).

Differentiation *is* legitimate when it is known that $f'(x)$ is continuous and its asymptotic expansion exists. This follows by integration (§8.2) of the assumed expansion of $f'(x)$, and use of the uniqueness property (§7.2).

Another set of circumstances in which differentiation is legitimate occurs when the given function $f(z)$ is an analytic function of the complex variable z. As a consequence of Theorem 4.2, *the asymptotic expansion of $f(z)$ may be differentiated any number of times in any sector that is properly interior to the original sector of validity and has the same vertex.*

8.4 The final operation we consider is *reversion*. This is possible when the variables are real or complex; for illustration we consider a case of the latter.

Let $\zeta(z)$ be holomorphic in a region containing a closed annular sector $\mathbf{S}$ with vertex at the origin and angle less than 2π, and suppose that

$$\zeta(z) \sim z + a_0 + \frac{a_1}{z} + \frac{a_2}{z^2} + \cdots \qquad (z \to \infty \text{ in } \mathbf{S}).$$

Also, let $\mathbf{S}_1$ and $\mathbf{S}_2$ be closed annular sectors with vertices at the origin, $\mathbf{S}_1$ being properly interior to $\mathbf{S}$, and $\mathbf{S}_2$ being properly interior to $\mathbf{S}_1$. Theorem 6.1 shows that when $\zeta \in \mathbf{S}_2$ there is a unique corresponding point z in $\mathbf{S}_1$ (provided that $|\zeta|$ is sufficiently large), and

$$z = \{1 + o(1)\}\zeta \qquad (\zeta \to \infty \text{ in } \mathbf{S}_2).$$

Beginning with this approximation and repeatedly resubstituting in the right-hand side of

$$z = \zeta - a_0 - \frac{a_1}{z} - \frac{a_2}{z^2} - \cdots - \frac{a_{n-1}}{z^{n-1}} + O\left(\frac{1}{z^n}\right),$$

n being an arbitrary integer, we see that there exists a representation of the form

$$z = \zeta - b_0 - \frac{b_1}{\zeta} - \frac{b_2}{\zeta^2} - \cdots - \frac{b_{n-1}}{\zeta^{n-1}} + O\left(\frac{1}{\zeta^n}\right) \qquad (\zeta \to \infty \text{ in } \mathbf{S}_2),$$

† Bromwich (1926, p. 345).

where the coefficients b_s are polynomials in the a_s which are independent of n. This is the required result.

The first four coefficients may be verified to be†

$$b_0 = a_0, \qquad b_1 = a_1, \qquad b_2 = a_0 a_1 + a_2, \qquad b_3 = a_0^2 a_1 + a_1^2 + 2a_0 a_2 + a_3.$$

Ex. 8.1 Let K_n and L_n be the nth implied constants in the asymptotic expansions given in §8.1 for $f(z)$ and $1/f(z)$ respectively, and m the infimum of $|f(z)|$ in **F**. Show that

$$L_n \leqslant m^{-1} \sum_{s=0}^{n-1} |k_s| K_{n-s} \qquad (n \geqslant 1).$$

Ex. 8.2 (*Substitution of asymptotic expansions*) Let

$$f \equiv f(z) \sim \sum_{s=0}^{\infty} f_s z^{-s} \qquad (z \to \infty \text{ in } \mathbf{F}),$$

$$z \equiv z(t) \sim t + \sum_{s=0}^{\infty} b_s t^{-s} \qquad (t \to \infty \text{ in } \mathbf{T}).$$

Show that if the z map of **T** is included in **F**, then f can be expanded in the form

$$f \sim \sum_{s=0}^{\infty} c_s t^{-s} \qquad (t \to \infty \text{ in } \mathbf{T}),$$

where $c_0 = f_0$, $c_1 = f_1$, $c_2 = f_2 - f_1 b_0$, $c_3 = f_3 - 2f_2 b_0 + f_1(b_0^2 - b_1)$.

Ex. 8.3 In the notation of §8.1 assume that $f_0 = 1$. Prove that

$$\ln\{f(z)\} \sim \sum_{s=1}^{\infty} \frac{l_s}{z^s} \qquad (z \to \infty \text{ in } \mathbf{F}),$$

where $l_1 = f_1$ and

$$sl_s = sf_s - (s-1) f_1 l_{s-1} - (s-2) f_2 l_{s-2} - \cdots - f_{s-1} l_1 \qquad (s \geqslant 2).$$

Ex. 8.4 In the notation of §8.1 show that if $f_0 = 1$ and v is a real or complex constant, then

$$\{f(z)\}^v \sim \sum_{s=0}^{\infty} \frac{p_s}{z^s} \qquad (z \to \infty \text{ in } \mathbf{F}),$$

where $p_0 = 1$ and

$$sp_s = (v-s+1) f_1 p_{s-1} + (2v-s+2) f_2 p_{s-2} + \cdots + \{(s-1)v - 1\} f_{s-1} p_1 + svf_s p_0.$$

9 Functions Having Prescribed Asymptotic Expansions

9.1 Let $a_0, a_1, a_2, \ldots$ be an infinite sequence of arbitrary numbers, real or complex, and **R** an unbounded region. Under what conditions does there exist a function having the formal series

$$a_0 + \frac{a_1}{z} + \frac{a_2}{z^2} + \cdots \tag{9.01}$$

† For any s, sb_s is the coefficient of z^{-1} in the asymptotic expansion of $\{\zeta(z)\}^s$ in descending powers of z. This is a consequence of Lagrange's formula for the reversion of power series; see, for example, Copson (1935, §6.23).

as its asymptotic expansion when $z \to \infty$ in $\mathbf{R}$? Somewhat surprisingly, the answer is *none*.

Consider the function

$$f(z) = \sum_{s=0}^{\nu(|z|)} \frac{a_s}{z^s}, \tag{9.02}$$

where $\nu(|z|)$ is the largest integer fulfilling

$$|a_0| + |a_1| + \cdots + |a_{\nu(|z|)}| + \nu(|z|) \leqslant |z|. \tag{9.03}$$

Clearly $\nu(|z|)$ is a nondecreasing function of $|z|$. Let n be an arbitrary positive integer, and

$$z_n = |a_0| + |a_1| + \cdots + |a_{n+1}| + n + 1.$$

If $|z| \geqslant z_n$, then $\nu(|z|) \geqslant n+1$, $|z| > 1$, and

$$\left| f(z) - \sum_{s=0}^{n-1} \frac{a_s}{z^s} \right| = \left| \sum_{s=n}^{\nu(|z|)} \frac{a_s}{z^s} \right| \leqslant \frac{|a_n|}{|z|^n} + \frac{1}{|z|^{n+1}} \sum_{s=n+1}^{\nu(|z|)} |a_s|. \tag{9.04}$$

From (9.03) it can be seen that the right-hand side of (9.04) is bounded by $(|a_n|+1)/|z|^n$; hence (9.01) is the asymptotic expansion of $f(z)$ as $z \to \infty$ in *any* unbounded region.

This solution is not unique. For example, if we change the definition of $\nu(|z|)$ by replacing the right-hand side of (9.03) by $k|z|$, where k is any positive constant, then (9.02) again has (9.01) as its asymptotic expansion. The infinite class of functions having (9.01) as asymptotic expansion is called the *asymptotic sum* of this series in $\mathbf{R}$.

9.2 The function (9.02) is somewhat artificial in the sense that it is discontinuous on an infinite set of circles. We shall now construct an *analytic* function with the desired property. The only restriction is that the range of $\operatorname{ph} z$ is bounded.

We suppose $\mathbf{R}$ to be a closed annular sector $\mathbf{S}$ which, by preliminary translation and rotation of the z plane, can be taken as $|\operatorname{ph} z| \leqslant \sigma$, $|z| \geqslant a$. No restrictions are imposed on the positive numbers σ and a. We shall prove that a suitable function is given by

$$f(z) = \sum_{s=0}^{\infty} \frac{a_s e_s(z)}{z^s}, \tag{9.05}$$

where

$$e_s(z) = 1 - \exp(-z^\rho b^s/|a_s|),$$

ρ and b being any fixed numbers satisfying $0 < \rho < \pi/(2\sigma)$ and $0 < b < a$. If any one of the a_s vanishes, then the corresponding $e_s(z)$ is taken to be unity.

An immediate consequence of the definitions is that

$$|\operatorname{ph}(z^\rho)| = |\rho \operatorname{ph} z| \leqslant \rho\sigma < \tfrac{1}{2}\pi.$$

Therefore,

$$\left| \frac{a_s e_s(z)}{z^s} \right| \leqslant \lambda b^s |z|^{\rho - s} \leqslant \lambda |z|^\rho \left(\frac{b}{a}\right)^s, \tag{9.06}$$

where λ is the supremum of $|(1-e^{-t})/t|$ in the right half of the t plane. Clearly λ is finite. By Weierstrass' M-test the series of analytic functions (9.05) converges uniformly in any compact set in **S**.† Hence $f(z)$ is holomorphic within **S**.

To demonstrate that $f(z)$ has the desired asymptotic expansion, let n be an arbitrary positive integer. Then

$$f(z) - \sum_{s=0}^{n-1} \frac{a_s}{z^s} = -\sum_{s=0}^{n-1} \frac{a_s}{z^s} \exp\left(-\frac{z^\rho b^s}{|a_s|}\right) + \sum_{s=n}^{\infty} \frac{a_s e_s(z)}{z^s}.$$

In consequence of the first of (9.06) the infinite sum is $O(z^{\rho-n})$. The exponential factors in the finite sum on the right-hand side are all of smaller asymptotic order, hence

$$f(z) - \sum_{s=0}^{n-1} \frac{a_s}{z^s} = O\left(\frac{1}{z^{n-\rho}}\right) \qquad (z \to \infty \text{ in } \mathbf{S}).$$

Replacing n by $n+[\rho]+1$, we see that this O term can be strengthened into $O(1/z^n)$. This is the desired result.

Ex. 9.1 Let $\{a_s\}$ be an arbitrary sequence of real or complex numbers, and $\{\alpha_s\}$ an arbitrary sequence of positive numbers such that $\sum \alpha_s$ converges. Also, let the sequence $\{b_s\}$ be defined by $b_0 = a_0$, $b_1 = a_1$, and $b_s = a_s - c_s$ $(s \geqslant 2)$, where c_s is the coefficient of z^{-s} in the expansion of the rational function

$$\sum_{j=1}^{s-1} \frac{b_j \alpha_j}{|b_j| + \alpha_j z} \frac{1}{z^{j-1}}$$

in descending powers of z. Show that in the annular sector $|\operatorname{ph} z| \leqslant \frac{1}{2}\pi$, $|z| \geqslant 1$, the function

$$f(z) = b_0 + \sum_{s=1}^{\infty} \frac{b_s \alpha_s}{|b_s| + \alpha_s z} \frac{1}{z^{s-1}}$$

is holomorphic and

$$f(z) \sim a_0 + \frac{a_1}{z} + \frac{a_2}{z^2} + \cdots \qquad (z \to \infty).$$

10 Generalizations of Poincaré's Definition

10.1 The definition of an asymptotic expansion given in §7.1 may be extended in a number of ways.

In the first place, attention need not be confined to the point at infinity. Similar definitions can be constructed when the variable z tends to any finite point c, say, by replacing z by $(z-c)^{-1}$. Thus, let **R** be a given region having a limit point c (which need not belong to **R**). Suppose that for each fixed n

$$f(z) = a_0 + a_1(z-c) + a_2(z-c)^2 + \cdots + a_{n-1}(z-c)^{n-1} + O\{(z-c)^n\}$$

† *Compact* means bounded and closed.

as $z \to c$ in **R**. Then we write

$$f(z) \sim a_0 + a_1(z-c) + a_2(z-c)^2 + \cdots \qquad (z \to c \text{ in } \mathbf{R}). \tag{10.01}$$

The results of §§7 and 8 carry over straightforwardly to the new definitions.

The point c is called the *distinguished point* of the asymptotic expansion; compare §3.1. In treating first the case $c = \infty$, we have followed historical precedent, and also acknowledged that infinity is the natural distinguished point in many physical applications.

10.2 The next extension is to series other than power series. Again, let **R** be a given point set having c as a finite or infinite limit point. Suppose that $\{\phi_s(z)\}$, $s = 0, 1, \ldots$, is a sequence of functions defined in **R**, such that for every s

$$\phi_{s+1}(z) = o\{\phi_s(z)\} \qquad (z \to c \text{ in } \mathbf{R}). \tag{10.02}$$

Then $\{\phi_s(z)\}$ is said to be an *asymptotic sequence* or *scale*, and the statement

$$f(z) \sim \sum_{s=0}^{\infty} a_s \phi_s(z) \qquad (z \to c \text{ in } \mathbf{R}) \tag{10.03}$$

means that for each nonnegative integer n

$$f(z) = \sum_{s=0}^{n-1} a_s \phi_s(z) + O\{\phi_n(z)\} \qquad (z \to c \text{ in } \mathbf{R}).$$

Many of the properties of ordinary Poincaré expansions hold for expansions of the type (10.03). Exceptions include multiplication and division: it is not always possible to arrange the doubly infinite array $\phi_r(z)\phi_s(z)$ as a single scale.†

10.3 The definition just given is still insufficiently general in many circumstances. For example, the series

$$\frac{\cos x}{x} + \frac{\cos(2x)}{x^2} + \frac{\cos(3x)}{x^3} + \cdots$$

converges uniformly when $x \in [a, \infty)$, provided that $a > 1$, and its leading terms exhibit the essential behavior of its sum as $x \to \infty$. Yet it is excluded because the ratio of any consecutive pair of terms is unbounded as $x \to \infty$. Series of this kind are accommodated by the following definition.

Let $\{\phi_s(z)\}$ be a scale as $z \to c$ in **R**, and $f(z)$, $f_s(z)$, $s = 0, 1, \ldots$, functions such that for each nonnegative integer n

$$f(z) = \sum_{s=0}^{n-1} f_s(z) + O\{\phi_n(z)\} \qquad (z \to c \text{ in } \mathbf{R}). \tag{10.04}$$

Then we say that $\sum f_s(z)$ is a *generalized asymptotic expansion with respect to the scale* $\{\phi_s(z)\}$, and write

$$f(z) \sim \sum_{s=0}^{\infty} f_s(z); \qquad \{\phi_s(z)\} \text{ as } z \to c \text{ in } \mathbf{R}.$$

† Conditions which permit multiplication have been given by Erdélyi (1956a, §1.5).

If $f(z)$, $f_s(z)$, and (possibly) $\phi_s(z)$ are functions of a parameter (or set of parameters) u, and the o and O terms in (10.02) and (10.04) are uniform with respect to u in a point set $\mathbf{U}$, then the generalized expansion is said to hold *uniformly* with respect to u in $\mathbf{U}$.

Great caution needs to be exercised in the manipulation of generalized asymptotic expansions because only a few properties of Poincaré expansions carry over. For example, for a given region $\mathbf{R}$, distinguished point c, and scale $\{\phi_s(z)\}$, a function $f(z)$ has either no generalized expansion or an infinity of such expansions: we have only to rearrange any one of them by including arbitrary multiples of later terms in earlier ones. In consequence, there is no analogue of formula (7.04) for constructing successive terms.

Next, efficacy cannot be judged merely by reference to scale. Suppose, for example, that

$$f(x) \sim \sum_{s=0}^{\infty} \frac{a_s}{x^s}; \qquad \{x^{-s}\} \text{ as } x \to \infty. \tag{10.05}$$

(In other words, we have an ordinary Poincaré expansion.) Simple regrouping of terms produces

$$f(x) \sim \sum_{s=0}^{\infty} \left(\frac{a_{2s}}{x^{2s}} + \frac{a_{2s+1}}{x^{2s+1}} \right); \qquad \{x^{-2s}\} \text{ as } x \to \infty. \tag{10.06}$$

Yet it can hardly be said that (10.06) is more powerful than (10.05), even though its scale diminishes at twice the rate.

Lastly, the definition admits expansions that have no conceivable value, in an analytical or numerical sense, concerning the functions they represent. An example is supplied by

$$\frac{\sin x}{x} \sim \sum_{s=1}^{\infty} \frac{s!\,e^{-(s+1)x/(2s)}}{(\ln x)^s}; \qquad \{(\ln x)^{-s}\} \text{ as } x \to \infty. \tag{10.07}$$

Ex. 10.1 Let $\mathbf{S}$ and $\mathbf{S}_\delta$ denote the sectors $\alpha < \operatorname{ph} z < \beta$ and $\alpha+\delta \leqslant \operatorname{ph} z \leqslant \beta-\delta$, respectively. Show that if $f(z)$ is holomorphic within the intersection of $\mathbf{S}$ with a neighborhood of $z = 0$, and

$$f(z) \sim a_0 + a_1 z + a_2 z^2 + \cdots$$

as $z \to 0$ in $\mathbf{S}_\delta$ for every δ such that $0 < \delta < \frac{1}{2}(\beta-\alpha)$, then $f^{(n)}(z) \to n!\,a_n$ as $z \to 0$ in $\mathbf{S}_\delta$.

Ex. 10.2 By use of Taylor's theorem prove the following converse of Exercise 10.1. Suppose that $f(z)$ is holomorphic within $\mathbf{S}$ for all sufficiently small $|z|$, and, for each n, $\lim\{f^{(n)}(z)\}$ exists uniformly with respect to $\operatorname{ph} z$ as $z \to 0$ in $\mathbf{S}_\delta$. Denoting this limit by $n!\,a_n$, prove that

$$f(z) \sim a_0 + a_1 z + a_2 z^2 + \cdots \qquad (z \to 0 \text{ in } \mathbf{S}_\delta).$$

Ex. 10.3 Let λ be a real constant exceeding unity. With the aid of the preceding exercise and Abel's theorem on the continuity of power series,† prove that

$$\sum_{s=0}^{\infty} \frac{z^s}{\lambda^{\sqrt{s}}} \sim \sum_{n=0}^{\infty} \left\{ \sum_{s=n}^{\infty} \binom{s}{n} \frac{1}{\lambda^{\sqrt{s}}} \right\} (z-1)^n$$

as $z \to 1$ between any two chords of the unit circle that meet at $z = 1$. [Davis, 1953.]

† See, for example, Titchmarsh (1939, §7.61).

Ex. 10.4 Let x be a real variable and $\{\phi_s(x)\}$ a sequence of positive continuous functions that form a scale as x tends to a finite point c. Show that the integrals $\int_c^x \phi_s(t)\,dt$ form a scale as $x \to c$, and that if $f(x)$ is a continuous function having an expansion

$$f(x) \sim \sum a_s \phi_s(x) \qquad (x \to c),$$

then

$$\int_c^x f(t)\,dt \sim \sum a_s \int_c^x \phi_s(t)\,dt \qquad (x \to c).$$

11 Error Analysis; Variational Operator

11.1 In this chapter we have seen how the Poincaré definition supplied an effective analytical meaning to the manipulation of a wide class of formal power series. The definition opened up a new branch of analysis, which has undergone continual development and application since Poincaré's day.

The importance and success of this theory (and its later generalizations) are beyond question, but there is an important drawback: the theory is strictly existential. There is no dependence on, nor information given about, the numerical values of the implied constants. For this reason, following van der Corput (1956), we call the Poincaré theory *pure asymptotics*, to distinguish it from the wider term *asymptotics* which is used to cover all aspects of the development and use of asymptotic approximations and expansions.

In this book we shall be concerned with both pure asymptotics and error analysis. In deriving implied constants frequent use will be made of the *variational operator* $\mathscr{V}$, which we now proceed to define and discuss.

11.2 In the theory of real variables the *variation*, or more fully *total variation*, of a function $f(x)$ over a finite or infinite interval (a, b), is the supremum of

$$\sum_{s=0}^{n-1} |f(x_{s+1}) - f(x_s)|$$

for unbounded n and all possible modes of subdivision

$$x_0 < x_1 < x_2 < \cdots < x_n,$$

with x_0 and x_n in the closure of (a, b). When this supremum is finite $f(x)$ is said to be of *bounded variation* in (a, b), and we denote the supremum by $\mathscr{V}_{x=a,b}\{f(x)\}$, $\mathscr{V}_{a,b}(f)$, or even $\mathscr{V}(f)$, when there is no ambiguity.

11.3 In the case of a compact interval $[a, b]$ one possible mode of subdivision is given by $n = 1$, $x_0 = a$, and $x_1 = b$. Hence

$$\mathscr{V}_{a,b}(f) \geqslant |f(b) - f(a)|.$$

Equality holds when $f(x)$ is monotonic over $[a, b]$:

$$\mathscr{V}_{a,b}(f) = |f(b) - f(a)|. \qquad (11.01)$$

The last relation affords a simple method for calculating the variation of a

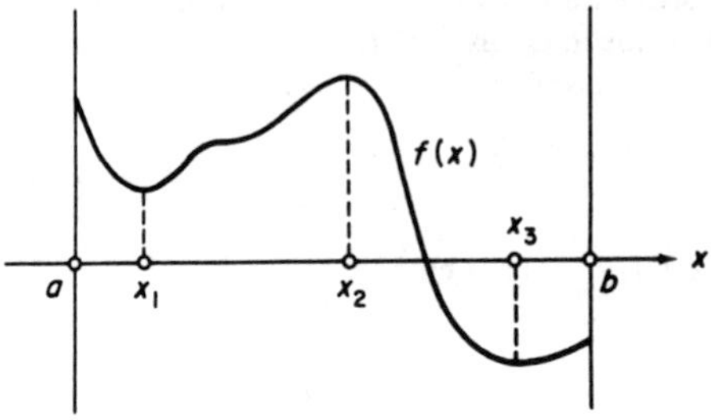

Fig. 11.1 Variation of a continuous function.

continuous function with a finite number of maxima and minima: we subdivide $[a, b]$ at the maxima and minima and apply (11.01) to each subrange. For example, in the case of the function depicted in Fig. 11.1, we see that

$$\begin{aligned}\mathscr{V}_{a,b}(f) &= \{f(a)-f(x_1)\} + \{f(x_2)-f(x_1)\} + \{f(x_2)-f(x_3)\} + \{f(b)-f(x_3)\} \\ &= f(a) - 2f(x_1) + 2f(x_2) - 2f(x_3) + f(b).\end{aligned}$$

When $f(x)$ is *continuously differentiable* in $[a, b]$ application of the mean-value theorem gives

$$\sum_{s=0}^{n-1} |f(x_{s+1})-f(x_s)| = \sum_{s=0}^{n-1} (x_{s+1}-x_s)|f'(\xi_s)| \qquad (x_s < \xi_s < x_{s+1}).$$

Continuity of $f'(x)$ implies that of $|f'(x)|$. Hence from Riemann's definition of an integral

$$\mathscr{V}_{a,b}(f) = \int_a^b |f'(x)|\, dx. \tag{11.02}$$

11.4 Suppose now that the interval (a, b) is finite or infinite, $f(x)$ is continuous in the closure of (a, b), $f'(x)$ is continuous within (a, b), and $|f'(x)|$ is integrable over (a, b). Using the subdivision points of §11.2 and the result of §11.3, we have

$$\mathscr{V}_{a,b}(f) \geqslant \mathscr{V}_{x_1, x_{n-1}}(f) = \int_{x_1}^{x_{n-1}} |f'(x)|\, dx.$$

Since x_1 and x_{n-1} are arbitrary points in (a, b) this result implies

$$\mathscr{V}_{a,b}(f) \geqslant \int_a^b |f'(x)|\, dx. \tag{11.03}$$

We also have

$$\sum_{s=0}^{n-1} |f(x_{s+1})-f(x_s)| = \sum_{s=0}^{n-1} \left| \int_{x_s}^{x_{s+1}} f'(x)\, dx \right| \leqslant \int_a^b |f'(x)|\, dx,$$

implying that (11.03) holds with the $\geqslant$ sign reversed. Therefore (11.02) again applies.

11.5 So far it has been assumed that $f(x)$ is real. If $f(x)$ is a complex function of the real variable x, then its variation is *defined* by (11.02) whenever this integral converges.

Suppose, for example, that $f(z)$ is a holomorphic function of z in a complex domain **D**.† Suppose also that **D** contains a *path* (or *contour*) $\mathscr{P}$, that is, a finite chain of regular (or smooth) arcs each having an equation of the form

$$z = z(\tau) \qquad (\alpha < \tau < \beta),$$

in which τ is the arc parameter and $z'(\tau)$ is continuous and nonvanishing in the closure of (α, β). Then

$$\mathscr{V}_{\mathscr{P}}(f) = \sum \int_{\alpha}^{\beta} |f'\{z(\tau)\} z'(\tau)| \, d\tau.$$

For a given pair of endpoints, the variation of $f(z)$ obviously depends on the path selected, quite unlike the integral of $f(z)$.

Ex. 11.1 Show that

$$\mathscr{V}(f+g) \leqslant \mathscr{V}(f) + \mathscr{V}(g), \qquad \mathscr{V}(f) \geqslant \mathscr{V}(|f|).$$

Show also that equality holds in the second relation when f is real and continuous.

Ex. 11.2 Evaluate

(i) $\mathscr{V}_{0,1}\{\sin^2(n\pi x)\}$, where n is an integer.
(ii) $\mathscr{V}_{-1,1}(f)$, where f is the step function defined by $f = 0$ $(x < 0)$, $f = \frac{1}{2}$ $(x = 0)$, and $f = 1$ $(x > 0)$.
(iii) $\mathscr{V}_{-\infty,\infty}(\operatorname{sech} x)$.
(iv) $\mathscr{V}_{x=0,\infty}\{\int_0^x (t-1) e^{-t} dt\}$.

Ex. 11.3 Evaluate $\mathscr{V}_{-1,1}(e^{iz})$: (i) along the join of -1 and 1; (ii) around the other three sides of the square having vertices at -1, 1, $1+2i$, $-1+2i$; (iii) around the path conjugate to (ii).

Ex. 11.4 In the notation of §11.5 let $\mathscr{P}$ be subdivided at the points $z_0, z_1, z_2, \ldots, z_n$, arranged in order. Show that

$$\mathscr{V}_{\mathscr{P}}(f) = \sup \sum_{s=0}^{n-1} |f(z_{s+1}) - f(z_s)|,$$

for all n and all possible modes of subdivision, provided that $z''(\tau)$ is continuous on each arc of $\mathscr{P}$.

Historical Notes and Additional References

§1.4 Historical details in this subsection were obtained from Bromwich (1926, §104). Further information is contained in this reference.

§§4–6 For further results concerning integration and differentiation of asymptotic and order relations, and the asymptotic solution of transcendental equations see de Bruijn (1961), Berg (1968), Dieudonné (1968), and Riekstiņš (1968). The result given by Theorem 6.1 may not have been stated quite so explicitly before.

§9 The constructions in §9.1, §9.2, and Exercise 9.1 are due to van der Corput (1956, Theorem 4.1), Ritt (1916), and Carleman (1926, Chapter 5), respectively. Accounts of further constructions have

† That is, an open point set any two members of which can be connected either by a finite chain of overlapping disks belonging to the set, or, equivalently, by a polygonal arc lying in the set. When the boundary points are added the domain is said to be *closed*; but unless specified otherwise a domain is assumed to be open.

been given by Davis (1953) and Pittnauer (1969). For discussions of the uniqueness problem see Watson (1911) and Davis (1957). Although such results are of great theoretical interest, practical applications are rare.

§10.3 This generalization is due to Schmidt (1937). For further generalizations of the definition of an asymptotic expansion see Erdélyi and Wyman (1963) and Riekstiņš (1966). The example (10.07) is taken from the last reference.

2

INTRODUCTION TO SPECIAL FUNCTIONS

1 The Gamma Function

1.1 The *Gamma function* originated as the solution of an interpolation problem for the factorial function. Can a function $\Gamma(x)$ be found which has continuous derivatives of all orders in $[1, \infty)$, and the properties $\Gamma(1) = 1$, $\Gamma(x+1) = x\Gamma(x)$? The answer is affirmative; indeed supplementary conditions are needed to make $\Gamma(x)$ unique. We shall not pursue the formulation of these conditions, because a simpler starting point for our purpose is *Euler's integral*†

$$\Gamma(z) = \int_0^\infty e^{-t} t^{z-1}\, dt \qquad (\operatorname{Re} z > 0), \tag{1.01}$$

in which the path of integration is the real axis and t^{z-1} has its principal value.

If δ and Δ are arbitrary positive constants and $\delta \leqslant \operatorname{Re} z \leqslant \Delta$, then

$$|t^{z-1}| \leqslant t^{\delta-1} \quad (0 < t \leqslant 1), \qquad |t^{z-1}| \leqslant t^{\Delta-1} \quad (t \geqslant 1).$$

Hence by Weierstrass' M-test the integral (1.01) converges uniformly with respect to z in this strip. That $\Gamma(z)$ is holomorphic in the half-plane $\operatorname{Re} z > 0$ is a consequence of this result and the following theorem.

Theorem 1.1‡ *Let t be a real variable ranging over a finite or infinite interval* (a, b) *and z a complex variable ranging over a domain* **D**. *Assume that the function $f(z, t)$ satisfies the following conditions:*

(i) *$f(z, t)$ is a continuous function of both variables.*
(ii) *For each fixed value of t, $f(z, t)$ is a holomorphic function of z.*
(iii) *The integral*

$$F(z) = \int_a^b f(z, t)\, dt$$

converges uniformly at both limits in any compact set in **D**.

† More fully, *Euler's integral of the second kind. Euler's integral of the first kind* is given by (1.11) below.

‡ This is an extension to complex variables of a standard theorem concerning differentiation of an infinite integral with respect to a parameter; for proofs see, for example, Levinson and Redheffer (1970, Chapter 6) or Copson (1935, §5.51).

Then $F(z)$ is holomorphic in **D**, *and its derivatives of all orders may be found by differentiating under the sign of integration.*

1.2 When $z = n$, a positive integer, (1.01) can be evaluated by repeated partial integrations. This gives

$$\Gamma(n) = (n-1)! \qquad (n = 1, 2, \ldots). \tag{1.02}$$

But for general values of z the integral cannot be evaluated in closed form in terms of elementary functions.

A single partial integration of (1.01) produces the *fundamental recurrence formula*

$$\Gamma(z+1) = z\Gamma(z). \tag{1.03}$$

This formula is invaluable for numerical purposes, and it also enables $\Gamma(z)$ to be continued analytically strip by strip into the left half-plane. The only points at which $\Gamma(z)$ remains undefined are $0, -1, -2, \ldots$. These are the singularities of $\Gamma(z)$.

To determine the nature of the singularities we have from Taylor's theorem

$$\Gamma(z+1) = 1 + zf(z),$$

where $f(z)$ is holomorphic in the neighborhood of $z = 0$. Hence

$$\Gamma(z) = \frac{1}{z}\Gamma(z+1) = \frac{1}{z} + f(z).$$

Thus $z = 0$ is a simple pole of residue 1. More generally, if n is any positive integer, then with the aid of the Binomial theorem we see that

$$\Gamma(z-n) = \frac{1+zf(z)}{z(z-1)\cdots(z-n)} = \frac{(-)^n}{n!z}\{1+zf(z)\}\{1+zg(z)\},$$

where $g(z)$ is analytic at $z = 0$. Therefore *the only singularities of* $\Gamma(z)$ *are simple poles at* $z = 0, -1, -2, \ldots$, *the residue at* $z = -n$ *being* $(-1)^n/n!$.

1.3 An alternative definition of $\Gamma(z)$, which is not restricted to the half-plane $\operatorname{Re} z > 0$, can be derived from (1.01) in the following way. We have

$$\lim_{n\to\infty}\left(1 - \frac{t}{n}\right)^n = e^{-t}.$$

This suggests that we consider the limiting behavior of the integral

$$\Gamma_n(z) = \int_0^n \left(1 - \frac{t}{n}\right)^n t^{z-1}\,dt \qquad (\operatorname{Re} z > 0),$$

as $n \to \infty$, z being fixed.

First, we evaluate $\Gamma_n(z)$ in the case when n is a positive integer. Repeated partial integrations produce

$$\Gamma_n(z) = \frac{1}{z}\frac{n-1}{(z+1)n}\frac{n-2}{(z+2)n}\cdots\frac{1}{(z+n-1)n}\int_0^n t^{z+n-1}\,dt = \frac{n!n^z}{z(z+1)\cdots(z+n)}. \tag{1.04}$$

Next, we prove that the limit of $\Gamma_n(z)$ as $n \to \infty$ is $\Gamma(z)$. Write

$$\Gamma(z) - \Gamma_n(z) = I_1 + I_2 + I_3,$$

where

$$I_1 = \int_n^\infty e^{-t} t^{z-1}\,dt, \qquad I_2 = \int_0^{n/2} \left\{ e^{-t} - \left(1 - \frac{t}{n}\right)^n \right\} t^{z-1}\,dt,$$

$$I_3 = \int_{n/2}^{n} \left\{ e^{-t} - \left(1 - \frac{t}{n}\right)^n \right\} t^{z-1}\,dt.$$

Clearly $I_1 \to 0$ as $n \to \infty$. For I_2 and I_3 we have, when $t \in [0, n)$,

$$\ln\left\{\left(1 - \frac{t}{n}\right)^n\right\} = n \ln\left(1 - \frac{t}{n}\right) = -t - T,$$

where

$$T = \frac{t^2}{2n} + \frac{t^3}{3n^2} + \frac{t^4}{4n^3} + \cdots.$$

Hence

$$\left(1 - \frac{t}{n}\right)^n = e^{-t-T} \leqslant e^{-t},$$

since $T \geqslant 0$. Accordingly,

$$|I_3| \leqslant \int_{n/2}^{n} e^{-t} t^{\operatorname{Re} z - 1}\,dt \to 0 \qquad (n \to \infty).$$

For I_2, $t/n \leqslant \frac{1}{2}$. Hence $T \leqslant ct^2/n$, where

$$c = \frac{1}{2 \cdot 1} + \frac{1}{3 \cdot 2} + \frac{1}{4 \cdot 2^2} + \cdots$$

and is finite. In consequence

$$0 \leqslant e^{-t} - \left(1 - \frac{t}{n}\right)^n = e^{-t}(1 - e^{-T}) \leqslant e^{-t} T \leqslant e^{-t} \frac{ct^2}{n},$$

and

$$|I_2| \leqslant \frac{c}{n} \int_0^{n/2} e^{-t} t^{\operatorname{Re} z + 1}\,dt \to 0 \qquad (n \to \infty).$$

Thus we have *Euler's limit formula*

$$\Gamma(z) = \lim_{n \to \infty} \frac{n!\,n^z}{z(z+1)(z+2)\cdots(z+n)}. \tag{1.05}$$

The condition $\operatorname{Re} z > 0$ assumed in the proof can be eased to $z \neq 0, -1, -2, \ldots,$ by use of the recurrence formula (1.03) in the following way. If $\operatorname{Re} z \in (-m, -m+1]$, where m is an arbitrary fixed positive integer, then

$$\Gamma(z) = \frac{\Gamma(z+m)}{z(z+1)\cdots(z+m-1)} = \frac{1}{z(z+1)\cdots(z+m-1)} \lim_{n\to\infty} \frac{(n-m)!(n-m)^{z+m}}{(z+m)(z+m+1)\cdots(z+n)}$$

$$= \lim_{n\to\infty} \frac{n!\,n^z}{z(z+1)\cdots(z+n)}.$$

1.4 In order to cast Euler's limit formula into the standard, or canonical, form of an infinite product, we need the following:

Lemma 1.1 *The sequence of numbers*

$$u_n = 1 + \frac{1}{2} + \frac{1}{3} + \cdots + \frac{1}{n} - \ln n \qquad (n = 1, 2, 3, \ldots)$$

tends to a finite limit as $n \to \infty$.

Since t^{-1} is decreasing, we have for $n \geqslant 2$

$$\frac{1}{2} + \frac{1}{3} + \cdots + \frac{1}{n} < \int_1^n \frac{dt}{t} < 1 + \frac{1}{2} + \frac{1}{3} + \cdots + \frac{1}{n-1}.$$

Therefore $1/n < u_n < 1$. Next,

$$u_{n+1} - u_n = \frac{1}{n+1} + \ln\left(1 - \frac{1}{n+1}\right) < 0.$$

Accordingly, $\{u_n\}$ is a sequence of decreasing positive numbers, and the lemma follows.

The limiting value of u_n is called *Euler's constant* and is usually denoted by γ. From the proof just given it is seen that $0 \leqslant \gamma < 1$. Numerical computations give, to ten decimal places,

$$\gamma = 0.57721\ 56649.$$

Now assume, temporarily, that $z \neq 0, -1, -2, \ldots$. Then from (1.04) we have identically

$$\frac{1}{\Gamma_n(z)} = z \exp\left\{z\left(1 + \frac{1}{2} + \frac{1}{3} + \cdots + \frac{1}{n} - \ln n\right)\right\} \prod_{s=1}^{n} \left\{\left(1 + \frac{z}{s}\right) e^{-z/s}\right\}.$$

Letting $n \to \infty$, we obtain the required infinite product in the form

$$\frac{1}{\Gamma(z)} = z e^{\gamma z} \prod_{s=1}^{\infty} \left\{\left(1 + \frac{z}{s}\right) e^{-z/s}\right\}. \qquad (1.06)$$

This result also holds when z is zero or a negative integer because both sides vanish at these points.

By taking logarithms it is easily seen that the right-hand side of (1.06) converges uniformly in any compact domain that excludes $z = 0, -1, -2, \ldots$. It therefore represents a holomorphic function in this domain. We have already shown that at the exceptional points $\Gamma(z)$ has simple poles, hence $1/\Gamma(z)$ is holomorphic in their neighborhoods. Therefore $1/\Gamma(z)$ *is an entire function*. As a corollary, $\Gamma(z)$ *has no zeros*.

1.5 Two important identities which are easy to verify by means of Euler's limit formula are

$$\Gamma(z)\Gamma(1-z) = \frac{\pi}{\sin \pi z} \qquad (z \neq 0, \pm 1, \pm 2, \ldots), \tag{1.07}$$

and

$$\Gamma(2z) = \frac{2^{2z-1}}{\pi^{1/2}}\Gamma(z)\Gamma(z+\tfrac{1}{2}) \qquad (2z \neq 0, -1, -2, \ldots). \tag{1.08}$$

In the case of (1.07), we have

$$\frac{1}{\Gamma(z)\Gamma(1-z)} = \lim_{n\to\infty}\left\{\frac{z(z+1)\cdots(z+n)}{n!\,n^z}\,\frac{(1-z)(2-z)\cdots(n+1-z)}{n!\,n^{1-z}}\right\}$$

$$= z\prod_{s=1}^{\infty}\left(1-\frac{z^2}{s^2}\right) = \frac{\sin \pi z}{\pi}.$$

As an immediate deduction

$$\Gamma(\tfrac{1}{2}) = \pi^{1/2}, \tag{1.09}$$

the possibility $-\pi^{1/2}$ being ruled out by reference to (1.01) or (1.05).

Next, in the case of (1.08) we have

$$\frac{2^{2z}\Gamma(z)\Gamma(z+\frac{1}{2})}{\Gamma(2z)} = \lim_{n\to\infty}\left\{2^{2z}\frac{n!\,n^z}{z(z+1)\cdots(z+n)}\,\frac{n!\,n^{z+(1/2)}}{(z+\frac{1}{2})(z+\frac{3}{2})\cdots(z+n+\frac{1}{2})}\right.$$

$$\left.\times\,\frac{2z(2z+1)\cdots(2z+2n)}{(2n)!\,(2n)^{2z}}\right\}$$

$$= \lim_{n\to\infty}\left\{\frac{(n!)^2\,2^{2n+1}}{(2n)!\,n^{1/2}}\right\}.$$

The last quantity is independent of z, and must be finite since the left-hand side exists. The value $2\pi^{1/2}$ is found by setting $z = \frac{1}{2}$ on the left and referring to (1.09). The relation (1.08) then follows.

Equation (1.07) is called the *reflection formula* and equation (1.08) the *duplication* or *multiplication formula*. The reflection formula enables properties of the Gamma function of negative argument (or, more generally, argument with negative real part) to be obtained readily from those for positive argument (or argument with positive real part). A generalization of the duplication formula is given in Chapter 8, Exercise 4.1.

The graph of $\Gamma(x)$, for real values of x, is indicated in Fig. 1.1.

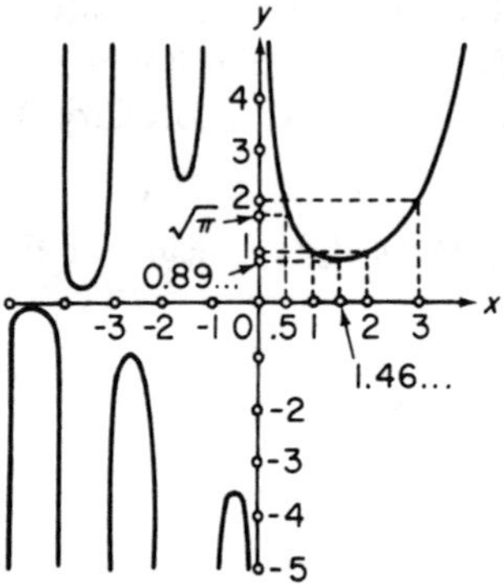

Fig. 1.1 Gamma function. $y = \Gamma(x)$.

1.6 The next formula concerns the product of two Gamma functions $\Gamma(p)\Gamma(q)$. At first we suppose that $p \geqslant 1$ and $q \geqslant 1$. From (1.01)

$$\begin{aligned}\Gamma(p)\Gamma(q) &= \lim_{R\to\infty}\left\{\left(\int_0^R e^{-y}y^{p-1}\,dy\right)\left(\int_0^R e^{-x}x^{q-1}\,dx\right)\right\}\\ &= \lim_{R\to\infty}\iint_{S_R} e^{-x-y}x^{q-1}y^{p-1}\,dx\,dy,\end{aligned}$$

where S_R denotes the square $x, y \in [0, R]$. The repeated integral equals the double integral since the integrand is continuous in both variables. Now let T_R denote the triangle bounded by the axes and the line $x+y = R$. Clearly,

$$\iint_{S_{R/2}} < \iint_{T_R} < \iint_{S_R}.$$

Since the integrals over $S_{R/2}$ and S_R have the same limiting value, it follows that

$$\Gamma(p)\Gamma(q) = \lim_{R\to\infty}\iint_{T_R} e^{-x-y}x^{q-1}y^{p-1}\,dx\,dy.$$

We transform to new variables u and v, given by

$$x + y = u, \qquad y = uv.$$

In the x, y plane the lines of constant u parallel the hypotenuse of T_R. And since $y/x = v/(1-v)$, lines of constant v are rays through the origin. The Jacobian $\partial(x,y)/\partial(u,v)$ equals u. Hence the transformation yields

$$\Gamma(p)\Gamma(q) = \lim_{R\to\infty}\left\{\left(\int_0^R e^{-u}u^{p+q-1}\,du\right)\left(\int_0^1 v^{p-1}(1-v)^{q-1}\,dv\right)\right\},$$

that is,

$$\Gamma(p)\Gamma(q) = \Gamma(p+q)\int_0^1 v^{p-1}(1-v)^{q-1}\,dv. \tag{1.10}$$

This is the required formula. The restrictions $p \geqslant 1$ and $q \geqslant 1$ may now be eased in the following way. The left-hand side of (1.10) is holomorphic in p when $\operatorname{Re} p > 0$ and holomorphic in q when $\operatorname{Re} q > 0$. Reference to Theorem 1.1 shows that the same is true of the right-hand side. Hence by analytic continuation with respect to p, and then q, the regions of validity of (1.10) are extended to $\operatorname{Re} p > 0$ and $\operatorname{Re} q > 0$.

The integral

$$\mathrm{B}(p,q) = \int_0^1 v^{p-1}(1-v)^{q-1}\,dv \qquad (\operatorname{Re} p > 0, \quad \operatorname{Re} q > 0) \tag{1.11}$$

is called the *Beta function*. In this notation (1.10) becomes

$$\mathrm{B}(p,q) = \frac{\Gamma(p)\Gamma(q)}{\Gamma(p+q)}.$$

By confining the proof of the required formula to positive real values of the parameters and then appealing to analytic continuation, we avoided possible complications in handling the case of complex parameters directly. This powerful artifice is of frequent use in establishing transformations for special functions.

1.7 The final formula for the Gamma function in this section is an integral representation valid for *unrestricted* z. It is constructed by using a loop contour in the complex plane instead of the straight-line path of (1.01). The idea is due to Hankel (1864) and is applicable to many similar integrals.

Consider

$$I(z) = \int_{-\infty}^{(0+)} e^t t^{-z}\,dt,$$

where the notation means that the path begins at $t = -\infty$, encircles $t = 0$ once in a positive sense, and returns to its starting point; see Fig. 1.2. We suppose that the branch of t^{-z} takes its principal value at the point (or points) where the contour crosses the positive real axis, and is continuous elsewhere. For a given choice of path, the integral converges uniformly with respect to z in any compact set, by the M-test. By taking the arc parameter of the path as integration variable and applying Theorem 1.1 it is seen that $I(z)$ is an entire function of z.

Let r be any positive number. Then by Cauchy's theorem the path can be deformed into the two sides of the interval $(-\infty, -r]$, together with the circle $|t| = r$; see

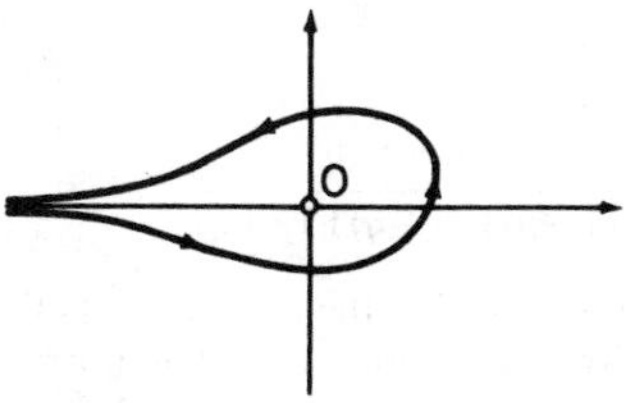

Fig. 1.2 t plane. Contour for Hankel's loop integral.

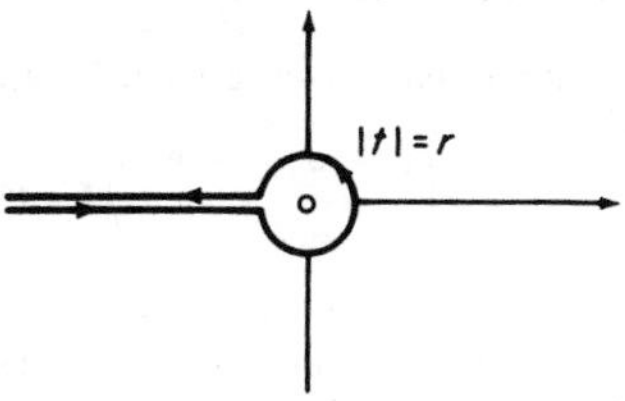

Fig. 1.3 t plane.

Fig. 1.3. Suppose temporarily that z is fixed and $\operatorname{Re} z < 1$. Then as $r \to 0$, the contribution to the integral from the circle vanishes. On the lower side of the negative real axis $\operatorname{ph} t = -\pi$, and on the upper side $\operatorname{ph} t = \pi$. Writing $\tau = |t|$, we obtain

$$I(z) = -\int_{\infty}^{0} e^{-\tau}\tau^{-z}e^{i\pi z}\,d\tau - \int_{0}^{\infty} e^{-\tau}\tau^{-z}e^{-i\pi z}\,d\tau = 2i\sin(\pi z)\Gamma(1-z) = 2\pi i/\Gamma(z);$$

compare (1.07). On returning to the original path we have

$$\frac{1}{\Gamma(z)} = \frac{1}{2\pi i}\int_{-\infty}^{(0+)} e^{t}t^{-z}\,dt. \tag{1.12}$$

This is *Hankel's loop integral*. Analytic continuation removes the temporary restriction on $\operatorname{Re} z$, provided that the branch of t^{-z} is chosen in the manner specified in the second paragraph of this subsection.

Ex. 1.1 Show that when $\operatorname{Re} \nu > 0$, $\mu > 0$, and $\operatorname{Re} z > 0$,

$$\int_0^\infty \exp(-zt^\mu)\,t^{\nu-1}\,dt = \frac{1}{\mu}\Gamma\left(\frac{\nu}{\mu}\right)\frac{1}{z^{\nu/\mu}},$$

where fractional powers have their principal values.

Ex. 1.2 If y is real and nonzero show that

$$|\Gamma(iy)| = \left(\frac{\pi}{y\sinh \pi y}\right)^{1/2}.$$

Ex. 1.3 When $\operatorname{Re} p > 0$ and $\operatorname{Re} q > 0$ show that

$$\mathrm{B}(p,q) = 2\int_0^{\pi/2} \sin^{2p-1}\theta\cos^{2q-1}\theta\,d\theta = \int_0^\infty \frac{t^{p-1}\,dt}{(1+t)^{p+q}}.$$

Ex. 1.4 If x and y are real show that

$$\left|\frac{\Gamma(x)}{\Gamma(x+iy)}\right|^2 = \prod_{s=0}^{\infty}\left\{1 + \frac{y^2}{(x+s)^2}\right\} \qquad (x \neq 0, -1, -2, \ldots),$$

and thence that $|\Gamma(x+iy)| \leqslant |\Gamma(x)|$.

Ex. 1.5 Prove that

$$\prod_{s=1}^{\infty}\frac{s(a+b+s)}{(a+s)(b+s)} = \frac{\Gamma(a+1)\Gamma(b+1)}{\Gamma(a+b+1)},$$

provided that neither a nor b is a negative integer.

Ex. 1.6 Show that for unrestricted p and q

$$\int_a^{(1+,0+,1-,0-)} v^{p-1}(1-v)^{q-1}\,dv = -\frac{4\pi^2 e^{\pi i(p+q)}}{\Gamma(1-p)\Gamma(1-q)\Gamma(p+q)}.$$

Here a is any point of the interval $(0,1)$, and the notation means that the integration path begins at a, encircles $v = 1$ once in the positive sense and returns to a without encircling $v = 0$, then encircles $v = 0$ once in the positive sense and returns to a without encircling $v = 1$, and so on. The factors in the integrand are assumed to be continuous on the path and take their principal values at the beginning. [Pochhammer, 1890.]

2 The Psi Function

2.1 The logarithmic derivative of the Gamma function is usually denoted by

$$\psi(z) = \Gamma'(z)/\Gamma(z).$$

Most of its properties follow straightforwardly from corresponding properties of the Gamma function. For example, the only singularities of $\psi(z)$ are simple poles of residue -1 at the points $z = 0, -1, -2, \ldots$.

Sometimes $\psi(z)$ is called the *Digamma function*, and its successive derivatives $\psi'(z)$, $\psi''(z), \ldots$, *the Trigamma function*, *Tetragamma function*, and so on.

The graph of $\psi(x)$, for real values of x, is indicated in Fig. 2.1.

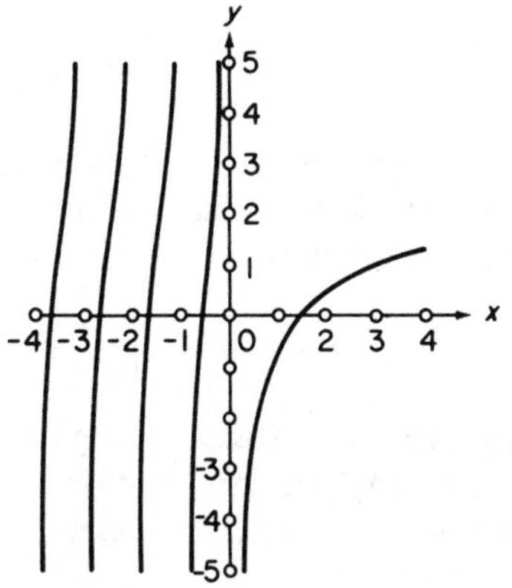

Fig. 2.1 Psi function. $y = \psi(x)$.

Ex. 2.1 Show that unless $z = 0, -1, -2, \ldots$,

$$\psi(z) = \psi(z+1) - (1/z) = \psi(1-z) - \pi \cot \pi z = \tfrac{1}{2}\psi(\tfrac{1}{2}z) + \tfrac{1}{2}\psi(\tfrac{1}{2}z+\tfrac{1}{2}) + \ln 2.$$

Ex. 2.2 Show that

$$\psi(z) = -\gamma - \frac{1}{z} + \sum_{s=1}^{\infty}\left(\frac{1}{s} - \frac{1}{s+z}\right) \qquad (z \neq 0, -1, -2, \ldots),$$

and thence that

$$\Gamma'(1) = \psi(1) = -\gamma, \qquad \psi(n) = -\gamma + \sum_{s=1}^{n-1}\frac{1}{s} \qquad (n = 2, 3, \ldots).$$

Ex. 2.3 From the preceding exercises derive $\psi(\frac{1}{2}) = -\gamma - 2\ln 2$.

Ex. 2.4 Prove that

$$\psi'(z) = \sum_{s=0}^{\infty}\frac{1}{(s+z)^2} \qquad (z \neq 0, -1, -2, \ldots).$$

Deduce that when z is real and positive $\Gamma(z)$ has a single minimum, which lies between 1 and 2.

Ex. 2.5 If y is real, show that

$$\sum_{s=1}^{\infty}\frac{y}{s^2+y^2} = \operatorname{Im}\{\psi(1+iy)\}.$$

Ex. 2.6 Verify that each of the following expressions equals γ:

$$-\int_0^\infty e^{-t}\ln t\,dt,\qquad \int_0^1 (1-e^{-t})\frac{dt}{t}-\int_1^\infty e^{-t}\frac{dt}{t},\qquad \int_0^\infty\left(\frac{e^{-t}}{1-e^{-t}}-\frac{e^{-t}}{t}\right)dt.$$

Deduce that $\gamma > 0$.

Ex. 2.7 By means of Exercises 2.2 and 2.6 establish Gauss's formula†

$$\psi(z)=\int_0^\infty\left(\frac{e^{-t}}{t}-\frac{e^{-zt}}{1-e^{-t}}\right)dt\qquad(\operatorname{Re} z>0).$$

3 Exponential, Logarithmic, Sine, and Cosine Integrals

3.1 The *exponential integral* is defined by

$$E_1(z)=\int_z^\infty \frac{e^{-t}}{t}\,dt. \tag{3.01}$$

The point $t=0$ is a pole of the integrand, hence $z=0$ is a branch point of $E_1(z)$. The principal branch is obtained by introducing a cut along the negative real axis.

An integral representation with a fixed path is provided by

$$E_1(z)=e^{-z}\int_0^\infty \frac{e^{-zt}}{1+t}\,dt\qquad(|\operatorname{ph} z|<\tfrac{1}{2}\pi). \tag{3.02}$$

This is easily proved by transforming variables when z is positive, and extending to $|\operatorname{ph} z|<\frac{1}{2}\pi$ by analytic continuation.

The *complementary exponential integral* is defined by

$$\operatorname{Ein}(z)=\int_0^z \frac{1-e^{-t}}{t}\,dt, \tag{3.03}$$

and is entire. By expanding the integrand in ascending powers of t and integrating term by term we obtain the Maclaurin expansion

$$\operatorname{Ein}(z)=\sum_{s=1}^\infty \frac{(-)^{s-1}}{s}\frac{z^s}{s!}. \tag{3.04}$$

The connection between $E_1(z)$ and $\operatorname{Ein}(z)$ is found by temporarily supposing that $z>0$ and rearranging (3.03) in the form

$$\operatorname{Ein}(z)=\int_0^1 \frac{1-e^{-t}}{t}\,dt+\ln z-\int_1^\infty \frac{e^{-t}}{t}\,dt+\int_z^\infty \frac{e^{-t}}{t}\,dt.$$

Referring to Exercise 2.6, we see that

$$\operatorname{Ein}(z)=E_1(z)+\ln z+\gamma. \tag{3.05}$$

Combination with (3.04) then yields

$$E_1(z)=-\ln z-\gamma+\sum_{s=1}^\infty \frac{(-)^{s-1}}{s}\frac{z^s}{s!}. \tag{3.06}$$

† Note that this integral representation involves only *single-valued* functions.

Analytic continuation immediately extends (3.05) and (3.06) to complex z. In both cases principal branches of $E_1(z)$ and $\ln z$ correspond.

3.2 When $z = x$ and is real, another notation often used for the exponential integral is given by

$$\operatorname{Ei}(x) = \int_{-\infty}^{x} \frac{e^t}{t}\,dt \qquad (x \neq 0), \tag{3.07}$$

it being understood that the integral takes its Cauchy principal value when x is positive.† The connection with the previous notation is given by

$$E_1(x) = -\operatorname{Ei}(-x), \qquad E_1(-x \pm i0) = -\operatorname{Ei}(x) \mp i\pi, \tag{3.08}$$

with $x > 0$ in both relations. These identities are obtained by replacing t by $-t$ and, in the case of the second one, using a contour with a vanishingly small indentation. The notation $E_1(-x+i0)$, for example, means the value of the principal branch of $E_1(-x)$ on the upper side of the cut.

A related function is the *logarithmic integral*, defined for positive x by

$$\operatorname{li}(x) = \int_0^x \frac{dt}{\ln t} \qquad (x \neq 1), \tag{3.09}$$

the Cauchy principal value being taken when $x > 1$. By transformation of integration variable we find that

$$\operatorname{li}(x) = \operatorname{Ei}(\ln x) \qquad (0 < x < 1 \quad \text{or} \quad 1 < x < \infty). \tag{3.10}$$

3.3 The *sine integrals* are defined by

$$\operatorname{Si}(z) = \int_0^z \frac{\sin t}{t}\,dt, \qquad \operatorname{si}(z) = -\int_z^\infty \frac{\sin t}{t}\,dt. \tag{3.11}$$

Each is entire. To relate them we need the following result:

Lemma 3.1

$$\int_0^\infty \frac{\sin t}{t}\,dt = \tfrac{1}{2}\pi. \tag{3.12}$$

This formula can be established by integrating e^{it}/t around the contour of Fig. 3.1, as follows. On the small semicircle $t = re^{i\theta}$, $\pi \geq \theta \geq 0$, we have

$$\int \frac{e^{it}}{t}\,dt = i\int_\pi^0 \exp(ire^{i\theta})\,d\theta \to -i\pi \qquad (r \to 0).$$

Fig. 3.1 t plane.

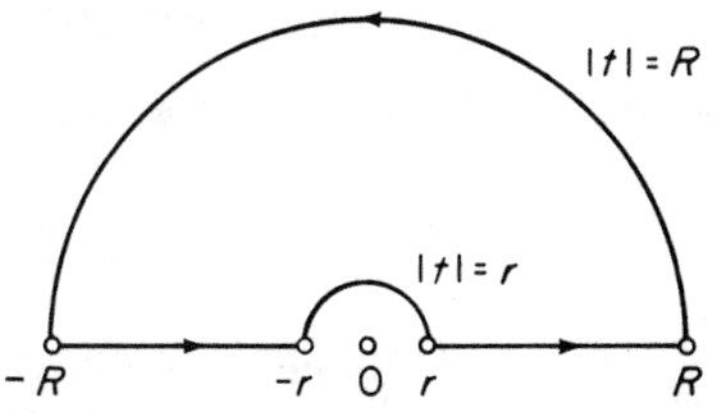

† That is, $\lim_{\delta \to 0+}\left(\int_{-\infty}^{-\delta} + \int_{\delta}^{x}\right)$.

On the large semicircle $t = Re^{i\theta}$, $0 \leqslant \theta \leqslant \pi$, we utilize *Jordan's inequality*:

$$\sin\theta \geqslant 2\theta/\pi \qquad (0 \leqslant \theta \leqslant \tfrac{1}{2}\pi). \tag{3.13}$$

Thus

$$\left|\int \frac{e^{it}}{t}\,dt\right| = \left|\int_0^{\pi} \exp(iRe^{i\theta})\,d\theta\right| \leqslant 2\int_0^{\pi/2} e^{-R\sin\theta}\,d\theta$$

$$\leqslant 2\int_0^{\pi/2} e^{-2R\theta/\pi}\,d\theta = \frac{\pi}{R}(1-e^{-R}) \to 0 \qquad (R\to\infty).$$

Equation (3.12) is obtained on using Cauchy's theorem and taking imaginary parts.

From (3.11) and (3.12) we derive

$$\mathrm{Si}(z) = \tfrac{1}{2}\pi + \mathrm{si}(z). \tag{3.14}$$

$\mathrm{Si}(z)$ may be expressed in terms of the complementary exponential integral by changing the integration variable in (3.11) from t to it; thus

$$2i\,\mathrm{Si}(z) = \mathrm{Ein}(iz) - \mathrm{Ein}(-iz). \tag{3.15}$$

Then by using (3.05) and (3.14), we obtain

$$2i\,\mathrm{si}(z) = E_1(iz) - E_1(-iz). \tag{3.16}$$

In the last relation the branches of $E_1(iz)$ and $E_1(-iz)$ take their principal values when z is positive, and are continuous elsewhere.

3.4 The notations generally employed for the corresponding *cosine integrals* are

$$\mathrm{Ci}(z) = -\int_z^{\infty} \frac{\cos t}{t}\,dt, \qquad \mathrm{Cin}(z) = \int_0^z \frac{1-\cos t}{t}\,dt. \tag{3.17}$$

$\mathrm{Ci}(z)$ has a branch point at $z = 0$; the principal branch is obtained by introducing a cut along the negative real axis. $\mathrm{Cin}(z)$ is entire.

From (3.01)

$$E_1(-iz) = \int_z^{i\infty} \frac{e^{it}}{t}\,dt = \int_z^{\infty} \frac{e^{it}}{t}\,dt,$$

the deformation of the path at infinity being justifiable as in the proof of Lemma 3.1. A similar result holds for $E_1(iz)$, whence

$$2\,\mathrm{Ci}(z) = -E_1(iz) - E_1(-iz). \tag{3.18}$$

This corresponds to (3.16). For the complementary functions we obtain, on replacing t in (3.03) by $\pm it$,

$$2\,\mathrm{Cin}(z) = \mathrm{Ein}(iz) + \mathrm{Ein}(-iz). \tag{3.19}$$

Addition of the last two equations and use of (3.05) connects the two cosine integrals:

$$\mathrm{Ci}(z) + \mathrm{Cin}(z) = \ln z + \gamma. \tag{3.20}$$

Again, principal branches of $\mathrm{Ci}(z)$ and $\ln z$ correspond.

Ex. 3.1 Show that

$$\mathrm{Si}(z) = \sum_{s=0}^{\infty} \frac{(-)^s}{2s+1} \frac{z^{2s+1}}{(2s+1)!}, \qquad \mathrm{Ci}(z) = \ln z + \gamma + \sum_{s=1}^{\infty} \frac{(-)^s}{2s} \frac{z^{2s}}{(2s)!}.$$

Ex. 3.2 Show that $\int_0^{\pi/2} \exp(-ze^{it})\, dt = -\mathrm{si}(z) - i\{\mathrm{Ci}(z) + E_1(z)\}$.

Ex. 3.3 If a is real and b is positive, prove that

$$\int_0^1 \frac{(1-e^{-at})\cos bt}{t}\, dt = \tfrac{1}{2}\ln\left(1 + \frac{a^2}{b^2}\right) + \mathrm{Ci}(b) + \mathrm{Re}\{E_1(a+ib)\}.$$

Ex. 3.4 Verify the following *Laplace transforms* when $\mathrm{Re}\, p > 0$:

$$\int_0^{\infty} e^{-pt}\, \mathrm{si}(t)\, dt = -\frac{\tan^{-1} p}{p}, \qquad \int_0^{\infty} e^{-pt}\, \mathrm{Ci}(t)\, dt = -\frac{\ln(1+p^2)}{2p}.$$

Ex. 3.5 The *generalized exponential integral* is defined by

$$E_n(z) = \int_1^{\infty} \frac{e^{-zt}}{t^n}\, dt \qquad (n = 1, 2, \ldots),$$

when $\mathrm{Re}\, z > 0$, and by analytic continuation elsewhere. Show that the only singularity of $E_n(z)$ is a branch point at $z = 0$.

Prove also that

$$nE_{n+1}(z) = e^{-z} - zE_n(z),$$

and

$$E_n(z) = \frac{(-z)^{n-1}}{(n-1)!}\{-\ln z + \psi(n)\} + \sum_{s=0}^{\infty}{}' \frac{(-z)^s}{s!(n-s-1)},$$

where the prime on the last sum signifies that the term $s = n-1$ is omitted.

Ex. 3.6 With the notation of the preceding exercise show that

$$E_n(z) = \int_z^{\infty} E_{n-1}(t)\, dt = \cdots = \int_z^{\infty} \cdots \int_t^{\infty} \frac{e^{-t}}{t} (dt)^n,$$

and hence that

$$E_n(z) = \frac{e^{-z}}{(n-1)!} \int_0^{\infty} \frac{e^{-t} t^{n-1}}{z+t}\, dt \qquad (|\mathrm{ph}\, z| < \pi).$$

4 Error Functions, Dawson's Integral, and Fresnel Integrals

4.1 The *error function* and *complementary error function* are important in probability theory and heat-conduction problems. They are defined respectively by

$$\mathrm{erf}\, z = \frac{2}{\pi^{1/2}} \int_0^z e^{-t^2}\, dt, \qquad \mathrm{erfc}\, z = \frac{2}{\pi^{1/2}} \int_z^{\infty} e^{-t^2}\, dt. \tag{4.01}$$

Each is entire. The factor $2/\pi^{1/2}$, that is, $2/\Gamma(\frac{1}{2})$, is introduced to simplify their connection formula:

$$\mathrm{erf}\, z + \mathrm{erfc}\, z = 1. \tag{4.02}$$

The Maclaurin expansion of erf z is given by

$$\operatorname{erf} z = \frac{2}{\pi^{1/2}} \sum_{s=0}^{\infty} \frac{(-)^s}{s!} \frac{z^{2s+1}}{2s+1}. \tag{4.03}$$

A related integral with positive exponential integrand when z is real is *Dawson's integral*

$$F(z) = e^{-z^2} \int_0^z e^{t^2}\, dt. \tag{4.04}$$

It is easily verified that

$$F(z) = \frac{\pi^{1/2}}{2i} e^{-z^2} \operatorname{erf}(iz). \tag{4.05}$$

4.2 Corresponding integrals of oscillatory type (when the variables are real) are the *Fresnel integrals*

$$C(z) = \int_0^z \cos(\tfrac{1}{2}\pi t^2)\, dt, \qquad S(z) = \int_0^z \sin(\tfrac{1}{2}\pi t^2)\, dt. \tag{4.06}$$

They, too, are entire. In terms of the error function,

$$C(z) + iS(z) = \tfrac{1}{2}(1+i) \operatorname{erf}\{\tfrac{1}{2}\pi^{1/2}(1-i)z\}. \tag{4.07}$$

Ex. 4.1 If $a > 0$ show that

$$\int_0^\infty \exp(-at^2) \sin(bt)\, dt = \frac{1}{a^{1/2}} F\left(\frac{b}{2a^{1/2}}\right),$$

where F is defined by (4.04).

Ex. 4.2 Let a and b be positive and

$$I = \int_0^\infty \exp(-at^2)(t^2+b^2)^{-1}\, dt.$$

By considering $d\{\exp(-ab^2) I\}/da$ prove that $I = \frac{1}{2}\pi b^{-1} \exp(ab^2) \operatorname{erfc}(ba^{1/2})$.

Ex. 4.3 Show that $C(\infty) = S(\infty) = \frac{1}{2}$.

Ex. 4.4† Let

$$f(x) = \int_0^\infty \frac{\exp(-u^2)}{u+x}\, du \qquad (x > 0).$$

Prove that

$$f(x) = -\ln x - \tfrac{1}{2}\gamma + o(1) \qquad (x \to 0),$$

and

$$\frac{d}{dx}\left\{\exp(x^2) f(x) - \pi^{1/2} \int_0^x \exp(u^2)\, du\right\} = -\frac{\exp(x^2)}{x}.$$

Hence establish that in terms of Dawson's integral and the exponential integral,

$$f(x) = \pi^{1/2} F(x) - \tfrac{1}{2} \exp(-x^2) \operatorname{Ei}(x^2).$$

† These results are due to Goodwin and Staton (1948) and Ritchie (1950), with a correction by Erdélyi (1950).

5 Incomplete Gamma Functions

5.1 All of the functions introduced in §§3 and 4 can be regarded as special cases of the *incomplete Gamma function*

$$\gamma(\alpha, z) = \int_0^z e^{-t} t^{\alpha-1}\, dt \qquad (\operatorname{Re}\alpha > 0), \tag{5.01}$$

or its complement $\Gamma(\alpha, z)$, defined in the next subsection. Clearly $\gamma(\alpha, z)$ is an analytic function of z, the only possible singularity being a branch point at the origin. The principal branch is obtained by introducing a cut along the negative real t axis, and requiring $t^{\alpha-1}$ to have its principal value.

If $\operatorname{Re}\alpha \geqslant 1$, then by uniform convergence we may expand e^{-t} in ascending powers of t and integrate term by term. In this way we obtain the following expansion, valid for all z:

$$\gamma(\alpha, z) = z^{\alpha} \sum_{s=0}^{\infty} (-)^s \frac{z^s}{s!\,(\alpha+s)}. \tag{5.02}$$

This enables $\gamma(\alpha, z)$ to be continued analytically with respect to α into the left half-plane, or with respect to z outside the principal phase range. Thus it is seen that when $z \neq 0$ the only singularities of $\gamma(\alpha, z)$ as a function of α are simple poles at $\alpha = 0, -1, -2, \ldots$. Also, if α is fixed, then the branch of $\gamma(\alpha, z)$ obtained after z encircles the origin m times is given by

$$\gamma(\alpha, ze^{2m\pi i}) = e^{2m\alpha\pi i}\gamma(\alpha, z) \qquad (\alpha \neq 0, -1, -2, \ldots). \tag{5.03}$$

5.2 The *complementary incomplete Gamma function*, or *Prym's function* as it is sometimes called, is defined by

$$\Gamma(\alpha, z) = \int_z^{\infty} e^{-t} t^{\alpha-1}\, dt, \tag{5.04}$$

there being no restriction on α. The principal branch is defined in the same way as for $\gamma(\alpha, z)$. Combination with (5.01) yields

$$\gamma(\alpha, z) + \Gamma(\alpha, z) = \Gamma(\alpha). \tag{5.05}$$

From (5.03) and (5.05) we derive

$$\Gamma(\alpha, ze^{2m\pi i}) = e^{2m\alpha\pi i}\Gamma(\alpha, z) + (1 - e^{2m\alpha\pi i})\Gamma(\alpha), \qquad (m = 0, \pm 1, \pm 2, \ldots). \tag{5.06}$$

Analytic continuation shows that this result also holds when α is zero or a negative integer, provided that the right-hand side is replaced by its limiting value.

Ex. 5.1 In the notation of §§3 and 4, show that

$$E_n(z) = z^{n-1}\Gamma(1-n, z), \qquad \operatorname{erf} z = \pi^{-1/2}\gamma(\tfrac{1}{2}, z^2), \qquad \operatorname{erfc} z = \pi^{-1/2}\Gamma(\tfrac{1}{2}, z^2).$$

Ex. 5.2 Show that $\gamma(\alpha, z)/\{z^{\alpha}\Gamma(\alpha)\}$ is entire in α and entire in z, and can be expanded in the form

$$e^{-z} \sum_{s=0}^{\infty} \frac{z^s}{\Gamma(\alpha+s+1)}.$$

Ex. 5.3 Show that

$$\partial^n\{z^{-\alpha}\Gamma(\alpha, z)\}/\partial z^n = (-)^n z^{-\alpha-n}\Gamma(\alpha+n, z).$$

6 Orthogonal Polynomials

6.1 Let (a, b) be a given finite or infinite interval, and $w(x)$ a function of x in (a, b) with the properties:

(i) *$w(x)$ is positive and continuous, except possibly at a finite set of points.*
(ii) $\int_a^b w(x)|x|^n \, dx < \infty$, $n = 0, 1, 2, \ldots$.

(In particular Condition (ii) implies that $w(x)$ is integrable over the given interval.) Then a set of real polynomials $\phi_n(x)$ of proper degree n,† $n = 0, 1, 2, \ldots$, is said to be *orthogonal over* (a, b) with *weight function* $w(x)$ if

$$\int_a^b w(x)\, \phi_n(x)\, \phi_s(x)\, dx = 0 \qquad (s \neq n). \tag{6.01}$$

Theorem 6.1 (i) *If the coefficient of x^n in $\phi_n(x)$ is prescribed for each n, then the set of orthogonal polynomials exists and is unique.*
(ii) *Each $\phi_n(x)$ is orthogonal to all polynomials of lower degree.*

Let $a_{n,n}$ $(\neq 0)$ denote the (prescribed) coefficient of x^n in $\phi_n(x)$. Assume that for a certain value of n the polynomials $\phi_0(x), \phi_1(x), \ldots, \phi_{n-1}(x)$ have been determined in such a way that they satisfy (6.01) among themselves—an assumption that is obviously valid in the case $n = 1$. Since each $\phi_s(x)$ is of proper degree s, any polynomial $\phi_n(x)$ of degree n with leading term $a_{n,n} x^n$ can be expressed in the form

$$\phi_n(x) = a_{n,n} x^n + b_{n,n-1} \phi_{n-1}(x) + b_{n,n-2} \phi_{n-2}(x) + \cdots + b_{n,0} \phi_0(x),$$

where the coefficients $b_{n,s}$ are independent of x. Application of the condition (6.01) with $s = 0, 1, \ldots, n-1$ in turn yields

$$a_{n,n} \int_a^b w(x)\, x^n \phi_s(x)\, dx + b_{n,s} \int_a^b w(x) \{\phi_s(x)\}^2 \, dx = 0.$$

Since $\int_a^b w(x) \{\phi_s(x)\}^2 \, dx$ cannot vanish, this determines $b_{n,s}$ finitely and uniquely. Part (i) of the theorem now follows by induction.

Part (ii) is easily proved by observing that any polynomial of degree $n-1$ or less can be expressed as a linear combination of $\phi_0(x), \phi_1(x), \ldots, \phi_{n-1}(x)$.

6.2 The specification of the $a_{n,n}$ is called the *normalization*. One method of normalization is to make each $a_{n,n}$ unity; another method sometimes used is implicitly given by

$$\int_a^b w(x)\, \phi_n(x)\, \phi_s(x)\, dx = \delta_{n,s}, \tag{6.02}$$

where $\delta_{n,s}$ is *Kronecker's delta symbol*, defined by

$$\delta_{n,s} = 0 \quad (n \neq s), \qquad \delta_{n,n} = 1.$$

A set of polynomials satisfying (6.02) is called *orthonormal*.

† That is, the degree of $\phi_n(x)$ is n and no less.

6.3 Theorem 6.2 *Each set of orthogonal polynomials satisfies a three-term recurrence relation of the form*

$$\phi_{n+1}(x) - (A_n x + B_n)\phi_n(x) + C_n \phi_{n-1}(x) = 0, \tag{6.03}$$

in which A_n, B_n, *and* C_n *are independent of* x.

To prove this result, we first choose A_n so that $\phi_{n+1}(x) - A_n x\phi_n(x)$ contains no term in x^{n+1}. Then we express

$$\phi_{n+1}(x) - A_n x\phi_n(x) = \sum_{s=0}^{n} c_{n,s}\phi_s(x).$$

The coefficients $c_{n,s}$ can be found by multiplying both sides of this equation by $w(x)\phi_s(x)$ and integrating from a to b. In consequence of (6.01) this yields

$$c_{n,s}\int_a^b w(x)\{\phi_s(x)\}^2\,dx = -A_n\int_a^b w(x)\,x\phi_s(x)\,\phi_n(x)\,dx.$$

Again, $x\phi_s(x)$ is a polynomial of degree $s+1$, and $\phi_n(x)$ is orthogonal to all polynomials of degree less than n. Hence all the $c_{n,s}$ vanish except possibly $c_{n,n-1}$ and $c_{n,n}$. This is the result stated with $B_n = c_{n,n}$ and $C_n = -c_{n,n-1}$.

6.4 Theorem 6.3 *The zeros of each member of a set of orthogonal polynomials are real, distinct, and lie in* (a,b).

Let $x_1, x_2, \ldots, x_m$, $0 \leqslant m \leqslant n$, be the distinct points in (a,b) at which $\phi_n(x)$ has a zero of odd multiplicity. Then in (a,b) the polynomial

$$\phi_n(x)(x-x_1)(x-x_2)\cdots(x-x_m)$$

has only zeros of even multiplicity. If $m < n$, then the orthogonal property shows that

$$\int_a^b w(x)\,\phi_n(x)(x-x_1)(x-x_2)\cdots(x-x_m)\,dx = 0,$$

which is a contradiction since the integrand does not change sign in (a,b). Therefore $m = n$. Moreover, since the total number of zeros is n, each x_s must be a simple zero. This completes the proof.

Ex. 6.1 In Theorem 6.1, show that the effect of renormalizing the $a_{n,n}$ is to multiply each $\phi_n(x)$ by a nonzero constant.

Show also that an orthonormal set is unique, except for signs.

Ex. 6.2 (*Gram–Schmidt orthonormalizing process*) Let $f_n(x)$, $n = 0, 1, \ldots$, be any set of polynomials in which $f_n(x)$ is of proper degree n. Define successively for $n = 0, 1, \ldots,$

$$\psi_n(x) = f_n(x) - \sum_{s=0}^{n-1}\left\{\int_a^b w(t)\,f_n(t)\,\phi_s(t)\,dt\right\}\phi_s(x), \qquad \phi_n(x) = \left[\int_a^b w(t)\{\psi_n(t)\}^2\,dt\right]^{-1/2}\psi_n(x).$$

Prove that the set $\phi_0(x), \phi_1(x), \ldots$ is orthonormal.

Ex. 6.3 Apply Theorem 6.2 to prove the *Christoffel–Darboux formula*

$$(x-y)\sum_{s=0}^{n}\frac{1}{h_s}\phi_s(x)\phi_s(y) = \frac{a_{n,n}}{h_n a_{n+1,n+1}}\{\phi_{n+1}(x)\phi_n(y)-\phi_n(x)\phi_{n+1}(y)\},$$

where $a_{n,n}$ is the coefficient of x^n in $\phi_n(x)$, and

$$h_n = \int_a^b w(x)\{\phi_n(x)\}^2\,dx.$$

Ex. 6.4 Let a and b be finite, $\{\phi_n(x)\}$ an orthonormal set, and $f(x)$ a given continuous function. Show that

$$\int_a^b w(x)\left\{f(x) - \sum_{s=0}^{n}\alpha_s\phi_s(x)\right\}^2 dx$$

is minimized by the choice $\alpha_s = \int_a^b w(x)f(x)\phi_s(x)\,dx$.

7 The Classical Orthogonal Polynomials

7.1 In this section we consider special families of orthogonal polynomials which are of importance in applied mathematics and numerical analysis. We again denote the interval under consideration by (a, b), the weight function by $w(x)$, and the highest term in $\phi_n(x)$ by $a_{n,n}x^n$.

Legendre polynomials $P_n(x)$. For these polynomials the interval is finite and the weight function the simplest possible:

$$a = -1, \qquad b = 1, \qquad w(x) = 1, \qquad a_{n,n} = (2n)!/\{2^n(n!)^2\}. \tag{7.01}$$

Jacobi polynomials $P_n^{(\alpha,\beta)}(x)$. These are generalizations of the Legendre polynomials:

$$a = -1, \qquad b = 1, \qquad w(x) = (1-x)^\alpha(1+x)^\beta, \qquad a_{n,n} = \frac{1}{2^n}\binom{2n+\alpha+\beta}{n}, \tag{7.02}$$

where α and β are real constants such that $\alpha > -1$ and $\beta > -1$. Thus

$$P_n(x) = P_n^{(0,0)}(x). \tag{7.03}$$

Laguerre polynomials $L_n^{(\alpha)}(x)$. For these, the range is infinite:

$$a = 0, \qquad b = \infty, \qquad w(x) = e^{-x}x^\alpha, \qquad a_{n,n} = (-1)^n/n!, \tag{7.04}$$

where α is a constant such that $\alpha > -1$. Sometimes $L_n^{(\alpha)}(x)$ is called the *generalized Laguerre polynomial*, the name Laguerre polynomial and notation $L_n(x)$ being reserved for $L_n^{(0)}(x)$.

Hermite polynomials $H_n(x)$. The range is doubly infinite, and the weight function an exponential that vanishes at both ends:

$$a = -\infty, \qquad b = \infty, \qquad w(x) = e^{-x^2}, \qquad a_{n,n} = 2^n. \tag{7.05}$$

7.2 Explicit expressions for the foregoing polynomials are supplied by *Rodrigues' formulas*:

$$P_n(x) = \frac{(-)^n}{2^n n!} \frac{d^n}{dx^n} \{(1-x^2)^n\}, \tag{7.06}$$

$$P_n^{(\alpha,\beta)}(x) = (-)^n \frac{(1-x)^{-\alpha}(1+x)^{-\beta}}{2^n n!} \frac{d^n}{dx^n} \{(1-x)^{n+\alpha}(1+x)^{n+\beta}\}, \tag{7.07}$$

$$L_n^{(\alpha)}(x) = \frac{e^x x^{-\alpha}}{n!} \frac{d^n}{dx^n} (e^{-x} x^{n+\alpha}), \tag{7.08}$$

and

$$H_n(x) = (-)^n e^{x^2} \frac{d^n}{dx^n} (e^{-x^2}). \tag{7.09}$$

That each of these expressions represents a polynomial of degree n is perceivable from Leibniz's theorem.

To prove (7.07), for example, let $\phi_n(x)$ denote the right-hand side and $\varpi(x)$ be any polynomial. Then by repeated partial integrations we arrive at

$$\int_{-1}^{1} (1-x)^\alpha (1+x)^\beta \phi_n(x) \varpi(x)\, dx = \frac{1}{2^n n!} \int_{-1}^{1} (1-x)^{n+\alpha}(1+x)^{n+\beta} \varpi^{(n)}(x)\, dx.$$

The last integral vanishes when the degree of $\varpi(x)$ is less than n. Therefore $\phi_n(x)$ satisfies the orthogonal relation of the Jacobi polynomials. By expansion in descending powers of x, it is seen that the coefficient of x^n in (7.07) is

$$\frac{1}{2^n} \binom{2n+\alpha+\beta}{n}.$$

Referring to (7.02) and Theorem 6.1 we see that (7.07) is established.

Formula (7.06) is a special case of (7.07), and formulas (7.08) and (7.09) may be verified in a similar manner. Formulas (7.07) and (7.08) can be used as definitions of $P_n^{(\alpha,\beta)}(x)$ and $L_n^{(\alpha)}(x)$ for values of α and β for which the orthogonal relations are inapplicable owing to divergence.

Another way of normalizing the classical polynomials would have been to specify the values of the constants

$$h_n = \int_a^b w(x) \{\phi_n(x)\}^2\, dx \tag{7.10}$$

and the signs of the $a_{n,n}$. Taking $\varpi(x) = P_n^{(\alpha,\beta)}(x)$ in the foregoing proof, we find that in the case of the Jacobi polynomials

$$h_n = \frac{a_{n,n}}{2^n} \int_{-1}^{1} (1-x)^{n+\alpha}(1+x)^{n+\beta}\, dx = a_{n,n} 2^{n+\alpha+\beta+1} \int_0^1 v^{n+\alpha}(1-v)^{n+\beta}\, dv$$

$$= \frac{2^{\alpha+\beta+1}}{2n+\alpha+\beta+1} \frac{\Gamma(n+\alpha+1)\Gamma(n+\beta+1)}{n!\Gamma(n+\alpha+\beta+1)}; \tag{7.11}$$

compare (7.02) and (1.10). In particular,

$$\int_{-1}^{1} \{P_n(x)\}^2\, dx = \frac{2}{2n+1}. \tag{7.12}$$

In a similar way

$$h_n = \Gamma(n+\alpha+1)/n! \quad \text{(Laguerre)}, \qquad h_n = \pi^{1/2} 2^n n! \quad \text{(Hermite)}. \tag{7.13}$$

Of the classical polynomials only the $L_n^{(0)}(x)$ comprise an orthonormal set.

7.3 In the remainder of this section we confine attention to the Legendre polynomials. Corresponding results for the other polynomials are stated as exercises at the end of the section.

The recurrence relation of type (6.03) can be determined by comparing coefficients. From (7.06) we see that the coefficients of x^n, x^{n-1}, and x^{n-2} in $P_n(x)$ are

$$\frac{(2n)!}{2^n (n!)^2}, \qquad 0, \qquad \text{and} \qquad -\frac{(2n-2)!}{2^n (n-2)!\,(n-1)!}, \tag{7.14}$$

respectively. Hence we derive

$$A_n = \frac{2n+1}{n+1}, \qquad B_n = 0, \qquad C_n = \frac{n}{n+1},$$

and

$$(n+1)\, P_{n+1}(x) - (2n+1)\, x P_n(x) + n P_{n-1}(x) = 0. \tag{7.15}$$

In addition to this second-order linear recurrence relation (or difference equation) $P_n(x)$ satisfies a second-order linear differential equation. The function

$$\frac{d}{dx}\{(1-x^2)\, P_n'(x)\} = (1-x^2)\, P_n''(x) - 2x P_n'(x) \tag{7.16}$$

is clearly a polynomial of degree n, and can therefore be expanded in the form

$$\sum_{s=0}^{n} c_{n,s} P_s(x). \tag{7.17}$$

To find the $c_{n,s}$, we multiply by $P_s(x)$, integrate from -1 to 1 and use (7.12). Then by two partial integrations we find that

$$\frac{2c_{n,s}}{2s+1} = \int_{-1}^{1} P_s(x) \frac{d}{dx}\{(1-x^2)\, P_n'(x)\}\, dx = \int_{-1}^{1} P_n(x) \frac{d}{dx}\{(1-x^2)\, P_s'(x)\}\, dx.$$

Again, since $P_n(x)$ is orthogonal to all polynomials of lower degree, it follows that

$$c_{n,s} = 0 \qquad (s < n).$$

To determine $c_{n,n}$ we compare coefficients of x^n in (7.16) and (7.17). This yields $-n(n+1)$. The desired differential equation is therefore

$$(1-x^2)\, P_n''(x) - 2x P_n'(x) + n(n+1)\, P_n(x) = 0. \tag{7.18}$$

7.4 Suppose that $G(x,h)$ is a function with a Maclaurin expansion of the form

$$G(x,h) = \sum_{n=0}^{\infty} \phi_n(x) h^n.$$

Then $G(x,h)$ is said to be a *generating function* for the set $\{\phi_n(x)\}$. In this concluding subsection we show how to construct a generating function for $\{P_n(x)\}$.

From Rodrigues' formula (7.06) and Cauchy's integral formula for the nth derivative of an analytic function, we immediately derive *Schläfli's integral*

$$P_n(x) = \frac{1}{2^{n+1}\pi i} \int_{\mathscr{C}} \frac{(t^2-1)^n}{(t-x)^{n+1}}\, dt, \tag{7.19}$$

in which $\mathscr{C}$ is any simple closed contour that encircles $t = x$; here x may be real or complex. For fixed $\mathscr{C}$ and sufficiently small $|h|$, the series

$$\sum_{n=0}^{\infty} \frac{(t^2-1)^n h^n}{2^{n+1}\pi i (t-x)^{n+1}}$$

converges uniformly with respect to $t \in \mathscr{C}$, by the M-test. Hence by integration and summation we obtain

$$\frac{1}{2\pi i} \int_{\mathscr{C}} \left\{1 - \frac{(t^2-1)h}{2(t-x)}\right\}^{-1} \frac{dt}{t-x} = \sum_{n=0}^{\infty} P_n(x) h^n = G(x,h),$$

and thence

$$G(x,h) = -\frac{1}{\pi i} \int_{\mathscr{C}} \frac{dt}{ht^2 - 2t + (2x-h)} = -\frac{1}{h\pi i} \int_{\mathscr{C}} \frac{dt}{(t-t_1)(t-t_2)},$$

where

$$t_1 = \{1 - (1-2xh+h^2)^{1/2}\}/h, \qquad t_2 = \{1 + (1-2xh+h^2)^{1/2}\}/h.$$

Clearly if $h \to 0$, then $t_1 \to x$ and $|t_2| \to \infty$. Hence for sufficiently small $|h|$, $\mathscr{C}$ contains t_1 but not t_2. The residue theorem yields

$$G(x,h) = -\frac{2}{h}\frac{1}{t_1 - t_2} = \frac{1}{(1-2xh+h^2)^{1/2}}.$$

Accordingly, the desired expansion is given by

$$\frac{1}{(1-2xh+h^2)^{1/2}} = \sum_{n=0}^{\infty} P_n(x) h^n, \tag{7.20}$$

provided that $|h|$ is sufficiently small and the chosen branch of the square root tends to 1 as $h \to 0$.

For $x \in [-1, 1]$ the singularities of the left-hand side of (7.20) both lie on the circle $|h| = 1$, hence in this case the radius of convergence of the series on the right-hand side is unity.

Ex. 7.1 Verify the following differential equations:

$$w'' - 2xw' + 2nw = 0, \qquad w = H_n(x); \qquad xw'' + (\alpha+1-x)w' + nw = 0, \qquad w = L_n^{(\alpha)}(x);$$

and

$$(1-x^2)w'' + \{(\beta-\alpha) - (\alpha+\beta+2)x\}w' + n(n+\alpha+\beta+1)w = 0, \qquad w = P_n^{(\alpha,\beta)}(x).$$

Ex. 7.2 Show that

$$P_n^{(\alpha,\beta)}(1) = \binom{n+\alpha}{n}, \qquad P_n^{(\alpha,\beta)}(-1) = (-)^n\binom{n+\beta}{n}.$$

Ex. 7.3 The *Chebyshev polynomials* $T_n(x)$ and $U_n(x)$ are defined by

$$T_n(x) = \cos n\theta, \qquad U_n(x) = \sin\{(n+1)\theta\}/\sin\theta,$$

where $\theta = \cos^{-1}x$. Show that

$$T_n(x) = \frac{2^{2n}(n!)^2}{(2n)!}P_n^{(-1/2,-1/2)}(x), \qquad U_n(x) = \frac{2^{2n}n!(n+1)!}{(2n+1)!}P_n^{(1/2,1/2)}(x).$$

Ex. 7.4 Show that

$$\sum_{n=0}^{\infty} H_n(x)\frac{h^n}{n!} = \exp(2xh-h^2).$$

Deduce that $H_n'(x) = 2nH_{n-1}(x)$, and

$$\sum_{s=0}^{n}\binom{n}{s}H_s(x)H_{n-s}(y) = 2^{n/2}H_n\left(\frac{x+y}{2^{1/2}}\right).$$

Ex. 7.5 Show that for $|h| < 1$

$$\sum_{n=0}^{\infty} L_n^{(\alpha)}(x)h^n = \frac{e^{-xh/(1-h)}}{(1-h)^{\alpha+1}}, \qquad \sum_{n=0}^{\infty} L_n^{(\alpha-n)}(x)h^n = (1+h)^{\alpha}e^{-xh}.$$

From the first expansion deduce that $dL_n^{(\alpha)}(x)/dx = -L_{n-1}^{(\alpha+1)}(x)$ when $n \geqslant 1$.

Ex. 7.6 Verify that for $n \geqslant 1$

$$H_{n+1}(x) - 2xH_n(x) + 2nH_{n-1}(x) = 0,$$

and

$$(n+1)L_{n+1}^{(\alpha)}(x) + (x-2n-\alpha-1)L_n^{(\alpha)}(x) + (n+\alpha)L_{n-1}^{(\alpha)}(x) = 0.$$

Ex. 7.7 Show that

$$H_{2n}(x) = (-)^n 2^{2n} n! L_n^{(-1/2)}(x^2), \qquad H_{2n+1}(x) = (-)^n 2^{2n+1} n! x L_n^{(1/2)}(x^2).$$

Ex. 7.8 Show that when $n \geqslant 1$,

$$xP_n'(x) - P_{n-1}'(x) = nP_n(x), \qquad n(n+1)\{P_{n+1}(x) - P_{n-1}(x)\} = (2n+1)(x^2-1)P_n'(x),$$

$$nxP_n(x) - nP_{n-1}(x) = (x^2-1)P_n'(x), \qquad nP_n(x) - nxP_{n-1}(x) = (x^2-1)P_{n-1}'(x).$$

Ex. 7.9 By taking the contour $\mathscr{C}$ in Schläfli's integral to be $|t-x| = |x^2-1|^{1/2}$ obtain *Laplace's integral*:

$$P_n(x) = \pi^{-1}\int_0^{\pi}\{x \pm (x^2-1)^{1/2}\cos\theta\}^n\,d\theta.$$

Deduce that if $x \in [-1, 1]$ then $|P_n(x)| \leqslant 1$, and, more generally, if $x = \cosh(\alpha + i\beta)$, where α and β are real, then $|P_n(x)| \leqslant e^{n|\alpha|}$. Thence show that the radius of convergence of the series (7.20) is at least $e^{-|\alpha|}$.

8 The Airy Integral

8.1 For real values of x the *Airy integral* is defined by

$$\mathrm{Ai}(x) = \frac{1}{\pi}\int_0^\infty \cos(\tfrac{1}{3}t^3 + xt)\,dt. \tag{8.01}$$

Although the integrand does not die away as $t \to \infty$, its increasingly rapid oscillations induce convergence of the integral. This can be confirmed by partial integration, as follows. We have

$$\int^t \cos(\tfrac{1}{3}t^3 + xt)\,dt = \frac{\sin(\tfrac{1}{3}t^3 + xt)}{t^2 + x} + 2\int^t \sin(\tfrac{1}{3}t^3 + xt)\frac{t\,dt}{(t^2+x)^2}.$$

As $t \to \infty$ the first term on the right-hand side vanishes, and the last integral converges absolutely.

When x lies off the real axis (8.01) diverges. To obtain the analytic continuation of $\mathrm{Ai}(x)$ into the complex plane we transform this integral into a contour integral, as follows. Set $t = v/i$. Then

$$\mathrm{Ai}(x) = \frac{1}{\pi i}\int_0^{i\infty} \cosh(\tfrac{1}{3}v^3 - xv)\,dv = \frac{1}{2\pi i}\int_{-i\infty}^{i\infty} \exp(\tfrac{1}{3}v^3 - xv)\,dv.$$

Assume temporarily that x is positive and consider

$$I(R) = \int_{iR}^{Re^{\pi i/6}} |\exp(\tfrac{1}{3}v^3 - xv)\,dv|,$$

where R is a large positive number, and the integration path is the shorter arc of the circle $|v| = R$. Substituting $v = iRe^{-i\theta/3}$ and applying Jordan's inequality (3.13) we derive

$$I(R) = \frac{R}{3}\int_0^\pi \exp(-\tfrac{1}{3}R^3\sin\theta - xR\sin\tfrac{1}{3}\theta)\,d\theta \leqslant \frac{R}{3}\int_0^\pi \exp(-\tfrac{1}{3}R^3\sin\theta)\,d\theta$$

$$\leqslant \frac{2R}{3}\int_0^{\pi/2} \exp\left(-\frac{2R^3\theta}{3\pi}\right)d\theta < \frac{\pi}{R^2}.$$

Hence $I(R)$ vanishes as $R \to \infty$. Clearly the same is true of the corresponding integral along the conjugate path.

Changing x into z and using Cauchy's theorem, we see that

$$\mathrm{Ai}(z) = \frac{1}{2\pi i}\int_{\mathscr{L}} \exp(\tfrac{1}{3}v^3 - zv)\,dv, \tag{8.02}$$

where $\mathscr{L}$ is any contour that begins at a point at infinity in the sector $-\frac{1}{2}\pi \leqslant \mathrm{ph}\,v \leqslant -\frac{1}{6}\pi$ and ends at infinity in the conjugate sector; see Fig. 8.1. This result has been established for positive z. However, if δ is an arbitrary small positive number and $\mathscr{L}$ begins in the sector $-\frac{1}{2}\pi + \delta \leqslant \mathrm{ph}\,v \leqslant -\frac{1}{6}\pi - \delta$ and ends at infinity in the conjugate sector, then at the extremities of $\mathscr{L}$ the factor $\exp(v^3/3)$ dominates

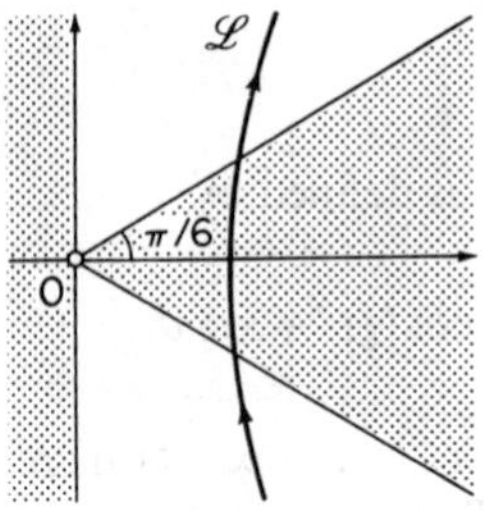

Fig. 8.1 v plane.

e^{-zv}, and (8.02) converges absolutely and uniformly in any compact z domain. Applying Theorem 1.1, with the t of this theorem taken to be the arc parameter of $\mathscr{L}$, we see that *with the contour chosen in this way*, (8.02) *supplies the analytic continuation of* Ai(z) *to the whole z plane; moreover*, Ai(z) *is entire*.

8.2 To obtain the Maclaurin expansion of Ai(z) we use the following general theorem concerning the integration of an infinite series over an infinite interval, or over an interval in which terms become infinite.

Theorem 8.1† *Let (a, b) be a given finite or infinite interval, and $u_1(t), u_2(t), u_3(t), \ldots$ be a sequence of real or complex functions which are continuous in (a, b) and have the properties:*

(i) $\sum_{s=1}^{\infty} u_s(t)$ *converges uniformly in any compact interval in (a, b).*
(ii) *At least one of the following quantities is finite:*

$$\int_a^b \left\{ \sum_{s=1}^{\infty} |u_s(t)| \right\} dt, \qquad \sum_{s=1}^{\infty} \int_a^b |u_s(t)|\, dt.$$

Then

$$\int_a^b \left\{ \sum_{s=1}^{\infty} u_s(t) \right\} dt = \sum_{s=1}^{\infty} \int_a^b u_s(t)\, dt.$$

Returning to (8.02) we take $\mathscr{L}$ to consist of the rays $\operatorname{ph} v = \pm\frac{1}{3}\pi$, and expand e^{-zv} in ascending powers of zv. Applying Theorem 8.1 and referring to the identities

$$\int_0^{\infty e^{\pm \pi i/3}} v^s \exp(\tfrac{1}{3}v^3)\, dv = 3^{(s-2)/3} e^{\pm(s+1)\pi i/3}\, \Gamma\!\left(\frac{s+1}{3}\right) \qquad (s = 0, 1, 2, \ldots),$$

obtained from (1.01) by means of the substitutions $v = (3t)^{1/3} e^{\pm \pi i/3}$, we arrive at

$$\operatorname{Ai}(z) = \operatorname{Ai}(0)\left(1 + \frac{1}{3!}z^3 + \frac{1\cdot 4}{6!}z^6 + \frac{1\cdot 4\cdot 7}{9!}z^9 + \cdots\right)$$
$$+ \operatorname{Ai}'(0)\left(z + \frac{2}{4!}z^4 + \frac{2\cdot 5}{7!}z^7 + \frac{2\cdot 5\cdot 8}{10!}z^{10} + \cdots\right), \qquad (8.03)$$

† This is the *dominated convergence theorem of Lebesgue* in the setting of Riemann integrals. For a proof see Bromwich (1926, §§175 and 176) or Titchmarsh (1939, §1.77).

where

$$\mathrm{Ai}(0) = \frac{\Gamma(\frac{1}{3})}{3^{1/6}2\pi} = \frac{1}{3^{2/3}\Gamma(\frac{2}{3})}, \qquad \mathrm{Ai}'(0) = -\frac{3^{1/6}\Gamma(\frac{2}{3})}{2\pi} = -\frac{1}{3^{1/3}\Gamma(\frac{1}{3})}. \tag{8.04}$$

8.3 One of the most important properties of $\mathrm{Ai}(z)$ is that it satisfies a second-order differential equation of particularly simple type. Referring to Theorem 1.1 and differentiating (8.02) under the sign of integration, we find that

$$\mathrm{Ai}''(z) - z\,\mathrm{Ai}(z) = \frac{1}{2\pi i}\int_{\mathscr{L}} (v^2 - z)\exp(\tfrac{1}{3}v^3 - zv)\,dv = \frac{1}{2\pi i}\Big[\exp(\tfrac{1}{3}v^3 - zv)\Big]_{\mathscr{L}}.$$

At the extremities of $\mathscr{L}$ the quantity in square brackets vanishes. Therefore the equation

$$d^2w/dz^2 = zw \tag{8.05}$$

is satisfied by $w = \mathrm{Ai}(z)$.

Equation (8.05) is unaffected when z is replaced by $ze^{\pm 2\pi i/3}$. Hence other solutions are $\mathrm{Ai}(ze^{2\pi i/3})$ and $\mathrm{Ai}(ze^{-2\pi i/3})$. We shall see in Chapter 5 that only two solutions can be independent, consequently a linear relation subsists between $\mathrm{Ai}(z)$, $\mathrm{Ai}(ze^{2\pi i/3})$, and $\mathrm{Ai}(ze^{-2\pi i/3})$. This can be found by integrating $\exp(\frac{1}{3}v^3 - zv)$ around a path in the v plane which begins at $\infty e^{-\pi i/3}$, passes to $\infty e^{\pi i/3}$, then to $-\infty$, and finally returns to $\infty e^{-\pi i/3}$. Application of Cauchy's theorem leads to the desired result

$$\mathrm{Ai}(z) + e^{2\pi i/3}\,\mathrm{Ai}(ze^{2\pi i/3}) + e^{-2\pi i/3}\,\mathrm{Ai}(ze^{-2\pi i/3}) = 0. \tag{8.06}$$

Applications and further properties of the Airy integral are given in Chapter 11. This chapter also introduces other solutions of (8.05).

Ex. 8.1 Show that $w = \mathrm{Ai}^2(z)$ satisfies $w''' - 4zw' - 2w = 0$.

9 The Bessel Function $J_\nu(z)$

9.1 For integer values of n and real or complex values of z, the function $J_n(z)$ is defined by *Bessel's integral*

$$J_n(z) = \frac{1}{\pi}\int_0^\pi \cos(n\theta - z\sin\theta)\,d\theta \qquad (n = 0, \pm 1, \pm 2, \ldots). \tag{9.01}$$

The variables n and z are called respectively the *order* and *argument* of $J_n(z)$. Theorem 1.1 shows that $J_n(z)$ is an entire function of z. To facilitate the evaluation of its Maclaurin coefficients we first construct a representation by a contour integral. Equation (9.01) may be rewritten

$$J_n(z) = \frac{1}{2\pi}\int_{-\pi}^{\pi} \exp(-in\theta + iz\sin\theta)\,d\theta. \tag{9.02}$$

Setting $h = e^{i\theta}$, we obtain

$$J_n(z) = \frac{1}{2\pi i}\int_{\mathscr{C}} \exp\{\tfrac{1}{2}z(h-h^{-1})\}\frac{dh}{h^{n+1}}, \tag{9.03}$$

where $\mathscr{C}$ is the unit circle. However, the only singularity of the integrand in the complex h plane is the origin, hence $\mathscr{C}$ may be deformed into any simple closed contour that encircles the origin.

Differentiating s times and setting $z = 0$, we see that $J_n^{(s)}(0)$ is the residue of $\{\frac{1}{2}(h-h^{-1})\}^s h^{-n-1}$ at $h = 0$. Suppose first that n is nonnegative. Then

$$J_n^{(s)}(0) = 0 \qquad (0 \leqslant s \leqslant n-1),$$

and

$$J_n^{(n+2s)}(0) = \frac{(-)^s}{2^{n+2s}}\binom{n+2s}{s}, \qquad J_n^{(n+2s+1)}(0) = 0 \qquad (s = 0, 1, 2, \ldots).$$

Accordingly,

$$J_n(z) = (\tfrac{1}{2}z)^n \sum_{s=0}^{\infty} \frac{(-)^s(\frac{1}{4}z^2)^s}{s!\,(n+s)!} \qquad (n = 0, 1, 2, \ldots). \tag{9.04}$$

The corresponding expansion when n is negative can be obtained in the same way, but it is simpler to refer to (9.01). Replacing θ by $\pi-\theta$, we immediately perceive that

$$J_{-n}(z) = (-)^n J_n(z). \tag{9.05}$$

9.2 A generating function and a differential equation for $J_n(z)$ may be derived as follows.

Referring to (9.03) and applying Laurent's theorem on the expansion of an analytic function in the neighborhood of an isolated essential singularity, we have

$$\exp\{\tfrac{1}{2}z(h-h^{-1})\} = \sum_{n=-\infty}^{\infty} J_n(z)h^n. \tag{9.06}$$

This is the required generating function. By Laurent's theorem the expansion converges for all values of h and z, other than $h = 0$.

Next, we differentiate (9.01) with respect to z. Writing

$$\Theta = n\theta - z\sin\theta$$

for brevity, we obtain

$$J_n'(z) = \frac{1}{\pi}\int_0^{\pi} \sin\theta \sin\Theta \, d\theta,$$

and

$$\{zJ_n'(z)\}' = -\frac{z}{\pi}\int_0^{\pi} \sin^2\theta \cos\Theta \, d\theta + \frac{1}{\pi}\int_0^{\pi} \sin\theta \sin\Theta \, d\theta.$$

Integration of the last term by parts produces

$$\{zJ_n'(z)\}' = -\frac{z}{\pi}\int_0^\pi \cos\Theta\, d\theta + \frac{n}{\pi}\int_0^\pi \cos\theta \cos\Theta\, d\theta.$$

Hence

$$z\{zJ_n'(z)\}' + (z^2-n^2)J_n(z) = \frac{n}{\pi}\int_0^\pi (z\cos\theta - n)\cos\Theta\, d\theta = \frac{n}{\pi}\Big[-\sin\Theta\Big]_0^\pi = 0. \tag{9.07}$$

Thus $w = J_n(z)$ satisfies

$$z^2w'' + zw' + (z^2-n^2)w = 0. \tag{9.08}$$

Equation (9.08) is *Bessel's equation.* It is of great importance in many physical problems.

9.3 When n is replaced by a general real or complex variable ν, we no longer define $J_\nu(z)$ by (9.01) because this integral does not satisfy Bessel's differential equation.† Instead, $J_\nu(z)$ is defined by the series

$$J_\nu(z) = (\tfrac{1}{2}z)^\nu \sum_{s=0}^\infty \frac{(-)^s(\tfrac{1}{4}z^2)^s}{s!\Gamma(\nu+s+1)}. \tag{9.09}$$

This obviously agrees with (9.04) when ν is zero or a positive integer. And it is not difficult to verify that it is consistent with the previous definition when ν is a negative integer, because in this event the first $-\nu$ terms of the series (9.09) vanish identically.

With the aid of (1.03) and the M-test it is easily seen that the sum in (9.09) converges uniformly in any compact sets in the planes of ν and z. Accordingly, $(\tfrac{1}{2}z)^{-\nu}J_\nu(z)$ is entire in z and entire in ν. Since $(\tfrac{1}{2}z)^\nu = \exp\{\nu\ln(\tfrac{1}{2}z)\}$, *the function $J_\nu(z)$ is an entire function of ν (except when $z=0$), and a many valued function of z (except when ν is zero or an integer).* The principal branch is obtained by taking the principal branch of $(\tfrac{1}{2}z)^\nu$ in (9.09); other branches are related by

$$J_\nu(ze^{m\pi i}) = e^{m\nu\pi i}J_\nu(z) \qquad (m = \text{integer}). \tag{9.10}$$

That the series (9.09) satisfies

$$\frac{d^2w}{dz^2} + \frac{1}{z}\frac{dw}{dz} + \left(1 - \frac{\nu^2}{z^2}\right)w = 0 \tag{9.11}$$

(compare (9.08)) is easily verifiable by term-by-term differentiation. Moreover, since this differential equation is unchanged when ν is replaced by $-\nu$, another solution is $w = J_{-\nu}(z)$.

9.4 A contour integral for $J_\nu(z)$ can be found by substituting Hankel's loop

† This can be seen from the analysis of §9.2: $\sin\Theta$ vanishes at $\theta = \pi$ only when n is an integer or zero.

integral (1.12) for the reciprocal of the Gamma function in (9.09). This gives

$$J_\nu(z) = \frac{(\frac{1}{2}z)^\nu}{2\pi i}\sum_{s=0}^{\infty}(-)^s\frac{(\frac{1}{4}z^2)^s}{s!}\int_{-\infty}^{(0+)} e^t t^{-\nu-s-1}\,dt.$$

Inverting the order of integration and summation—a procedure which is justifiable by taking the arc parameter as integration variable and referring to Theorem 8.1—we obtain

$$J_\nu(z) = \frac{(\frac{1}{2}z)^\nu}{2\pi i}\int_{-\infty}^{(0+)} \exp\left(t - \frac{z^2}{4t}\right)\frac{dt}{t^{\nu+1}}. \tag{9.12}$$

This is *Schläfli's integral for* $J_\nu(z)$; compare (7.19). As in (1.12) the branch of $t^{\nu+1}$ takes its principal value where the path crosses the positive real axis, and is continuous elsewhere.

If we suppose, temporarily, that z is positive and set $t = \frac{1}{2}zh$ in (9.12), then we find that

$$J_\nu(z) = \frac{1}{2\pi i}\int_{-\infty}^{(0+)} \exp\{\tfrac{1}{2}z(h-h^{-1})\}\frac{dh}{h^{\nu+1}}.$$

(We notice in passing that when ν is an integer the integrand is single valued and the integral reduces to (9.03).) Now set $h = e^\tau$. Then we obtain

$$J_\nu(z) = \frac{1}{2\pi i}\int_{\infty-\pi i}^{\infty+\pi i} e^{z\sinh\tau-\nu\tau}\,d\tau, \tag{9.13}$$

also due to Schläfli; the contour is indicated in Fig. 9.1. Analytic continuation immediately extends this result to $|\mathrm{ph}\, z| < \frac{1}{2}\pi$.

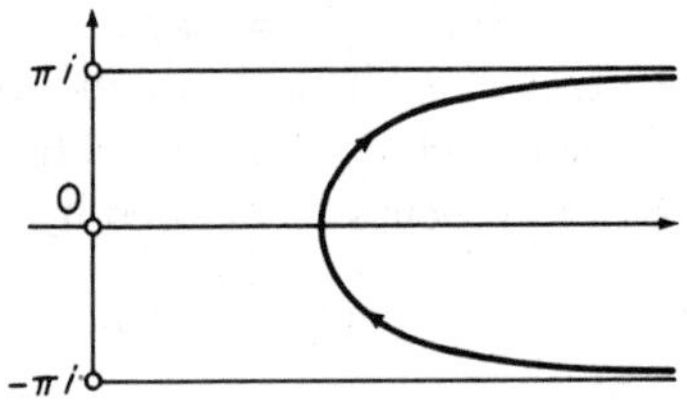

Fig. 9.1 τ plane.

9.5 Recurrence relations for the Bessel functions can be derived either from the series definition or from Schläfli's integrals. The latter is the more constructive approach. From (9.13) we have

$$\begin{aligned}\tfrac{1}{2}zJ_{\nu-1}(z) + \tfrac{1}{2}zJ_{\nu+1}(z) - \nu J_\nu(z) &= \frac{1}{2\pi i}\int_{\infty-\pi i}^{\infty+\pi i}(z\cosh\tau-\nu)\,e^{z\sinh\tau-\nu\tau}\,d\tau\\ &= (2\pi i)^{-1}\left[e^{z\sinh\tau-\nu\tau}\right]_{\infty-\pi i}^{\infty+\pi i} = 0;\end{aligned}$$

whence

$$J_{\nu-1}(z) + J_{\nu+1}(z) = (2\nu/z)J_\nu(z). \tag{9.14}$$

Although (9.13) holds only when $|\mathrm{ph}\, z| < \frac{1}{2}\pi$, analytic continuation removes this restriction from (9.14).

Similarly,

$$J'_\nu(z) = \frac{1}{2\pi i}\int_{\infty-\pi i}^{\infty+\pi i} \sinh\tau\, e^{z\sinh\tau - \nu\tau}\,d\tau;$$

whence

$$J_{\nu-1}(z) - J_{\nu+1}(z) = 2J'_\nu(z). \tag{9.15}$$

From (9.14) and (9.15) the further relations

$$J_{\nu+1}(z) = (\nu/z)J_\nu(z) - J'_\nu(z), \qquad J_{\nu-1}(z) = (\nu/z)J_\nu(z) + J'_\nu(z), \tag{9.16}$$

are easily found. In particular, $J'_0(z) = -J_1(z)$.

Ex. 9.1 From the generating function deduce that

$$1 = J_0(z) + 2J_2(z) + 2J_4(z) + 2J_6(z) + \cdots,$$
$$\cos z = J_0(z) - 2J_2(z) + 2J_4(z) - 2J_6(z) + \cdots,$$
$$\tfrac{1}{2}z\cos z = J_1(z) - 9J_3(z) + 25J_5(z) - 49J_7(z) + \cdots.$$

Ex. 9.2 Prove *Neumann's addition theorem* for integer orders n:

$$J_n(z_1+z_2) = \sum_{s=-\infty}^{\infty} J_s(z_1)J_{n-s}(z_2).$$

Deduce that $1 = J_0^2(z) + 2\sum_{s=1}^{\infty} J_s^2(z)$.

Ex. 9.3 Show that

$$J_{1/2}(z) = \left(\frac{2}{\pi z}\right)^{1/2}\sin z, \qquad J_{3/2}(z) = \left(\frac{2}{\pi z}\right)^{1/2}\left(\frac{\sin z}{z} - \cos z\right),$$

$$J_{-1/2}(z) = \left(\frac{2}{\pi z}\right)^{1/2}\cos z, \qquad J_{-3/2}(z) = -\left(\frac{2}{\pi z}\right)^{1/2}\left(\frac{\cos z}{z} + \sin z\right).$$

Ex. 9.4 Show that

$$\left(\frac{1}{z}\frac{d}{dz}\right)^s\{z^\nu J_\nu(z)\} = z^{\nu-s}J_{\nu-s}(z), \qquad \left(\frac{1}{z}\frac{d}{dz}\right)^s\{z^{-\nu}J_\nu(z)\} = (-)^s z^{-\nu-s}J_{\nu+s}(z).$$

Ex. 9.5 By expansion of the cosine factor in the integrand establish *Poisson's integral*:

$$J_\nu(z) = \frac{(\frac{1}{2}z)^\nu}{\pi^{1/2}\Gamma(\nu+\frac{1}{2})}\int_0^\pi \cos(z\cos\theta)\sin^{2\nu}\theta\,d\theta \qquad (\mathrm{Re}\,\nu > -\tfrac{1}{2}).$$

Verify directly that this integral satisfies Bessel's differential equation.

Ex. 9.6 From the preceding exercise deduce that

$$|J_\nu(z)| \leqslant |\tfrac{1}{2}z|^\nu e^{|\mathrm{Im}\, z|}/\Gamma(\nu+1) \qquad (\nu \geqslant -\tfrac{1}{2}),$$

and from (9.02) that

$$|J_n(z)| \leqslant e^{|\mathrm{Im}\, z|} \qquad (n = 0, \pm 1, \pm 2, \ldots).$$

Ex. 9.7 Show that for $\mathrm{Re}\,\nu > -1$

$$\int_0^z J_\nu(t)\,dt = 2\sum_{s=0}^{\infty} J_{\nu+2s+1}(z).$$

Using Exercise 9.3 and the notation of §4.2, deduce that

$$C(z) = \sum_{s=0}^{\infty} J_{2s+(1/2)}(\tfrac{1}{2}\pi z^2), \qquad S(z) = \sum_{s=0}^{\infty} J_{2s+(3/2)}(\tfrac{1}{2}\pi z^2).$$

Ex. 9.8 From the definition (9.09) deduce that if a, b, and $\nu+\frac{1}{2}$ are positive numbers and $b < a$, then

$$\int_0^{\infty} e^{-at} J_\nu(bt)\, t^\nu \, dt = \frac{\Gamma(\nu+\frac{1}{2})(2b)^\nu}{\pi^{1/2}(a^2+b^2)^{\nu+(1/2)}}.$$

Show also that the restriction $b < a$ can be removed by use of Exercise 9.6 and appeal to analytic continuation.

10 The Modified Bessel Function $I_\nu(z)$

10.1 The modified Bessel function $I_\nu(z)$ is defined for all values of ν and z, other than $z = 0$, by the series

$$I_\nu(z) = (\tfrac{1}{2}z)^\nu \sum_{s=0}^{\infty} \frac{(\frac{1}{4}z^2)^s}{s!\,\Gamma(\nu+s+1)}. \tag{10.01}$$

Like (9.09) this is a many valued function of z, unless ν is an integer or zero. The principal branch is obtained by assigning $(\frac{1}{2}z)^\nu$ its principal value.

Comparing (9.09) with (10.01), we see that

$$I_\nu(z) = e^{-\nu\pi i/2} J_\nu(iz), \tag{10.02}$$

where the branches have their principal values when $\operatorname{ph} z = 0$, and are continuous elsewhere.† In consequence, $I_\nu(z)$ is sometimes called the *Bessel function of imaginary argument.*

Most properties of I_ν can be deduced straightforwardly from those of J_ν by means of (10.02). For example, the *modified Bessel equation*

$$\frac{d^2w}{dz^2} + \frac{1}{z}\frac{dw}{dz} - \left(1 + \frac{\nu^2}{z^2}\right) w = 0 \tag{10.03}$$

is satisfied by $w = I_{\pm\nu}(z)$. Recurrence relations for the modified functions are

$$I_{\nu-1}(z) - I_{\nu+1}(z) = (2\nu/z) I_\nu(z), \qquad I_{\nu-1}(z) + I_{\nu+1}(z) = 2I_\nu'(z), \tag{10.04}$$

$$I_{\nu+1}(z) = -(\nu/z) I_\nu(z) + I_\nu'(z), \qquad I_{\nu-1}(z) = (\nu/z) I_\nu(z) + I_\nu'(z). \tag{10.05}$$

Further properties of $J_\nu(z)$, $I_\nu(z)$, and other solutions of the differential equations (9.11) and (10.03) are developed in Chapter 7.

Ex. 10.1 Show that when n is an integer,

$$I_n(z) = I_{-n}(z) = \pi^{-1} \int_0^{\pi} e^{z\cos\theta} \cos(n\theta)\, d\theta.$$

† It should be noticed that the cuts for the principal branches of $I_\nu(z)$ and $J_\nu(iz)$ are not the same; compare Exercise 10.2 below.

Ex. 10.2 When principal branches are used, show that

$$I_\nu(z) = e^{-\nu\pi i/2} J_\nu(iz) \qquad (-\pi < \operatorname{ph} z \leqslant \tfrac{1}{2}\pi), \qquad I_\nu(z) = e^{3\nu\pi i/2} J_\nu(iz) \qquad (\tfrac{1}{2}\pi < \operatorname{ph} z \leqslant \pi).$$

Ex. 10.3 Prove that

$$\exp\{\tfrac{1}{2}z(h+h^{-1})\} = \sum_{n=-\infty}^{\infty} I_n(z) h^n \qquad (h \neq 0).$$

Ex. 10.4 Show that with the transformations $\xi = \tfrac{2}{3}z^{3/2}$ and $W = z^{-1/2} w$, equation (8.05) becomes

$$\frac{d^2 W}{d\xi^2} + \frac{1}{\xi}\frac{dW}{d\xi} - \left(1 + \frac{1}{9\xi^2}\right) W = 0.$$

Show also that

$$\operatorname{Ai}(z) = \tfrac{1}{3}z^{1/2}\{I_{-1/3}(\xi) - I_{1/3}(\xi)\}, \qquad \operatorname{Ai}(-z) = \tfrac{1}{3}z^{1/2}\{J_{-1/3}(\xi) + J_{1/3}(\xi)\},$$

$$\operatorname{Ai}'(z) = \tfrac{1}{3}z\{I_{2/3}(\xi) - I_{-2/3}(\xi)\}, \qquad \operatorname{Ai}'(-z) = \tfrac{1}{3}z\{J_{2/3}(\xi) - J_{-2/3}(\xi)\},$$

where all functions take their principal values when $\operatorname{ph} z = 0$ and are related by continuity elsewhere.

Ex. 10.5 By means of Exercise 9.4 prove that

$$I_\nu(z) = \sum_{s=0}^{\infty} \frac{z^s}{s!} J_{\nu+s}(z), \qquad J_\nu(z) = \sum_{s=0}^{\infty} (-)^s \frac{z^s}{s!} I_{\nu+s}(z),$$

where the branches take their principal values when $\operatorname{ph} z = 0$.

Ex. 10.6 Show that solutions of the differential equation

$$x^4 w^{iv} + 2x^3 w''' - (1+2\nu^2)(x^2 w'' - xw') + (\nu^4 - 4\nu^2 + x^4) w = 0$$

are the *Kelvin functions* $\operatorname{ber}_\nu x$, $\operatorname{bei}_\nu x$, $\operatorname{ber}_{-\nu} x$, and $\operatorname{bei}_{-\nu} x$, defined by

$$\operatorname{ber}_\nu x \pm i \operatorname{bei}_\nu x = J_\nu(xe^{\pm 3\pi i/4}) = e^{\pm \nu\pi i/2} I_\nu(xe^{\pm \pi i/4}).$$

11 The Zeta Function

11.1 The *Zeta function* (of Riemann) is defined by the series

$$\zeta(z) = \sum_{s=1}^{\infty} \frac{1}{s^z} \tag{11.01}$$

when $\operatorname{Re} z > 1$, and by analytic continuation elsewhere. The series converges absolutely and uniformly in any compact domain within $\operatorname{Re} z > 1$, hence $\zeta(z)$ is holomorphic in this half-plane.

An integral representation for $\zeta(z)$ can be found by substituting Euler's integral for the Gamma function in the form

$$\frac{1}{s^z} = \frac{1}{\Gamma(z)} \int_0^\infty e^{-st} t^{z-1}\, dt \qquad (\operatorname{Re} z > 0).$$

When $\operatorname{Re} z > 1$ we are permitted, by Theorem 8.1, to invert the order of summation and integration. This gives

$$\zeta(z) = \frac{1}{\Gamma(z)} \int_0^\infty \frac{t^{z-1}}{e^t - 1}\, dt \qquad (\operatorname{Re} z > 1). \tag{11.02}$$

In many respects this integral resembles its parent (1.01).

11.2 The analytic continuation of $\zeta(z)$ to the region $\operatorname{Re} z \leqslant 1$ is obtainable by constructing a loop integral of Hankel's type. Consider

$$I(z) = \int_{-\infty}^{(0+)} \frac{t^{z-1}}{e^{-t}-1}\,dt,$$

where the contour does not enclose any of the points $\pm 2\pi i, \pm 4\pi i, \ldots$. By applying Theorem 1.1, taking the t of this theorem to be the arc parameter of the path, we readily see that $I(z)$ is entire. Following §1.7, we temporarily suppose that $\operatorname{Re} z > 1$ and collapse the path on the negative real axis, to obtain

$$I(z) = 2i\sin(\pi z)\int_0^{\infty} \frac{\tau^{z-1}}{e^{\tau}-1}\,d\tau = 2i\sin(\pi z)\,\Gamma(z)\,\zeta(z);$$

compare (11.02). Use of the reflection formula for the Gamma function immediately produces

$$\zeta(z) = \frac{\Gamma(1-z)}{2\pi i}\int_{-\infty}^{(0+)} \frac{t^{z-1}}{e^{-t}-1}\,dt. \tag{11.03}$$

This is the required formula. As in Hankel's integral the branch of the complex power takes its principal value where the contour crosses the positive real axis, and is defined by continuity elsewhere.

When $\operatorname{Re} z \leqslant 1$, formula (11.03) provides the required analytic continuation of $\zeta(z)$. Clearly the only possible singularities are the singularities of $\Gamma(1-z)$, that is, $z = 1, 2, \ldots$. Since we already know that $\zeta(z)$ is holomorphic when $\operatorname{Re} z > 1$ it remains to consider $z = 1$. By the residue theorem

$$\int_{-\infty}^{(0+)} \frac{dt}{e^{-t}-1} = -2\pi i.$$

Accordingly, *the only singularity of* $\zeta(z)$ *is a simple pole of residue* 1 *at* $z = 1$.

11.3 Can the integral (11.03) be evaluated for general values of z by deformation of the path? Apart from $t = 0$, the singularities of the integrand are simple poles at $t = \pm 2s\pi i$, $s = 1, 2, \ldots$. Let N be a large positive integer, and consider the integral

$$\int_{\mathscr{R}_N} \frac{t^{z-1}}{e^{-t}-1}\,dt, \tag{11.04}$$

where $\mathscr{R}_N$ is the perimeter of the rectangle with vertices $\pm N \pm (2N-1)\pi i$.† It is easily verified that

$$|e^{-t}-1| \geqslant 1 - e^{-N} \qquad (t \in \mathscr{R}_N).$$

Accordingly, if $\operatorname{Re} z < 0$, then (11.04) vanishes as $N \to \infty$. The residue of $t^{z-1}/(e^{-t}-1)$ at $t = \pm 2s\pi i$ is $-(\pm 2s\pi i)^{z-1}$. Applying the residue theorem and (11.03), we derive

$$\zeta(z) = \Gamma(1-z)\left\{\sum_{s=1}^{\infty} (2s\pi i)^{z-1} + \sum_{s=1}^{\infty} (-2s\pi i)^{z-1}\right\},$$

†The integrand is discontinuous at $t = -N$.

that is,

$$\zeta(z) = \Gamma(1-z)2^z\pi^{z-1}\cos\{\tfrac{1}{2}\pi(z-1)\}\zeta(1-z).$$

Again, analytic continuation extends this result to all z, other than $z = 1$.

Thus although deformation of the path does not lead to an actual evaluation of $\zeta(z)$, it supplies a valuable reflection formula. This formula is due to Riemann, and is more commonly quoted in the form

$$\zeta(1-z) = 2^{1-z}\pi^{-z}\cos(\tfrac{1}{2}\pi z)\Gamma(z)\zeta(z). \tag{11.05}$$

11.4 It was possible to evaluate the integral in (11.03) at $z = 1$ because the integrand is then a single-valued function of t and the residue theorem applies. Similar evaluations may be made for other integer values of z; compare Chapter 8, §1.5. For the time being we record the following special cases of (11.05), or its limiting form:

$$\zeta(-2m) = 0, \qquad \zeta(1-2m) = (-)^m 2^{1-2m}\pi^{-2m}(2m-1)!\,\zeta(2m) \qquad (m = 1, 2, 3, \ldots),$$

and

$$\zeta(0) = -\tfrac{1}{2}.$$

11.5 The final formula in this section is an infinite product due to Euler. Assume that $\operatorname{Re} z > 1$ and subtract from (11.01) the corresponding series for $2^{-z}\zeta(z)$. Then

$$\zeta(z)(1-2^{-z}) = \frac{1}{1^z} + \frac{1}{3^z} + \frac{1}{5^z} + \frac{1}{7^z} + \cdots.$$

Similarly,

$$\zeta(z)(1-2^{-z})(1-3^{-z}) = \sum \frac{1}{s^z},$$

where the sum is taken over all positive integers s, excluding multiples of 2 or 3.

Now let ϖ_s be the sth prime number, counting from $\varpi_1 = 2$. By continuing the previous argument, we see that

$$\zeta(z)\prod_{s=1}^{n}(1-\varpi_s^{-z}) = 1 + \sum \frac{1}{s^z},$$

where the last sum excludes terms for which $s = 1$ or a multiple of $\varpi_1, \varpi_2, \ldots, \varpi_n$. This sum is bounded in absolute value by

$$\sum_{s=\varpi_n+1}^{\infty} \frac{1}{s^{\operatorname{Re} z}},$$

and therefore vanishes as $n \to \infty$ (since $\varpi_n \to \infty$). Hence we obtain the required formula

$$\zeta(z)\prod_{s=1}^{\infty}(1-\varpi_s^{-z}) = 1 \qquad (\operatorname{Re} z > 1).$$

This relation is one of many important connections between the Zeta function and the theory of prime numbers.

Comparing the infinite products

$$\prod_{s=2}^{\infty}(1-s^{-z}), \qquad \prod_{s=1}^{\infty}(1-\varpi_s^{-z}) \qquad (\operatorname{Re} z > 1),$$

we note that the factors of the latter are a subset of those of the former. Since the former product is absolutely convergent, so is the latter. An immediate corollary is that $\zeta(z)$ *has no zeros in the half-plane* $\operatorname{Re} z > 1$. And by combining this result with the reflection formula (11.05), we see that *the only zeros of* $\zeta(z)$ *in the half-plane* $\operatorname{Re} z < 0$ *are* $-2, -4, -6, \ldots$.

In the remaining strip $0 \leqslant \operatorname{Re} z \leqslant 1$, the nature of the zeros of $\zeta(z)$ is not fully known. A famous, and still unproved, conjecture of Riemann is that they all lie on the midline $\operatorname{Re} z = \frac{1}{2}$. One of the many results which depend on this conjecture is the following formula for the number of primes $\pi(x)$ not exceeding x:

$$\operatorname{li}(x) - \pi(x) = O(x^{1/2} \ln x) \qquad (x \to \infty),$$

where $\operatorname{li}(x)$ is defined in §3.2.

Ex. 11.1 Show that when $\operatorname{Re} z > 0$

$$(1-2^{1-z})\zeta(z) = \frac{1}{1^z} - \frac{1}{2^z} + \frac{1}{3^z} - \frac{1}{4^z} + \cdots = \frac{1}{\Gamma(z)}\int_0^{\infty} \frac{t^{z-1}}{e^t+1}\,dt.$$

Ex. 11.2 With the aid of Exercise 2.6 show that

$$\int_{-\infty}^{(0+)} \frac{\ln t}{e^{-t}-1}\,dt = 0,$$

and thence that $\lim_{z\to 1}\{\zeta(z)-(z-1)^{-1}\} = \gamma$, $\zeta'(0) = -\frac{1}{2}\ln(2\pi)$.

Ex. 11.3 With the aid of Exercise 2.4 prove that

$$\ln\{\Gamma(z)\} = -\gamma(z-1) + \sum_{s=2}^{\infty} (-)^s \frac{\zeta(s)}{s}(z-1)^s \qquad (|z-1| < 1).$$

Historical Notes and Additional References

The material in this chapter is classical. Considerable use has been made of the books by Whittaker and Watson (1927), Copson (1935), B.M.P. (1953a, b), and N.B.S. (1964).

§1 (i) An excellent history of the Gamma function has been given by Davis (1959).

(ii) Euler's constant has been computed to 3566 decimal places by Sweeney (1963). Whether γ is an algebraic or transcendental number—that is, whether γ is, or is not, a root of a polynomial equation with integer coefficients—is an unsolved problem.

§§3–5 Collections of formulas for definite and indefinite integrals involving the exponential integral and error functions have been given by Geller and Ng (1969) and Ng and Geller (1969). Further properties of these functions, and the incomplete Gamma functions, are included in the book by Luke (1962).

§§6–7 The definitive treatise on orthogonal polynomials is that of Szegö (1967). The monograph of Hochstadt (1961) was helpful in preparing these sections.

§§8–10 For notes on the Airy integral and Bessel functions see pp. 277–278 and 433.

§11 Although the Zeta function was known to Euler, its more important properties awaited the researches of Riemann (1859). For further results see Titchmarsh (1951).

3

INTEGRALS OF A REAL VARIABLE

1 Integration by Parts

1.1 A simple and often effective way of deriving the asymptotic expansion of an integral containing a parameter consists of repeated integrations by parts. Each integration yields a new term in the expansion, and the error term is given explicitly as an integral, from which bounds or estimates may be derived.

Consider the incomplete Gamma function with real arguments α and x, x being positive. The convergent series expansion (5.02) of Chapter 2 is useful for computing $\gamma(\alpha, x)$ when x is small or moderate in size, but not when x is large owing to severe numerical cancellation among the terms. We therefore seek an asymptotic expansion, and for this purpose it is more convenient to work with the complementary function $\Gamma(\alpha, x)$.

Integration by parts of the definition (5.04) of Chapter 2 produces

$$\Gamma(\alpha, x) = e^{-x}x^{\alpha-1} + (\alpha-1)\,\Gamma(\alpha-1, x).$$

Repeated application of this result leads to

$$\Gamma(\alpha, x) = e^{-x}x^{\alpha-1}\left\{1 + \frac{\alpha-1}{x} + \frac{(\alpha-1)(\alpha-2)}{x^2} + \cdots + \frac{(\alpha-1)(\alpha-2)\cdots(\alpha-n+1)}{x^{n-1}}\right\} + \varepsilon_n(x), \tag{1.01}$$

where n is an arbitrary nonnegative integer, and

$$\varepsilon_n(x) = (\alpha-1)(\alpha-2)\cdots(\alpha-n)\int_x^\infty e^{-t}t^{\alpha-n-1}\,dt. \tag{1.02}$$

If $n \geqslant \alpha-1$, then $t^{\alpha-n-1} \leqslant x^{\alpha-n-1}$ and we immediately obtain

$$|\varepsilon_n(x)| \leqslant |(\alpha-1)(\alpha-2)\cdots(\alpha-n)|\,e^{-x}x^{\alpha-n-1}. \tag{1.03}$$

Accordingly, for fixed α and large x

$$\Gamma(\alpha, x) \sim e^{-x}x^{\alpha-1}\sum_{s=0}^{\infty}\frac{(\alpha-1)(\alpha-2)\cdots(\alpha-s)}{x^s}. \tag{1.04}$$

Moreover, the nth error term is bounded in absolute value by the $(n+1)$th term of the series and has the same sign, provided that $n \geqslant \alpha-1$.

For later use we record the special case

$$\Gamma(\alpha, x) \leqslant e^{-x}x^{\alpha-1} \qquad (\alpha \leqslant 1, \quad x>0). \tag{1.05}$$

1.2 If $n < \alpha-1$, then $\varepsilon_n(x)$ is *not* bounded in absolute value by the first neglected term in the series. This can be seen from the identity

$$\varepsilon_n(x) = (\alpha-1)(\alpha-2)\cdots(\alpha-n)e^{-x}x^{\alpha-n-1} + (\alpha-1)(\alpha-2)\cdots(\alpha-n-1)\int_x^\infty e^{-t}t^{\alpha-n-2}\,dt$$

obtained by partial integration of (1.02); both terms on the right-hand side are positive when $n<\alpha-1$. However, by continuing the process of expansion we see that the first $[\alpha]-n+1$ neglected terms of the series are nonnegative and $\varepsilon_n(x)$ is bounded by their sum.

Ex. 1.1 Show that (1.04) is uniform for α in a compact interval.

Ex. 1.2 Prove that

$$\operatorname{erfc} x \sim \frac{\exp(-x^2)}{\pi^{1/2}x}\sum_{s=0}^{\infty}(-)^s\frac{1\cdot3\cdots(2s-1)}{(2x^2)^s} \qquad (x\to\infty).$$

Show also that for $x\in(0,\infty)$ the error term does not exceed the first neglected term in the series in absolute value, and has the same sign.

Ex. 1.3 Show that for $x>0$ and $n=0,1,2,\ldots$

$$\operatorname{Ci}(x)+i\operatorname{Si}(x) = \frac{i\pi}{2}+\frac{e^{ix}}{ix}\left\{\sum_{s=0}^{n-1}\frac{s!}{(ix)^s}+\vartheta_n(x)\frac{n!}{(ix)^n}\right\},$$

where $|\vartheta_n(x)| \leqslant 2$.

Ex. 1.4 With the aid of Exercise 4.3 of Chapter 2, show that the asymptotic expansion of the Fresnel integrals can be represented in the form

$$C\left\{\left(\frac{2x}{\pi}\right)^{1/2}\right\}+iS\left\{\left(\frac{2x}{\pi}\right)^{1/2}\right\} \sim \frac{1+i}{2}-\frac{ie^{ix}}{(2\pi x)^{1/2}}\sum_{s=0}^{\infty}\frac{1\cdot3\cdots(2s-1)}{(2ix)^s} \qquad (x\to\infty).$$

Show also that when $x>0$ and $n\geqslant1$, the nth implied constant of this expansion does not exceed twice the absolute value of the coefficient of the $(n+1)$th term.

2 Laplace Integrals

2.1 A general type of integral amenable to the method of integration by parts is given by

$$I(x) = \int_0^\infty e^{-xt}q(t)\,dt, \tag{2.01}$$

in which the function $q(t)$ is independent of the positive parameter x. We assume here that $q(t)$ is infinitely differentiable in $[0, \infty)$, and that for each s

$$q^{(s)}(t) = O(e^{\sigma t}) \qquad (0 \leqslant t < \infty), \tag{2.02}$$

where σ is a real constant which is independent of s.

The integral (2.01) converges when $x > \sigma$. Repeated integrations by parts produce

$$I(x) = \frac{q(0)}{x} + \frac{q'(0)}{x^2} + \cdots + \frac{q^{(n-1)}(0)}{x^n} + \varepsilon_n(x), \tag{2.03}$$

where n is an arbitrary nonnegative integer, and

$$\varepsilon_n(x) = \frac{1}{x^n}\int_0^\infty e^{-xt} q^{(n)}(t)\,dt. \tag{2.04}$$

With the assumed conditions,

$$\varepsilon_n(x) = \frac{1}{x^n}\int_0^\infty e^{-xt} O(e^{\sigma t})\,dt = \frac{1}{x^n} O\left\{\int_0^\infty e^{-(x-\sigma)t}\,dt\right\} = O\left\{\frac{1}{x^n(x-\sigma)}\right\}.$$

Therefore

$$I(x) \sim \sum_{s=0}^{\infty} \frac{q^{(s)}(0)}{x^{s+1}} \qquad (x \to \infty). \tag{2.05}$$

Less restrictive conditions for the validity of this result are given in Exercise 3.3 below.

2.2 Should the maximum value of $|q^{(n)}(t)|$ be attained at $t = 0$, then (2.04) immediately gives

$$|\varepsilon_n(x)| \leqslant |q^{(n)}(0)|\,x^{-n-1}, \tag{2.06}$$

when $x > 0$. This situation obtains, for example, when $q(t)$ is an *alternating function*† in $[0, \infty)$, that is, when

$$(-)^s q^{(s)}(t) \geqslant 0 \qquad (t \geqslant 0, \quad s = 0, 1, 2, \ldots).$$

The result (2.06) can be regarded as a special case of the so-called *error test*. This simple test asserts that if consecutive error terms associated with a series expansion have opposite signs, then each error term is numerically less than the first neglected term of the series, and has the same sign. In the present case we have

$$\varepsilon_n(x) - \varepsilon_{n+1}(x) = q^{(n)}(0)\,x^{-n-1}.$$

Clearly if $\varepsilon_n(x)$ and $\varepsilon_{n+1}(x)$ have opposite signs, then (2.06) applies and $\varepsilon_n(x)$ has the same sign as $q^{(n)}(0)$. The test has wider applicability than asymptotic expansions;

† Also known as a *completely monotonic function*. Some general properties of these functions have been given by Widder (1941, Chapter 4) and van der Corput and Franklin (1951). See also Exercises 2.1–2.3 below.

it can be used, for example, for finite-difference expansions arising in numerical analysis.†

It needs to be stressed that the error test has to be applied to consecutive *error* terms and not *actual* terms of the series. If it is merely known that $q^{(n)}(0)$ and $q^{(n+1)}(0)$ are of opposite sign, then the relation

$$\varepsilon_n(x) = \frac{q^{(n)}(0)}{x^{n+1}} + \frac{q^{(n+1)}(0)}{x^{n+2}} + O\left(\frac{1}{x^{n+3}}\right)$$

shows that (2.06) is certainly true for all $x > X_n$, provided that X_n is taken to be sufficiently large. But an actual value for X_n is not available from this analysis.

2.3 When $|q^{(n)}(t)|$ is not majorized by $|q^{(n)}(0)|$ we may consider the obvious extension

$$|\varepsilon_n(x)| \leqslant C_n x^{-n-1} \qquad (x > 0), \tag{2.07}$$

where

$$C_n = \sup_{(0,\infty)} |q^{(n)}(t)|.$$

Very often, however, C_n is infinite or else so large compared with $|q^{(n)}(0)|$ that this bound grossly overestimates the actual error. In these cases it is preferable to seek a majorant of the form

$$|q^{(n)}(t)| \leqslant |q^{(n)}(0)| \, e^{\sigma_n t} \qquad (0 \leqslant t < \infty), \tag{2.08}$$

in which the quantity σ_n is independent of t. Substitution of this majorant in (2.04) leads to

$$|\varepsilon_n(x)| \leqslant \frac{|q^{(n)}(0)|}{x^n(x-\sigma_n)} \qquad (x > \max(\sigma_n, 0)). \tag{2.09}$$

The previous condition $x > \sigma$ is not needed here because repeated integrations of (2.08) show that for $s < n$ and t large, $q^{(s)}(t)$ is $O(e^{\sigma_n t})$, $O(t^{n-s})$, or $O(t^{n-s-1})$ according as σ_n is positive, zero, or negative. In any event, (2.03) is valid for $x > \max(\sigma_n, 0)$.

The best value of σ_n is evidently

$$\sigma_n = \sup_{(0,\infty)} \left\{ \frac{1}{t} \ln \left| \frac{q^{(n)}(t)}{q^{(n)}(0)} \right| \right\}. \tag{2.10}$$

With the condition (2.02) this supremum is finite unless $q^{(n)}(0) = 0$. In the latter event we proceed to a nonvanishing higher term of the series.

Unlike (2.07) the ratio of the bound (2.09) to the actual value of $|\varepsilon_n(x)|$ has the desirable property of tending to unity as $x \to \infty$. The need to compute the derivatives of $q(t)$ is sometimes a drawback. A later method (§9) avoids this difficulty.

2.4 As an illustration of the error bound of the preceding subsection, consider again the expansion of the incomplete Gamma function. If we set $t = x(1+\tau)$

† Steffensen (1927, §4).

and $q(\tau) = (1+\tau)^{\alpha-1}$, then (1.02) becomes

$$e^x x^{-\alpha} \varepsilon_n(x) = x^{-n} \int_0^\infty e^{-x\tau} q^{(n)}(\tau)\, d\tau \qquad (x > 0); \tag{2.11}$$

compare (2.04). From (2.10) we have

$$\sigma_n = \sup_{(0,\infty)} \left\{ (\alpha - n - 1) \frac{\ln(1+\tau)}{\tau} \right\}. \tag{2.12}$$

When $\alpha - n - 1 \leqslant 0$, it is immediately seen that this supremum is attained at $\tau = \infty$ and equals zero. This leads to the same result as §1.1.

In the case $\alpha - n - 1 > 0$ the content of the braces in (2.12) is positive. Since $\ln(1+\tau)$ and τ are equal in the limit at $\tau = 0$ and the latter function grows more quickly than the former, the supremum must be approached as $\tau \to 0$. Hence $\sigma_n = \alpha - n - 1$, and (2.09) and (2.11) lead to

$$|\varepsilon_n(x)| \leqslant \frac{(\alpha-1)(\alpha-2)\cdots(\alpha-n) e^{-x} x^{\alpha-n}}{x - \alpha + n + 1} \qquad (x > \alpha - n - 1 > 0). \tag{2.13}$$

Moreover, it follows from (2.11) that $\varepsilon_n(x)$ is positive in these circumstances. By comparison with §1.2 the bound (2.13) is a slightly weaker, but more concise result. For the particular case $n = 0$, we have

$$\Gamma(\alpha, x) \leqslant \frac{e^{-x} x^\alpha}{x - \alpha + 1} \qquad (\alpha > 1, \quad x > \alpha - 1); \tag{2.14}$$

compare (1.05).

Ex. 2.1 Show that the sum or product of two alternating functions is itself alternating.

Ex. 2.2 If $q(t) > 0$ and $q'(t)$ is alternating, show that $1/q(t)$ is alternating.

Ex. 2.3 If $q(t)$ is nonnegative and continuous for $t > 0$, and each of its *moments* $\int_0^\infty t^s q(t)\, dt$, $s = 0, 1, \ldots$, is finite, show that the function $I(x)$ defined by (2.01) is alternating in $(0, \infty)$.

Ex. 2.4 Prove that

$$\int_0^\infty e^{-x \sinh t}\, dt \sim \sum_{s=0}^\infty (-)^s \frac{1^2 \cdot 3^2 \cdot 5^2 \cdots (2s-1)^2}{x^{2s+1}} \qquad (x \to \infty).$$

Ex. 2.5 Prove that

$$\int_1^\infty \frac{dt}{t^2 (x + \ln t)^{1/3}} \sim \frac{1}{x^{1/3}} \sum_{s=0}^\infty (-)^s \frac{1 \cdot 4 \cdot 7 \cdots (3s-2)}{(3x)^s} \qquad (x \to \infty).$$

Show also that for all positive x the error term is less than the first neglected term, and has the same sign.

Ex. 2.6 Show that

$$\int_0^\infty \exp\{-xt + (1+t)^{1/2}\}\, dt = (e/x)\{1 + \delta(x)\},$$

where

$$0 < \delta(x) \leqslant \{2(x - \sigma)\}^{-1} \qquad (x > \sigma); \qquad \sigma = \sup_{(0,\infty)} [\{(1+t)^{1/2} - 1 - \tfrac{1}{2}\ln(1+t)\}/t].$$

Estimate σ numerically by calculating the last expression for $t = 0, 2, 4, 6, 8, 10, 15, 20, 25, \infty$.

3 Watson's Lemma

3.1 A direct way of producing the expansion (2.05) is to substitute the Maclaurin expansion

$$q(t) = q(0) + tq'(0) + t^2 \frac{q''(0)}{2!} + \cdots \tag{3.01}$$

for $q(t)$ in (2.01) and integrate term by term. Of course this does not constitute a proof; the expansion (3.01) may not even be valid throughout the interval of integration. But this formal process suggests a natural extension: can a similar asymptotic result be constructed by termwise integration in cases when the expansion of $q(t)$ near $t = 0$ is in terms of noninteger powers of t?

An affirmative answer was supplied by Watson (1918a). It transpires that it is immaterial whether the expansion of $q(t)$ ascends in regularly spaced powers, or whether the series converges or is merely asymptotic. The general result is illustrated sufficiently well by the following theorem, which is probably the most frequently used result for deriving asymptotic expansions.

3.2 Theorem 3.1 *Let $q(t)$ be a function of the positive real variable t, such that*

$$q(t) \sim \sum_{s=0}^{\infty} a_s t^{(s+\lambda-\mu)/\mu} \qquad (t \to 0), \tag{3.02}$$

where λ and μ are positive constants. Then

$$\int_0^\infty e^{-xt} q(t)\, dt \sim \sum_{s=0}^{\infty} \Gamma\left(\frac{s+\lambda}{\mu}\right) \frac{a_s}{x^{(s+\lambda)/\mu}} \qquad (x \to \infty), \tag{3.03}$$

provided that this integral converges throughout its range for all sufficiently large x.

We may say that the expansion (3.02) *induces* the expansion (3.03). Subject to convergence, the conditions permit $q(t)$ to have a finite number of discontinuities and infinities anywhere in the range of integration, including $t = 0$. Convergence of the integral at $t = 0$, for all x, is assured by (3.02).

This theorem cannot be proved by a straightforward application of the method of integration by parts. Instead, we proceed as follows. For each nonnegative integer n, define

$$\phi_n(t) = q(t) - \sum_{s=0}^{n-1} a_s t^{(s+\lambda-\mu)/\mu} \qquad (t > 0). \tag{3.04}$$

Multiplying both sides of this identity by e^{-xt} and integrating by use of Euler's integral, we obtain

$$\int_0^\infty e^{-xt} q(t)\, dt = \sum_{s=0}^{n-1} \Gamma\left(\frac{s+\lambda}{\mu}\right) \frac{a_s}{x^{(s+\lambda)/\mu}} + \int_0^\infty e^{-xt} \phi_n(t)\, dt. \tag{3.05}$$

The integral on the right-hand side exists for all sufficiently large x, because the same is true of the integral on the left (by hypothesis).

As $t \to 0$ we have $\phi_n(t) = O(t^{(n+\lambda-\mu)/\mu})$. This means that there exist positive numbers k_n and K_n, say, such that

$$|\phi_n(t)| \leqslant K_n t^{(n+\lambda-\mu)/\mu} \qquad (0 < t \leqslant k_n).$$

Accordingly,

$$\left|\int_0^{k_n} e^{-xt}\phi_n(t)\,dt\right| \leqslant K_n \int_0^{k_n} e^{-xt} t^{(n+\lambda-\mu)/\mu}\,dt < \Gamma\left(\frac{n+\lambda}{\mu}\right)\frac{K_n}{x^{(n+\lambda)/\mu}}. \tag{3.06}$$

For the contribution from the range $[k_n, \infty)$, let X be a value of x for which $\int_0^\infty e^{-xt}\phi_n(t)\,dt$ converges, and write

$$\Phi_n(t) = \int_{k_n}^t e^{-Xv}\phi_n(v)\,dv,$$

so that $\Phi_n(t)$ is continuous and bounded in $[k_n, \infty)$. Let L_n denote the supremum of $|\Phi_n(t)|$ in this range. When $x > X$, we find by partial integration

$$\int_{k_n}^\infty e^{-xt}\phi_n(t)\,dt = \int_{k_n}^\infty e^{-(x-X)t}e^{-Xt}\phi_n(t)\,dt = (x-X)\int_{k_n}^\infty e^{-(x-X)t}\Phi_n(t)\,dt. \tag{3.07}$$

Therefore

$$\left|\int_{k_n}^\infty e^{-xt}\phi_n(t)\,dt\right| \leqslant (x-X)L_n\int_{k_n}^\infty e^{-(x-X)t}\,dt = L_n e^{-(x-X)k_n}. \tag{3.08}$$

Combining (3.06) and (3.08), we immediately see that the integral on the right-hand side of (3.05) is $O(x^{-(n+\lambda)/\mu})$ as $x \to \infty$, and the theorem is proved.

Error bounds for the expansion (3.03) are given later in the chapter (§9).

Ex. 3.1 Prove that

$$\int_0^\infty e^{-x\cosh t}\,dt \sim \left(\frac{\pi}{2x}\right)^{1/2} e^{-x}\sum_{s=0}^\infty (-)^s \frac{1^2\cdot 3^2\cdot 5^2 \cdots (2s-1)^2}{s!(8x)^s} \qquad (x \to \infty).$$

Ex. 3.2 In the notation of Chapter 2, Exercise 3.5, show that

$$E_x(x) \sim e^{-x}\sum_{s=1}^\infty e_s x^{-s} \qquad (x \to \infty),$$

where $e_1 = \frac{1}{2}$, $e_2 = \frac{1}{8}$, $e_3 = -\frac{1}{32}$, and $e_4 = -\frac{1}{128}$. [Airey, 1937.]

Ex. 3.3 If the integral (2.01) converges for all sufficiently large x, show that a sufficient condition that (2.05) furnishes an asymptotic expansion to n terms is that $q^{(n)}(t)$ be continuous in the neighborhood of $t = 0$.

Ex. 3.4 Suppose that $q(t)$ satisfies the conditions of Theorem 3.1, except that it has a simple pole at an interior point of $(0, \infty)$. Prove that (3.03) applies, provided that the integral is interpreted as a Cauchy principal value.

4 The Riemann–Lebesgue Lemma

4.1 Suppose that in a neighborhood of a finite point d, the function $q(t)$ is continuous except possibly at d. Moreover, suppose that as $t \to d$ from the left the limiting value $q(d-)$ exists; similarly as $t \to d$ from the right $q(d+)$ exists. If $q(d-) \neq q(d+)$, then d is called a *jump discontinuity*. Alternatively, if $q(d-) = q(d+)$ but either $q(d) \neq q(d-)$ or $q(d)$ does not exist, then d is called a *removable discontinuity*. For example, at $t = 0$ the function defined by

$$q(t) = 0 \quad (t < 0), \qquad q(0) = \tfrac{1}{2}, \qquad q(t) = 1 \quad (t > 0),$$

has a jump discontinuity, and its derivative has a removable discontinuity.

A *simple discontinuity* is either a jump discontinuity or a removable discontinuity.

Next, suppose that $q(t)$ is continuous in a finite or infinite interval (a, b), save for a finite number of simple discontinuities. Then we say that $q(t)$ is *sectionally continuous* (or *piecewise continuous*) in (a, b). If, also, a is finite and $q(a+)$ exists, then we say that $q(t)$ is sectionally continuous in $[a, b)$; compare Fig. 4.1. Similarly for the intervals $(a, b]$ and $[a, b]$.

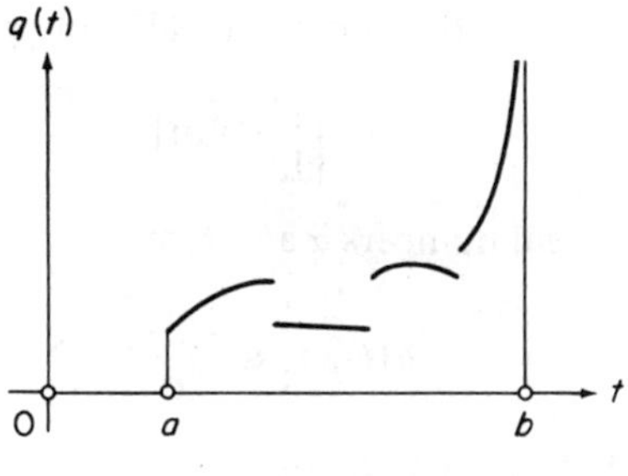

Fig. 4.1 Sectional continuity in $[a, b)$.

4.2 In developing the asymptotic theory of definite integrals of oscillatory functions, we shall refer frequently to the *Riemann–Lebesgue lemma*:

Theorem 4.1 (i) *Let $q(t)$ be sectionally continuous in a compact interval $[a, b]$. Then*

$$\int_a^b e^{ixt} q(t)\, dt = o(1) \qquad (x \to \infty). \tag{4.01}$$

(ii) *Let a be finite or $-\infty$, b be finite or $+\infty$, and $q(t)$ be continuous in (a, b) save possibly at a finite number of points. Then* (4.01) *again applies, provided that the integral converges uniformly at a, b, and the exceptional points, for all sufficiently large x.*

Two facets of this statement deserve attention. First, the result (ii) includes (i). Secondly, if the integral (4.01) converges absolutely, then necessarily it converges uniformly; on the other hand, it is readily verified by partial integration that the integral

$$\int_0^\infty \frac{e^{ixt}}{t^\delta}\,dt \qquad (0<\delta<1), \tag{4.02}$$

for example, converges uniformly at both limits for $x \geqslant X$ (>0), but does not converge absolutely at the upper limit.†

To prove (i) we observe that it is sufficient to establish the result when $q(t)$ is continuous in $[a,b]$; the extension to sectional continuity will follow by subdivision and summation. If $q(t)$ is continuous in $[a,b]$, then it is automatically uniformly continuous in this interval. This means that corresponding to an arbitrary positive number ε we can find a finite number of subdivision points t_s that satisfy

$$a \equiv t_0 < t_1 < t_2 < \cdots < t_n \equiv b$$

and

$$|q(t)-q(t_s)| < \frac{\varepsilon}{2(b-a)} \qquad (t_{s-1} \leqslant t \leqslant t_s)$$

for $s = 1, 2, \ldots, n$. Then

$$\int_a^b e^{ixt} q(t)\,dt = \sum_{s=1}^n q(t_s) \int_{t_{s-1}}^{t_s} e^{ixt}\,dt + \sum_{s=1}^n \int_{t_{s-1}}^{t_s} e^{ixt}\{q(t)-q(t_s)\}\,dt.$$

Let Q denote the maximum value of $|q(t)|$ in $[a,b]$. Since

$$\left|\int_\alpha^\beta e^{ixt}\,dt\right| = \left|\frac{e^{ix\beta}-e^{ix\alpha}}{ix}\right| \leqslant \frac{2}{x} \qquad (x>0)$$

for any real numbers α and β, we have

$$\left|\int_a^b e^{ixt} q(t)\,dt\right| \leqslant \frac{2Qn}{x} + \sum_{s=1}^n (t_s - t_{s-1}) \frac{\varepsilon}{2(b-a)} = \frac{2Qn}{x} + \frac{\varepsilon}{2} < \varepsilon,$$

provided that $x > 4Qn/\varepsilon$.

To prove (ii), let $d_1, d_2, \ldots, d_m$ be the interior points of (a,b), arranged in ascending order, at which $q(t)$ is discontinuous or infinite. The given conditions show that there exist finite points α_s and β_s such that

$$a < \alpha_0 < \beta_0 < d_1 < \alpha_1 < \beta_1 < d_2 < \cdots < d_m < \alpha_m < \beta_m < b$$

and each of the integrals

$$\int_a^{\alpha_0} e^{ixt} q(t)\,dt, \qquad \int_{\beta_m}^b e^{ixt} q(t)\,dt, \qquad \int_{\beta_s}^{\alpha_{s+1}} e^{ixt} q(t)\,dt \quad (s = 0, 1, \ldots, m-1),$$

is bounded in absolute value by ε for all $x \geqslant X$, assignable. To complete the proof the result (i) is applied to each of the intervals $[\alpha_s, \beta_s]$, $s = 0, 1, \ldots, m$.

Ex. 4.1 If $q(t)$ is continuous in $[0,\infty)$, $q'(t)$ is absolutely integrable over the same interval, and $q(t)$ vanishes as $t \to \infty$, prove that $\int_0^\infty e^{ixt} q(t)\,dt$ converges uniformly for all sufficiently large x.

† Integrals that converge, but do not converge absolutely, are sometimes called *conditionally convergent* or *improper*.

5 Fourier Integrals

5.1 A second type of integral to which the method of integration by parts may be directly applied is the finite Fourier integral

$$I(x) = \int_a^b e^{ixt} q(t)\, dt, \tag{5.01}$$

in which a, b, and $q(t)$ are independent of the positive parameter x.

If $q(t)$ is continuous and $q'(t)$ is absolutely integrable over $[a, b]$, then

$$I(x) = \frac{i}{x}\{e^{iax}q(a) - e^{ibx}q(b)\} + \varepsilon_1(x), \tag{5.02}$$

where

$$\varepsilon_1(x) = \frac{i}{x}\int_a^b e^{ixt} q'(t)\, dt. \tag{5.03}$$

The last integral is absolutely and uniformly convergent, hence by Theorem 4.1 we have $\varepsilon_1(x) = o(x^{-1})$ as $x \to \infty$.

Next, if all derivatives of $q(t)$ are continuous in $[a, b]$, then n integrations by parts yield

$$I(x) = \sum_{s=0}^{n-1} \left(\frac{i}{x}\right)^{s+1} \{e^{iax}q^{(s)}(a) - e^{ibx}q^{(s)}(b)\} + \varepsilon_n(x), \tag{5.04}$$

where

$$\varepsilon_n(x) = \left(\frac{i}{x}\right)^n \int_a^b e^{ixt} q^{(n)}(t)\, dt. \tag{5.05}$$

Again, $\varepsilon_n(x) = o(x^{-n})$ by the Riemann–Lebesgue lemma. Hence (5.04) furnishes an asymptotic expansion of $I(x)$ for large x.†

5.2 These results extend easily to an infinite range of integration. Suppose that all derivatives of $q(t)$ are continuous in $[a, \infty)$ and each of the integrals

$$\int_a^\infty e^{ixt} q^{(s)}(t)\, dt \qquad (s = 0, 1, \ldots)$$

converges uniformly for all sufficiently large x. Letting $b \to \infty$ in (5.02) and (5.03) we see that $e^{ibx}q(b)$ must tend to a constant limiting value, and since x can take more than one value it follows that $q(b) \to 0$ as $b \to \infty$. Application of Theorem 4.1 then shows that

$$I(x) = \frac{ie^{iax}}{x} q(a) + o\left(\frac{1}{x}\right) \qquad (x \to \infty).$$

This argument may be repeated successively for $n = 1, 2, \ldots$ in (5.04) and (5.05). In this way we establish

$$I(x) \sim \frac{ie^{iax}}{x} \sum_{s=0}^{\infty} q^{(s)}(a) \left(\frac{i}{x}\right)^s \qquad (x \to \infty).$$

† Compare Chapter 1, Exercise 7.1.

5.3 For a finite range of integration a simple bound for the error term (5.05) is given by

$$|\varepsilon_n(x)| \leqslant (b-a)\,Q_n x^{-n}, \qquad Q_n \equiv \max_{[a,b]} |q^{(n)}(t)|.$$

Often, however, this bound is a considerable overestimate and it is better to use

$$|\varepsilon_n(x)| \leqslant x^{-n} \mathcal{V}_{a,b}(q^{(n-1)}).$$

This form is also applicable when the range is infinite.

Ex. 5.1 By using (5.04) with $n = 3$, prove that

$$\int_0^\infty e^{ix\sinh t}\,dt = \frac{i}{x} + \varepsilon(x), \qquad \text{where } |\varepsilon(x)| \leqslant \left(2 + \frac{16}{25}\sqrt{\frac{2}{5}}\right)\frac{1}{x^3}.$$

Ex. 5.2 If $x > 0$ and n is any nonnegative integer, prove that

$$\left| \int_0^1 e^{-ixt} \ln(1+t)\,dt - \frac{i}{x} e^{-ix} \ln 2 - \sum_{s=0}^{n-1} s! \left(1 - \frac{e^{-ix}}{2^{s+1}}\right) \left(\frac{i}{x}\right)^{s+2} \right| \leqslant 2 \frac{n!}{x^{n+2}}.$$

6 Examples; Cases of Failure

6.1 There is especial need for care in appraising errors associated with the expansions derived in the preceding section. This is borne out by the following example.†

Consider

$$I(m) = \int_0^\pi \frac{\cos mt}{t^2+1}\,dt, \tag{6.01}$$

in which m is a large positive integer. Application of the analysis of §5.1 with $q(t) = (t^2+1)^{-1}$ yields

$$I(m) \sim (-)^m \sum_{s=0}^\infty (-)^s \frac{q^{(2s+1)}(\pi)}{m^{2s+2}} \qquad (m \to \infty), \tag{6.02}$$

since $q^{(2s+1)}(0) = 0$. The first three odd derivatives of $q(t)$ are

$$q'(t) = -\frac{2t}{(t^2+1)^2}, \qquad q^{(3)}(t) = -\frac{24(t^3-t)}{(t^2+1)^4}, \qquad q^{(5)}(t) = -\frac{240(3t^5-10t^3+3t)}{(t^2+1)^6},$$

from which it may be verified that

$$q'(\pi) = -0.05318, \qquad q^{(3)}(\pi) = -0.04791, \qquad q^{(5)}(\pi) = -0.08985,$$

correct to five decimal places. Accordingly, for $m = 10$ the first three terms of (6.02) contribute

$$-0.0005318 + 0.0000048 - 0.0000001 = -0.0005271. \tag{6.03}$$

† Olver (1964a).

But this plausible answer is quite incorrect, because direct numerical quadrature of the given integral (6.01) informs us that to seven decimal places

$$I(10) = -0.0004558. \tag{6.04}$$

The discrepancy is entirely attributable to neglect of the error term. When (6.02) is truncated at the term for which $s = n-1$, the error is given by

$$\varepsilon_{2n}(m) = \frac{(-)^n}{m^{2n}} \int_0^\pi \cos(mt)\, q^{(2n)}(t)\, dt = \frac{(-)^{n+1}}{m^{2n+1}} \int_0^\pi \sin(mt)\, q^{(2n+1)}(t)\, dt.$$

Accordingly,

$$|\varepsilon_{2n}(m)| \leqslant \mathscr{V}_{0,\pi}(q^{(2n)})/m^{2n+1}. \tag{6.05}$$

For $n = 2$,

$$q^{(4)}(t) = 24(5t^4 - 10t^2 + 1)/(t^2+1)^5.$$

The stationary points of this function are the zeros of $q^{(5)}(t)$. Those in the interval of variation are $t = 0$, $1/\sqrt{3}$, and $\sqrt{3}$, and computations yield

$$q^{(4)}(0) = 24.00, \qquad q^{(4)}(1/\sqrt{3}) = -10.12, \qquad q^{(4)}(\sqrt{3}) = 0.38, \qquad q^{(4)}(\pi) = 0.06.$$

Hence $\mathscr{V}_{0,\pi}(q^{(4)}) = 44.94$, and (6.05) becomes

$$|\varepsilon_4(10)| \leqslant 0.00045.$$

The size of this bound warns us that the series (6.03) may be grossly in error (although the actual error lies well within the bound).

6.2 A substantial improvement in the asymptotic expansion (6.02) is attainable by use of the identity

$$\int_0^\infty \frac{\cos mt}{t^2+1}\, dt = \frac{1}{2}\pi e^{-m}, \tag{6.06}$$

which is easily verifiable by contour integration. Addition to (6.01) produces

$$I(m) = \frac{1}{2}\pi e^{-m} - \int_\pi^\infty \frac{\cos mt}{t^2+1}\, dt.$$

Applying the method of §5 to the last integral, we obtain

$$I(m) = \frac{1}{2}\pi e^{-m} + (-)^m \sum_{s=0}^{n-1} (-)^s \frac{q^{(2s+1)}(\pi)}{m^{2s+2}} + \eta_{2n}(m), \tag{6.07}$$

where the new error term is bounded by

$$|\eta_{2n}(m)| \leqslant \mathscr{V}_{\pi,\infty}(q^{(2n)})/m^{2n+1}. \tag{6.08}$$

The representation (6.07) differs from (6.02) by the presence of the term $\frac{1}{2}\pi e^{-m}$. For $m = 10$, this term has the value 0.0000713, which is *exactly* the discrepancy between (6.03) and the correct value (6.04). This success is largely confirmed by evaluation of the error bound (6.08) for $n = 2$: the derivative $q^{(4)}(t)$ has no stationary points in (π, ∞); accordingly

$$\mathscr{V}_{\pi,\infty}(q^{(4)}) = q^{(4)}(\pi) = 0.06,$$

and (6.08) becomes

$$|\eta_4(10)| \leqslant 0.0000006.$$

The reason the error term $\varepsilon_{2n}(m)$ is generally much larger than the corresponding $\eta_{2n}(m)$ is that the derivatives of $q(t)$ are considerably larger in the interval $(0, \pi)$ than they are in (π, ∞). In turn this is traceable to the fact that in the complex plane the singularities of $q(t)$ at $t = \pm i$ are closer to the interval $(0, \pi)$ than they are to (π, ∞).

Two important lessons emerge from the foregoing example. First, the numerical use of an asymptotic expansion without investigation of its error terms may lead to disastrously wrong answers. Secondly, inclusion of terms that are exponentially small compared with other terms in the expansion may improve numerical results substantially, *even though the terms are negligible in Poincaré's sense*.

6.3 The last two subsections furnish an example of partial failure of the method of integration by parts to produce a satisfactory asymptotic representation of a Fourier integral. Complete failure can occur in the following way.† Let

$$I(m) = \int_a^b \cos(mt)\, q(t)\, dt,$$

in which a and b are multiples of π, and all odd derivatives of $q(t)$ vanish at a and b. The integral (6.06), for example, is effectively of this type. Application of the method of §5 yields

$$I(m) \sim \frac{0}{m} + \frac{0}{m^2} + \frac{0}{m^3} + \cdots \qquad (m \to \infty).$$

This result is valid, but useless for numerical and most analytical purposes. Similarly for the corresponding integral with $\cos(mt)$ replaced by $\sin(mt)$ and vanishing even derivatives of $q(t)$ at a and b.

As in §6.2 it may be necessary to resort to methods of contour integration to obtain a satisfactory approximation to $I(m)$ in these circumstances.

6.4 As an example consider

$$I(x) = \int_{-\infty}^{\infty} \frac{t \sin(xt)}{t^2 + \alpha^2} h(t)\, dt,$$

in which α is a positive constant and x a large positive parameter. Suppose that the function $h(t)$ is real when t is real and holomorphic in a domain containing the strip $|\operatorname{Im} t| \leqslant \beta$, where $\beta > \alpha$. Suppose also that

$$h(t) = O(t^{-\delta}) \qquad (\operatorname{Re} t \to \pm\infty) \tag{6.09}$$

† Pointed out to the author by L. Maximon.

uniformly with respect to $\operatorname{Im} t$ in the strip, where $\delta > 0$. Then we have the kind of failure discussed in the previous subsection, because the function $th(t)(t^2+\alpha^2)^{-1}$ is uniformly $O(t^{-1-\delta})$ as $\operatorname{Re} t \to \pm\infty$, and therefore its derivatives all vanish as $t \to \pm\infty$; compare Chapter 1, Exercise 4.7.

Application of the residue theorem to the boundary of the upper half of the strip gives

$$\int_{-\infty}^{\infty} \frac{te^{ixt}}{t^2+\alpha^2} h(t)\, dt = \pi i e^{-\alpha x} h(i\alpha) + \varepsilon(x),$$

where

$$\varepsilon(x) = \int_{\mathscr{L}} \frac{te^{ixt}}{t^2+\alpha^2} h(t)\, dt,$$

$\mathscr{L}$ being the line defined parametrically by $t = i\beta+\tau$, $-\infty < \tau < \infty$. Clearly

$$|\varepsilon(x)| \leqslant e^{-\beta x} \int_{\mathscr{L}} \left| \frac{th(t)}{t^2+\alpha^2}\, dt \right| = O(e^{-\beta x}),$$

since the integral is necessarily finite with the condition (6.09). Combination of these results gives an explicit representation

$$I(x) = \pi e^{-\alpha x} \operatorname{Re}\{h(i\alpha)\} + O(e^{-\beta x}) \qquad (x \to \infty).$$

Ex. 6.1 Let m be a positive integer. By expressing the integral

$$I(m) = \int_0^{\pi} \frac{\sin(mt)}{\sinh t}\, dt$$

in the form

$$\operatorname{Si}(m\pi) + \int_0^{\pi} \left(\frac{1}{\sinh t} - \frac{1}{t} \right) \sin(mt)\, dt,$$

show that

$$I(m) \sim \frac{1}{2}\pi + \frac{(-)^{m-1}}{m} \sum_{s=0}^{\infty} (-)^s \frac{h_{2s}}{m^{2s}} \qquad (m \to \infty),$$

where h_{2s} is the value of the $2s$th derivative of $\operatorname{csch} t$ at $t = \pi$.

Show also that a more accurate representation is given by

$$I(m) \sim \frac{1}{2}\pi - \frac{\pi}{e^{m\pi}+1} + \frac{(-)^{m-1}}{m} \sum_{s=0}^{\infty} (-)^s \frac{h_{2s}}{m^{2s}}.$$

Ex. 6.2 If α and ρ are positive constants and x is positive, show that

$$\int_0^{\infty} \frac{t \exp(-\rho^2 t^2) \sin(xt)}{t^2+\alpha^2}\, dt = \frac{1}{2}\pi \exp(\alpha^2\rho^2 - \alpha x) + \varepsilon(x),$$

where

$$|\varepsilon(x)| \leqslant \pi^{1/2}\beta \exp(-\beta x + \rho^2\beta^2)/\{2\rho(\beta^2-\alpha^2)\},$$

β being any number exceeding α. By allowing β to depend on x deduce that

$$\varepsilon(x) = O\{x^{-1} e^{-x^2/(4\rho^2)}\} \qquad (x \to \infty).$$

7 Laplace's Method

7.1 Consider the generalization of the integral of §2 given by

$$I(x) = \int_a^b e^{-xp(t)} q(t)\, dt, \tag{7.01}$$

in which a, b, $p(t)$, and $q(t)$ are independent of the positive parameter x. Either a or b or both may be infinite. The following powerful method for approximating $I(x)$ originated with Laplace (1820). The peak value of the factor $e^{-xp(t)}$ occurs at the point $t = t_0$, say, at which $p(t)$ is a minimum. When x is large, this peak is very sharp, and the graph of the integrand suggests that the overwhelming contribution to the integral comes from the neighborhood of t_0. Accordingly, we replace $p(t)$ and $q(t)$ by the leading terms in their series expansions in ascending powers of $t - t_0$, and then, as appropriate, extend the integration limits to $-\infty$ or $+\infty$. The resulting integral is explicitly evaluable and yields the required approximation.

Suppose, for example, that $t_0 = a$, $p'(a) > 0$, and $q(a) \neq 0$. Then Laplace's procedure is expressed by

$$I(x) \risingdotseq \int_a^b e^{-x\{p(a)+(t-a)p'(a)\}} q(a)\, dt$$

$$\risingdotseq q(a) e^{-xp(a)} \int_a^\infty e^{-x(t-a)p'(a)}\, dt = \frac{q(a) e^{-xp(a)}}{xp'(a)}. \tag{7.02}$$

Another common case arises when $p(t)$ has a simple minimum at an interior point t_0 of (a, b) and $q(t_0) \neq 0$. Then

$$I(x) \risingdotseq \int_a^b \exp[-x\{p(t_0) + \tfrac{1}{2}(t-t_0)^2 p''(t_0)\}]\, q(t_0)\, dt$$

$$\risingdotseq q(t_0) e^{-xp(t_0)} \int_{-\infty}^{\infty} \exp\{-\tfrac{1}{2}x(t-t_0)^2 p''(t_0)\}\, dt = q(t_0) e^{-xp(t_0)} \left\{\frac{2\pi}{xp''(t_0)}\right\}^{1/2}. \tag{7.03}$$

It should be observed that in constructing these approximations the assumption that only the neighborhood of the peak is of importance is used twice: first, when $p(t)$ and $q(t)$ are replaced by the leading terms of their expansions in powers of $t - t_0$; secondly, when b is replaced by ∞ and, in the case of (7.03), a is replaced by $-\infty$.

7.2 The foregoing analysis is heuristic. With precisely formulated conditions on $p(t)$ and $q(t)$ we shall prove that the Laplace approximation is asymptotic to the given integral as $x \to \infty$. Without loss of generality it may be supposed that a is finite and the minimum of $p(t)$ occurs at $t = a$: in other cases the integration range can be subdivided at the minima and maxima of $p(t)$, and the sign of t reversed where necessary.

We suppose that the limits a and b are independent of x, a being finite and b $(> a)$

finite or infinite. The functions $p(t)$ and $q(t)$ are independent of x, $p(t)$ being real and $q(t)$ either real or complex. In addition:

(i) *$p(t) > p(a)$ when $t \in (a,b)$, and for every $c \in (a,b)$ the infimum of $p(t)-p(a)$ in $[c,b)$ is positive.*†

(ii) *$p'(t)$ and $q(t)$ are continuous in a neighborhood of a, except possibly at a.*

(iii) *As $t \to a$ from the right*

$$p(t)-p(a) \sim P(t-a)^{\mu}, \qquad q(t) \sim Q(t-a)^{\lambda-1},$$

and the first of these relations is differentiable. Here P, μ, and λ are positive constants (integers or otherwise), and Q is a nonzero real or complex constant.

(iv)

$$I(x) \equiv \int_a^b e^{-xp(t)} q(t)\, dt \tag{7.04}$$

converges absolutely throughout its range for all sufficiently large x.

Theorem 7.1‡ *With the conditions of this subsection*

$$I(x) \sim \frac{Q}{\mu}\Gamma\left(\frac{\lambda}{\mu}\right)\frac{e^{-xp(a)}}{(Px)^{\lambda/\mu}} \qquad (x \to \infty). \tag{7.05}$$

The proof follows.

7.3 Conditions (ii) and (iii) show that a number k can be found which is close enough to a to ensure that in $(a,k]$, $p'(t)$ is continuous and positive and $q(t)$ is continuous. Since $p(t)$ is increasing in (a,k) we may take

$$v = p(t) - p(a)$$

as new integration variable in this interval. Then v and t are continuous functions of each other and

$$e^{xp(a)} \int_a^k e^{-xp(t)} q(t)\, dt = \int_0^{\kappa} e^{-xv} f(v)\, dv, \tag{7.06}$$

where

$$\kappa = p(k) - p(a), \qquad f(v) = q(t)\frac{dt}{dv} = \frac{q(t)}{p'(t)}. \tag{7.07}$$

Clearly κ is finite and positive, and $f(v)$ is continuous when $v \in (0,\kappa]$.

Since $v \sim P(t-a)^{\mu}$ as $t \to a$, we have§

$$t - a \sim (v/P)^{1/\mu} \qquad (v \to 0+),$$

and hence

$$f(v) \sim \frac{Qv^{(\lambda/\mu)-1}}{\mu P^{\lambda/\mu}} \qquad (v \to 0+). \tag{7.08}$$

† In other words, the minimum of $p(t)$ is approached only at a.

‡ Erdélyi (1956a, §2.4).

§ Compare Chapter 1, Theorem 5.1.

In consequence of this relation we rearrange the integral (7.06) in the form

$$\int_0^\kappa e^{-xv} f(v)\,dv = \frac{Q}{\mu P^{\lambda/\mu}}\left\{\int_0^\infty e^{-xv} v^{(\lambda/\mu)-1}\,dv - \varepsilon_1(x)\right\} + \varepsilon_2(x), \tag{7.09}$$

where

$$\varepsilon_1(x) = \int_\kappa^\infty e^{-xv} v^{(\lambda/\mu)-1}\,dv, \qquad \varepsilon_2(x) = \int_0^\kappa e^{-xv}\left\{f(v) - \frac{Qv^{(\lambda/\mu)-1}}{\mu P^{\lambda/\mu}}\right\} dv.$$

The first term on the right-hand side of (7.09) is evaluable by use of Euler's integral and immediately yields the required approximation (7.05).

Secondly, given an arbitrary positive number ε we make κ small enough (by choosing k sufficiently close to a) to ensure that

$$\left| f(v) - \frac{Qv^{(\lambda/\mu)-1}}{\mu P^{\lambda/\mu}} \right| < \varepsilon \frac{|Q|\, v^{(\lambda/\mu)-1}}{\mu P^{\lambda/\mu}} \qquad (0 < v \le \kappa);$$

compare (7.08). Then by use again of Euler's integral we derive

$$|\varepsilon_2(x)| < \varepsilon \frac{|Q|}{\mu} \Gamma\left(\frac{\lambda}{\mu}\right) \frac{1}{(Px)^{\lambda/\mu}}. \tag{7.10}$$

Thirdly, in the notation of the incomplete Gamma function we have

$$\varepsilon_1(x) = \frac{1}{x^{\lambda/\mu}} \Gamma\left(\frac{\lambda}{\mu}, \kappa x\right) = O\left(\frac{e^{-\kappa x}}{x}\right) \tag{7.11}$$

for large x; compare (1.04).

Lastly, let X be a value of x for which $I(x)$ is absolutely convergent and write

$$\eta \equiv \inf_{[k,b)} \{p(t) - p(a)\}. \tag{7.12}$$

In consequence of Condition (i) η is positive. Restricting $x \geqslant X$, we have

$$\begin{aligned} xp(t) - xp(a) &= (x-X)\{p(t)-p(a)\} + X\{p(t)-p(a)\} \\ &\geqslant (x-X)\eta + Xp(t) - Xp(a), \end{aligned}$$

and hence

$$\left| e^{xp(a)} \int_k^b e^{-xp(t)} q(t)\,dt \right| \leqslant e^{-(x-X)\eta + Xp(a)} \int_k^b e^{-Xp(t)} |q(t)|\,dt. \tag{7.13}$$

The proof of Theorem 7.1 is completed by making x large enough to guarantee that the right-hand sides of (7.11) and (7.13) are both bounded by $\varepsilon x^{-\lambda/\mu}$; this is always possible since κ and η are positive.

7.4 An example is supplied by the modified Bessel function of integer order, given by

$$I_n(x) = \frac{1}{\pi} \int_0^\pi e^{x\cos t} \cos(nt)\,dt;$$

compare Chapter 2, Exercise 10.1. In the notation of §7.2

$$p(t) = -\cos t, \qquad q(t) = \pi^{-1}\cos(nt).$$

Clearly $p(t)$ is increasing for $0 < t < \pi$, and Conditions (i) and (ii) are satisfied. Condition (iv) does not apply. As $t \to 0$

$$p(t) = -1 + \tfrac{1}{2}t^2 + O(t^4), \qquad q(t) = \pi^{-1} + O(t^2).$$

Hence $p(a) = -1$, $P = \frac{1}{2}$, $\mu = 2$, $Q = \pi^{-1}$, and $\lambda = 1$. Accordingly, Theorem 7.1 gives

$$I_n(x) \sim (2\pi x)^{-1/2}e^x \qquad (x \to \infty, \quad n \text{ fixed}).$$

Higher terms in this approximation are given in Exercise 8.5 below (for $n = 0$) and Chapter 7, §8.2 (for general n).

7.5 A harder example is provided by†

$$I(x) = \int_0^\infty e^{xt-(t-1)\ln t}\,dt.$$

We note first that the obvious choice $p(t) = -t$ is unfruitful, because $-t$ has no minimum in the integration range. We therefore consider the peak value of the *whole* integrand. This occurs where

$$x - 1 - \ln t + (1/t) = 0.$$

For large x, the relevant root of this equation is given by

$$t \sim e^{x-1} = \xi,$$

say. To apply our theory, the location of the peak needs to be independent of x. Therefore we take $\tau = t/\xi$ as new integration variable, so that

$$I(x) = \xi^2 \int_0^\infty e^{-\xi p(\tau)}q(\tau)\,d\tau, \tag{7.14}$$

where

$$p(\tau) = \tau(\ln\tau - 1), \qquad q(\tau) = \tau.$$

The only minimum of $p(\tau)$ is at $\tau = 1$. Expansions in powers of $\tau - 1$ are given by

$$p(\tau) = -1 + \tfrac{1}{2}(\tau-1)^2 - \tfrac{1}{6}(\tau-1)^3 + \cdots, \qquad q(\tau) = 1 + (\tau - 1).$$

Accordingly, in the notation of §7.2, $p(a) = -1$, $P = \frac{1}{2}$, $\mu = 2$, $Q = 1$, and $\lambda = 1$. Hence (7.05) yields

$$\int_1^\infty e^{-\xi p(\tau)}q(\tau)\,d\tau \sim \left(\frac{\pi}{2\xi}\right)^{1/2} e^{\xi}.$$

On replacing τ by $2 - \tau$, we see that the same asymptotic approximation holds for

† Based on an example of Evgrafov (1961, p. 27).

the corresponding integral over the range $0 \leqslant \tau \leqslant 1$. Substitution of these results in (7.14) and restoration of the original variable x leads to the required result

$$I(x) \sim (2\pi)^{1/2} e^{3(x-1)/2} \exp(e^{x-1}) \qquad (x \to \infty). \tag{7.15}$$

The reader is advised to understand thoroughly the preliminary steps in this example, since they recur often with other examples and other methods. First, an equation was set up for the abscissa t of the peak value of the whole integrand. Secondly, this transcendental equation was solved asymptotically for large x, giving $t = \xi(x)$, say. Thirdly, a new integration variable $\tau = t/\xi(x)$ was introduced with the object of making the (approximate) location of the new peak independent of the parameter x.

Ex. 7.1 Using the integral given in Chapter 2, Exercise 7.9, show that for fixed positive α and large n the Legendre polynomial $P_n(\cosh\alpha)$ is approximated by $(2\pi n \sinh\alpha)^{-1/2} e^{n\alpha+(\alpha/2)}$.

Ex. 7.2† Let $\mathbf{A}_\nu(x) = \int_0^\infty e^{-\nu t - x\sinh t}\,dt$. Show that

$$\mathbf{A}_\nu(x) \sim 1/x \qquad (x \to \infty, \quad \nu \text{ fixed}),$$

and

$$\mathbf{A}_\nu(a\nu) \sim 1/(a\nu + \nu) \qquad (\nu \to \infty, \quad a \text{ fixed and nonnegative}).$$

Show also that if a is fixed and $\nu \to \infty$, then $\mathbf{A}_{-\nu}(a\nu)$ is asymptotic to

$$\frac{1}{a\nu - \nu}, \quad \left(\frac{2}{9}\right)^{1/3} \Gamma\left(\frac{1}{3}\right) \frac{1}{\nu^{1/3}}, \quad \text{or} \quad \left(\frac{2\pi}{\nu}\right)^{1/2} \left(\frac{1+(1-a^2)^{1/2}}{a}\right)^\nu \frac{\exp\{-\nu(1-a^2)^{1/2}\}}{(1-a^2)^{1/4}},$$

according as $a > 1$, $a = 1$, or $0 < a < 1$.

Ex. 7.3 Let α and β be constants such that $0 < \alpha < 1$ and $\beta > 0$. Show that for large positive values of x

$$\int_0^\infty \exp(-t - xt^\alpha)\, t^{\beta-1}\, dt \sim \frac{\Gamma(\beta/\alpha)}{\alpha x^{\beta/\alpha}},$$

and

$$\int_0^\infty \exp(-t + xt^\alpha)\, t^{\beta-1}\, dt \sim \left(\frac{2\pi}{1-\alpha}\right)^{1/2} (\alpha x)^{(2\beta-1)/(2-2\alpha)} \exp\{(1-\alpha)(\alpha^\alpha x)^{1/(1-\alpha)}\}.$$

[Bakhoom, 1933.]

Ex. 7.4 Show that

$$\int_0^\infty t^x e^{-t} \ln t\, dt \sim (2\pi)^{1/2} e^{-x} x^{x+(1/2)} \ln x \qquad (x \to \infty).$$

Ex. 7.5 With the conditions of §7.2, assume that as $t \to a+$

$$p'(t) = \mu P(t-a)^{\mu-1} + O\{(t-a)^{\mu_1-1}\}, \qquad q(t) = Q(t-a)^{\lambda-1} + O\{(t-a)^{\lambda_1-1}\},$$

where $\mu_1 > \mu$ and $\lambda_1 > \lambda$. Prove that the relative error in (7.05) is $O(x^{-\varpi/\mu})$, where

$$\varpi = \min(\lambda_1 - \lambda, \mu_1 - \mu).$$

Ex. 7.6 Assume that $p'(t)$ is continuous and $p(t)$ has a finite number of maxima and minima in (a, b). Using the method of proof of §3.2 show that Condition §7.2(iv) of Theorem 7.1 may be replaced by: *$I(x)$ converges for at least one value of x.*

† This integral is related to the so-called *Anger function*; compare Exercise 13.3 below and also Chapter 9, §12.

8 Asymptotic Expansions by Laplace's Method; Gamma Function of Large Argument

8.1 Theorem 7.1 confirms the prediction of §7.1 that in a wide range of circumstances the asymptotic form of the integral (7.01) for large x depends solely on the behavior of the integrand near the minimum of $p(t)$. An extension of the analysis enables an asymptotic expansion to be developed for $I(x)$ in descending powers of x. We assume that $p(t)$ and $q(t)$ can be expanded in series of ascending powers of $t-a$ in the neighborhood of a. As in the case of Watson's lemma, it matters not whether these series are convergent or merely asymptotic, or whether the powers of $t-a$ are integers. The procedure is adequately illustrated by the following case.

Assume that

$$p(t) \sim p(a) + \sum_{s=0}^{\infty} p_s(t-a)^{s+\mu}, \tag{8.01}$$

and

$$q(t) \sim \sum_{s=0}^{\infty} q_s(t-a)^{s+\lambda-1}, \tag{8.02}$$

as $t \to a$ from the right, where μ and λ are again positive constants.† Without loss of generality we may suppose that $p_0 \neq 0$ and $q_0 \neq 0$. Since $t = a$ is to be a minimum of $p(t)$, p_0 is necessarily positive. Assume also that (8.01) can be differentiated, that is,

$$p'(t) \sim \sum_{s=0}^{\infty} (s+\mu)p_s(t-a)^{s+\mu-1} \qquad (t \to a+). \tag{8.03}$$

By substituting (8.01) in the equation

$$v = p(t) - p(a)$$

and reverting in the manner of Chapter 1, §8.4, we arrive at an expansion of the form

$$t - a \sim \sum_{s=1}^{\infty} c_s v^{s/\mu} \qquad (v \to 0+). \tag{8.04}$$

The first three coefficients may be verified to be

$$c_1 = \frac{1}{p_0^{1/\mu}}, \qquad c_2 = -\frac{p_1}{\mu p_0^{1+(2/\mu)}}, \qquad c_3 = \frac{(\mu+3)p_1^2 - 2\mu p_0 p_2}{2\mu^2 p_0^{2+(3/\mu)}}. \tag{8.05}$$

Substitution of this result in (8.02) and (8.03) and use of the equation

$$f(v) = q(t)\frac{dt}{dv} = \frac{q(t)}{p'(t)} \tag{8.06}$$

† Actually λ can be complex without complication, provided that $\operatorname{Re}\lambda > 0$.

(compare (7.07)), then yields

$$f(v) \sim \sum_{s=0}^{\infty} a_s v^{(s+\lambda-\mu)/\mu} \qquad (v \to 0+), \tag{8.07}$$

where the a_s are expressible in terms of p_s and q_s. In particular,

$$a_0 = \frac{q_0}{\mu p_0^{\lambda/\mu}}, \qquad a_1 = \left\{\frac{q_1}{\mu} - \frac{(\lambda+1)p_1 q_0}{\mu^2 p_0}\right\} \frac{1}{p_0^{(\lambda+1)/\mu}},$$

and

$$a_2 = \left[\frac{q_2}{\mu} - \frac{(\lambda+2)p_1 q_1}{\mu^2 p_0} + \{(\lambda+\mu+2)p_1^2 - 2\mu p_0 p_2\} \frac{(\lambda+2)q_0}{2\mu^3 p_0^2}\right] \frac{1}{p_0^{(\lambda+2)/\mu}}.$$

(In the common case $q(t) = 1$, we have $\lambda = 1$ and $a_s = (s+1)c_{s+1}/\mu$.)

8.2 Theorem 8.1† *Let Conditions* (i), (ii), *and* (iv) *of* §7.2 *be satisfied and the expansions* (8.01), (8.02), *and* (8.03) *hold. Then*

$$\int_a^b e^{-xp(t)} q(t)\, dt \sim e^{-xp(a)} \sum_{s=0}^{\infty} \Gamma\left(\frac{s+\lambda}{\mu}\right) \frac{a_s}{x^{(s+\lambda)/\mu}} \qquad (x \to \infty), \tag{8.08}$$

where the coefficients a_s are defined in §8.1.

This result is proved in a similar manner to Theorem 7.1. We again suppose that k is a point on the right of a close enough to ensure that $p'(t)$ is continuous and positive and $q(t)$ is continuous in $(a, k]$, and write $\kappa = p(k) - p(a)$. We proceed from (7.06), as follows. For each positive integer n, let the remainder coefficient $f_n(v)$ be defined by $f_n(0) = a_n$ and

$$f(v) = \sum_{s=0}^{n-1} a_s v^{(s+\lambda-\mu)/\mu} + v^{(n+\lambda-\mu)/\mu} f_n(v) \qquad (v > 0). \tag{8.09}$$

Corresponding to (7.09) we have

$$\int_0^\kappa e^{-xv} f(v)\, dv = \sum_{s=0}^{n-1} \Gamma\left(\frac{s+\lambda}{\mu}\right) \frac{a_s}{x^{(s+\lambda)/\mu}} - \varepsilon_{n,1}(x) + \varepsilon_{n,2}(x), \tag{8.10}$$

where

$$\varepsilon_{n,1}(x) = \sum_{s=0}^{n-1} \Gamma\left(\frac{s+\lambda}{\mu}, \kappa x\right) \frac{a_s}{x^{(s+\lambda)/\mu}}, \tag{8.11}$$

and

$$\varepsilon_{n,2}(x) = \int_0^\kappa e^{-xv} v^{(n+\lambda-\mu)/\mu} f_n(v)\, dv. \tag{8.12}$$

From (1.04) it is seen that for large x

$$\varepsilon_{n,1}(x) = O(e^{-\kappa x}/x).$$

† Erdélyi (1956a, §2.4). Theorem 3.1 corresponds to the special case obtained by taking $a = 0$, $b = \infty$, $p(t) = t^\mu$, and then replacing t^μ by t.

Also, since κ is finite and $f_n(v)$ is continuous in $[0, \kappa]$, it follows that

$$\varepsilon_{n,2}(x) = \int_0^\kappa e^{-xv} v^{(n+\lambda-\mu)/\mu} O(1)\, dv = O\left(\frac{1}{x^{(n+\lambda)/\mu}}\right).$$

Accordingly, the contribution to $I(x)$ from the integration range (a, k) has the stated asymptotic expansion. For the remaining range (k, b), the bound (7.13) again applies, and the asymptotic expansion is unaffected. This completes the proof.

8.3 An important illustration is provided by Euler's integral, in the form

$$\Gamma(x) = x^{-1} \int_0^\infty e^{-w} w^x\, dw \qquad (x > 0).$$

The integrand is zero at $w = 0$, increases to a maximum at $w = x$, then decreases steadily back to zero as $w \to \infty$. The location of the maximum is made independent of x on taking w/x as new integration variable, but because the notation simplifies slightly with the maximum at the origin, we set $w = x(1+t)$. This gives

$$\Gamma(x) = e^{-x} x^x \int_{-1}^\infty e^{-xt} (1+t)^x\, dt = e^{-x} x^x \int_{-1}^\infty e^{-xp(t)}\, dt, \tag{8.13}$$

where

$$p(t) = t - \ln(1+t).$$

Subdivision at the minimum of $p(t)$ produces

$$e^x x^{-x} \Gamma(x) = \int_0^\infty e^{-xp(t)}\, dt + \int_0^1 e^{-xp(-t)}\, dt. \tag{8.14}$$

Since $p'(t) = t/(1+t)$, and

$$p(t) = \tfrac{1}{2}t^2 - \tfrac{1}{3}t^3 + \tfrac{1}{4}t^4 - \cdots \qquad (-1 < t < 1),$$

it is easily seen that the conditions of Theorem 8.1 are satisfied by each integral in (8.14). With $v = p(t)$, reversion of the last expansion yields, for the first integral,

$$t = 2^{1/2} v^{1/2} + \frac{2}{3} v + \frac{2^{1/2}}{18} v^{3/2} - \frac{2}{135} v^2 + \frac{2^{1/2}}{1080} v^{5/2} + \cdots,$$

this expansion converging for sufficiently small v. Thence we derive

$$f(v) \equiv \frac{dt}{dv} = a_0 v^{-1/2} + a_1 + a_2 v^{1/2} + \cdots, \tag{8.15}$$

where, for example,

$$a_0 = \frac{2^{1/2}}{2}, \qquad a_1 = \frac{2}{3}, \qquad a_2 = \frac{2^{1/2}}{12}, \qquad a_3 = -\frac{4}{135}, \qquad a_4 = \frac{2^{1/2}}{432}.$$

From (8.08) we find that

$$\int_0^\infty e^{-xp(t)}\, dt \sim \sum_{s=0}^\infty \Gamma\left(\frac{s+1}{2}\right) \frac{a_s}{x^{(s+1)/2}}.$$

Similarly,

$$\int_0^1 e^{-xp(-t)}\,dt \sim \sum_{s=0}^{\infty} (-)^s \Gamma\left(\frac{s+1}{2}\right) \frac{a_s}{x^{(s+1)/2}}.$$

Substitution of these series in (8.14) yields the required result:

$$\Gamma(x) \sim e^{-x}x^x\left(\frac{2\pi}{x}\right)^{1/2}\left(1+\frac{1}{12x}+\frac{1}{288x^2}+\cdots\right) \qquad (x\to\infty). \qquad (8.16)$$

The leading term in this expansion is often known as *Stirling's formula*. No general expression is available for the coefficients.†

An alternative way of expanding $\Gamma(x)$ asymptotically for large x, complete with error bounds, will be given in Chapter 8, §4.

Ex. 8.1 Assume $p^{iv}(t)$ and $q''(t)$ are continuous in (a,b), the minimum of $p(t)$ is attained at an interior point t_0, and $p(t)$ is bounded away from $p(t_0)$ as $t\to a$ or b. Show that

$$\int_a^b e^{-xp(t)}q(t)\,dt = q(t_0)e^{-xp(t_0)}\left\{\frac{2\pi}{xp''(t_0)}\right\}^{1/2}\left\{1+O\left(\frac{1}{x}\right)\right\} \qquad (x\to\infty),$$

provided that $p''(t_0)$ and $q(t_0)$ are nonzero and the integral converges absolutely for all sufficiently large x.

Ex. 8.2 Using the preceding exercise deduce that the relative error in (7.15) is $O(e^{-x})$.

Ex. 8.3 Show that the coefficients a_s of §8.3 satisfy

$$a_0a_s+\frac{1}{2}a_1a_{s-1}+\frac{1}{3}a_2a_{s-2}+\cdots+\frac{1}{s+1}a_sa_0=\frac{1}{s}a_{s-1} \qquad (s\geqslant 1).$$

Ex. 8.4 Show that

$$\int_0^{\pi^2/4} e^{x\cos\sqrt{t}}\,dt \sim e^x\left(\frac{2}{x}+\frac{2}{3x^2}+\frac{8}{15x^3}+\cdots\right) \qquad (x\to\infty).$$

Does this result still hold if the integration limits are changed to (a) 0 and π^2, (b) 0 and $4\pi^2$?

Ex. 8.5 In the notation of §7.4, show that

$$I_0(x) \sim \frac{e^x}{(2\pi x)^{1/2}}\sum_{s=0}^{\infty}\frac{1^2\cdot 3^2\cdot 5^2\cdots(2s-1)^2}{s!(8x)^s} \qquad (x\to\infty).$$

Ex. 8.6 Prove that

$$\int_0^\infty \frac{t^{v-1}}{\Gamma(v)}\,dv \sim \frac{1}{t}\sum_{s=0}^{\infty}(-)^s\frac{\mathrm{Rg}^{(s+1)}(0)}{(\ln t)^{s+2}} \qquad (t\to 0+),$$

where $\mathrm{Rg}(v)=1/\Gamma(v)$.

Ex. 8.7‡ By using Stirling's formula, show that for fixed nonnegative α

$$\int_\alpha^\infty \frac{t^{v-1}}{\Gamma(v)}\,dv \sim e^t \qquad (t\to\infty).$$

† The first twenty-one have been given by Wrench (1968), together with approximate values of the next ten.

‡ Higher approximations to this integral are derived in Chapter 8, §11.4.

*9 Error Bounds for Watson's Lemma and Laplace's Method

9.1 In the case of Theorem 3.1 a natural way of extending the error analysis of §2.3 is to introduce a number σ_n such that the function $\phi_n(t)$ defined by (3.04) is majorized by

$$|\phi_n(t)| \leqslant |a_n|\, t^{(n+\lambda-\mu)/\mu} e^{\sigma_n t} \qquad (0 < t < \infty). \tag{9.01}$$

The error term in (3.05) is then bounded by

$$\left|\int_0^\infty e^{-xt}\phi_n(t)\,dt\right| \leqslant \Gamma\left(\frac{n+\lambda}{\mu}\right)\frac{|a_n|}{(x-\sigma_n)^{(n+\lambda)/\mu}} \qquad (x > \max(\sigma_n, 0)).^{\dagger} \tag{9.02}$$

The best value of σ_n is given by

$$\sigma_n = \sup_{(0,\infty)} \{\psi_n(t)\}, \tag{9.03}$$

where

$$\psi_n(t) = \frac{1}{t}\ln\left|\frac{\phi_n(t)}{a_n t^{(n+\lambda-\mu)/\mu}}\right| = \frac{1}{t}\ln\left|\frac{q(t) - \sum_{s=0}^{n-1} a_s t^{(s+\lambda-\mu)/\mu}}{a_n t^{(n+\lambda-\mu)/\mu}}\right|.$$

Like (2.09), the bound (9.02) enjoys the property of being asymptotic to the absolute value of the actual error when $x \to \infty$.

The preceding approach fails when σ_n is infinite. This obviously happens when $a_n = 0$, in which event we would simply proceed to a higher value of n. If $a_n \neq 0$, then the commonest way failure occurs is for the function $\psi_n(t)$ to tend to $+\infty$ as $t \to 0+$. For small t, we have from (3.02)

$$\phi_n(t) \sim a_n t^{(n+\lambda-\mu)/\mu} + a_{n+1} t^{(n+1+\lambda-\mu)/\mu} + a_{n+2} t^{(n+2+\lambda-\mu)/\mu} + \cdots.$$

Therefore

$$\psi_n(t) \sim \frac{a_{n+1}}{a_n} t^{(1/\mu)-1} + \left(\frac{a_{n+2}}{a_n} - \frac{a_{n+1}^2}{2a_n^2}\right) t^{(2/\mu)-1} + \cdots.$$

If $\mu > 1$, then $t^{(1/\mu)-1} \to \infty$. No problem arises if a_{n+1} and a_n have opposite signs because the right-hand side tends to $-\infty$ as $t \to 0$. But if $\mu > 1$ and a_{n+1}/a_n is positive, then $\sigma_n = \infty$.

9.2 A simple way of overcoming the difficulty is to modify the majorant (9.01) by the inclusion of an arbitrary factor M exceeding unity; $M = 2$ would be a realistic choice in many circumstances. Then in place of (9.02) we derive

$$\left|\int_0^\infty e^{-xt}\phi_n(t)\,dt\right| \leqslant \Gamma\left(\frac{n+\lambda}{\mu}\right)\frac{M|a_n|}{(x-\hat{\sigma}_n)^{(n+\lambda)/\mu}} \qquad (x > \max(\hat{\sigma}_n, 0)), \tag{9.04}$$

where

$$\hat{\sigma}_n = \sup_{(0,\infty)}\left\{\frac{1}{t}\ln\left|\frac{\phi_n(t)}{M a_n t^{(n+\lambda-\mu)/\mu}}\right|\right\}. \tag{9.05}$$

† The condition $x > 0$ is needed for the validity of (3.05).

This bound generally succeeds because as $t \to 0$ the content of the braces in the last equation tends to $-\infty$.

A variation of this procedure is to take $M = M_n$, where

$$M_n = \sup_{(0,\infty)} |\phi_n(t)/\{a_n t^{(n+\lambda-\mu)/\mu}\}|.$$

Then $\hat{\sigma}_n = 0$, which means that the ratio of the bound (9.04) to the absolute value of the first neglected term in the asymptotic expansion is M_n, independently of x. In practice, however, M_n may turn out to be infinite or unacceptably large.

9.3 Another approach is to let m be the largest integer such that $m < \mu$, and a_{n+j+1} the first member of the set $a_{n+1}, a_{n+2}, \ldots, a_{n+m}$ that has opposite sign to a_n, or, if no such member exists, let $j = m$. Define

$$\rho_n = \sup_{(0,\infty)} \left\{ \frac{1}{t} \ln \left| \frac{\phi_n(t) t^{-(n+\lambda-\mu)/\mu}}{a_n + a_{n+1} t^{1/\mu} + \cdots + a_{n+j} t^{j/\mu}} \right| \right\}. \tag{9.06}$$

Then

$$|\phi_n(t)| \leqslant |a_n t^{(n+\lambda-\mu)/\mu} + a_{n+1} t^{(n+1+\lambda-\mu)/\mu} + \cdots + a_{n+j} t^{(n+j+\lambda-\mu)/\mu}| \, e^{\rho_n t},$$

and

$$\left| \int_0^\infty e^{-xt} \phi_n(t)\, dt \right| \leqslant \sum_{s=n}^{n+j} \Gamma\left(\frac{s+\lambda}{\mu}\right) \frac{|a_s|}{(x-\rho_n)^{(s+\lambda)/\mu}}. \tag{9.07}$$

This bound is successful because as $t \to 0$,

$$\frac{1}{t} \ln \left| \frac{\phi_n(t) t^{-(n+\lambda-\mu)/\mu}}{a_n + a_{n+1} t^{1/\mu} + \cdots + a_{n+j} t^{j/\mu}} \right| = \frac{a_{n+j+1}}{a_n} t^{(j+1-\mu)/\mu} + O(t^{(j+2-\mu)/\mu}),$$

and tends to $-\infty$ if $j \leqslant m-1$, or is bounded if $j = m$. Moreover, $a_n + a_{n+1} t^{1/\mu} + \cdots + a_{n+j} t^{j/\mu}$ cannot vanish when $t \in (0, \infty)$.

The advantage of (9.07) is that its ratio to the absolute value of the actual error tends to unity as $x \to \infty$, unlike (9.04). Disadvantages are increased complexity and the need to evaluate coefficients beyond a_n.

9.4 In the case of Theorem 8.1, it is seen from the proof that the nth truncation error of the expansion (8.08) can be expressed

$$\int_a^b e^{-xp(t)} q(t)\, dt - e^{-xp(a)} \sum_{s=0}^{n-1} \Gamma\left(\frac{s+\lambda}{\mu}\right) \frac{a_s}{x^{(s+\lambda)/\mu}}$$

$$= -e^{-xp(a)} \varepsilon_{n,1}(x) + e^{-xp(a)} \varepsilon_{n,2}(x) + \int_k^b e^{-xp(t)} q(t)\, dt, \tag{9.08}$$

where k is a number in $(a, b]$ satisfying the criteria of §8.2, and $\varepsilon_{n,1}(x)$ and $\varepsilon_{n,2}(x)$ are defined by (7.07), (8.09), (8.11), and (8.12), with $v = p(t) - p(a)$.

The first error term in (9.08) is absent if $k = b$ and $p(b) = \infty$, for then $\kappa = \infty$.†

† The requirement in the proof of Theorem 8.1 that k and κ be finite does not apply to (9.08).

In other cases, we have from (1.05) and (2.14)

$$\Gamma(\alpha, x) \leqslant \frac{e^{-x}x^{\alpha}}{x-\max(\alpha-1,0)} \qquad (x > \max(\alpha-1,0)).$$

Substituting in (8.11) by means of this inequality, we obtain

$$|e^{-xp(a)}\varepsilon_{n,1}(x)| \leqslant \frac{e^{-xp(k)}}{\kappa x-\alpha_n}\sum_{s=0}^{n-1}|a_s|\,\kappa^{(s+\lambda)/\mu} \qquad (x > \alpha_n/\kappa), \tag{9.09}$$

where $\kappa = p(k)-p(a)$ as before, and

$$\alpha_n = \max\{(n+\lambda-\mu-1)/\mu, 0\}. \tag{9.10}$$

The second error term, $e^{-xp(a)}\varepsilon_{n,2}(x)$, can be bounded by methods similar to those of §§9.1 to 9.3. The role of t is now played by v, and $\phi_n(t)$ is replaced by $v^{(n+\lambda-\mu)/\mu}f_n(v)$; the essential difference is that the suprema in (9.03), (9.05), and (9.06) are evaluated over the range $0 < v < \kappa$ instead of $0 < t < \infty$. The bounds (9.02), (9.04), and (9.07) apply unchanged to $|\varepsilon_{n,2}(x)|$.

For the tail, the inequality (7.13) can be used, the integral on the right-hand side being found numerically for a suitably chosen value of X. Alternatively, as in §10.1 below, it may be possible to majorize $-p(t)$ and $|q(t)|$ by simple functions and evaluate the resulting integral analytically. Because the contribution of the tail is exponentially small compared with $e^{-xp(a)}\varepsilon_{n,2}(x)$ a crude bound is often acceptable.

9.5 Some of the complications in bounding $|\varepsilon_{n,2}(x)|$ can be circumvented in the following common case. Suppose that $p(t)$ and $q(t)$ have Taylor-series expansions at all points of (a,b), $p(t)$ has a simple minimum at an interior point of (a,b), and $q(t)$ does not vanish at this minimum. Without loss of generality, we may assume that (i) the minimum is located at $t=0$; (ii) $p(0)=p'(0)=0$; (iii) the integration range is truncated in such a way that $p'(t)/t$ is positive for $a<t<b$, and $p(a) = p(b) = \kappa$, say.

As before, the range $(0,b)$ is treated by taking a new integration variable $v = p(t)$. Then

$$\int_0^b e^{-xp(t)}q(t)\,dt = \int_0^{\kappa} e^{-xv}f(v)\,dv,$$

where

$$f(v) = \frac{q(t)}{p'(t)} = \sum_{s=0}^{\infty} a_s v^{(s-1)/2},$$

this expansion converging for all sufficiently small v; compare (8.07) with $\mu=2$ and $\lambda=1$.

Similarly,

$$\int_a^0 e^{-xp(t)}q(t)\,dt = \int_0^{\kappa} e^{-xv}\hat{f}(v)\,dv,$$

where

$$\hat{f}(v) = -\frac{q(t)}{p'(t)} = \sum_{s=0}^{\infty}(-)^s a_s v^{(s-1)/2}.$$

Hence

$$\int_a^b e^{-xp(t)} q(t)\, dt = \int_0^\kappa e^{-xv} F(v)\, dv,$$

where for small v

$$F(v) = 2 \sum_{s=0}^{\infty} a_{2s} v^{s-(1/2)}.$$

Since the last expansion ascends in powers of v and not $v^{1/2}$, an error bound of the type (9.02) can be constructed with a finite value of the exponent σ_n.

Ex. 9.1 Show that for $x > 0$

$$\int_0^\infty e^{-x\cosh t}\, dt = \left(\frac{\pi}{2x}\right)^{1/2} e^{-x}\{1 - \vartheta(x)\},$$

where $0 < \vartheta(x) < (8x)^{-1}$.

Ex. 9.2 Show that

$$\int_{-\infty}^{\infty} \exp(-xt^2) \ln(1+t+t^2)\, dt = \frac{\pi^{1/2}}{4}\left\{\frac{1}{x^{3/2}} + \frac{3}{4x^{5/2}} - \frac{5}{2x^{7/2}} + \varepsilon(x)\right\},$$

where

$$0 < \varepsilon(x) < \frac{105}{32(x-\frac{4}{5})^{9/2}} \qquad (x > \tfrac{4}{5}).$$

[Olver, 1968.]

Ex. 9.3 If each of the integrals (or moments)

$$M_s = \int_0^\infty t^s f(t)\, dt \qquad (s = 0, 1, 2, \ldots)$$

is finite, prove that for large positive x the asymptotic expansion of the *Stieltjes transform*

$$\mathscr{S}(x) = \int_0^\infty \frac{f(t)}{t+x}\, dt$$

is given by

$$\mathscr{S}(x) = \sum_{s=0}^{n-1} (-)^s M_s x^{-s-1} + \varepsilon_n(x),$$

where n is an arbitrary positive integer or zero, and

$$|\varepsilon_n(x)| \leqslant x^{-n-1} \sup_{(0,\infty)} \left|\int_0^t v^n f(v)\, dv\right|.$$

*10 Examples

10.1 Consider the asymptotic expansion given in Exercise 8.5 for the function

$$I_0(x) = \frac{1}{\pi}\int_0^\pi e^{x\cos t}\, dt.$$

In the notation of §9.4, $p(t) \equiv -\cos t$ increases steadily from a minimum at $t = 0$ to a maximum at $t = \pi$. We cannot take $k = \pi$, however, since $p'(t)$ vanishes at this point. The "best" value of k is not easily specified, but the choice is not critical. For simplicity, take k to be the midpoint $\frac{1}{2}\pi$.

When $\frac{1}{2}\pi \leqslant t \leqslant \pi$ we have, from Jordan's inequality, $\cos t \leqslant 1-(2t/\pi)$. Accordingly, a bound for the tail of the integral is supplied by

$$\frac{1}{\pi}\int_{\pi/2}^{\pi} e^{x\cos t}\,dt \leqslant \frac{e^x}{\pi}\int_{\pi/2}^{\pi} e^{-2tx/\pi}\,dt < \frac{1}{2x}. \tag{10.01}$$

Next, in the notation of §§7 and 8, we have $a = 0$, $p_0 = \frac{1}{2}$, $\mu = 2$, $\lambda = 1$, $\kappa = 1$, $v = 1-\cos t = 2\sin^2(\frac{1}{2}t)$, and

$$f(v) = \frac{1}{\pi \sin t} = \frac{1}{\pi(2v-v^2)^{1/2}} = \sum_{s=0}^{\infty} a_s v^{(s-1)/2} \qquad (0 < v < 2),$$

where

$$a_{2s} = \frac{1\cdot 3\cdots(2s-1)}{\pi 2^{2s+(1/2)} s!}, \qquad a_{2s+1} = 0.$$

Because the a_s of odd suffix vanish, we apply the results of §9 with n replaced by $2n$. From (9.10) we derive $\alpha_{2n} = n-1$ $(n \geqslant 1)$. Hence (9.09) yields

$$|\varepsilon_{2n,1}(x)| \leqslant \frac{e^{-x}}{x-n+1}\sum_{s=0}^{2n-1} a_s < \frac{e^{-x}}{(x-n+1)\pi} \qquad (x > n-1 \geqslant 0). \tag{10.02}$$

Next,

$$f_{2n}(v) = \frac{1}{v^{n-(1/2)}}\left\{\frac{1}{\pi(2v-v^2)^{1/2}} - \sum_{s=0}^{n-1} a_{2s} v^{s-(1/2)}\right\}$$
$$= a_{2n} + a_{2n+2}v + a_{2n+4}v^2 + \cdots \qquad (0 \leqslant v < 2).$$

Since no term in $v^{1/2}$ is present in the last expansion, the methods of §§9.1 and 9.3 lead to the same bound for $\varepsilon_{2n,2}(x)$, given by

$$|\varepsilon_{2n,2}(x)| < \frac{\Gamma(n+\frac{1}{2})a_{2n}}{(x-\sigma_{2n})^{n+(1/2)}} \qquad (x > \sigma_{2n}), \tag{10.03}$$

where

$$\sigma_{2n} = \sup_{(0,1)}\left\{\frac{1}{v}\ln\left|\frac{f_{2n}(v)}{a_{2n}}\right|\right\}. \tag{10.04}$$

The aggregate of (10.01), (10.02), and (10.03) furnishes the desired bounds for the error terms in the expansion

$$I_0(x) = e^x\left\{\sum_{s=0}^{n-1}\frac{\Gamma(s+\frac{1}{2})a_{2s}}{x^{s+(1/2)}} - \varepsilon_{2n,1}(x) + \varepsilon_{2n,2}(x)\right\} + \frac{1}{\pi}\int_{\pi/2}^{\pi} e^{x\cos t}\,dt.$$

The requisite values of σ_{2n} may be obtained by numerical computation from (10.04).

The first three are found to be

$$\sigma_0 = 0.35, \qquad \sigma_2 = 0.50, \qquad \sigma_4 = 0.56,$$

to two decimal places.†

An alternative way of deriving the asymptotic expansion of $I_0(x)$ complete with error bounds, is included in Chapter 7, especially §8.2 and Exercise 13.2.

10.2‡ As a second application, consider

$$S(m) = \frac{2}{\pi}\int_0^\infty \left(\frac{\sin t}{t}\right)^m dt, \tag{10.05}$$

in which m is a positive integer. Methods of contour integration, for example, yield the closed form

$$S(m) = \frac{m}{2^{m-1}} \sum_{s=0}^{[(m-1)/2]} \frac{(-)^s (m-2s)^{m-1}}{s!\,(m-s)!}, \tag{10.06}$$

but the numerical evaluation of this sum is cumbersome for large m and we seek instead an asymptotic representation.

The function $\sin t/t$ has an infinite sequence of alternating peaks and troughs, located at the successive nonnegative roots $0, t_1, t_2, t_3, \ldots$, of the equation

$$\tan t = t.$$

Only the first of these lies in $[0, \pi]$, and for this interval we define a new integration variable τ by

$$\tau = \ln\left(\frac{t}{\sin t}\right), \qquad \frac{dt}{d\tau} = \frac{t \sin t}{\sin t - t\cos t}.$$

As t increases from 0 to π, τ increases monotonically from 0 to ∞. Hence

$$S_0(m) \equiv \frac{2}{\pi}\int_0^\pi \left(\frac{\sin t}{t}\right)^m dt = \frac{2}{\pi}\int_0^\infty e^{-m\tau}\frac{dt}{d\tau}\,d\tau. \tag{10.07}$$

For small t and τ we find, by expansion and reversion,

$$t = (6\tau)^{1/2}\left(1 - \tfrac{1}{10}\tau - \tfrac{13}{4200}\tau^2 + \tfrac{9}{14000}\tau^3 + \cdots\right).$$

Application of Watson's lemma immediately produces

$$S_0(m) \sim \left(\frac{6}{\pi m}\right)^{1/2} \sum_{s=0}^\infty \frac{h_s}{m^s} \qquad (m \to \infty), \tag{10.08}$$

where

$$h_0 = 1, \qquad h_1 = -\tfrac{3}{20}, \qquad h_2 = -\tfrac{13}{1120}, \qquad h_3 = \tfrac{27}{3200}, \qquad \ldots.$$

† Analytical tests for locating the supremum in the expression (9.03) for the exponent σ_n have been developed by Olver (1968). For the present example, these tests establish that the supremum in (10.04) occurs at $v = 1$. In consequence, the computation of σ_{2n} reduces to the evaluation of $\ln\{(\pi^{-1} - a_0 - a_2 - \cdots - a_{2n-2})/a_{2n}\}$.

‡ The analysis in §§10.2 and 10.3 is based on that of Medhurst and Roberts (1965).

Now consider the interval $[s\pi, (s+1)\pi]$, where s is any positive integer. We have

$$s\pi < t_s < (s+\tfrac{1}{2})\pi.$$

Hence

$$\left|\frac{2}{\pi}\int_{s\pi}^{(s+1)\pi}\left(\frac{\sin t}{t}\right)^m dt\right| \leqslant 2\left|\frac{\sin t_s}{t_s}\right|^m = \frac{2}{(1+t_s^2)^{m/2}} < \frac{2}{(s\pi)^m}.$$

Summation produces

$$\frac{2}{\pi}\int_{\pi}^{\infty}\left(\frac{\sin t}{t}\right)^m dt \leqslant \frac{2}{\pi^m}\sum_{s=1}^{\infty}\frac{1}{s^m} \qquad (m \geqslant 2).$$

Since this is $O(\pi^{-m})$ for large m, the desired expansion is given by

$$S(m) \sim \left(\frac{6}{\pi m}\right)^{1/2}\sum_{s=0}^{\infty}\frac{h_s}{m^s} \qquad (m \to \infty). \tag{10.09}$$

10.3 Numerical results obtained from the last series are somewhat disappointing. For example, with $m = 4$ the fourth partial sum gives

$$0.6910(1-0.0375-0.0007+0.0001) = 0.6647, \tag{10.10}$$

to four decimal places, compared with the exact value $S(4) = \frac{2}{3}$ obtained from (10.06). Thus the absolute error is about 20 or 30 times the last term retained.

The smooth behavior of the function $dt/d\tau$ in (10.07) suggests that the discrepancy does not arise from the error term associated with (10.08). A more likely source is the neglect of the contribution from the remaining part of the integration range, especially as the value of the original integrand $(\sin t/t)^m$ equals 0.0022 when $m = 4$ and $t = t_1 = 4.4934\ldots$.

Consider the interval $[\pi, 2\pi]$. Application of the methods of §§4 and 5 yields

$$S_1(m) \equiv \frac{2}{\pi}\int_{\pi}^{2\pi}\left(\frac{\sin t}{t}\right)^m dt \sim 2(\cos t_1)^m\left(\frac{2}{\pi m}\right)^{1/2}\sum_{s=0}^{\infty}\frac{k_s}{m^s} \qquad (m \to \infty), \tag{10.11}$$

where†

$$k_0 = 1, \qquad k_1 = -\frac{1}{4} - \frac{1}{6t_1^2} = -0.2583\ldots.$$

For $m = 4$, the numerical form of this expansion is

$$0.0018(1-0.0646+\cdots) = 0.0017.$$

Adding this result to (10.10) we obtain 0.6664, which is much closer to the correct value. Even closer agreement could be achieved by inclusion of the approximate contribution $2(\cos t_2)^m\{2/(\pi m)\}^{1/2}$ from the next interval $[2\pi, 3\pi]$. Accordingly, this example furnishes another illustration of the numerical importance of exponentially small terms in an asymptotic expansion.

† In the cited reference k_1 is incorrectly given as $-\frac{1}{4}-t_1^{-2}$.

10.4 The conclusions of the preceding subsection can be supported by strict error analyses on the lines of §9. The full form of (10.08) may be expressed

$$S_0(m) = \left(\frac{6}{\pi m}\right)^{1/2}\left\{\sum_{s=0}^{n-1}\frac{h_s}{m^s} + \varepsilon_n(m)\right\} \qquad (n = 1, 2, \ldots),$$

where

$$|\varepsilon_n(m)| \leqslant \frac{|h_n|\, m^{1/2}}{(m-\rho_n)^{n+(1/2)}} \qquad (m > \rho_n),$$

and

$$\rho_n = \sup_{t\in(0,\pi)}\left[\frac{1}{\tau}\ln\left|\frac{1}{l_n\tau^{n-(1/2)}}\left\{\left(\frac{2}{3}\right)^{1/2}\frac{dt}{d\tau} - \sum_{s=0}^{n-1} l_s\tau^{s-(1/2)}\right\}\right|\right],$$

with $l_s = \Gamma(\frac{1}{2})h_s/\Gamma(s+\frac{1}{2})$. Numerical calculation gives $\rho_3 = 0.45\ldots$. In consequence, the value for $S_0(4)$ obtained by summing the first three terms in (10.08) namely $0.6646\ldots$, is correct to within ± 0.00014.

A similar result for (10.11) is given by

$$S_1(m) = 2(\cos t_1)^m\left(\frac{2}{\pi m}\right)^{1/2}\left\{\sum_{s=0}^{n-1}\frac{k_s}{m^s} + \eta_n(m)\right\},$$

where

$$|\eta_n(m)| \leqslant \frac{2|k_n|\, m^{1/2}}{(m-\hat{\sigma}_n)^{n+(1/2)}} \qquad (m > \hat{\sigma}_n),$$

$\hat{\sigma}_n$ being defined by formula (9.05) with $M = 2$. Again, by numerical calculation we find that $\hat{\sigma}_1$ vanishes to two decimal places, from which we conclude that $S_1(4)$ equals $2(\cos t_1)^4(2\pi)^{-1/2}$, that is, $0.0018\ldots$, correct to within ± 0.00023.

11 The Method of Stationary Phase

11.1 Consider the integrals

$$\int_a^b \cos\{xp(t)\}\, q(t)\, dt, \qquad \int_a^b \sin\{xp(t)\}\, q(t)\, dt,$$

in which a, b, $p(t)$, and $q(t)$ are independent of the parameter x. For large x, the integrands oscillate rapidly and cancel themselves over most of the range. Cancellation does not occur, however, in the neighborhoods of the following points: (i) the endpoints a and b (when finite), owing to lack of symmetry; (ii) zeros of $p'(t)$, because $p(t)$ changes relatively slowly near these "stationary points." Kelvin's *method of stationary phase* stems from these somewhat vague ideas.†

† Also called the *method of critical points*.

Both integrals are covered simultaneously by combining them into

$$I(x) = \int_a^b e^{ixp(t)} q(t)\, dt. \tag{11.01}$$

In the neighborhood of $t = a$, the new integrand is approximately

$$\exp[ix\{p(a) + (t-a)p'(a)\}]\,q(a).$$

An indefinite integral of this function is

$$\frac{\exp[ix\{p(a) + (t-a)p'(a)\}]\,q(a)}{ixp'(a)}, \tag{11.02}$$

provided that $p'(a) \neq 0$. The lower limit $t = a$ contributes

$$-e^{ixp(a)} q(a)/\{ixp'(a)\} \tag{11.03}$$

to the value of $I(x)$. As t recedes from a the real and imaginary parts of (11.02) oscillate about the mean value zero, accordingly it is reasonable to neglect other contributions from (11.02). Similar reasoning suggests that the upper limit $x = b$ asymptotically contributes

$$e^{ixp(b)} q(b)/\{ixp'(b)\}. \tag{11.04}$$

11.2 Next, if $t_0 \in (a, b)$ is a stationary point of $p(t)$, then near this point the integrand is approximately

$$\exp[ix\{p(t_0) + \tfrac{1}{2}(t-t_0)^2 p''(t_0)\}]\,q(t_0),$$

provided that $p''(t_0)$ and $q(t_0)$ are nonzero. On integrating this function we pursue our belief that only the neighborhood of t_0 matters, by extending the limits to $-\infty$ and $+\infty$. The resulting integral is then explicitly evaluable. We have†

$$\int_{-\infty}^{\infty} \exp(\pm iyt^2)\, dt = e^{\pm\pi i/4}\left(\frac{\pi}{y}\right)^{1/2} \qquad (y > 0).$$

Hence the contribution to $I(x)$ from the neighborhood of t_0 is expected to be

$$e^{\pm\pi i/4} q(t_0) \exp\{ixp(t_0)\} \left|\frac{2\pi}{xp''(t_0)}\right|^{1/2}, \tag{11.05}$$

where the upper or lower sign is taken according as $xp''(t_0)$ is positive or negative. It should be noticed, incidentally, that (11.05) is of a larger order of magnitude than (11.03) and (11.04).

Similar results can be found for (i) stationary points of higher order, that is, points at which the lowest nonvanishing derivative of $p(t)$ is of order higher than 2; (ii) some cases with $q(t_0) = 0$.

The approximate value of $I(x)$ for large x is obtained by summing expressions of the form (11.05) over the various stationary points in the range of integration and adding the contributions (11.03) and (11.04) from the endpoints. This approach is,

† Compare §12.1 below.

of course, heuristic, but in following sections we shall place the method on a firm foundation.

The similarity of the approximations (11.03) and (11.05) to (7.02) and (7.03) attracts attention. From the standpoint of complex-variable theory (Chapter 4), Laplace's method and the method of stationary phase can be regarded as special cases of the same general procedure. This is reflected in the analysis: the proofs of §13 below resemble those of §7 in many ways.

11.3 The case in which stationary points are absent is an exercise in integration by parts. Since $p'(t)$ is of constant sign in $[a, b]$ we may take $v = p(t)$ as new integration variable. Then (11.01) becomes

$$I(x) = \int_{p(a)}^{p(b)} e^{ixv} f(v)\, dv,$$

where $f(v) = q(t)/p'(t)$. This is a Fourier integral, and the asymptotic analysis of §5 is directly applicable. In particular, if $f(v)$ is continuous and $f'(v)$ is sectionally continuous, that is, if $p'(t)$ and $q(t)$ are continuous and $p''(t)$ and $q'(t)$ are sectionally continuous in $[a, b]$, then

$$I(x) = \frac{ie^{ixp(a)}q(a)}{xp'(a)} - \frac{ie^{ixp(b)}q(b)}{xp'(b)} + o\left(\frac{1}{x}\right) \qquad (x \to \infty).$$

This confirms the predictions of §11.1 in this case.

In other cases, the range of integration can be subdivided in such a way that the only stationary point in each subrange is located at one of the endpoints, and without loss of generality we may suppose that this is the left endpoint. Before proceeding to these cases we establish a number of preliminary results.

12 Preliminary Lemmas

12.1 Lemma 12.1

$$\int_0^\infty e^{ixv} v^{\alpha - 1}\, dv = \frac{e^{\alpha\pi i/2}\Gamma(\alpha)}{x^\alpha} \qquad (0 < \alpha < 1, \quad x > 0). \tag{12.01}$$

The restriction $\alpha \in (0, 1)$ is needed since the integral diverges at its lower limit when $\alpha \leqslant 0$ and at its upper limit when $\alpha \geqslant 1$. The result is proved by integrating $e^{ixv}v^{\alpha-1}$ around the contour indicated in Fig. 12.1, and then letting $r \to 0$ and $R \to \infty$. Details are straightforward and left to the reader.

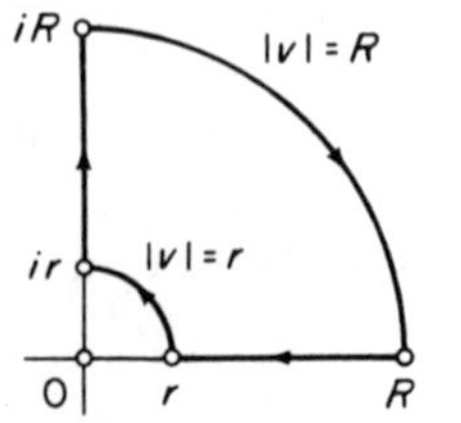

Fig. 12.1 v plane.

12.2 Lemma 12.2 *If α and κ are fixed numbers such that $\alpha < 1$ and $\kappa > 0$, then*

$$\int_\kappa^\infty e^{ixv} v^{\alpha-1}\, dv = O\left(\frac{1}{x}\right) \qquad (x \to \infty). \tag{12.02}$$

This result is another exercise in integration by parts:

$$\left|\int_\kappa^\infty e^{ixv} v^{\alpha-1}\, dv\right| = \left|\left[\frac{e^{ixv}}{ix} v^{\alpha-1}\right]_\kappa^\infty - \frac{\alpha-1}{ix}\int_\kappa^\infty e^{ixv} v^{\alpha-2}\, dv\right|$$
$$\leqslant \frac{\kappa^{\alpha-1}}{x} + \frac{1-\alpha}{x}\int_\kappa^\infty v^{\alpha-2}\, dv = \frac{2\kappa^{\alpha-1}}{x}.$$

12.3 Lemma 12.3 *For the integral*

$$\Phi(x) = \int_0^\infty e^{ixv} \phi(v)\, dv \tag{12.03}$$

assume that:

(i) *$\phi(v)$ is sectionally continuous and $\phi'(v)$ has at most a finite number of discontinuities and infinities in the interval* $(0, \infty)$.
(ii) $\phi(v) = o(v^{\alpha-1})$ *and* $\phi'(v) = o(v^{\alpha-2})$ *as* $v \to 0+$, *where α is a constant in the interval* $(0, 1)$.
(iii) $\mathscr{V}_{\kappa,\infty}(\phi)$ *is finite for each positive constant κ.*
(iv) $\phi(v) \to 0$ *as* $v \to \infty$.

Then the integral (12.03) *converges uniformly for $x \geqslant X$, where X is any positive constant, and*

$$\Phi(x) = o(x^{-\alpha}) \qquad (x \to \infty). \tag{12.04}$$

This is an extension of the Riemann–Lebesgue lemma. Condition (ii) shows that the given integral converges at its lower limit absolutely and uniformly for all real x. Next, if v_1 and v_2 are any two numbers exceeding the affixes of all discontinuities and infinities of $\phi(v)$ and $\phi'(v)$, then by partial integration

$$\left|\int_{v_1}^{v_2} e^{ixv}\phi(v)\, dv\right| = \left|\frac{\exp(ixv_2)\,\phi(v_2) - \exp(ixv_1)\,\phi(v_1)}{ix} - \frac{1}{ix}\int_{v_1}^{v_2} e^{ixv}\phi'(v)\, dv\right|$$
$$\leqslant x^{-1}\{|\phi(v_2)| + |\phi(v_1)| + \mathscr{V}_{v_1,v_2}(\phi)\}.$$

From this inequality and Conditions (iii) and (iv) it follows that (12.03) converges uniformly at its upper limit for $x \geqslant X$.

It remains to establish (12.04). Given an arbitrary positive number ε, Conditions (i) and (ii) show that there exists a finite positive number κ such that in $(0, \kappa]$ the functions $\phi(v)$ and $\phi'(v)$ are continuous, and

$$|\phi(v)| < \varepsilon v^{\alpha-1}, \qquad |\phi'(v)| < \varepsilon v^{\alpha-2}. \tag{12.05}$$

Assume that $x \geqslant 1/\kappa$ and subdivide the integration range at $v = 1/x$. Then

$$\left|\int_0^{1/x} e^{ixv}\phi(v)\, dv\right| < \int_0^{1/x} \varepsilon v^{\alpha-1}\, dv = \frac{\varepsilon}{\alpha x^\alpha}.$$

Using Condition (iv), we find on integration by parts

$$\int_{1/x}^{\infty} e^{ixv}\phi(v)\,dv = \sum_{s=1}^{m} \frac{\exp(ixd_s)}{ix}\{\phi(d_s-) - \phi(d_s+)\} - \frac{e^i}{ix}\phi\left(\frac{1}{x}\right)$$

$$-\frac{1}{ix}\int_{1/x}^{\kappa} e^{ixv}\phi'(v)\,dv - \frac{1}{ix}\int_{\kappa}^{\infty} e^{ixv}\phi'(v)\,dv, \qquad (12.06)$$

where $d_1, d_2, \ldots, d_m$ are the discontinuities of $\phi(v)$. The sum is $O(x^{-1})$ for large x. The inequalities (12.05) show that the next term on the right-hand side is bounded in absolute value by $\varepsilon x^{-\alpha}$, and also that

$$\left|\frac{1}{ix}\int_{1/x}^{\kappa} e^{ixv}\phi'(v)\,dv\right| < \frac{1}{x}\int_{1/x}^{\kappa} \varepsilon v^{\alpha-2}\,dv < \frac{\varepsilon}{(1-\alpha)x^{\alpha}}.$$

Lastly, from (11.03) of Chapter 1, we see that

$$\left|\frac{1}{ix}\int_{\kappa}^{\infty} e^{ixv}\phi'(v)\,dv\right| \leqslant \frac{\mathscr{V}_{\kappa,\infty}(\phi)}{x} = O\left(\frac{1}{x}\right) \qquad (x\to\infty).$$

The proof of Lemma 12.3 is completed by combining the foregoing results.

Ex. 12.1 Show that

$$\int_{\kappa}^{\infty} e^{ixv}v^{\alpha-1}\,dv = e^{\alpha\pi i/2}x^{-\alpha}\Gamma(\alpha, -i\kappa x) \qquad (\alpha < 1),$$

and

$$\int_{0}^{\kappa} e^{ixv}v^{\alpha-1}\,dv = e^{\alpha\pi i/2}x^{-\alpha}\gamma(\alpha, -i\kappa x) \qquad (\alpha > 0),$$

where the incomplete Gamma functions take their principal values.

Ex. 12.2 Show that Lemma 12.3 remains valid when the o symbols are replaced throughout by O symbols.

13 Asymptotic Nature of the Stationary Phase Approximation

13.1 As in §7.2, we suppose that in the integral

$$I(x) = \int_{a}^{b} e^{ixp(t)}q(t)\,dt \qquad (13.01)$$

the limits a and b are independent of x, a being finite and b $(>a)$ finite or infinite. The functions $p(t)$ and $q(t)$ are independent of x, $p(t)$ being real and $q(t)$ either real or complex. In accordance with the closing paragraph of §11.3, we assume that in the closure of (a, b), the only possible point at which $p'(t)$ vanishes is a. Without loss of generality both x and $p'(t)$ are taken to be positive; cases in which one of these quantities is negative can be handled by changing the sign of i throughout. We shall use the notation $p(b) \equiv \lim\{p(t)\}$ as $t \to b-$ when this limit exists, otherwise $p(b) = \infty$. Corresponding to Conditions (i) to (iv) of §7.2, we require:

(i) *In* (a, b), *the functions* $p'(t)$ *and* $q(t)$ *are continuous,* $p'(t) > 0$, *and* $p''(t)$ *and* $q'(t)$ *have at most a finite number of discontinuities and infinities.*

(ii) *As* $t \to a+$

$$p(t) - p(a) \sim P(t-a)^{\mu}, \qquad q(t) \sim Q(t-a)^{\lambda-1}, \tag{13.02}$$

the first of these relations being differentiable. Here P, μ, *and* λ *are positive constants, and* Q *is a real or complex constant.*

(iii) $\mathscr{V}_{k,b}\{q(t)/p'(t)\}$ *is finite for each* $k \in (a, b)$.

(iv) *As* $t \to b-$, $q(t)/p'(t)$ *tends to a finite limit, and this limit is zero when* $p(b) = \infty$.

Condition (ii) immediately shows that the integral (13.01) converges at its lower limit absolutely and uniformly for all real x. Next, by partial integration

$$\int e^{ixp(t)} q(t)\, dt = \frac{e^{ixp(t)}}{ix} \frac{q(t)}{p'(t)} - \frac{1}{ix} \int e^{ixp(t)} \frac{d}{dt}\left\{\frac{q(t)}{p'(t)}\right\} dt. \tag{13.03}$$

Using Conditions (iii) and (iv), we see that (13.01) converges at its upper limit; moreover in the case $p(b) = \infty$ the convergence is uniform for all sufficiently large x.

With the foregoing conditions, the nature of the asymptotic approximation to $I(x)$ for large x depends on the sign of $\lambda - \mu$. When $\lambda < \mu$ the contribution from the endpoint a dominates, when $\lambda > \mu$ the contribution from b dominates, and when $\lambda = \mu$ the contributions from a and b are equally important. The commonest case in physical applications is $\lambda < \mu$, and we begin with this.

13.2 Theorem 13.1 *In addition to the conditions of* §13.1, *assume that* $\lambda < \mu$, *the first of* (13.02) *is twice differentiable, and the second of* (13.02) *is differentiable.*† *Then*

$$I(x) \sim e^{\lambda\pi i/(2\mu)} \frac{Q}{\mu} \Gamma\left(\frac{\lambda}{\mu}\right) \frac{e^{ixp(a)}}{(Px)^{\lambda/\mu}} \qquad (x \to \infty). \tag{13.04}$$

To prove this result we take a new integration variable $v = p(t) - p(a)$. In consequence of Condition (i), the relationship between t and v is one to one. Denote

$$\beta = p(b) - p(a), \qquad f(v) = q(t)/p'(t). \tag{13.05}$$

Then

$$I(x) = e^{ixp(a)} \int_0^{\beta} e^{ixv} f(v)\, dv.$$

As in §7.3, Condition (ii) implies that

$$f(v) \sim \frac{Qv^{(\lambda/\mu)-1}}{\mu P^{\lambda/\mu}} \qquad (v \to 0+).$$

Moreover in the present case this relation can be differentiated. We now express

$$\int_0^{\beta} e^{ixv} f(v)\, dv = \frac{Q}{\mu P^{\lambda/\mu}} \left\{ \int_0^{\infty} e^{ixv} v^{(\lambda/\mu)-1}\, dv - \varepsilon_1(x) \right\} + \varepsilon_2(x), \tag{13.06}$$

† When $\mu = 1$ this is to be interpreted as $p'(t) \to P$ and $p''(t) = o\{(t-a)^{-1}\}$. Similarly, $q'(t) = o\{(t-a)^{-1}\}$ in the case $\lambda = 1$.

where

$$\varepsilon_1(x) = \int_\beta^\infty e^{ixv} v^{(\lambda/\mu)-1}\, dv, \qquad \varepsilon_2(x) = \int_0^\infty e^{ixv}\phi(v)\, dv,$$

and

$$\phi(v) = f(v) - \frac{Qv^{(\lambda/\mu)-1}}{\mu P^{\lambda/\mu}} \quad \text{or} \quad 0, \tag{13.07}$$

according as v lies inside or outside the interval $(0, \beta)$.

The first term on the right-hand side of (13.06) is evaluable by means of Lemma 12.1 and yields the required approximation (13.04).

Next, Lemma 12.2 shows that

$$\varepsilon_1(x) = O(x^{-1}) \qquad (x \to \infty).$$

For the remaining error term, it is readily verified by reference to the given conditions that the function $\phi(v)$, defined by (13.07) and (13.05), satisfies the conditions of Lemma 12.3 with $\alpha = \lambda/\mu$. Therefore

$$\varepsilon_2(x) = o(x^{-\lambda/\mu}) \qquad (x \to \infty).$$

Since $\lambda/\mu < 1$, the estimate $O(x^{-1})$ for $\varepsilon_1(x)$ may be absorbed in the estimate $o(x^{-\lambda/\mu})$ for $\varepsilon_2(x)$, and the proof of Theorem 13.1 is complete.

13.3 Theorem 13.2 *In addition to the conditions of* §13.1, *assume that* $\lambda \geqslant \mu$ *and* $\mathscr{V}_{a,b}\{q(t)/p'(t)\} < \infty$. *Then*

$$I(x) = -\lim_{t\to a+}\left\{\frac{q(t)}{p'(t)}\right\}\frac{e^{ixp(a)}}{ix} + \lim_{t\to b-}\left\{\frac{q(t)e^{ixp(t)}}{p'(t)}\right\}\frac{1}{ix} + \varepsilon(x), \tag{13.08}$$

where $\varepsilon(x) = o(x^{-1})$ *as* $x \to \infty$.

The existence of both limits on the right-hand side of (13.08) when $\lambda \geqslant \mu$ is a consequence of Conditions (ii) and (iv) of §13.1. Equation (13.03) yields the following integral for the error term:

$$\varepsilon(x) = -\frac{e^{ixp(a)}}{ix}\int_0^\beta e^{ixv} f'(v)\, dv,$$

where β and $f(v)$ are defined by (13.05). The given conditions show that this integral converges absolutely and uniformly throughout its range; accordingly the Riemann–Lebesgue lemma immediately yields the desired result $\varepsilon(x) = x^{-1}o(1)$.

It should be observed that if $\lambda > \mu$ and $p(b) = \infty$, then both limits in (13.08) are zero. In this case the theorem furnishes only an order of magnitude and not an asymptotic estimate (compare §6.3).

13.4 An illustrative example is provided by the Airy integral of negative argument:

$$\mathrm{Ai}(-x) = \frac{1}{\pi}\int_0^\infty \cos\left(\tfrac{1}{3}w^3 - xw\right) dw \qquad (x > 0).$$

The stationary points of the integrand satisfy $w^2 - x = 0$, giving $w = x^{1/2}$ or $-x^{1/2}$, the former of which lies in the range of integration. Substitution of $w = x^{1/2}(1+t)$ yields

$$\mathrm{Ai}(-x) = \frac{x^{1/2}}{\pi}\int_{-1}^{\infty} \cos\left\{x^{3/2}\left(-\frac{2}{3}+t^2+\frac{1}{3}t^3\right)\right\} dt. \tag{13.09}$$

In the notation of §13.1, replace x by $x^{3/2}$ and take

$$a = 0, \qquad b = \infty, \qquad p(t) = -\tfrac{2}{3}+t^2+\tfrac{1}{3}t^3, \qquad q(t) = 1.$$

Then $p(a) = -\frac{2}{3}$, $P = 1$, $\mu = 2$, and $Q = \lambda = 1$. Clearly as $t \to \infty$, the quotient $q(t)/p'(t)$ vanishes and its variation converges. Thus Conditions (i) to (iv) of §13.1 are all satisfied.

The appropriate theorem is Theorem 13.1, and we derive

$$\int_0^{\infty} \exp\{ix^{3/2}p(t)\}\, dt \sim \tfrac{1}{2}\pi^{1/2}e^{\pi i/4}x^{-3/4}\exp(-\tfrac{2}{3}ix^{3/2}).$$

On changing the sign of t and again using Theorem 13.1, it is seen that the same approximation holds for $\int_{-1}^{0}$. Taking real parts and substituting in (13.09), we arrive at the desired result:

$$\mathrm{Ai}(-x) = \pi^{-1/2}x^{-1/4}\cos(\tfrac{2}{3}x^{3/2}-\tfrac{1}{4}\pi) + o(x^{-1/4}) \qquad (x\to\infty).$$

Harder problems may need preliminary transformations of the kind outlined for Laplace's method in the closing paragraph of §7.5.

Ex. 13.1 Show that

$$\int_0^{\pi/2} t\sin(x\cos t)\, dt = x^{-1}(\tfrac{1}{2}\pi-\cos x) + o(x^{-1}) \qquad (x\to\pm\infty).$$

Ex. 13.2 The functions of Anger and H. F. Weber are respectively defined by

$$\mathbf{J}_\nu(x) = \frac{1}{\pi}\int_0^{\pi}\cos(\nu\theta - x\sin\theta)\, d\theta, \qquad \mathbf{E}_\nu(x) = \frac{1}{\pi}\int_0^{\pi}\sin(\nu\theta - x\sin\theta)\, d\theta.$$

Prove that when ν is real and fixed, and x is large and positive

$$\mathbf{J}_\nu(x) + i\mathbf{E}_\nu(x) \sim 2^{1/2}(\pi x)^{-1/2}\exp\{i(\tfrac{1}{2}\nu\pi+\tfrac{1}{4}\pi - x)\},$$

$$\mathbf{J}_{\nu x}(x) + i\mathbf{E}_{\nu x}(x) = \frac{i}{(\nu-1)\pi x} - \frac{i\exp(i\nu\pi x)}{(\nu+1)\pi x} + o\left(\frac{1}{x}\right) \qquad (|\nu| > 1),$$

$$\mathbf{J}_{\nu x}(x) + i\mathbf{E}_{\nu x}(x) \sim 2^{1/2}(\pi x\sin\alpha)^{-1/2}\exp\{i(x\alpha\cos\alpha - x\sin\alpha + \tfrac{1}{4}\pi)\}$$
$$(|\nu| < 1, \quad \alpha \equiv \cos^{-1}\nu),$$

$$\mathbf{J}_x(x) \sim 2^{-2/3}3^{-1/6}\pi^{-1}\Gamma(\tfrac{1}{3})x^{-1/3}, \qquad \mathbf{E}_x(x) \sim 6^{-2/3}\pi^{-1}\Gamma(\tfrac{1}{3})x^{-1/3},$$

and

$$\mathbf{J}_{-x}(x) + i\mathbf{E}_{-x}(x) \sim 2^{1/3}3^{-2/3}\pi^{-1}\Gamma(\tfrac{1}{3})x^{-1/3}\exp\{i\pi(\tfrac{1}{6}-x)\}.$$

Ex. 13.3 By use of equation (9.13) of Chapter 2, prove that with the notations of Exercises 7.2 and 13.2

$$\mathbf{J}_\nu(x) = J_\nu(x) + \pi^{-1}\sin(\nu\pi)\mathbf{A}_\nu(x).$$

Ex. 13.4 Show that for large positive x

$$\int_1^{\infty}(1-e^{1-t})e^{ixt(1-\ln t)}\, dt \sim -(i/x)e^{ix}.$$

Ex. 13.5 Show that for large positive x

$$\int_0^\infty t \exp\{it^2(\ln t - x)\}\, dt \sim (\pi/e)^{1/2} \exp(x - \tfrac{1}{2}ie^{2x-1} + \tfrac{1}{4}\pi i).$$

*14 Asymptotic Expansions by the Method of Stationary Phase

14.1 Suppose that $p(t)$ is increasing in (a, b), and in the neighborhood of $t = a$ both $p(t)$ and $q(t)$ can be expanded in ascending powers of $t - a$. In §§7 and 8 we saw that an asymptotic expansion of the integral

$$\int_a^b e^{-xp(t)} q(t)\, dt \tag{14.01}$$

for large x could be constructed by transforming it into the form

$$e^{-xp(a)} \int_0^{p(b)-p(a)} e^{-xv} f(v)\, dv,$$

expanding $f(v)$ in ascending powers of v, and integrating formally term by term over the interval $(0, \infty)$.

There is an analogous procedure for the oscillatory integral

$$\int_a^b e^{ixp(t)} q(t)\, dt.$$

Compared with (14.01), however, two major complications arise. First, direct integration of the terms in the expansion of $f(v)$ over the interval $(0, \infty)$ is permissible only for the first few. This is because $\int_0^\infty e^{ixv} v^{\alpha-1}\, dv$ diverges when $\alpha \geqslant 1$. Secondly, the upper limit b contributes to the final asymptotic expansion when $p(b)$ is finite, whether or not $t = b$ is a stationary point.

Treatments of the problem will be found in the references Erdélyi (1956a, §2.9), Lyness (1971b), and Olver (1974); the last two include methods for estimating and bounding the error terms. Often alternative methods, such as those of Chapters 4 and 7, are available in applications and may provide an easier way of calculating higher terms and error bounds.

Historical Notes and Additional References

§3 Wyman and Wong (1969) have pointed out that Watson's result can be regarded as a special case of an earlier theorem of Barnes (1906). The present form is due to Doetsch (1955, p. 45).

§4 (i) A common misconception—possibly stemming from present-day emphasis on the Lebesgue theory of integration—is that the Riemann–Lebesgue lemma applies only to absolutely convergent integrals. Uniform convergence suffices. Moreover, the name *improper* for an integral which converges but does not converge absolutely is only really justified in the context of Lebesgue theory.

(ii) An extension of the Riemann–Lebesgue lemma has been given by Bleistein, Handelsman, and Lew (1972).

§6 It is interesting to note that in an appendix to a paper published many years before Poincaré's definition of an asymptotic expansion, Stokes (1857) observed that numerical results obtained from an asymptotic expansion of the Airy integral were greatly improved by including exponentially small terms.

§§7–9 Following Laplace, contributors to the theory of the Laplace approximation include Burkhardt (1914), Pólya and Szegö (1925), Widder (1941, Chapter 7), and Erdélyi (1956a, §2.4). §§7 to 9 are based on the last reference and Olver (1968). Extensions of Laplace's method are considered in Chapters 4 and 9.

§§9.2–9.3 A third way of attacking the difficulty is to employ a majorant of the form

$$|\phi_n(t)| \leqslant |a_n| t^{(n+\lambda-\mu)/\mu} \exp(\hat{\rho}_n t^{1/\mu}).$$

This has been discussed by Olver (1968) for the case $\mu = 2$, and by D. S. Jones (1972) for $\mu \geqslant 2$.

§§11–14 The method of stationary phase originated in the interference principle of water waves. It was used by Stokes (1850) in investigating the Airy integral (§13.4), and formulated in more general terms by Kelvin (1887). Further advances in the theory are due to Poincaré (1904), Watson (1918b), van der Corput (1934, 1936), Erdélyi (1955), D. S. Jones (1966), and Cīrulis (1969). Dieudonné (1968, p. 135) gives Theorem 13.1 in the case $p(b) < \infty$, but the present version is somewhat more general than previous results pertaining to the first approximation. Recently, work on diffraction and other problems has caused the method to be extended to multiple integrals. This topic is outside the scope of the present book; accounts and references have been given by Boin (1965), Chako (1965), Fedoryuk (1970), de Kok (1971), and Bleistein and Handelsman (1974).

Some mystery was attached to the method of stationary phase in its infancy, more perhaps than to other results in asymptotic analysis. To some extent this attitude persists. The method is frequently regarded as weak, suited only to the derivation of the first term of an asymptotic expansion, and either best avoided or regarded as a special case of complex-variable procedures (thereby requiring the functions $p(t)$ and $q(t)$ to be analytic). This view is not well founded. In essential respects the method of stationary phase resembles Laplace's method. The main differences are heavier differentiability requirements on the given functions $p(t)$ and $q(t)$, harder proofs, and weaker forms of error bound.

4

CONTOUR INTEGRALS

1 Laplace Integrals with a Complex Parameter

1.1 The theory of Chapter 3, §2 is easily extended to the integral

$$I(z) = \int_0^\infty e^{-zt} q(t)\, dt, \tag{1.01}$$

in which z is a complex parameter. We again suppose that $q(t)$ is a real or complex function that is infinitely differentiable in $[0, \infty)$ and has the property

$$|q^{(s)}(t)| \leqslant A_s e^{\sigma t} \qquad (t \geqslant 0), \tag{1.02}$$

where A_s and σ are real constants, σ being independent of s. Without loss of generality it may be assumed that $\sigma \geqslant 0$.

The principal results are given by

$$I(z) = \frac{q(0)}{z} + \frac{q'(0)}{z^2} + \cdots + \frac{q^{(n-1)}(0)}{z^n} + \varepsilon_n(z) \qquad (\operatorname{Re} z > \sigma),$$

where n is an arbitrary positive integer or zero,

$$\varepsilon_n(z) = \frac{1}{z^n} \int_0^\infty e^{-zt} q^{(n)}(t)\, dt, \tag{1.03}$$

and

$$|\varepsilon_n(z)| \leqslant \frac{A_n}{|z|^n (\operatorname{Re} z - \sigma)}. \tag{1.04}$$

Suppose that z is confined to the annular sector

$$|\operatorname{ph} z| \leqslant \tfrac{1}{2}\pi - \delta, \qquad |z| > \sigma \csc \delta, \tag{1.05}$$

where δ is a constant in $(0, \frac{1}{2}\pi)$. Then $\operatorname{Re} z \geqslant |z| \sin \delta > \sigma$, and

$$|\varepsilon_n(z)| \leqslant \frac{A_n}{|z|^n (|z| \sin \delta - \sigma)}$$

Accordingly, as $z \to \infty$ in (1.05) we have

$$I(z) \sim \sum_{s=0}^{\infty} \frac{q^{(s)}(0)}{z^{s+1}}. \tag{1.06}$$

Provided that $q^{(n)}(0) \neq 0$, a useful form of the bound (1.04) for the nth error term of the last expansion is

$$|\varepsilon_n(z)| \leqslant \frac{|q^{(n)}(0)|}{|z|^n(\operatorname{Re} z - \sigma_n)} \qquad (\operatorname{Re} z > \max(\sigma_n, 0)), \tag{1.07}$$

where

$$\sigma_n = \sup_{(0,\infty)} \left\{ \frac{1}{t} \ln \left| \frac{q^{(n)}(t)}{q^{(n)}(0)} \right| \right\}. \tag{1.08}$$

As in the case of real variables (Chapter 3, Exercise 3.3) the assumed restrictions on $q(t)$ can be eased somewhat without invalidating the expansion (1.06). In these more general circumstances, however, the relations (1.03) and (1.07) are inapplicable.

1.2 Next, suppose that as a function of the complex variable t, $q(t)$ is holomorphic in a domain which includes the sector $\mathbf{S}$: $\alpha_1 \leqslant \operatorname{ph} t \leqslant \alpha_2$.† We require $\mathbf{S}$ to contain $\operatorname{ph} t = 0$ in its interior, so that $\alpha_1 < 0$ and $\alpha_2 > 0$. We suppose further that

$$|q(t)| \leqslant Ae^{\sigma|t|} \qquad (t \in \mathbf{S}), \tag{1.09}$$

where A and σ are nonnegative constants.

Let δ be any positive number satisfying $\alpha_1 + \delta \leqslant 0 \leqslant \alpha_2 - \delta$. Then the method of Chapter 1, §4.3 shows that

$$|q^{(s)}(t)| \leqslant A_s e^{\sigma|t|} \qquad (\alpha_1 + \delta \leqslant \operatorname{ph} t \leqslant \alpha_2 - \delta), \tag{1.10}$$

where A_s is independent of t. In particular, when $\operatorname{ph} t = 0$ the conditions of §1.1 are satisfied and (1.06) applies. In the present circumstances, the region of validity of this asymptotic expansion can be extended in the following way.

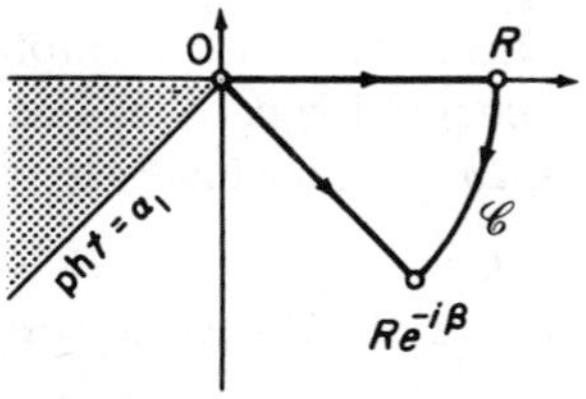

Fig. 1.1 t plane.

Let R be an arbitrary positive number and β be an arbitrary angle in the interval $0 \leqslant \beta \leqslant \min(-\alpha_1 - \delta, \frac{1}{2}\pi)$. By Cauchy's theorem

$$\int_0^R e^{-zt} q^{(n)}(t)\, dt = \int_0^{Re^{-i\beta}} e^{-zt} q^{(n)}(t)\, dt - \int_{\mathscr{C}} e^{-zt} q^{(n)}(t)\, dt, \tag{1.11}$$

† If $\alpha_2 - \alpha_1 \geqslant 2\pi$, then $\mathbf{S}$ extends to more than one Riemann sheet.

where $\mathscr{C}$ is the arc with parametric equation

$$t = Re^{-i\chi} \qquad (0 \leqslant \chi \leqslant \beta);$$

see Fig. 1.1. Write $\theta \equiv \operatorname{ph} z$ and assume that $0 < \delta < \frac{1}{4}\pi$ and $\delta \leqslant \theta \leqslant \frac{1}{2}\pi - \delta$. Then $|\theta - \chi| \leqslant \frac{1}{2}\pi - \delta$; hence on $\mathscr{C}$

$$\operatorname{Re}(zt) = |z|\,R\cos(\theta - \chi) \geqslant |z|\,R\sin\delta.$$

Accordingly, using (1.10) with $s = n$ we have

$$\left|\int_{\mathscr{C}} e^{-zt} q^{(n)}(t)\,dt\right| \leqslant A_n R\beta \exp(-|z|\,R\sin\delta + \sigma R)$$

and therefore vanishes as $R \to \infty$, provided that $|z| > \sigma \csc\delta$.

We have therefore shown that

$$\varepsilon_n(z) = \frac{1}{z^n}\int_0^{\infty e^{-i\beta}} e^{-zt} q^{(n)}(t)\,dt, \tag{1.12}$$

with the conditions $\delta \leqslant \operatorname{ph} z \leqslant \frac{1}{2}\pi - \delta$ and $|z| > \sigma \csc\delta$. It is seen from (1.10), however, that the last integral defines a holomorphic function of z in a region that includes the annular sector

$$|\operatorname{ph}(ze^{-i\beta})| \leqslant \tfrac{1}{2}\pi - \delta, \qquad |z| > \sigma\csc\delta; \tag{1.13}$$

compare Theorem 1.1 of Chapter 2. Therefore (1.12) represents the analytic continuation of $\varepsilon_n(z)$ within this region. In particular, with $n = 0$ we have the analytic continuation of the original integral $I(z)$.

1.3 From (1.10) and (1.12) it follows that $\varepsilon_n(z) = O(z^{-n-1})$, and thence that the expansion (1.06) is valid in the sector $|\operatorname{ph}(ze^{-i\beta})| \leqslant \frac{1}{2}\pi - \delta$, provided that $I(z)$ is interpreted as the analytic continuation of the original integral.

If $\alpha_1 \geqslant -\frac{1}{2}\pi - \delta$, then the largest value we may assign to β is $-\alpha_1 - \delta$. This extends the sector of validity from $|\operatorname{ph} z| \leqslant \frac{1}{2}\pi - \delta$ to $-\frac{1}{2}\pi + \delta \leqslant \operatorname{ph} z \leqslant -\alpha_1 + \frac{1}{2}\pi - 2\delta$. Alternatively, if $\alpha_1 < -\frac{1}{2}\pi - \delta$, then we can set $\beta = \frac{1}{2}\pi$: the extended region of validity becomes $-\frac{1}{2}\pi + \delta \leqslant \operatorname{ph} z \leqslant \pi - \delta$. And in this event further rotations of the integration path may be made in the negative angular sense. Each is less than or equal to $\frac{1}{2}\pi$, and the maximum permissible total rotation is $\beta = -\alpha_1 - \delta$.

In a similar way the integration path may be rotated through a positive angle up to $\alpha_2 - \delta$. On replacing 2δ by δ, we finally have:

Theorem 1.1 *Let $I(z)$ denote $\int_0^\infty e^{-zt} q(t)\,dt$ or the analytic continuation of this integral. With the conditions of the opening paragraph of* §1.2

$$I(z) \sim \sum_{s=0}^{\infty} \frac{q^{(s)}(0)}{z^{s+1}} \tag{1.14}$$

as $z \to \infty$ in the sector $-\alpha_2 - \frac{1}{2}\pi + \delta \leqslant \operatorname{ph} z \leqslant -\alpha_1 + \frac{1}{2}\pi - \delta$, where $\delta > 0$.

When $\alpha_2 - \alpha_1 > \pi$ the expansion (1.14) holds in a sector of angle exceeding 2π. In this event Theorem 7.2 of Chapter 1 shows that either (1.14) converges for all sufficiently large $|z|$, or $I(z)$ has a branch point at infinity.

1.4 The corresponding extension of the error bound (1.07) is given by

$$|\varepsilon_n(z)| \leqslant \frac{|q^{(n)}(0)|}{|z|^n\{\mathrm{Re}(ze^{-i\beta})-\sigma_n(\beta)\}}, \tag{1.15}$$

where n is an arbitrary positive integer or zero, β is an arbitrary angle in the interval $(-\alpha_2, -\alpha_1)$,

$$\sigma_n(\beta) = \sup_{\mathrm{ph}\, t = -\beta}\left\{\frac{1}{|t|}\ln\left|\frac{q^{(n)}(t)}{q^{(n)}(0)}\right|\right\}, \tag{1.16}$$

and z restricted by

$$|\mathrm{ph}(ze^{-i\beta})| < \tfrac{1}{2}\pi, \qquad \mathrm{Re}(ze^{-i\beta}) > \max\{\sigma_n(\beta), 0\}. \tag{1.17}$$

For a prescribed value of z the magnitude of the bound (1.15) depends on the value assigned to β. With $\mathrm{ph}\, z$ again denoted by θ, the ratio of the absolute value of the first neglected term in the series (1.14) to the right-hand side of (1.15) is $\cos(\theta-\beta)-|z|^{-1}\sigma_n(\beta)$. For large $|z|$, this is approximately $\cos(\theta-\beta)$. When $-\alpha_2 < \theta < -\alpha_1$, we can set $\beta = \theta$, in which event the ratio is approximately unity for large $|z|$, which is ideal.

If θ lies in either of the remaining intervals $[-\alpha_1, -\alpha_1+\frac{1}{2}\pi)$ and $(-\alpha_2-\frac{1}{2}\pi, -\alpha_2]$, then β has to differ from θ. As θ approaches $-\alpha_1+\frac{1}{2}\pi$ or $-\alpha_2-\frac{1}{2}\pi$, the bound (1.15) exceeds the absolute value of the first neglected term by an increasingly large factor. This warns us that the direct use of the asymptotic expansion near the boundaries of its region of validity may lead to grave inaccuracies and should be avoided.

Ex. 1.1 With the conditions of §1.1, show that (1.06) is also valid in the half-plane $\mathrm{Re}\, z \geqslant \sigma+\delta$.

Ex. 1.2 Let $I(x)$ denote the analytic continuation of the integral

$$\int_1^\infty \frac{dt}{t^2(x+\ln t)^{1/3}}$$

from $\mathrm{ph}\, x = 0$ into the complex plane. What is the region of validity of the asymptotic expansion for $I(x)$ given in Chapter 3, Exercise 2.5?

Ex. 1.3 Show that the number $\sigma_n(\beta)$ defined by (1.16) satisfies

$$\lambda_n \cos(\mu_n-\beta) \leqslant \sigma_n(\beta) \leqslant \sup_{\mathrm{ph}\, t=-\beta}\left|\frac{q^{(n+1)}(t)}{q^{(n)}(t)}\right|,$$

where $\lambda_n e^{i\mu_n} = q^{(n+1)}(0)/q^{(n)}(0)$. [Olver, 1965c.]

2 Incomplete Gamma Functions of Complex Argument

2.1 Let us apply the foregoing theory to the integral

$$\Gamma(\alpha, z) = e^{-z}z^\alpha\int_0^\infty e^{-zt}(1+t)^{\alpha-1}\,dt \qquad (|\mathrm{ph}\, z| < \tfrac{1}{2}\pi), \tag{2.01}$$

in which all functions have their principal values. This expression is derivable from Chapter 2, (5.04) by simple transformation of integration variable when z is positive; the extension to $|\mathrm{ph}\, z| < \frac{1}{2}\pi$ follows by analytic continuation.

In the notation of §1, we have

$$q(t) = (1+t)^{\alpha-1}, \qquad q^{(s)}(t) = (\alpha-1)(\alpha-2)\cdots(\alpha-s)(1+t)^{\alpha-s-1}.$$

Except when α is a positive integer—in which event (2.01) is evaluable in terms of elementary functions—$q(t)$ has a singularity at $t=-1$. We therefore take $\alpha_1 = -\pi+\delta$ and $\alpha_2 = \pi-\delta$. Clearly the condition (1.09) is satisfiable with σ either zero or an assignable positive number. On replacing 2δ by δ, Theorem 1.1 immediately gives the expansion

$$\Gamma(\alpha, z) = e^{-z}z^{\alpha-1}\left\{\sum_{s=0}^{n-1} \frac{(\alpha-1)(\alpha-2)\cdots(\alpha-s)}{z^s} + \varepsilon_n(z)\right\} \qquad (n = 0, 1, 2, \ldots), \tag{2.02}$$

where $\varepsilon_n(z) = O(z^{-n})$ as $z \to \infty$ in the sector $|\mathrm{ph}\, z| \leqslant \frac{3}{2}\pi-\delta$, α being kept fixed.

2.2 In evaluating bounds for $\varepsilon_n(z)$, *we make the simplifying assumption that* α *is real*. The definition (1.16) yields

$$\sigma_n(\beta) = \sup_{\mathrm{ph}\, t = -\beta}\left(\frac{\alpha-n-1}{|t|}\ln|1+t|\right). \tag{2.03}$$

And from (1.15) and (1.17) we obtain

$$|\varepsilon_n(z)| \leqslant \frac{|(\alpha-1)(\alpha-2)\cdots(\alpha-n)|}{|z|\cos(\theta-\beta)-\sigma_n(\beta)}\frac{1}{|z|^{n-1}}, \tag{2.04}$$

where $\beta \in (-\pi, \pi)$ is arbitrary, $\theta \equiv \mathrm{ph}\, z$, and z is restricted by

$$|\theta-\beta| < \tfrac{1}{2}\pi, \qquad |z|\cos(\theta-\beta) > \sigma_n(\beta).$$

The essential problem in the error analysis is to evaluate or bound $\sigma_n(\beta)$.

2.3 Suppose first that $n \leqslant \alpha-1$ (which can only happen when $\alpha \geqslant 1$). From (2.03) we obtain

$$\sigma_n(\beta) = (\alpha-n-1)\sup_{\tau\in(0,\infty)}\left\{\frac{\ln(1+2\tau\cos\beta+\tau^2)}{2\tau}\right\}. \tag{2.05}$$

For positive τ and real β,

$$\frac{\ln(1+2\tau\cos\beta+\tau^2)}{2\tau} \leqslant \frac{\ln(1+\tau)}{\tau} < 1.$$

Hence $\sigma_n(\beta) \leqslant \alpha-n-1$. With $\beta=\theta$ the inequality (2.04) yields

$$|\varepsilon_n(z)| \leqslant \frac{(\alpha-1)(\alpha-2)\cdots(\alpha-n)}{|z|-(\alpha-n-1)}\frac{1}{|z|^{n-1}}, \tag{2.06}$$

valid when $-\pi < \theta < \pi$ and $|z| > \alpha-n-1$. Since $\varepsilon_n(z)$ is a continuous function of θ in $(-\frac{3}{2}\pi, \frac{3}{2}\pi)$ and the right-hand side of (2.06) is independent of θ, the first of these restrictions can be eased to $-\pi \leqslant \theta \leqslant \pi$. Thus sufficient conditions for the validity of (2.06) are

$$|z| > \alpha - n - 1 \geqslant 0, \qquad |\mathrm{ph}\, z| \leqslant \pi. \tag{2.07}$$

Next, by taking β different from θ we obtain bounds for $|\varepsilon_n(z)|$ which apply to the sectors $\pi < |\theta| < \frac{3}{2}\pi$. These bounds become increasingly large as θ approaches $\pm\frac{3}{2}\pi$, but this is of only theoretical interest, since in practice the continuation formula (5.06) of Chapter 2 would be used to compute $\Gamma(\alpha, z)$ outside the range $\mathrm{ph}\, z \in [-\pi, \pi]$.

2.4 Now suppose that $n \geqslant \alpha - 1$. Instead of (2.05) we have

$$\sigma_n(\beta) = (n-\alpha+1)\,\sigma(\beta),$$

where

$$\sigma(\beta) = \sup_{\mathrm{ph}\, t = -\beta} \left\{ -\frac{\ln|1+t|}{|t|} \right\} = \sup_{\tau \in (0,\infty)} \left\{ -\frac{1}{2\tau} \ln(1+2\tau\cos\beta+\tau^2) \right\}. \tag{2.08}$$

Clearly if $|\beta| \leqslant \frac{1}{2}\pi$, then $\sigma(\beta) = 0$. Setting $\beta = \theta$, we derive

$$|\varepsilon_n(z)| \leqslant \frac{|(\alpha-1)(\alpha-2)\cdots(\alpha-n)|}{|z|^n} \qquad (n \geqslant \alpha-1, \quad |\theta| \leqslant \tfrac{1}{2}\pi). \tag{2.09}$$

In other words, in these circumstances the error is bounded by the absolute value of the first neglected term in the expansion.

When $\frac{1}{2}\pi \leqslant \theta < \pi$, we may set $\beta = \frac{1}{2}\pi$. This gives

$$|\varepsilon_n(z)| \leqslant \frac{|(\alpha-1)(\alpha-2)\cdots(\alpha-n)|}{|z|^n \sin\theta} \qquad (n \geqslant \alpha-1, \quad \tfrac{1}{2}\pi \leqslant \theta < \pi). \tag{2.10}$$

With $\sin\theta$ replaced by $|\sin\theta|$ this result also holds for $-\pi < \theta \leqslant -\frac{1}{2}\pi$. Alternatively, we may again set $\beta = \theta$. This produces

$$|\varepsilon_n(z)| \leqslant \frac{|(\alpha-1)(\alpha-2)\cdots(\alpha-n)|}{|z| - (n-\alpha+1)\,\sigma(\theta)} \frac{1}{|z|^{n-1}} \tag{2.11}$$

when $n \geqslant \alpha-1$, $\frac{1}{2}\pi \leqslant |\theta| < \pi$, $|z| > (n-\alpha+1)\,\sigma(\theta)$. The value of $\sigma(\theta)$ is numerically calculable from its definition (2.08).† The right-hand side of (2.11) is asymptotic to $|(\alpha-1)\cdots(\alpha-n)z^{-n}|$ as $|z| \to \infty$. Hence (2.11) is a better bound than (2.10) when $|z|$ is sufficiently large, in fact when

$$|z| > (n-\alpha+1)\,\sigma(\theta)/(1-|\sin\theta|).$$

The opposite is true when $|z|$ is of moderate size; indeed, (2.11) is unavailable when $|z| \leqslant (n-\alpha+1)\,\sigma(\theta)$.

Both (2.10) and (2.11) fail as θ approaches $\pm\pi$, because $\sin\theta$ vanishes and $\sigma(\theta)$ becomes infinite, but useful bounds for this region can be obtained by taking other values of β in (2.04). For example, $\beta = \frac{3}{4}\pi$ gives acceptable bounds when

$$|z|\cos(\theta - \tfrac{3}{4}\pi) > (n-\alpha+1)\,\sigma(\tfrac{3}{4}\pi) \qquad (n \geqslant \alpha-1);$$

† Or it can be replaced by the upper bound given in Exercise 2.3 below.

in particular, this includes the upper side of the negative real axis to the left of the point $-2^{1/2}(n-\alpha+1)\sigma(\frac{3}{4}\pi)$.

Ex. 2.1 In Chapter 3, Exercise 1.2 an asymptotic expansion for erfc x was given. What is its region of validity in the complex plane?

Ex. 2.2 For the generalized exponential integral (Chapter 2, Exercise 3.5) prove that for fixed n and large z in $|\text{ph}\, z| \leqslant \frac{3}{2}\pi-\delta\,(<\frac{3}{2}\pi)$

$$E_n(z) \sim \frac{e^{-z}}{z}\sum_{s=0}^{\infty}(-)^s\frac{n(n+1)\cdots(n+s-1)}{z^s}.$$

Ex. 2.3 Show that when $\frac{1}{2}\pi < |\beta| < \pi$ the number $\sigma(\beta)$ defined by (2.08) satisfies

$$|\sec\beta|\ln(|\csc\beta|) \leqslant \sigma(\beta) \leqslant 2|\sec\beta|\ln(|\csc\beta|).$$

[Olver, 1965c.]

3 Watson's Lemma

3.1 When z is a complex parameter the integral

$$I(z) = \int_0^\infty e^{-zt}q(t)\,dt \tag{3.01}$$

is known in the operational calculus as the *Laplace transform* of $q(t)$. It is often denoted by $\mathscr{L}(q)$ or $\bar{q}(z)$, and in operational work the symbol z is commonly replaced by p; thus

$$\mathscr{L}(q) = \int_0^\infty e^{-pt}q(t)\,dt.$$

If (3.01) converges for a certain value of z, then it is reasonable to expect convergence when the exponential factor in the integrand decays at a faster rate:

Theorem 3.1† *Let $q(t)$ be a real or complex function of the positive real variable t with a finite number of discontinuities and infinities. If the integral* (3.01) *converges throughout its range for $z=z_0$, then it also converges when* $\text{Re}\, z > \text{Re}\, z_0$.

The proof is similar to analysis in Chapter 3, §3.2. Let

$$Q(t) = \int_0^t e^{-z_0 v}q(v)\,dv,$$

so that $Q(t)$ is continuous and bounded in $[0,\infty)$. If $\text{Re}\, z > \text{Re}\, z_0$, then

$$\int_0^\infty e^{-zt}q(t)\,dt = (z-z_0)\int_0^\infty e^{-(z-z_0)t}Q(t)\,dt.$$

Since $|Q(t)|$ is bounded, the right-hand integral converges (absolutely), hence the integral on the left converges.

3.2 As a consequence of Theorem 3.1 there are three possibilities concerning the convergence of $I(z)$ in the complex plane: (a) $I(z)$ converges for all z; (b) $I(z)$

† Doetsch (1950, pp. 35 and 549).

diverges for all z; (c) there exists a number ζ such that $I(z)$ converges when $\operatorname{Re} z > \zeta$ and diverges when $\operatorname{Re} z < \zeta$. The number ζ is called the *abscissa of convergence* of $I(z)$ and, conventionally, we write $\zeta = -\infty$ in Case (a) and $\zeta = +\infty$ in Case (b). Moreover, on replacing $q(t)$ by $|q(t)|$ in the analysis it is seen that there is also an *abscissa of absolute convergence* ζ_A, say. Evidently $\zeta \leqslant \zeta_A$.

3.3 Theorem 3.2 *Assume that:*

(i) *$q(t)$ is a real or complex function of the positive real variable t with a finite number of discontinuities and infinities.*

(ii) *As $t \to 0+$*

$$q(t) \sim \sum_{s=0}^{\infty} a_s t^{(s+\lambda-\mu)/\mu}, \tag{3.02}$$

where μ is a positive constant and λ is a real or complex constant such that $\operatorname{Re} \lambda > 0$.

(iii) *The abscissa of convergence of the integral* (3.01) *is not* $+\infty$.

Then

$$I(z) \sim \sum_{s=0}^{\infty} \Gamma\left(\frac{s+\lambda}{\mu}\right) \frac{a_s}{z^{(s+\lambda)/\mu}} \tag{3.03}$$

as $z \to \infty$ in the sector $|\operatorname{ph} z| \leqslant \frac{1}{2}\pi - \delta$ $(<\frac{1}{2}\pi)$, where $z^{(s+\lambda)/\mu}$ has its principal value.

The proof of this theorem parallels that of Theorem 3.1 of Chapter 3. As before, write

$$\phi_n(t) = q(t) - \sum_{s=0}^{n-1} a_s t^{(s+\lambda-\mu)/\mu}, \tag{3.04}$$

and define k_n and K_n to be positive numbers such that

$$|\phi_n(t)| \leqslant K_n t^{(n+\operatorname{Re}\lambda-\mu)/\mu} \qquad (0 < t \leqslant k_n).$$

Then

$$\left|\int_0^{k_n} e^{-zt} \phi_n(t)\, dt\right| < \Gamma\left(\frac{n+\operatorname{Re}\lambda}{\mu}\right) \frac{K_n}{(\operatorname{Re} z)^{(n+\operatorname{Re}\lambda)/\mu}} = O\left(\frac{1}{z^{(n+\lambda)/\mu}}\right) \tag{3.05}$$

as $z \to \infty$ in $|\operatorname{ph} z| \leqslant \frac{1}{2}\pi - \delta$.

Again, if

$$L_n \equiv \sup_{t \in [k_n, \infty)} \left|\int_{k_n}^{t} e^{-Xv} \phi_n(v)\, dv\right|,$$

where X is a positive real value of z for which $I(z)$ converges, then for $\operatorname{Re} z > X$ we have

$$\left|\int_{k_n}^{\infty} e^{-zt} \phi_n(t)\, dt\right| \leqslant \frac{|z - X|}{\operatorname{Re} z - X} L_n \exp\{-(\operatorname{Re} z - X) k_n\}.$$

Because $|z| \leqslant (\operatorname{Re} z) \csc \delta$ this is $O\{\exp(-k_n |z| \sin \delta)\}$ as $z \to \infty$ in $|\operatorname{ph} z| \leqslant \frac{1}{2}\pi - \delta$. Combination of this estimate with (3.05) yields (3.03).

3.4 Theorem 3.3 *Assume that:*

(i) $q(t)$ *is holomorphic within the sector* $\mathbf{S}$: $\alpha_1 < \operatorname{ph} t < \alpha_2$, *where* $\alpha_1 < 0$ *and* $\alpha_2 > 0$.

(ii) *For each* $\delta \in (0, \frac{1}{2}\alpha_2 - \frac{1}{2}\alpha_1)$ *the expansion* (3.02) *holds as* $t \to 0$ *in the sector* $\mathbf{S}_\delta$: $\alpha_1 + \delta \leqslant \operatorname{ph} t \leqslant \alpha_2 - \delta$. *Again,* $\mu > 0$ *and* $\operatorname{Re} \lambda > 0$.

(iii) $q(t) = O(e^{\sigma|t|})$ *as* $t \to \infty$ *in* $\mathbf{S}_\delta$, *where* σ *is an assignable constant.*

Then if $I(z)$ *denotes the integral* (3.01) *or its analytic continuation, the expansion* (3.03) *holds in the sector* $-\alpha_2 - \frac{1}{2}\pi + \delta \leqslant \operatorname{ph} z \leqslant -\alpha_1 + \frac{1}{2}\pi - \delta$.

In this result the branches of $t^{(s+\lambda-\mu)/\mu}$ and $z^{(s+\lambda)/\mu}$ have their principal values on the positive real axis and are defined by continuity elsewhere.

This extension of Theorem 3.2 is established by rotation of the path of integration as in §§1.2 and 1.3.† Let β be any number in the interval $[-\alpha_2 + \delta, -\alpha_1 - \delta]$. Then

$$\int_0^{\infty e^{-i\beta}} e^{-zt} q(t)\, dt \tag{3.06}$$

represents the analytic continuation of $I(z)$ in the annular sector (1.13). We now apply Theorem 3.2 with $te^{i\beta}$ and $ze^{-i\beta}$ playing the roles of t and z, respectively, and subsequently replace 2δ by δ.

The reader may notice that, as in the case of Theorem 3.2, Condition (iii) could be eased to weaker (but more complicated) convergence conditions by use of partial integration.

***3.5** In the case of Theorem 3.3, the nth error term of the expansion (3.03) is given by

$$\varepsilon_n(z) = \int_0^{\infty e^{-i\beta}} e^{-zt} \phi_n(t)\, dt \qquad (|\operatorname{ph}(ze^{-i\beta})| < \tfrac{1}{2}\pi),$$

where $\phi_n(t)$ is defined by (3.04) and β is any number in $(-\alpha_2, -\alpha_1)$. Accordingly,

$$|\varepsilon_n(z)| \leqslant \Gamma\left(\frac{n+\lambda_{\mathrm{R}}}{\mu}\right) \frac{\exp(\lambda_{\mathrm{I}}\beta/\mu)|a_n|}{\{\operatorname{Re}(ze^{-i\beta}) - \sigma_n(\beta)\}^{(n+\lambda_{\mathrm{R}})/\mu}}, \tag{3.07}$$

where $\lambda_{\mathrm{R}} = \operatorname{Re}\lambda$, $\lambda_{\mathrm{I}} = \operatorname{Im}\lambda$, and

$$\sigma_n(\beta) = \sup_{\operatorname{ph} t = -\beta} \left\{ \frac{1}{|t|} \ln \left| \frac{\phi_n(t)}{a_n t^{(n+\lambda-\mu)/\mu}} \right| \right\}.$$

The bound (3.07) is valid when z lies in the annular sector

$$|\operatorname{ph}(ze^{-i\beta})| < \tfrac{1}{2}\pi, \qquad \operatorname{Re}(ze^{-i\beta}) > \max\{\sigma_n(\beta), 0\}.$$

When $\sigma_n(\beta)$ is infinite, modifications of this result on the lines of §§9.2 and 9.3 of Chapter 3 may be made.

It will be observed that Theorem 1.1 belongs to the special case $\lambda = \mu = 1$ of Theorem 3.3. The forms of the error bound associated with the two theorems are quite different, however. The exponent $\sigma_n(\beta)$ of §1.4 is defined in terms of the nth

† The function $q(t)$ is not analytic at $t = 0$, but because $q(t) = O(t^{(\lambda/\mu)-1})$ as $t \to 0$ in $\mathbf{S}_\delta$, the rotation is justified.

derivative of $q(t)$; this is not the case in the present section. Furthermore, for large $|z|$ the overestimation factor associated with (1.15) is approximately $\sec(\theta-\beta)$, compared with $\sec^{n+1}(\theta-\beta)$ for (3.07) (in the case $\lambda = \mu = 1$). The former bound is therefore sharper when $\beta \neq \theta$, $n \geqslant 1$, and $|z|$ is sufficiently large.

Ex. 3.1 By applying Cauchy's theorem to the rectangle with vertices 0, T, $T+\frac{1}{2}\pi i$, and $\frac{1}{2}\pi i$, and letting $T \to +\infty$, prove that the abscissa of convergence of the Laplace transform of $q(t) = \exp(ie^t)$ differs from its abscissa of absolute convergence.

Ex. 3.2 Show that $\int_0^\infty \exp(-z^2 t)\ln(1+t^{1/2})\,dt$ and its analytic continuation share the asymptotic expansion

$$\sum_{s=1}^{\infty} (-)^{s-1}\frac{\Gamma(\frac{1}{2}s)}{2z^{s+2}}$$

as $z \to \infty$ in the sector $|\mathrm{ph}\, z| \leqslant \frac{5}{4}\pi - \delta$ $(< \frac{5}{4}\pi)$.

Ex. 3.3 If α is a positive constant, show that in the sector $|\mathrm{ph}\, z| \leqslant \frac{3}{2}\pi - \delta$ $(< \frac{3}{2}\pi)$ the analytic continuation of the integral $\int_0^\infty \exp\{-z\exp(t^\alpha)\}\,dt$ is approximated by

$$\Gamma\left(1+\frac{1}{\alpha}\right)\frac{e^{-z}}{z^{1/\alpha}}\left\{1 - \frac{1+\alpha}{2\alpha^2 z} + O\left(\frac{1}{z^2}\right)\right\} \qquad (z \to \infty).$$

Ex. 3.4† In Theorem 3.2 suppose that Condition (ii) is replaced by

$$q(t) \sim \sum_{s=0}^{\infty} q_s\{2\sinh(\tfrac{1}{2}t)\}^{2s} \qquad (t \to 0+).$$

Show that for each positive integer n

$$I(z) = \sum_{s=0}^{n-1} \frac{(2s)!\,q_s}{(z-s)(z-s+1)\cdots(z+s)} + O\left(\frac{1}{z^{2n+1}}\right)$$

as $z \to \infty$ in the sector $|\mathrm{ph}\, z| \leqslant \frac{1}{2}\pi - \delta$ $(< \frac{1}{2}\pi)$.

***Ex. 3.5** For the Goodwin–Staton integral of Chapter 2, Exercise 4.4, prove that

$$\int_0^\infty \frac{\exp(-u^2)}{u+z}\,du = \sum_{s=0}^{n-1} (-)^s \frac{\Gamma(\frac{1}{2}s+\frac{1}{2})}{2z^{s+1}} + \varepsilon_n(z) \qquad (n = 0, 1, \ldots),$$

where $\varepsilon_n(z) = O(z^{-n-1})$ as $z \to \infty$ in the sector $|\mathrm{ph}\, z| \leqslant \frac{5}{4}\pi - \delta$ $(< \frac{5}{4}\pi)$.

Show also that:

(i) If $|\mathrm{ph}\, z| \leqslant \frac{1}{4}\pi$, then $|\varepsilon_n(z)| \leqslant \frac{1}{2}\Gamma(\frac{1}{2}n+\frac{1}{2})|z|^{-n-1}$.

(ii) If $\frac{1}{4}\pi \leqslant |\mathrm{ph}\, z| \leqslant \frac{3}{4}\pi$, then $|\varepsilon_n(z)| \leqslant \frac{1}{2}\Gamma(\frac{1}{2}n+\frac{1}{2})|z|^{-n-1} + \frac{1}{2}\Gamma(\frac{1}{2}n+1)|z|^{-n-2}$.

(iii) If β is an arbitrary number such that $\frac{3}{4}\pi \leqslant |\beta| < \pi$ and z is restricted by

$$|\mathrm{ph}(ze^{-i\beta})| < \tfrac{1}{4}\pi, \qquad \mathrm{Re}(z^2 e^{-2i\beta}) > \sigma(2|\beta|-\pi),$$

where $\sigma(\beta)$ is defined by (2.08), then

$$|\varepsilon_n(z)| \leqslant \frac{\Gamma(\frac{1}{2}n+\frac{1}{2})}{2\{\mathrm{Re}(z^2 e^{-2i\beta}) - \sigma(2|\beta|-\pi)\}^{(n+1)/2}} + \frac{\Gamma(\frac{1}{2}n+1)}{2\{\mathrm{Re}(z^2 e^{-2i\beta}) - \sigma(2|\beta|-\pi)\}^{(n+2)/2}}.$$

***Ex. 3.6** With the notation and conditions of Theorem 3.3 let $g(t) = \mu t^{\mu-\lambda} q(t^\mu)$, and assume that $g(t)$ is holomorphic in the neighborhood of the origin. Show that the error term $\varepsilon_n(z)$ of §3.5 is given by

$$\varepsilon_n(z) = \frac{1}{(n-1)!}\int_0^{\infty e^{-i\beta/\mu}} g^{(n)}(v)\,dv \int_v^{\infty e^{-i\beta/\mu}} (t-v)^{n-1} t^{\lambda-1}\exp(-zt^\mu)\,dt,$$

provided that $-\alpha_2 < \beta < -\alpha_1$, $|\mathrm{ph}(ze^{-i\beta})| < \frac{1}{2}\pi$, and $\mathrm{Re}(ze^{-i\beta}) > \sigma$.

† Further results of this kind have been given by Erdélyi (1946, 1961).

***Ex. 3.7** By rotating the path of integration of the inner integral in the preceding exercise to parallel $\operatorname{ph} t = -(\operatorname{ph} z)/\mu$, show that if $\mu = 2$, $\lambda = 1$, $n \geqslant 1$, $\theta = \operatorname{ph} z$, and

$$|g^{(n)}(v)| \leqslant G_n \exp\{\gamma_n(\beta)|v|^2\} \qquad (\operatorname{ph} v = -\tfrac{1}{2}\beta),$$

then

$$|\varepsilon_n(z)| \leqslant \frac{G_n \Gamma(\frac{1}{2}n+\frac{1}{2})}{2|z|^{(n+1)/2} n!} \frac{|z|\cos(\frac{1}{2}\theta - \frac{1}{2}\beta)}{|z|\cos(\theta-\beta) - \gamma_n(\beta)},$$

whenever the denominator is positive.

4 Airy Integral of Complex Argument; Compound Asymptotic Expansions

4.1 To determine the asymptotic behavior of $\operatorname{Ai}(z)$ for large $|z|$ we follow Copson (1963) and use the representation

$$\operatorname{Ai}(z) = \frac{\exp(-\frac{2}{3}z^{3/2})}{2\pi} \int_0^\infty \exp(-z^{1/2}t) \cos(\tfrac{1}{3}t^{3/2}) t^{-1/2}\, dt \qquad (|\operatorname{ph} z| < \pi), \tag{4.01}$$

in which fractional powers take their principal values. This integral may be obtained as follows. For positive $z = x$, say, we have from Chapter 2, §8.1

$$\operatorname{Ai}(x) = \frac{1}{2\pi i} \int_{-i\infty}^{i\infty} \exp(\tfrac{1}{3}v^3 - xv)\, dv.$$

The analysis accompanying this formula shows that the path may be translated to pass through the point $v = x^{1/2}$.† Setting $v = x^{1/2} + it^{1/2}$ on the upper half of the new path and $v = x^{1/2} - it^{1/2}$ on the lower half, we obtain (4.01) with $z = x$. The extension to $|\operatorname{ph} z| < \pi$ follows by analytic continuation.

Application of Theorem 3.2 to (4.01)—with $\lambda = \frac{1}{6}$, $\mu = \frac{1}{3}$, and the role of z played by $z^{1/2}$—yields the required expansion

$$\operatorname{Ai}(z) \sim \frac{e^{-\xi}}{2\pi^{1/2} z^{1/4}} \sum_{s=0}^{\infty} (-)^s \frac{u_s}{\xi^s} \tag{4.02}$$

as $z \to \infty$ in the sector $|\operatorname{ph} z| \leqslant \pi - \delta$ $(<\pi)$. Here $\xi = \frac{2}{3}z^{3/2}$, $u_0 = 1$,

$$u_s = \frac{2^s}{3^{3s}(2s)!} \frac{\Gamma(3s+\frac{1}{2})}{\Gamma(\frac{1}{2})} = \frac{(2s+1)(2s+3)(2s+5)\cdots(6s-1)}{(216)^s s!} \qquad (s \geqslant 1), \tag{4.03}$$

and fractional powers of z take their principal values.

To bound the error terms, we have from Taylor's theorem

$$\left| \cos\tau - \sum_{s=0}^{n-1} (-)^s \frac{\tau^{2s}}{(2s)!} \right| \leqslant \frac{\tau^{2n}}{(2n)!} \qquad (\tau \text{ real}, \quad n = 0, 1, 2, \ldots).$$

† Motivations for these transformations are clarified in §7 below.

Putting $\tau = \frac{1}{3}t^{3/2}$, we conclude that the ratio of the nth error term of (4.02) to the nth term of the series does not exceed $\{\sec(\frac{1}{2}\operatorname{ph} z)\}^{3n+(1/2)}$ in absolute value. In the case of positive z this means that each error term is bounded in absolute value by the first neglected term of the expansion.

An alternative way of deriving (4.02) is mentioned in Chapter 11, §8.1. It yields better error bounds when z is complex.

4.2 The sector of validity of (4.02) cannot be extended by use of Theorem 3.3 because the convergence condition (iii) is violated off the real t axis. To derive an asymptotic expansion for $\operatorname{Ai}(z)$ that is uniformly valid in a region embracing the negative real axis we employ the identity

$$\operatorname{Ai}(-z) = e^{\pi i/3}\operatorname{Ai}(ze^{\pi i/3}) + e^{-\pi i/3}\operatorname{Ai}(ze^{-\pi i/3}), \tag{4.04}$$

obtained from Chapter 2, (8.06).

Let l be an arbitrary positive integer and δ an arbitrary constant in $(0, \frac{2}{3}\pi)$. Truncating the expansion (4.02) at its lth term and replacing z by $ze^{\pi i/3}$, we derive

$$e^{\pi i/3}\operatorname{Ai}(ze^{\pi i/3}) = \frac{e^{\pi i/4}e^{-i\xi}}{2\pi^{1/2}z^{1/4}}\left\{\sum_{s=0}^{l-1} i^s\frac{u_s}{\xi^s} + \varepsilon_l^{(1)}(\xi)\right\},$$

where

$$\varepsilon_l^{(1)}(\xi) = O(\xi^{-l}) \quad \text{as} \quad z \to \infty \quad \text{in} \quad \operatorname{ph} z \in [-\tfrac{4}{3}\pi + \delta, \tfrac{2}{3}\pi - \delta].$$

The corresponding expansion for $e^{-\pi i/3}\operatorname{Ai}(ze^{-\pi i/3})$ is obtained by replacing i by $-i$ and $\varepsilon_l^{(1)}(\xi)$ by an error term $\varepsilon_l^{(2)}(\xi)$ with the property

$$\varepsilon_l^{(2)}(\xi) = O(\xi^{-l}) \quad \text{as} \quad z \to \infty \quad \text{in} \quad \operatorname{ph} z \in [-\tfrac{2}{3}\pi + \delta, \tfrac{4}{3}\pi - \delta].$$

Substituting these results in (4.04) and rearranging, we find that

$$\operatorname{Ai}(-z) = \frac{1}{\pi^{1/2}z^{1/4}}\left[\cos\left(\xi - \frac{1}{4}\pi\right)\left\{\sum_{s=0}^{[\frac{1}{2}l-\frac{1}{2}]} (-)^s\frac{u_{2s}}{\xi^{2s}} + \eta_l^{(1)}(\xi)\right\}\right.$$
$$\left. + \sin\left(\xi - \frac{1}{4}\pi\right)\left\{\sum_{s=0}^{[\frac{1}{2}l-1]} (-)^s\frac{u_{2s+1}}{\xi^{2s+1}} + \eta_l^{(2)}(\xi)\right\}\right], \tag{4.05}$$

where

$$2\eta_l^{(1)}(\xi) = \varepsilon_l^{(1)}(\xi) + \varepsilon_l^{(2)}(\xi), \qquad 2i\eta_l^{(2)}(\xi) = \varepsilon_l^{(1)}(\xi) - \varepsilon_l^{(2)}(\xi).$$

Clearly $\eta_l^{(1)}(\xi)$ and $\eta_l^{(2)}(\xi)$ are both $O(\xi^{-l})$ as $z \to \infty$ in $|\operatorname{ph} z| \leqslant \frac{2}{3}\pi - \delta$.

On replacing l by $2m$ and $2n+1$ in turn, we see that

$$\operatorname{Ai}(-z) = \frac{1}{\pi^{1/2}z^{1/4}}\left[\cos\left(\xi - \frac{1}{4}\pi\right)\left\{\sum_{s=0}^{m-1} (-)^s\frac{u_{2s}}{\xi^{2s}} + O\left(\frac{1}{\xi^{2m}}\right)\right\}\right.$$
$$\left. + \sin\left(\xi - \frac{1}{4}\pi\right)\left\{\sum_{s=0}^{n-1} (-)^s\frac{u_{2s+1}}{\xi^{2s+1}} + O\left(\frac{1}{\xi^{2n+1}}\right)\right\}\right], \tag{4.06}$$

where m and n are arbitrary positive integers, or zero. An expansion of this form will be called a *compound asymptotic expansion*. It is characterized by having two or more error terms, none of which is absorbable in the others.

With an extension of the meaning of the ~ sign, we write

$$\mathrm{Ai}(-z) \sim \frac{1}{\pi^{1/2}z^{1/4}}\left\{\cos\left(\xi - \frac{1}{4}\pi\right)\sum_{s=0}^{\infty}(-)^s\frac{u_{2s}}{\xi^{2s}} + \sin\left(\xi - \frac{1}{4}\pi\right)\sum_{s=0}^{\infty}(-)^s\frac{u_{2s+1}}{\xi^{2s+1}}\right\} \tag{4.07}$$

as $z \to \infty$ in $|\mathrm{ph}\, z| \leqslant \frac{2}{3}\pi - \delta$, ξ and u_s being defined as in §4.1, and fractional powers taking their principal values. For $\mathrm{ph}\, z = 0$, the leading term in (4.07) was found in Chapter 3, §13.4 by the method of stationary phase.

The reader will observe that the content of the square brackets in (4.05) can also be rearranged as a generalized asymptotic expansion $\sum \cos(\xi - \frac{1}{4}\pi - \frac{1}{2}s\pi) u_s \xi^{-s}$, with scale $e^{|\mathrm{Im}\,\xi|}\xi^{-s}$; compare Chapter 1, §10.3.

4.3 An immediate deduction from (4.07) is that on the negative real axis $\mathrm{Ai}(z)$ changes sign infinitely often, and therefore has a sequence of zeros with limit point at $z = -\infty$. Since the right-hand side of (4.02) is nonvanishing for all sufficiently large $|z|$, the sector of validity of (4.02) cannot be extended beyond $|\mathrm{ph}\, z| < \pi$. For similar reasons, $|\mathrm{ph}\, z| < \frac{2}{3}\pi$ is the maximal region of validity of (4.07).

Ex. 4.1 Verify that the expansions (4.02) and (4.07) agree in their common regions of validity, except for the presence of terms which are exponentially small (for large $|z|$) compared with the main series.

5 Ratio of Two Gamma Functions; Watson's Lemma for Loop Integrals

5.1 The asymptotic expansion of $\Gamma(z+a)/\Gamma(z+b)$ for fixed a and b and large z can be found by deriving the asymptotic expansions of $\Gamma(z+a)$ and $\Gamma(z+b)$ from Chapter 3, (8.16) in the case of real variables, or Chapter 8, §4 in the case of complex variables, and dividing the results. We shall illustrate the methods of the present chapter by obtaining the required expansion directly from the Beta-function integral. Throughout it is supposed that a and b are real or complex constants.

From Chapter 2, (1.10), we have

$$\frac{\Gamma(z+a)\Gamma(b-a)}{\Gamma(z+b)} = \int_0^1 v^{z+a-1}(1-v)^{b-a-1}\,dv \qquad (\mathrm{Re}(z+a) > 0, \quad \mathrm{Re}(b-a) > 0),$$

fractional powers taking their principal values. Substituting $v = e^{-t}$, we obtain

$$\frac{\Gamma(z+a)}{\Gamma(z+b)} = \frac{1}{\Gamma(b-a)}\int_0^{\infty} e^{-zt}q(t)\,dt \tag{5.01}$$

valid with the same restrictions, where

$$q(t) = e^{-at}(1-e^{-t})^{b-a-1}.$$

The expansion of $q(t)$ in ascending powers of t has the form

$$q(t) = \sum_{s=0}^{\infty} (-)^s q_s(a,b)\, t^{s+b-a-1} \qquad (|t| < 2\pi),$$

and the conditions of Theorem 3.3 are met with the choices $\alpha_1 = -\frac{1}{2}\pi$, $\alpha_2 = \frac{1}{2}\pi$, $\lambda = b-a$, $\mu = 1$, and $\sigma = |a|$. Application of the theorem gives the desired result

$$\frac{\Gamma(z+a)}{\Gamma(z+b)} \sim z^{a-b} \sum_{s=0}^{\infty} \frac{G_s(a,b)}{z^s}, \tag{5.02}$$

as $z \to \infty$ in the sector $|\mathrm{ph}\, z| \leqslant \pi - \delta$ $(<\pi)$, where

$$G_s(a,b) = (a-b)(a-b-1)\cdots(a-b-s+1)\, q_s(a,b).$$

The first three coefficients are easily verified to be

$$G_0(a,b) = 1, \qquad G_1(a,b) = \tfrac{1}{2}(a-b)(a+b-1),$$
$$G_2(a,b) = \tfrac{1}{24}(a-b)(a-b-1)\{3(a+b)^2 - 7a - 5b + 2\}.$$

5.2 The expansion (5.02) has been established with the restriction $\mathrm{Re}(b-a) > 0$. This can be removed in the following way. Let n be an arbitrary positive integer and $\phi_n(t)$ be defined for all positive t by

$$q(t) = \sum_{s=0}^{n-1} (-)^s q_s(a,b)\, t^{s+b-a-1} + \phi_n(t), \tag{5.03}$$

so that

$$\phi_n(t) = \sum_{s=n}^{\infty} (-)^s q_s(a,b)\, t^{s+b-a-1} \qquad (|t| < 2\pi). \tag{5.04}$$

Substituting in (5.01) by means of (5.03), we obtain

$$\frac{\Gamma(z+a)}{\Gamma(z+b)} = z^{a-b} \sum_{s=0}^{n-1} \frac{G_s(a,b)}{z^s} + I_n(a,b,z), \tag{5.05}$$

where

$$I_n(a,b,z) = \frac{1}{\Gamma(b-a)} \int_0^{\infty} e^{-zt} \phi_n(t)\, dt. \tag{5.06}$$

The conditions assumed in establishing equation (5.05) are

$$\mathrm{Re}(z+a) > 0, \qquad \mathrm{Re}\, z > 0, \qquad \mathrm{Re}(b-a) > 0.$$

From (5.03) and (5.04), however, it can be seen that $I_n(a,b,z)$ still converges at both limits if the last condition is relaxed to $\mathrm{Re}(n+b-a) > 0$. Noting that $G_s(a,b)$ is a polynomial, we see by analytic continuation with respect to b that (5.05) holds with the new condition. Then applying Theorem 3.3 to (5.06), and bearing in mind that the integer n is arbitrary, we conclude that *the expansion* (5.02) *holds without restriction on a or b*.

The artifice employed in the present subsection is frequently useful in asymptotic (and numerical) analysis. It is sometimes called the *method of extraction of the singular part*. In essence, we subtract poles or other troublesome singularities from a given function, evaluate their contribution analytically, and then employ the general asymptotic (or numerical) method under consideration to determine the contribution from the remainder.

5.3 Following Tricomi and Erdélyi (1951) we may generalize the foregoing analysis into a useful result known as *Watson's lemma for loop integrals*. Consider

$$I(z) = \frac{1}{2\pi i}\int_{-\infty}^{(0+)} e^{zt} q(t)\, dt, \tag{5.07}$$

where the path comprises the lower and upper sides of the real axis to the left of the point $-d$, say, together with the circle $|t| = d$. Assume that $q(t)$ is holomorphic, but not necessarily single valued, in an annulus $0 < |t| < d'$, where $d' > d$, and also that $q(t)$ is continuous on the path of integration. Then by analysis similar to §§3.1 and 3.2 we can prove that (5.07) possesses an abscissa of convergence. Next, we have:

Theorem 5.1 *Assume the conditions of this subsection, and also that*

$$q(t) \sim \sum_{s=0}^{\infty} a_s t^{(s+\lambda-\mu)/\mu} \tag{5.08}$$

as $t \to 0$ *in* $|\mathrm{ph}\, t| \leqslant \pi$, *where* μ *is a positive constant and* λ *is an unrestricted real or complex constant. Then*

$$I(z) \sim \sum_{s=0}^{\infty} \left\{\Gamma\left(\frac{\mu-\lambda-s}{\mu}\right)\right\}^{-1} \frac{a_s}{z^{(s+\lambda)/\mu}} \tag{5.09}$$

as $z \to \infty$ *in the sector* $|\mathrm{ph}\, z| \leqslant \frac{1}{2}\pi - \delta$ $(<\frac{1}{2}\pi)$.

In this result all fractional powers have their principal values. As in the case of Watson's lemma it is assumed that the abscissa of convergence of (5.07) is finite or $-\infty$, otherwise (5.09) is meaningless.

To prove the theorem, we again write

$$q(t) = \sum_{s=0}^{n-1} a_s t^{(s+\lambda-\mu)/\mu} + \phi_n(t) \qquad (n = 0, 1, \ldots).$$

Substituting the sum in (5.07) and integrating termwise by use of Hankel's loop integral for the reciprocal of the Gamma function (Chapter 2, (1.12)) we immediately obtain the first n terms in (5.09). Next, from (5.08) we see that $\phi_n(t)$ is $O(t^{(n+\lambda-\mu)/\mu})$ as $t \to 0$. As long as n is large enough to ensure that $n + \mathrm{Re}\,\lambda$ is positive, we may collapse the loop integral for $e^{zt}\phi_n(t)$ onto the two sides of the negative real axis, to obtain

$$\frac{1}{2\pi i}\int_{-\infty}^{(0+)} e^{zt}\phi_n(t)\, dt = \int_0^{\infty} e^{-z\tau} Q_n(\tau)\, d\tau, \tag{5.10}$$

where

$$Q_n(\tau) = \{\phi_n(\tau e^{-\pi i}) - \phi_n(\tau e^{\pi i})\}/(2\pi i)$$

$$\sim -\frac{1}{\pi}\sum_{s=0}^{\infty} \sin\left(\frac{s+n+\lambda-\mu}{\mu}\pi\right) a_{s+n}\,\tau^{(s+n+\lambda-\mu)/\mu} \qquad (\tau \to 0+).$$

The proof is completed by applying Theorem 3.2 to (5.10) and using the reflection formula for the Gamma function.

The reader will perceive that the essential difference of Theorem 5.1 from Theorem 3.2 is that λ is no longer restricted to the right half-plane. The reader should also notice that if the region in which $q(t)$ is holomorphic and has the expansion (5.08) includes the sector $\alpha_1 < \mathrm{ph}(-t) < \alpha_2$, where $\alpha_1 < 0$ and $\alpha_2 > 0$, and if, also, $q(t)$ is $O(e^{\sigma|t|})$ as $t \to \infty$ in this sector, then by use of Theorem 3.3 the expansion (5.09) for $I(z)$ (or its analytic continuation) is extendible to the sector $-\alpha_2 - \frac{1}{2}\pi + \delta \leqslant \mathrm{ph}\, z \leqslant -\alpha_1 + \frac{1}{2}\pi - \delta$, where $\delta > 0$.

6 Laplace's Method for Contour Integrals

6.1 Consider the integral

$$I(z) = \int_a^b e^{-zp(t)} q(t)\, dt, \tag{6.01}$$

in which the path $\mathscr{P}$, say, is a contour in the complex plane, $p(t)$ and $q(t)$ are analytic functions of t, and z is a real or complex parameter. By analogy with the real-variable theory of Chapter 3, §7 we might expect that when $|z|$ is large the main contribution to $I(z)$ comes from the neighborhood of the point $t = t_0$, say, at which $\mathrm{Re}\{zp(t)\}$ attains its minimum value. We shall see that this conjecture is correct when t_0 is an endpoint of $\mathscr{P}$, but is generally false when t_0 is an interior point of $\mathscr{P}$. In the latter event a deformation of the path is necessary before an asymptotic approximation can be obtained. In the present section we treat the former case.

It is convenient to introduce the following notation. Let t_1 and t_2 be any two points of $\mathscr{P}$. The part of $\mathscr{P}$ lying between t_1 and t_2 will be denoted by $(t_1, t_2)_{\mathscr{P}}$ when t_1 and t_2 are both excluded, and by $[t_1, t_2]_{\mathscr{P}}$ when t_1 and t_2 are both included. Similarly for $(t_1, t_2]_{\mathscr{P}}$ and $[t_1, t_2)_{\mathscr{P}}$. We also denote

$$\omega = \text{angle of slope of } \mathscr{P} \text{ at } a = \lim\{\mathrm{ph}(t-a)\} \qquad (t \to a \text{ along } \mathscr{P}). \tag{6.02}$$

Assumptions

(i) *$p(t)$ and $q(t)$ are independent of z, and single valued and holomorphic in a domain* **T**.

(ii) *$\mathscr{P}$ is independent of z, a is finite, b is finite or infinite, and $(a, b)_{\mathscr{P}}$ lies within* **T**.†

† Thus either a or b or both may be boundary points of **T**.

(iii) *In the neighborhood of a, the functions $p(t)$ and $q(t)$ can be expanded in convergent series of the form*

$$p(t) = p(a) + \sum_{s=0}^{\infty} p_s (t-a)^{s+\mu}, \qquad q(t) = \sum_{s=0}^{\infty} q_s (t-a)^{s+\lambda-1},$$

where $p_0 \neq 0$, μ is real and positive, and $\operatorname{Re}\lambda > 0$. When μ or λ is not an integer—and this can only happen when a is a boundary point of **T**—*the branches of $(t-a)^\mu$ and $(t-a)^\lambda$ are determined by the relations*

$$(t-a)^\mu \sim |t-a|^\mu e^{i\mu\omega}, \qquad (t-a)^\lambda \sim |t-a|^\lambda e^{i\lambda\omega},$$

as $t \to a$ along $\mathscr{P}$, and by continuity elsewhere on $\mathscr{P}$.

(iv) *z ranges along a ray or over an annular sector given by $\theta_1 \le \theta \le \theta_2$ and $|z| \ge Z$, where $\theta \equiv \operatorname{ph} z$, $\theta_2 - \theta_1 < \pi$, and $Z > 0$. $I(Ze^{i\theta})$ converges at b absolutely and uniformly with respect to θ.*

(v) $\operatorname{Re}\{e^{i\theta}p(t) - e^{i\theta}p(a)\}$ *is positive when $t \in (a,b)_{\mathscr{P}}$, and is bounded away from zero uniformly with respect to θ as $t \to b$ along $\mathscr{P}$.*

Remark. Neither ω nor θ need be confined to the principal range $(-\pi, \pi]$, provided that consistency is maintained.

6.2 Throughout the analysis great care is needed in specifying the branches of the many-valued functions which appear. With this in mind we introduce the following convention: *the value of $\omega_0 \equiv \operatorname{ph} p_0$ is not necessarily the principal one, but is chosen to satisfy*

$$|\omega_0 + \theta + \mu\omega| \leqslant \tfrac{1}{2}\pi, \tag{6.03}$$

and this branch of $\operatorname{ph} p_0$ is used in constructing all fractional powers of p_0 which occur. For example, $p_0^{1/\mu}$ means $\exp\{(\ln|p_0| + i\omega_0)/\mu\}$. Since

$$e^{i\theta}p(t) - e^{i\theta}p(a) \sim e^{i\theta}p_0(t-a)^\mu$$

as $t \to a$ along $\mathscr{P}$ (Condition (iii)), and $e^{i\theta}p(t) - e^{i\theta}p(a)$ has nonnegative real part (Condition (v)), it is always possible to choose ω_0 uniquely in this way. Moreover, because θ is restricted to an interval of length less than π, the value of ω_0 satisfying (6.03) is independent of θ.

We introduce new variables v and w by the equations

$$w^\mu = v = p(t) - p(a). \tag{6.04}$$

The branches of $\operatorname{ph} v$ and $\operatorname{ph} w$ are determined by

$$\operatorname{ph} v,\ \mu \operatorname{ph} w \to \omega_0 + \mu\omega \qquad (t \to a \text{ along } \mathscr{P}), \tag{6.05}$$

and by continuity elsewhere. Again, it is to be understood that these branches of $\operatorname{ph} v$ and $\operatorname{ph} w$ are to be used in constructing all fractional powers of v and w. Since v and w cannot vanish on $(a,b)_{\mathscr{P}}$ (Condition (v)), the branches are specified uniquely on $\mathscr{P}$; furthermore, $\operatorname{ph} v = \mu \operatorname{ph} w$ at every point of $\mathscr{P}$. From (6.03), (6.05), and Condition (v), it follows that

$$|\theta + \operatorname{ph} v| < \tfrac{1}{2}\pi \qquad (t \in (a,b)_{\mathscr{P}}). \tag{6.06}$$

Accordingly, v is confined to a single Riemann sheet as t ranges over $\mathscr{P}$.

For small $|t-a|$, Condition (iii) and the Binomial theorem yield

$$w = p_0^{1/\mu}(t-a)\left\{1 + \frac{p_1}{\mu p_0}(t-a) + \cdots\right\}.$$

Thus w is a single-valued holomorphic function of t in a neighborhood of a, and dw/dt is nonzero at a. Application of the inversion theorem for analytic functions† shows that for all sufficiently small values of the positive number ρ, the disk $|t-a| < \rho$ is mapped conformally on a domain $\mathbf{W}$ containing $w = 0$. Moreover, if $w \in \mathbf{W}$ then $t-a$ can be expanded in a convergent series

$$t - a = \sum_{s=1}^{\infty} c_s w^s = \sum_{s=1}^{\infty} c_s v^{s/\mu},$$

in which the coefficients c_s are expressible in terms of the p_s; compare Chapter 3, (8.05).

Let k be a point of $(a, b)_{\wp}$ chosen independently of z and sufficiently close to a to ensure that the disk $|w| \leq |p(k) - p(a)|^{1/\mu}$ is contained in $\mathbf{W}$. Then $[a, k]_{\wp}$ may be deformed to make its w map a straight line. Transformation to the variable v gives

$$\int_a^k e^{-zp(t)} q(t)\, dt = e^{-zp(a)} \int_0^{\kappa} e^{-zv} f(v)\, dv, \tag{6.07}$$

where

$$\kappa = p(k) - p(a), \qquad f(v) = q(t)\frac{dt}{dv} = \frac{q(t)}{p'(t)}, \tag{6.08}$$

and the path for the integral on the right-hand side of (6.07) also is a straight line.

For small $|v|, f(v)$ has a convergent expansion of the form

$$f(v) = \sum_{s=0}^{\infty} a_s v^{(s+\lambda-\mu)/\mu}, \tag{6.09}$$

in which the coefficients a_s are related to p_s and q_s in exactly the same way as in Chapter 3, §8.1; for example, $a_0 = q_0/(\mu p_0^{\lambda/\mu})$.

6.3 Following the real-variable approach, we define $f_n(v)$, $n = 0, 1, 2, \ldots$, by the relations $f_n(0) = a_n$ and

$$f(v) = \sum_{s=0}^{n-1} a_s v^{(s+\lambda-\mu)/\mu} + v^{(n+\lambda-\mu)/\mu} f_n(v) \qquad (v \neq 0). \tag{6.10}$$

Then $f_n(v)$ is $O(1)$ as $v \to 0$. The integral on the right-hand side of (6.07) is rearranged in the form

$$\int_0^{\kappa} e^{-zv} f(v)\, dv = \sum_{s=0}^{n-1} \Gamma\left(\frac{s+\lambda}{\mu}\right) \frac{a_s}{z^{(s+\lambda)/\mu}} - \varepsilon_{n,1}(z) + \varepsilon_{n,2}(z), \tag{6.11}$$

† See, for example, Levinson and Redheffer (1970, p. 308) or Copson (1935, §6.22).

where

$$\varepsilon_{n,1}(z) = \sum_{s=0}^{n-1} \Gamma\left(\frac{s+\lambda}{\mu}, \kappa z\right) \frac{a_s}{z^{(s+\lambda)/\mu}}, \tag{6.12}$$

$$\varepsilon_{n,2}(z) = \int_0^\kappa e^{-zv} v^{(n+\lambda-\mu)/\mu} f_n(v)\, dv. \tag{6.13}$$

Because $|\theta + \mathrm{ph}\,\kappa| < \frac{1}{2}\pi$ (compare (6.06)), the branch of $z^{(s+\lambda)/\mu}$ in (6.11) and (6.12) is $\exp\{(s+\lambda)(\ln|z| + i\theta)/\mu\}$, and each incomplete Gamma function in (6.12) takes its principal value.

Application of (2.02) immediately shows that

$$\varepsilon_{n,1}(z) = O(e^{-\kappa z}/z) \qquad (|z| \to \infty), \tag{6.14}$$

uniformly with respect to θ.

For $\varepsilon_{n,2}(z)$, the substitution $v = \kappa\tau$ produces

$$\varepsilon_{n,2}(z) = \int_0^1 e^{-z\kappa\tau} \tau^{(n+\lambda-\mu)/\mu} O(1)\, d\tau.$$

In consequence of Condition (v) and the fact that θ is restricted to a closed interval, we have

$$\mathrm{Re}(z\kappa) = |z|\, \mathrm{Re}\{e^{i\theta} p(k) - e^{i\theta} p(a)\} \geqslant |z|\, \eta_k, \tag{6.15}$$

where η_k is independent of z and positive. Hence

$$\varepsilon_{n,2}(z) = O(z^{-(n+\mathrm{Re}\,\lambda)/\mu}) = O(z^{-(n+\lambda)/\mu}),$$

uniformly with respect to θ.

Combination of the results of this subsection with (6.07) leads to

$$\int_a^k e^{-zp(t)} q(t)\, dt = e^{-zp(a)} \left\{ \sum_{s=0}^{n-1} \Gamma\left(\frac{s+\lambda}{\mu}\right) \frac{a_s}{z^{(s+\lambda)/\mu}} + O\left(\frac{1}{z^{(n+\lambda)/\mu}}\right) \right\}, \tag{6.16}$$

uniformly with respect to θ as $|z| \to \infty$.

6.4 It remains to consider the tail of the integral, that is, the contribution from $(k, b)_{\mathscr{P}}$. From Condition (v) it follows that

$$\mathrm{Re}\{e^{i\theta} p(t) - e^{i\theta} p(a)\} \geqslant \eta > 0, \qquad (t \in [k, b)_{\mathscr{P}}), \tag{6.17}$$

where η is independent of θ. Accordingly,

$$\begin{aligned} \mathrm{Re}\{zp(t) - zp(a)\} &= \{(|z| - Z) + Z\}\, \mathrm{Re}\{e^{i\theta} p(t) - e^{i\theta} p(a)\} \\ &\geqslant (|z| - Z)\eta + \mathrm{Re}\{Z e^{i\theta} p(t)\} - \mathrm{Re}\{Z e^{i\theta} p(a)\}, \end{aligned}$$

and

$$\left| \int_k^b e^{-zp(t)} q(t)\, dt \right| \leqslant |e^{-zp(a)}|\, e^{(Z-|z|)\eta} |\exp\{Z e^{i\theta} p(a)\}| \int_k^b |\exp\{-Z e^{i\theta} p(t)\}\, q(t)\, dt|. \tag{6.18}$$

Condition (iv) shows that the last quantity is $e^{-zp(a)}O(e^{-|z|\eta})$, uniformly with respect to θ. Therefore the asymptotic expansion (6.16) is unaffected by addition of the tail.

6.5 We have established the following fundamental result:

Theorem 6.1 *With the assumptions of* §6.1,

$$\int_a^b e^{-zp(t)}q(t)\,dt \sim e^{-zp(a)}\sum_{s=0}^{\infty}\Gamma\left(\frac{s+\lambda}{\mu}\right)\frac{a_s}{z^{(s+\lambda)/\mu}} \tag{6.19}$$

as $z\to\infty$ *in the sector* $\theta_1 \leqslant \operatorname{ph} z \leqslant \theta_2$. *Here the coefficients* a_s *are determined by the procedure of* §6.2, *and the branch of* $z^{(s+\lambda)/\mu}$ *is* $\exp\{(s+\lambda)(\ln|z|+i\theta)/\mu\}$.

As in the case of real variables, it will be noticed that in the more important respects Watson's lemma (Theorem 3.2) is a particular case of the present theorem.

Ex. 6.1 If $\mathscr{L}$ denotes the straight line joining $t=0$ and $t=\pi(1+i)$, $(1+t)^{ix}$ has its principal value, and x is large and positive, show that

$$\int_{\mathscr{L}}(1+t)^{ix}\exp(ixe^t)\,dt = e^{ix}\left\{\frac{i}{2x}+\frac{3i}{16x^3}+\frac{5}{32x^4}+O\left(\frac{1}{x^5}\right)\right\}.$$

Ex. 6.2 Let $\mathscr{S}$ denote the semicircle in the upper half of the t plane which begins at $t=1$ and ends at $t=-1$. Show that

$$\int_{\mathscr{S}} e^{z(t-\ln t)}\,dt = e^z\left\{\left(\frac{\pi}{2}\right)^{1/2}\frac{i}{z^{1/2}}-\frac{2}{3z}-\left(\frac{\pi}{2}\right)^{1/2}\frac{i}{12z^{3/2}}+O\left(\frac{1}{z^2}\right)\right\}$$

as $z\to\infty$ in the sector $-\frac{1}{2}\pi+\delta\leqslant \operatorname{ph} z\leqslant \tan^{-1}(2/\pi)-\delta$, where δ is an arbitrarily small positive number, and $\ln t$ and the powers of z have their principal values.

7 Saddle Points

7.1 Consider now the integral

$$I(z) = \int_a^b e^{-zp(t)}q(t)\,dt \tag{7.01}$$

in cases when the minimum value of $\operatorname{Re}\{zp(t)\}$ on the path occurs at an interior point t_0. For simplicity, assume that $\theta\ (\equiv \operatorname{ph} z)$ is fixed, so that t_0 is independent of z.

The path may be subdivided at t_0, giving

$$I(z) = \int_{t_0}^b e^{-zp(t)}q(t)\,dt - \int_{t_0}^a e^{-zp(t)}q(t)\,dt. \tag{7.02}$$

In the neighborhood of t_0 the functions $p(t)$ and $q(t)$ have Taylor-series expansions

$$p(t) = p(t_0)+(t-t_0)p'(t_0)+(t-t_0)^2\frac{p''(t_0)}{2!}+\cdots, \tag{7.03}$$

$$q(t) = q(t_0)+(t-t_0)q'(t_0)+(t-t_0)^2\frac{q''(t_0)}{2!}+\cdots. \tag{7.04}$$

For large $|z|$, the asymptotic expansion of each integral on the right-hand side of (7.02) is obtainable by application of Theorem 6.1, the roles of the series in Condition (iii) of §6.1 being played by (7.03) and (7.04). However, if $p'(t_0) \neq 0$ then the condition that $\operatorname{Re}\{e^{i\theta}p(t)\}$ has a minimum at t_0 gives

$$\cos(\omega_0+\theta+\omega) = 0,$$

where, again, $\omega_0 = \operatorname{ph}\{p'(t_0)\}$ and ω is the angle of slope of the path at t_0. Since the two values of ω differ by π, $\omega_0+\theta+\omega$ is $\frac{1}{2}\pi$ for one integral and $-\frac{1}{2}\pi$ for the other; compare (6.03). The values of ω_0 and the coefficients a_s are exactly the same in the two cases. In consequence, the asymptotic expansions of the integrals are the same, and all that remains on substitution in (7.02) is an error term $O\{z^{-n}e^{-zp(t_0)}\}$, n being an arbitrary positive integer.†

On the other hand, if $p'(t_0) = 0$ then the μ of Condition (iii) of §6.1 is an integer such that $\mu \geqslant 2$. Thus $\mu\omega$ differs by $\mu\pi$ for the two integrals, causing the values of ω_0 that satisfy (6.03) to differ by $\mu\pi$ if μ is even, or $(\mu\pm 1)\pi$ if μ is odd. In consequence, *different* branches are used for $p_0^{1/\mu}$ in constructing the coefficients a_s, and the asymptotic expansions no longer cancel on substitution in (7.02).

7.2 The last observation provides the clue for handling cases in which $p'(t_0) \neq 0$. We try to deform the path of integration in such a way that the minimum of $\operatorname{Re}\{e^{i\theta}p(t)\}$ occurs either at one of the endpoints, or at a point at which $p'(t)$ vanishes. If this is successful, then the asymptotic expansion of $I(z)$ is found with one or two applications of Theorem 6.1.

Thus the points at which $p'(t) = 0$ are of great importance. For reasons given later (§10.3) they are called *saddle points*. The task of locating the saddle points is generally fairly easy, but the construction of a path on which $\operatorname{Re}\{e^{i\theta}p(t)\}$ attains its minimum at an endpoint or saddle point may be troublesome. An intelligent guess is sometimes successful, especially when the parameter z is real. Failing this, it may be necessary to make a partial study of the conformal mapping between the planes of t and v, where $v = p(t)-p(a)$ and a is the endpoint or saddle point. Once the map $\mathbf{V}$ of the original domain $\mathbf{T}$ has been constructed, it is easy to ascertain whether the point $p(b)-p(a)$ can be joined to the origin by a path $\mathcal{Q}$ lying entirely in the intersection of $\mathbf{V}$ with the sector $|\operatorname{ph}(e^{i\theta}v)| < \frac{1}{2}\pi$. An admissible $\mathcal{P}$ is the t map of $\mathcal{Q}$, *but its actual location need not be determined*: an existence demonstration is sufficient for the purpose of applying Theorem 6.1.

If a is a saddle point of order $\mu-1$, that is, if

$$p'(a) = p''(a) = \cdots = p^{(\mu-1)}(a) = 0, \qquad p^{(\mu)}(a) \neq 0,$$

then the neighborhood of a is mapped on μ Riemann sheets in the v plane. Fortunately, however, the full neighborhood of a need not be considered, because $\mathcal{Q}$ is confined to half a sheet.

7.3 The most common case in practice is for the integral (7.01) to have a simple saddle point, that is, a saddle point of order unity, at an interior point t_0 of the

† This situation has no analogue with real variables: when $p(t)$ is real and continuously differentiable, it cannot attain a minimum at an interior point t_0 of (a,b), unless $p'(t_0) = 0$.

integration path. Since some simplifications then become available, we state in full the result of combining the contributions from $(t_0, b)_{\mathscr{P}}$ and $(a, t_0)_{\mathscr{P}}$.

Assumptions

(i) *$p(t)$ and $q(t)$ are independent of z, and single valued and holomorphic in a domain* **T**.

(ii) *The integration path $\mathscr{P}$ is independent of z. The endpoints a and b of $\mathscr{P}$ are finite or infinite, and $(a,b)_{\mathscr{P}}$ lies within* **T**.

(iii) *$p'(t)$ has a simple zero at an interior point t_0 of $\mathscr{P}$.*

(iv) *z ranges along a ray or over an annular sector given by $\theta_1 \le \theta \le \theta_2$ and $|z| \ge Z$, where $\theta \equiv \operatorname{ph} z$, $\theta_2 - \theta_1 < \pi$, and $Z > 0$. $I(Ze^{i\theta})$ converges at a and b absolutely and uniformly with respect to θ.*

(v) $\operatorname{Re}\{e^{i\theta}p(t) - e^{i\theta}p(t_0)\}$ *is positive on $(a,b)_{\mathscr{P}}$, except at t_0, and is bounded away from zero uniformly with respect to θ as $t \to a$ or b along $\mathscr{P}$.*

Theorem 7.1 *With the foregoing assumptions,*

$$\int_a^b e^{-zp(t)} q(t)\, dt \sim 2e^{-zp(t_0)} \sum_{s=0}^{\infty} \Gamma\left(s + \frac{1}{2}\right) \frac{a_{2s}}{z^{s+(1/2)}} \tag{7.05}$$

as $z \to \infty$ in the sector $\theta_1 \leqslant \operatorname{ph} z \leqslant \theta_2$.

Formulas for the first two coefficients are

$$a_0 = \frac{q}{(2p'')^{1/2}}, \qquad a_2 = \left\{2q'' - \frac{2p'''q'}{p''} + \left(\frac{5p'''^2}{6p''^2} - \frac{p^{iv}}{2p''}\right)q\right\}\frac{1}{(2p'')^{3/2}}, \tag{7.06}$$

where p, q, and their derivatives are evaluated at $t = t_0$. In forming $(2p'')^{1/2}$ and $(2p'')^{3/2}$, the branch of $\omega_0 \equiv \operatorname{ph}\{p''(t_0)\}$ must satisfy

$$|\omega_0 + \theta + 2\omega| \leqslant \tfrac{1}{2}\pi, \tag{7.07}$$

where ω is the limiting value of $\operatorname{ph}(t - t_0)$ as $t \to t_0$ along $(t_0, b)_{\mathscr{P}}$.

8 Examples

8.1 Schläfli's integral for the Legendre polynomial of degree n is given by Chapter 2, (7.19). It may be cast into the form

$$P_n(\cos\alpha) = \frac{1}{2^{n+1}\pi i} \int_{\mathscr{C}} e^{-np(t)} q(t)\, dt,$$

in which $\mathscr{C}$ is a simple closed contour encircling the point $t = \cos\alpha$, and

$$p(t) = \ln\left(\frac{t - \cos\alpha}{t^2 - 1}\right), \qquad q(t) = \frac{1}{t - \cos\alpha},$$

the branch of the logarithm being real when $t \in (1, \infty)$. Let us seek an asymptotic approximation for $P_n(\cos\alpha)$ when n is large and α is a fixed point in the interval

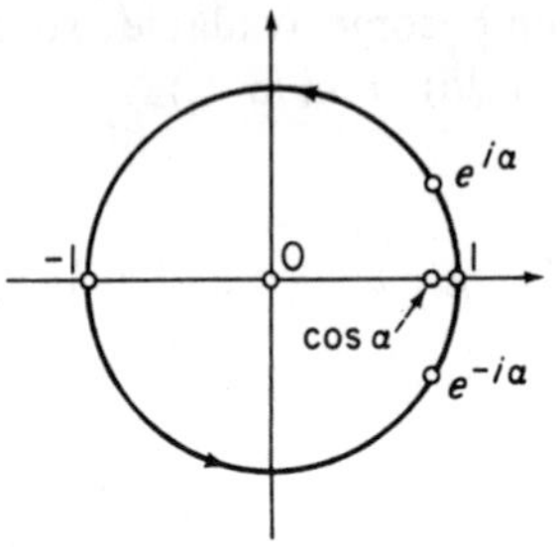

Fig. 8.1 Integration path for $P_n(\cos\alpha)$.

$(0, \pi)$. Since

$$p'(t) = -\frac{t^2-2t\cos\alpha+1}{(t^2-1)(t-\cos\alpha)}$$

the saddle points are located at $e^{i\alpha}$ and $e^{-i\alpha}$. In accordance with §7.2, we deform $\mathscr{C}$ to pass through these points. One possibility is the unit circle; see Fig. 8.1. Since $p(t)$ is real on part of the real axis it takes conjugate values at conjugate values of t.† The same holds for $q(t)$. Hence

$$P_n(\cos\alpha) = \frac{1}{2^n\pi}\operatorname{Im}\left\{\int_{\mathscr{S}} e^{-np(t)}q(t)\,dt\right\},$$

where $\mathscr{S}$ is the semicircle in the upper half-plane extending from 1 to -1.

The saddle point at $t = e^{i\alpha}$ is simple, hence Theorem 7.1 is relevant. Setting $t = e^{i\tau}$, we find that

$$\operatorname{Re}\{p(t)-p(e^{i\alpha})\} = \ln\left|\frac{e^{i\tau}-\cos\alpha}{e^{2i\tau}-1}\right| + \ln 2 = \frac{1}{2}\ln\left\{1+\left(\frac{\cos\tau-\cos\alpha}{\sin\tau}\right)^2\right\}.$$

This is positive in the interval $0 < \tau < \pi$, except at $\tau = \alpha$. Because $\theta = 0$ in the present case the key condition (v) is satisfied. The remaining conditions (i)–(iv) of §7.3 are also satisfied. Noting that $p(e^{i\alpha}) = -\ln 2 - i\alpha$,‡ we derive

$$\int_{\mathscr{S}} e^{-np(t)}q(t)\,dt \sim 2^{n+1}e^{in\alpha}\sum_{s=0}^{\infty}\Gamma\left(s+\frac{1}{2}\right)\frac{a_{2s}}{n^{s+(1/2)}} \qquad (n\to\infty).$$

To evaluate a_0, we have

$$p''(e^{i\alpha}) = ie^{-i\alpha}\csc\alpha, \qquad q(e^{i\alpha}) = -i\csc\alpha.$$

With $\omega = \frac{1}{2}\pi+\alpha$, the appropriate choice of branch of $\omega_0 \equiv \operatorname{ph}\{p''(e^{i\alpha})\}$ is $-\alpha-\frac{3}{2}\pi$; compare (7.07) with $\theta = 0$. Hence the first of (7.06) gives

$$a_0 = (2\sin\alpha)^{-1/2}e^{(2\alpha+\pi)i/4}.$$

† This is an application of *Schwarz's principle of symmetry*; see, for example, Levinson and Redheffer (1970, p. 318).

‡ The choice of branch of the logarithm here is immaterial.

Accordingly, as a first approximation

$$P_n(\cos\alpha) = \left(\frac{2}{\pi n \sin\alpha}\right)^{1/2} \sin\left(n\alpha + \frac{1}{2}\alpha + \frac{1}{4}\pi\right) + O\left(\frac{1}{n^{3/2}}\right). \tag{8.01}$$

The value of a_2 can be calculated from the second of (7.06), and higher coefficients found by the general procedure of §6.2. We shall not pursue these calculations, however, because a more convenient form of expansion will be derived in Chapter 8, §§10.1 and 10.2.

8.2 A second example is furnished by the integral

$$I(x) = \int_{-\infty}^{\infty} \frac{\exp(-t^2)}{t^{2x}}\, dt, \tag{8.02}$$

in which x is large and positive, the path of integration passes above the origin, and t^{2x} is continuous and takes its principal value as $t \to +\infty$.

The natural choice $p(t) = 2 \ln t$, $q(t) = \exp(-t^2)$ produces no saddle points. Recalling the approach of Chapter 3, §7.5, we seek instead the points at which the derivative of the *whole* integrand vanishes. This produces the equation

$$2t \exp(-t^2)t^{-2x} + 2x \exp(-t^2)t^{-2x-1} = 0,$$

the roots of which are $t = \pm i\sqrt{x}$. Because our theory applies only when the saddle points are independent of the parameter, we replace the integration variable t by $t\sqrt{x}$, giving

$$I(x) = \frac{1}{x^{x-(1/2)}} \int_{-\infty}^{\infty} e^{-xp(t)}\, dt, \tag{8.03}$$

where

$$p(t) = t^2 + 2 \ln t. \tag{8.04}$$

The new saddle points are $t = \pm i$, and both are simple. As a possible path consider the straight line through i parallel to the real axis, the minor deformations at $t = \pm\infty$ being easily justified by Cauchy's theorem. Setting $t = i + \tau$ and noting that on the new path the logarithm in (8.04) takes its principal value, we obtain

$$p(t) = -1 + 2i\tau + \tau^2 + 2 \ln i + 2 \ln(1 - i\tau),$$

and thence

$$\operatorname{Re}\{p(t)\} = -1 + \tau^2 + \ln(1+\tau^2).$$

This quantity attains its minimum at $\tau = 0$; hence Condition (v) of §7.3 is satisfied, as are the other four conditions. Again, in the notation of §7.3 we have

$$t_0 = i, \qquad p(i) = -1 + i\pi, \qquad p''(i) = 4, \qquad p'''(i) = 4i, \qquad p^{iv}(i) = -12.$$

Equations (7.06) yield $a_0 = 1/(2\sqrt{2})$ and $a_2 = 1/(24\sqrt{2})$, and applying Theorem 7.1 to (8.03) we immediately obtain the required result:

$$I(x) \sim e^{-x\pi i}\left(\frac{\pi}{2}\right)^{1/2}\left(\frac{e}{x}\right)^{x}\left(1 + \frac{1}{24x} + \cdots\right) \qquad (x \to \infty).$$

This result can be verified by means of Chapter 3, (8.16): by taking $-t^2$ as new integration variable in (8.02) and using Hankel's loop integral (Chapter 2, (1.12)), we find that

$$I(x) = \pi e^{-x\pi i}/\Gamma(x+\tfrac{1}{2}).$$

Ex. 8.1 Establish the result of Chapter 3, Exercise 7.1, by the method of §8.1, using as integration path the circle having $[e^{-\alpha}, e^{\alpha}]$ as diameter.

Ex. 8.2 Let $f(x) = \int_{-\infty}^{\infty} \exp\{-2xt^2-(4x/t)\}\,dt$, where x is positive and the integration path passes above the origin. By employing a path comprising the segments of the real axis outside the unit circle, together with the upper half of this circle, show that

$$f(x) = \pi^{1/2}(6x)^{-1/2}e^{3x+3ix\sqrt{3}}\{1+O(x^{-1})\} \qquad (x\to\infty). \qquad \text{[Lauwerier, 1966.]}$$

Ex. 8.3 If the path of integration is the imaginary axis and the integrand has its principal value, prove that for large positive x

$$\int_{-i\infty}^{i\infty} t^{t^2}\exp(-xt^2)\,dt \sim i\pi^{1/2}\exp(-\tfrac{1}{2}e^{2x-1}),$$

with a relative error $O(e^{-2x})$.

Ex. 8.4 In the integral

$$I(z) = \int_{-\infty}^{\infty} \exp\{-z(t^2-2it)\}\operatorname{csch}(1+t^2)\,dt,$$

the saddle point $t = i$ coincides with a pole of the integrand. By extracting the singular part (§5.2), show that

$$I(z) = e^{-z}\left\{\frac{1}{2}\pi + \frac{\pi^{1/2}}{4z^{1/2}} - \frac{11}{96}\frac{\pi^{1/2}}{z^{3/2}} + O\left(\frac{1}{z^{5/2}}\right)\right\},$$

as $z\to\infty$ in the sector $|\operatorname{ph} z| \leqslant \frac{1}{2}\pi-\delta$ $(<\frac{1}{2}\pi)$. Is this region of validity maximal?

9 Bessel Functions of Large Argument and Order

9.1 In this section the theory of §§6 and 7 is applied to derive two important expansions for the Bessel function $J_\nu(z)$. Our starting point is the contour integral (9.13) of Chapter 2. On replacing τ by $-t$ this becomes

$$J_\nu(z) = -\frac{1}{2\pi i}\int_{-\infty+\pi i}^{-\infty-\pi i} e^{-z\sinh t+\nu t}\,dt \qquad (|\operatorname{ph} z| < \tfrac{1}{2}\pi). \tag{9.01}$$

In the first case it is supposed that ν and z are real or complex numbers, ν being fixed and $|z|$ large. The saddle points are located at the zeros of $\cosh t$, that is, at $t = \pm\frac{1}{2}\pi i, \pm\frac{3}{2}\pi i, \ldots$. The integration path can be deformed to pass through any number of these points, but it is not obvious how to choose a path on which $\operatorname{Re}(z\sinh t)$ attains its minimum at one or more of the saddle points. Accordingly, we follow the suggestion made in §7.2 and begin by mapping the strip $0 < \operatorname{Im} t < \pi$ (which contains one of the saddle points) on the plane of

$$v = \sinh t - i.$$

The map is quickly determined by the following considerations:

(a) The positive real t axis corresponds to the line segment $\operatorname{Im} v = -1$, $\operatorname{Re} v \geqslant 0$.
(b) $v \sim \frac{1}{2}e^t$ as $\operatorname{Re} t \to +\infty$.
(c) Increasing t by πi changes the sign of $v + i$.
(d) dv/dt is real on the imaginary axis and changes sign at $t = \frac{1}{2}\pi i$.
(e) $v \sim \frac{1}{2}i(t - \frac{1}{2}\pi i)^2$ as $t \to \frac{1}{2}\pi i$.
(f) Images in the imaginary axis correspond.

Corresponding points in the two planes are indicated in Figs. 9.1, 9.2, and 9.3. The v map has two sheets, passage from Fig. 9.2 to 9.3 taking place across the dotted line segment DG.

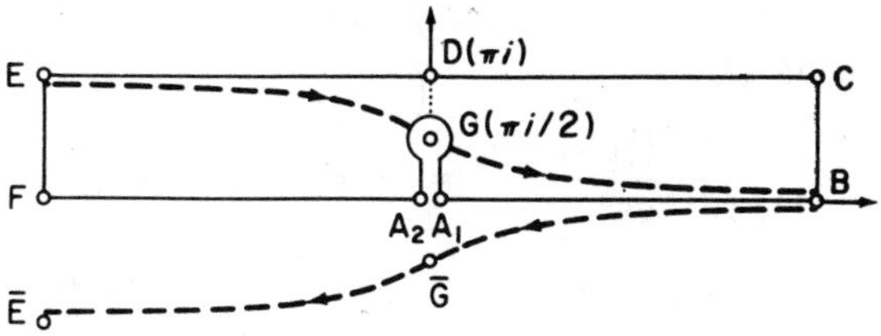

Fig. 9.1 t plane.

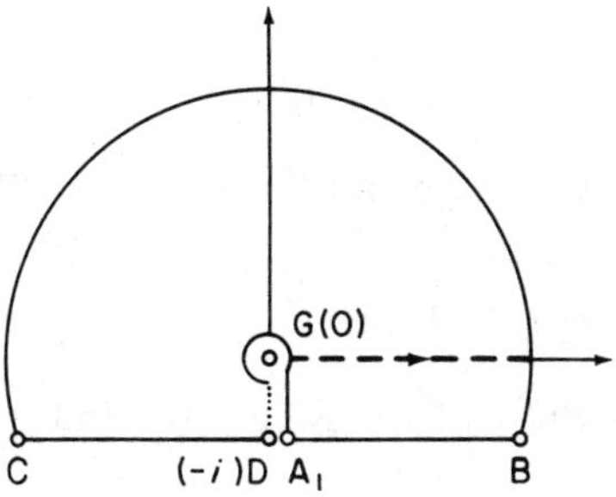

Fig. 9.2 v plane (i).

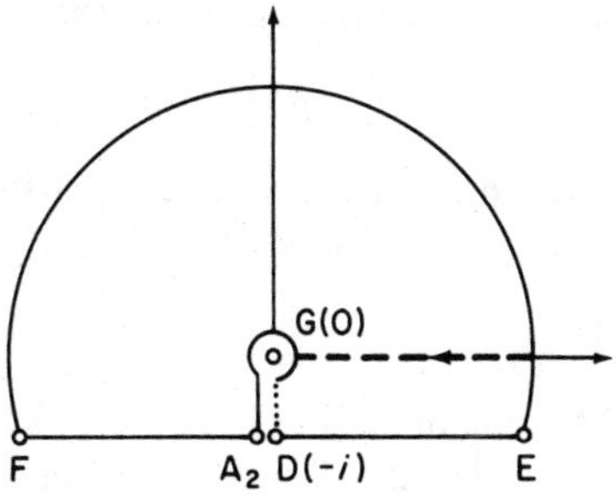

Fig. 9.3 v plane (ii).

The positive real axes of Figs. 9.2 and 9.3 map onto the broken curves GB and GE indicated in Fig. 9.1. As a possible path for the integral (9.01) we try the whole curve EGB, together with the conjugate curve $B\bar{G}\bar{E}$ (also depicted in Fig. 9.1). Obviously $\operatorname{Re} v$ attains its minimum on EGB at G. And if δ is an arbitrary small positive number and $\theta \equiv \operatorname{ph} z$ is restricted by $|\theta| \leqslant \frac{1}{2}\pi - \delta$, then Condition (v) of §7.3 is satisfied. It is easily seen that the other four conditions are satisfied, and Theorem 7.1 immediately gives

$$\int_{-\infty+\pi i}^{\infty} e^{-z\sinh t + vt}\,dt \sim 2e^{-iz} \sum_{s=0}^{\infty} \Gamma\left(s + \frac{1}{2}\right) \frac{a_{2s}}{z^{s+(1/2)}} \tag{9.02}$$

as $z \to \infty$ in $|\operatorname{ph} z| \leqslant \frac{1}{2}\pi - \delta$.

9.2 The next task is to evaluate the coefficients a_{2s}. This is an exercise in trigonometric series. From (6.08) and (6.09), with $\lambda = 1$ and $\mu = 2$, we have

$$\frac{e^{vt}}{\cosh t} = \sum_{s=0}^{\infty} a_s v^{(s-1)/2}. \tag{9.03}$$

Set $t = \frac{1}{2}\pi i + \tau$, so that $v = 2i \sinh^2(\frac{1}{2}\tau)$. The relation (7.07) is satisfied with $\omega = -\frac{1}{4}\pi$, $\omega_0 = \frac{1}{2}\pi$, and $|\mathrm{ph}\, z| < \frac{1}{2}\pi$; hence $\mathrm{ph}\, \tau = 0$ corresponds to $\mathrm{ph}\, v = \frac{1}{2}\pi$. Accordingly, the correct choice of branches in (9.03) leads to

$$\frac{e^{(2\nu-1)\pi i/4} e^{\nu\tau}}{2^{1/2}\cosh(\frac{1}{2}\tau)} = \sum_{s=0}^{\infty} a_s e^{s\pi i/4} 2^{s/2} \sinh^s(\tfrac{1}{2}\tau).$$

Since only the a_s of even suffix are needed, we replace τ by $-\tau$ and take the mean of the two expansions; thus

$$\frac{e^{(2\nu-1)\pi i/4} \cosh(\nu\tau)}{2^{1/2}\cosh(\frac{1}{2}\tau)} = \sum_{s=0}^{\infty} a_{2s}(2i)^s \sinh^{2s}(\tfrac{1}{2}\tau). \tag{9.04}$$

If we denote $\sinh(\frac{1}{2}\tau)$ by y and the left-hand side of (9.04) by $F(y)$, then from Taylor's theorem

$$a_{2s} = F^{(2s)}(0)/\{(2i)^s(2s)!\}. \tag{9.05}$$

Direct differentiation shows that

$$(1+y^2)F''(y) + 3yF'(y) + (1-4\nu^2)F(y) = 0.$$

Again, differentiating this equation $2s-2$ times by Leibniz's theorem and setting $y = 0$, we find that

$$F^{(2s)}(0) = \{4\nu^2 - (2s-1)^2\} F^{(2s-2)}(0). \tag{9.06}$$

The value of a_0 is obtainable by setting $\tau = 0$ in (9.04). Then using (9.05) and (9.06) we arrive at the desired general expression

$$a_{2s} = \frac{(4\nu^2-1^2)(4\nu^2-3^2)\cdots\{4\nu^2-(2s-1)^2\}}{(2s)!(2i)^s} \frac{e^{(2\nu-1)\pi i/4}}{2^{1/2}}.$$

Returning to (9.02) and using the duplication formula for the Gamma function, we find that

$$\int_{-\infty+\pi i}^{\infty} e^{-z\sinh t + \nu t}\, dt \sim \left(\frac{2\pi}{z}\right)^{1/2} \exp\left\{i\left(\frac{1}{2}\nu\pi - \frac{1}{4}\pi - z\right)\right\} \sum_{s=0}^{\infty} \frac{A_s(\nu)}{(iz)^s}, \tag{9.07}$$

where

$$A_s(\nu) = \frac{(4\nu^2-1^2)(4\nu^2-3^2)\cdots\{4\nu^2-(2s-1)^2\}}{s!\, 8^s}. \tag{9.08}$$

The corresponding expansion for the integral along the path from $-\infty - \pi i$ to

∞ is obtained from (9.07) by changing the sign of i. Substituting these results in (9.01) we obtain the required asymptotic expansion, in compound form, given by

$$J_\nu(z) \sim \left(\frac{2}{\pi z}\right)^{1/2}\left[\cos\left(z-\frac{1}{2}\nu\pi-\frac{1}{4}\pi\right)\sum_{s=0}^{\infty}(-)^s\frac{A_{2s}(\nu)}{z^{2s}} - \sin\left(z-\frac{1}{2}\nu\pi-\frac{1}{4}\pi\right)\sum_{s=0}^{\infty}(-)^s\frac{A_{2s+1}(\nu)}{z^{2s+1}}\right] \tag{9.09}$$

as $z\to\infty$ in the sector $|\mathrm{ph}\,z|\leqslant\frac{1}{2}\pi-\delta$ $(<\frac{1}{2}\pi)$. This expansion is due to Hankel (1869).

9.3 The region of validity of (9.09) can be extended by taking new paths of integration, as in §1.2 and the proof of Theorem 3.3. By use of Cauchy's theorem, the path EGB in Fig. 9.1 may be deformed into the path whose map is the ray $\mathrm{ph}\,v=-\beta$, provided that $\beta\in(-\frac{3}{2}\pi,\frac{1}{2}\pi)$. The last restriction is needed because, as a function of v, t has singularities on the rays $\mathrm{ph}\,v=\frac{3}{2}\pi$ and $\mathrm{ph}\,v=-\frac{1}{2}\pi$. For each admissible β, the integral along the new path is the analytic continuation, in the sector $|\mathrm{ph}(ze^{-i\beta})|<\frac{1}{2}\pi$, of the integral on the left-hand side of (9.07).

On the new path the conditions of §7.3 are satisfied, provided that $\theta\in[-\frac{1}{2}\pi+\beta+\delta,\ \frac{1}{2}\pi+\beta-\delta]$. In consequence, the right-hand side of (9.07) furnishes the asymptotic expansion of the analytic continuation of the integral on the left, provided that $-2\pi+\delta\leqslant\mathrm{ph}\,z\leqslant\pi-\delta$. The corresponding extension for the integral along the path $B\bar{G}\bar{E}$ is given by $-\pi+\delta\leqslant\mathrm{ph}\,z\leqslant2\pi-\delta$. Therefore (9.09) *is valid in the intersection of these sectors, that is, in* $|\mathrm{ph}\,z|\leqslant\pi-\delta$.

9.4 The second combination of variables we consider for $J_\nu(z)$ is given by $z=\nu\,\mathrm{sech}\,\alpha$, where α and ν are both real and positive, α being fixed, and ν being large. Changing the sign of t in (9.01), we have

$$J_\nu(\nu\,\mathrm{sech}\,\alpha)=\frac{1}{2\pi i}\int_{\infty-\pi i}^{\infty+\pi i}e^{-\nu p(t)}\,dt,$$

where

$$p(t)=t-\mathrm{sech}\,\alpha\,\sinh t. \tag{9.10}$$

The saddle points are now the roots of $\cosh t=\cosh\alpha$, and are therefore given by $t=\pm\alpha,\ \pm\alpha\pm2\pi i,\ \pm\alpha\pm4\pi i,\dots$. The most promising is α, and as a possible path we consider that indicated in Fig. 9.4.

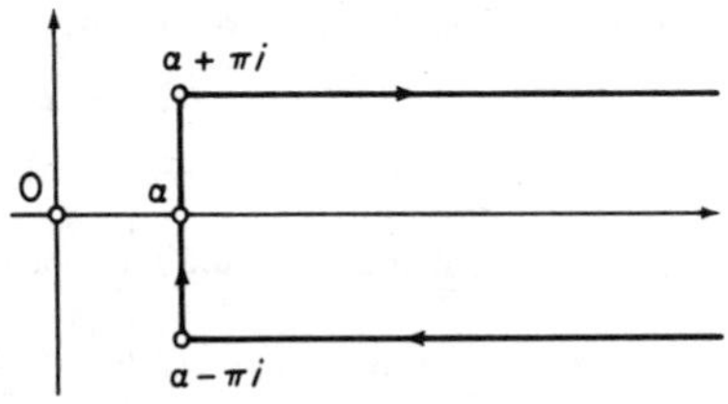

Fig. 9.4 Integration path for $J_\nu(\nu\,\mathrm{sech}\,\alpha)$.

On the vertical segment $t = \alpha + i\tau$, $-\pi \leqslant \tau \leqslant \pi$, we have

$$\operatorname{Re}\{p(t)\} = \alpha - \tanh\alpha\cos\tau > \alpha - \tanh\alpha \qquad (\tau \neq 0).$$

On the horizontal parts $t = \alpha \pm \pi i + \tau$, $0 \leqslant \tau < \infty$, we have

$$\operatorname{Re}\{p(t)\} = \alpha + \tau + \operatorname{sech}\alpha\sinh(\alpha+\tau) \geqslant \alpha + \tanh\alpha.$$

Therefore $\operatorname{Re}\{p(t)\}$ attains its minimum on the path at α, as required by Condition (v) of §7.3. The other four conditions are also satisfied, and applying Theorem 7.1 we obtain

$$J_\nu(\nu\operatorname{sech}\alpha) \sim \frac{e^{-\nu(\alpha-\tanh\alpha)}}{\pi i}\sum_{s=0}^{\infty}\Gamma\left(s+\frac{1}{2}\right)\frac{a_{2s}}{\nu^{s+(1/2)}} \qquad (\nu\to\infty).$$

Unlike (9.09), an explicit general expression for the coefficients is unavailable. The first two are easily found from (7.06), however. Differentiation of (9.10) gives

$$p''(\alpha) = p^{iv}(\alpha) = -\tanh\alpha, \qquad p'''(\alpha) = -1.$$

Since $\omega = \frac{1}{2}\pi$, the correct choice of branch for the powers of $p''(\alpha)$ is given by $\operatorname{ph}\{p''(\alpha)\} = -\pi$. In consequence,

$$a_0 = (\tfrac{1}{2}\coth\alpha)^{1/2}i, \qquad a_2 = (\tfrac{1}{2}-\tfrac{5}{6}\coth^2\alpha)(\tfrac{1}{2}\coth\alpha)^{3/2}i,$$

and

$$J_\nu(\nu\operatorname{sech}\alpha) \sim \frac{e^{-\nu(\alpha-\tanh\alpha)}}{(2\pi\nu\tanh\alpha)^{1/2}}\left\{1+\left(\frac{1}{8}\coth\alpha-\frac{5}{24}\coth^3\alpha\right)\frac{1}{\nu}+\cdots\right\}. \tag{9.11}$$

This expansion is due to Debye (1909). Higher terms have been given, for example, by B.A. (1952); they are obtainable more easily from differential-equation theory (Chapter 10, §7.3) than by the present method.

9.5 In the analysis of §9.4 it was possible to avoid the need for conformal mapping because a suitable path was easily guessed. When ν or α is complex, however, conformal mapping is almost unavoidable; compare the next exercise.

Ex. 9.1 Construct the map of the half-strip $0 < \operatorname{Im} t < 2\pi$, $\operatorname{Re} t > 0$ on the plane of

$$t - \operatorname{sech}\alpha\sinh t - \alpha + \tanh\alpha,$$

where α is fixed and positive. Thence show that the expansion (9.11) is valid for $|\operatorname{ph}\nu| \leqslant \pi - \delta$ $(< \pi)$.

Ex. 9.2 By use of Theorem 6.1 prove that

$$J_\nu(\nu) \sim 2^{1/3}/\{3^{2/3}\Gamma(\tfrac{2}{3})\nu^{1/3}\},$$

for large $|\nu|$ in the sector $|\operatorname{ph}\nu| \leqslant \pi - \delta$ $(< \pi)$.

Ex. 9.3 Show that when α is a fixed number in the interval $(0, \frac{1}{2}\pi)$ and ν is large and positive,

$$\int_0^\infty \exp(-\nu\sec\alpha\cosh t)\cos(\nu t)\,dt = \left(\frac{\pi}{2\nu\tan\alpha}\right)^{1/2}\exp\{\nu(\alpha-\tan\alpha-\tfrac{1}{2}\pi)\}\left\{1+O\left(\frac{1}{\nu}\right)\right\}.$$

*10 Error Bounds for Laplace's Method; the Method of Steepest Descents

10.1 Using the notation of §6 assume, for the moment, that the whole of the integration path $\mathscr{P}$ can be deformed to make its v map lie along the real axis, so that on $\mathscr{P}$

$$\operatorname{Im}\{p(t)\} = \text{constant} = \operatorname{Im}\{p(a)\}. \tag{10.01}$$

With v as integration variable the original integral (6.01) may be decomposed, as in §6, into

$$\int_a^b e^{-zp(t)}q(t)\,dt = e^{-zp(a)}\left\{\sum_{s=0}^{n-1}\Gamma\left(\frac{s+\lambda}{\mu}\right)\frac{a_s}{z^{(s+\lambda)/\mu}} - \varepsilon_{n,1}(z) + \varepsilon_{n,2}(z)\right\}, \tag{10.02}$$

where

$$\varepsilon_{n,1}(z) = \sum_{s=0}^{n-1}\Gamma\left\{\frac{s+\lambda}{\mu},\ zp(b)-zp(a)\right\}\frac{a_s}{z^{(s+\lambda)/\mu}}, \tag{10.03}$$

and

$$\varepsilon_{n,2}(z) = \int_0^{p(b)-p(a)} e^{-zv}v^{(n+\lambda-\mu)/\mu}f_n(v)\,dv. \tag{10.04}$$

For simplicity, *attention will be confined to the case of real* λ. In (10.03) $|\operatorname{ph}\{zp(b)-zp(a)\}| < \frac{1}{2}\pi$, and each incomplete Gamma function takes its principal value. From (2.02), (2.06), and (2.09), with $n = 0$, we have

$$|\Gamma(\alpha,\zeta)| \leqslant \frac{|e^{-\zeta}\zeta^{\alpha}|}{|\zeta|-\alpha_0} \qquad (|\operatorname{ph}\zeta| \leqslant \tfrac{1}{2}\pi,\quad |\zeta| > \alpha_0),$$

where $\alpha_0 = \max(\alpha-1, 0)$. This inequality enables $|\varepsilon_{n,1}(z)|$ to be bounded in a realistic way.

For the other error term, assume that $a_n \neq 0$ and

$$|f_n(v)| \leqslant |a_n|\,e^{\sigma_n v} \qquad (0 \leqslant v < p(b)-p(a)).$$

Then

$$|\varepsilon_{n,2}(z)| \leqslant \Gamma\left(\frac{n+\lambda}{\mu}\right)\frac{|a_n|}{(|z|\cos\theta-\sigma_n)^{(n+\lambda)/\mu}} \qquad (|\theta| < \tfrac{1}{2}\pi,\quad |z|\cos\theta > \sigma_n). \tag{10.05}$$

In terms of the original variables the best value of σ_n is given by

$$\sigma_n = \sup_{t\in\mathscr{P}}\left[\frac{1}{|p(t)-p(a)|}\ln\left|\frac{\{q(t)/p'(t)\} - \sum a_s\{p(t)-p(a)\}^{(s+\lambda-\mu)/\mu}}{a_n\{p(t)-p(a)\}^{(n+\lambda-\mu)/\mu}}\right|\right], \tag{10.06}$$

the summation extending from $s = 0$ to $s = n-1$.

More generally, suppose that β is an arbitrary real number, the v map of $\mathscr{P}$ lies along the ray $\operatorname{ph} v = -\beta$, and $|\operatorname{ph}(ze^{-i\beta})| < \frac{1}{2}\pi$. Then (10.03) again applies with the principal value of each incomplete Gamma function. In place of (10.05), however, we have

$$|\varepsilon_{n,2}(z)| \leqslant \Gamma\left(\frac{n+\lambda}{\mu}\right)\frac{|a_n|}{\{|z|\cos(\theta-\beta)-\sigma_n\}^{(n+\lambda)/\mu}} \qquad (|\theta-\beta|<\tfrac{1}{2}\pi,\quad |z|\cos(\theta-\beta)>\sigma_n). \tag{10.07}$$

Here σ_n is defined by (10.06), and now depends on β. Cases in which a_n vanishes or σ_n is infinite can be handled by modifications on the lines of Chapter 3, §§9.1 to 9.3.†

10.2 Now suppose that a and b *cannot* be linked by a path having an equation of the form

$$\operatorname{ph}\{p(t)-p(a)\} = -\beta. \tag{10.08}$$

In this event, we proceed from a along a path of type (10.08) until a conveniently chosen point k is reached.‡ The journey from k to b is completed by any convenient path lying in $\mathbf{T}$ along which $\operatorname{Re}\{e^{i\theta}p(t)-e^{i\theta}p(a)\}$ is positive. The integral over $(a,k)_{\mathscr{P}}$ has the same asymptotic expansion as the integral over $(a,b)_{\mathscr{P}}$, and its error terms can be bounded by the methods just given. The contribution from $(k,b)_{\mathscr{P}}$ is bounded by an inequality of the form (6.18), in which η can be taken as the largest number fulfilling (6.17). Since (6.18) is exponentially small compared with the bound for $|e^{-zp(a)}\varepsilon_{n,2}(z)|$, the choices of k and the path from k to b are not crucial.

An illustrative example of the foregoing methods for constructing error bounds has been given by Olver (1970a, §7). Error bounds for the expansions of §§8 and 9 will be derived by other methods in later chapters.

10.3 The curves defined by equation (10.01), or more generally (10.08), have an interesting geometrical interpretation. If a is not a saddle point, then the theory of conformal mapping shows that in the neighborhood of a equation (10.01) defines a regular arc passing through a; see Fig. 10.1(i). Alternatively, if a is a saddle point and $\mu-1$ is its order (§7.2), then μ regular arcs pass through a on which (10.01) is satisfied, and adjacent arcs intersect at angle π/μ. Figures 10.1(ii) and (iii) illustrate the cases $\mu = 2$ and $\mu = 3$, respectively.

Consider the surface $|e^{p(t)}|$ plotted against the real and imaginary parts of t. In consequence of the maximum-modulus theorem there can be no peaks or hollows on this surface. If $p'(a) = 0$, then the tangent plane at a is horizontal. If, in addition, $p''(a) \neq 0$, then the surface is shaped like a saddle in the neighborhood of a. Hence the name *saddle point* or *col*. Deformation of a path in the t plane to pass through a saddle point is equivalent to crossing a mountain ridge via a pass.

The surface loci of constant $\operatorname{Im}\{p(t)\}$ are paths of *steepest descent* or *steepest ascent* through a. This can be seen as follows. Let the real and imaginary parts of $p(t)$ be denoted by

$$p(t) = p_{\mathrm{R}}(t) + ip_{\mathrm{I}}(t), \tag{10.09}$$

† See also D. S. Jones (1972).

‡ §6.2 shows that this is always possible. In the present context, however, k need not satisfy the criteria of that subsection.

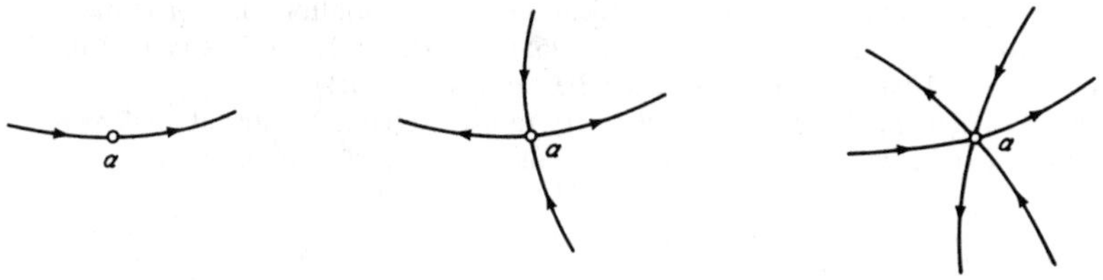

(i) $p'(a) \neq 0$ (ii) Simple zero of $p(t)$ at a (iii) Double zero of $p(t)$ at a

Fig. 10.1 t plane: curves of constant $\operatorname{Im}\{p(t)\}$. Arrows typify directions in which $\operatorname{Re}\{p(t)\}$ increases.

and let the equation of an arbitrary path in the t plane passing through a be $t = t(\tau)$, where τ is the arc parameter. If $\rho_R(\tau) \equiv p_R\{t(\tau)\}$ and $\rho_I(\tau) \equiv p_I\{t(\tau)\}$, then

$$\frac{d}{d\tau}|e^{p(t)}| = \frac{d}{d\tau}e^{p_R(t)} = \rho_R'(\tau)\, e^{p_R(t)}.$$

For given t, this is a maximum or minimum according as $\rho_R'(\tau)$ is a maximum or minimum. Differentiation of (10.09) yields

$$p'(t)\, t'(\tau) = \rho_R'(\tau) + i\rho_I'(\tau).$$

Since $|t'(\tau)| = 1$, it follows that

$$\{\rho_R'(\tau)\}^2 = |p'(t)|^2 - \{\rho_I'(\tau)\}^2.$$

Hence $|\rho_R'(\tau)|$ is greatest when $\rho_I'(\tau) = 0$. If the last equation is fulfilled everywhere on the path, then $p_I(t)$ is constant. In other words, (10.01) holds.

A common misunderstanding, encouraged by the name *method of steepest descents*, is that deformation of the original integration path into paths of constant $\operatorname{Im}\{p(t)\}$ is an *essential* step in the asymptotic analysis of integrals of the form (6.01). As we have seen, it suffices that the minimum value of $\operatorname{Re}\{e^{i\theta}p(t)\}$ be attained either at an endpoint of the path or at a saddle point. Sound reasons for attaching importance to paths of steepest descent are that they help in finding maximum regions of validity in the complex plane (compare §9.3), and in constructing explicit error bounds (§§10.1 and 10.2).

Historical Notes and Additional References

§2 Error bounds for the asymptotic expansion of the incomplete Gamma function can also be obtained by specializing results for Whittaker functions given by Olver (1965d); compare Chapter 7, Exercise 10.2 and equation (11.03).

§8.1 This analysis is based on that of Szegö (1967, §8.71) and Copson (1965, §37).

§10.3 A strong motivation for using paths of steepest descent is to facilitate application of Watson's lemma; see, for example, an illuminating discussion by Ursell (1970). This approach was sketched in a posthumous paper of Riemann (1863) and developed more fully in the researches of Debye on Bessel functions of large order (§9.4). Because of the difficulty of constructing steepest paths exactly, writers have often modified the paths in applications. These variations are sometimes called the *saddle-point method* (de Bruijn, 1961; Copson, 1965). Theorem 6.1, which is taken from Wyman (1964) and Olver (1970a), unifies and extends the various approaches.

It is worth commenting that it is not necessary to use descending paths at all. The method of stationary phase (Chapter 3) in effect uses paths along which $|e^{p(t)}|$ is constant.

5

DIFFERENTIAL EQUATIONS WITH REGULAR SINGULARITIES; HYPERGEOMETRIC AND LEGENDRE FUNCTIONS

1 Existence Theorems for Linear Differential Equations: Real Variables

1.1 Several of the special functions introduced in Chapter 2 were shown to satisfy differential equations of the form

$$\frac{d^2w}{dx^2}+f(x)\frac{dw}{dx}+g(x)\,w=0. \tag{1.01}$$

Other special functions of importance will be defined later as solutions of equations of the same type. At this stage it behooves us to study the existence and nature of solutions of (1.01) in general terms.

Although much of the subsequent analysis carries over straightforwardly to the general homogeneous linear differential equation of arbitrary order n, given by

$$\frac{d^nw}{dx^n}+f_{n-1}(x)\frac{d^{n-1}w}{dx^{n-1}}+\cdots+f_0(x)\,w=0,$$

in the interests of clarity and relevance to special functions we confine attention for the most part to $n=2$. This is the lowest value of n for which the equation has nontrivial solutions. In the case $n=1$ the general solution is easily verified to be

$$w=\exp\left\{-\int f_0(x)\,dx\right\}. \tag{1.02}$$

1.2 Theorem 1.1 *Let $f(x)$ and $g(x)$ be continuous in a finite or infinite interval (a,b). Then the differential equation* (1.01) *has an infinity of solutions which are twice continuously differentiable in (a,b). If the values of w and dw/dx are prescribed at any point, then the solution is unique.*

This is a well-known result from differential equation theory. We give the proof in full, however, since similar analysis will be used for more difficult problems.

At $x=x_0$, say, let a_0 and a_1 be arbitrarily prescribed values of w and dw/dx, respectively. We construct a sequence of functions $h_s(x)$, $s=0,1,2,\ldots$, defined by $h_0(x)=0$ and

$$h_s''(x)=-f(x)\,h_{s-1}'(x)-g(x)\,h_{s-1}(x),\qquad h_s(x_0)=a_0,\qquad h_s'(x_0)=a_1, \tag{1.03}$$

when $s \geqslant 1$. Thus, for example,

$$h_1'(x) = a_1, \qquad h_1(x) = a_1(x-x_0) + a_0. \tag{1.04}$$

The proof of the theorem consists in showing that when $s \to \infty$ the limit of the sequence exists, is twice differentiable, and satisfies (1.01).

Integration of (1.03) with respect to x yields

$$h_s'(x) = -\int_{x_0}^{x} \{f(t)h_{s-1}'(t)+g(t)h_{s-1}(t)\}\,dt + a_1. \tag{1.05}$$

Integrating again, using the method of parts, we obtain

$$h_s(x) = -\int_{x_0}^{x} (x-t)\{f(t)h_{s-1}'(t)+g(t)h_{s-1}(t)\}\,dt + a_1(x-x_0) + a_0. \tag{1.06}$$

When $s \geqslant 1$, subtraction gives

$$h_{s+1}'(x) - h_s'(x) = -\int_{x_0}^{x} [f(t)\{h_s'(t)-h_{s-1}'(t)\} + g(t)\{h_s(t)-h_{s-1}(t)\}]\,dt,$$

and

$$h_{s+1}(x) - h_s(x) = -\int_{x_0}^{x} (x-t)[f(t)\{h_s'(t)-h_{s-1}'(t)\} + g(t)\{h_s(t)-h_{s-1}(t)\}]\,dt.$$

Let $[\alpha, \beta]$ be any compact interval contained in (a, b) which itself contains x_0. From (1.04) and the assumed conditions it is seen that there exist finite constants H and K such that in $[\alpha, \beta]$

$$|h_1(x)| \leqslant H, \qquad |h_1'(x)| \leqslant H, \qquad |f(x)|+|g(x)| \leqslant K.$$

Therefore

$$|h_2'(x)-h_1'(x)| \leqslant HK|x-x_0|, \qquad |h_2(x)-h_1(x)| \leqslant (\beta-\alpha)HK|x-x_0|.$$

And by means of induction it is readily verified that

$$|h_{s+1}'(x)-h_s'(x)|,\ |h_{s+1}(x)-h_s(x)| \leqslant HK^sL^s|x-x_0|^s/s! \qquad (s \geqslant 0), \tag{1.07}$$

where $L = \max(\beta-\alpha, 1)$. Hence by the M-test each series

$$k(x) = \sum_{s=0}^{\infty} \{h_{s+1}'(x)-h_s'(x)\}, \qquad h(x) = \sum_{s=0}^{\infty} \{h_{s+1}(x)-h_s(x)\},$$

converges uniformly in $[\alpha, \beta]$. Therefore $k(x)$ is continuous, $h(x)$ is differentiable, and $k(x) = h'(x)$.

Next, from (1.03) we obtain

$$h_{s+1}''(x) - h_s''(x) = -f(x)\{h_s'(x)-h_{s-1}'(x)\} - g(x)\{h_s(x)-h_{s-1}(x)\} \qquad (s \geqslant 1). \tag{1.08}$$

Hence

$$\sum_{s=0}^{\infty} \{h_{s+1}''(x)-h_s''(x)\}$$

converges uniformly. Its sum is therefore continuous and equal to $h''(x)$.

Summing each side of equation (1.08) from $s = 1$ to $s = \infty$, we see that $h(x)$ satisfies the given differential equation (1.01) in $[\alpha, \beta]$. Moreover, it fulfills the conditions

$$h(x_0) = a_0, \qquad h'(x_0) = a_1. \tag{1.09}$$

1.3 Since β can be chosen arbitrarily close to b, and α arbitrarily close to a,† it remains to prove that $h(x)$ is the *only* twice continuously differentiable solution which satisfies (1.09). The difference $l(x)$, say, between $h(x)$ and any other solution which meets the requirements has the initial values $l(x_0) = l'(x_0) = 0$. By integration of (1.01) we derive

$$l'(x) = -\int_{x_0}^{x} \{f(t)\,l'(t) + g(t)\,l(t)\}\, dt,$$

and

$$l(x) = -\int_{x_0}^{x} (x-t)\{f(t)\,l'(t) + g(t)\,l(t)\}\, dt;$$

compare (1.05) and (1.06).

Let H now denote the least number such that $|l(x)| \leqslant H$ and $|l'(x)| \leqslant H$ when $x \in [\alpha, \beta]$; H is finite since $l(x)$ and $l'(x)$ are continuous, by hypothesis. Successive resubstitutions on the right-hand sides of the last two equations yield

$$|l(x)|,\ |l'(x)| \leqslant HK^sL^s|x - x_0|^s/s!,$$

where K and L are defined as before, and s is an arbitrary positive integer. Letting $s \to \infty$ we see that $l(x)$ and $l'(x)$ are both identically zero. This completes the proof of Theorem 1.1.

The foregoing method for constructing a solution of (1.01) is called *Picard's method of successive approximations*, although, of course, there is nothing approximate about the final solution. Extensions of Theorem 1.1 are stated in Exercises 1.1 and 1.2 below.

1.4 Let $w_1(x)$ and $w_2(x)$ be a pair of solutions of (1.01) with the property that any other solution can be expressed in the form

$$w(x) = Aw_1(x) + Bw_2(x),$$

where A and B are constants. Then $w_1(x)$ and $w_2(x)$ are said to comprise a *fundamental pair*. An example is furnished by the solutions satisfying the conditions

$$w_1(x_0) = 1, \qquad w_1'(x_0) = 0, \qquad w_2(x_0) = 0, \qquad w_2'(x_0) = 1,$$

at any chosen point x_0 of (a, b). Clearly in this case $A = w(x_0)$ and $B = w'(x_0)$.

Theorem 1.2 *Let $f(x)$ and $g(x)$ be continuous in (a, b), and $w_1(x)$ and $w_2(x)$ be solutions of* (1.01). *Then the following three statements are equivalent:*

† When $b = \infty$ this means "β can be chosen arbitrarily large"; similarly when $a = -\infty$.

(i) *$w_1(x)$ and $w_2(x)$ are a fundamental pair.*
(ii) *The* Wronskian

$$\mathscr{W}\{w_1(x), w_2(x)\} \equiv w_1(x)\, w_2'(x) - w_2(x)\, w_1'(x)$$

does not vanish at any interior point of (a, b).†
(iii) *$w_1(x)$ and $w_2(x)$ are linearly independent, that is, the only constants A and B such that*

$$Aw_1(x) + Bw_2(x) = 0$$

identically in (a, b) are $A = 0$ and $B = 0$.

To establish this result, we begin with the identity

$$\frac{d}{dx}\mathscr{W}\{w_1(x), w_2(x)\} = -f(x)\,\mathscr{W}\{w_1(x), w_2(x)\},$$

obtained by differentiation and use of (1.01). Integration gives

$$\mathscr{W}\{w_1(x), w_2(x)\} = Ce^{-\int f(x)\,dx}, \tag{1.10}$$

where C is independent of x. Accordingly, the Wronskian either vanishes for all x within (a, b), or it does not vanish at all.

Suppose first that (i) holds. Then for any point x_0 of (a, b) and any prescribed values of $w(x_0)$ and $w'(x_0)$, numbers A and B can be found such that

$$w(x_0) = Aw_1(x_0) + Bw_2(x_0), \qquad w'(x_0) = Aw_1'(x_0) + Bw_2'(x_0).$$

From elementary linear algebra it is known that this is possible if, and only if, $w_1(x_0)\, w_2'(x_0) - w_2(x_0)\, w_1'(x_0)$ is nonzero. That is, (i) implies (ii) and, conversely, (ii) implies (i).

Next, assume that (ii) holds. Then the only numbers A and B satisfying

$$Aw_1(x_0) + Bw_2(x_0) = 0, \qquad Aw_1'(x_0) + Bw_2'(x_0) = 0,$$

are $A = B = 0$. That is, (ii) implies (iii).

Lastly, assume that (iii) holds and $\mathscr{W}(w_1, w_2) = 0$. Clearly the solution

$$w(x) = w_2(x_0)\, w_1(x) - w_1(x_0)\, w_2(x)$$

satisfies $w(x_0) = w'(x_0) = 0$. Hence, by §1.3, $w(x) \equiv 0$, and hence, by (iii), $w_1(x_0) = w_2(x_0) = 0$. Similarly by considering the solution $w_2'(x_0)\, w_1(x) - w_1'(x_0)\, w_2(x)$ we see that $w_1'(x_0) = w_2'(x_0) = 0$. Again, by use of §1.3 it follows that $w_1(x) \equiv 0$ and $w_2(x) \equiv 0$. This contradicts (iii), however; hence the assumption $\mathscr{W}(w_1, w_2) = 0$ is false. That is, (iii) implies (ii). This completes the proof.

Equation (1.10) is called *Abel's identity*. An immediate corollary is that when $f(x) = 0$, that is, *when the differential equation has no term in the first derivative, the Wronskian of any pair of solutions is constant.*

† The possibility that the Wronskian vanishes as x tends to either of the endpoints a and b is not excluded, however.

Ex. 1.1 (*Existence theorem for inhomogeneous equations*) Show that Theorem 1.1 remains valid when the right-hand side of equation (1.01) is replaced by a given function of x which is continuous in (a, b).

Ex. 1.2 Let a and b be finite or infinite, and suppose that $f(x)$ and $g(x)$ are continuous in (a, b) save at a finite point set $\mathbf{X}$, and $|f(x)|$ and $|g(x)|$ are integrable over (a, b). Show that there is a unique function $w(x)$ with the following properties in the closure of (a, b): (i) $w'(x)$ is continuous; (ii) $w''(x)$ is continuous, except when $x \in \mathbf{X}$; (iii) $w(x)$ satisfies (1.01) except when $x \in \mathbf{X}$; (iv) $w(x_0)$ and $w'(x_0)$ are prescribed, where x_0 is any point in the closure of (a, b), including $\mathbf{X}$.

2 Equations Containing a Real or Complex Parameter

2.1 Many of the differential equations satisfied by the special functions contain one or more parameters, and we often need to know how the solutions behave as the parameters vary.

Theorem 2.1 *In the equation*

$$\frac{d^2w}{dx^2} + f(u, x)\frac{dw}{dx} + g(u, x)\, w = 0 \tag{2.01}$$

let u and x range over the finite rectangle $\mathbf{R}$: $u_0 \leqslant u \leqslant u_1$, $\alpha \leqslant x \leqslant \beta$, and assume that $f(u, x)$ and $g(u, x)$ are continuous in $\mathbf{R}$. Assume also that x_0 is a fixed point in $[\alpha, \beta]$ and that the values of w and $\partial w/\partial x$ at x_0 are prescribed continuous functions of u. Then the solution w and its partial derivatives $\partial w/\partial x$ and $\partial^2 w/\partial x^2$ are continuous in $\mathbf{R}$.

If, in addition, $\partial f/\partial u$ and $\partial g/\partial u$ are continuous in $\mathbf{R}$, and the values of $\partial w/\partial u$ and $\partial^2 w/(\partial u\, \partial x)$ at $x = x_0$ are continuous functions of u, then $\partial w/\partial u$, $\partial^2 w/(\partial u\, \partial x)$, and $\partial^3 w/(\partial u\, \partial x^2)$ are continuous in $\mathbf{R}$.

In this statement "continuous in $\mathbf{R}$" means, as usual, continuous functions simultaneously of both variables in $\mathbf{R}$. The theorem is a special case of general results in differential-equation theory.†

For the proof, we reexamine the steps of the analysis given in §1.2, bearing in mind that the functions $h_s(x) = h_s(u, x)$ now depend on u. From (1.04) and the given conditions it is immediately seen that $h_1(u, x)$ and $\partial h_1(u, x)/\partial x$ are continuous in $\mathbf{R}$. Write

$$H_s(u, x) = f(u, x)\{\partial h_s(u, x)/\partial x\} + g(u, x)\, h_s(u, x),$$

and let δu and δx be arbitrary changes in u and x, respectively. From (1.05) we have

$$\frac{\partial h_2(u+\delta u, x+\delta x)}{\partial x} - \frac{\partial h_2(u, x)}{\partial x}$$

$$= -\int_{x_0}^{x} \{H_1(u+\delta u, t) - H_1(u, t)\}\, dt - \int_{x}^{x+\delta x} H_1(u+\delta u, t)\, dt + a_1(u+\delta u) - a_1(u), \tag{2.02}$$

† Hartman (1964, Chapter V).

where $a_1(u)$ is the prescribed value of $\partial w/\partial x$ at x_0. Since f, g, h_1, $\partial h_1/\partial x$, and a_1 are continuous in **R**, they are automatically uniformly continuous there. Therefore the right-hand side of (2.02) is numerically less than an arbitrarily assigned positive number ε whenever $|\delta u|$ and $|\delta x|$ are both sufficiently small. Accordingly, $\partial h_2/\partial x$ is continuous. Similarly for h_2. And by similar analysis and induction we see that $\partial h_s/\partial x$ and h_s are continuous for $s = 3, 4, \ldots$.

In the remaining part of the analysis of §1.2, the numbers H, K, and L are assignable independently of u. In consequence, the series

$$\sum (h_{s+1} - h_s), \qquad \sum \left(\frac{\partial h_{s+1}}{\partial x} - \frac{\partial h_s}{\partial x}\right) \tag{2.03}$$

converge uniformly with respect to both variables. Their respective sums w and $\partial w/\partial x$ are therefore continuous in **R**. From this result and the differential equation (2.01) it follows that $\partial^2 w/\partial x^2$ is continuous. This completes the proof of the first part of the theorem.

For the second part, we observe that in consequence of the given conditions both $\partial h_1/\partial u$ and $\partial^2 h_1/(\partial u\, \partial x)$ are continuous, and thence that in the case $s = 2$ the integrals (1.05) and (1.06) may be differentiated under the sign of integration.† Then, as in the above analysis of $\partial h_2/\partial x$ and h_2, it follows that $\partial^2 h_2/(\partial u\, \partial x)$ and $\partial h_2/\partial u$ are continuous in **R**. Repetition of this argument establishes that $\partial h_s/\partial u$ and $\partial^2 h_s/(\partial u\, \partial x)$ are continuous for $s = 3, 4, \ldots$. From the u-differentiated forms of (1.05) and (1.06) it follows, by similar analysis to §1.2, that the series

$$\sum \left(\frac{\partial h_{s+1}}{\partial u} - \frac{\partial h_s}{\partial u}\right), \qquad \sum \left(\frac{\partial^2 h_{s+1}}{\partial u\, \partial x} - \frac{\partial^2 h_s}{\partial u\, \partial x}\right)$$

converge uniformly in **R**. Accordingly, $\partial w/\partial u$ and $\partial^2 w/(\partial u\, \partial x)$ are continuous. For $\partial^3 w/(\partial u\, \partial x^2)$ we merely refer to the u-differentiated form of (2.01). This completes the proof.

2.2 In the case when u is a complex variable (x still being real), holomorphicity of the coefficients of the differential equation implies holomorphicity of the solutions, provided that the initial values are holomorphic:

Theorem 2.2 *Assume that:*

(i) *$f(u, x)$ and $g(u, x)$ are continuous functions of both variables when u ranges over a domain* **U** *and x ranges over a compact interval $[\alpha, \beta]$.*

(ii) *For each x in $[\alpha, \beta]$, $f(u, x)$ and $g(u, x)$ are holomorphic functions of u.*

(iii) *The values of w and $\partial w/\partial x$ at a fixed point x_0 in $[\alpha, \beta]$ are holomorphic functions of u.*

Then for each $x \in [\alpha, \beta]$ the solution $w(u, x)$ of (2.01) *and its first two partial x derivatives are holomorphic functions of u.*

A straightforward extension of the proof of the first part of Theorem 2.1 shows that $\partial h_s(u, x)/\partial x$ and $h_s(u, x)$ are continuous functions of u and x for each s. We now

† Apostol (1957, pp. 219 and 220).

apply Theorem 1.1 of Chapter 2 to the integrals (1.05) and (1.06). By induction, it is seen that $\partial h_s(u,x)/\partial x$ and $h_s(u,x)$ are holomorphic in u for $s = 1, 2, \ldots$. Again, as in §2.1, the series (2.03) converge uniformly with respect to u and x in compact sets. This establishes the holomorphicity of w and $\partial w/\partial x$. For $\partial^2 w/\partial x^2$, we again refer to (2.01).

Ex. 2.1 Show that the first part of Theorem 2.1 can be extended to permit the initial point x_0 to depend on u, provided that x_0 and the values of w and $\partial w/\partial x$ at x_0 are continuous functions of u.

3 Existence Theorems for Linear Differential Equations: Complex Variables

3.1 Theorem 3.1† *Let $f(z)$ and $g(z)$ be holomorphic in a simply connected domain* $\mathbf{Z}$. *Then the equation*

$$\frac{d^2w}{dz^2} + f(z)\frac{dw}{dz} + g(z)w = 0 \tag{3.01}$$

has an infinity of solutions which are holomorphic in $\mathbf{Z}$. *If the values of w and dw/dz are prescribed at any point, then the solution is unique.*

The proof is an adaptation of that of Theorem 1.1. We first suppose that $\mathbf{Z}$ is a disk $|z-a| < r$, and that z_0 is a point of $\mathbf{Z}$ at which the values

$$a_0 = w(z_0), \qquad a_1 = w'(z_0), \tag{3.02}$$

are prescribed.

The sequence $h_s(z)$, $s = 0, 1, 2, \ldots$, is defined as before, with z replacing x and the integration paths taken to be straight lines. Suppose that $z \in \mathbf{Z}_1$, where $\mathbf{Z}_1$ is the closed disk $|z-a| \leqslant \rho$, ρ being any number such that $|z_0 - a| < \rho < r$. Then bounds H and K exist such that

$$|h_1(z)| \leqslant H, \qquad |h_1'(z)| \leqslant H, \qquad |f(z)| + |g(z)| \leqslant K,$$

when $z \in \mathbf{Z}_1$. Corresponding to (1.07) we have

$$|h_{s+1}'(z) - h_s'(z)|,\ |h_{s+1}(z) - h_s(z)| \leqslant HK^sL^s|z-z_0|^s/s!,$$

where $L = \max(2\rho, 1)$. Accordingly,

$$h(z) = \sum_{s=0}^{\infty} \{h_{s+1}(z) - h_s(z)\} \tag{3.03}$$

is a series of holomorphic functions; it converges uniformly in $\mathbf{Z}_1$ and therefore in any compact set in $\mathbf{Z}$, since ρ may be chosen arbitrarily close to r. The sum $h(z)$ is therefore holomorphic in $\mathbf{Z}$, and the series may be differentiated term by term any number of times. In consequence, $h(z)$ satisfies (3.01). Uniqueness is established as in §1.3, or by observing that in consequence of (3.01) and (3.02) *all* derivatives of the solution are prescribed at z_0.

† Fuchs (1866).

To complete the proof of Theorem 3.1, we recall that because **Z** is a domain any two points can be connected by a finite chain of overlapping disks lying within **Z**. We merely apply the result just obtained to each disk in turn. The condition that **Z** be simply connected is needed to ensure that the solution obtained by the continuation process is single valued.†

3.2 The definitions of a fundamental pair of solutions, Wronskian relation, and linear independence, as well as the result expressed by Theorem 1.2, all carry over straightforwardly to the complex plane.

The series (3.03) is called the *Liouville–Neumann expansion* of the solution of the differential equation. It is important in existence proofs, but for computational and other purposes preference is usually given to other forms of expansion, for example, Taylor series. Let r be the distance of the nearest of the singularities of $f(z)$ and $g(z)$ from $z = z_0$, and

$$f(z) = \sum_{s=0}^{\infty} f_s(z-z_0)^s, \qquad g(z) = \sum_{s=0}^{\infty} g_s(z-z_0)^s,$$

be the expansions of $f(z)$ and $g(z)$ within $|z-z_0|<r$. Theorem 3.1 shows that all holomorphic solutions of (3.01) can be expanded in series of the form

$$w(z) = \sum_{s=0}^{\infty} a_s(z-z_0)^s, \tag{3.04}$$

also convergent within $|z-z_0|<r$. Substituting in (3.01) and equating coefficients, we find that a_0 and a_1 may be prescribed arbitrarily (as we expect); higher coefficients are then determined recursively by

$$-s(s-1)\,a_s = (s-1)\,f_0\,a_{s-1} + (s-2)\,f_1\,a_{s-2} + \cdots$$
$$+ f_{s-2}\,a_1 + g_0\,a_{s-2} + g_1\,a_{s-3} + \cdots + g_{s-2}\,a_0 \qquad (s \geqslant 2).$$

3.3 Consider again the case in which the differential equation contains a parameter:

Theorem 3.2 *In the equation*

$$\frac{d^2w}{dz^2} + f(u,z)\frac{dw}{dz} + g(u,z)\,w = 0 \tag{3.05}$$

assume that u and z range over fixed, but not necessarily bounded, complex domains **U** *and* **Z**, *respectively, and*

(i) *$f(u,z)$ and $g(u,z)$ are continuous functions of both variables.*
(ii) *For each u, $f(u,z)$ and $g(u,z)$ are holomorphic functions of z.*
(iii) *For each z, $f(u,z)$ and $g(u,z)$ are holomorphic functions of u.*
(iv) *The values of w and $\partial w/\partial z$ at a fixed point z_0 in* **Z** *are holomorphic functions of u.*

Then at each point z of **Z** *the solution $w(u,z)$ of* (3.04) *and its first two partial z derivatives are holomorphic functions of u.*

† This is the *monodromy theorem*; see Levinson and Redheffer (1970, p. 402).

This result is provable by application of Theorem 2.2, as follows. The initial point z_0 is joined to z by a path $\mathscr{P}$ which lies in $\mathbf{Z}$ and has an equation of the form $t = t(\tau)$, where t is a typical point of the path, and τ is the arc parameter. On $\mathscr{P}$, w is a complex function of the real variable τ which satisfies the equation

$$\frac{d^2w}{d\tau^2} + \left[t'(\tau) f\{u, t(\tau)\} - \frac{t''(\tau)}{t'(\tau)}\right]\frac{dw}{d\tau} + \{t'(\tau)\}^2 g\{u, t(\tau)\}\, w = 0. \tag{3.06}$$

Now suppose that $\mathscr{P}$ can be chosen in such a way that: (a) $t''(\tau)$ *is continuous*; (b) $t'(\tau)$ *does not vanish*. Then the coefficients of $dw/d\tau$ and w in (3.06) are continuous; consequently from Theorem 2.2 it follows that each of the three functions

$$w, \qquad \frac{dw}{d\tau} \equiv t'(\tau)\frac{dw}{dt}, \qquad \frac{d^2w}{d\tau^2} \equiv \{t'(\tau)\}^2\frac{d^2w}{dt^2} + t''(\tau)\frac{dw}{dt}$$

is holomorphic in u at all points of $\mathscr{P}$, including, in particular, $t = z$.

Conditions (a) and (b) are certainly fulfilled when $\mathscr{P}$ is a straight line. But in any event $\mathscr{P}$ can always be chosen to consist of a finite chain of line segments. On each segment w satisfies an equation of the form (3.06). At the beginning of each segment the values of w and $\{t'(\tau)\}^{-1}\, dw/d\tau$ are the same as at the end of the previous segment, and therefore holomorphic in u.† Application of Theorem 2.2 to each segment in turn establishes Theorem 3.2.

Another way of completing the proof is suggested by Exercise 3.4 below.

3.4 Conditions (a) and (b) of §3.3 demand more than that $\mathscr{P}$ be a regular arc, for in this case (a) would be replaced by "$t'(\tau)$ is continuous"; compare Chapter 1, §11.5. We define paths satisfying (a) and (b) to be R_2 *arcs*. By analogy, regular arcs can be called R_1 *arcs*. Similarly, a path on which all derivatives of $t(\tau)$ are continuous and $t'(\tau)$ is nonvanishing is said to be an R_∞ *arc*. All paths normally used in complex-variable theory, consisting of straight lines, circular arcs, parabolic arcs, and so on, are chains of R_∞ arcs, and *a fortiori* chains of R_2 arcs.

Ex. 3.1 Show that the equation $(\cosh z)\, w'' + w = 0$ has a fundamental pair of solutions whose Maclaurin expansions begin

$$1 - \tfrac{1}{2}z^2 + \tfrac{1}{12}z^4 - \tfrac{13}{720}z^6 + \cdots, \qquad z - \tfrac{1}{6}z^3 + \tfrac{1}{30}z^5 - \tfrac{13}{1680}z^7 + \cdots,$$

and check the coefficients by use of the Wronskian relation.

What is the radius of convergence of each series?

Ex. 3.2 Show that throughout the z plane *Weber's differential equation*

$$d^2w/dz^2 = (\tfrac{1}{4}z^2 + a)\, w$$

has independent solutions

$$w_1 = \sum_{s=0}^{\infty} a_{2s}\frac{z^{2s}}{(2s)!}, \qquad w_2 = \sum_{s=0}^{\infty} a_{2s+1}\frac{z^{2s+1}}{(2s+1)!},$$

in which $a_0 = a_1 = 1$, $a_2 = a_3 = a$, and

$$a_{s+2} = a a_s + \tfrac{1}{4}s(s-1)\, a_{s-2} \qquad (s \geqslant 2).$$

† The values of $dw/d\tau$ at the junction differ, however.

Show also that

$$w_1 = \exp(\mp\tfrac{1}{4}z^2) \sum_{s=0}^{\infty} (\tfrac{1}{2}a \pm \tfrac{1}{4})(\tfrac{1}{2}a \pm \tfrac{5}{4}) \cdots (\tfrac{1}{2}a \pm s \mp \tfrac{3}{4}) \frac{2^s z^{2s}}{(2s)!},$$

$$w_2 = \exp(\mp\tfrac{1}{4}z^2) \sum_{s=0}^{\infty} (\tfrac{1}{2}a \pm \tfrac{3}{4})(\tfrac{1}{2}a \pm \tfrac{7}{4}) \cdots (\tfrac{1}{2}a \pm s \mp \tfrac{1}{4}) \frac{2^s z^{2s+1}}{(2s+1)!},$$

where either the upper or the lower signs are taken consistently throughout.

Ex. 3.3 In the notation of §3.2 let F and G denote respectively the maximum moduli of $f(z)$ and $g(z)$ on the circle $|z-z_0| = \rho$, where ρ is any number less than r. Also let K be the greater of F and $G\rho$. By use of Cauchy's formula and induction verify that $|a_s| \leqslant b_s$, $s = 0, 1, 2, \ldots$, where $b_0 = |a_0|$, $b_1 = |a_1|$, and

$$s(s-1)\,b_s = K\{sb_{s-1} + (s-1)\,b_{s-2}\,\rho^{-1} + (s-2)\,b_{s-3}\,\rho^{-2} + \cdots + b_0\,\rho^{-s+1}\},$$

when $s \geqslant 2$. Deduce that

$$s(s-1)\,b_s - (s-1)(s-2)\,b_{s-1}\,\rho^{-1} = Ksb_{s-1} \qquad (s \geqslant 3),$$

and thence prove directly that the radius of convergence of the series (3.04) is at least r.†

Ex. 3.4 Show that any two points of a domain can be connected by a single R_2 arc lying in the domain.

4 Classification of Singularities; Nature of the Solutions in the Neighborhood of a Regular Singularity

4.1 If the functions $f(z)$ and $g(z)$ are both analytic at $z = z_0$, then this point is said to be an *ordinary point* of the differential equation

$$\frac{d^2w}{dz^2} + f(z)\frac{dw}{dz} + g(z)\,w = 0. \tag{4.01}$$

If $z = z_0$ is not an ordinary point, but both $(z-z_0)\,f(z)$ and $(z-z_0)^2 g(z)$ are analytic there, then z_0 is said to be a *regular singularity*, or *singularity of the first kind*.

Lastly, if z_0 is neither an ordinary point nor a regular singularity, then it is said to be an *irregular singularity*, or *singularity of the second kind*. When the singularities of $f(z)$ and $g(z)$ at z_0 are no worse than poles z_0 is said to be a *singularity of rank* $l-1$, where l is the least integer such that both $(z-z_0)^l f(z)$ and $(z-z_0)^{2l} g(z)$ are analytic. Thus a regular singularity is of rank zero. If either $f(z)$ or $g(z)$ has an essential singularity at z_0, then the rank may be said to be infinite.

In §3 we showed that in the neighborhood of an ordinary point the differential equation has linearly independent pairs of holomorphic solutions. In the present section and §5 we construct convergent series solutions in the neighborhood of a regular singularity. In general this cannot be done for an irregular singularity; treatment of this more difficult case is deferred until Chapters 6 and 7.

4.2 Without loss of generality the regular singularity can be taken at the origin. Thus we assume that in a neighborhood $|z| < r$ there exist convergent series

† This is Cauchy's method of proof of the existence of solutions.

expansions

$$zf(z) = \sum_{s=0}^{\infty} f_s z^s, \qquad z^2 g(z) = \sum_{s=0}^{\infty} g_s z^s, \tag{4.02}$$

in which at least one of the coefficients f_0, g_0, and g_1 is nonzero.

The form that we may reasonably expect the solutions to take can be found by approximating $f(z)$ and $g(z)$ by means of the leading terms in (4.02); thus

$$\frac{d^2w}{dz^2} + \frac{f_0}{z}\frac{dw}{dz} + \frac{g_0}{z^2}w = 0.$$

Exact solutions of this equation are given by $w = z^\alpha$, where α is a root of the quadratic equation

$$\alpha(\alpha-1) + f_0\alpha + g_0 = 0. \tag{4.03}$$

Accordingly, as a possible solution of (4.01) we try the series

$$w(z) = z^\alpha \sum_{s=0}^{\infty} a_s z^s, \tag{4.04}$$

in which α is a root of the *indicial equation* (4.03). The two possible values of α are called the *exponents* or *indices* of the singularity. Substituting in the given differential equation by means of (4.02) and (4.04) and formally equating coefficients of $z^{\alpha+s-2}$, we derive

$$Q(\alpha+s)a_s = -\sum_{j=0}^{s-1} \{(\alpha+j)f_{s-j} + g_{s-j}\}a_j \qquad (s = 1, 2, \ldots), \tag{4.05}$$

where $Q(\alpha)$ denotes the left-hand side of (4.03). Equation (4.05) determines $a_1, a_2, \ldots$ recursively in terms of an arbitrarily assigned (nonzero) value of a_0. The procedure runs into difficulty if, and only if, $Q(\alpha+s)$ vanishes for a positive integer value of s.

Accordingly, when the roots of the indicial equation are distinct and do not differ by an integer, two series of the form (4.04) can be found that formally satisfy the differential equation. In other cases only one solution of this type is available, unless the right-hand side of (4.05) vanishes at the same value of the positive integer s for which $Q(\alpha+s) = 0$.

4.3 Theorem 4.1† *With the notation and conditions of* §4.2, *the series* (4.04) *converges and defines a solution of the differential equation* (4.01) *when* $|z| < r$, *provided that the other exponent is not of the form* $\alpha + s$, *where* s *is a positive integer.*

Let ρ be any number less than r, and K denote the greater of

$$\max_{|z|=\rho} |zf(z)|, \qquad \max_{|z|=\rho} |z^2 g(z)|.$$

Then Cauchy's formula yields the following inequalities for the coefficients in the

† Frobenius (1873). The proof should be compared with Cauchy's method sketched in Exercise 3.3.

series (4.02):

$$|f_s| \leqslant K\rho^{-s}, \qquad |g_s| \leqslant K\rho^{-s}.$$

Next, let β denote the second exponent and $n \equiv [|\alpha-\beta|]$. Define b_s by $b_s = |a_s|$ when $s = 0, 1, \ldots, n$, and by

$$s(s-|\alpha-\beta|)\,b_s = K\sum_{j=0}^{s-1}(|\alpha|+j+1)\,b_j\,\rho^{j-s}, \tag{4.06}$$

when $s \geqslant n+1$. Then by using (4.05) and the identity $Q(\alpha+s) = s(s+\alpha-\beta)$, it may be verified by induction that $|a_s| \leqslant b_s$.

In (4.06) if we replace s by $s-1$ and combine the two equations, then we find that the majorizing coefficients b_s also satisfy the simpler recurrence relation

$$\rho s(s-|\alpha-\beta|)\,b_s - (s-1)(s-1-|\alpha-\beta|)\,b_{s-1} = K(|\alpha|+s)\,b_{s-1} \qquad (s \geqslant n+2).$$

Dividing this by $s^2 b_s$ and letting $s \to \infty$, we find that

$$b_{s-1}/b_s \to \rho,$$

which means that the radius of convergence of the series $\sum b_s z^s$ is ρ. Therefore, by the comparison test, the radius of convergence of the series (4.04) is at least ρ. Since ρ can be arbitrarily close to r, this radius of convergence is at least r. Well-known properties of power series now confirm that the processes of substitution and termwise differentiation used in §4.2 are justified, and hence that the series (4.04) is a solution of (4.01) within $|z| < r$. This completes the proof.

If α is a nonnegative integer, then the solution with exponent α is analytic at $z = 0$. When α is a negative integer the solution has a pole, and when α is nonintegral there is a branch point. Again, provided that the exponent difference is not an integer the theorem can be applied twice and the solutions obtained comprise a fundamental pair, at least one of which has a branch point at the singularity.

Ex. 4.1 Find independent series solutions of the equation

$$z^2(z-1)\,w'' + (\tfrac{3}{2}z-1)\,zw' + (z-1)\,w = 0$$

(i) in the neighborhood of $z = 0$; (ii) in the neighborhood of $z = 1$.

5 Second Solution When the Exponents Differ by an Integer or Zero

5.1 Suppose that α and β are the roots of the indicial equation (4.03) and $\alpha-\beta = n$, where n is a positive integer or zero. Theorem 4.1 furnishes the solution

$$w_1(z) = z^\alpha \sum_{s=0}^{\infty} a_s z^s. \tag{5.01}$$

To find an independent second solution we use a standard substitution for depressing the order of a differential equation having a known solution, given by

$$w(z) = w_1(z)\,v(z).$$

Then

$$v''(z) + \left\{2\frac{w_1'(z)}{w_1(z)} + f(z)\right\}v'(z) = 0.$$

Regarding this as a first-order differential equation in $v'(z)$ and referring to (1.02), we obtain

$$v(z) = \int \frac{1}{\{w_1(z)\}^2} \exp\left\{-\int f(z)\,dz\right\} dz.$$

5.2 What is the nature of the solution $w_2(z) \equiv w_1(z)v(z)$ in the neighborhood of $z = 0$? From (4.02) and (5.01) it is seen that

$$\frac{1}{\{w_1(z)\}^2} \exp\left\{-\int f(z)\,dz\right\} = \frac{1}{z^{2\alpha}(a_0 + a_1 z + \cdots)^2} \exp(-f_0 \ln z - f_1 z - \tfrac{1}{2} f_2 z^2 - \cdots),$$

and from (4.03) we have $f_0 = 1 - \alpha - \beta = 1 + n - 2\alpha$. Hence

$$\frac{1}{\{w_1(z)\}^2} \exp\left\{-\int f(z)\,dz\right\} = \frac{\phi(z)}{z^{n+1}},$$

where $\phi(z)$ is analytic at $z = 0$. Let the Maclaurin expansion of $\phi(z)$ be denoted by

$$\phi(z) = \sum_{s=0}^{\infty} \phi_s z^s,$$

where the ϕ_s are expressible in terms of the a_s and f_s; in particular, $\phi_0 = 1/a_0^2$. Integrating $z^{-n-1}\phi(z)$ and multiplying the result by $w_1(z)$, we obtain

$$w_2(z) = w_1(z)\left\{-\sum_{s=0}^{n-1} \frac{\phi_s}{(n-s)z^{n-s}} + \phi_n \ln z + \sum_{s=n+1}^{\infty} \frac{\phi_s z^{s-n}}{s-n}\right\}. \tag{5.02}$$

When $n = 0$, that is, when the exponents coincide, (5.02) has the form

$$w_2(z) = \phi_0 w_1(z) \ln z + z^{\alpha+1} \sum_{s=0}^{\infty} b_s z^s. \tag{5.03}$$

Since ϕ_0 does not vanish, $w_2(z)$ has a logarithmic branch point at the singularity and

$$w_2(z) \sim (z^\alpha \ln z)/a_0 \qquad (z \to 0).$$

Alternatively, when n is a positive integer (5.02) takes the form

$$w_2(z) = \phi_n w_1(z) \ln z + z^\beta \sum_{s=0}^{\infty} c_s z^s. \tag{5.04}$$

The leading coefficient in the last sum is

$$c_0 = -a_0 \phi_0 / n = -1/(n a_0),$$

and is always nonzero. Thus

$$w_2(z) \sim -z^\beta/(n a_0) \qquad (z \to 0).$$

It may happen that $\phi_n = 0$, in which event the logarithmic term in (5.04) is absent.†

Since the only possible singularities of $w_1(z)$ and $w_2(z)$ are the singularities of $f(z)$ and $g(z)$, the radius of convergence of the series (5.03) and (5.04) is not less than the distance from the origin of the nearest singularity of $zf(z)$ and $z^2 g(z)$.

Having established the *form* of the second solution, we would not normally use the foregoing construction to evaluate the coefficients. *Generally it is easier to substitute* (5.03) *or* (5.04) *directly in the original differential equation and equate coefficients.* Since the second solution is undetermined to the extent of an arbitrary constant factor, in the case $n > 0$ the value of c_0 can be assigned arbitrarily, in advance. In this way ϕ_n is determined automatically.

5.3 If the coefficients in the differential equation are functions of a parameter u and the exponent difference of the singularity is an integer or zero for a critical value u_0, say, of u, then another way of constructing the series expansion for a second solution when $u = u_0$ is to determine the limiting value of the quotient

$$\{w_2(u,z) - w_1(u,z)\}/(u-u_0). \tag{5.05}$$

Here $w_1(u,z)$ and $w_2(u,z)$ are solutions obtained by the method of §4 which are linearly independent when $u \neq u_0$ and coincide when $u = u_0$. For real variables, the limiting process is justifiable in the following way.

Write

$$\phi(u,x) = w_2(u,x) - w_1(u,x).$$

With the conditions of Theorem 2.1, $[\partial\phi(u,x)/\partial u]_{u=u_0}$ exists and equals the limiting value of (5.05) as $u \to u_0$. Differentiation of the original differential equation (2.01) with respect to u yields

$$\frac{\partial^3\phi}{\partial u\,\partial x^2} + \frac{\partial f}{\partial u}\frac{\partial\phi}{\partial x} + f\frac{\partial^2\phi}{\partial u\,\partial x} + \frac{\partial g}{\partial u}\phi + g\frac{\partial\phi}{\partial u} = 0. \tag{5.06}$$

Again, provided that the conditions of Theorem 2.1 are satisfied all partial derivatives appearing in this equation are continuous functions of both variables. Since this is also true of $\partial^2\phi/\partial x^2$ we have‡

$$\frac{\partial^2\phi}{\partial u\,\partial x} = \frac{\partial^2\phi}{\partial x\,\partial u}, \qquad \frac{\partial^3\phi}{\partial u\,\partial x^2} = \frac{\partial^3\phi}{\partial x^2\,\partial u}.$$

Let $u \to u_0$. By hypothesis, $\phi(u,x)$ and $\partial\phi(u,x)/\partial x$ both vanish. Accordingly, (5.06) reduces to (2.01) with $w = [\partial\phi/\partial u]_{u=u_0}$. This is the required result.

For complex z, we extend the series solution obtained by the real-variable procedure by analytic continuation.

This method is due to Frobenius (1873). When applicable it usually furnishes the easiest way of calculating the series for the second solution. Illustrations are given later in this chapter and also in Chapter 7.

† This occurs in the situation mentioned in the closing sentence of §4.2.

‡ Apostol (1957, p. 121).

Ex. 5.1 Show that within the unit disk the equation $z(z-1)w''+(2z-1)w'+\frac{1}{4}w=0$ has independent solutions

$$\sum_{s=0}^{\infty} a_s z^s, \qquad \left(\sum_{s=0}^{\infty} a_s z^s\right)\ln z + 4\sum_{s=1}^{\infty}\{\psi(2s+1)-\psi(s+1)\}a_s z^s,$$

where ψ is the logarithmic derivative of the Gamma function, and

$$a_s = 1^2\cdot 3^2\cdots(2s-1)^2/\{2^2\cdot 4^2\cdots(2s)^2\}.$$ [Whittaker and Watson, 1927.]

6 Large Values of the Independent Variable

6.1 To discuss solutions in the neighborhood of the point at infinity, we make the transformation $z=1/t$. Equation (4.01) becomes

$$\frac{d^2w}{dt^2}+p(t)\frac{dw}{dt}+q(t)w=0, \tag{6.01}$$

where

$$p(t)=\frac{2}{t}-\frac{1}{t^2}f\left(\frac{1}{t}\right), \qquad q(t)=\frac{1}{t^4}g\left(\frac{1}{t}\right).$$

The singularity of (4.01) at $z=\infty$ is classified according to the nature of the singularity of (6.01) at $t=0$.

Thus *infinity is an ordinary point of* (4.01) *if $p(t)$ and $q(t)$ are analytic at $t=0$, that is, if $2z-z^2f(z)$ and $z^4g(z)$ are analytic at infinity.* In this case all analytic solutions can be expanded in series of the form

$$\sum_{s=0}^{\infty} a_s z^{-s}$$

which converge for sufficiently large $|z|$.

Next, *infinity is a regular singularity of* (4.01) *if $t^{-1}f(t^{-1})$ and $t^{-2}g(t^{-1})$ are analytic at $t=0$, that is, if $f(z)$ and $g(z)$ can be expanded in convergent series of the form*

$$f(z)=\frac{1}{z}\sum_{s=0}^{\infty}\frac{f_s}{z^s}, \qquad g(z)=\frac{1}{z^2}\sum_{s=0}^{\infty}\frac{g_s}{z^s},$$

when $|z|$ is large. In this case there exists at least one solution of the form

$$w(z)=\frac{1}{z^\alpha}\sum_{s=0}^{\infty}\frac{a_s}{z^s}.$$

The number α is again termed the *exponent* of the solution or singularity. It satisfies the equation

$$\alpha(\alpha+1)-f_0\alpha+g_0=0;$$

compare (4.03) and (4.04).

Lastly, *if either of the functions* $zf(z)$ *and* $z^2g(z)$ *is singular at infinity, then* $z = \infty$ *is an irregular singularity of the differential equation.* The rank is $m+1$, where m is the least nonnegative integer such that $z^{-m}f(z)$ and $z^{-2m}g(z)$ are analytic at infinity.

Ex. 6.1 For each of the following equations what is the nature of the singularity at infinity? Evaluate the exponents or rank, as appropriate.

$$(z^2+1)^{1/2}w'' = w' + w, \qquad w'' + (\sin z)\,w' + (\cos z)\,w = 0,$$

$$\frac{d}{dz}\left\{(z^4+2z^2)\frac{dw}{dz}\right\} + (z^2+1)\,w = 0.$$

Ex. 6.2 Construct independent series solutions of the equation $(1-z^2)w'' - 2zw' + 12w = 0$ valid outside the unit disk.

7 Numerically Satisfactory Solutions

7.1 In §1.4 we saw that all twice continuously differentiable solutions of a second-order homogeneous linear differential equation can be expressed as a linear combination of a fundamental pair of solutions. In numerical and physical applications, however, knowledge of a fundamental pair of solutions may not determine all the other solutions in an adequate manner. Consider, for example, the equation

$$d^2w/dz^2 = w.$$

This has the general solution

$$w = Ae^z + Be^{-z}, \tag{7.01}$$

in which A and B are arbitrary constants. Another representation of the general solution is afforded by

$$w = A\cosh z + B\sinh z. \tag{7.02}$$

Given numerical tables of e^z and e^{-z} having a certain number of significant figures, we can evaluate the expression (7.01) to an almost constant precision for any chosen values of A and B. This precision may not be attainable, however, if we use instead comparable tables of $\cosh z$ and $\sinh z$. When A and $-B$ are equal, or very nearly equal, severe cancellation takes place between the terms on the right-hand side of (7.02) for large positive values of $\operatorname{Re} z$. Similarly in the case when A and B are equal and $\operatorname{Re} z$ is large and negative.

For this reason e^z and e^{-z} are said to comprise a *numerically satisfactory*† pair of solutions in the neighborhood of infinity. The pair $\cosh z$ and $\sinh z$ are not numerically satisfactory in this region, even though they are linearly independent.

7.2 In the foregoing example the point at infinity is an irregular singularity of the differential equation. Similar considerations apply to regular singularities. Indeed, it is easily seen that *in the neighborhood of a regular singularity one member of a*

† J. C. P. Miller (1950).

numerically satisfactory pair of solutions has to be the solution constructed by the methods of §§4–6 *from the exponent of largest real part, or, in the case of equal exponents, the solution not containing a logarithmic term in its expansion.* This solution, which is undetermined to the extent of an arbitrary constant factor, is called the *recessive* solution at the singularity.† Any solution which is linearly independent of the recessive solution is said to be *dominant* at the singularity, because the ratio of its magnitude to that of the recessive solution tends to infinity as the singularity is approached.

The distinction between recession and dominance is also important in the *identification* of solutions of the differential equation. If α and β are the exponents at a finite singularity z_0, say, and $\operatorname{Re}\alpha > \operatorname{Re}\beta$, then it is clear that the condition

$$w \sim (z-z_0)^{\alpha} \qquad (z \to z_0) \tag{7.03}$$

specifies the solution uniquely. On the other hand, there is an infinite number of solutions which satisfy the condition

$$w \sim (z-z_0)^{\beta} \qquad (z \to z_0),$$

because the addition of an arbitrary multiple of the recessive solution does not change the overall asymptotic behavior.

Similarly when $\alpha = \beta$ the condition (7.03) again specifies w uniquely, but not the condition

$$w \sim (z-z_0)^{\alpha} \ln(z-z_0) \qquad (z \to z_0);$$

compare (5.03).

The one case excluded from the foregoing discussion occurs when $\alpha \neq \beta$ but $\operatorname{Re}\alpha = \operatorname{Re}\beta$. Neither the series solution constructed from α nor that constructed from β dominates the other, and the two solutions comprise a numerically satisfactory pair in the neighborhood of z_0.

Similar considerations apply when the singularity is located at infinity.

7.3 Recession and dominance are tied to the singularity under consideration. A solution that is recessive at one singularity may well be dominant at others; indeed, it generally is.

In a region containing two regular singularities z_1 and z_2, say, a numerically satisfactory pair of solutions would consist of one which is recessive at z_1 and dominant at z_2, and another which is recessive at z_2 and dominant at z_1. If, by chance, the *same* solution is recessive at z_1 and z_2, then it could be paired with any independent solution, since the latter is necessarily dominant at z_1 and z_2.

In a region containing n ($\geqslant 3$) regular singularities it is not possible, as a rule, to select a *single* pair that is numerically satisfactory throughout the entire region. Altogether, there are n recessive solutions, and some explicit knowledge of each is needed in order to have a satisfactory basis for constructing all possible solutions of the differential equation.

† Other adjectives in use are *subdominant, distinguished,* and *minimal.*

8 The Hypergeometric Equation

8.1 The differential equation

$$z(1-z)\frac{d^2w}{dz^2}+\{c-(a+b+1)z\}\frac{dw}{dz}-abw=0, \tag{8.01}$$

in which a, b, and c are real or complex parameters, is called the *hypergeometric equation*. Its only singularities are 0, 1, and ∞; each is easily seen to be regular, the corresponding exponent pairs being $(0,1-c)$, $(0,c-a-b)$, and (a,b), respectively.

The importance of equation (8.01) stems in part from the following theorem, the proof of which is the theme of this section.†

Theorem 8.1 *Any homogeneous linear differential equation of the second order whose singularities—including the point at infinity—are regular and not more than three in number, is transformable into the hypergeometric equation.*

8.2 We first construct the second-order equation

$$\frac{d^2w}{dz^2}+f(z)\frac{dw}{dz}+g(z)w=0,$$

having regular singularities at given distinct finite points ξ, η, and ζ, with arbitrarily assigned exponent pairs (α_1,α_2), (β_1,β_2), and (γ_1,γ_2), respectively.‡

Since the only possible singularities (including infinity) of $f(z)$ and $g(z)$ are poles, these functions are rational.§ Therefore

$$f(z)=\frac{F(z)}{(z-\xi)(z-\eta)(z-\zeta)},\qquad g(z)=\frac{G(z)}{(z-\xi)^2(z-\eta)^2(z-\zeta)^2},$$

where $F(z)$ and $G(z)$ are polynomials. If infinity is to be an ordinary point then, as we observed in §6, $2z-z^2f(z)$ and $z^4g(z)$ must be analytic there. Accordingly, both $F(z)$ and $G(z)$ are quadratics, the coefficient of z^2 in the former being 2. Thus

$$f(z)=\frac{A}{z-\xi}+\frac{B}{z-\eta}+\frac{C}{z-\zeta},$$

and

$$(z-\xi)(z-\eta)(z-\zeta)g(z)=\frac{D}{z-\xi}+\frac{E}{z-\eta}+\frac{F}{z-\zeta},$$

where

$$A+B+C=2. \tag{8.02}$$

To express the constants A, B, C, D, E, and F in terms of the assumed exponents,

† Compare also Exercises 8.1 and 8.2 below.

‡ Equations with less than three singularities are automatically included by allowing the choice $(0,1)$ for one or more of the exponent pairs.

§ This is a consequence of Laurent's theorem. See Copson (1935, §4.56).

we see from the indicial equation at ξ, namely,

$$\alpha(\alpha-1) + A\alpha + D(\xi-\eta)^{-1}(\xi-\zeta)^{-1} = 0,$$

that

$$A = 1 - \alpha_1 - \alpha_2, \qquad D = (\xi-\eta)(\xi-\zeta)\alpha_1\alpha_2.$$

Similarly,

$$B = 1 - \beta_1 - \beta_2, \qquad E = (\eta-\zeta)(\eta-\xi)\beta_1\beta_2,$$

$$C = 1 - \gamma_1 - \gamma_2, \qquad F = (\zeta-\xi)(\zeta-\eta)\gamma_1\gamma_2.$$

In consequence of (8.02) the six exponents cannot be chosen independently; they have to satisy

$$\alpha_1 + \alpha_2 + \beta_1 + \beta_2 + \gamma_1 + \gamma_2 = 1. \tag{8.03}$$

The desired differential equation then takes the form

$$\frac{d^2w}{dz^2} + \left(\frac{1-\alpha_1-\alpha_2}{z-\xi} + \frac{1-\beta_1-\beta_2}{z-\eta} + \frac{1-\gamma_1-\gamma_2}{z-\zeta}\right)\frac{dw}{dz}$$
$$-\left\{\frac{\alpha_1\alpha_2}{(z-\xi)(\eta-\zeta)} + \frac{\beta_1\beta_2}{(z-\eta)(\zeta-\xi)} + \frac{\gamma_1\gamma_2}{(z-\zeta)(\xi-\eta)}\right\}\frac{(\xi-\eta)(\eta-\zeta)(\zeta-\xi)}{(z-\xi)(z-\eta)(z-\zeta)}w = 0. \tag{8.04}$$

This is called the *Papperitz* or *Riemann* equation.

In a notation due to Riemann, equation (8.04) is represented by the array

$$w = P\left\{\begin{matrix} \xi & \eta & \zeta & \\ \alpha_1 & \beta_1 & \gamma_1 & z \\ \alpha_2 & \beta_2 & \gamma_2 & \end{matrix}\right\}.$$

The singularities appear in the first row, the order being immaterial. The corresponding exponents appear in columns below them, again the order of each pair being immaterial.

With the same method it is verifiable that the explicit form of

$$w = P\left\{\begin{matrix} \xi & \infty & \zeta & \\ \alpha_1 & \beta_1 & \gamma_1 & z \\ \alpha_2 & \beta_2 & \gamma_2 & \end{matrix}\right\},$$

that is, the differential equation having regular singularities at ξ, ζ, and the point at infinity, is given by

$$\frac{d^2w}{dz^2} + \left(\frac{1-\alpha_1-\alpha_2}{z-\xi} + \frac{1-\gamma_1-\gamma_2}{z-\zeta}\right)\frac{dw}{dz}$$
$$+\left\{\frac{\alpha_1\alpha_2(\xi-\zeta)}{z-\xi} + \beta_1\beta_2 + \frac{\gamma_1\gamma_2(\zeta-\xi)}{z-\zeta}\right\}\frac{w}{(z-\xi)(z-\zeta)} = 0, \tag{8.05}$$

provided that condition (8.03) is again satisfied. Not surprisingly, (8.05) is the limiting form of (8.04) as $\eta \to \infty$.

8.3 We now transform (8.04) by taking new variables

$$t = \frac{(\zeta-\eta)(z-\xi)}{(\zeta-\xi)(z-\eta)}, \qquad W = t^{-\alpha_1}(1-t)^{-\gamma_1}w. \tag{8.06}$$

The first of these relations is a fractional linear transformation which maps the z plane in a one-to-one manner onto the t plane.

The differential equation in W and t is, again, second order and linear. Its only singularities are the points corresponding to $z = \xi$, η, and ζ, that is, $t = 0$, ∞, and 1, respectively. From the opening paragraph of §6.1 it follows that the new singularities are regular (or possibly ordinary points), and from the second of (8.06) it is seen that the new exponent pairs are

$$(0,\ \alpha_2-\alpha_1), \qquad (\beta_1+\alpha_1+\gamma_1,\ \beta_2+\alpha_1+\gamma_1), \qquad (0,\ \gamma_2-\gamma_1),$$

respectively. The analysis of §8.2 shows that the differential equation is uniquely determined by the affixes of the singularities and the values of (five of) the exponents. Hence from (8.05) we can immediately write the new equation:

$$\frac{d^2W}{dt^2}+\left(\frac{1-\alpha_2+\alpha_1}{t}+\frac{1-\gamma_2+\gamma_1}{t-1}\right)\frac{dW}{dt}+\frac{(\alpha_1+\beta_1+\gamma_1)(\alpha_1+\beta_2+\gamma_1)}{t(t-1)}W = 0. \tag{8.07}$$

In consequence of (8.03) this equation is of the form (8.01) with

$$a = \alpha_1+\beta_1+\gamma_1, \qquad b = \alpha_1+\beta_2+\gamma_1, \qquad c = 1+\alpha_1-\alpha_2.$$

The foregoing analysis covers the case of three finite singularities. In a similar way the differential equation (8.05) can be transformed into (8.07) and thence into (8.01). This completes the proof of Theorem 8.1.

Ex. 8.1 Show that there is no second-order homogeneous linear differential equation which is entirely free from singularities.

Ex. 8.2 Show that any homogeneous linear differential equation of the second order having no irregular singularities and one or two regular singularities can be solved in closed form in terms of elementary functions.

Ex. 8.3 If $\beta_1+\beta_2+\gamma_1+\gamma_2 = \frac{1}{2}$, prove that

$$P\begin{Bmatrix} 0 & \infty & 1 & \\ 0 & \beta_1 & \gamma_1 & z^2 \\ \frac{1}{2} & \beta_2 & \gamma_2 & \end{Bmatrix} = P\begin{Bmatrix} -1 & \infty & 1 & \\ \gamma_1 & 2\beta_1 & \gamma_1 & z \\ \gamma_2 & 2\beta_2 & \gamma_2 & \end{Bmatrix}. \qquad \text{[Riemann, 1857.]}$$

Ex. 8.4 Show that the most general homogeneous linear differential equation of the second order having regular singularities at the distinct points $\xi_1, \xi_2, \ldots, \xi_n$, and no other singularities, is given by

$$\frac{d^2w}{dz^2}+\left\{\sum\frac{1-\alpha_s-\beta_s}{z-\xi_s}\right\}\frac{dw}{dz}+\left\{\sum\frac{\alpha_s\beta_s}{(z-\xi_s)^2}+\sum\frac{\lambda_s}{z-\xi_s}\right\}w = 0,$$

where the constants α_s, β_s, and λ_s satisfy

$$\sum(\alpha_s+\beta_s) = n-2, \qquad \sum\lambda_s = \sum(\lambda_s\xi_s+\alpha_s\beta_s) = \sum(\lambda_s\xi_s^2+2\alpha_s\beta_s\xi_s) = 0,$$

all summations being from $s=1$ to $s=n$. [Klein, 1894.]

9 The Hypergeometric Function

9.1 Series solutions of equation (8.01) valid in the neighborhoods of $z=0$, 1, or ∞ can be constructed by direct application of the methods of §§4–6. In particular, corresponding to the exponent 0 at $z=0$ the solution assuming the value unity at $z=0$ is found to be

$$F(a,b;c;z) = \sum_{s=0}^{\infty} \frac{a(a+1)\cdots(a+s-1)\,b(b+1)\cdots(b+s-1)}{c(c+1)\cdots(c+s-1)}\frac{z^s}{s!}, \tag{9.01}$$

provided that c is not zero or a negative integer. This series evidently converges when $|z|<1$—as we expect—and is known as the *hypergeometric series*. Its sum $F(a,b;c;z)$ is the *hypergeometric function*.

$F(a,b;c;z)$ is the standard notation for the principal solution of the hypergeometric equation, but it is more convenient to develop the theory in terms of the function

$$\mathbf{F}(a,b;c;z) = F(a,b;c;z)/\Gamma(c), \tag{9.02}$$

because this leads to fewer restrictions and simpler formulas. Most of the results obtained will be restated in the F notation. From (9.01) and (9.02), we have

$$\mathbf{F}(a,b;c;z) = \sum_{s=0}^{\infty} \frac{(a)_s(b)_s}{\Gamma(c+s)}\frac{z^s}{s!} \qquad (|z|<1), \tag{9.03}$$

where, for brevity, we have used *Pochhammer's notation* $(a)_0=1$, and

$$(a)_s = a(a+1)(a+2)\cdots(a+s-1) \qquad (s=1,2,\ldots). \tag{9.04}$$

Unlike $F(a,b;c;z)$, the function $\mathbf{F}(a,b;c;z)$ exists and satisfies (8.01) for *all* values of a, b, and c; from (9.03) it is easily verified that when n is a positive integer or zero

$$\begin{aligned}\mathbf{F}(a,b;-n;z) &= (a)_{n+1}(b)_{n+1}z^{n+1}\mathbf{F}(a+n+1,\ b+n+1;\ n+2;\ z)\\ &= (a)_{n+1}(b)_{n+1}z^{n+1}F(a+n+1,\ b+n+1;\ n+2;\ z)/(n+1)!.\end{aligned} \tag{9.05}$$

Consequently at these exceptional values $\mathbf{F}(a,b;c;z)$ corresponds to the exponent $1-c$ and not 0.

Outside the disk $|z|<1$ the function $\mathbf{F}(a,b;c;z)$ is defined by analytic continuation. The theory of §§4–6 shows that *if the z plane is cut along the real axis from* 1 *to* $+\infty$, *then the only possible singularities of* $\mathbf{F}(a,b;c;z)$ *are branch points* (*or poles*) *at* $z=1$ *and* $z=\infty$. The cut restricts $\mathbf{F}(a,b;c;z)$ to its *principal branch*. Other branches are obtained by analytic continuation across the cut; in their case $z=0$ is generally a singularity.

9.2 We may also regard $\mathbf{F}(a,b;c;z)$ as a function of a, b, or c:

Theorem 9.1 *If z is fixed and does not have any of the values 0, 1, or ∞, then each branch of $\mathbf{F}(a,b;c;z)$ is an entire function of each of the parameters a, b, and c.*

For the principal branch with $|z|<1$ this result is verifiable from the definition (9.03): the M-test shows that this series converges uniformly in any bounded region of the complex a,b,c space. The extension to $|z|\geqslant 1$ and other branches is immediately achieved by means of Theorem 3.2; any point within the unit disk, other than the origin, may be taken as z_0 in Condition (iv) of this theorem. The points $z=0$, 1, and ∞ are excluded in the statement of the final result, because $\mathbf{F}(a,b;c;z)$ may not exist there.†

9.3 Many well-known functions are expressible in the notation of the hypergeometric function. For example, the principal branch of $(1-z)^{-a}$ is also the principal branch of $\mathbf{F}(a,1;1;z)$. Other examples are stated in Exercises 9.1, 9.2, and 10.1 below.

The particular case $a=1$ of $(1-z)^{-a}$, given by

$$1+z+z^2+\cdots = \mathbf{F}(1,1;1;z),$$

indicates the origin of the name *hypergeometric*.

9.4 An integral representation for $\mathbf{F}(a,b;c;z)$ can be found by use of the Beta-function integral of Chapter 2, §1.6. Assume that

$$\operatorname{Re} c > \operatorname{Re} b > 0, \qquad |z| < 1. \tag{9.06}$$

Using Pochhammer's symbol (9.04), we have

$$\begin{aligned}\mathbf{F}(a,b;c;z) &= \frac{1}{\Gamma(b)}\sum_{s=0}^{\infty} z^s \frac{(a)_s}{s!}\frac{\Gamma(b+s)}{\Gamma(c+s)} \\ &= \frac{1}{\Gamma(b)\Gamma(c-b)}\sum_{s=0}^{\infty} z^s \frac{(a)_s}{s!}\int_0^1 t^{b+s-1}(1-t)^{c-b-1}\,dt,\end{aligned} \tag{9.07}$$

where t^{b+s-1} and $(1-t)^{c-b-1}$ both assume their principal values.

Because $|z|<1$, the M-test shows that the series

$$\sum_{s=0}^{\infty} \frac{(a)_s}{s!} z^s t^{b+s-1}(1-t)^{c-b-1} \tag{9.08}$$

converges uniformly in any compact t interval within $(0,1)$. Using the conditions (9.06) and appealing to Theorem 8.1 of Chapter 2, we see that the order of summation and integration in (9.07) may be interchanged.‡ This produces the desired

† For the principal branch, $z=0$ need not be excluded since $\mathbf{F}(a,b;c;0)=1/\Gamma(c)$.

‡ A variation on the proof which avoids the need for the dominated convergence theorem is to restrict $\operatorname{Re} b \geqslant 1$ and $\operatorname{Re}(c-b)\geqslant 1$. The series (9.08) then converges uniformly in $[0,1]$ and may therefore be integrated term by term. The extension of the final result to $\operatorname{Re} c > \operatorname{Re} b > 0$ is achieved by analytic continuation with respect to b and c.

result

$$\mathbf{F}(a,b;c;z) = \frac{1}{\Gamma(b)\Gamma(c-b)}\int_0^1 t^{b-1}(1-t)^{c-b-1}(1-zt)^{-a}\,dt. \tag{9.09}$$

Equation (9.09) (which is due to Euler) has been established on the assumption that $|z|<1$. But as a function of z the integral on the right-hand side converges uniformly in any compact domain which excludes all points of the interval $[1,\infty)$. Hence *when* $\operatorname{Re} c > \operatorname{Re} b > 0$ *the integral* (9.09) *furnishes the principal value of* $\mathbf{F}(a,b;c;z)$, *except along the cut* $1 \leqslant z < \infty$. All powers in the integrand are assigned their principal values.

By further analytic continuation we see that the lower side of the cut can be included in the region of validity of (9.09) when $\operatorname{Re} a < 1$, but not otherwise.

9.5 What is the sum of the hypergeometric series at the singularity $z=1$? From Chapter 4, §5 we have

$$\frac{\Gamma(a+s)\Gamma(b+s)}{\Gamma(c+s)s!} \sim \frac{1}{s^{c-a-b+1}} \qquad (s\to\infty).$$

Hence the sum $\mathbf{F}(a,b;c;1)$ certainly exists when $\operatorname{Re}(c-a-b)>0$.

Suppose, temporarily, that $\operatorname{Re} c > \operatorname{Re} b > 0$ and $\operatorname{Re} a \leqslant 0$. Letting $z\to 1$ from within the unit circle, we find that the right-hand side of (9.09) tends to

$$\frac{1}{\Gamma(b)\Gamma(c-b)}\int_0^1 t^{b-1}(1-t)^{c-a-b-1}\,dt,$$

that is, $\Gamma(c-a-b)/\{\Gamma(c-a)\Gamma(c-b)\}$. By Abel's theorem on the continuity of power series† this expression equals the sum of the series at $z=1$:

$$\mathbf{F}(a,b;c;1) = \frac{\Gamma(c-a-b)}{\Gamma(c-a)\Gamma(c-b)}. \tag{9.10}$$

Again, analytic continuation with respect to a, c, and b in turn shows that (9.10) *is valid when* $\operatorname{Re}(c-a-b)>0$, *with no other restrictions*. This important formula is due to Gauss, and is more usually quoted in the form

$$F(a,b;c;1) = \frac{\Gamma(c)\Gamma(c-a-b)}{\Gamma(c-a)\Gamma(c-b)}, \tag{9.11}$$

with the added condition $c \neq 0, -1, -2, \ldots$.

Ex. 9.1 Show that when $|z|<1$

$$\ln(1+z) = zF(1,1;2;-z), \qquad \ln\{(1+z)/(1-z)\} = 2zF(\tfrac{1}{2},1;\tfrac{3}{2};z^2),$$

$$\sin^{-1}z = zF(\tfrac{1}{2},\tfrac{1}{2};\tfrac{3}{2};z^2), \qquad \tan^{-1}z = zF(\tfrac{1}{2},1;\tfrac{3}{2};-z^2).$$

Ex. 9.2 Show that when $|k|<1$ the *elliptic integrals*

$$K(k^2) = \int_0^1 \frac{dt}{\{(1-t^2)(1-k^2t^2)\}^{1/2}}, \qquad E(k^2) = \int_0^1 \frac{(1-k^2t^2)^{1/2}}{(1-t^2)^{1/2}}\,dt,$$

can be expressed as $K(k^2) = \frac{1}{2}\pi F(\frac{1}{2},\frac{1}{2};1;k^2)$, $E(k^2) = \frac{1}{2}\pi F(-\frac{1}{2},\frac{1}{2};1;k^2)$.

† Titchmarsh (1939, §7.61).

Ex. 9.3 Show that

$$(\partial/\partial z)^n \mathbf{F}(a,b;c;z) = (a)_n (b)_n \mathbf{F}(a+n,b+n;c+n;z),$$
$$(\partial/\partial z)^n \{z^{a+n-1}\mathbf{F}(a,b;c;z)\} = (a)_n z^{a-1}\mathbf{F}(a+n,b;c;z).$$

Ex. 9.4† Verify that

$$(c-a)\mathbf{F}(a-1,b;c;z) + \{2a-c+(b-a)z\}\mathbf{F}(a,b;c;z) + a(z-1)\mathbf{F}(a+1,b;c;z) = 0,$$
$$(z-1)\mathbf{F}(a,b;c-1;z) + \{c-1-(2c-a-b-1)z\}\mathbf{F}(a,b;c;z) + (c-a)(c-b)z\mathbf{F}(a,b;c+1;z) = 0.$$

Ex. 9.5‡ Assume that z is any point of the complex plane not in the interval $[1,\infty)$, and write

$$I = \int_\alpha^{(1+,0+,1-,0-)} t^{b-1}(1-t)^{c-b-1}(1-zt)^{-a}\,dt.$$

The integration path begins at an arbitrary point α of the interval $(0,1)$, encircles the interval $(\alpha,1]$ once in the positive sense, returns to α, then encircles $[0,\alpha)$ once in the positive sense, returns to α, and so on. The point $1/z$ is exterior to all loops. Assume also that the factors in the integrand are continuous on the path and take their principal values at the starting point. Prove Pochhammer's result that the principal branch of $\mathbf{F}(a,b;c;z)$ is given by

$$\mathbf{F}(a,b;c;z) = -e^{-c\pi i}\Gamma(1-b)\Gamma(1+b-c)I/(4\pi^2),$$

provided that neither b nor $c-b$ is a positive integer.

Can this result be extended to other branches of $\mathbf{F}(a,b;c;z)$?

Ex. 9.6§ Let a, b, and z be fixed, and $z \notin [1,\infty)$. By applying the methods of Chapter 4, §§3 and 5.2 to (9.09) show that

$$\mathbf{F}(a,b;c;z) \sim \sum_{s=0}^{\infty} \frac{(b)_s q_s}{\Gamma(c-b)\, c^{s+b}},$$

as $c\to\infty$ in the sector $|\mathrm{ph}\, c| \leqslant \frac{1}{2}\pi-\delta$ $(<\frac{1}{2}\pi)$, where $q_0 = 1$ and higher coefficients are defined by the expansion

$$e^\tau(e^\tau-1)^{b-1}(1-z+ze^{-\tau})^{-a} = \sum_{s=0}^{\infty} q_s \tau^{s+b-1}.$$

Show also that when $\mathrm{Re}\, z \leqslant \frac{1}{2}$ the region of validity can be increased to $|\mathrm{ph}\, c| \leqslant \pi-\delta$ $(<\pi)$.

Ex. 9.7 Let a, b, c, and z be fixed, and $z \in (-\infty, 1)$. Show that

$$\mathbf{F}(a+\lambda, b+\lambda; c+\lambda; z) \sim \frac{(1-z)^{c-a-b-\lambda}}{\Gamma(b+\lambda)} \sum_{s=0}^{\infty} \frac{(c-b)_s q_s}{\lambda^{c-b+s}},$$

as $\lambda\to\infty$ in the sector $|\mathrm{ph}\,\lambda| \leqslant \frac{1}{2}\pi-\delta$ $(<\frac{1}{2}\pi)$, where $q_0 = 1$ and higher coefficients are defined by the expansion

$$e^{-b\tau}(1-e^{-\tau})^{c-b-1}(1-z+ze^{-\tau})^{a-c} = \sum_{s=0}^{\infty} q_s \tau^{s+c-b-1}.$$

By application of Theorem 6.1 of Chapter 4, show also that this result can be extended to complex z, with the conditions $\mathrm{Re}\, z \leqslant 1$, $z \neq 1$, and $\mathrm{ph}\,\lambda = 0$.

† These identities are two of Gauss's fifteen linear relations which connect $\mathbf{F}(a,b;c;z)$ with two *contiguous* hypergeometric functions, that is, functions obtained from $\mathbf{F}(a,b;c;z)$ by increasing or decreasing one of the parameters by unity.

‡ Compare Chapter 2, Exercise 1.6.

§ In Exercises 9.6 and 9.7 all functions take their principal values. Further results of this kind have been given by Watson (1918c) and Luke (1969a, Chapter VII). There is an error on p. 299 of Watson's paper: $\log(1-x^{-1})$ should be replaced by $-\log(1-x^{-1})$. This affects the regions of validity.

10 Other Solutions of the Hypergeometric Equation

10.1 In §9.4 we derived an integral formula for $\mathbf{F}(a,b;c;z)$ that furnished the analytic continuation of this function in the z plane cut along the interval $[1,\infty)$, subject to certain restrictions on the parameters. In the present section we construct further analytic continuations by expressing $\mathbf{F}(a,b;c;z)$ in terms of other solutions of the hypergeometric equation:

$$z(1-z)(d^2w/dz^2)+\{c-(a+b+1)z\}(dw/dz)-abw = 0. \qquad (10.01)$$

First, we consider the full solution of this equation in the neighborhood of the origin.

The solution $\mathbf{F}(a,b;c;z)$ corresponds to the exponent 0, provided that $c \neq 0,-1,-2,\ldots$. The method of §4 shows that the solution corresponding to the other exponent at $z=0$ is $z^{1-c}F(1+a-c,1+b-c;2-c;z)$, provided that, now, $c \neq 2,3,4,\ldots$. Again, it is sometimes more convenient to adopt as second solution

$$\mathbf{G}(a,b;c;z) = z^{1-c}\mathbf{F}(1+a-c,1+b-c;2-c;z),$$

since this exists for all c.

When c is not an integer or zero, the limiting forms of $\mathbf{F}$, $\mathbf{G}$, and their derivatives as $z\to 0$ are supplied by

$$\mathbf{F}(a,b;c;z) \to \frac{1}{\Gamma(c)}, \qquad \frac{\partial}{\partial z}\mathbf{F}(a,b;c;z) \to \frac{ab}{\Gamma(c+1)},$$

$$\mathbf{G}(a,b;c;z) \sim \frac{z^{1-c}}{\Gamma(2-c)}, \qquad \frac{\partial}{\partial z}\mathbf{G}(a,b;c;z) \sim \frac{z^{-c}}{\Gamma(1-c)}.$$

Accordingly, the Wronskian of $\mathbf{F}(a,b;c;z)$ and $\mathbf{G}(a,b;c;z)$ is given by

$$\mathscr{W}(\mathbf{F},\mathbf{G}) = \frac{\sin(\pi c)}{\pi}z^{-c}(1-z)^{c-a-b-1};$$

compare (1.10). Analytic continuation immediately extends this identity to all values of c. From this result and Theorem 1.2 it is seen that $\mathbf{F}$ and $\mathbf{G}$ are linearly independent, except when c is an integer or zero. In these exceptional cases an independent series solution, involving a logarithm, can be constructed by Frobenius' method (§5.3); see Exercise 10.3 below.

In the terminology of §7, both $\mathbf{F}(a,b;c;z)$ and $\mathbf{G}(a,b;c;z)$ are recessive at $z=0$ when c is an integer or zero. In other cases, $\mathbf{F}(a,b;c;z)$ is recessive and $\mathbf{G}(a,b;c;z)$ is dominant when $\operatorname{Re} c>1$; these roles are reversed when $\operatorname{Re} c<1$; neither solution dominates the other when $\operatorname{Re} c = 1$.

10.2 In Riemann's notation the hypergeometric equation (10.01) becomes

$$w = P\left\{\begin{matrix} 0 & 1 & \infty & \\ 0 & 0 & a & z \\ 1-c & c-a-b & b & \end{matrix}\right\}. \qquad (10.02)$$

The transformation $w=(1-z)^{\rho}W$ decreases the exponents at the singularity 1 by ρ, and increases the exponents at ∞ by the same amount. If we set $\rho=c-a-b$, then the new equation again has a zero exponent at 1:

$$W = P\left\{\begin{matrix} 0 & 1 & \infty & \\ 0 & a+b-c & c-b & z \\ 1-c & 0 & c-a & \end{matrix}\right\}.$$

When $\operatorname{Re} c > 1$, the recessive solution of the last equation at the origin is

$$W = \mathbf{F}(c-a, c-b; c; z).$$

Its ratio to the corresponding recessive solution of (10.02) is proportional to $(1-z)^{a+b-c}$, and the proportionality constant is derivable by setting $z=0$. Thus we obtain

$$\mathbf{F}(a,b;c;z) = (1-z)^{c-a-b}\mathbf{F}(c-a,c-b;c;z). \tag{10.03}$$

In this equation principal branches correspond; the only cut needed is the interval $[1,\infty)$. Moreover, by analytic continuation with respect to c the restriction $\operatorname{Re} c > 1$ is removed.

10.3 Now consider the transformations

$$w = (1-z)^{-a}W, \qquad t = z/(z-1).$$

The first of these alters the exponents at 1 and ∞; in particular, it reduces one of the exponents at ∞ to zero. The second transformation interchanges the singularities at 1 and ∞. The new equation is therefore

$$W = P\left\{\begin{matrix} 0 & \infty & 1 & \\ 0 & a & 0 & t \\ 1-c & c-b & b-a & \end{matrix}\right\}. \tag{10.04}$$

Again, the recessive solution of (10.02) at $z=0$ has to be a multiple of the recessive solution of (10.04) at $t=0$. Hence we derive

$$\mathbf{F}(a,b;c;z) = (1-z)^{-a}\mathbf{F}\left(a, c-b; c; \frac{z}{z-1}\right), \tag{10.05}$$

again without restrictions on the parameters.

In a similar way, or by use of (10.03), we have

$$\mathbf{F}(a,b;c;z) = (1-z)^{-b}\mathbf{F}\left(b, c-a; c; \frac{z}{z-1}\right). \tag{10.06}$$

As z ranges from 1 to $+\infty$, $z/(z-1)$ ranges from $+\infty$ to 1, hence in each of the last two equations principal branches correspond. The hypergeometric series for the functions on the right-hand sides converge when $|z/(z-1)|<1$, that is, when

$\operatorname{Re} z < \frac{1}{2}$. Accordingly, these relations supply the analytic continuation of $\mathbf{F}(a,b;c;z)$ into this half-plane.

When $c \neq 0,-1,-2,\ldots$, the symbol $\mathbf{F}$ in (10.03), (10.05), and (10.06) may be replaced throughout by F.

10.4 Next, consider series solutions of the hypergeometric equation in the neighborhood of the singularity $z = 1$. Either by direct use of the method of §4 or, more simply, by applying the transformation $z = 1-t$ we see that these solutions are given by

$$\mathbf{F}(a,b;1+a+b-c;1-z), \tag{10.07}$$

and

$$(1-z)^{c-a-b}\mathbf{F}(c-a,c-b;1+c-a-b;1-z). \tag{10.08}$$

They are independent, except when $a+b-c$ is an integer or zero.

Since the principal branch of $\mathbf{F}(a,b;c;z)$ necessitates a cut along the real axis from $z = 1$ to $z = +\infty$, the principal branches of the $\mathbf{F}$ functions in (10.07) and (10.08) necessitate a cut from $z = 0$ to $z = -\infty$. If we further assume that $(1-z)^{c-a-b}$ has its principal value, then we also need a cut from 1 to $+\infty$.

In the doubly cut plane the three solutions $\mathbf{F}(a,b;c;z)$, (10.07), and (10.08) are connected by a relation of the form

$$\begin{aligned}\mathbf{F}(a,b;c;z) &= A\mathbf{F}(a,b;1+a+b-c;1-z)\\ &\quad + B(1-z)^{c-a-b}\mathbf{F}(c-a,c-b;1+c-a-b;1-z).\end{aligned}$$

To determine the coefficients A and B, assume temporarily that

$$\operatorname{Re}(a+b) < \operatorname{Re} c < 1, \tag{10.09}$$

so that each of the series

$$\mathbf{F}(a,b;c;1), \qquad \mathbf{F}(a,b;1+a+b-c;1), \qquad \mathbf{F}(c-a,c-b;1+c-a-b;1),$$

converges; compare §9.5.

Letting $z \to 1-$, and using (9.10) and Abel's theorem on the continuity of power series, we derive

$$A = \Gamma(1+a+b-c)\,\mathbf{F}(a,b;c;1) = \frac{\pi}{\sin\{\pi(c-a-b)\}\,\Gamma(c-a)\,\Gamma(c-b)}. \tag{10.10}$$

Similarly, on letting $z \to 0+$, we obtain

$$1/\Gamma(c) = A\mathbf{F}(a,b;1+a+b-c;1) + B\mathbf{F}(c-a,c-b;1+c-a-b;1).$$

Substituting by means of (9.10) and (10.10), and again using the reflection formula for the Gamma function, we arrive at

$$B = -\frac{\pi}{\sin\{\pi(c-a-b)\}\,\Gamma(a)\,\Gamma(b)}.$$

Accordingly, the desired *connection formula* is given by

$$\frac{\sin\{\pi(c-a-b)\}}{\pi}\mathbf{F}(a,b;c;z) = \frac{1}{\Gamma(c-a)\Gamma(c-b)}\mathbf{F}(a,b;1+a+b-c;1-z)$$
$$-\frac{(1-z)^{c-a-b}}{\Gamma(a)\Gamma(b)}\mathbf{F}(c-a,c-b;1+c-a-b;1-z), \tag{10.11}$$

each function having its principal value in the z plane cut along $(-\infty,0]$ and $[1,\infty)$. The conditions (10.09) may now be removed by appealing to analytic continuation.

Except when $a+b-c$ is an integer or zero, equation (10.11) confirms that $\mathbf{F}(a,b;c;z)$ has a branch point at $z=1$. In the F notation (10.11) becomes

$$F(a,b;c;z) = \frac{\Gamma(c)\Gamma(c-a-b)}{\Gamma(c-a)\Gamma(c-b)}F(a,b;1+a+b-c;1-z)$$
$$+\frac{\Gamma(c)\Gamma(a+b-c)}{\Gamma(a)\Gamma(b)}(1-z)^{c-a-b}F(c-a,c-b;1+c-a-b;1-z), \tag{10.12}$$

provided that $a+b-c$ is not an integer or zero, and c is not a negative integer or zero.

10.5 In (10.11) set $z=(t-1)/t$. Then using (10.05), we obtain

$$\frac{\sin\{\pi(c-a-b)\}}{\pi}t^a\mathbf{F}(a,c-b;c;1-t)$$
$$=\frac{1}{\Gamma(c-a)\Gamma(c-b)}\mathbf{F}(a,b;1+a+b-c;t^{-1})$$
$$-\frac{t^{a+b-c}}{\Gamma(a)\Gamma(b)}\mathbf{F}(c-a,c-b;1+c-a-b;t^{-1}).$$

Replacement of b by $1+a-c$, c by $1+a+b-c$, and t by z produces

$$\frac{\sin\{\pi(b-a)\}}{\pi}\mathbf{F}(a,b;1+a+b-c;1-z)$$
$$=\frac{z^{-a}}{\Gamma(b)\Gamma(1+b-c)}\mathbf{F}(a,1+a-c;1+a-b;z^{-1})$$
$$-\frac{z^{-b}}{\Gamma(a)\Gamma(1+a-c)}\mathbf{F}(b,1+b-c;1+b-a;z^{-1}). \tag{10.13}$$

This formula connects a series solution of (10.01) at $z=1$ with series solutions at $z=\infty$. It is valid without restriction on the parameters, and principal branches correspond; in aggregate these branches introduce a cut along $(-\infty,1]$.

10.6 The last formula we seek in this section connects $\mathbf{F}(a,b;c;z)$ with series solutions at $z=\infty$:

$$\mathbf{F}(a,b;c;z) = A(-z)^{-a}\mathbf{F}(a,1+a-c;1+a-b;z^{-1}) + B(-z)^{-b}\mathbf{F}(b,1+b-c;1+b-a;z^{-1}). \tag{10.14}$$

The necessary cut for principal values now extends from 0 to $+\infty$.

To evaluate the constants A and B, replace c and z in (10.13) by $1+a+b-c$ and $1-z$ respectively, and then expand the right-hand side in descending powers of z. The result has the form

$$\frac{\sin\{\pi(b-a)\}}{\pi}\mathbf{F}(a,b;c;z) = \frac{(-z)^{-a}}{\Gamma(b)\Gamma(c-a)\Gamma(1+a-b)}\left(1+\frac{\lambda_1}{z}+\frac{\lambda_2}{z^2}+\cdots\right) - \frac{(-z)^{-b}}{\Gamma(a)\Gamma(c-b)\Gamma(1+b-a)}\left(1+\frac{\mu_1}{z}+\frac{\mu_2}{z^2}+\cdots\right),$$

where the coefficients λ_s and μ_s are independent of z. Comparing this with the expansion of the right-hand side of (10.14) in descending powers of z, we immediately obtain the values of A and B, and thence

$$\frac{\sin\{\pi(b-a)\}}{\pi}\mathbf{F}(a,b;c;z) = \frac{(-z)^{-a}}{\Gamma(b)\Gamma(c-a)}\mathbf{F}(a,1+a-c;1+a-b;z^{-1}) - \frac{(-z)^{-b}}{\Gamma(a)\Gamma(c-b)}\mathbf{F}(b,1+b-c;1+b-a;z^{-1}). \tag{10.15}$$

Again, analytic continuation removes all restrictions from the parameters in the final result.

In the F notation

$$F(a,b;c;z) = \frac{\Gamma(c)\Gamma(b-a)}{\Gamma(b)\Gamma(c-a)}(-z)^{-a}F(a,1+a-c;1+a-b;z^{-1}) + \frac{\Gamma(c)\Gamma(a-b)}{\Gamma(a)\Gamma(c-b)}(-z)^{-b}F(b,1+b-c;1+b-a;z^{-1}), \tag{10.16}$$

provided that $c \neq 0,-1,-2,\ldots$ and $a-b$ is not an integer or zero.

An alternative derivation of this result based upon a contour integral representation of $F(a,b;c;z)$ is given in Chapter 8, §6.3.

Ex. 10.1 Show that the Jacobi polynomials can be expressed in the forms

$$P_n^{(\alpha,\beta)}(x) = \binom{n+\alpha}{n}F(-n,\alpha+\beta+n+1;\alpha+1;\tfrac{1}{2}-\tfrac{1}{2}x)$$
$$= (-)^n\binom{n+\beta}{n}F(-n,\alpha+\beta+n+1;\beta+1;\tfrac{1}{2}+\tfrac{1}{2}x).$$

Ex. 10.2† Show that $\mathbf{F}(a,b;a+b+\tfrac{1}{2};4z-4z^2) = \mathbf{F}(2a,2b;a+b+\tfrac{1}{2};z)$.

† This is an example of several possible *quadratic transformations* of the hypergeometric function.

Ex. 10.3† Let m be any positive integer. By using the method of §5.3 and considering the limiting value of

$$\frac{1}{c-1+m}\left\{\frac{\mathbf{F}(a,b;c;z)}{\Gamma(1-a)\Gamma(1-b)}-\frac{\mathbf{G}(a,b;c;z)}{\Gamma(c-a)\Gamma(c-b)}\right\}$$

as $c \to 1-m$, prove that a second solution of the hypergeometric equation in the case $c = 1-m$ is given by

$$z^m\left\{\sum_{s=1}^{m}(-)^{s-1}\lambda_{m,-s}\frac{(s-1)!}{z^s}+\lambda_{m,0}F(a+m,b+m;1+m;z)\ln z+\sum_{s=0}^{\infty}\lambda_{m,s}\mu_{m,s}\frac{z^s}{s!}\right\},$$

where

$$\lambda_{m,s} = 1/\{\Gamma(1-a-m-s)\Gamma(1-b-m-s)(m+s)!\},$$
$$\mu_{m,s} = \psi(1-a-m-s)+\psi(1-b-m-s)-\psi(1+m+s)-\psi(1+s).$$

11 Generalized Hypergeometric Functions

11.1 In terms of the operator

$$\vartheta = z\,d/dz$$

the hypergeometric equation (10.01) becomes

$$\vartheta(\vartheta+c-1)\,w = z(\vartheta+a)(\vartheta+b)\,w. \tag{11.01}$$

The *generalized hypergeometric equation* is defined by

$$\vartheta(\vartheta+c_1-1)(\vartheta+c_2-1)\cdots(\vartheta+c_q-1)\,w = z(\vartheta+a_1)(\vartheta+a_2)\cdots(\vartheta+a_p)\,w, \tag{11.02}$$

where the c_s and a_s are constants. This is a linear differential equation of order $\max(p,q+1)$. Employing Pochhammer's notation (§9.1), we easily find that the solution of exponent zero at the origin is

$${}_pF_q(a_1,a_2,\ldots,a_p;c_1,c_2,\ldots,c_q;z) \equiv \sum_{s=0}^{\infty}\frac{(a_1)_s(a_2)_s\cdots(a_p)_s}{(c_1)_s(c_2)_s\cdots(c_q)_s}\frac{z^s}{s!}, \tag{11.03}$$

provided that none of the c_s is a negative integer or zero and the series converges. For brevity, this function is denoted by ${}_pF_q(z)$.

When $p \leqslant q$, the series (11.03) converges for all z and ${}_pF_q(z)$ is entire. In Chapter 7 we consider the case $p = q = 1$ in detail.

When $p = q+1$ the radius of convergence of (11.03) is unity. Outside the unit disk ${}_pF_q(z)$ has to be defined by analytic continuation. In the present notation the function $F(a,b;c;z)$ discussed in preceding sections becomes ${}_2F_1(a,b;c;z)$.

Lastly, when $p > q+1$ the generalized hypergeometric series (11.03) diverges for nonzero z, unless one of the parameters $a_1, a_2, \ldots, a_p$ happens to be zero or a negative integer. Except in these cases the series fails to define a solution of the differential equation.‡

† This result is simpler than one often quoted (N.B.S., 1964, eq. 15.5.21).
‡ The origin is an irregular singularity.

Ex. 11.1 If w satisfies the differential equation $w'' + fw' + gw = 0$, show that the product of any two solutions satisfies

$$W''' + 3fW'' + (2f^2 + f' + 4g)W' + (4fg + 2g')W = 0.$$

Thence verify the identity

$$\{F(a, b; a+b+\tfrac{1}{2}; z)\}^2 = {}_3F_2(2a, a+b, 2b; a+b+\tfrac{1}{2}, 2a+2b; z),$$

provided that $2a+2b$ is not zero or a negative integer. [Clausen, 1828.]

12 The Associated Legendre Equation

12.1 In Chapter 2, §7.3, it was shown that the Legendre polynomial $P_n(z)$ is a solution of *Legendre's equation*

$$(1-z^2)\frac{d^2w}{dz^2} - 2z\frac{dw}{dz} + n(n+1)w = 0. \tag{12.01}$$

This is a special case of the *associated Legendre equation*

$$(1-z^2)\frac{d^2w}{dz^2} - 2z\frac{dw}{dz} + \left\{\nu(\nu+1) - \frac{\mu^2}{1-z^2}\right\}w = 0, \tag{12.02}$$

which is of importance in various branches of applied mathematics, particularly the solution of Laplace's equation in spherical polar or spheroidal coordinates.

In most applications the parameters ν and μ are integers, but in much of the analysis we allow them to range over the whole complex plane. In this way the powerful tool of analytic continuation can be used to establish fundamental formulas in a simple manner.

We first observe that the differential equation (12.02) is unchanged on replacing μ by $-\mu$, ν by $-\nu-1$, or z by $-z$. Therefore from the standpoint of representing the general solution in a satisfactory way, it suffices to construct a numerically satisfactory set of solutions (§7) for the half-plane $\operatorname{Re} z \geqslant 0$ when $\operatorname{Re}\mu \geqslant 0$ and $\operatorname{Re}\nu \geqslant -\frac{1}{2}$. Although it would be inconvenient to restrict the variable and parameters in exactly this way, our primary objective will be to cover these regions satisfactorily.

12.2 The singularities of equation (12.02) are located at $z = 1, -1$, and ∞, and each is easily seen to be regular. In Riemann's notation, (12.02) becomes

$$w = P\left\{\begin{matrix} 1 & \infty & -1 & \\ \frac{1}{2}\mu & \nu+1 & \frac{1}{2}\mu & z \\ -\frac{1}{2}\mu & -\nu & -\frac{1}{2}\mu & \end{matrix}\right\}. \tag{12.03}$$

From §12.1 it follows that important solutions of this equation are (i) the solution which is recessive at $z = 1$ when $\operatorname{Re}\mu > 0$ or $\mu = 0$; (ii) the solution which is recessive at $z = \infty$ when $\operatorname{Re}\nu > -\frac{1}{2}$ or $\nu = -\frac{1}{2}$. These solutions are denoted respectively by $P_\nu^{-\mu}(z)$ and $\mathbf{Q}_\nu^\mu(z)$, subject to the choice of suitable normalizing factors, as follows.

The transformation of (12.03) into hypergeometric form is expressed by

$$w = \frac{(z-1)^{\mu/2}}{(z+1)^{\mu/2}} P \begin{Bmatrix} 0 & \infty & 1 & \\ 0 & \nu+1 & \mu & \dfrac{1-z}{2} \\ -\mu & -\nu & 0 & \end{Bmatrix}.$$

$P_\nu^{-\mu}(z)$ is defined to be the solution

$$P_\nu^{-\mu}(z) = \frac{(z-1)^{\mu/2}}{(z+1)^{\mu/2}} \mathbf{F}(\nu+1,-\nu;\mu+1;\tfrac{1}{2}-\tfrac{1}{2}z). \tag{12.04}$$

The choice of branches is discussed below. From (10.03) this definition is seen to be equivalent to

$$P_\nu^{-\mu}(z) = 2^{-\mu}(z-1)^{\mu/2}(z+1)^{\mu/2}\mathbf{F}(\mu-\nu,\nu+\mu+1;\mu+1;\tfrac{1}{2}-\tfrac{1}{2}z). \tag{12.05}$$

Next, from the solution

$$(-z)^{-a}\mathbf{F}(a,1+a-c;1+a-b;z^{-1})$$

of the hypergeometric equation (§10.6), we frame the definition

$$\mathbf{Q}_\nu^\mu(z) = 2^\nu\Gamma(\nu+1)\frac{(z-1)^{(\mu/2)-\nu-1}}{(z+1)^{\mu/2}}\mathbf{F}\left(\nu+1,\nu-\mu+1;2\nu+2;\frac{2}{1-z}\right); \tag{12.06}$$

equivalently,

$$\mathbf{Q}_\nu^\mu(z) = 2^\nu\Gamma(\nu+1)\frac{(z+1)^{\mu/2}}{(z-1)^{(\mu/2)+\nu+1}}\mathbf{F}\left(\nu+1,\nu+\mu+1;2\nu+2;\frac{2}{1-z}\right). \tag{12.07}$$

The factors 2^ν and $\Gamma(\nu+1)$ are introduced as a matter of convenience; without the latter the function $\mathbf{Q}_\nu^\mu(z)$ would have the undesirable property of vanishing identically when ν is a negative integer; compare (9.05). As a consequence of Theorem 3.2, the right-hand side of (12.06) or (12.07) tends to a finite limit as ν tends to a negative integer, and the limiting value satisfies (12.02).†

Both $P_\nu^{-\mu}(z)$ and $\mathbf{Q}_\nu^\mu(z)$ exist for all values of ν, μ, and z, except possibly the singular points $z = \pm 1$ and ∞. As functions of z they are many valued with branch points at $z = \pm 1$ and ∞. The principal branches of both solutions are obtained by introducing a cut along the real axis from $z = -\infty$ to $z = 1$, and assigning the principal value to each function appearing in (12.04) to (12.07).

It should be noticed that with the z plane cut in this manner the ratio of the principal values of $(z-1)^{\mu/2}$ and $(z+1)^{\mu/2}$ in (12.04) can be replaced by the principal value of $\{(z-1)/(z+1)\}^{\mu/2}$, since $\mathrm{ph}(z-1)$ and $\mathrm{ph}(z+1)$ have the same sign. On the other hand, if the factors $(z-1)^{\mu/2}(z+1)^{\mu/2}$ in (12.05) are combined into $(z^2-1)^{\mu/2}$, then for the principal value of $P_\nu^{-\mu}(z)$ *the correct choice of branch of* $(z^2-1)^{\mu/2}$ *is positive when* $z > 1$ *and continuous in the* z *plane cut along the interval* $(-\infty, 1]$. The reader will easily verify that in the left half-plane this is not the principal branch of $(z^2-1)^{\mu/2}$.

† In applying Theorem 3.2 the point z_0 of Condition (iv) is taken to be any fixed finite point in the annulus $|z-1| > 2$.

Wherever noninteger powers of z^2-1 occur in the remainder of this section, or in §§13 and 14, it is intended that the branch be chosen in this manner.

For fixed z (again other than ± 1 *or* ∞*) each branch of* $P_\nu^{-\mu}(z)$ *or* $\mathbf{Q}_\nu^\mu(z)$ *is an entire function of each of the parameters* ν *and* μ. This follows from the corresponding property of the **F** function (Theorem 9.1) and, in the case of $\mathbf{Q}_\nu^\mu(z)$, Theorem 3.2.

The motivating properties (i) and (ii) stated at the beginning of this subsection are easily recovered from the definitions (12.04) and (12.06). They are expressed by

$$P_\nu^{-\mu}(z) \sim \frac{(z-1)^{\mu/2}}{2^{\mu/2}\Gamma(\mu+1)} \qquad (z\to 1, \quad \mu \neq -1,-2,-3,\ldots), \tag{12.08}$$

and

$$\mathbf{Q}_\nu^\mu(z) \sim \frac{\pi^{1/2}}{2^{\nu+1}\Gamma(\nu+\frac{3}{2})z^{\nu+1}} \qquad (z\to\infty, \quad \nu\neq -\tfrac{3}{2}, -\tfrac{5}{2}, -\tfrac{7}{2},\ldots), \tag{12.09}$$

principal values of both sides corresponding in each case.

12.3 To ascertain whether $P_\nu^{-\mu}(z)$ and $\mathbf{Q}_\nu^\mu(z)$ comprise a numerically satisfactory pair of solutions of the associated Legendre equation in the right half of the z plane, we need to know the behavior of the former as $z\to\infty$ and the latter as $z\to 1$. As a preliminary step we specialize the connection formulas developed in §10 for the hypergeometric functions.

Because the associated Legendre equation is unchanged on replacing μ by $-\mu$, or ν by $-\nu-1$, each of the eight functions $P_\nu^{\pm\mu}(z)$, $P_{-\nu-1}^{\pm\mu}(z)$, $\mathbf{Q}_\nu^{\pm\mu}(z)$, $\mathbf{Q}_{-\nu-1}^{\pm\mu}(z)$ is a solution. Only four of these solutions are distinct, however, since from (12.04), (12.06), and (12.07) it is immediately verifiable that

$$\begin{aligned} P_{-\nu-1}^{-\mu}(z) &= P_\nu^{-\mu}(z), & P_{-\nu-1}^{\mu}(z) &= P_\nu^{\mu}(z), \\ \mathbf{Q}_\nu^{-\mu}(z) &= \mathbf{Q}_\nu^{\mu}(z), & \mathbf{Q}_{-\nu-1}^{-\mu}(z) &= \mathbf{Q}_{-\nu-1}^{\mu}(z). \end{aligned} \tag{12.10}$$

The first connection formula is obtained from (10.15) by taking $a=\nu+1$, $b=\nu+\mu+1$, $c=2\nu+2$, and replacing z by $2/(1-z)$. This yields

$$\frac{2\sin(\mu\pi)}{\pi}\mathbf{Q}_\nu^\mu(z) = \frac{P_\nu^\mu(z)}{\Gamma(\nu+\mu+1)} - \frac{P_\nu^{-\mu}(z)}{\Gamma(\nu-\mu+1)}. \tag{12.11}$$

Next, in (10.15) we substitute $a=\nu+1$, $b=-\nu$, $c=\mu+1$, and replace z by $(1-z)/2$. Then using (12.10) we arrive at

$$\cos(\nu\pi)P_\nu^{-\mu}(z) = \frac{\mathbf{Q}_{-\nu-1}^\mu(z)}{\Gamma(\nu+\mu+1)} - \frac{\mathbf{Q}_\nu^\mu(z)}{\Gamma(\mu-\nu)}. \tag{12.12}$$

From these two formulas and (12.10) the remaining connection formulas easily follow:

$$\frac{2\sin(\mu\pi)}{\pi}\mathbf{Q}_{-\nu-1}^\mu(z) = \frac{P_\nu^\mu(z)}{\Gamma(\mu-\nu)} - \frac{P_\nu^{-\mu}(z)}{\Gamma(-\nu-\mu)}, \tag{12.13}$$

and

$$\cos(\nu\pi)P_\nu^{\mu}(z) = \frac{\mathbf{Q}_{-\nu-1}^\mu(z)}{\Gamma(\nu-\mu+1)} - \frac{\mathbf{Q}_\nu^\mu(z)}{\Gamma(-\nu-\mu)}. \tag{12.14}$$

12.4 We now establish the main result in this section concerning the associated Legendre equation:

Theorem 12.1 *When* $\operatorname{Re}\nu \geqslant -\frac{1}{2}$, $\operatorname{Re}\mu \geqslant 0$, *and* z *ranges over the right half-plane, the principal values of* $P_\nu^{-\mu}(z)$ *and* $\mathbf{Q}_\nu^\mu(z)$ *comprise a numerically satisfactory pair of solutions, in the sense of* §7.

Differentiation of (12.05) yields

$$\frac{dP_\nu^{-\mu}(z)}{dz} \sim \frac{(z-1)^{(\mu/2)-1}}{2^{(\mu/2)+1}\Gamma(\mu)} \qquad (z \to 1, \quad \mu \neq 0, -1, -2, \ldots).$$

From this result and (12.08) it is seen that

$$\mathscr{W}\{P_\nu^{-\mu}(z), P_\nu^\mu(z)\} \sim -\frac{\sin(\mu\pi)}{\pi(z-1)} \qquad (z \to 1, \quad \mu \text{ nonintegral}).$$

From (1.10) it is known that the Wronskian of any pair of solutions of the associated Legendre equation is of the form $C/(z^2-1)$, where C is independent of z. Hence

$$\mathscr{W}\{P_\nu^{-\mu}(z), P_\nu^\mu(z)\} = -\frac{2\sin(\mu\pi)}{\pi(z^2-1)}, \tag{12.15}$$

analytic continuation removing all restrictions on μ.

Substituting in the last relation for $P_\nu^\mu(z)$ by means of (12.11), we derive

$$\mathscr{W}\{P_\nu^{-\mu}(z), \mathbf{Q}_\nu^\mu(z)\} = -\frac{1}{\Gamma(\nu+\mu+1)(z^2-1)}. \tag{12.16}$$

Hence from Theorem 1.2 $P_\nu^{-\mu}(z)$ and $\mathbf{Q}_\nu^\mu(z)$ are linearly dependent if, and only if, $\nu+\mu$ is a negative integer—a case which is irrelevant in the present theorem.

If $\operatorname{Re}\mu > 0$ or $\mu = 0$, then $P_\nu^{-\mu}(z)$ is recessive at $z = 1$. Hence in these circumstances $\mathbf{Q}_\nu^\mu(z)$ must be dominant. Similarly, if $\operatorname{Re}\nu > -\frac{1}{2}$ or $\nu = -\frac{1}{2}$, then at infinity $\mathbf{Q}_\nu^\mu(z)$ is recessive and $P_\nu^{-\mu}(z)$ is dominant. Two cases remain: (i) $\operatorname{Re}\mu = 0$ and $\operatorname{Im}\mu \neq 0$; (ii) $\operatorname{Re}\nu = -\frac{1}{2}$ and $\operatorname{Im}\nu \neq 0$. In (i) neither recessive nor dominant solutions exist at $z = 1$, and in (ii) neither recessive nor dominant solutions exist at $z = \infty$. Since $P_\nu^{-\mu}(z)$ and $\mathbf{Q}_\nu^\mu(z)$ are linearly independent in these circumstances they again comprise a numerically satisfactory pair (§7.2). This completes the proof.

12.5 The importance of Theorem 12.1 is that for the purpose of representing the general solution of the associated Legendre equation by numerical tables, computational algorithms, or, as we shall develop in Chapter 12, asymptotic expansions for large values of the parameters, we need concentrate only on $P_\nu^{-\mu}(z)$ and $\mathbf{Q}_\nu^\mu(z)$ with

$$\operatorname{Re}\nu \geqslant -\tfrac{1}{2}, \qquad \operatorname{Re}\mu \geqslant 0, \qquad \operatorname{Re}z \geqslant 0. \tag{12.17}$$

For other combinations of the parameters and variable, connection formulas can be relied upon to provide corresponding representations in a satisfactory way.

Perhaps it needs emphasizing that when conditions (12.17) are violated, $P_\nu^{-\mu}(z)$ and $\mathbf{Q}_\nu^\mu(z)$ are no longer a satisfactory pair, as a rule, regardless of whether or not they are linearly independent. For example, if $\operatorname{Re}\mu < 0$ and neither μ nor $\nu-\mu$ is a negative integer, then both $P_\nu^{-\mu}(z)$ and $\mathbf{Q}_\nu^\mu(z)$ are dominant at $z = 1$. This is because the recessive solution in these circumstances is $P_\nu^\mu(z)$, and from (12.15) and (12.16) (with μ replaced by $-\mu$) it is seen that both $P_\nu^{-\mu}(z)$ and $\mathbf{Q}_\nu^\mu(z)$ are linearly independent of $P_\nu^\mu(z)$.

12.6 It is of interest to determine the actual limiting forms of $P_\nu^{-\mu}(z)$ and $\mathbf{Q}_\nu^\mu(z)$ as $z \to \infty$ and $z \to 1$, respectively.

From (12.09) and (12.12) we deduce that

$$P_\nu^{-\mu}(z) \sim \frac{\Gamma(\nu+\frac{1}{2})}{\pi^{1/2}\Gamma(\nu+\mu+1)}(2z)^\nu \qquad (z \to \infty), \tag{12.18}$$

provided that $\operatorname{Re}\nu > -\frac{1}{2}$, $\nu+\mu$ is not a negative integer, and $\nu+\frac{1}{2}$ is not a positive integer. The last of these restrictions may be removed by appeal to Cauchy's formula

$$\phi(n-\tfrac{1}{2}, z) = \frac{1}{2\pi i}\int_{\mathscr{C}} \frac{\phi(\nu, z)}{\nu-n+\frac{1}{2}}\, d\nu,$$

in which n is a positive integer,

$$\phi(\nu, z) \equiv \frac{\pi^{1/2}\Gamma(\nu+\mu+1)}{\Gamma(\nu+\frac{1}{2})(2z)^\nu} P_\nu^{-\mu}(z),$$

and $\mathscr{C}$ is the circle $|\nu-n+\frac{1}{2}| = \delta$, δ being arbitrary. By hypothesis, $n+\frac{1}{2}+\mu$ is not a negative integer or zero; hence $\mathscr{C}$ contains no singularity of $\Gamma(\nu+\mu+1)$ when δ is sufficiently small. From (12.18) it follows that on $\mathscr{C}$, $\phi(\nu, z) \to 1$ as $z \to \infty$; moreover, it is easily seen that the approach to this limit is uniform with respect to ν. Therefore $\phi(n-\frac{1}{2}, z) \to 1$ when $z \to \infty$, as asserted.

Next, in the case $\nu = -\frac{1}{2}$ we find from (12.12) by expanding in powers of $\nu+\frac{1}{2}$

$$P_{-1/2}^{-\mu}(z) = -\frac{2}{\pi\Gamma(\mu+\frac{1}{2})}\left\{\left[\frac{\partial \mathbf{Q}_\nu^\mu(z)}{\partial\nu}\right]_{\nu=-1/2} + \psi(\mu+\tfrac{1}{2})\,\mathbf{Q}_{-1/2}^\mu(z)\right\}, \tag{12.19}$$

the limiting form of the right-hand side being taken when $\mu-\frac{1}{2}$ is a negative integer. The right-hand side of (12.06) can be expanded as a convergent series of powers of $2/(1-z)$. Differentiating the dominant terms with respect to ν, setting $\nu = -\frac{1}{2}$, and substituting the result in (12.19), we arrive at

$$P_{-1/2}^{-\mu}(z) \sim \frac{1}{\Gamma(\mu+\frac{1}{2})}\left(\frac{2}{\pi z}\right)^{1/2} \ln z \qquad (z \to \infty, \quad \mu \neq -\tfrac{1}{2}, -\tfrac{3}{2}, -\tfrac{5}{2}, \ldots). \tag{12.20}$$

In a similar way it may be verified that

$$\mathbf{Q}_\nu^\mu(z) \sim \frac{2^{(\mu/2)-1}\Gamma(\mu)}{\Gamma(\nu+\mu+1)}\frac{1}{(z-1)^{\mu/2}} \qquad (z \to 1, \quad \operatorname{Re}\mu > 0, \quad \nu+\mu \neq -1, -2, -3, \ldots), \tag{12.21}$$

$$\mathbf{Q}_\nu^0(z) = \frac{1}{\Gamma(\nu+1)}\left\{\left[\frac{\partial P_\nu^\mu(z)}{\partial\mu}\right]_{\mu=0} - \psi(\nu+1)\,P_\nu^0(z)\right\}, \tag{12.22}$$

and

$$\mathbf{Q}_\nu^0(z) \sim -\frac{\ln(z-1)}{2\Gamma(\nu+1)} \qquad (z \to 1, \quad \nu \neq -1, -2, -3, \ldots). \tag{12.23}$$

Ex. 12.1 Prove that

$$\mathbf{Q}_\nu^\mu(z) = \pi^{1/2} 2^{-\nu-1} z^{-\nu-\mu-1} (z^2-1)^{\mu/2} \mathbf{F}(\tfrac{1}{2}\nu+\tfrac{1}{2}\mu+1, \tfrac{1}{2}\nu+\tfrac{1}{2}\mu+\tfrac{1}{2}; \nu+\tfrac{3}{2}; z^{-2}).$$

Ex. 12.2 Prove *Whipple's formula*

$$\mathbf{Q}_\nu^\mu(z) = (\tfrac{1}{2}\pi)^{1/2} (z^2-1)^{-1/4} P_{-\mu-(1/2)}^{-\nu-(1/2)}\{z(z^2-1)^{-1/2}\}.$$

Ex. 12.3 Verify that

$$P_\nu^{-1/2}(\cosh\zeta) = \left(\frac{2}{\pi \sinh\zeta}\right)^{1/2} \frac{\sinh\{(\nu+\frac{1}{2})\zeta\}}{\nu+\frac{1}{2}},$$

$$\mathbf{Q}_\nu^{1/2}(\cosh\zeta) = \left(\frac{\pi}{2\sinh\zeta}\right)^{1/2} \frac{\exp\{-(\nu+\frac{1}{2})\zeta\}}{\Gamma(\nu+\frac{3}{2})}, \qquad P_\nu^{1/2}(\cosh\zeta) = \left(\frac{2}{\pi\sinh\zeta}\right)^{1/2} \cosh\{(\nu+\tfrac{1}{2})\zeta\}.$$

13 Legendre Functions of General Degree and Order

13.1 When $\nu = n$, a positive integer, and $\mu = 0$, equation (12.04) becomes

$$P_n^0(z) = F(n+1, -n; 1; \tfrac{1}{2}-\tfrac{1}{2}z).$$

This is a polynomial of degree n in z, which takes the value 1 at $z = 1$ and has $(n+1)_n/(2^n n!)$ as coefficient of z^n. Since the associated Legendre equation (12.02) reduces to Legendre's equation (12.01) in these circumstances, and recessive solutions are unique apart from a normalizing factor, it follows that

$$P_n^0(z) = P_n(z),$$

where $P_n(z)$ is the Legendre polynomial defined in Chapter 2, §7 (compare especially (7.14)). Because of this identity, ν is sometimes referred to as the *degree* of $P_\nu^\mu(z)$; μ is called the *order*.

Many of the properties of $P_n(z)$ given in Chapter 2 are capable of extension to the functions $P_\nu^\mu(z)$ and $\mathbf{Q}_\nu^\mu(z)$. We begin with generalizations of Schläfli's integral.

13.2 Theorem 13.1 *When z does not lie in the interval $(-\infty, -1]$ the principal value of $P_\nu^{-\mu}(z)$ is given by*

$$P_\nu^{-\mu}(z) = \frac{e^{\mu\pi i}\Gamma(-\nu)}{2^{\nu+1}\pi i \Gamma(\mu-\nu)} (z^2-1)^{\mu/2} \int_\infty^{(1+,z+)} \frac{(t^2-1)^\nu}{(t-z)^{\nu+\mu+1}} \, dt \qquad (\operatorname{Re}\mu > \operatorname{Re}\nu), \tag{13.01}$$

$$P_\nu^{-\mu}(z) = \frac{2^\nu e^{\mu\pi i}\Gamma(\nu+1)}{\pi i \Gamma(\nu+\mu+1)} (z^2-1)^{\mu/2} \int_\infty^{(1+,z+)} \frac{(t-z)^{\nu-\mu}}{(t^2-1)^{\nu+1}} \, dt \qquad (\operatorname{Re}\nu + \operatorname{Re}\mu > -1). \tag{13.02}$$

The path for both integrals is a single closed loop which begins at infinity on the positive real axis, encircles the points $t = 1$ and $t = z$ once in the positive sense, and

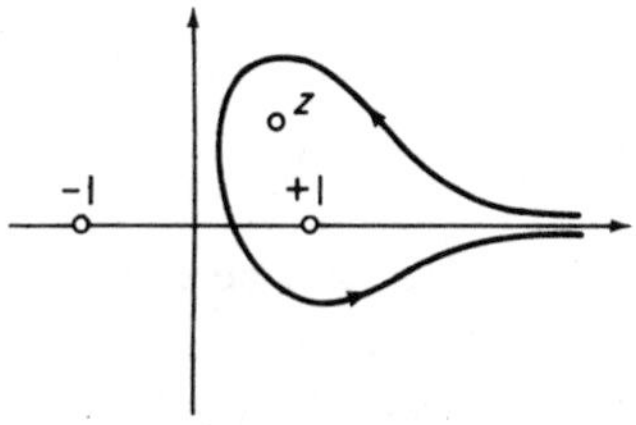

Fig. 13.1 t plane. Path for $P_\nu^{-\mu}(z)$.

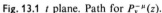

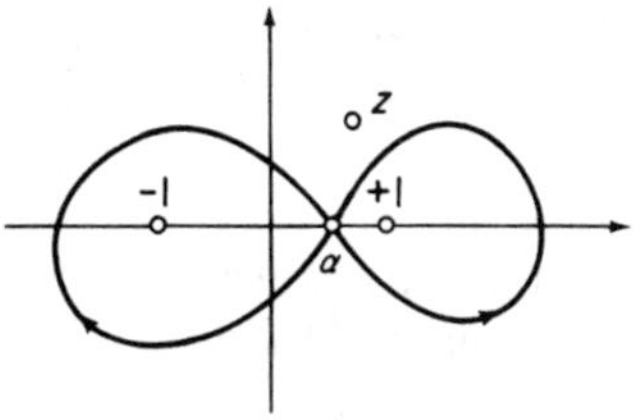

Fig. 13.2 t plane. Path for $\mathbf{Q}_\nu^\mu(z)$.

returns to its starting point without intersecting itself or the interval $(-\infty, -1]$. *The branches of the numerators and denominators of the integrands are continuous on the path and take their principal values in the neighborhood of the starting point. The branch of* $(z^2-1)^{\mu/2}$ *is determined as in* §12.2.†

The integration path is depicted in Fig. 13.1.

We first observe that it suffices to prove either (13.01) or (13.02); the other representation follows immediately from the identity $P_{-\nu-1}^{-\mu}(z) = P_\nu^{-\mu}(z)$.

The differential equation satisfied by $w = (z^2-1)^{-\mu/2} P_\nu^{-\mu}(z)$ is found to be

$$(z^2-1)\frac{d^2w}{dz^2} + 2(\mu+1)z\frac{dw}{dz} - (\nu-\mu)(\nu+\mu+1)w = 0. \tag{13.03}$$

Let us substitute for w by means of a contour integral of the form

$$I(z) = \int_{\mathscr{P}} \frac{(t^2-1)^\nu}{(t-z)^{\nu+\mu+1}}\,dt. \tag{13.04}$$

We have

$$(z^2-1)I''(z) + 2(\mu+1)zI'(z) - (\nu-\mu)(\nu+\mu+1)I(z) = (\nu+\mu+1)J(z),$$

where

$$J(z) = \int_{\mathscr{P}} \frac{(t^2-1)^\nu}{(t-z)^{\nu+\mu+3}} \{(z^2-1)(\nu+\mu+2) + 2(\mu+1)z(t-z) - (\nu-\mu)(t-z)^2\}\,dt$$

$$= \int_{\mathscr{P}} \frac{(t^2-1)^\nu}{(t-z)^{\nu+\mu+3}} \{(\nu+\mu+2)(t^2-1) - 2(\nu+1)t(t-z)\}\,dt$$

$$= \left[-\frac{(t^2-1)^{\nu+1}}{(t-z)^{\nu+\mu+2}}\right]_{\mathscr{P}}.$$

Thus $I(z)$ satisfies (13.03) when the content of the square brackets has the same value at the two ends of $\mathscr{P}$. This condition is fulfilled by the loop integral on the right of (13.01), since the integral converges at the extremities of the path when $\operatorname{Re}\mu > \operatorname{Re}\nu$, and the content of the square brackets vanishes there. Accordingly, the right-hand side of (13.01) is a solution of the associated Legendre equation.

† When ν is a nonnegative integer the right-hand side of (13.01) is to be replaced by its limiting value; see Exercise 13.4 below. Similarly for (13.02) when ν is a negative integer.

Next, the asymptotic form of (13.01) as $z \to 1$ is $A(z-1)^{\mu/2}$, where

$$A = \frac{2^{\mu/2} e^{\mu\pi i} \Gamma(-\nu)}{2^{\nu+1} \pi i \Gamma(\mu-\nu)} \int_{\infty}^{(1+)} \frac{(t+1)^{\nu}}{(t-1)^{\mu+1}} dt.$$

With the temporary added condition $\operatorname{Re}\mu < 0$, this integral can be evaluated by collapsing the path onto the two sides of the interval $[1, \infty)$; thus

$$\int_{\infty}^{(1+)} \frac{(t+1)^{\nu}}{(t-1)^{\mu+1}} dt = (e^{-2\mu\pi i} - 1) \int_{1}^{\infty} \frac{(t+1)^{\nu}}{(t-1)^{\mu+1}} dt = \frac{2^{\nu-\mu+1} \pi i \Gamma(\mu-\nu)}{e^{\mu\pi i} \Gamma(\mu+1) \Gamma(-\nu)},$$

the last step being completed by means of the substitution $t = (2-\tau)/\tau$, followed by use of the Beta-function integral and the reflection formula for the Gamma function. Hence

$$A = 1/\{2^{\mu/2} \Gamma(\mu+1)\}.$$

The condition $\operatorname{Re}\mu < 0$ may now be removed by analytic continuation with respect to μ and ν, provided that we still have $\operatorname{Re}\mu > \operatorname{Re}\nu$.

Now assume that $\operatorname{Re}\mu > 0$. Then the right-hand side of (13.01) is recessive at $z = 1$. It also has the same normalizing factor as $P_{\nu}^{-\mu}(z)$; compare (12.08). Accordingly, the two solutions are identical. Thus (13.01) is proved when $\operatorname{Re}\mu$ exceeds $\max(\operatorname{Re}\nu, 0)$, and hence—again by analytic continuation with respect to μ—when $\operatorname{Re}\mu > \operatorname{Re}\nu$. This establishes the theorem.

13.3 Recurrence relations with respect to ν or μ or both of these parameters can be found with the aid of Theorem 13.1.

Write

$$A_{\nu,\mu} = \frac{e^{-\mu\pi i} \Gamma(-\nu)}{2^{\nu+1} \pi i \Gamma(-\nu-\mu)}, \tag{13.05}$$

and

$$\hat{P}_{\nu}^{\mu}(z) = (z^2-1)^{\mu/2} P_{\nu}^{\mu}(z), \tag{13.06}$$

so that from (13.01), with μ replaced by $-\mu$, we have

$$\hat{P}_{\nu}^{\mu}(z) = A_{\nu,\mu} \int_{\mathscr{P}} \frac{(t^2-1)^{\nu}}{(t-z)^{\nu-\mu+1}} dt \qquad (\operatorname{Re}\nu + \operatorname{Re}\mu < 0), \tag{13.07}$$

where $\mathscr{P}$ now denotes the path used in (13.01). Then

$$\frac{d}{dz} \hat{P}_{\nu}^{\mu}(z) = (\nu-\mu+1) A_{\nu,\mu} \int_{\mathscr{P}} \frac{(t^2-1)^{\nu}}{(t-z)^{\nu-\mu+2}} dt = (\nu-\mu+1)(\nu+\mu) \hat{P}_{\nu}^{\mu-1}(z). \tag{13.08}$$

Two applications of this formula produce

$$\frac{d^2}{dz^2} \hat{P}_{\nu}^{\mu}(z) = (\nu-\mu+1)(\nu+\mu)(\nu-\mu+2)(\nu+\mu-1) \hat{P}_{\nu}^{\mu-2}(z). \tag{13.09}$$

The differential equation satisfied by $\hat{P}_{\nu}^{\mu}(z)$ is found from (13.03) by changing the sign of μ. Substituting therein by means of (13.08) and (13.09), we obtain

$$(z^2-1)(\nu-\mu+2)(\nu+\mu-1) \hat{P}_{\nu}^{\mu-2}(z) - 2(\mu-1) z \hat{P}_{\nu}^{\mu-1}(z) - \hat{P}_{\nu}^{\mu}(z) = 0.$$

Then changing μ into $\mu+2$ and using (13.06) we arrive at the first of the desired relations:

$$P_\nu^{\mu+2}(z)+2(\mu+1)z(z^2-1)^{-1/2}P_\nu^{\mu+1}(z)-(\nu-\mu)(\nu+\mu+1)P_\nu^\mu(z)=0. \tag{13.10}$$

Since $P_\nu^\mu(z)$ is entire in ν and μ, all restrictions on the parameters assumed in the proof are removable by analytic continuation; this is also true of the other recurrence relations derived below.

For the next formula we employ partial integration:

$$\hat{P}_{\nu+1}^\mu(z)=A_{\nu+1,\mu}\int_{\mathscr{P}}\frac{(t^2-1)^{\nu+1}}{(t-z)^{\nu-\mu+2}}\,dt$$

$$=\frac{2(\nu+1)A_{\nu+1,\mu}}{\nu-\mu+1}\int_{\mathscr{P}}\frac{t(t^2-1)^\nu}{(t-z)^{\nu-\mu+1}}\,dt=\frac{2(\nu+1)A_{\nu+1,\mu}}{\nu-\mu+1}\left\{\frac{\hat{P}_\nu^{\mu+1}(z)}{A_{\nu,\mu+1}}+z\frac{\hat{P}_\nu^\mu(z)}{A_{\nu,\mu}}\right\};$$

whence

$$(z^2-1)^{1/2}P_\nu^{\mu+1}(z)=(\nu-\mu+1)P_{\nu+1}^\mu(z)-(\nu+\mu+1)zP_\nu^\mu(z). \tag{13.11}$$

Other recurrence relations involving functions obtained from $P_\nu^\mu(z)$ by increasing or decreasing the parameters ν and μ by unity can be found by combination of (13.10) and (13.11). Each can be regarded as a special case of Gauss's relations between contiguous hypergeometric functions.† For example, to construct the ν-wise recurrence relation, we have from (13.11)

$$(z^2-1)^{1/2}P_\nu^{\mu+2}(z)=(\nu-\mu)P_{\nu+1}^{\mu+1}(z)-(\nu+\mu+2)zP_\nu^{\mu+1}(z).$$

Again,

$$\begin{aligned}(z^2-1)P_\nu^{\mu+2}(z)&=(\nu-\mu)\{(\nu-\mu+2)P_{\nu+2}^\mu(z)-(\nu+\mu+2)zP_{\nu+1}^\mu(z)\}\\&\quad-(\nu+\mu+2)z\{(\nu-\mu+1)P_{\nu+1}^\mu(z)-(\nu+\mu+1)zP_\nu^\mu(z)\}\\&=(\nu-\mu)(\nu-\mu+2)P_{\nu+2}^\mu(z)-(\nu+\mu+2)(2\nu-2\mu+1)zP_{\nu+1}^\mu(z)\\&\quad+(\nu+\mu+1)(\nu+\mu+2)z^2P_\nu^\mu(z).\end{aligned}$$

Substitution in (13.10) by means of this result and (13.11) leads to the desired equation, given by

$$(\nu-\mu+2)P_{\nu+2}^\mu(z)-(2\nu+3)zP_{\nu+1}^\mu(z)+(\nu+\mu+1)P_\nu^\mu(z)=0. \tag{13.12}$$

13.4 A contour integral for $\mathbf{Q}_\nu^\mu(z)$ similar to (13.01) and (13.02) can be constructed by selecting a different integration path.

Theorem 13.2 *When z does not lie on the cut $(-\infty,1]$ the principal value of $\mathbf{Q}_\nu^\mu(z)$ is given by*

$$\mathbf{Q}_\nu^\mu(z)=\frac{e^{-\nu\pi i}\Gamma(-\nu)}{2^{\nu+2}\pi i}(z^2-1)^{\mu/2}\int_a^{(1+,-1-)}\frac{(1-t^2)^\nu}{(z-t)^{\nu+\mu+1}}\,dt. \tag{13.13}$$

† Exercise 9.4.

The integration path begins at an arbitrary point a of the interval $(-1,1)$, encircles the interval $(a,1]$ once in the positive sense, returns to a, then encircles $[-1,a)$ once in the negative sense, again returning to a. The point z is exterior to both loops. The branches of the numerator and denominator of the integrand are continuous on the path and take their principal values at the starting point. The branch of $(z^2-1)^{\mu/2}$ is determined as in §12.2.†

The integration path is the "figure of eight" depicted in Fig. 13.2.

The proof parallels that of Theorem 13.1. With the chosen path, the branch of $(z-t)^{\nu+\mu+1}$ has the same value at the beginning and end. The phase of the numerator $(1-t^2)^\nu$ increases by $2\nu\pi$ on encircling $t=1$, and decreases by the same amount on encircling $t=-1$ in the opposite sense. Thus $(1-t^2)^\nu$ assumes the same value at the extremities of the path. In consequence, the right-hand side of (13.13) satisfies the associated Legendre equation.

For large z the integration path can be fixed. Then $(z-t)^{\nu+\mu+1}$ is asymptotic to the principal value of $z^{\nu+\mu+1}$ as $z\to\infty$ in the sector $|\mathrm{ph}\, z|\leqslant\pi-\delta$ $(<\pi)$, uniformly with respect to t on the path. Therefore the right member of (13.13) is asymptotic to $Bz^{-\nu-1}$, where

$$B = \frac{e^{-\nu\pi i}\Gamma(-\nu)}{2^{\nu+2}\pi i}\int_a^{(1+,-1-)}(1-t^2)^\nu\, dt.$$

When $\mathrm{Re}\,\nu>-1$ we can evaluate B by collapsing the path onto the interval $[-1,1]$ in the usual manner; thus

$$B = \pi^{1/2}/\{2^{\nu+1}\Gamma(\nu+\tfrac{3}{2})\}.$$

Comparison with (12.09) establishes (13.13) in the recessive circumstances $\mathrm{Re}\,\nu>-\frac{1}{2}$. The proof of the theorem is completed by analytic continuation.

13.5 Although $\mathbf{Q}_\nu^\mu(z)$ is the most satisfactory companion to $P_\nu^{-\mu}(z)$ in the analytic theory of the associated Legendre equation, it is not the second solution used in most applications. This is defined by

$$Q_\nu^\mu(z) = e^{\mu\pi i}\Gamma(\nu+\mu+1)\,\mathbf{Q}_\nu^\mu(z), \tag{13.14}$$

provided that $\nu+\mu$ is not a negative integer. When this condition is violated $Q_\nu^\mu(z)$ does not exist, as a rule. From the identity $\mathbf{Q}_\nu^{-\mu}(z)=\mathbf{Q}_\nu^\mu(z)$ we derive

$$Q_\nu^{-\mu}(z) = e^{-2\mu\pi i}\{\Gamma(\nu-\mu+1)/\Gamma(\nu+\mu+1)\}\,Q_\nu^\mu(z). \tag{13.15}$$

And from (12.07) and (13.13) (with μ replaced by $-\mu$) we have

$$Q_\nu^\mu(z) = \frac{\pi^{1/2}e^{\mu\pi i}\Gamma(\nu+\mu+1)}{2^{\nu+1}\Gamma(\nu+\frac{3}{2})}\,\frac{(z+1)^{\mu/2}}{(z-1)^{(\mu/2)+\nu+1}}\,F\left(\nu+1,\nu+\mu+1;2\nu+2;\frac{2}{1-z}\right) \tag{13.16}$$

$$= \frac{e^{(\mu-\nu)\pi i}\Gamma(-\nu)\Gamma(\nu+\mu+1)}{2^{\nu+2}\pi i}(z^2-1)^{-\mu/2}\int_a^{(1+,-1-)}\frac{(1-t^2)^\nu}{(z-t)^{\nu-\mu+1}}\,dt. \tag{13.17}$$

† Again if $\nu=0,1,2,\ldots$ the right-hand side of (13.13) is to be replaced by its limiting value; see Exercise 13.4 below.

The importance of $Q_\nu^\mu(z)$ stems from the fact that it obeys the same recurrence relations as $P_\nu^\mu(z)$. This can be seen as follows. In (13.17), on replacing $1-t^2$ and $z-t$ by $e^{\pm\pi i}(t^2-1)$ and $e^{\pm\pi i}(t-z)$, respectively, we have

$$\hat{Q}_\nu^\mu(z) \equiv (z^2-1)^{\mu/2} Q_\nu^\mu(z) = B_{\nu,\mu} \int_a^{(1+,-1-)} \frac{(t^2-1)^\nu}{(t-z)^{\nu-\mu+1}}\, dt, \tag{13.18}$$

where

$$B_{\nu,\mu} = e^{(\mu-\nu)\pi i}\Gamma(-\nu)\Gamma(\nu+\mu+1)e^{\pm\nu\pi i}e^{\mp(\nu-\mu+1)\pi i}/(2^{\nu+2}\pi i).$$

Whether the ambiguous signs be + or −, it follows that

$$\frac{B_{\nu,\mu}}{B_{\nu+1,\mu}} = \frac{2(\nu+1)}{\nu+\mu+1} = \frac{A_{\nu,\mu}}{A_{\nu+1,\mu}}, \qquad \frac{B_{\nu,\mu}}{B_{\nu,\mu+1}} = \frac{1}{\nu+\mu+1} = \frac{A_{\nu,\mu}}{A_{\nu,\mu+1}},$$

where $A_{\nu,\mu}$ is defined by (13.05). In consequence of these identities, on starting with (13.18) in place of (13.07) and retracing the analysis of §13.3 we are bound to arrive at (13.10), (13.11), and (13.12) with the symbol P replaced throughout by Q.

Ex. 13.1 Let s be an arbitrary positive or negative integer, and $P_\nu^{-\mu}(ze^{s\pi i})$ and $Q_\nu^\mu(ze^{s\pi i})$ denote the branches of the Legendre functions obtained from the principal branches by making $\frac{1}{2}s$ circuits, in the positive sense, of the ellipse having ± 1 as foci and passing through z. Similarly, let $P_{\nu,s}^{-\mu}(z)$ and $Q_{\nu,s}^\mu(z)$ denote the branches obtained from the principal branches by encircling the point 1 (but not the point −1) s times in the positive sense. With the aid of Exercise 12.1 show that

$$Q_\nu^\mu(ze^{s\pi i}) = (-)^s e^{-s\nu\pi i} Q_\nu^\mu(z), \qquad P_{\nu,s}^{-\mu}(z) = e^{s\mu\pi i} P_\nu^{-\mu}(z).$$

Thence from the connection formulas of §12.3 derive

$$P_\nu^{-\mu}(ze^{s\pi i}) = e^{s\nu\pi i} P_\nu^{-\mu}(z) + \frac{2\sin\{(\nu+\frac{1}{2})s\pi\}}{\cos(\nu\pi)} \frac{e^{(1-s)\pi i/2} e^{-\mu\pi i}}{\Gamma(\nu+\mu+1)\Gamma(\mu-\nu)} Q_\nu^\mu(z),$$

$$\frac{Q_{\nu,s}^\mu(z)}{\Gamma(\nu+\mu+1)} = e^{-s\mu\pi i}\frac{Q_\nu^\mu(z)}{\Gamma(\nu+\mu+1)} - \frac{\pi i e^{\mu\pi i}\sin(s\mu\pi)}{\sin(\mu\pi)\Gamma(\nu-\mu+1)} P_\nu^{-\mu}(z).$$

Ex. 13.2 Show that

$$P_{\nu+1}^\mu(z) - (\nu+\mu)(z^2-1)^{1/2} P_\nu^{\mu-1}(z) - zP_\nu^\mu(z) = 0,$$

$$P_{\nu+1}^\mu(z) - (2\nu+1)(z^2-1)^{1/2} P_\nu^{\mu-1}(z) - P_{\nu-1}^\mu(z) = 0,$$

$$(z^2-1)\, dP_\nu^\mu(z)/dz = (\nu-\mu+1)P_{\nu+1}^\mu(z) - (\nu+1)zP_\nu^\mu(z).$$

Ex. 13.3 By deforming the integration path in (13.01), show that unless $z \in (-\infty, -1]$

$$P_\nu^{-\mu}(z) = \frac{(z^2-1)^{\mu/2}}{2^\nu\Gamma(\mu-\nu)\Gamma(\nu+1)} \int_0^\infty \frac{(\sinh\tau)^{2\nu+1}}{(z+\cosh\tau)^{\nu+\mu+1}}\, d\tau \qquad (\operatorname{Re}\mu > \operatorname{Re}\nu > -1).$$

Ex. 13.4 Deduce from (13.01) and (13.13) that when $\nu = n$, a positive integer or zero,

$$P_n^{-\mu}(z) = \frac{(-)^n e^{\mu\pi i}(z^2-1)^{\mu/2}}{2^{n+1} n!\,\pi i\Gamma(\mu-n)} \int_\infty^{(1+,z+)} \frac{(t^2-1)^n}{(t-z)^{n+\mu+1}} \ln\left(\frac{t-z}{t^2-1}\right) dt \qquad (\operatorname{Re}\mu > n),$$

and

$$\mathbf{Q}_n^\mu(z) = \frac{(z^2-1)^{\mu/2}}{2^{n+2} n!\,\pi i} \int_a^{(1+,-1-)} \frac{(1-t^2)^n}{(z-t)^{n+\mu+1}} \ln\left(\frac{z-t}{1-t^2}\right) dt,$$

where the logarithms are continuous on the integration paths and take their principal values in the neighborhoods of the starting points.

14 Legendre Functions of Integer Degree and Order

14.1 When ν and μ are nonnegative integers it is customary to replace them by the symbols n and m, respectively. This case is especially important in physical applications. In defining the branches of $(z^2-1)^{\pm m/2}$ as in §12.2, we observe that the segment of the cut from $-\infty$ to -1 is now unnecessary: *the chosen branches of* $(z^2-1)^{\pm m/2}$ *are positive when* $z>1$ *and continuous in the z plane cut along the interval* $[-1,1]$.

From equations (9.05) and (12.04), and the differentiation formula for the hypergeometric function stated in Exercise 9.3, we perceive that

$$P_n^m(z) = \frac{(z+1)^{m/2}}{(z-1)^{m/2}}\mathbf{F}(n+1,-n;1-m;\tfrac{1}{2}-\tfrac{1}{2}z) = (z^2-1)^{m/2}\frac{d^m}{dz^m}P_n(z), \tag{14.01}$$

since $P_n^0(z) = P_n(z)$. Immediate consequences of this important formula are (i) when $m>n$, $P_n^m(z)=0$; (ii) when $m<n$ and m is even, $P_n^m(z)$ is a polynomial of degree n; (iii) when $m<n$ and m is odd, the only cut needed for the principal branch of $P_n^m(z)$ is the interval $[-1,1]$.

From (14.01) and Rodrigues' formula (Chapter 2, (7.06)) we derive

$$P_n^m(z) = \frac{(z^2-1)^{m/2}}{2^n n!}\frac{d^{n+m}}{dz^{n+m}}(z^2-1)^n, \tag{14.02}$$

and thence by Cauchy's formula

$$P_n^m(z) = \frac{(n+m)!}{2^{n+1}n!}\frac{(z^2-1)^{m/2}}{\pi i}\int_{\mathscr{C}}\frac{(t^2-1)^n}{(t-z)^{n+m+1}}\,dt, \tag{14.03}$$

where $\mathscr{C}$ is a simple closed contour surrounding $t=z$.

Another integral of Schläfli's type can be found from (13.02). When $\nu=n$ and $\mu=m\ (\leqslant n)$ the integrand is single valued and free from singularity at $t=z$. Hence the loop path can be replaced by a simple closed contour $\mathscr{C}'$ which encircles $t=1$ but not $t=-1$:

$$P_n^{-m}(z) = (-)^m\frac{2^n n!}{(n+m)!}\frac{(z^2-1)^{m/2}}{\pi i}\int_{\mathscr{C}'}\frac{(t-z)^{n-m}}{(t^2-1)^{n+1}}\,dt.$$

Then by use of the relation

$$(n-m)!\,P_n^m(z) = (n+m)!\,P_n^{-m}(z) \qquad (n\geqslant m), \tag{14.04}$$

obtained from (12.11), we derive the desired result

$$P_n^m(z) = (-)^m\frac{2^n n!}{(n-m)!}\frac{(z^2-1)^{m/2}}{\pi i}\int_{\mathscr{C}'}\frac{(t-z)^{n-m}}{(t^2-1)^{n+1}}\,dt \qquad (n\geqslant m). \tag{14.05}$$

When $z\neq\pm1$ the contour $\mathscr{C}$ in (14.03) can be taken to be the circle

$$t = z+(z^2-1)^{1/2}e^{i\theta} \qquad (-\pi\leqslant\theta\leqslant\pi). \tag{14.06}$$

Then

$$t^2-1 = 2(z^2-1)^{1/2}e^{i\theta}\{z+(z^2-1)^{1/2}\cos\theta\},$$

and we obtain the representation

$$P_n^m(z) = \frac{(n+m)!}{n!\pi} \int_0^\pi \{z + (z^2-1)^{1/2}\cos\theta\}^n \cos(m\theta)\, d\theta; \qquad (14.07)$$

compare Chapter 2, Exercise 7.9. The restrictions $z \neq \pm 1$ are now removed by continuity.

When $\operatorname{Re} z > 0$, it is easily verified that the circle (14.06) contains $t = 1$ but not $t = -1$. Taking $\mathscr{C}'$ to be this circle we derive from (14.05)

$$P_n^m(z) = \frac{(-)^m n!}{(n-m)!\pi} \int_0^\pi \frac{\cos(m\theta)\, d\theta}{\{z + (z^2-1)^{1/2}\cos\theta\}^{n+1}} \qquad (\operatorname{Re} z > 0). \qquad (14.08)$$

In both (14.07) and (14.08) $P_n^m(z)$ has its principal value.

14.2 As in the case of $P_n^m(z)$, the index in the second solution is usually suppressed when $m = 0$. Thus from (13.16) we have

$$Q_n(z) \equiv Q_n^0(z) = \frac{\pi^{1/2} n!}{2^{n+1}\Gamma(n+\frac{3}{2})} \frac{1}{(z-1)^{n+1}} F\left(n+1, n+1; 2n+2; \frac{2}{1-z}\right).$$

Again, the only cut needed for the principal value is the interval $[-1, 1]$.

Corresponding to (14.01) we have

$$Q_n^m(z) = (z^2-1)^{m/2} \frac{d^m}{dz^m} Q_n(z). \qquad (14.09)$$

To prove this result we differentiate Legendre's equation (12.01) m times by use of Leibniz's theorem. In this way it is seen that if w satisfies Legendre's equation, then $v \equiv d^m w/dz^m$ satisfies

$$(1-z^2)\frac{d^2 v}{dz^2} - 2(m+1)z\frac{dv}{dz} + (n-m)(n+m+1)v = 0.$$

By making the further substitution $u = (z^2-1)^{m/2} v$ we find that u satisfies the associated Legendre equation (12.02); compare (13.03) with $\nu = n$ and $\mu = m$. In particular, this means that the right-hand side of (14.09) satisfies the associated Legendre equation. By inspection, this solution is recessive at $z = \infty$; therefore it must be a multiple of $Q_n^m(z)$. That the multiple is unity is settled by reference to (13.16).

A closed expression for $Q_n(z)$ in terms of $P_n(z)$ is derivable as follows. The **F** function in (12.04), with $\nu = n$, is expanded as a finite series of powers of $z-1$ and the resulting expression for $P_n^{-\mu}(z)$ differentiated with respect to μ. Then using (12.22) we obtain

$$Q_n(z) = \tfrac{1}{2}P_n(z)\ln\left(\frac{z+1}{z-1}\right) - \sum_{s=0}^{n-1} \frac{(n+s)!}{(n-s)!(s!)^2 2^s}\{\psi(n+1) - \psi(s+1)\}(z-1)^s. \qquad (14.10)$$

Principal values of $Q_n(z)$ and the logarithm correspond.

Provided that z does not lie on the cut from -1 to 1, an integral for the principal branch of $Q_n^m(z)$ analogous to (14.03) can be found from (13.13) by collapsing the integration path onto the cut and subsequently setting $\nu = n$ and $\mu = m$:

$$Q_n^m(z) = \frac{(-)^m(n+m)!}{2^{n+1}n!}(z^2-1)^{m/2}\int_{-1}^{1}\frac{(1-t^2)^n}{(z-t)^{n+m+1}}\,dt \qquad (z \notin [-1,1]). \tag{14.11}$$

It may be noted in passing that unlike (14.03) and (14.05) this formula remains valid when n and m are replaced throughout by ν and μ, provided that the integral converges, that is, provided that $\operatorname{Re}\nu > -1$.

Now suppose, temporarily, that $z > 1$. Substituting

$$t = z - (z^2-1)^{1/2}e^{\theta}$$

in (14.11), we find that

$$Q_n^m(z) = (-)^m\frac{(n+m)!}{n!}\int_0^{\zeta}\{z-(z^2-1)^{1/2}\cosh\theta\}^n\cosh(m\theta)\,d\theta, \tag{14.12}$$

where

$$\zeta = \frac{1}{2}\ln\left(\frac{z+1}{z-1}\right) = \coth^{-1}z.$$

The temporary restriction is removable by analytic continuation: equation (14.12) holds for complex z, provided that $Q_n^m(z)$ has its principal value and the branches of $(z^2-1)^{1/2}$ and ζ are continuous in the cut plane.

14.3 An integral for $Q_n^m(z)$ in terms of $P_n^m(z)$ (with principal branches in each case) can be found from their Wronskian formula.† From (12.16), (13.14), and (14.04) we derive

$$P_n^m(z)\,Q_n^{m\prime}(z) - Q_n^m(z)\,P_n^{m\prime}(z) = (-)^{m-1}\frac{(n+m)!}{(n-m)!}\frac{1}{z^2-1}\cdot \tag{14.13}$$

By repeated applications of Rolle's theorem, we see from (14.02) that the zeros of $P_n^m(z)$ all lie in the interval $[-1,1]$. Hence on dividing (14.13) throughout by $\{P_n^m(z)\}^2$ and integrating, we find that

$$Q_n^m(z) = (-)^m P_n^m(z)\frac{(n+m)!}{(n-m)!}\int_z^{\infty}\frac{dt}{(t^2-1)\{P_n^m(t)\}^2} \qquad (n \geqslant m), \tag{14.14}$$

provided that the path does not intersect the cut $[-1,1]$.

Again, provided that z does not lie on the cut, an integral for $Q_n^m(z)$ involving $P_n^m(z)$ in a different way may be found by means of Cauchy's integral formula. For simplicity, restrict m to be zero. Then

$$Q_n(z) = \frac{1}{2\pi i}\int_{\mathscr{C}_1+\mathscr{C}_2}\frac{Q_n(t)}{t-z}\,dt,$$

† This is, in effect, the construction of a second solution of a differential equation when one solution is known; compare §5.1.

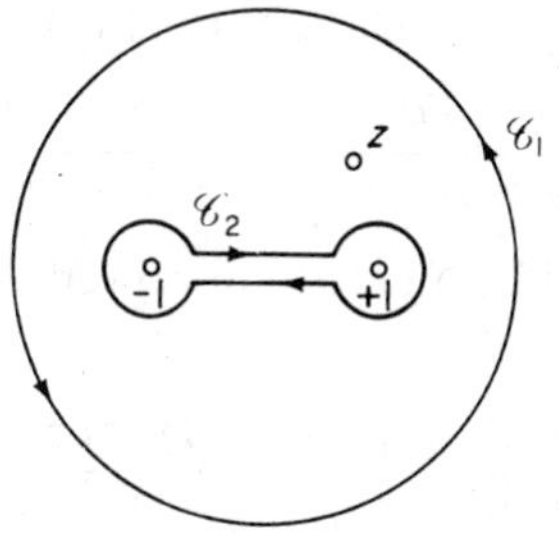

Fig. 14.1 t plane.

where $\mathscr{C}_1$ is a large circle and $\mathscr{C}_2$ is a closed contour within $\mathscr{C}_1$ which itself contains the interval $[-1, 1]$ but not the point z; see Fig. 14.1. The contribution from $\mathscr{C}_1$ vanishes as the radius of $\mathscr{C}_1$ tends to infinity; compare (12.09). Then collapsing $\mathscr{C}_2$ onto the two sides of the interval $[-1, 1]$, we find that

$$Q_n(z) = \frac{1}{2\pi i}\int_{-1}^{1} \frac{Q_n(t-i0) - Q_n(t+i0)}{z-t}\,dt.$$

By encircling the logarithmic singularity of $Q_n(t)$ at $t = 1$, we derive from (14.10)

$$Q_n(t-i0) - Q_n(t+i0) = \pi i P_n(t) \qquad (-1 < t < 1).$$

Thus we have *Neumann's integral*:

$$Q_n(z) = \frac{1}{2}\int_{-1}^{1} \frac{P_n(t)}{z-t}\,dt \qquad (z \notin [-1, 1]). \tag{14.15}$$

14.4 The last result to be established in this section is the so-called *addition theorem* for Legendre polynomials.

Theorem 14.1 *Let z, z_1, z_2, and ϕ be real or complex numbers such that*

$$z = z_1 z_2 - (z_1^2-1)^{1/2}(z_2^2-1)^{1/2}\cos\phi, \tag{14.16}$$

the branches of the square roots being chosen in accordance with §14.1. *Then*

$$P_n(z) = P_n(z_1)P_n(z_2) + 2\sum_{m=1}^{n} (-)^m \frac{(n-m)!}{(n+m)!} P_n^m(z_1)P_n^m(z_2)\cos(m\phi). \tag{14.17}$$

In the proof it is adequate to consider real values of z_1, z_2, and ϕ, with $z_1 > 1$ and $z_2 > 1$: the extension to complex variables follows by analytic continuation.

The theorem is based upon the identity

$$\frac{1}{2\pi}\int_{-\pi}^{\pi} \frac{d\theta}{z_2 + (z_2^2-1)^{1/2}\cos\theta - h\{z_1 + (z_1^2-1)^{1/2}\cos(\phi-\theta)\}} = \frac{1}{(1-2zh+h^2)^{1/2}}, \tag{14.18}$$

valid when $|h|$ is sufficiently small, which itself is derived from the following easily verified identity:

Lemma 14.1 *If a, b, and c are real and* $a > (b^2+c^2)^{1/2}$, *then*

$$\int_{-\pi}^{\pi} \frac{d\theta}{a+b\cos\theta+c\sin\theta} = \frac{2\pi}{(a^2-b^2-c^2)^{1/2}}.$$

Expanding the left-hand side of (14.18) in powers of h and referring to equation (7.20) of Chapter 2, we see that

$$P_n(z) = \frac{1}{2\pi}\int_{-\pi}^{\pi} \frac{\{z_1+(z_1^2-1)^{1/2}\cos(\phi-\theta)\}^n}{\{z_2+(z_2^2-1)^{1/2}\cos\theta\}^{n+1}}\,d\theta. \tag{14.19}$$

From (14.16) it is seen that $P_n(z)$ is a polynomial in $\cos\phi$ of degree n, and hence capable of expansion in the form

$$P_n(z) = \tfrac{1}{2}\alpha_0 + \sum_{m=1}^{n} \alpha_m \cos(m\phi),$$

where the α_m are independent of ϕ. This is the required form of expansion. Since it is a Fourier cosine series in ϕ, the coefficients are given by

$$\alpha_m = \pi^{-1}\int_{-\pi}^{\pi} P_n(z)\cos(m\phi)\,d\phi \qquad (m = 0, 1, \dots, n).$$

Substituting in the last integral by means of (14.19) and inverting the order of integration, we find that

$$\alpha_m = \frac{1}{2\pi^2}\int_{-\pi}^{\pi} \frac{I_{n,m}(\theta)\,d\theta}{\{z_2+(z_2^2-1)^{1/2}\cos\theta\}^{n+1}}, \tag{14.20}$$

where

$$I_{n,m}(\theta) = \int_{-\pi}^{\pi} \{z_1+(z_1^2-1)^{1/2}\cos(\phi-\theta)\}^n \cos(m\phi)\,d\phi$$

$$= \int_{-\pi}^{\pi} \{z_1+(z_1^2-1)^{1/2}\cos\chi\}^n \cos(m\theta+m\chi)\,d\chi.$$

If $\cos(m\theta+m\chi)$ is replaced by $\cos m\theta\cos m\chi - \sin m\theta\sin m\chi$, then the sine terms make no contribution to $I_{n,m}(\theta)$ because the other factor in the integrand is even in χ. The contribution from the cosine terms is evaluable by use of (14.07); thus

$$I_{n,m}(\theta) = \frac{n!\,2\pi}{(n+m)!}P_n^m(z_1)\cos(m\theta).$$

Substituting in (14.20) and applying (14.08), we obtain

$$\alpha_m = 2(-)^m\frac{(n-m)!}{(n+m)!}P_n^m(z_1)\,P_n^m(z_2).$$

This completes the proof.

The addition theorem can be generalized to the case in which n is replaced by the real or complex variable ν; in this event the sum in (14.17) is taken from $m = 1$ to $m = \infty$, and the factorials are replaced by Gamma functions.†

† Hobson (1931, §220).

Ex. 14.1 By making the substitution

$$\{z+(z^2-1)^{1/2}\cosh\phi\}\{z-(z^2-1)^{1/2}\cosh\theta\} = 1$$

in (14.12), with $m = 0$, derive *Heine's integral*

$$Q_n(z) = \int_0^\infty \frac{d\phi}{\{z+(z^2-1)^{1/2}\cosh\phi\}^{n+1}}.$$

Ex. 14.2† From the preceding exercise, deduce that if $z = \cosh(\alpha+i\beta)$ and α and β are real, then

$$|Q_n(z)| \leqslant e^{-(n-1)|\alpha|}Q_0(\cosh 2\alpha) \qquad (n \geqslant 1).$$

Ex. 14.3 Either by referring to Rodrigues' formula for the Chebyshev polynomial $U_{m-1}(x)$ (Chapter 2, (7.07) and Exercise 7.3), or by use of induction, prove *Jacobi's lemma*

$$\frac{d^{m-1}\sin^{2m-1}\theta}{d(\cos\theta)^{m-1}} = \frac{(-)^{m-1}}{m}\frac{(2m)!}{2^m m!}\sin m\theta.$$

Thence by repeated integrations by parts deduce from (14.07) that

$$P_n^m(z) = \frac{2^m m!(n+m)!}{(2m)!(n-m)!}\frac{(z^2-1)^{m/2}}{\pi}\int_0^\pi \{z+(z^2-1)^{1/2}\cos\theta\}^{n-m}\sin^{2m}\theta\, d\theta \qquad (n \geqslant m).$$

Ex. 14.4 From the preceding exercise deduce that when $\zeta > 0$

$$P_n^m(\cosh\zeta) = \frac{2^{2m+(1/2)}m!(n+m)!}{\pi(2m)!(n-m)!\sinh^m\zeta}\int_0^\zeta (\cosh\zeta-\cosh t)^{m-(1/2)}\cosh\{(n+\tfrac{1}{2})t\}\, dt.$$

Ex. 14.5 From Neumann's integral and the expansion (7.20) of Chapter 2 show that if $z > 1$ and h is positive and sufficiently small, then

$$\sum_{n=0}^\infty Q_n(z)h^n = \frac{1}{(1-2zh+h^2)^{1/2}}\ln\left\{\frac{z-h+(1-2zh+h^2)^{1/2}}{(z^2-1)^{1/2}}\right\}.$$

With the aid of Exercise 7.9 of Chapter 2 show that when $|h| < 1$ the sum on the left-hand side converges uniformly in any compact z domain not intersecting the cut $[-1,1]$, and thence extend the expansion to complex z.

15 Ferrers Functions

15.1 When ν and μ are real the principal branches of $P_\nu^{-\mu}(z)$ and $\mathbf{Q}_\nu^\mu(z)$ are real on the part of the real axis between 1 and ∞. On the cut from $-\infty$ to 1 there are two possible values for each function, depending whether the cut is approached from the upper or lower side. Replacing z by x, we denote these values by $P_\nu^{-\mu}(x+i0)$, $P_\nu^{-\mu}(x-i0)$, $\mathbf{Q}_\nu^\mu(x+i0)$, and $\mathbf{Q}_\nu^\mu(x-i0)$. None of these functions is real, as a rule. Because the associated Legendre equation is real in these circumstances, however, it is desirable to have real standard solutions. For the interval $-\infty < x \leqslant -1$ the obvious choice is $P_\nu^{-\mu}(-x)$ and $\mathbf{Q}_\nu^\mu(-x)$ or $e^{-\mu\pi i}Q_\nu^\mu(-x)$. To cover the remaining interval $-1 \leqslant x \leqslant 1$ we introduce the following solutions, called *Ferrers functions*:

$$\mathrm{P}_\nu^\mu(x) = e^{\mu\pi i/2}P_\nu^\mu(x+i0) = e^{-\mu\pi i/2}P_\nu^\mu(x-i0), \tag{15.01}$$

$$\begin{aligned}\mathrm{Q}_\nu^\mu(x) &= \tfrac{1}{2}\Gamma(\nu+\mu+1)\{e^{-\mu\pi i/2}\mathbf{Q}_\nu^\mu(x+i0)+e^{\mu\pi i/2}\mathbf{Q}_\nu^\mu(x-i0)\}\\ &= \tfrac{1}{2}e^{-3\mu\pi i/2}Q_\nu^\mu(x+i0)+\tfrac{1}{2}e^{-\mu\pi i/2}Q_\nu^\mu(x-i0).\end{aligned} \tag{15.02}$$

† Compare Exercise 7.9 of Chapter 2.

These equations define $\mathrm{P}_\nu^\mu(x)$ and $\mathrm{Q}_\nu^\mu(x)$ for all combinations of ν and μ, except $\nu+\mu = -1, -2, -3, \ldots$. Clearly, $\mathrm{P}_\nu^{-\mu}(x)$ and $\mathrm{Q}_\nu^{-\mu}(x)$ are further solutions. Again, when the index μ is zero it is customarily omitted; thus $\mathrm{P}_n(x) = P_n(x)$, when n is a nonnegative integer.

That the two definitions (15.01) of $\mathrm{P}_\nu^\mu(x)$ are consistent can be seen by encircling the singularity at $x = 1$ and referring to (12.04) (with μ replaced by $-\mu$). This analysis also shows that

$$\mathrm{P}_\nu^\mu(x) = \left(\frac{1+x}{1-x}\right)^{\mu/2} \mathbf{F}(\nu+1, -\nu; 1-\mu; \tfrac{1}{2}-\tfrac{1}{2}x). \tag{15.03}$$

Equation (15.03) can be used to extend the definition of $\mathrm{P}_\nu^\mu(x)$ to complex values of ν, μ, and x: cuts are introduced along the x intervals $(-\infty, -1]$ and $[1, \infty)$.

The corresponding expression for the other Ferrers function is derivable from (15.03) and the connection formula

$$\frac{2\sin(\mu\pi)}{\pi}\mathrm{Q}_\nu^\mu(x) = \cos(\mu\pi)\,\mathrm{P}_\nu^\mu(x) - \frac{\Gamma(\nu+\mu+1)}{\Gamma(\nu-\mu+1)}\mathrm{P}_\nu^{-\mu}(x), \tag{15.04}$$

which is itself obtained from the foregoing relations and (12.11). This gives

$$\frac{2\sin(\mu\pi)}{\pi}\mathrm{Q}_\nu^\mu(x) = \cos(\mu\pi)\left(\frac{1+x}{1-x}\right)^{\mu/2} \mathbf{F}(\nu+1, -\nu; 1-\mu; \tfrac{1}{2}-\tfrac{1}{2}x)$$
$$- \frac{\Gamma(\nu+\mu+1)}{\Gamma(\nu-\mu+1)}\left(\frac{1-x}{1+x}\right)^{\mu/2} \mathbf{F}(\nu+1, -\nu; 1+\mu; \tfrac{1}{2}-\tfrac{1}{2}x). \tag{15.05}$$

For real values of ν and μ such that $\nu \geqslant -\frac{1}{2}$ and $\mu \geqslant 0$,† the limiting forms of $\mathrm{P}_\nu^{\pm\mu}(x)$ and $\mathrm{Q}_\nu^{\pm\mu}(x)$, as x tends to the singularity 1 from the left, are derivable from (15.03), (12.21), and (12.23). They are given by

$$\mathrm{P}_\nu^\mu(x) \sim \frac{1}{\Gamma(1-\mu)}\left(\frac{2}{1-x}\right)^{\mu/2}, \qquad \mathrm{P}_\nu^{-\mu}(x) \sim \frac{1}{\Gamma(1+\mu)}\left(\frac{1-x}{2}\right)^{\mu/2}, \tag{15.06}$$

$$\mathrm{Q}_\nu^\mu(x) \sim \tfrac{1}{2}\cos(\mu\pi)\,\Gamma(\mu)\left(\frac{2}{1-x}\right)^{\mu/2}, \qquad \mathrm{Q}_\nu^{-\mu}(x) \sim \frac{\Gamma(\mu)\,\Gamma(\nu-\mu+1)}{2\Gamma(\nu+\mu+1)}\left(\frac{2}{1-x}\right)^{\mu/2}, \tag{15.07}$$

$$\mathrm{P}_\nu(x) \to 1, \qquad \mathrm{Q}_\nu(x) \sim \tfrac{1}{2}\ln\left(\frac{1}{1-x}\right), \tag{15.08}$$

provided that the Gamma functions are finite and $\cos(\mu\pi)$ is nonzero.

Inspection of these limiting forms indicates that no single pair of the solutions $\mathrm{P}_\nu^{\pm\mu}(x)$, $\mathrm{Q}_\nu^{\pm\mu}(x)$ is numerically satisfactory in the neighborhood of $x = 1$ for all nonnegative values of $\nu+\frac{1}{2}$ and μ. In the case when ν and μ are nonnegative integers, however, $\mathrm{P}_n^{-m}(x)$ and $\mathrm{Q}_n^m(x)$ are satisfactory.

† Compare §12.1.

15.2 Both $\mathrm{P}_\nu^\mu(x)$ and $\mathrm{Q}_\nu^\mu(x)$ are analytic at $x=0$, and therefore capable of expansion in Maclaurin series. These series are needed in a later chapter; they may be derived in the following way.

The differential equation satisfied by $(1-x^2)^{\mu/2}\mathrm{P}_\nu^\mu(x)$ and $(1-x^2)^{\mu/2}\mathrm{Q}_\nu^\mu(x)$ is given by

$$(1-x^2)\frac{d^2w}{dx^2}+2(\mu-1)x\frac{dw}{dx}+(\nu+\mu)(\nu-\mu+1)w = 0;$$

compare (13.03). The method of §3.2 yields even and odd solutions:

$$w_1 = F(-\tfrac{1}{2}\nu-\tfrac{1}{2}\mu,\tfrac{1}{2}\nu-\tfrac{1}{2}\mu+\tfrac{1}{2};\tfrac{1}{2};x^2), \qquad w_2 = xF(-\tfrac{1}{2}\nu-\tfrac{1}{2}\mu+\tfrac{1}{2},\tfrac{1}{2}\nu-\tfrac{1}{2}\mu+1;\tfrac{3}{2};x^2). \tag{15.09}$$

Accordingly,

$$(1-x^2)^{\mu/2}\mathrm{P}_\nu^\mu(x) = A_1w_1+A_2w_2, \tag{15.10}$$

$$(1-x^2)^{\mu/2}\mathrm{Q}_\nu^\mu(x) = B_1w_1+B_2w_2, \tag{15.11}$$

where A_1, A_2, B_1, and B_2 are independent of x.

A real-variable method for determining A_1, A_2, B_1, and B_2, would be to let $x\to 1$ in (15.10) and (15.11), and their x-differentiated forms, and use Gauss's formula (9.11) for F functions of argument 1. Instead, we use a less laborious method based upon the limiting forms of the solutions as x tends to $\pm i\infty$.

Set $x=i\xi$ and assume temporarily that $\operatorname{Re}\nu>-\tfrac{1}{2}$, and none of μ, 2ν, $\nu\pm\mu$ is an integer or zero. Letting $\xi\to\infty$ we obtain from (15.03), (15.09), and (10.16)

$$(1-x^2)^{\mu/2}\mathrm{P}_\nu^\mu(x) \sim \frac{\Gamma(2\nu+1)e^{(\nu+\mu)\pi i/2}\xi^{\nu+\mu}}{2^\nu\Gamma(\nu+1)\Gamma(\nu-\mu+1)},$$

and

$$w_1 \sim \frac{\pi^{1/2}\Gamma(\nu+\tfrac{1}{2})\xi^{\nu+\mu}}{\Gamma(\tfrac{1}{2}\nu+\tfrac{1}{2}\mu+\tfrac{1}{2})\Gamma(\tfrac{1}{2}\nu-\tfrac{1}{2}\mu+\tfrac{1}{2})}, \qquad w_2 \sim i\frac{\pi^{1/2}\Gamma(\nu+\tfrac{1}{2})\xi^{\nu+\mu}}{2\Gamma(\tfrac{1}{2}\nu+\tfrac{1}{2}\mu+1)\Gamma(\tfrac{1}{2}\nu-\tfrac{1}{2}\mu+1)}.$$

Hence

$$\frac{2^\nu e^{(\nu+\mu)\pi i/2}}{\pi\Gamma(\nu-\mu+1)} = \frac{A_1}{\Gamma(\tfrac{1}{2}\nu+\tfrac{1}{2}\mu+\tfrac{1}{2})\Gamma(\tfrac{1}{2}\nu-\tfrac{1}{2}\mu+\tfrac{1}{2})}+\frac{iA_2}{2\Gamma(\tfrac{1}{2}\nu+\tfrac{1}{2}\mu+1)\Gamma(\tfrac{1}{2}\nu-\tfrac{1}{2}\mu+1)}.$$

Similarly, by setting $x=-i\xi$ and letting $\xi\to\infty$, we obtain the same equation with the sign of i changed throughout. Solution of these two equations yields

$$A_1 = \frac{2^\mu\pi^{1/2}}{\Gamma(\tfrac{1}{2}\nu-\tfrac{1}{2}\mu+1)\Gamma(-\tfrac{1}{2}\nu-\tfrac{1}{2}\mu+\tfrac{1}{2})}, \qquad A_2 = -\frac{2^{\mu+1}\pi^{1/2}}{\Gamma(\tfrac{1}{2}\nu-\tfrac{1}{2}\mu+\tfrac{1}{2})\Gamma(-\tfrac{1}{2}\nu-\tfrac{1}{2}\mu)}, \tag{15.12}$$

all restrictions on the parameters being removable by analytic continuation.

To determine the coefficients in (15.11) we apply the connection formula (15.04). From (15.10) and (10.03) we derive

$$(1-x^2)^{\mu/2}\mathrm{P}_\nu^{-\mu}(x) = \hat{A}_1w_1+\hat{A}_2w_2,$$

where $\hat{A}_1$ and $\hat{A}_2$ are obtained from A_1 and A_2, respectively, by replacing μ in (15.12) by $-\mu$. Accordingly,

$$B_1 = \frac{\pi\Gamma(\nu+\mu+1)}{2\sin(\mu\pi)}\left\{\frac{\cos(\mu\pi)A_1}{\Gamma(\nu+\mu+1)} - \frac{\hat{A}_1}{\Gamma(\nu-\mu+1)}\right\},$$

and

$$B_2 = \frac{\pi\Gamma(\nu+\mu+1)}{2\sin(\mu\pi)}\left\{\frac{\cos(\mu\pi)A_2}{\Gamma(\nu+\mu+1)} - \frac{\hat{A}_2}{\Gamma(\nu-\mu+1)}\right\}.$$

Substituting by means of (15.12) and carrying out some reduction, we find that†

$$\begin{aligned} B_1 &= -2^{\mu-1}\pi^{1/2}\sin\{(\tfrac{1}{2}\nu+\tfrac{1}{2}\mu)\pi\}\,\Gamma(\tfrac{1}{2}\nu+\tfrac{1}{2}\mu+\tfrac{1}{2})/\Gamma(\tfrac{1}{2}\nu-\tfrac{1}{2}\mu+1), \\ B_2 &= 2^{\mu}\pi^{1/2}\cos\{(\tfrac{1}{2}\nu+\tfrac{1}{2}\mu)\pi\}\,\Gamma(\tfrac{1}{2}\nu+\tfrac{1}{2}\mu+1)/\Gamma(\tfrac{1}{2}\nu-\tfrac{1}{2}\mu+\tfrac{1}{2}), \end{aligned} \qquad (15.13)$$

provided that $\nu+\mu \neq -1, -2, -3, \ldots$.

Ex. 15.1 Show that

$$Q_0(x) = \frac{1}{2}\ln\left(\frac{1+x}{1-x}\right), \qquad Q_n(x) = \frac{1}{2}P_n(x)\ln\left(\frac{1+x}{1-x}\right) - W_{n-1}(x) \qquad (n \geqslant 1),$$

where $W_n(x)$ is a polynomial of degree n, and

$$W_0(x) = 1, \qquad W_1(x) = \tfrac{3}{2}x, \qquad W_2(x) = \tfrac{5}{2}x^2 - \tfrac{2}{3}.$$

Ex. 15.2 Show that if n and m are positive integers, then

$$P_n^m(x) = (-)^m(1-x^2)^{m/2}\,P_n^{(m)}(x), \qquad Q_n^m(x) = (-)^m(1-x^2)^{m/2}\,Q_n^{(m)}(x),$$

and

$$P_n^{-m}(x) = (1-x^2)^{-m/2}\int_x^1\int_x^1\cdots\int_x^1 P_n(x)\,(dx)^m.$$

Ex. 15.3 From (15.06) and (15.07), derive the Wronskians

$$\mathscr{W}\{P_\nu^{-\mu}(x), Q_\nu^\mu(x)\} = \frac{\cos(\mu\pi)}{1-x^2}, \qquad \mathscr{W}\{P_\nu^\mu(x), Q_\nu^\mu(x)\} = \frac{\Gamma(\nu+\mu+1)}{\Gamma(\nu-\mu+1)}\frac{1}{1-x^2}.$$

Confirm these results by evaluation at $x = 0$.

Ex. 15.4 Prove that when $x \in (-1,1)$

$$\begin{aligned} P_\nu^\mu(-x) &= \cos\{(\nu+\mu)\pi\}\,P_\nu^\mu(x) - (2/\pi)\sin\{(\nu+\mu)\pi\}\,Q_\nu^\mu(x), \\ Q_\nu^\mu(-x) &= -\tfrac{1}{2}\pi\sin\{(\nu+\mu)\pi\}\,P_\nu^\mu(x) - \cos\{(\nu+\mu)\pi\}\,Q_\nu^\mu(x). \end{aligned}$$

Ex. 15.5 Show that

$$\int\left\{(\nu-\nu')(\nu+\nu'+1) + \frac{\mu'^2-\mu^2}{1-x^2}\right\}P_\nu^\mu(x)\,P_{\nu'}^{\mu'}(x)\,dx = (1-x^2)\,\mathscr{W}\{P_\nu^\mu(x), P_{\nu'}^{\mu'}(x)\}.$$

With the aid of (14.04) and the preceding exercise deduce that if l, m, and n are nonnegative integers, then

$$\int_{-1}^1 P_l^m(x)\,P_n^m(x)\,dx = \delta_{l,n}\frac{(n+m)!}{(n-m)!\,(n+\frac{1}{2})},$$

† There is a sign error in the version of these formulas appearing in B.M.P. (1953a, p. 144).

and

$$\int_{-1}^{1} \frac{\mathrm{P}_n^l(x)\,\mathrm{P}_n^m(x)}{1-x^2}\,dx = \delta_{l,m}\frac{(n+m)!}{(n-m)!\,m} \qquad (m>0).$$

Ex. 15.6† Let μ and x be fixed, x being real and positive. From (15.03) deduce that if ν tends to infinity through a sequence of positive values, then $\nu^{\mu}\,\mathrm{P}_\nu^{-\mu}\{\cos(x/\nu)\}$ tends to $J_\mu(x)$.

Historical Notes and Additional References

Almost all of this chapter is classical material. Heavy use has been made of the books by B.M.P. (1953a) and Whittaker and Watson (1927). The present derivation of the properties of the hypergeometric and Legendre functions places somewhat more emphasis on the analytic theory of the defining differential equations, particularly holomorphicity with respect to the parameters (Theorem 3.2).

The new notations **F** and **Q** for solutions of the hypergeometric and associated Legendre equations have been introduced with some reluctance. With the present approach, however, there are considerable advantages in working with solutions which are entire in each parameter. Moreover, many formulas involving F or Q simplify when expressed in terms of **F** or **Q**.

§§1–6 For extensions to linear differential equations of higher order and systems of equations see, for example, Ince (1927) or Hartman (1964). Each of these references also includes much historical information.

§§9–11 For further properties of, and references to, the hypergeometric function, and, especially, generalized hypergeometric functions, see B.M.P. (1953a), Carathéodory (1960), Slater (1966), and Luke (1969a, b).

§§12–15 The classical reference on Legendre functions is Hobson (1931). Other comprehensive treatises include those of Snow (1952), Robin (1957, 1958, 1959), and MacRobert (1967).

The integral representations of Theorem 13.1 for $P_\nu^{-\mu}(z)$ are more restrictive with respect to the parameters than the loop integral used as definition by Hobson (1931, §118). Their advantage is to bring out the recessive property of $P_\nu^{-\mu}(z)$ at $z = 1$ in a direct manner. The integral representation of Theorem 13.2 for $\mathbf{Q}_\nu^\mu(z)$ is essentially Hobson's definition (Hobson, 1931, §125).

§15.1 Ferrers (1877) considered the associated Legendre equation in the case when ν and μ are nonnegative integers and discussed in detail only one solution, which he denoted by $T_\nu^{(\mu)}(x)$. In the present notation $T_\nu^{(\mu)}(x) = (-)^\mu \mathrm{P}_\nu^\mu(x)$.

† More general relations of this kind are established in Chapter 12.

6

THE LIOUVILLE–GREEN APPROXIMATION

1 The Liouville Transformation

1.1 In this chapter we begin the study of the approximation of solutions of differential equations of the form

$$d^2w/dx^2 = f(x)w, \tag{1.01}$$

in which x is a real or complex variable, and $f(x)$ a prescribed function. All homogeneous linear differential equations of the second order can be put in this form by appropriate change of dependent or independent variable.

The simplest approximation is obtained by assuming that $f(x)$ may be treated as constant. This yields

$$w \fallingdotseq Ae^{x\surd\{f(x)\}} + Be^{-x\surd\{f(x)\}}, \tag{1.02}$$

where A and B are arbitrary constants. The assumption is reasonable if $f(x)$ is continuous and the interval or domain under consideration is sufficiently small and does not contain a zero. In other words, (1.02) furnishes a guide to the *local behavior* of the solutions. In particular, in an interval in which $f(x)$ is real, positive, and slowly varying, the solutions of (1.01) may be expected to be *exponential* in character, that is, expressible as a linear combination of two solutions whose magnitudes change monotonically, one increasing and the other decreasing. Similarly, in an interval in which $f(x)$ is negative the solutions of (1.01) may be expected to be *trigonometric* (or *oscillatory*) in character. In succeeding sections, it will be seen that these inferences are correct, in general.†

1.2 For most purposes, the approximation (1.02) is too crude. We seek to improve it by preliminary transformation of (1.01) into a differential equation of the same type, but with $f(x)$ replaced by a function that varies more slowly.

Theorem 1.1 *Let w satisfy equation* (1.01), $\xi(x)$ *be any thrice-differentiable function of x, and*

$$W = \{\xi'(x)\}^{1/2}w. \tag{1.03}$$

† A noteworthy exception is provided by $f(x) = \alpha(\alpha-1)/x^2$, where $x > 0$ and α is a constant such that $0 < \alpha < 1$. Although $f(x)$ is negative, the solutions $w = Ax^{\alpha} + Bx^{1-\alpha}$, when $\alpha \neq \frac{1}{2}$, or $w = x^{1/2}(A + B\ln x)$, when $\alpha = \frac{1}{2}$, do not oscillate.

Then W satisfies

$$\frac{d^2W}{d\xi^2} = \left\{\dot{x}^2 f(x) + \dot{x}^{1/2}\frac{d^2}{d\xi^2}(\dot{x}^{-1/2})\right\} W, \tag{1.04}$$

where dots signify differentiations with respect to ξ.

This result is verifiable by direct substitutions. With ξ as independent variable (1.01) transforms into

$$\frac{d^2w}{d\xi^2} - \frac{\ddot{x}}{\dot{x}}\frac{dw}{d\xi} = \dot{x}^2 f(x) w.$$

The term in the first derivative is then removed by taking the new dependent variable (1.03). This yields (1.04).

The transformation supplied by the theorem is known as the *Liouville transformation*. The second term in the coefficient of W in (1.04) is often expressed in the form

$$\dot{x}^{1/2}\frac{d^2}{d\xi^2}(\dot{x}^{-1/2}) = -\frac{1}{2}\{x, \xi\},$$

where $\{x, \xi\}$ is the *Schwarzian derivative*

$$\{x, \xi\} = \frac{\dddot{x}}{\dot{x}} - \frac{3}{2}\left(\frac{\ddot{x}}{\dot{x}}\right)^2.$$

1.3 For a given function $f(x)$, it is no less difficult to arrange that the coefficient of W in (1.04) be constant, than to solve exactly the original differential equation (1.01). We compromise by choosing $\xi(x)$ so that the term $\dot{x}^2 f(x)$ is a constant, which we take to be unity without loss of generality; thus

$$\xi(x) = \int f^{1/2}(x)\, dx. \tag{1.05}$$

Provided that $f(x)$ is twice differentiable, the Schwarzian derivative can be evaluated, and equation (1.04) becomes

$$d^2W/d\xi^2 = (1+\phi)W, \tag{1.06}$$

where

$$\phi = \frac{4f(x)f''(x) - 5f'^2(x)}{16 f^3(x)} = -\frac{1}{f^{3/4}}\frac{d^2}{dx^2}\left(\frac{1}{f^{1/4}}\right). \tag{1.07}$$

So far, the analysis is exact. If, now, ϕ is neglected, then independent solutions of (1.06) are $e^{\pm\xi}$. Restoring the original variables, and noting that $\xi'(x) = f^{1/2}(x)$, we obtain

$$w \doteq A f^{-1/4} e^{\int f^{1/2} dx} + B f^{-1/4} e^{-\int f^{1/2} dx}, \tag{1.08}$$

where, again, A and B are arbitrary constants. This is the *Liouville–Green* (*LG*) *approximation*† for the general solution of (1.01). The expressions $f^{-1/4}\exp(\int f^{1/2}\, dx)$ and $f^{-1/4}\exp(-\int f^{1/2}\, dx)$ are the *LG functions*.

† Also called the *WKB approximation*. See p. 228.

Obviously the accuracy of (1.08) relates to the magnitude of the neglected function ϕ in the region under consideration. We investigate this dependence rigorously in the next section. At this stage we merely observe that we expect $|\phi|$ to be small, and therefore the approximation to be successful, when $|f^{-1/4}|$ is sufficiently small or slowly varying. This includes the situation in which the simpler approximation (1.02) is applicable.

We notice immediately an important case of failure: intervals or domains containing zeros of f. Clearly ϕ becomes infinite and the approximation fails at these points. Zeros of f are called *turning points* or *transition points* of the differential equation (1.01). The reason for these names is that when the variables are real and the zero is simple (or, more generally, of odd order), it separates an interval in which the solutions are of exponential type from one in which they oscillate.

Throughout the present chapter we suppose that all regions under consideration are free from turning points. The approximation of solutions in regions containing turning points is considered in Chapter 11.

1.4 Another formal way† of deriving the LG approximation is to use the Riccati equation

$$v' + v^2 = f,$$

obtained from (1.01) by the substitution $w = \exp(\int v\,dx)$. To solve this equation, we first ignore the term v' to obtain $v \risingdotseq \pm f^{1/2} = v_1$, say. As a second approximation,

$$v \risingdotseq \pm(f - v_1')^{1/2} = \pm f^{1/2}\left(1 \mp \frac{f'}{2f^{3/2}}\right)^{1/2} \risingdotseq \pm f^{1/2} - \frac{f'}{4f},$$

provided that $|f'| \ll 2|f|^{3/2}$. Integration of the last expression immediately yields (1.08).

1.5 The transformation given by (1.03) and (1.05) can also be applied to the differential equation

$$d^2w/dx^2 = \{f(x) + g(x)\}\,w. \tag{1.09}$$

It yields

$$\frac{d^2W}{d\xi^2} = \left(1 + \phi + \frac{g}{f}\right)W, \tag{1.10}$$

where ϕ is given by (1.07). Again, if $|\phi| \ll 1$ and $|g| \ll |f|$ in the region of interest, then we have hopes that (1.08) approximates the solutions of (1.09).

Of course we can regard the coefficient $f(x)+g(x)$ in (1.09) as a single function of x and use (1.08) with $f+g$ in place of f. When the coefficient of w is separated into two parts, however, a better approximation can result‡; alternatively, the evaluation of the integral in (1.08) may be eased. These advantages will become clearer in §§4 and 5 below.

† Jeffreys and Jeffreys (1956, §17.122).
‡ Jeffreys (1924) appears to have been the first to point this out.

Ex. 1.1 By considering successive Liouville transformations prove *Cayley's identity*

$$\{x,\zeta\} = (d\xi/d\zeta)^2\{x,\xi\} + \{\xi,\zeta\}.$$

Deduce that

$$\{x,\xi\} = -(dx/d\xi)^2\{\xi,x\}.$$

Ex. 1.2 If p is twice differentiable and q is differentiable, show that the equation

$$\frac{d^2W}{dx^2} + q\frac{dW}{dx} + \left\{\frac{1}{2}\frac{dq}{dx} + \frac{1}{4}q^2 - p - p^{1/4}\frac{d^2}{dx^2}(p^{-1/4})\right\}W = 0,$$

is satisfied exactly by

$$W = p^{-1/4}\exp\left(\pm\int p^{1/2}\,dx - \tfrac{1}{2}\int q\,dx\right).$$

Ex. 1.3 Show that the approximation (1.08) is exact if, and only if, $f = (ax+b)^{-4}$, where a and b are constants.

Ex.1.4 Given the equation $d^2w/dx^2 = \alpha(\alpha-1)x^{-2}w$ in which α is a large positive constant, show that with $g(x) = 0$ the ratio of the LG functions to the corresponding exact solutions is approximately unity when $-8\alpha \ll \ln x \ll 8\alpha$. Show also that with $g(x) = -\frac{1}{4}x^{-2}$ the LG functions are exact solutions.

2 Error Bounds: Real Variables

2.1 The analysis leading to (1.08) is purely formal. The key assumption is that the solutions of the differential equation (1.06), or more generally (1.10), do not differ significantly from those of the simpler equation $d^2W/d\xi^2 = W$. The following theorem provides a rigorous justification in the case of solutions of exponential type. A second theorem (§2.4) covers the oscillatory case.

Theorem 2.1 *In a given finite or infinite interval (a_1, a_2), let $f(x)$ be a positive, real, twice continuously differentiable function, $g(x)$ a continuous real or complex function, and*

$$F(x) = \int\left\{\frac{1}{f^{1/4}}\frac{d^2}{dx^2}\left(\frac{1}{f^{1/4}}\right) - \frac{g}{f^{1/2}}\right\}dx. \tag{2.01}$$

Then in this interval the differential equation

$$d^2w/dx^2 = \{f(x)+g(x)\}w \tag{2.02}$$

has twice continuously differentiable solutions

$$\begin{aligned} w_1(x) &= f^{-1/4}(x)\exp\left\{\int f^{1/2}(x)\,dx\right\}\{1+\varepsilon_1(x)\}, \\ w_2(x) &= f^{-1/4}(x)\exp\left\{-\int f^{1/2}(x)\,dx\right\}\{1+\varepsilon_2(x)\}, \end{aligned} \tag{2.03}$$

such that

$$|\varepsilon_j(x)|,\ \tfrac{1}{2}f^{-1/2}(x)|\varepsilon_j'(x)| \leqslant \exp\{\tfrac{1}{2}\mathscr{V}_{a_j,x}(F)\} - 1 \qquad (j=1,2), \tag{2.04}$$

provided that $\mathscr{V}_{a_j,x}(F) < \infty$. If $g(x)$ is real, then the solutions are real.

The integral (2.01) will be called the *error-control function* for the solutions (2.03). It suffices to establish the theorem for the case $j = 1$; the corresponding result for $j = 2$ then follows on replacing x in (2.02) by $-x$.

2.2 We begin the proof of Theorem 2.1 by applying the transformations (1.05) and $w = f^{-1/4}(x)W$. Equation (2.02) becomes

$$d^2W/d\xi^2 = \{1+\psi(\xi)\}\,W, \tag{2.05}$$

where

$$\psi(\xi) = \frac{g}{f} - \frac{1}{f^{3/4}}\frac{d^2}{dx^2}\left(\frac{1}{f^{1/4}}\right); \tag{2.06}$$

compare (1.07) and (1.10). The choice of integration constant in (1.05) is immaterial: it merely affects the final solution by a constant factor. Since f is positive, ξ is an increasing function of x. Let $\xi = \alpha_1$ and α_2 correspond to $x = a_1$ and a_2, respectively; then with the assumed conditions $\psi(\xi)$ is continuous in (α_1, α_2).

In (2.05) we substitute

$$W(\xi) = e^{\xi}\{1+h(\xi)\}, \tag{2.07}$$

and obtain

$$h''(\xi) + 2h'(\xi) - \psi(\xi)\,h(\xi) = \psi(\xi). \tag{2.08}$$

To solve this inhomogeneous differential equation for $h(\xi)$, the term $\psi(\xi)\,h(\xi)$ is regarded as a correction and transferred to the right. Applying the method of variation of parameters (or constants), we find that

$$h(\xi) = \tfrac{1}{2}\int_{\alpha_1}^{\xi} \{1-e^{2(v-\xi)}\}\,\psi(v)\,\{1+h(v)\}\,dv. \tag{2.09}$$

Conversely, it is easily verified by differentiation that any twice-differentiable solution of this Volterra integral equation satisfies (2.08).

Equation (2.09) is solvable by the method of successive approximations used in Chapter 5, §1. At first, we assume α_1 is finite and $\psi(\xi)$ is continuous at α_1. We define a sequence $h_s(\xi)$, $s = 0, 1, \ldots$, by $h_0(\xi) = 0$ and

$$h_s(\xi) = \tfrac{1}{2}\int_{\alpha_1}^{\xi} \{1-e^{2(v-\xi)}\}\,\psi(v)\,\{1+h_{s-1}(v)\}\,dv \qquad (s \geqslant 1); \tag{2.10}$$

in particular,

$$h_1(\xi) = \tfrac{1}{2}\int_{\alpha_1}^{\xi} \{1-e^{2(v-\xi)}\}\,\psi(v)\,dv. \tag{2.11}$$

Since $\xi - v \geqslant 0$, we have

$$0 \leqslant 1 - e^{2(v-\xi)} < 1. \tag{2.12}$$

Hence† $|h_1(\xi)| \leqslant \tfrac{1}{2}\Psi(\xi)$, where

$$\Psi(\xi) = \int_{\alpha_1}^{\xi} |\psi(v)|\,dv.$$

† Equality occurs when $\xi = \alpha_1$.

Now suppose that for a particular value of s

$$|h_s(\xi)-h_{s-1}(\xi)| \leqslant \Psi^s(\xi)/(s!\,2^s), \tag{2.13}$$

as indeed is the case when $s = 1$. From (2.10) we have

$$h_{s+1}(\xi) - h_s(\xi) = \tfrac{1}{2}\int_{\alpha_1}^{\xi} \{1-e^{2(v-\xi)}\}\,\psi(v)\{h_s(v)-h_{s-1}(v)\}\,dv \qquad (s \geqslant 1). \tag{2.14}$$

Hence

$$|h_{s+1}(\xi)-h_s(\xi)| \leqslant \frac{1}{s!\,2^{s+1}}\int_{\alpha_1}^{\xi} |\psi(v)|\,\Psi^s(v)\,dv = \frac{\Psi^{s+1}(\xi)}{(s+1)!\,2^{s+1}}.$$

Therefore by induction (2.13) holds for all s. Since $\Psi(\xi)$ is bounded when ξ is finite, the series

$$h(\xi) = \sum_{s=0}^{\infty} \{h_{s+1}(\xi)-h_s(\xi)\} \tag{2.15}$$

converges uniformly in any compact ξ interval. That $h(\xi)$ satisfies the integral equation (2.09) follows by summation of (2.14) and use of (2.11).

To prove that $h(\xi)$ is twice differentiable, it suffices to show that the series $\sum\{h''_{s+1}(\xi)-h''_s(\xi)\}$ is uniformly convergent. By differentiation of (2.11) and (2.14) we have

$$h'_1(\xi) = \int_{\alpha_1}^{\xi} e^{2(v-\xi)}\psi(v)\,dv, \qquad h'_{s+1}(\xi) - h'_s(\xi) = \int_{\alpha_1}^{\xi} e^{2(v-\xi)}\psi(v)\{h_s(v)-h_{s-1}(v)\}\,dv. \tag{2.16}$$

Substituting by means of (2.13) and the bound $|e^{2(v-\xi)}| \leqslant 1$, we obtain

$$|h'_{s+1}(\xi)-h'_s(\xi)| \leqslant \Psi^{s+1}(\xi)/\{(s+1)!\,2^s\} \qquad (s = 0, 1, \ldots). \tag{2.17}$$

This establishes the uniform convergence of $\sum\{h'_{s+1}(\xi)-h'_s(\xi)\}$ in any compact interval. For the second differentiation we use

$$h''_1(\xi) = -2h'_1(\xi) + \psi(\xi),$$

$$h''_{s+1}(\xi) - h''_s(\xi) = -2\{h'_{s+1}(\xi)-h'_s(\xi)\} + \psi(\xi)\{h_s(\xi)-h_{s-1}(\xi)\}.$$

Summarizing so far, we have shown that equation (2.05) is satisfied by the function (2.07) with $h(\xi)$ given by (2.15). Summation of (2.13) and (2.17) produces

$$|h(\xi)|,\ \tfrac{1}{2}|h'(\xi)| \leqslant e^{\Psi(\xi)/2} - 1. \tag{2.18}$$

On transforming back to the variable x by means of the differential relation $d\xi = f^{1/2}\,dx$, we find $-\int \psi(\xi)\,d\xi$ becomes the error-control function $F(x)$. Therefore $\Psi(\xi) = \mathcal{V}_{a_1,x}(F)$, and the inequalities (2.18) transform into the desired bounds (2.04).

2.3 It remains to consider the cases (i) α_1 finite and $\psi(\xi)$ discontinuous at $\xi = \alpha_1$, (ii) $\alpha_1 = -\infty$. The principal way in which the analysis is affected is that the functions $h_s(\xi)$ are now defined in terms of infinite integrals. By hypothesis, however, $\int_{\alpha_1}^{\xi}|\psi(v)|\,dv$ converges, and this ensures the (absolute) convergence of all integrals appearing in the analysis. That the series (2.15) satisfies (2.09) is established by use of the dominated convergence theorem (Chapter 2, §8.2). The rest of the proof is unchanged.

It should be noted that the bounds (2.04) show that $w_1(x)$ satisfies the conditions

$$\varepsilon_1(x) \to 0, \qquad f^{-1/2}(x)\,\varepsilon_1'(x) \to 0 \qquad (x \to a_1+). \tag{2.19}$$

Similarly for the second solution.

2.4 The corresponding theorem for equations with oscillatory type solutions is as follows:

Theorem 2.2 *Assume the conditions of Theorem* 2.1, *and also that a is an arbitrary finite or infinite point in the closure of (a_1, a_2). Then in (a_1, a_2) the differential equation*

$$d^2w/dx^2 = \{-f(x)+g(x)\}\,w \tag{2.20}$$

has twice continuously differentiable solutions

$$\begin{aligned} w_1(x) &= f^{-1/4}(x)\exp\left\{i\int f^{1/2}(x)\,dx\right\}\{1+\varepsilon_1(x)\},\\ w_2(x) &= f^{-1/4}(x)\exp\left\{-i\int f^{1/2}(x)\,dx\right\}\{1+\varepsilon_2(x)\}, \end{aligned} \tag{2.21}$$

such that

$$|\varepsilon_j(x)|,\ f^{-1/2}(x)\,|\varepsilon_j'(x)| \leqslant \exp\{\mathscr{V}_{a,x}(F)\} - 1 \qquad (j = 1, 2), \tag{2.22}$$

provided that $\mathscr{V}_{a,x}(F) < \infty$. If $g(x)$ is real, then the solutions $w_1(x)$ and $w_2(x)$ are complex conjugates.

The proof is similar. The integral equation corresponding to (2.09) is

$$h(\xi) = \frac{1}{2i}\int_\alpha^\xi \{1-e^{2i(v-\xi)}\}\,\psi(v)\{1+h(v)\}\,dv, \tag{2.23}$$

where α is the value of ξ at $x = a$. The absence of the coefficient $\frac{1}{2}$ from the variation in (2.22), compared with (2.04), stems from the fact that the best bound for the kernel in (2.23) is given by $|1-e^{2i(v-\xi)}| \leqslant 2$.

The choice of reference point a governs the initial conditions satisfied by the solutions:

$$\varepsilon_j(x) \to 0, \qquad f^{-1/2}(x)\,\varepsilon_j'(x) \to 0 \qquad (x \to a, \quad j = 1, 2).$$

Similar freedom of choice is unavailable for Theorem 2.1, essentially because (2.12) does not hold when $\xi < v$.

Ex. 2.1 If the continuity conditions on $f''(x)$ and $g(x)$ are relaxed to sectional continuity, $f(x)$ and $f'(x)$ still being continuous, show that Theorems 2.1 and 2.2 apply except that the second derivatives of the solutions are discontinuous.

Ex. 2.2 Show that in the case of Theorem 2.1, $\frac{1}{2}|\varepsilon_j(x)+(-)^{j-1}f^{-1/2}(x)\,\varepsilon_j'(x)|$ is bounded by the right-hand side of (2.04), and in the case of Theorem 2.2, $|\varepsilon_j(x)+(-)^j if^{-1/2}(x)\,\varepsilon_j'(x)|$ is bounded by the right of (2.22).†

† These results are useful for the derivative of $f^{1/4}(x)\,w_j(x)$.

Ex. 2.3 Let a and b be arbitrary positive numbers. Show that in $[a, \infty)$ the equation

$$w''(x) = (e^{2x} + 2ib\cos x)\,w(x)$$

has solutions $\rho_1(x)\exp(-\frac{1}{2}x + e^x)$ and $\rho_2(x)\exp(-\frac{1}{2}x - e^x)$, where

$$|\rho_1(x) - 1| \leqslant \exp\{(b^2 + \tfrac{1}{64})^{1/2}(e^{-a} - e^{-x})\} - 1, \qquad |\rho_2(x) - 1| \leqslant \exp\{(b^2 + \tfrac{1}{64})^{1/2} e^{-x}\} - 1.$$

Ex. 2.4 If $w''(x) = (1 + \frac{1}{10}x^{-3})\,w(x)$, $w(1) = 1$, and $w'(1) = 0$, show that

$$w(2) = \{w_2(2)\,w_1'(1) - w_1(2)\,w_2'(1)\}/\{w_2(1)\,w_1'(1) - w_1(1)\,w_2'(1)\},$$

where $w_1(x)$, $w_2(x)$ are given by (2.03), with $a_1 = 1$, $a_2 = 2$. Hence compute the approximate value of $w(2)$, and estimate the maximum error in your result.

Ex. 2.5 With the aid of Exercise 2.2 show that for real $g(x)$ equation (2.20) has the general solution

$$w(x) = Af^{-1/4}(x)\left[\sin\left\{\int f^{1/2}(x)\,dx + \delta\right\} + \varepsilon(x)\right],$$

in which A and δ are arbitrary constants, and

$$|\varepsilon(x)|,\ f^{-1/2}(x)\,|\varepsilon'(x)| \leqslant \exp\{\mathscr{V}_{a,x}(F)\} - 1.$$

Deduce that if a_1 and a_2 are finite and the values of $w(a_1)$ and $w(a_2)$ are prescribed (a *boundary value problem*), then $w(x)$ is given by

$$\left\{\frac{f(a_2)}{f(x)}\right\}^{1/4} \frac{\sin\{\int_{a_1}^{x} f^{1/2}(t)\,dt\} + \varepsilon_1(x)}{\sin c + \varepsilon_1(a_2)}\, w(a_2) + \left\{\frac{f(a_1)}{f(x)}\right\}^{1/4} \frac{\sin\{\int_{x}^{a_2} f^{1/2}(t)\,dt\} + \varepsilon_2(x)}{\sin c + \varepsilon_2(a_1)}\, w(a_1),$$

where $c = \int_{a_1}^{a_2} f^{1/2}(t)\,dt$ and $|\varepsilon_j(x)| \leqslant \exp\{\mathscr{V}_{a_j,x}(F)\} - 1$.

3 Asymptotic Properties with Respect to the Independent Variable

3.1 From (2.19) we have the following information concerning the behavior, at the endpoint a_1, of the solution $w_1(x)$ introduced in Theorem 2.1:

$$w_1(x) \sim f^{-1/4}\exp\left(\int f^{1/2}\,dx\right) \qquad (x \to a_1+). \tag{3.01}$$

Similarly,

$$w_2(x) \sim f^{-1/4}\exp\left(-\int f^{1/2}\,dx\right) \qquad (x \to a_2-). \tag{3.02}$$

These results are valid whether or not a_1 and a_2 are finite, and also whether or not f and $|g|$ are bounded as a_1 and a_2 are approached: *it suffices that the error-control function $F(x)$ is of bounded variation in (a_1, a_2).*

In the interesting situation in which $\int f^{1/2}\,dx$ is unbounded as x approaches an endpoint, it is natural to enquire whether there exist solutions $w_3(x)$ and $w_4(x)$, say, with the complementary properties

$$w_3(x) \sim f^{-1/4}\exp\left(\int f^{1/2}\,dx\right) \qquad (x \to a_2-), \tag{3.03}$$

$$w_4(x) \sim f^{-1/4}\exp\left(-\int f^{1/2}\,dx\right) \qquad (x \to a_1+). \tag{3.04}$$

To resolve this question we first consider the behavior of $w_1(x)$ at a_2.

3.2 Theorem 3.1 *In addition to the conditions of Theorem* 2.1 *assume that* $\mathscr{V}_{a_1,a_2}(F) < \infty$ *and also that* $\int f^{1/2}\,dx \to \infty$ *as* $x \to a_2-$. *Then*

$$\varepsilon_1(x) \to a \text{ constant}, \qquad f^{-1/2}(x)\varepsilon_1'(x) \to 0 \qquad (x \to a_2-). \tag{3.05}$$

From Theorem 2.1 we know that $|\varepsilon_1(x)|$ is bounded throughout (a_1, a_2). The message of the present theorem is that there is no possibility of undamped oscillation in $\varepsilon_1(x)$ as x tends to a_2. The proof follows.

With the given conditions we have $\alpha_2 = \infty$; compare (1.05). Corresponding to an arbitrary small positive number η there exists $\gamma \in (\alpha_1, \infty)$, such that

$$\int_\gamma^\infty |\psi(v)|\,dv = \eta.$$

Assume that $\xi \geqslant \gamma$. Then by subdividing the integration range in the first of (2.16) at γ, we see that

$$|h_1'(\xi)| \leqslant \int_{\alpha_1}^{\gamma} e^{2(\gamma-\xi)}|\psi(v)|\,dv + \int_\gamma^\xi |\psi(v)|\,dv \leqslant e^{2(\gamma-\xi)}\Psi(\gamma) + \eta.$$

Similarly, from (2.13) and the second of (2.16) we derive

$$|h_{s+1}'(\xi) - h_s'(\xi)| \leqslant \frac{e^{2(\gamma-\xi)}\Psi^{s+1}(\gamma)}{(s+1)!\,2^s} + \frac{\Psi^s(\infty)}{s!\,2^s}\eta \qquad (s \geqslant 1).$$

Summation yields

$$|h'(\xi)| \leqslant 2e^{2(\gamma-\xi)}\{e^{\Psi(\gamma)/2} - 1\} + e^{\Psi(\infty)/2}\eta.$$

The first term on the right vanishes when $\xi \to \infty$. And since η is arbitrary this implies that $h'(\xi) \to 0$ as $\xi \to \infty$, which is equivalent to the second of (3.05).

Next, from (2.11), (2.14), (2.15), and (2.16), we have

$$h(\xi) = \tfrac{1}{2}\sum_{s=0}^{\infty} l_s(\xi) - \tfrac{1}{2}h'(\xi), \tag{3.06}$$

where

$$l_0(\xi) = \int_{\alpha_1}^\xi \psi(v)\,dv, \qquad l_s(\xi) = \int_{\alpha_1}^\xi \psi(v)\{h_s(v) - h_{s-1}(v)\}\,dv \qquad (s \geqslant 1). \tag{3.07}$$

Again, if $\xi \geqslant \gamma$ we deduce from (2.13) and the fact that $\Psi(\xi)$ is an increasing function

$$|l_s(\xi) - l_s(\gamma)| \leqslant \frac{\Psi^s(\xi)}{s!\,2^s}\int_\gamma^\xi |\psi(v)|\,dv \qquad (s \geqslant 0).$$

Therefore

$$|h(\xi) - h(\gamma)| \leqslant \tfrac{1}{2}e^{\Psi(\xi)/2}\{\Psi(\xi) - \Psi(\gamma)\} + \tfrac{1}{2}|h'(\xi) - h'(\gamma)|. \tag{3.08}$$

The right-hand side vanishes as ξ and γ tend to infinity independently, hence $h(\xi)$ tends to a constant limiting value. This establishes the first of (3.05), and completes the proof of Theorem 3.1. By symmetry, there is a similar result concerning $w_2(x)$ at a_1.

3.3 Before leaving the proof of Theorem 3.1, we indicate how to obtain information concerning the *manner* of approach of $\varepsilon_1(x)$ to its limit, $\varepsilon_1(a_2)$, say, as $x \to a_2-$. Letting $\xi \to \infty$ in (3.08), and then replacing γ by ξ, we obtain

$$|\varepsilon_1(x) - \varepsilon_1(a_2)| = |h(\xi) - h(\infty)| \leqslant \tfrac{1}{2}e^{\Psi(\infty)/2}\{\Psi(\infty) - \Psi(\xi)\} + \tfrac{1}{2}|h'(\xi)|. \tag{3.09}$$

From (2.13) and (2.16)

$$|h_1'(\xi)| \leqslant \int_{\alpha_1}^{\xi} e^{2(v-\xi)}|\psi(v)|\,dv, \qquad |h_{s+1}'(\xi) - h_s'(\xi)| \leqslant \frac{\Psi^s(\infty)}{s!\,2^s}\int_{\alpha_1}^{\xi} e^{2(v-\xi)}|\psi(v)|\,dv.$$

Summation and substitution in (3.09) gives

$$|\varepsilon_1(x) - \varepsilon_1(a_2)| \leqslant \tfrac{1}{2}e^{\Psi(\infty)/2}\left\{\int_{\xi}^{\infty}|\psi(v)|\,dv + \int_{\alpha_1}^{\xi} e^{2(v-\xi)}|\psi(v)|\,dv\right\}. \tag{3.10}$$

Further progress depends on the nature of $\psi(v)$ as $v \to \infty$: an illustration is provided in §4.1 below.

3.4 We return to the questions posed in §3.1. Again, let $\varepsilon_1(a_2)$ denote the limiting value of $\varepsilon_1(x)$ as $x \to a_2-$. Then from (2.03) we have

$$w_1(x) \sim \{1 + \varepsilon_1(a_2)\} f^{-1/4} \exp\left(\int f^{1/2}\,dx\right) \qquad (x \to a_2-), \tag{3.11}$$

provided that $\varepsilon_1(a_2) \neq -1$. The actual value of $\varepsilon_1(a_2)$ is not supplied by our theory, but a bound is given by (2.04) with $j = 1$ and $x = a_2$. From the standpoint of investigating the asymptotic behavior of solutions of the differential equation at a_2, we may replace a_1 by any convenient point $\hat{a}_1$, say, of (a_1, a_2). This change of course affects $w_1(x)$ and $\varepsilon_1(a_2)$, but (3.11) still holds. By making $\hat{a}_1$ sufficiently close to a_2, $\mathcal{V}_{\hat{a}_1, a_2}(F)$ can be made arbitrarily small, ensuring that $1 + \varepsilon_1(a_2)$ does not vanish. Then division of both sides of (3.11) by $1 + \varepsilon_1(a_2)$ establishes that there *does* exist a solution $w_3(x)$ with the property (3.03).

Furthermore, since the choice of $\hat{a}_1$ in the foregoing construction is arbitrary (to some extent) the solution $w_3(x)$ is not unique. Thus the situation at the endpoints is analogous to that encountered in Chapter 5, §7. At a_2 the solution $w_3(x)$ characterized by (3.03) is *dominant* but *not unique*, whereas the solution characterized by (3.02) is *recessive* and *unique*. Similarly for the solutions $w_1(x)$ and $w_4(x)$ at a_1.

3.5 Different conclusions obtain in the case of Theorem 2.2. For example, if $\int f^{1/2}\,dx \to \infty$ as $x \to a_2-$, then in general the error terms $\varepsilon_1(x)$ and $\varepsilon_2(x)$ oscillate as $x \to a_2-$. Furthermore, the solution with either of the properties

$$w(x) \sim f^{-1/4} \exp\left(\pm i \int f^{1/2}\,dx\right) \qquad (x \to a_2-) \tag{3.12}$$

is unique. Both of these statements may be verified by expressing the general solution of the differential equation as a linear combination of solutions furnished by Theorem 2.2.

3.6 As an example, consider solutions of the equation

$$w'' = (x+\ln x)w \tag{3.13}$$

as $x \to \infty$. We cannot take $f = x$ and $g = \ln x$, because $\int gf^{-1/2}\,dx$ would diverge at $a_2 \equiv \infty$. Accordingly, we set $f = x + \ln x$ and $g = 0$. Then it is easily seen that for large x, $f^{-1/4}(f^{-1/4})''$ is $O(x^{-5/2})$. Hence $\mathscr{V}(F)$ converges at ∞, and asymptotic solutions of (3.13) are

$$(x+\ln x)^{-1/4}\exp\left\{\pm\int(x+\ln x)^{1/2}\,dx\right\}.$$

This result can be simplified. We have, again for large x,

$$(x+\ln x)^{1/2} = x^{1/2} + \tfrac{1}{2}x^{-1/2}\ln x + O\{x^{-3/2}(\ln x)^2\}.$$

Hence

$$\int(x+\ln x)^{1/2}\,dx = \tfrac{2}{3}x^{3/2} + x^{1/2}\ln x - 2x^{1/2} + \text{constant} + o(1).$$

Accordingly, equation (3.13) has a unique solution $w_2(x)$ such that

$$w_2(x) \sim x^{-(1/4)-\sqrt{x}}\exp(2x^{1/2} - \tfrac{2}{3}x^{3/2}) \qquad (x\to\infty),$$

and a nonunique solution $w_3(x)$ such that

$$w_3(x) \sim x^{-(1/4)+\sqrt{x}}\exp(\tfrac{2}{3}x^{3/2} - 2x^{1/2}) \qquad (x\to\infty).$$

Ex. 3.1 Assume the conditions of Theorem 2.1, and also that $\mathscr{V}_{a_1,a_2}(F) < \infty$, $\int f^{1/2}dx \to \infty$ as $x \to a_2-$, and $\int f^{1/2}\,dx \to -\infty$ as $x \to a_1+$. By considering the Wronskian with respect to ξ of $f^{1/4}w_1$ and $f^{1/4}w_2$, prove that $\varepsilon_1(a_2) = \varepsilon_2(a_1)$.

Ex. 3.2 Show that the equation $w'' - \frac{1}{2}w' + (\frac{1}{16} + x - e^x)w = 0$ has solutions of the form $\{1 + O(xe^{-x/2})\}\exp(-2e^{x/2})$ and $\{1 + o(1)\}\exp(2e^{x/2})$ as $x \to \infty$.

Ex. 3.3 Show that the equation $w'' + (2x^{-3} + x^{-4})w = 0$ has a pair of conjugate solutions of the form $x^{1\mp i}e^{\pm i/x}\{1 + \frac{1}{2}(\pm i - 1)x + O(x^2)\}$ as $x \to 0+$.

4 Convergence of $\mathscr{V}(F)$ at a Singularity

4.1 If a_2 is finite, then sufficient conditions for $\mathscr{V}(F)$ to be bounded at a_2 are given by

$$f(x) \sim \frac{c}{(a_2-x)^{2\alpha+2}}, \qquad g(x) = O\left\{\frac{1}{(a_2-x)^{\alpha-\beta+2}}\right\} \qquad (x \to a_2-), \tag{4.01}$$

where c, α, and β are positive constants, provided that the first of these relations is twice differentiable. For then

$$f^{-1/4}(f^{-1/4})'' = O\{(a_2-x)^{\alpha-1}\}, \qquad gf^{-1/2} = O\{(a_2-x)^{\beta-1}\}; \tag{4.02}$$

compare (1.07). Accordingly, $F'(x) = O\{(a_2-x)^{\delta-1}\}$, where $\delta = \min(\alpha, \beta)$. Since $\delta > 0$ we have $\mathscr{V}_{x,a_2}(F) < \infty$, enabling Theorems 2.1, 2.2, and 3.1 to be applied.

In the case of Theorems 2.1 and 3.1, more refined information concerning the limiting behavior of $\varepsilon_1(x)$ and $\varepsilon_2(x)$ at a_2 is derivable as follows. From (4.02) $\mathcal{V}_{x,a_2}(F) = O\{(a_2-x)^\delta\}$. Hence

$$\varepsilon_2(x) = O\{(a_2-x)^\delta\} \qquad (x \to a_2-); \tag{4.03}$$

compare (2.04). Next, from (1.05), (2.06), and (4.02) we see that

$$\xi \sim \frac{c^{1/2}}{\alpha(a_2-x)^\alpha}, \qquad \psi(\xi) = O\left\{\frac{1}{\xi^{1+(\delta/\alpha)}}\right\} \qquad (x \to a_2-). \tag{4.04}$$

Therefore in (3.10) the first integral within the braces is $O(\xi^{-\delta/\alpha})$ for large ξ. And by subdividing the interval (α_1, ξ) at $\frac{1}{2}\xi$, we see that the second integral is bounded by

$$e^{-\xi}\int_{\alpha_1}^{\xi/2} |\psi(v)|\,dv + O\left(\frac{1}{\xi^{1+(\delta/\alpha)}}\right)\int_{\xi/2}^{\xi} e^{2(v-\xi)}\,dv,$$

that is, $O(\xi^{-1-(\delta/\alpha)})$. Hence the whole of the right-hand side of (3.10) is $O(\xi^{-\delta/\alpha})$. Accordingly,

$$\varepsilon_1(x) - \varepsilon_1(a_2) = O\{(a_2-x)^\delta\} \qquad (x \to a_2-). \tag{4.05}$$

Relations (4.03) and (4.05) are the desired refinements.

It will be noted that the conditions (4.01) include the case when the differential equation (2.02) or (2.20) has an irregular singularity at a_2 of arbitrary rank α; compare Chapter 5, §4.1.†

4.2 In a similar way, if $a_2 = \infty$ then sufficient conditions for $\mathcal{V}_{x,\infty}(F) < \infty$ are given by

$$f(x) \sim cx^{2\alpha-2}, \qquad g(x) = O(x^{\alpha-\beta-2}) \qquad (x \to \infty), \tag{4.06}$$

where c, α, and β are positive constants. Again, the first of these relations has to be twice differentiable: when $\alpha = \frac{3}{2}$ we interpret this as $f'(x) \to c$ and $f''(x) = O(x^{-1})$; when $\alpha = 1$ we require $f'(x) = O(x^{-1})$ and $f''(x) = O(x^{-2})$. The conditions include the case of an irregular singularity at infinity of arbitrary rank α.

Corresponding to (4.03) and (4.05), we derive, by similar analysis,

$$\varepsilon_2(x) = O(x^{-\delta}), \qquad \varepsilon_1(x) - \varepsilon_1(\infty) = O(x^{-\delta}) \qquad (x \to \infty), \tag{4.07}$$

where, again, $\delta = \min(\alpha, \beta)$.

4.3 Success with irregular singularities suggests we enquire whether the LG approximation holds at regular singularities. By definition (Chapter 5, §4.1), a_2 is a regular singularity of the equation

$$d^2w/dx^2 = q(x)w \tag{4.08}$$

if $q(x)$ can be expanded in a series of the form

$$q(x) = \frac{1}{(a_2-x)^2}\sum_{s=0}^{\infty} q_s(a_2-x)^s,$$

† The symbols f and g are now being used differently.

convergent in a neighborhood of a_2. Equation (4.08) can be expressed in the standard form of Theorem 2.1 or Theorem 2.2 by arbitrarily partitioning $q(x) = \pm f(x) + g(x)$, with

$$f(x) = \frac{1}{(a_2-x)^2}\sum_{s=0}^{\infty} f_s(a_2-x)^s, \qquad g(x) = \frac{1}{(a_2-x)^2}\sum_{s=0}^{\infty} g_s(a_2-x)^s, \quad (4.09)$$

and $\pm f_s + g_s = q_s$. We stipulate that the f_s are real and $f_0 \geqslant 0$ (because $f(x)$ must be positive).

Suppose first that $f_0 \neq 0$. Then for sufficiently small $|a_2 - x|$ we have

$$f^{-1/4}(f^{-1/4})'' - gf^{-1/2} = \frac{1}{a_2-x}\sum_{s=0}^{\infty} c_s(a_2-x)^s,$$

where the coefficients c_s depend on f_s and g_s; in particular, $c_0 = -(\tfrac{1}{4}+g_0)f_0^{-1/2}$. For $\mathscr{V}_{x,a_2}(F)$ to converge, it is clearly necessary and sufficient that $c_0 = 0$. Except when $q_0 = -\tfrac{1}{4}$ this can be arranged by taking $g_0 = -\tfrac{1}{4}$ and $f_0 = |q_0 + \tfrac{1}{4}|$. If $q_0 > -\tfrac{1}{4}$, then relations (3.02) and (3.11) apply; alternatively if $q_0 < -\tfrac{1}{4}$, then (3.12) applies.

In the exceptional case $q_0 = -\tfrac{1}{4}$, f and g cannot be chosen in such a way that $\mathscr{V}_{x,a_2}(F) < \infty$. For suppose that f_r $(r \geqslant 1)$ is the first nonvanishing coefficient in the expansion (4.09) of $f(x)$. Since $g_0 = -\tfrac{1}{4}$, we have

$$f^{-1/4}(f^{-1/4})'' - gf^{-1/2} \sim \tfrac{1}{16} r^2 f_r^{-1/2}(a_2-x)^{-(r/2)-1} \qquad (x \to a_2-).$$

Hence $\mathscr{V}_{x,a_2}(F) = \infty$. Complications also arise in the theory of Chapter 5, §§4 and 5 when $q_0 = -\tfrac{1}{4}$, because the indicial equation has equal roots.

Similar analysis and conclusions apply when a_2 is a regular singular point which is located at $+\infty$: details are left to the reader.

4.4 The main results of §3 and the present section may be summarized by the following statement. *With proper choice of f and g, the LG functions provide asymptotic representations of dominant and recessive solutions in the neighborhood of an irregular singularity of arbitrary rank, and also in the neighborhood of a regular singularity not having equal exponents.*

Ex. 4.1 Prove that for large positive x the equation $w'' - x^3w' + x^{-2}w = 0$ has independent solutions of the forms $1 + O(x^{-4})$ and $x^{-3}\exp(\tfrac{1}{4}x^4)\{1 + O(x^{-4})\}$.

Ex. 4.2 If $q(x)$ is continuous in $(0, b)$ and $\int_0^b x|q(x)|\,dx < \infty$, establish that the equation $w'' = q(x)w$ has solutions of the form $1 + o(1)$ and $x + o(x)$ as $x \to 0+$.

Ex. 4.3 Let $f(x)$ be analytic at a finite point a, having a zero of any order there, and $g(x)$ be bounded as $x \to a$. Show that $\mathscr{V}(F)$ diverges at a.

***Ex. 4.4** If $f > 0$, f'' is continuous, $g = 0$, and $\int_x^\infty |f^{-3/2}f''|\,dx < \infty$, show that $\mathscr{V}_{x,\infty}(F) < \infty$ and $\int_x^\infty f^{1/2}\,dx = \infty$. [Coppel, 1965.]

***Ex. 4.5** If $f > 0$, f'' is continuous, $g = 0$, $\mathscr{V}_{x,\infty}(F) < \infty$, and $\int_x^\infty f^{1/2}\,dx < \infty$, deduce from Exercise 4.4 that $\int_x^\infty f^{-5/2}f'^2\,dx = \infty$, and thence that $f^{-3/2}f' \to -\infty$ as $x \to \infty$.

From these results and the identity $(f^{-1/4})' = \text{constant} - \int_x^\infty f^{-1/4}(f^{-1/4})''f^{1/4}\,dx$, show that $f \sim dx^{-4}$ and $f' \sim -4dx^{-5}$ as $x \to \infty$, where d is a positive constant. [Coppel, 1965.]

5 Asymptotic Properties with Respect to Parameters

5.1 Consider the equation

$$d^2w/dx^2 = \{u^2f(x)+g(x)\}w \tag{5.01}$$

in which u is a positive parameter, and the functions $f(x)$ and $g(x)$ are independent of u. Equations of this form are satisfied, for example, by several of the special functions of Chapters 2 and 5. We again suppose that in a given interval (a_1, a_2), $f(x)$ is positive, and $f''(x)$ and $g(x)$ are continuous. Applying Theorem 2.1 and discarding an irrelevant factor $u^{-1/2}$, we see that equation (5.01) has solutions

$$w_j(u,x) = f^{-1/4}(x)\exp\{(-)^{j-1}u\int f^{1/2}(x)\,dx\}\{1+\varepsilon_j(u,x)\} \qquad (j=1,2),$$

such that

$$|\varepsilon_j(u,x)|,\ \frac{|\varepsilon_j'(u,x)|}{2uf^{1/2}(x)} \leqslant \exp\left\{\frac{\mathscr{V}_{a_j,x}(F)}{2u}\right\} - 1. \tag{5.02}$$

Here primes denote partial differentiations with respect to x, and $F(x)$ is again defined by (2.01). Since $F(x)$ is independent of u, the right-hand side of (5.02) is $O(u^{-1})$ for large u and fixed x. Moreover, *if* $\mathscr{V}_{a_1,a_2}(F) < \infty$, *then this* O *term is uniform with respect to* x, *because* $\mathscr{V}_{a_j,x}(F) \leqslant \mathscr{V}_{a_1,a_2}(F)$. Asymptotically,

$$w_j(u,x) \sim f^{-1/4}\exp\{(-)^{j-1}u\int f^{1/2}\,dx\} \qquad (u\to\infty), \tag{5.03}$$

uniformly in (a_1, a_2).

The original importance of the LG approximation stemmed from this property and the analogous result obtained when Theorem 2.2 is applied to the equation

$$d^2w/dx^2 = \{-u^2f(x)+g(x)\}w. \tag{5.04}$$

We obtained (5.03) as an immediate consequence of the error bounds supplied by Theorem 2.1. Furthermore, as we saw in §4, these bounds reveal an asymptotic property of the approximation in the neighborhood of a singularity of the differential equation. On account of this *double* asymptotic feature the LG approximation is a remarkably powerful tool for approximating solutions of linear second-order differential equations.

5.2 By how much do the error bounds (5.02) overestimate the *actual* errors? A partial answer is found by determining the asymptotic forms of the $\varepsilon_j(u,x)$ as $u\to\infty$. Using hats to distinguish the symbols in the present case from the corresponding symbols in §2, we have

$$\hat{\xi} = u\xi,\ \hat{\alpha}_j = u\alpha_j,\ \hat{\psi}(\hat{\xi}) = u^{-2}\psi(\xi),\ \hat{\Psi}(\hat{\xi}) = u^{-1}\Psi(\xi) = u^{-1}\mathscr{V}_{a_1,x}(F).$$

By separating the first term in the expansion (2.15) and using (2.11), we see that

$$\varepsilon_1(u,x) \equiv \hat{\varepsilon}_1(x) = (2u)^{-1}\int_{\alpha_1}^{\xi}\psi(v)\,dv - \theta_1(u,x) + \theta_2(u,x), \tag{5.05}$$

where

$$\theta_1(u,x) = (2u)^{-1}\int_{\alpha_1}^{\xi} e^{2u(v-\xi)}\psi(v)\,dv, \qquad \theta_2(u,x) = \sum_{s=1}^{\infty}\{\hat{h}_{s+1}(\xi) - \hat{h}_s(\xi)\}.$$

Because $\psi(v)$ is continuous in (α_1, α_2), Laplace's method (Chapter 3, §7) shows that

$$\theta_1(u,x) = O(u^{-2}) \qquad (u \to \infty),$$

except possibly when $x = a_2$. Also, from (2.13)

$$|\theta_2(u,x)| \leqslant \sum_{s=2}^{\infty} \frac{\{\mathscr{V}_{a_1,x}(F)\}^s}{s!\,(2u)^s}.$$

Substitution of these results in (5.05) gives

$$\varepsilon_1(u,x) = -(2u)^{-1}\{F(x) - F(a_1)\} + O(u^{-2}) \tag{5.06}$$

as $u \to \infty$. This is the required result.

The asymptotic form of the bound (5.02) for $|\varepsilon_1(u,x)|$ is

$$(2u)^{-1}\mathscr{V}_{a_1,x}(F) + O(u^{-2}).$$

Obviously this is related closely to (5.06). Indeed, to within $O(u^{-2})$ it is the *same* as the modulus of (5.06) in the case when F is monotonic in the interval (a_1, x). In these circumstances the error bound is particularly realistic.

5.3 The differential equation may have a singularity at either, or both, of the endpoints without invalidating the uniform validity of (5.03), provided that $\mathscr{V}(F)$ converges at both endpoints. This is the case, for example, when $f(x)$ and $g(x)$ satisfy conditions (4.01), when a_2 is finite, or (4.06), when $a_2 = \infty$.

Next, consider a regular singularity at a finite endpoint a_2, say. For small $|a_2 - x|$ the functions $f(x)$ and $g(x)$ can be expanded in convergent power series

$$f(x) = \frac{1}{(a_2-x)^2}\sum_{s=0}^{\infty} f_s(a_2-x)^s, \qquad g(x) = \frac{1}{(a_2-x)^2}\sum_{s=0}^{\infty} g_s(a_2-x)^s,$$

in which the f_s are real, and $f_0 \geqslant 0$. Suppose first that $f_0 \neq 0$. As in §4.3, we can show that $\mathscr{V}(F)$ converges at a_2 when $g_0 = -\frac{1}{4}$. When $g_0 \neq -\frac{1}{4}$, we can arrange for a convergent variation by adopting a new parameter

$$\hat{u} \equiv \{u^2 \pm f_0^{-1}(\tfrac{1}{4} + g_0)\}^{1/2},$$

the upper sign applying to equation (5.01), and the lower sign to (5.04). In terms of $\hat{u}$, the differential equation becomes

$$d^2w/dx^2 = \{\pm\hat{u}^2 f(x) + \hat{g}(x)\}\,w,$$

where

$$\hat{g}(x) = g(x) - f_0^{-1}(\tfrac{1}{4} + g_0)\,f(x).$$

In the expansion of $\hat{g}(x)$ in ascending powers of $a_2 - x$, the coefficient of $(a_2 - x)^{-2}$ is $-\frac{1}{4}$; hence the variation of the new error-control function converges at a_2.

Now suppose that $f_0 = 0$ but $f_1 \neq 0$. In the neighborhood of a_2

$$f^{-1/4}(f^{-1/4})'' - gf^{-1/2} = (a_2-x)^{-3/2} \sum_{s=0}^{\infty} c_s(a_2-x)^s,$$

where $c_0 = -(g_0+\frac{3}{16})f_1^{-1/2}$. Accordingly, $\mathscr{V}(F)$ converges if and only if $g_0 = -\frac{3}{16}$. This time, however, we are unable to treat cases in which $g_0 \neq -\frac{3}{16}$ by simple redefinition of the parameter. We defer these more difficult cases until Chapter 12.

When the singularities of $f(x)$ and $g(x)$ are poles the results of this subsection can be summarized as follows. At a finite point a, let $f(x)$ have a pole of order m, and $g(x)$ a pole of order n, with the understanding that $n = 0$ signifies $g(x)$ is analytic.

(i) *If* $m > 2$ *and* $0 \leqslant n < \frac{1}{2}m+1$, *then* $\mathscr{V}(F)$ *converges at* a.

(ii) *If* $m = 2$ *and* $n = 0$, 1, *or* 2, *then by redefinition of the parameter* u *it can be arranged for* $\mathscr{V}(F)$ *to converge at* a.

(iii) *If* $m = 1$, *then* $\mathscr{V}(F)$ *diverges at* a, *except in the special case given by* $g(x) \sim -\frac{3}{16}(x-a)^{-2}$ *as* $x \to a$.

Similar conclusions hold when $f(x)$ and $g(x)$ are singular at the point at infinity.

5.4 We can also cope with certain differential equations in which the parameter u appears in other ways. Consider the more general equation

$$d^2w/dx^2 = \{u^2 f(u,x) + g(u,x)\}\, w. \tag{5.07}$$

It is readily seen from Theorem 2.1 that there exist solutions $w_j(u,x)$ of (5.07) which satisfy (5.03) uniformly with respect to x, provided that the following conditions are fulfilled for $x \in (a_1, a_2)$ and all sufficiently large positive u:

(i) $f(u,x) > 0$.

(ii) $\partial^2 f(u,x)/\partial x^2$ *and* $g(u,x)$ *are continuous functions of* x.

(iii) $\mathscr{V}_{a_1,a_2}(F) = o(u)$ *as* $u \to \infty$.

The points a_1 and a_2 may depend on u.

Included in (5.07), for example, are equations of the form

$$d^2w/dx^2 = \{u^2 f_0(x) + u f_1(x) + f_2(x)\}\, w,$$

in which the functions $f_s(x)$ are independent of u. Obviously we may take

$$f(u,x) = f_0(x) + u^{-1} f_1(x) + u^{-2} f_2(x), \qquad g(u,x) = 0,$$

but other choices may be preferable, for example,

$$f(u,x) = f_0(x) + u^{-1} f_1(x), \qquad g(u,x) = f_2(x),$$

or

$$f(u,x) = \left\{ f_0^{1/2}(x) + \frac{f_1(x)}{2u f_0^{1/2}(x)} \right\}^2, \qquad g(u,x) = f_2(x) - \frac{f_1^2(x)}{4 f_0(x)},$$

the last version having the advantage that it simplifies the evaluation of $\int f^{1/2}(u,x)\, dx$. Of course, the magnitude of the error bound is affected by the actual choice of $f(u,x)$ and $g(u,x)$. It might happen that $\mathscr{V}_{a_1,a_2}(F)$ converges for one choice but not with others. Clearly the convergent choice would then be preferable.†

Ex. 5.1 Show that the error bound obtained when Theorem 2.2 is applied to equation (5.04) overestimates the actual value of $|\varepsilon_j(u,x)|$ by a factor of approximately 2 when u is large and F is monotonic in (a,x).

Ex. 5.2 Suppose that in the neighborhood of the origin

$$f(x) = x\sum_{s=0}^{\infty} f_s x^s, \qquad g(x) = \frac{1}{x^2}\sum_{s=0}^{\infty} g_s x^s,$$

where $f_0 \neq 0$. Show that $\mathscr{V}(F)$ converges at $x=0$ if and only if $g_0 = \frac{5}{16}$ and $g_1 = f_1/(8f_0)$.

Ex. 5.3 By constructing the differential equation for $(x^2-1)^{1/2}Q_n^m(x)$ prove that if m is fixed and n is large and positive, then the Legendre function of the second kind is given by

$$Q_n^m(\cosh t) = \pi^{1/2}e^{m\pi i}n^{m-(1/2)}(2\sinh t)^{-1/2}e^{-\{n+(1/2)\}t}\{1+O(n^{-1})\},$$

uniformly in the t interval $[\delta,\infty)$, where δ is any positive constant.

Ex. 5.4 If a, x, and u are positive, a being fixed and u large, show that in $[a,\infty)$ the equation $d^2w/dx^2 = (u^4x^2+u^2x^4)w$ has solutions uniformly of the form

$$\{1+O(u^{-2})\}x^{-1/2}(u^2+x^2)^{-1/4}\exp\{\pm\tfrac{1}{3}u(u^2+x^2)^{3/2}\}.$$

Show also that the x interval may be extended to $[au^{-1/2},\infty)$, provided that the uniform error term $O(u^{-2})$ is changed to $O(u^{-1})$.

Ex. 5.5 Show that for positive x and u the equation

$$\frac{d^2w}{dx^2} = \left(1+\frac{\cos u}{ux^{3/4}}\right)w$$

has a solution of the form $\{1+\varepsilon(u,x)\}\exp(-x-2u^{-1}x^{1/4}\cos u)$, where (i) $\varepsilon(u,x)=O(x^{-1/2})$ as $x\to\infty$, u fixed; (ii) $\varepsilon(u,x)=O(u^{-1}\cos u)$ as $u\to\infty$ uniformly for $x\in[a,\infty)$, a being any positive constant.

6 Example: Parabolic Cylinder Functions of Large Order

6.1 The differential equation for the *Weber parabolic cylinder functions* is

$$d^2w/dx^2 = (\tfrac{1}{4}x^2+a)w, \tag{6.01}$$

a being a parameter. The only singularity is at infinity, and is irregular and of rank 2. Accordingly, asymptotic solutions for fixed a and large x are derivable from the LG approximation. The choice $f=\frac{1}{4}x^2$, $g=a$, is inappropriate (unless $a=0$) because the corresponding error-control function F diverges at infinity. Instead, we take $f=\frac{1}{4}x^2+a$, $g=0$; then $f^{-1/4}(f^{-1/4})''$ is asymptotic to $\frac{3}{2}x^{-3}$ and $\mathscr{V}(F)<\infty$. From the theory of §3 there exist solutions of (6.01) that are asymptotic to $f^{-1/4}e^{\pm\xi}$ as $x\to\infty$, where

$$\xi = \int(\tfrac{1}{4}x^2+a)^{1/2}\,dx.$$

†An interesting example has been given by Jeffreys (1953, §3.3).

For large x,

$$\xi = \tfrac{1}{4}x^2 + a \ln x + \text{constant} + O(x^{-2}).$$

Hence the asymptotic forms of the solutions reduce to constant multiples of $x^{a-(1/2)}e^{x^2/4}$ and $x^{-a-(1/2)}e^{-x^2/4}$.

The principal solution $U(a, x)$ is specified (completely) by the condition

$$U(a,x) \sim x^{-a-(1/2)}e^{-x^2/4} \qquad (x \to +\infty). \tag{6.02}$$

Like all solutions, it is entire in x. In an older notation, due to Whittaker, $U(a, x)$ is denoted by $D_{-a-(1/2)}(x)$.

6.2 How does $U(a, x)$ behave as $a \to +\infty$? If we apply the theory of §5 with $u^2 = a$, $f = 1$, and $g = \frac{1}{4}x^2$, then the resulting $\mathscr{V}(F)$ diverges at infinity. Hence this approach produces asymptotic approximations for large a which are valid only in compact x intervals.

To derive an approximation which is uniformly valid for *unbounded* real x, we again take $f = \frac{1}{4}x^2 + a$. The variables are separated in a convenient way by setting $a = \frac{1}{2}u$ and $x = (2u)^{1/2}t$. Equation (6.01) becomes

$$d^2w/dt^2 = u^2(t^2+1)w.$$

From §5.1, the solution which is recessive at $t = +\infty$ is given by

$$w(u,t) = (t^2+1)^{-1/4}\exp\{-u\hat{\xi}(t)\}\{1+\varepsilon(u,t)\},$$

where

$$\hat{\xi}(t) = \int (t^2+1)^{1/2}\,dt = \tfrac{1}{2}t(t^2+1)^{1/2} + \tfrac{1}{2}\ln\{t+(t^2+1)^{1/2}\}. \tag{6.03}$$

The error term satisfies

$$|\varepsilon(u,t)| \leqslant \exp\{(2u)^{-1}\mathscr{V}_{t,\infty}(F)\} - 1, \tag{6.04}$$

in which

$$F(t) = \int (t^2+1)^{-1/4}\{(t^2+1)^{-1/4}\}''\,dt = \int \frac{3t^2-2}{4(t^2+1)^{5/2}}\,dt = -\frac{t^3+6t}{12(t^2+1)^{3/2}}. \tag{6.05}$$

For fixed u and large t, we have

$$\hat{\xi}(t) = \tfrac{1}{2}t^2 + \tfrac{1}{2}\ln(2t) + \tfrac{1}{4} + O(t^{-2}), \qquad F(t) = -\tfrac{1}{12} + O(t^{-2}),$$
$$\varepsilon(u,t) = O(t^{-2}).$$

Hence

$$w(u,t) = 2^{-u/2}e^{-u/4}t^{-(u+1)/2}e^{-ut^2/2}\{1+O(t^{-2})\}.$$

Since $U(\frac{1}{2}u, \sqrt{2u}\,t)$ is recessive in the same circumstances, it is a multiple of $w(u,t)$. The actual value of the multiple is easily found by comparison with (6.02); in this way we arrive at the desired result, given by

$$U(\tfrac{1}{2}u, \sqrt{2u}\,t) = 2^{(u-1)/4}e^{u/4}u^{-(u+1)/4}(t^2+1)^{-1/4}\exp\{-u\hat{\xi}(t)\}\{1+\varepsilon(u,t)\}. \tag{6.06}$$

6.3 Relations (6.04) and (6.06) hold for positive u and all real t, or on returning to the original variables, positive a and all real x. For fixed u (not necessarily large) and large positive t, we have $\varepsilon(u,t) = O(t^{-2})$. On the other hand, since $\mathscr{V}_{-\infty,\infty}(F) < \infty$ we have $\varepsilon(u,t) = O(u^{-1})$ for large u, uniformly with respect to t. These results illustrate the doubly asymptotic nature of the LG approximation.

The uniform error bound is evaluable, as follows. From (6.05) it is seen that the stationary points of $F(t)$ are $t = \pm\sqrt{\frac{2}{3}}$. We find that

$$F(-\infty) = \tfrac{1}{12}, \qquad F(-\sqrt{\tfrac{2}{3}}) = \tfrac{1}{3}\sqrt{\tfrac{2}{3}}, \qquad F(\sqrt{\tfrac{2}{3}}) = -\tfrac{1}{3}\sqrt{\tfrac{2}{3}}, \qquad F(\infty) = -\tfrac{1}{12}.$$

Hence $\mathscr{V}_{-\infty,\infty}(F) = \frac{4}{3}\sqrt{\frac{2}{3}} - \frac{1}{6} = 0.67\ldots$, giving

$$|\varepsilon(u,t)| \leqslant \exp\{(0.33\ldots)/u\} - 1.$$

In particular, with $\varepsilon(u,t)$ neglected the LG approximation for $U(\frac{1}{2}u, \sqrt{2u}\,t)$ is correct to within 10% when $u > 3.6$, that is, $a > 1.8$. This low value of the "large" parameter illustrates the powerful nature of the approximation; it is by no means untypical.

Ex. 6.1 By differentiation under the sign of integration verify that the integral

$$\int_0^\infty \exp(-xt - \tfrac{1}{2}t^2)\, t^{a-(1/2)}\, dt \qquad (a > -\tfrac{1}{2})$$

satisfies the same differential equation as $\exp(\frac{1}{4}x^2)\,U(a,x)$. By considering the asymptotic form of the integral for large x deduce that

$$U(a,x) = \frac{\exp(-\frac{1}{4}x^2)}{\Gamma(a+\frac{1}{2})} \int_0^\infty \exp(-xt - \tfrac{1}{2}t^2)\, t^{a-(1/2)}\, dt \qquad (a > -\tfrac{1}{2}).$$

7 A Special Extension

7.1 Let $g(x)$ have a simple pole at $x = 0$, u again denote a large positive parameter, and w satisfy

$$d^2w/dx^2 = \{-u^2 + g(x)\}\,w. \tag{7.01}$$

With $f(x) = u^2$ the error-control function for this equation is $-u^{-1}\int g\,dx$, and is infinite at $x = 0$. Accordingly, Theorem 2.2 yields no information in the neighborhood of this point. This is to be expected; the theory of Chapter 5, §5 informs us that the general solution of (7.01) has a logarithmic singularity at $x = 0$; therefore it cannot be represented adequately by the general solution of

$$d^2w/dx^2 = -u^2 w. \tag{7.02}$$

However, the *recessive* solution of (7.01) at the origin is free from singularity, and can be approximated uniformly for large u by the solution of (7.02) which vanishes at $x = 0$. Because of applications in scattering theory and the intrinsic interest of the problem, details follow.†

† See also Chapter 12, §6, and p. 479. Similar results can be found for the equation $d^2w/dx^2 = \{u^2 + g(x)\}\,w$.

Fig. 7.1 sm(x).

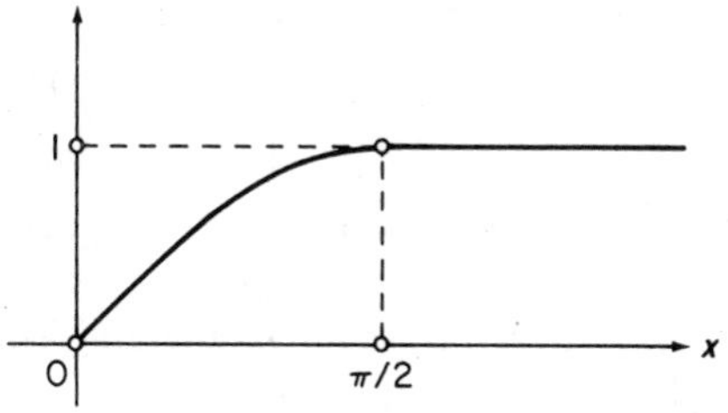

7.2 Since we shall give explicit error bounds, $g(x)$ is allowed to depend on u in the main theorem; thus

$$d^2w/dx^2 = \{-u^2+g(u,x)\}\, w. \tag{7.03}$$

We assume x ranges over a finite or infinite interval $(0, b)$, and do not restrict the singularity of $g(u,x)$ at $x = 0$ to be a simple pole. We also introduce the majorant

$$\mathrm{sm}(x) = \max_{0 \leqslant t \leqslant x} |\sin t|. \tag{7.04}$$

Obviously $\mathrm{sm}(x)$ is a nondecreasing function; see Fig. 7.1.

Theorem 7.1 *Assume that* $g(u,x)$ *is a continuous real or complex function of* x *in* $(0, b)$ *and the integral*

$$G(u,x) \equiv \frac{1}{u}\int_0^x \mathrm{sm}(ut)\,|g(u,t)|\, dt \tag{7.05}$$

converges at its lower limit. Then equation (7.03) *has a solution* $w(u,x)$ *that is continuously differentiable in* $[0, b)$, *twice continuously differentiable in* $(0, b)$, *and is given by*

$$w(u,x) = \sin(ux) + \varepsilon(u,x), \tag{7.06}$$

where

$$|\varepsilon(u,x)| \leqslant \mathrm{sm}(ux)[\exp\{G(u,x)\}-1]. \tag{7.07}$$

The proof of this result is a refinement of the proof of Theorems 2.1 and 2.2. The integral equation for $\varepsilon(u,x)$ is found to be

$$\varepsilon(u,x) = \frac{1}{u}\int_0^x \sin\{u(x-t)\}\, g(u,t)\{\sin(ut)+\varepsilon(u,t)\}\, dt.$$

Solution by successive approximations and use of the bound $|\sin\{u(x-t)\}| \leqslant 1$ would lead to a result equivalent to Theorem 2.2. The desired inequality (7.07) is obtained by use instead of the sharper bound

$$|\sin\{u(x-t)\}| \leqslant \mathrm{sm}(ux) \qquad (0 \leqslant t \leqslant x).$$

Details are left to the reader.

7.3 In scattering problems $g(u,x) \equiv g(x)$ is independent of u, $b = \infty$, and $g(x)$ is absolutely integrable at ∞. The theory of §3.5 (with $a_2 = \infty$) shows that $w(u,x)$

can also be expressed in the form

$$w(u,x) = (1+\rho)\sin(ux+\delta) + o(1) \qquad (x\to\infty), \tag{7.08}$$

where ρ and δ are independent of x, with $1+\rho>0$ and $-\pi<\delta\leqslant\pi$. Physically, u^2 represents the *energy* of the scattering particle, $g(x)$ the *potential*, and δ the *phase shift*.†

Bounds for ρ and δ can be derived from the uniform bound for $\varepsilon(u,x)$, as follows. Combination of (7.06) with (7.08) yields

$$\varepsilon(u,x) + o(1) = (1+\rho)\sin(ux+\delta) - \sin(ux) = \sigma\sin(ux+\eta), \tag{7.09}$$

where σ and η are related to ρ and δ by

$$(1+\rho)\cos\delta - 1 = \sigma\cos\eta, \qquad (1+\rho)\sin\delta = \sigma\sin\eta, \tag{7.10}$$

σ being nonnegative. Letting $x\to\infty$ through a sequence of values for which $ux+\eta$ is an odd integer multiple of $\frac{1}{2}\pi$, and using (7.07), (7.09), and the fact that $\operatorname{sm}(ux)=1$ when $ux\geqslant\frac{1}{2}\pi$, we obtain

$$\sigma \leqslant e^{G(u,\infty)} - 1. \tag{7.11}$$

To express ρ and δ in terms of σ, we have from (7.10)

$$(1+\rho)e^{i\delta} = 1 + \sigma e^{i\eta}.$$

If $\sigma\leqslant 1$, then by elementary geometry and Jordan's inequality, we derive

$$|\rho|\leqslant\sigma, \qquad |\delta|\leqslant\sin^{-1}\sigma\leqslant\tfrac{1}{2}\pi\sigma.$$

Substitution of (7.11) yields the desired bounds

$$|\rho|,\ 2|\delta|/\pi \leqslant e^{G(u,\infty)} - 1, \tag{7.12}$$

provided that the right-hand side does not exceed unity.

7.4 The asymptotic forms, for large u, of $\varepsilon(u,x)$, ρ, and δ depend on the behavior of $g(x)$ at $x=0$. This can be seen by subdividing the integration range for $G(u,\infty)$ at $\pi/(2u)$ and k, where k is any constant exceeding $\pi/(2u)$.

In the case mentioned in §7.1, for example, $g(x)$ has a simple pole at $x=0$. With K denoting the maximum value of $|tg(t)|$ in $[0,k]$, we have

$$G(u,\infty) \leqslant \frac{K}{u}\int_0^{\pi/(2u)} \frac{\sin(ut)}{t}\,dt + \frac{K}{u}\int_{\pi/(2u)}^{k}\frac{dt}{t} + \frac{1}{u}\int_k^\infty |g(t)|\,dt = O\left(\frac{\ln u}{u}\right).$$

Since $G(u,x)\leqslant G(u,\infty)$, (7.07) gives

$$\varepsilon(u,x) = \operatorname{sm}(ux)\,O(u^{-1}\ln u) \qquad (u\to\infty),$$

uniformly for $x\in[0,\infty)$. And from (7.12) it is seen that ρ and δ are both $O(u^{-1}\ln u)$. Strict bounds are given in Exercises 7.2 and 7.3 below.

Ex. 7.1 Let $g(u,x)\equiv g(x)$ be absolutely integrable at b and $g(x)=O(x^{-1-\beta})$ as $x\to 0+$, where $\beta\in(0,1)$. Show that the $\varepsilon(u,x)$ of Theorem 7.1 is $\operatorname{sm}(ux)\,O(u^{\beta-1})$ as $u\to\infty$, uniformly in $[0,b)$.

† Calogero (1967, Chapter 2).

Ex. 7.2 Let $g(x)$ be absolutely integrable at ∞ and have a simple pole of residue r at $x=0$. Show that

$$uG(u,\infty) = |r|\ln u + c + u^{-1}l(u),$$

where

$$c = |r|\{\ln(2k/\pi)+\mathrm{Si}(\tfrac{1}{2}\pi)\} + \int_0^k \{|g(t)|-t^{-1}|r|\}\,dt + \int_k^\infty |g(t)|\,dt,$$

k being any positive number, and

$$|l(u)| \leqslant (1+\tfrac{1}{2}\pi) \max_{0\leqslant t\leqslant \pi/(2u)} \big||g(t)|-t^{-1}|r|\big|.$$

Show also that c is independent of k.

Ex. 7.3 Let $g(x)$ be the *Yukawa potential* $\mu e^{-mx}/x$ in which μ and m are constants, m being positive. By means of the preceding exercise and Chapter 2, §3.1 show that each solution of equation (7.01) that vanishes at the origin can be expressed in the form (7.08) with

$$|\delta| \leqslant \frac{1}{2}\pi \exp\left[\frac{|\mu|}{u}\left\{\ln\left(\frac{2u}{\pi m}\right) + \mathrm{Si}\left(\frac{1}{2}\pi\right) - \gamma + \left(1+\frac{1}{2}\pi\right)\frac{m}{u}\right\}\right] - \frac{1}{2}\pi \qquad (\gamma = \text{Euler's constant}),$$

provided that this bound does not exceed $\tfrac{1}{2}\pi$.

*8 Zeros

8.1 Consider the differential equation

$$d^2w/dx^2 + \{f(x)-g(x)\}\,w = 0, \tag{8.01}$$

in which $f(x)$ and $g(x)$ satisfy the conditions of Theorems 2.1 and 2.2. Assume also that $g(x)$ is real and

$$\xi(x) \equiv \int f^{1/2}(x)\,dx \to \infty \qquad (x\to a_2-).$$

As indicated in Exercise 2.5, the general solution can be expressed

$$w(x) = Af^{-1/4}(x)[\sin\{\xi(x)+\delta\} + \varepsilon(x)], \tag{8.02}$$

where A and δ are constants whose values specify the particular solution under consideration, and

$$|\varepsilon(x)| \leqslant \exp\{\mathcal{V}_{x,a_2}(F)\} - 1 \qquad (a_1 < x < a_2).$$

The condition $w(x)=0$ yields

$$\xi(x) = n\pi - \delta + (-)^{n-1}\sin^{-1}\{\varepsilon(x)\},$$

where n is an arbitrary integer. As $x\to a_2-$, we have $\varepsilon(x) = o(1)$; hence

$$\xi(x) = n\pi - \delta + o(1) \qquad (n\to\infty).$$

Accordingly, in the neighborhood of a_2 the zeros of $w(x)$ are given by

$$x = X\{n\pi-\delta+o(1)\} \qquad (n\to\infty), \tag{8.03}$$

where $X(\xi)$ is the inverse function to $\xi(x)$. By use of the mean-value theorem this result can also be stated as

$$x = X(n\pi-\delta) + o(1)\,X'\{n\pi-\delta+o(1)\} \qquad (n\to\infty). \tag{8.04}$$

In a similar way, if the differential equation contains a large positive parameter u in the form

$$d^2w/dx^2 + \{u^2 f(x) - g(x)\}\,w = 0$$

and $\mathscr{V}_{a_1,a_2}(F) < \infty$, then the zeros of $w(u,x)$ in (a_1,a_2) are uniformly given by

$$x = X\left\{\frac{n\pi-\delta(u)}{u}\right\} + O\left(\frac{1}{u^2}\right) X'\left\{\frac{n\pi-\delta(u)}{u} + O\left(\frac{1}{u^2}\right)\right\} \qquad (u\to\infty). \tag{8.05}$$

Here $\delta(u)$ depends on the boundary conditions satisfied by $w(u,x)$, and n is any integer such that $u^{-1}\{n\pi-\delta(u)\}+O(u^{-2})$ lies in the ξ interval corresponding to (a_1,a_2).

8.2 Further progress with (8.04) and (8.05) depends on properties of $X(\xi)$. Suppose, for example, that $a_2=\infty$, and $f(x)$ and $g(x)$ satisfy conditions (4.06). Then

$$\xi(x) \sim c^{1/2}x^{\alpha}/\alpha \quad (x\to\infty), \qquad X(\xi) \sim (\alpha c^{-1/2}\xi)^{1/\alpha} \quad (\xi\to\infty),$$

and

$$X'(\xi) = \frac{1}{f^{1/2}(x)} \sim \frac{x^{1-\alpha}}{c^{1/2}} \sim \frac{(\alpha\xi)^{(1-\alpha)/\alpha}}{c^{1/(2\alpha)}}, \qquad \frac{X'(\xi)}{X(\xi)} \sim \frac{1}{\alpha\xi}.$$

Therefore $X'\{n\pi-\delta+o(1)\} \sim X'(n\pi-\delta)$ as $n\to\infty$, and substitution in (8.04) produces†

$$x = X(n\pi-\delta)\{1+o(n^{-1})\} \qquad (n\to\infty).$$

8.3 An error bound for the asymptotic approximation (8.03) can be constructed by the following method. Let b be the least number in the closure of (a_1,a_2) such that

$$\mathscr{V}_{x,a_2}(F) < \ln 2 \qquad (b<x<a_2),$$

and write

$$\sigma(x) = \exp\{\mathscr{V}_{x,a_2}(F)\} - 1, \qquad \theta(x) = \sin^{-1}\{\varepsilon(x)\}.$$

Then in (b,a_2) we have $|\varepsilon(x)| \leqslant \sigma(x) < 1$ and $|\theta(x)| < \frac{1}{2}\pi$. The equation for the zeros of the function (8.02) becomes

$$\varpi(x) \equiv \xi(x) - n\pi + \delta + (-)^n\theta(x) = 0. \tag{8.06}$$

If n is large enough to ensure that

$$X(n\pi-\delta-\tfrac{1}{2}\pi) > b, \tag{8.07}$$

then

$$\varpi\{X(n\pi-\delta-\tfrac{1}{2}\pi)\} = -\tfrac{1}{2}\pi + (-)^n\theta\{X(n\pi-\delta-\tfrac{1}{2}\pi)\} < 0,$$

† A simplification is $x = (\alpha c^{-1/2}n\pi)^{1/\alpha}\{1+o(1)\}$, but this is too crude because it does not separate the zeros.

and

$$\varpi\{X(n\pi-\delta+\tfrac{1}{2}\pi)\} = \tfrac{1}{2}\pi + (-)^n\theta\{X(n\pi-\delta+\tfrac{1}{2}\pi)\} > 0.$$

Therefore there is at least one zero in the interval

$$X(n\pi-\delta-\tfrac{1}{2}\pi) < x < X(n\pi-\delta+\tfrac{1}{2}\pi).$$

To delimit this zero in a shorter interval, denote it by

$$x = X(n\pi-\delta+\eta). \tag{8.08}$$

Then η is numerically less than $\frac{1}{2}\pi$ and satisfies

$$\eta = (-)^{n-1}\theta\{X(n\pi-\delta+\eta)\}.$$

By Jordan's inequality, $|\theta(x)| \leqslant \frac{1}{2}\pi|\varepsilon(x)| \leqslant \frac{1}{2}\pi\sigma(x)$. Hence

$$|\eta| \leqslant \tfrac{1}{2}\pi \exp\{\mathscr{V}_{X(n\pi-\delta-\frac{1}{2}\pi),a_2}(F)\} - \tfrac{1}{2}\pi. \tag{8.09}$$

In summary, if n fulfills (8.07), then the function (8.02) has a zero of the form (8.08) with η bounded by (8.09).

8.4 The analysis just given does not preclude the possibility of there being *more* than one zero fulfilling (8.09). To resolve this question we investigate the sign of $\varpi'(x)$. From (8.06)

$$\varpi'(x) = \xi'(x) + (-)^n\theta'(x) = f^{1/2}(x)\{1+(-)^n\theta'(x)f^{-1/2}(x)\}.$$

Now $\theta'(x) = \varepsilon'(x)\{1-\varepsilon^2(x)\}^{-1/2}$, and from Theorem 2.2 $|\varepsilon'(x)| \leqslant f^{1/2}(x)\sigma(x)$. If $x > b$, then $\sigma(x) < 1$ and therefore

$$|\theta'(x)|f^{-1/2}(x) \leqslant \sigma(x)\{1-\sigma^2(x)\}^{-1/2}.$$

As a function of σ, $\sigma(1-\sigma^2)^{-1/2}$ increases monotonically from zero at $\sigma = 0$, to unity at $\sigma = 2^{-1/2}$. Let $\hat{b}$ be the least number in the closure of (b, a_2) for which

$$\mathscr{V}_{x,a_2}(F) < \ln(1+2^{-1/2}) \qquad (\hat{b} < x < a_2).$$

Then $\varpi'(x) > 0$ in $(\hat{b}, a_2)$. Thus if n is large enough to ensure that

$$X(n\pi-\delta-\tfrac{1}{2}\pi) > \hat{b},$$

then *exactly* one zero (8.08) fulfills (8.09).

8.5 Similar analysis yields the following result for the approximation (8.05). Let

$$u > \mathscr{V}_{a_1,a_2}(F)/\ln(1+2^{-1/2}),$$

and n be such that $X\{u^{-1}(n\pi-\delta-\frac{1}{2}\pi)\} \in (a_1, a_2)$. Then $w(u, x)$ has exactly one zero of the form $X\{u^{-1}(n\pi-\delta+\eta)\}$, where

$$|\eta| \leqslant \tfrac{1}{2}\pi \exp[u^{-1}\mathscr{V}_{X\{u^{-1}(n\pi-\delta-\frac{1}{2}\pi)\},a_2}(F)] - \tfrac{1}{2}\pi. \tag{8.10}$$

It will be observed that the bound (8.10) vanishes as $u \to \infty$ or as $n \to \infty$, again reflecting the doubly asymptotic nature of the LG approximation.

8.6 The bounds (8.09) and (8.10) apply to the variable ξ. In applications, the error in the corresponding value of x can be bounded by use of special properties of the function $X(\xi)$, or by use of a result of the following type, the proof of which is left as an exercise for the reader.

Lemma 8.1 *In a finite or infinite* ξ *interval* (ξ_1, ξ_2), *assume that* $X(\xi)$ *is positive,* $X'(\xi)$ *is continuous, and* $|X'(\xi)/X(\xi)| \leqslant K$. *Then for any numbers* ξ *and* δ *such that* ξ *and* $\xi+\delta$ *lie in* (ξ_1, ξ_2) *and* $|\delta| < 1/K$, *we have*

$$(1-K|\delta|)\,X(\xi) \leqslant X(\xi+\delta) \leqslant X(\xi)/(1-K|\delta|).$$

Ex. 8.1 From (8.04) deduce that if a_2 is finite and $f(x)$ and $g(x)$ satisfy (4.01), then the zeros of $w(x)$ in the neighborhood of a_2 are given by

$$a_2 - x = \{a_2 - X(n\pi - \delta)\}\{1 + o(n^{-1})\} \qquad (n \to \infty).$$

Ex. 8.2 Let m be fixed and positive. By taking $(\frac{1}{2}x/m)^{2m}$ as new independent variable in Bessel's equation $x^2 w'' + xw' + (x^2 - m^2)w = 0$, show that the zeros of each solution are of the form $x = n\pi - \delta + o(1)$, where n is a large positive integer and δ is an arbitrary constant.

Show also that if $n\pi > \delta + \frac{1}{2}\pi + \{|m^2 - \frac{1}{4}|/\ln(1+2^{-1/2})\}$, then there is exactly one zero such that

$$|x - n\pi + \delta| \leqslant \tfrac{1}{2}\pi \exp\{|m^2 - \tfrac{1}{4}|/(n\pi - \delta - \tfrac{1}{2}\pi)\} - \tfrac{1}{2}\pi.$$

Ex. 8.3 Using (8.05) show that for positive values of the parameter u the equation

$$w'' + u^2(x^2+1)w = 0$$

has a solution whose real zeros are $T\{(n\pi - \delta)/u\} + \eta(u,n)$, where δ is an arbitrary constant, $n = 0, \pm 1, \pm 2, \ldots, T(\hat{\xi})$ is the inverse function to the $\hat{\xi}(t)$ of (6.03), and $\eta(u,n) = u^{-3/2}(u+|n|)^{-1/2}O(1)$ as $u \to \infty$ uniformly with respect to unbounded n.

In the case of the positive zeros use (8.10) to prove the stronger result

$$\eta(u,n) = u^{-1/2}(u+n)^{-3/2}O(1).$$

Ex. 8.4 In the notation of §8.3 show that at a zero of $w(x)$,

$$w'(x) = (-)^n A f^{1/4}(x)(1+\tau),$$

where $-\rho-\rho^2 \leqslant \tau \leqslant \rho$ and $\frac{1}{2}\pi\rho$ denotes the right-hand side of (8.09).

*9 Eigenvalue Problems

9.1 Consider the equation

$$d^2w/dx^2 + \{u^2 f(x) - g(x)\}w = 0 \tag{9.01}$$

in a finite interval $a_1 \leqslant x \leqslant a_2$ in which $f(x)$ and $g(x)$ satisfy the conditions of Theorems 2.1 and 2.2, and in addition $g(x)$ is real, and at the endpoints $f''(x)$ and $g(x)$ are continuous and $f(x)$ is nonzero. Is there a solution $w(u,x)$ that satisfies the boundary conditions $w(u,a_1) = w(u,a_2) = 0$ without being identically zero? The answer is affirmative only for certain special values of the positive parameter u, called the *eigenvalues* of the system. The corresponding solutions are called the *eigensolutions*; they are arbitrary to the extent of a factor which is independent of x. Asymptotic approximations for the large eigenvalues can be found in the following way.

From Theorem 2.2 the general solution of (9.01) is expressible as

$$w(u,x) = A(u)\,f^{-1/4}(x)\left[\sin\left\{u\int_{a_1}^{x} f^{1/2}(t)\,dt + \delta(u)\right\} + \varepsilon(u,x)\right], \tag{9.02}$$

where $A(u)$ and $\delta(u)$ are independent of x, and

$$|\varepsilon(u,x)| \leqslant \exp\{u^{-1}\mathscr{V}_{a_1,x}(F)\} - 1, \tag{9.03}$$

$F(x)$ again being given by (2.01). By hypothesis, $\mathscr{V}_{a_1,a_2}(F)$ is finite; hence $\varepsilon(u,x)$ is $O(u^{-1})$ for large u, uniformly with respect to x.

At $x = a_1$ we have $\varepsilon(u,x) = 0$. Therefore $\sin\{\delta(u)\} = 0$. Without loss of generality we may take $\delta(u) = 0$ since any other multiple of π merely affects $w(u,x)$ by a factor ± 1. The other boundary condition demands that

$$\sin(uc) + \varepsilon(u,a_2) = 0; \qquad c \equiv \int_{a_1}^{a_2} f^{1/2}(t)\,dt. \tag{9.04}$$

Since $\varepsilon(u,a_2) = O(u^{-1})$, we derive, as in Chapter 1, §5,

$$u = n\pi c^{-1} + O(n^{-1}) \qquad (n \to \infty), \tag{9.05}$$

where n is a positive integer. This is the required approximation for the eigenvalues.

9.2 To obtain bounds for the O term in (9.05) we introduce the notations

$$d = \mathscr{V}_{a_1,a_2}(F), \qquad \theta(u) = \sin^{-1}\{\varepsilon(u,a_2)\}. \tag{9.06}$$

From (9.03)

$$|\varepsilon(u,a_2)| \leqslant e^{d/u} - 1. \tag{9.07}$$

Hence if $u > d/\ln 2$, we have $|\varepsilon(u,a_2)| < 1$ and thence, by Jordan's inequality,

$$|\theta(u)| \leqslant \tfrac{1}{2}\pi(e^{d/u} - 1). \tag{9.08}$$

The equation for the eigenvalues is given by

$$\varpi(u) \equiv uc - n\pi + (-)^n\theta(u) = 0. \tag{9.09}$$

Let n be large enough to ensure that $(n-\tfrac{1}{2})\pi c^{-1}$ exceeds $d/\ln 2$. Then

$$\varpi\{(n-\tfrac{1}{2})\pi c^{-1}\} = -\tfrac{1}{2}\pi + (-)^n\theta\{(n-\tfrac{1}{2})\pi c^{-1}\} < 0;$$

whereas

$$\varpi\{(n+\tfrac{1}{2})\pi c^{-1}\} = \tfrac{1}{2}\pi + (-)^n\theta\{(n+\tfrac{1}{2})\pi c^{-1}\} > 0.$$

In consequence of Theorem 2.1 of Chapter 5, $\theta(u)$ is a continuous function of u. Therefore at least one eigenvalue satisfies $(n-\tfrac{1}{2})\pi c^{-1} < u < (n+\tfrac{1}{2})\pi c^{-1}$. To delimit it in a shorter interval, write

$$u = (n+\nu)\pi c^{-1}, \tag{9.10}$$

where $|\nu| < \tfrac{1}{2}$. Then from (9.09) we have

$$\nu\pi = (-)^{n-1}\theta\{(n+\nu)\pi c^{-1}\}.$$

Hence from (9.08)

$$|\nu| \leqslant \frac{1}{2}\exp\left\{\frac{cd}{(n-\frac{1}{2})\pi}\right\} - \frac{1}{2}. \tag{9.11}$$

Relations (9.10) and (9.11) comprise the required formulation of the eigenconditions. They are valid when $n > \frac{1}{2} + \{cd/(\pi \ln 2)\}$. The corresponding eigensolution is

$$f^{-1/4}(x)\left[\sin\left\{\frac{(n+\nu)\pi}{c}\int_{a_1}^{x} f^{1/2}(t)\,dt\right\} + \varepsilon_n(x)\right],$$

where

$$|\varepsilon_n(x)| \leqslant \exp\left\{\frac{c\mathscr{V}_{a_1,x}(F)}{(n-\frac{1}{2})\pi}\right\} - 1.$$

The eigensolution may also be expressed with a_1 replaced by a_2 in both places.

9.3 To rule out the possibility of there being more than one eigenvalue of the form (9.10) with $|\nu| < \frac{1}{2}$ we investigate the sign of $\varpi'(u)$; compare §8.4. With the assumed conditions we know from Theorem 2.2 that for any chosen point a of $[a_1, a_2]$, equation (9.01) has solutions

$$w_j(u,x) = f^{-1/4}(x)\exp\left\{(-)^{j-1} iu\int f^{1/2}(x)\,dx\right\}\{1+\varepsilon_j(u,x)\} \qquad (j=1,2),$$

such that

$$|\varepsilon_j(u,x)|,\ \frac{1}{uf^{1/2}(x)}\left|\frac{\partial\varepsilon_j(u,x)}{\partial x}\right| \leqslant \exp\left\{\frac{\mathscr{V}_{a,x}(F)}{u}\right\} - 1. \tag{9.12}$$

We now need information concerning the u derivatives of the error terms.

Theorem 9.1 *With the conditions of* §9.1 $\varepsilon_j(u,x)$, $\partial\varepsilon_j/\partial x$, *and* $\partial\varepsilon_j/\partial u$ *are continuous functions of* u *and* x *when* $u > 0$ *and* $x \in [a_1, a_2]$, *and*

$$\left|\frac{\partial\varepsilon_j(u,x)}{\partial u}\right| \leqslant \left[\frac{\mathscr{V}_{a,x}(I)}{u} + \frac{\{1+\mathscr{V}_{a,x}(I)\}\mathscr{V}_{a,x}(F)}{u^2}\right]\exp\left\{\frac{\mathscr{V}_{a,x}(F)}{u}\right\}. \tag{9.13}$$

Here $F(x)$ *is defined by* (2.01), *and*

$$I(x) = \int f^{1/2}(x)\mathscr{V}_{a,x}(F)\,dx. \tag{9.14}$$

This result is provable by a straightforward extension of the proofs of Theorems 2.1 and 2.2. Details are left as an exercise for the reader.

To apply Theorem 9.1 to the present problem, take $a = a_1$. Then the error term of §§9.1 and 9.2 is related to the error terms of the theorem by

$$2i\varepsilon(u,x) = \exp\left\{iu\int_{a_1}^{x} f^{1/2}(t)\,dt\right\}\varepsilon_1(u,x) - \exp\left\{-iu\int_{a_1}^{x} f^{1/2}(t)\,dt\right\}\varepsilon_2(u,x).$$

Using (9.12) and (9.13), we derive

$$\left|\frac{\partial\varepsilon(u,a_2)}{\partial u}\right| \leqslant \left\{\frac{d_1}{u} + \frac{(1+d_1)d}{u^2}\right\}e^{d/u} + c(e^{d/u}-1),$$

where c and d are defined by (9.04) and (9.06), and

$$d_1 = \int_{a_1}^{a_2} f^{1/2}(t)\,\mathscr{V}_{a_1,t}(F)\,dt.$$

Differentiation of (9.09) and the second of (9.06) yields

$$\varpi'(u) = c + (-)^n\theta'(u), \qquad \theta'(u) = \{1-\varepsilon^2(u,a_2)\}^{-1/2}\{\partial\varepsilon(u,a_2)/\partial u\}. \quad (9.15)$$

If $u > d/\ln 2$ then $e^{d/u} < 2$, and

$$|\theta'(u)| \leqslant \frac{1}{(2e^{-d/u}-1)^{1/2}}\left\{\frac{d_1}{u} + \frac{(1+d_1)d}{u^2} + c(1-e^{-d/u})\right\} \equiv \rho(u), \qquad (9.16)$$

say. The function $\rho(u)$ decreases strictly from infinity at $u = d/\ln 2$ to zero at $u = \infty$. Let $u = u_0$ be the root of $\rho(u) = c$ in this range. Then from the first of (9.15) we see that $\varpi'(u) > 0$ when $u > u_0$. Therefore there is *exactly* one eigenvalue of the form (9.10) with $|\nu| < \frac{1}{2}$, provided that $n > \frac{1}{2} + \pi^{-1}cu_0$.

By symmetry, d_1 may be replaced by

$$d_2 \equiv \int_{a_1}^{a_2} f^{1/2}(t)\,\mathscr{V}_{a_2,t}(F)\,dt$$

in the expression for $\rho(u)$. This would be advantageous when $d_2 < d_1$.

Ex. 9.1 Let b be any number such that $1 < b < 1+2^{-1/2}$. In the notation of §9.3, show that $\rho(u) < c$ when u exceeds both of the numbers

$$\frac{d}{\ln b}, \qquad \frac{d_2 + (1+d_2)\ln b}{c(2b^{-1}-1)^{1/2} - c(1-b^{-1})}.$$

Ex. 9.2 By taking $b = \frac{3}{2}$ in the preceding exercise, show that if n is any integer exceeding 1, then exactly one eigenvalue u of the differential system

$$w'' + u^2x^4w = 0, \qquad w(1) = w(2) = 0,$$

lies between the numbers $(3n\pi/7) \pm (3\pi/14)[\exp\{49/(36\pi n - 18\pi)\} - 1]$.

Ex. 9.3 Let κ be a constant in the interval $[0, \frac{1}{3}]$, and $\eta = 3\kappa\pi/\{2(1-\kappa)\}$. Show that for each integer n exceeding $(\eta/\ln 2) + \frac{1}{2}$ there is at least one number ν such that $|\nu| \leqslant \frac{1}{2}\exp\{\eta/(n-\frac{1}{2})\} - \frac{1}{2}$ and the differential equation

$$\frac{d^2w}{d\theta^2} + \left\{(n+\nu)^2 - \frac{3\kappa(\kappa - 3\kappa\cos^2\theta - 2\cos\theta)}{4(1+\kappa\cos\theta)^2}\right\}w = 0$$

has a nontrivial periodic solution that is an odd function of θ.

10 Theorems on Singular Integral Equations

10.1 The proofs of Theorems 2.1 and 2.2 may be adapted to other types of approximate solutions of linear differential equations. For second-order equations the steps used are as follows.

(a) Construction of a (Volterra) integral equation for the error term by the method of variation of parameters.

(b) Construction of a uniformly convergent series—the Liouville–Neumann expansion—for the solution $h(\xi)$, say, of the integral equation by the method of successive approximations.

(c) Verification that $h(\xi)$ is twice differentiable by construction of similar series for $h'(\xi)$ and $h''(\xi)$.

(d) Derivation of bounds for $|h(\xi)|$ and $|h'(\xi)|$ by majorizing the Liouville–Neumann expansion.

It would be tedious to carry out each of these steps from first principles in subsequent work. We now establish two general theorems which eliminate (b), (c), and (d) in most of the problems we shall encounter.

10.2 The standard form of integral equation is taken to be

$$h(\xi) = \int_\alpha^\xi \mathsf{K}(\xi, v)\{\phi(v)J(v)+\psi_0(v)h(v)+\psi_1(v)h'(v)\}\,dv. \tag{10.01}$$

For equation (2.09), for example, we would have

$$\mathsf{K}(\xi, v) = \tfrac{1}{2}\{1-e^{2(v-\xi)}\}, \qquad J(v) = 1, \qquad \phi(v) = \psi_0(v) = \psi(v), \qquad \psi_1(v) = 0.$$

Assumptions are as follows:

(i) The path of integration lies along a given path $\mathscr{P}$ comprising a finite chain of R_2 arcs in the complex plane. Either, or both, of the endpoints α and β, say, may be at infinity. (In real-variable problems, of course, $\mathscr{P}$ would consist of a segment of the real axis.)

(ii) The real or complex functions $J(v)$, $\phi(v)$, $\psi_0(v)$, and $\psi_1(v)$ are continuous when $v \in (\alpha, \beta)_\mathscr{P}$, save for a finite number of discontinuities and infinities.†

(iii) The real or complex kernel $\mathsf{K}(\xi, v)$ and its first two partial ξ derivatives are continuous functions of both variables when ξ, $v \in (\alpha, \beta)_\mathscr{P}$, including the arc junctions. Here, and in what follows, *all differentiations with respect to ξ are performed along $\mathscr{P}$.*

(iv) $\mathsf{K}(\xi, \xi) = 0$.

(v) When $\xi \in (\alpha, \beta)_\mathscr{P}$ and $v \in (\alpha, \xi]_\mathscr{P}$

$$|\mathsf{K}(\xi, v)| \leqslant P_0(\xi)Q(v), \qquad \left|\frac{\partial \mathsf{K}(\xi, v)}{\partial \xi}\right| \leqslant P_1(\xi)Q(v), \qquad \left|\frac{\partial^2 \mathsf{K}(\xi, v)}{\partial \xi^2}\right| \leqslant P_2(\xi)Q(v),$$

where the $P_j(\xi)$ and $Q(v)$ are continuous real functions, the $P_j(\xi)$ being positive.

(vi) When $\xi \in (\alpha, \beta)_\mathscr{P}$, the following integrals converge

$$\Phi(\xi) = \int_\alpha^\xi |\phi(v)\,dv|, \qquad \Psi_0(\xi) = \int_\alpha^\xi |\psi_0(v)\,dv|, \qquad \Psi_1(\xi) = \int_\alpha^\xi |\psi_1(v)\,dv|,$$

and the following suprema are finite

$$\kappa \equiv \sup\{Q(\xi)|J(\xi)|\}, \qquad \kappa_0 \equiv \sup\{P_0(\xi)Q(\xi)\}, \qquad \kappa_1 \equiv \sup\{P_1(\xi)Q(\xi)\},$$

except that κ_1 need not exist when $\psi_1(v) \equiv 0$.

† As in Chapter 4, §6.1, $(\alpha, \beta)_\mathscr{P}$ denotes the part of $\mathscr{P}$ lying *between* α and β.

Theorem 10.1 *With the foregoing conditions, equation* (10.01) *has a unique solution* $h(\xi)$ *which is continuously differentiable in* $(\alpha, \beta)_{\mathscr{P}}$ *and satisfies*

$$h(\xi)/P_0(\xi) \to 0, \qquad h'(\xi)/P_1(\xi) \to 0 \qquad (\xi \to \alpha \text{ along } \mathscr{P}). \tag{10.02}$$

Furthermore,†

$$\frac{|h(\xi)|}{P_0(\xi)}, \frac{|h'(\xi)|}{P_1(\xi)} \leqslant \kappa\Phi(\xi) \exp\{\kappa_0 \Psi_0(\xi) + \kappa_1 \Psi_1(\xi)\}, \tag{10.03}$$

and $h''(\xi)$ *is continuous except at the discontinuities (if any) of* $\phi(\xi) J(\xi)$, $\psi_0(\xi)$, *and* $\psi_1(\xi)$.

10.3 Theorem 10.1 is proved in a similar way to earlier theorems. We define a sequence $\{h_s(\xi)\}$ by $h_0(\xi) = 0$,

$$h_1(\xi) = \int_\alpha^\xi \mathsf{K}(\xi, v) \phi(v) J(v)\, dv, \tag{10.04}$$

and

$$h_{s+1}(\xi) - h_s(\xi) = \int_\alpha^\xi \mathsf{K}(\xi, v)[\psi_0(v)\{h_s(v) - h_{s-1}(v)\} + \psi_1(v)\{h_s'(v) - h_{s-1}'(v)\}]\, dv \qquad (s \geqslant 1). \tag{10.05}$$

Using Conditions (v) and (vi), we derive

$$|h_1(\xi)| \leqslant P_0(\xi) \int_\alpha^\xi Q(v)|\phi(v) J(v)\, dv| \leqslant \kappa P_0(\xi)\Phi(\xi).$$

Accordingly, if ξ_1 and ξ_2 are any fixed points in $(\alpha, \beta)_{\mathscr{P}}$, then the integral (10.04) converges throughout its range uniformly for $\xi \in [\xi_1, \xi_2]_{\mathscr{P}}$. Combination of this result with Conditions (ii) and (iii) shows that $h_1(\xi)$ is continuous in $(\alpha, \beta)_{\mathscr{P}}$.‡

Next, by differentiation of (10.04)§ and use of Condition (iv), we have

$$h_1'(\xi) = \int_\alpha^\xi \frac{\partial \mathsf{K}(\xi, v)}{\partial \xi} \phi(v) J(v)\, dv.$$

Hence by similar arguments $h_1'(\xi)$ is continuous and bounded by

$$|h_1'(\xi)| \leqslant \kappa P_1(\xi)\Phi(\xi).$$

Starting from these results and using equation (10.05) and its differentiated form, we may verify by induction that every $h_s(\xi)$ is continuously differentiable, and

$$\frac{|h_{s+1}(\xi) - h_s(\xi)|}{P_0(\xi)}, \frac{|h_{s+1}'(\xi) - h_s'(\xi)|}{P_1(\xi)} \leqslant \kappa\Phi(\xi) \frac{\{\kappa_0 \Psi_0(\xi) + \kappa_1 \Psi_1(\xi)\}^s}{s!} \qquad (s \geqslant 0). \tag{10.06}$$

The required solution is

$$h(\xi) = \sum_{s=0}^\infty \{h_{s+1}(\xi) - h_s(\xi)\}.$$

† The term $\kappa_1 \Psi_1(\xi)$ is to be omitted from (10.03) when $\psi_1(v) \equiv 0$.
‡ Apostol (1957, p. 441).
§ Apostol (1957, pp. 220 and 442).

That this sum is continuously differentiable follows by summing (10.06) and applying the M-test for uniform convergence in $[\xi_1, \xi_2]_{\mathscr{P}}$. This summation also shows that $h(\xi)$ does indeed satisfy (10.01), and yields the desired bounds (10.03). Relations (10.02) immediately follow since $\Phi(\xi)$, $\Psi_0(\xi)$, and $\Psi_1(\xi)$ all vanish as $\xi \to \alpha$. And the stated property of $h''(\xi)$ is verifiable by a second differentiation of (10.04) and (10.05).

To complete the proof of the theorem we have to establish that $h(\xi)$ is unique. This is effected by analysis similar to that of §1.3 of Chapter 5. Details are left to the reader.

10.4 The bounds for $h(\xi)$ and $h'(\xi)$ can be sharpened in the following common case:

Theorem 10.2 *Assume the conditions of* §10.2, *and also that* $\phi(v) = \psi_0(v)$, $\psi_1(v) = 0$. *Then the solution* $h(\xi)$ *given by Theorem* 10.1 *satisfies*

$$\frac{|h(\xi)|}{P_0(\xi)}, \frac{|h'(\xi)|}{P_1(\xi)} \leqslant \frac{\kappa}{\kappa_0}[\exp\{\kappa_0 \Phi(\xi)\} - 1]. \tag{10.07}$$

The modifications to the proof are straightforward, and again left as an exercise for the reader.

Ex. 10.1 Show how to apply Theorem 10.2 to the proofs of Theorems 2.1 and 2.2.

11 Error Bounds: Complex Variables

11.1 We turn now to the approximate solution of the differential equation

$$d^2w/dz^2 = \{f(z) + g(z)\}\, w \tag{11.01}$$

in a complex domain $\mathbf{D}$ in which $f(z)$ and $g(z)$ are holomorphic and $f(z)$ does not vanish. We suppose, temporarily, that $\mathbf{D}$ is simply connected, ensuring that the solutions of (11.01) are single valued (Chapter 5, §3.1).

The transformation $\xi = \int f^{1/2}(z)\, dz$ maps $\mathbf{D}$ on a domain $\mathbf{\Delta}$, say. The mapping is free from singularity, since $d\xi/dz$ is nonvanishing, and can be made one-to-one by supposing (if necessary) that $\mathbf{\Delta}$ comprises several Riemann sheets. The function $\psi(\xi)$ defined by (2.06) (with x replaced by z) is holomorphic in $\mathbf{\Delta}$. The analysis of §2.2 is reproducible until (2.12) is reached; thus we again have

$$h''(\xi) + 2h'(\xi) = \psi(\xi)\{1 + h(\xi)\}, \tag{11.02}$$

and

$$h(\xi) = \tfrac{1}{2}\int_{\alpha_1}^{\xi} \{1 - e^{2(v-\xi)}\}\, \psi(v)\{1 + h(v)\}\, dv. \tag{11.03}$$

In order to bound the kernel when the variables are complex, we suppose that the integral equation (11.03) is solved along a given path $\mathscr{Q}$ comprising a finite chain of

R_2 arcs in the complex plane, and $\operatorname{Re} v$ is nondecreasing as v moves along $\mathscr{Q}$ away from the initial point α_1. Then

$$|e^{2(v-\xi)}| \leqslant 1, \qquad |1-e^{2(v-\xi)}| \leqslant 2, \qquad \text{when} \quad v \in (\alpha_1, \xi]_{\mathscr{Q}}. \tag{11.04}$$

Applying the theory of §10 with $\alpha = \alpha_1$, $\mathsf{K}(\xi, v) = \frac{1}{2}\{1-e^{2(v-\xi)}\}$, $\partial \mathsf{K}/\partial \xi = e^{2(v-\xi)}$, $P_0(\xi) = P_1(\xi) = Q(v) = 1$, $J(v) = 1$, $\phi(v) = \psi_0(v) = \psi(v)$, and $\psi_1(v) = 0$, we deduce from Theorems 10.1 and 10.2 that equation (11.03) has a solution which is continuously differentiable along $\mathscr{Q}$ and bounded by

$$|h(\xi)|, \ |h'(\xi)| \leqslant e^{\Psi(\xi)} - 1,$$

where

$$\Psi(\xi) = \int_{\alpha_1}^{\xi} |\psi(v)\, dv|$$

evaluated along $\mathscr{Q}$.

11.2 To complete the analysis we must show that $h(\xi)$ also satisfies the differential equation (11.02) in the complex plane. The direct approach is troublesome, because the admissible points ξ need not comprise a domain.† Instead we proceed as follows.

Suppose first that α_1 is a given finite point of $\mathbf{\Delta}$. From Chapter 5, Theorem 3.1 we know that with prescribed initial conditions each holomorphic solution $W(\xi)$ of (2.05) is unique. With (2.07) this implies that in $\mathbf{\Delta}$ there is a unique holomorphic function $\hat{h}(\xi)$, say, which satisfies (11.02) and the conditions $\hat{h}(\alpha_1) = \hat{h}'(\alpha_1) = 0$. Variation of parameters shows that $\hat{h}(\xi)$ also satisfies (11.03), and since by Theorem 10.1 the solution of (11.03) is unique, it follows that $\hat{h}(\xi) = h(\xi)$ along $\mathscr{Q}$.

Alternatively, let α_1 be the point at infinity on a given R_2 arc $\mathscr{M}_1$, say. If $\hat{h}(\xi)$ is the solution of (11.02) which satisfies $\hat{h}(\gamma) = h(\gamma)$ and $\hat{h}'(\gamma) = h'(\gamma)$, where γ is any designated finite point of $\mathscr{Q}$, then $\hat{h}(\xi) = h(\xi)$ everywhere on $\mathscr{Q}$. To prove this assertion, we have from (11.02), by variation of parameters and use of the conditions at $\xi = \gamma$,

$$\hat{h}(\xi) = \tfrac{1}{2}\int_{\gamma}^{\xi} \{1-e^{2(v-\xi)}\}\, \psi(v)\{1+\hat{h}(v)\}\, dv + \tfrac{1}{2}\int_{\alpha_1}^{\gamma} \{1-e^{2(v-\xi)}\}\, \psi(v)\{1+h(v)\}\, dv.$$

Subtraction of (11.03) yields

$$\hat{h}(\xi) - h(\xi) = \tfrac{1}{2}\int_{\gamma}^{\xi} \{1-e^{2(v-\xi)}\}\, \psi(v)\{\hat{h}(v)-h(v)\}\, dv.$$

Regarding this as an integral equation for $\hat{h}(\xi)-h(\xi)$ and applying Theorem 10.1 with the role of α played by γ, we deduce that $\hat{h}(\xi) = h(\xi)$.‡ To ensure that $\hat{h}(\xi)$ is the same solution of (11.02) for all paths $\mathscr{Q}$, we stipulate that these paths coincide with $\mathscr{M}_1$ in the neighborhood of α_1.

† Compare Exercise 11.2 below.

‡ We have $P_0(\xi) = P_1(\xi) = Q(v) = 1$, or $P_0(\xi) = P_1(\xi) = |e^{-2\xi}|$ and $Q(v) = |e^{2v}|$, depending on which side of γ the point ξ happens to be, but because $\phi(v) = J(v) = 0$ the conclusion that $\hat{h}(\xi) - h(\xi)$ is zero applies in both cases.

11.3 Collecting together these results and similar results for a second solution of the differential equation, and transforming back to the original variable z, we arrive at the following:

Theorem 11.1 *With the conditions stated in the opening paragraph of* §11.1, *equation* (11.01) *has solutions* $w_j(z)$, $j = 1, 2$, *holomorphic in* $\mathbf{D}$, *and depending on arbitrary reference points* a_1, a_2, *such that*

$$w_j(z) = f^{-1/4}(z) \exp\{(-)^{j-1}\xi(z)\}\{1+\varepsilon_j(z)\}, \tag{11.05}$$

where

$$\xi(z) = \int f^{1/2}(z)\, dz, \tag{11.06}$$

and

$$|\varepsilon_j(z)|,\ |f^{-1/2}(z)\varepsilon_j'(z)| \leqslant \exp\{\mathscr{V}_{a_j,z}(F)\} - 1, \tag{11.07}$$

provided that $z \in \mathbf{H}_j(a_j)$ *(defined below).*

In this theorem the error-control function is again

$$F(z) = \int \left\{ \frac{1}{f^{1/4}} \frac{d^2}{dz^2}\left(\frac{1}{f^{1/4}}\right) - \frac{g}{f^{1/2}} \right\} dz,$$

and the branches of the fractional powers of $f(z)$ must be continuous in $\mathbf{D}$, that of $f^{1/2}(z)$ being the square of $f^{1/4}(z)$. Each region of validity $\mathbf{H}_j(a_j)$ comprises the z point set for which there exists a path $\mathscr{P}_j$ in $\mathbf{D}$ linking z with a_j, and having the properties:

(i) $\mathscr{P}_j$ *consists of a finite chain of* R_2 *arcs.*

(ii) *As* t *passes along* $\mathscr{P}_j$ *from* a_j *to* z, $\operatorname{Re}\{\xi(t)\}$ *is nondecreasing if* $j = 1$ *or nonincreasing if* $j = 2$.

The variation of F in (11.07) is evaluated along $\mathscr{P}_j$. Finally, the point a_j may be at infinity on a curve $\mathscr{L}_j$, provided that $\mathscr{P}_j$ coincides with $\mathscr{L}_j$ in the neighborhood of a_j, and $\mathscr{V}(F)$ converges.

11.4 If the condition that $\mathbf{D}$ be simply connected is relaxed, then the solutions of (11.01) are many-valued functions. In this case each branch of $w_j(z)$ satisfies (11.05) and (11.07) whenever a path $\mathscr{P}_j$ can be found in $\mathbf{D}$ fulfilling Conditions (i) and (ii). Again, each fractional power of $f(z)$ must be continuous along $\mathscr{P}_j$.

We call a path $\mathscr{P}_j$ which fulfills Conditions (i) and (ii) a *ξ-progressive path.* The map of $\mathscr{P}_j$ in the ξ plane is called simply a *progressive path.*

We shall refer to (ii) as the *monotonicity condition* governing the regions of validity. Suppose, for example, that the only finite singularity of $\psi(\xi)$ is $\xi = 0$, and the variation of the error-control function converges at infinity. We may take $\mathbf{\Delta}$ to consist of the whole ξ plane, with the origin deleted, and to begin with render $\mathbf{\Delta}$ simply connected by introducing a cut along the negative real axis. Set $\alpha_1 = -\infty + i\delta$, where $\delta\ (\geqslant 0)$ is arbitrary. Then the ξ map $\mathbf{K}_1(\alpha_1)$, say, of $\mathbf{H}_1(a_1)$ comprises the sector $-\frac{1}{2}\pi < \operatorname{ph}\xi \leqslant \pi$; see Fig. 11.1. Points in the remaining quadrant cannot be joined to α_1 without violating the monotonicity condition. The ξ regions excluded in this

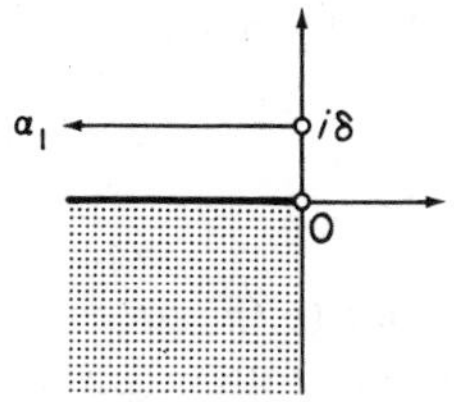

Fig. 11.1 $\mathbf{K}_1(-\infty+i\delta)$.

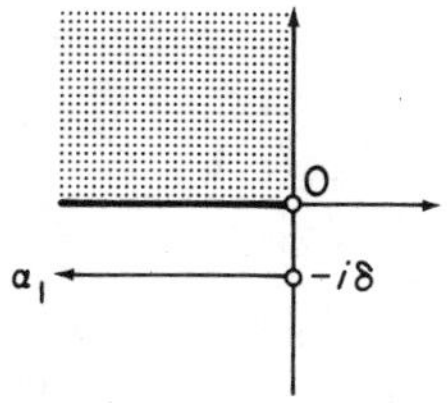

Fig. 11.2 $\mathbf{K}_1(-\infty-i\delta)$.

way and their z maps are called *shadow zones*.† Although the solution $w_1(z)$ exists and is holomorphic in the shadow zone, the bounds (11.07) do not apply there.

A solution in the shadow zone can be constructed by taking $\alpha_1 = -\infty - i\delta$, as indicated in Fig. 11.2. This has the same form (11.05) as the previous solution, but the error terms $\varepsilon_1(z)$ and regions of validity $\mathbf{H}_1(a_1)$ are quite different in the two cases.

Although the choice of the negative real axis as boundary for $\mathbf{\Delta}$ simplified the narrative in this example, it restricted the regions $\mathbf{K}_1(\alpha_1)$ (and their z maps) unnecessarily. When $\alpha_1 = -\infty + i\delta$, the region $\mathbf{K}_1(\alpha_1)$ can be extended by rotating the cut in the positive sense until it coincides with the positive imaginary axis. The total region of validity is then $-\frac{1}{2}\pi < \operatorname{ph}\xi < \frac{5}{2}\pi$. Further extension is precluded by the monotonicity condition. Similarly when $\alpha_1 = -\infty - i\delta$ the maximal $\mathbf{K}_1(\alpha_1)$ is $-\frac{5}{2}\pi < \operatorname{ph}\xi < \frac{1}{2}\pi$.

Ex. 11.1 Let $\xi = \pm 1$ be the only finite singularities of $\psi(\xi)$, and $\mathscr{V}(F)$ converge at infinity. Using all necessary Riemann sheets sketch the maximal region $\mathbf{K}_1(-\infty)$.

Ex. 11.2 Let a_j be at infinity and Conditions (i) and (ii) of §11.3 be replaced by the stronger conditions: (i) the ξ map of $\mathscr{P}_j$ is a polygonal arc; (ii) as t passes along $\mathscr{P}_j$ from a_j to z, $\operatorname{Re}\{\xi(t)\}$ is strictly increasing if $j=1$ or strictly decreasing if $j=2$. Show that $\mathbf{H}_j(a_j)$ is a domain.

[Thorne, 1960.]

12 Asymptotic Properties for Complex Variables

12.1 Asymptotic properties of the LG approximation with respect to the independent variable established in §§3 and 4 carry over to complex variables. If $\operatorname{Re}\xi \to -\infty$ as $z \to a_1$ and $\operatorname{Re}\xi \to +\infty$ as $z \to a_2$, then $w_1(z)$ is recessive at a_1 and $w_2(z)$ is recessive at a_2. As we noted in Chapter 5, §7.3, the construction of a numerically satisfactory set of solutions may necessitate the use of more than two reference points a_1 and a_2; compare Exercise 12.1 below.

Corresponding to Theorem 3.1, we have:

Theorem 12.1 *Let $\mathscr{L}$ be a finite or infinite ξ-progressive path in* $\mathbf{D}$ *and a_1, a_2 its endpoints. Assume that along $\mathscr{L}$, F is of bounded variation,* $\operatorname{Re}\xi \to -\infty$ *as $z \to a_1$, and* $\operatorname{Re}\xi \to +\infty$ *as $z \to a_2$. Then*

† This name is due to Cherry (1950a).

(i) $\varepsilon_1(z) \to$ *a constant* $\varepsilon_1(a_2)$, *say, and* $f^{-1/2}(z)\varepsilon_1'(z) \to 0$, *as* $z \to a_2$.
(ii) $\varepsilon_2(z) \to$ *a constant* $\varepsilon_2(a_1)$, *say, and* $f^{-1/2}(z)\varepsilon_2'(z) \to 0$, *as* $z \to a_1$.
(iii) $\varepsilon_1(a_2) = \varepsilon_2(a_1)$.
(iv) $|\varepsilon_1(a_2)| \leqslant \frac{1}{2}[\exp\{\mathscr{V}_{\mathscr{L}}(F)\} - 1]$.

The proof of Parts (i) and (ii) of this theorem is similar to the proof of Theorem 3.1. Part (iii) is proved as indicated in Exercise 3.1. Part (iv) follows on summing the inequalities

$$|l_s(\xi)| \leqslant \Psi^{s+1}(\xi)/(s+1)! \qquad (s = 0, 1, \ldots)$$

obtained from (3.07), and then letting $\xi \to \alpha_2$ in (3.06).

It is noteworthy that the bound (iv) is twice as sharp as the limiting form of (11.07).

12.2 As in §5, uniform asymptotic properties with respect to parameters derive naturally from the error bounds of Theorem 11.1. An added feature in the complex case is that the regions of validity $\mathbf{H}_j(a_j)$ depend strongly on the parameter u. In the case of equation (5.01), for example, we have $\xi = u\int f^{1/2}(z)\,dz$. If u is complex, then the ξ map of $\mathbf{D}$ rotates about the origin as ph u varies. Therefore a path in the z plane may be ξ-progressive for some values of ph u but not for others, causing the shadow zones to vary with ph u.

Ex. 12.1 Let m be a positive integer, j an integer or zero, and δ ($< 3\pi$) a positive constant. Show that the solution of the equation $d^2w/dz^2 = z^{m-2}w$ which is recessive at infinity along the ray ph $z = 2j\pi/m$ is given by

$$w(z) = \{1 + O(z^{-m/2})\}\, z^{(2-m)/4} \exp\{(-)^{j+1} 2z^{m/2}/m\}$$

as $z \to \infty$ in the sector $|m \operatorname{ph} z - 2j\pi| \leqslant 3\pi - \delta$.

How many of these solutions are needed to comprise a numerically satisfactory set in the neighborhood of infinity?

Ex. 12.2 If $f(z) = \frac{1}{4}u^2z^{-1}$ and $g(z) = z^{-1/2}(z+1)^{-3/2}$, where $u = |u|e^{i\omega}$ is a complex parameter, show that the boundaries of the maximal regions $\mathbf{H}_j(\infty e^{-2i\omega})$ lie along the ray ph $z = \pi - 2\omega$ and the parabola

$$\{(x+1)\sin 2\omega + y\cos 2\omega\}^2 = 4\sin\omega\{(x+1)\sin\omega + y\cos\omega\},$$

where x and y are the real and imaginary parts of z, respectively.

13 Choice of Progressive Paths

13.1 A new feature introduced by complex variables is the choice of ξ-progressive paths $\mathscr{P}_j$. For each pair of points z and a_j, the most effective use of Theorem 11.1 requires $\mathscr{P}_j$ to be determined in $\mathbf{D}$ in such a way that the total variation of the error-control function $F(z)$ along $\mathscr{P}_j$ is minimized, subject to fulfillment of the monotonicity condition.

For general $\mathbf{D}$ and $F(z)$, a solution of this minimization problem is unavailable. In applications, we select paths which fulfill the monotonicity condition and keep well away from singularities of F, including turning points of the differential equation. The resulting variations may not be minimal, but often are sufficiently small to provide satisfactory error bounds.

In this section we show how to choose actual minimizing paths in the special case $f(z) = 1$ and $g(z) = az^{-a-1}$, where a is a fixed positive number. Here **D** comprises the z plane with the origin deleted, $F(z) = z^{-a}$, and $\xi = z$. By symmetry, it suffices to consider the case $j = 2$. Taking a_2 to be the point at infinity on the positive real axis, we have

$$\mathscr{V}_{z,\infty}(F) = \mathscr{V}_{z,\infty}(t^{-a}) = a\int_z^\infty \left|\frac{dt}{t^{a+1}}\right|,$$

with $\operatorname{Re} t$ nondecreasing on the path. Then $\mathbf{H}_2(\infty)$ is the sector $|\operatorname{ph} z| < \frac{3}{2}\pi$; compare §11.4. We write $\theta = \operatorname{ph} z$, and study in turn the cases $|\theta| \leqslant \frac{1}{2}\pi$, $\frac{1}{2}\pi < |\theta| \leqslant \pi$, and $\pi < |\theta| < \frac{3}{2}\pi$. First we establish the following:

Lemma 13.1 *Let $\mathscr{L}$ be any doubly infinite straight line in the complex plane, and a a positive constant. Then*

$$\mathscr{V}_{\mathscr{L}}(t^{-a}) = 2\chi(a)d^{-a}, \tag{13.01}$$

where d is the shortest distance from the origin to $\mathscr{L}$, and

$$\chi(a) = \pi^{1/2}\Gamma(\tfrac{1}{2}a+1)/\Gamma(\tfrac{1}{2}a+\tfrac{1}{2}). \tag{13.02}$$

To prove this result, let z be the nearest point of $\mathscr{L}$ to $t = 0$, so that $|z| = d$. The parametric equation of $\mathscr{L}$ can be expressed

$$t = z + i\tau z \qquad (-\infty < \tau < \infty).$$

Hence

$$\mathscr{V}_{\mathscr{L}}(t^{-a}) = a\int_{-\infty}^{\infty}\left|\frac{iz\,d\tau}{(z+i\tau z)^{a+1}}\right| = \frac{2a}{d^a}\int_0^\infty \frac{d\tau}{(1+\tau^2)^{(a+1)/2}}.$$

Equation (13.01) follows on replacing τ^2 by t and referring to Exercise 1.3 of Chapter 2.

For reference, two-decimal values of $\chi(a)$ for the first ten integer values of a are as follows:

$\chi(1) = 1.57$, $\chi(2) = 2.00$, $\chi(3) = 2.36$, $\chi(4) = 2.67$, $\chi(5) = 2.95$,

$\chi(6) = 3.20$, $\chi(7) = 3.44$, $\chi(8) = 3.66$, $\chi(9) = 3.87$, $\chi(10) = 4.06$.

The results of Chapter 2, §2 and Chapter 4, §5 show that $\chi(a)$ is increasing in $(0, \infty)$ and $\chi(a) \sim (\frac{1}{2}\pi a)^{1/2}$ as $a \to \infty$.

13.2 (i) $|\theta| \leqslant \frac{1}{2}\pi$. Consider the path indicated in Fig. 13.1, consisting of part of the positive real axis, a circular arc of radius R $(>|z|)$ centered at the origin, and the line segment

$$t = z + \tau e^{i\theta} \qquad (0 \leqslant \tau \leqslant R - |z|).$$

It is readily seen that as $R \to \infty$ the contributions to the variation from the real axis and circular arc both vanish, and we obtain

$$\mathscr{V}_{z,\infty}(t^{-a}) = \int_0^\infty \frac{a\,d\tau}{|z+\tau e^{i\theta}|^{a+1}} = \int_0^\infty \frac{a\,d\tau}{(|z|+\tau)^{a+1}} = \frac{1}{|z|^a}. \tag{13.03}$$

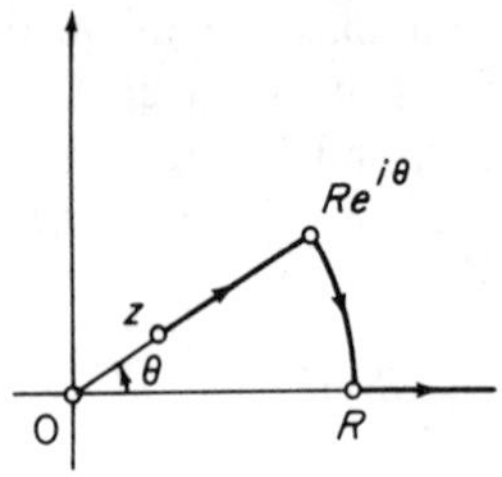

Fig. 13.1 $-\frac{1}{2}\pi \leqslant \theta \leqslant \frac{1}{2}\pi$.

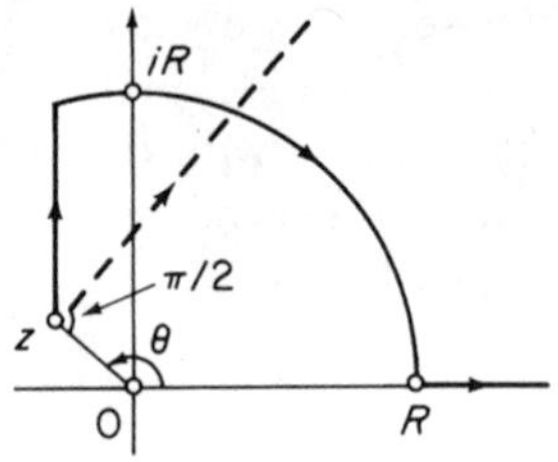

Fig. 13.2 $\frac{1}{2}\pi < \theta \leqslant \pi$.

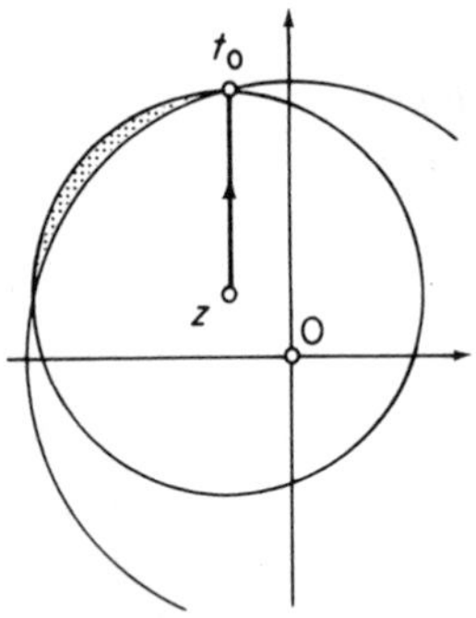

Fig. 13.3 $\frac{1}{2}\pi < \theta \leqslant \pi$.

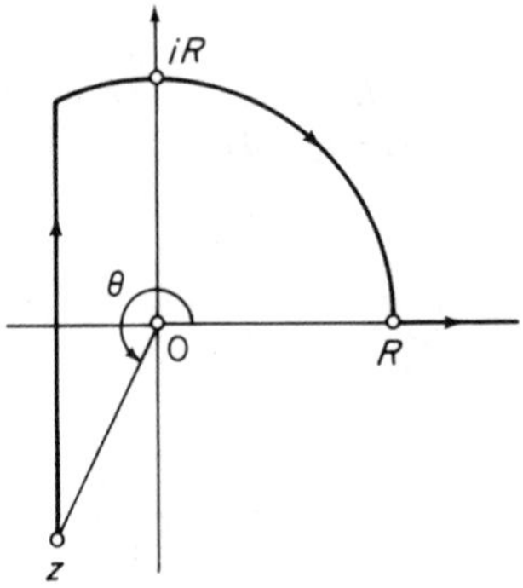

Fig. 13.4 $\pi < \theta < \frac{3}{2}\pi$.

Since this equals the modulus of the difference between the values of t^{-a} at the extremities of the path, no other path can yield a smaller variation.†

13.3 (ii) $\frac{1}{2}\pi < |\theta| \leqslant \pi$. Consider the path indicated by the heavy continuous line in Fig. 13.2 when θ is positive, or the conjugate path when θ is negative. Again, as the radius R of the circular arc tends to infinity the contributions from this arc and the real axis both vanish, and we obtain

$$\mathscr{V}_{z,\infty}(t^{-a}) = \int_0^\infty \frac{a\,d\tau}{|z+i\tau|^{a+1}} = \int_0^\infty \frac{a\,d\tau}{\{x^2+(|y|+\tau)^2\}^{(a+1)/2}}, \tag{13.04}$$

where $x+iy = z$.

The variation is minimized by this choice. To see this, we travel a prescribed distance τ along any admissible path from z, arriving at t, say. On the nominated path t is at $t_0 = z+i\tau$; for any other path t lies within or on the circle centered at z and passing through t_0, as shown in Fig. 13.3. Clearly, $|t| > |t_0|$ only when t lies within the shaded lune bounded by this circle and the circular arc $|t| = |t_0|$. No path can be admitted to this lune however, because $\operatorname{Re} t < \operatorname{Re} z$ in its interior. Hence $|t| \leqslant |t_0|$, confirming that the variation has been minimized.

† Compare Chapter 1, Exercise 11.4. Strictly speaking (13.03) is not an actual variation along an admissible path, but the infimum of a set of variations. This distinction is unimportant for the purpose of obtaining error bounds, and we shall not dwell on it again.

For integer a, the integral (13.04) can be evaluated in terms of elementary functions. For example,

$$\mathscr{V}_{z,\infty}(t^{-1}) = \frac{1}{|x|}\tan^{-1}\left|\frac{x}{y}\right| \quad (x \neq 0); \qquad \mathscr{V}_{z,\infty}(t^{-1}) = \frac{1}{|y|} \quad (x = 0).$$

But to avoid excessive complication in the general case we replace the content of the braces in (13.04) by the lower bound $x^2+y^2+\tau^2$. Evaluation yields

$$\mathscr{V}_{z,\infty}(t^{-a}) \leqslant \chi(a)|z|^{-a}, \tag{13.05}$$

where $\chi(a)$ is defined by (13.02). From Lemma 13.1 it is seen that this slightly weaker result is equivalent to using the broken-line path of Figure 13.2.

13.4 (iii) $\pi < |\theta| < \frac{3}{2}\pi$. The minimizing path is the limiting form of the path indicated in Fig. 13.4 when the radius R of the arc tends to infinity. To verify this, let any other path intersect the negative real axis at $t = l$. If $l = x$, then the result follows immediately from §13.3. If $l \in (x, 0)$, then for each positive number τ we compare points in the disk $|t-l| \leqslant \tau$ with $t_0 \equiv x - i\tau$. Again $|t| \leqslant |t_0|$ except within an inadmissible lune.

On letting $R \to \infty$ and using Lemma 13.1, we obtain

$$\mathscr{V}_{z,\infty}(t^{-a}) \leqslant 2\chi(a)|\operatorname{Re} z|^{-a}. \tag{13.06}$$

We notice that if $|z|$ is fixed and $\operatorname{ph} z \to \pm\frac{3}{2}\pi$, then the path moves toward the origin, causing $\mathscr{V}_{z,\infty}(t^{-a}) \to \infty$. This is to be expected because the boundaries of the region of validity $\mathbf{H}_2(\infty)$ are being approached.

Ex. 13.1 Show that a solution of the equation $d^2w/dz^2 = (z^2 - \frac{1}{4}z^{-2})w$ is

$$z^{-1/2}\exp(-\tfrac{1}{2}z^2)\{1+\varepsilon(z)\},$$

where $|\varepsilon(z)|$ is bounded by $\exp(\frac{1}{2}|z|^{-2})-1$, $\exp(\frac{1}{4}\pi|z|^{-2})-1$, or $\exp(\frac{1}{2}\pi|\operatorname{Re} z^2|^{-1})-1$, according as $|\operatorname{ph} z|$ lies in the interval $[0, \frac{1}{4}\pi]$, $(\frac{1}{4}\pi, \frac{1}{2}\pi]$, or $(\frac{1}{2}\pi, \frac{3}{4}\pi)$.

Ex. 13.2 Let $\mathscr{A}$ be an infinite R_1 arc such that $\mathscr{V}_{\mathscr{A}}(t^{-1}) < \infty$, and a a constant such that $a > 1$. Show that $\mathscr{V}_{\mathscr{A}}(t^{-a}) < \infty$.

Ex. 13.3 Let $t = t(\sigma)$ be an infinite R_1 arc, σ being the arc parameter. Show that if $|t(\sigma)|^{-1}$ is $O(\sigma^{-\alpha})$ as $\sigma \to \infty$, where $\alpha > \frac{1}{2}$, then $\mathscr{V}(t^{-1})$ converges along the arc.

Show also that these conditions are met by any parabolic arc.

Ex. 13.4 Show that on the path $t = 1+\tau+i\tau\sin\tau$ $(0 \leqslant \tau < \infty)$, $\mathscr{V}(t^{-a})$ converges if $a > 1$ and diverges if $a = 1$.

Ex. 13.5 From the definition of $\operatorname{Ei}(x)$ given in Chapter 2, §3.2, deduce that

$$\operatorname{Ei}(x) = -\frac{1}{2}\left(\int_x^{x+i\infty} + \int_x^{x-i\infty}\right)\frac{e^t}{t}\,dt \qquad (x > 0).$$

Thence by partial integrations prove that in the asymptotic expansion

$$\operatorname{Ei}(x) \sim e^x \sum_{s=0}^{\infty} s!\,x^{-s-1} \qquad (x \to +\infty)$$

the ratio of the nth error term to the $(n+1)$th term cannot exceed $1+\chi(n+1)$ in absolute value.

Historical Notes and Additional References

This chapter is based on the reference Olver (1961). The original material has been considerably expanded, particularly concerning the doubly asymptotic nature of the LG approximation. The availability of an explicit error bound has enabled much existing theory to be unified and simplified.

The approximation (1.08) was used independently by Liouville (1837) and Green (1837). Watson (1944, §1.4) noted that essentially the same procedure was used in a special case by Carlini in 1817. Theoretical physicists often refer to (1.08) as the WKB (or BKW) approximation in recognition of the papers by Wentzel (1926), Kramers (1926), and Brillouin (1926). However, the contribution of these authors was not the construction of the approximation (which was already known), but the determination of connection formulas for linking exponential and oscillatory LG approximations across a turning point on the real axis.† In recent usage, J is sometimes added to the initials to acknowledge that the approximate connection formulas of Wentzel, Kramers, and Brillouin had been discovered previously by Jeffreys (1924). And Jeffreys (1953) has pointed out that he himself had been anticipated by Gans (1915) and (to a lesser extent) Rayleigh (1912). Accordingly, following Jeffreys it seems best to associate the approximation (1.08) with the names of Liouville and Green, as we have done, and to reserve the initials WKBJ for the connection formula problem.

Further historical information may be found in the papers by Pike (1964a) and McHugh (1971).

§1 Liouville (1837) used only the special form of the transformation given in §1.3. Langer (1931, 1935) was the first to exploit the more general form for the purpose of constructing uniform asymptotic approximations.

§6 The notation $U(a, x)$ is due to J. C. P. Miller (1955); extensive properties and tables of parabolic cylinder functions are to be found in this reference. An extension of (6.06) into an asymptotic expansion in descending powers of u has been given by Olver (1959).

§8 Some further results and references concerning error bounds for asymptotic approximations of zeros have been given by Hethcote (1970b).

§9 For further asymptotic analyses of eigenvalues see Fix (1967), Cohn (1967), and Natterer (1969).

§10 Erdélyi (1964) seems to have been the first to study systematically singular integral equations arising in the asymptotic solution of ordinary differential equations. The present theorems resemble his results.

† This problem is studied in Chapter 13.

7

DIFFERENTIAL EQUATIONS WITH IRREGULAR SINGULARITIES; BESSEL AND CONFLUENT HYPERGEOMETRIC FUNCTIONS

1 Formal Series Solutions

1.1 In the preceding chapter we saw that in the neighborhood of an irregular singularity, the solutions of a linear second-order differential equation are asymptotically represented by the LG functions. The opening sections of the present chapter show how to extend these approximations into asymptotic expansions. The method applies to a singularity of any finite rank, but to simplify the text attention is restricted to the commonest case in applications, that is, unit rank.

As in Chapter 5, the standard form of differential equation is taken to be

$$\frac{d^2w}{dz^2}+f(z)\frac{dw}{dz}+g(z)w = 0. \tag{1.01}$$

Without loss of generality we may suppose that the singularity is located at infinity. This means that there exists an annulus $|z| > a$ in which $f(z)$ and $g(z)$ can be expanded in convergent power series of the form

$$f(z) = \sum_{s=0}^{\infty}\frac{f_s}{z^s}, \qquad g(z) = \sum_{s=0}^{\infty}\frac{g_s}{z^s}. \tag{1.02}$$

Not all of the coefficients f_0, g_0, and g_1 vanish, otherwise the singularity would be regular.

The term in the first derivative is removed from (1.01) by the substitution

$$w = \exp\left\{-\tfrac{1}{2}\int f(z)\,dz\right\}y. \tag{1.03}$$

Thus

$$d^2y/dz^2 = q(z)y, \tag{1.04}$$

where

$$q(z) = \tfrac{1}{4}f^2(z)+\tfrac{1}{2}f'(z)-g(z).$$

When $|z| > a$, expansion gives

$$q(z) = (\tfrac{1}{4}f_0^2-g_0)+(\tfrac{1}{2}f_0f_1-g_1)z^{-1}+\cdots. \tag{1.05}$$

Sections 3 and 12 of Chapter 6 show that with appropriate restrictions equation (1.04) has solutions with the properties

$$y \sim q^{-1/4}(z) \exp\left\{\pm \int q^{1/2}(z)\, dz\right\}$$

as $z \to \infty$. By use of (1.05) these representations simplify into

$$y \sim (\text{constant}) \times \exp\{\pm(\rho z + \sigma \ln z)\}, \tag{1.06}$$

where

$$\rho = (\tfrac{1}{4}f_0^2 - g_0)^{1/2}, \qquad \sigma = (\tfrac{1}{4}f_0 f_1 - \tfrac{1}{2}g_1)/\rho.$$

The forms (1.06) hold unless $\rho = 0$; this exceptional case is treated in §1.3 below.

Returning to the original differential equation, we have from (1.03) and (1.06)

$$w \sim (\text{constant}) \times \exp(\lambda z + \mu \ln z), \tag{1.07}$$

where

$$\lambda = \pm\rho - \tfrac{1}{2}f_0, \qquad \mu = \pm\sigma - \tfrac{1}{2}f_1.$$

1.2 Since the coefficients $f(z)$ and $g(z)$ have expansions in descending powers of z, it is natural to try to extend (1.07) into formal series solutions of the form

$$w = e^{\lambda z} z^{\mu} \sum_{s=0}^{\infty} \frac{a_s}{z^s}. \tag{1.08}$$

Substituting this expansion and (1.02) in (1.01) and equating coefficients, we obtain

$$\lambda^2 + f_0 \lambda + g_0 = 0, \tag{1.09}$$

$$(f_0 + 2\lambda)\mu = -(f_1 \lambda + g_1), \tag{1.10}$$

and

$$(f_0 + 2\lambda) s a_s = (s-\mu)(s-1-\mu) a_{s-1} + \{\lambda f_2 + g_2 - (s-1-\mu) f_1\} a_{s-1} + \{\lambda f_3 + g_3 - (s-2-\mu) f_2\} a_{s-2} + \cdots + \{\lambda f_{s+1} + g_{s+1} + \mu f_s\} a_0. \tag{1.11}$$

The first of these equations yields two possible values

$$\lambda_1, \lambda_2 = -\tfrac{1}{2}f_0 \pm (\tfrac{1}{4}f_0^2 - g_0)^{1/2}$$

for λ. Equation (1.10) determines corresponding values μ_1, μ_2, of μ. These findings are easily verified to be consistent with §1.1.

The values of a_0, say $a_{0,1}$ and $a_{0,2}$ in the two cases, may be assigned arbitrarily. Higher coefficients $a_{s,1}$ and $a_{s,2}$ are then determined recursively by (1.11). The process fails if and only if $f_0 + 2\lambda = 0$: this is the excepted case $f_0^2 = 4g_0$.

The discovery that a differential equation can be satisfied in the neighborhood of an irregular singularity by a series of the form (1.08) was made by Thomé. This kind of expansion is sometimes called a *normal series* or *normal solution* to distinguish it from expansions of Laurent type for w, although the actual choice of name (like "regular singularity" and "irregular singularity") has little to commend it. Equation (1.09) is called the *characteristic equation*, and its roots the *characteristic values* of the singularity.

1.3 In the case $f_0^2 = 4g_0$, the analysis of §1.1 can be modified to yield similar asymptotic forms for the solutions. An alternative procedure, which leads to the same results, is the transformation of Fabry†:

$$w = e^{-f_0 z/2} W, \qquad t = z^{1/2}.$$

This gives

$$\frac{d^2W}{dt^2} + F(t)\frac{dW}{dt} + G(t)W = 0, \tag{1.12}$$

where

$$F(t) = 2tf(t^2) - 2tf_0 - t^{-1}, \qquad G(t) = t^2\{4g(t^2)+f_0^2-2f_0 f(t^2)\}.$$

Equation (1.12) has the same form as (1.01). For $|t| > a^{1/2}$ its coefficients may be expanded in series

$$F(t) = \frac{2f_1-1}{t} + \frac{2f_2}{t^3} + \cdots, \qquad G(t) = (4g_1-2f_0 f_1) + \frac{4g_2-2f_0 f_2}{t^2} + \cdots.$$

If $4g_1 = 2f_0 f_1$, then (1.12) has a regular singularity at $t = \infty$, and therefore admits of solutions in convergent power series. Alternatively, if $4g_1 \neq 2f_0 f_1$, then (1.12) has an irregular singularity at infinity with *unequal* characteristic values $\pm(2f_0 f_1 - 4g_1)^{1/2}$; compare (1.09). Therefore we can construct formal series expansions for W of the form (1.08), with z replaced by t. Thus the Fabry transformation obviates the need for a special theory.‡

Restoration of the original variables in the case $4g_1 \neq 2f_0 f_1$ yields series solutions of the form

$$w = \exp\{-\tfrac{1}{2}f_0 z \pm (2f_0 f_1 - 4g_1)^{1/2} z^{1/2}\}\, z^{(1-2f_1)/4} \sum_{s=0}^{\infty} \frac{\hat{a}_s}{z^{s/2}}.$$

Again, the coefficients $\hat{a}_s$ may be found by direct substitution in the original differential equation. Expansions of this kind, involving fractional powers of z, are called *subnormal solutions*.

Ex. 1.1 The differential equation $w'' = (z^2 + z^{-6})w$ has a singularity of rank 2 at infinity. Show that it may be transformed into an equation in which the corresponding singularity is of rank 1.

Ex. 1.2 Solve exactly

$$z\frac{d^2w}{dz^2} + 2\frac{dw}{dz} - \left(\frac{1}{4} + \frac{5}{16z}\right)w = 0. \qquad \text{[Ince, 1927.]}$$

† Ince (1927, §17.53).

‡ This contrasts agreeably with the difficulties caused at a regular singularity by coincidence of the exponents (Chapter 5, §5).

Ex. 1.3 Construct the subnormal solutions at infinity of the equation

$$\frac{d^2 w}{dz^2} + \left\{\frac{2}{z} - \frac{L(L+1)}{z^2}\right\} w = 0,$$

in which L is a constant. [Curtis, 1964.]

2 Asymptotic Nature of the Formal Series

2.1 The analysis of §1 is purely formal. If it transpired that the expansion (1.08) converges for all sufficiently large $|z|$, then the process of termwise differentiation would be valid and the series would define a solution of the differential equation. That this is not the usual state of affairs can be seen as follows. When all terms beyond the first are neglected on the right of (1.11), we have

$$a_s/a_{s-1} \sim s/(f_0+2\lambda) \qquad (s \to \infty).$$

This implies (1.08) diverges. Hence only in cases in which the first term on the right of (1.11) is largely cancelled by the contribution of other terms—as, for example, in Exercise 1.2—is there any possibility of convergence.

The most that can be hoped of (1.08), in general, is that it provides the asymptotic expansion of a solution in a certain region of the z plane. It is reasonable, moreover, to expect this region to be symmetric with respect to the direction of strongest recession as $z \to \infty$. Since the ratio of the leading terms of the formal solutions is $e^{(\lambda_1-\lambda_2)z} z^{\mu_1-\mu_2} a_{0,1}/a_{0,2}$, this direction is given by $\operatorname{ph}\{(\lambda_2-\lambda_1)z\} = 0$ for the first solution, and $\operatorname{ph}\{(\lambda_1-\lambda_2)z\} = 0$ for the second solution.

Theorem 2.1 *Let $f(z)$ and $g(z)$ be analytic functions of the complex variable z having convergent series expansions*

$$f(z) = \sum_{s=0}^{\infty} \frac{f_s}{z^s}, \qquad g(z) = \sum_{s=0}^{\infty} \frac{g_s}{z^s}, \tag{2.01}$$

in the annulus **A**: $|z| > a$, *with* $f_0^2 \neq 4g_0$. *Then the equation*

$$\frac{d^2 w}{dz^2} + f(z)\frac{dw}{dz} + g(z)\, w = 0 \tag{2.02}$$

has unique solutions $w_j(z)$, $j = 1, 2$, such that in the respective intersections of **A** *with the sectors*†

$$|\operatorname{ph}\{(\lambda_2-\lambda_1)z\}| \leqslant \pi \quad (j=1), \qquad |\operatorname{ph}\{(\lambda_1-\lambda_2)z\}| \leqslant \pi \quad (j=2), \tag{2.03}$$

$w_j(z)$ is holomorphic and

$$w_j(z) \sim e^{\lambda_j z} z^{\mu_j} \sum_{s=0}^{\infty} \frac{a_{s,j}}{z^s} \qquad (z \to \infty). \tag{2.04}$$

† In effect cuts are introduced in **A**. The regions are not maximal; see Theorem 2.2 below.

In this theorem λ_j, μ_j, and $a_{s,j}$ are defined as in §1.2. Any branch of z^{μ_j} may be used, provided that it is continuous throughout the appropriate sector (2.03). The proof follows.

2.2 Let the solution of (2.02) be denoted by

$$w(z) = L_n(z) + \varepsilon_n(z),$$

where $L_n(z)$ is the nth partial sum

$$L_n(z) = e^{\lambda_1 z} z^{\mu_1} \sum_{s=0}^{n-1} \frac{a_{s,1}}{z^s}, \tag{2.05}$$

and $\varepsilon_n(z)$ the corresponding error term. If $L_n(z)$ is substituted for w in the left-hand side of (2.02), then the coefficient of $e^{\lambda_1 z} z^{\mu_1 - s}$ vanishes for $s = 0, 1, \ldots, n$, in consequence of (1.09) to (1.11). Accordingly,

$$L_n''(z) + f(z) L_n'(z) + g(z) L_n(z) = e^{\lambda_1 z} z^{\mu_1} R_n(z), \tag{2.06}$$

where $R_n(z) = O(z^{-n-1})$ as $z \to \infty$. Therefore

$$\varepsilon_n''(z) + f(z)\varepsilon_n'(z) + g(z)\varepsilon_n(z) = -e^{\lambda_1 z} z^{\mu_1} R_n(z). \tag{2.07}$$

To solve the last equation, let b be an arbitrary constant exceeding a, and z lie in the closed annulus $\mathbf{B}$: $|z| \geqslant b$. Then

$$|R_n(z)| \leqslant B_n |z|^{-n-1}, \tag{2.08}$$

where B_n is assignable. On the left-hand side of (2.07) we retain the dominant terms in the expansions of $f(z)$ and $g(z)$; the rest are transferred to the right. Thus

$$\varepsilon_n''(z) + f_0 \varepsilon_n'(z) + g_0 \varepsilon_n(z) = -e^{\lambda_1 z} z^{\mu_1} R_n(z) - \{g(z) - g_0\}\varepsilon_n(z) - \{f(z) - f_0\}\varepsilon_n'(z). \tag{2.09}$$

Variation of parameters yields the equivalent integral equation

$$\varepsilon_n(z) = \int_z^{\infty e^{-i\omega}} \mathsf{K}(z,t)[e^{\lambda_1 t} t^{\mu_1} R_n(t) + \{g(t) - g_0\}\varepsilon_n(t) + \{f(t) - f_0\}\varepsilon_n'(t)]\, dt, \tag{2.10}$$

where

$$\mathsf{K}(z,t) = \{e^{\lambda_1(z-t)} - e^{\lambda_2(z-t)}\}/(\lambda_1 - \lambda_2).$$

The direction of the upper limit is at our disposal. We prescribe it to be that of strongest recession of the wanted solution, given by

$$\omega = \mathrm{ph}(\lambda_2 - \lambda_1).$$

2.3 Provided that $n > \operatorname{Re}\mu_1 \equiv m_1$, say, $z \in \mathbf{B}$, and $|\mathrm{ph}(ze^{i\omega})| \leqslant \pi$, equation (2.10) is solvable by the method of successive approximations used in earlier chapters. We express

$$\varepsilon_n(z) = \sum_{s=0}^{\infty} \{h_{s+1}(z) - h_s(z)\}, \tag{2.11}$$

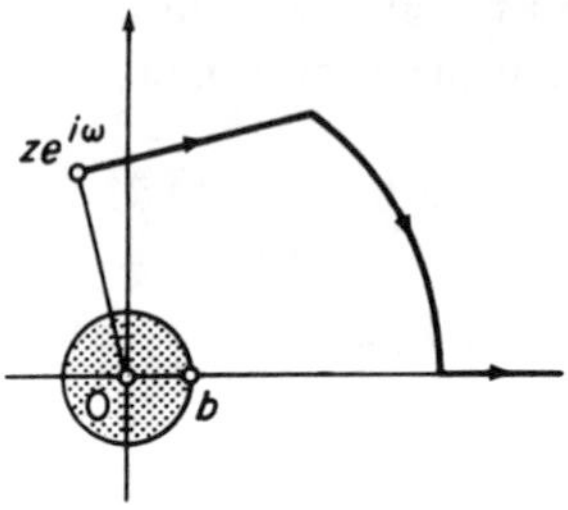

Fig. 2.1 $te^{i\omega}$ plane. $\omega = \mathrm{ph}(\lambda_2 - \lambda_1)$.

where the sequence $\{h_s(z)\}$ is defined by $h_0(z) = 0$ and

$$h_{s+1}(z) = \int_z^{\infty e^{-i\omega}} \mathsf{K}(z,t)[e^{\lambda_1 t} t^{\mu_1} R_n(t) + \{g(t) - g_0\} h_s(t) + \{f(t) - f_0\} h_s'(t)]\, dt \tag{2.12}$$

when $s \geqslant 0$. The integration path is chosen so that its map in the $te^{i\omega}$ plane consists of (i) a straight line segment through $ze^{i\omega}$ perpendicular to the join of $ze^{i\omega}$ and the origin; (ii) an arc of a large circle centered at the origin; (iii) part of the real axis: see Fig. 2.1.

On this path $\mathrm{Re}\{(\lambda_2 - \lambda_1)t\}$ is nondecreasing, hence

$$|\mathsf{K}(z,t)| \leqslant \frac{2|e^{\lambda_1(z-t)}|}{|\lambda_1 - \lambda_2|}, \qquad \left|\frac{\partial \mathsf{K}(z,t)}{\partial z}\right| \leqslant \frac{(|\lambda_1| + |\lambda_2|)|e^{\lambda_1(z-t)}|}{|\lambda_1 - \lambda_2|}.$$

Since $|\mathrm{ph}(te^{i\omega})| \leqslant \pi$, we also have

$$|t^{\mu_1}| \leqslant M|t|^{m_1}; \qquad M \equiv \exp\{(\pi + |\omega|)|\mathrm{Im}\,\mu_1|\}.$$

Taking $s = 0$ in (2.12) and its z-differentiated form, substituting by means of the bounds just obtained and (2.08), and letting the radius of the circular arc in Fig. 2.1 tend to infinity, we find that

$$\frac{|h_1(z)|}{2}, \quad \frac{|h_1'(z)|}{|\lambda_1| + |\lambda_2|} \leqslant \frac{MB_n}{|\lambda_1 - \lambda_2|} \frac{\chi(n - m_1)}{n - m_1} \frac{|e^{\lambda_1 z}|}{|z|^{n - m_1}},$$

where χ is the function introduced in Lemma 13.1 of Chapter 6. Beginning with this result, we may verify by induction that

$$\frac{|h_{s+1}(z) - h_s(z)|}{2}, \quad \frac{|h_{s+1}'(z) - h_s'(z)|}{|\lambda_1| + |\lambda_2|} \leqslant \frac{MB_n \beta^s}{|\lambda_1 - \lambda_2|^{s+1}} \left\{\frac{\chi(n - m_1)}{n - m_1}\right\}^{s+1} \frac{|e^{\lambda_1 z}|}{|z|^{n - m_1}} \tag{2.13}$$

for $s = 0, 1, \ldots$, where

$$\beta = \sup_{t \in \mathbf{B}} [|t| \{2|g(t) - g_0| + (|\lambda_1| + |\lambda_2|)|f(t) - f_0|\}]$$

and is finite; compare (2.01).

Now suppose that n is large enough to ensure that

$$|\lambda_1 - \lambda_2|(n - m_1) > \beta\chi(n - m_1); \tag{2.14}$$

this is always possible because

$$\chi(n - m_1) \sim (\tfrac{1}{2}\pi n)^{1/2} \qquad (n \to \infty).$$

Then the series (2.11) converges uniformly in any compact set in the intersection of **B** and the cut plane $|\mathrm{ph}(ze^{i\omega})| \leqslant \pi$. Termwise differentiation is therefore legitimate, and from (2.12) it is seen that the sum is an analytic function which satisfies the integral equation (2.10) and therefore the differential equations (2.07) and (2.09).

Summation of (2.13) shows that

$$\varepsilon_n(z),\ \varepsilon_n'(z) = O(e^{\lambda_1 z} z^{m_1 - n}) \qquad (z \to \infty).$$

Therefore for all sufficiently large values of n, equation (2.02) has an analytic solution $w_{n,1}(z)$ with the property

$$w_{n,1}(z) = e^{\lambda_1 z} z^{\mu_1} \left\{ \sum_{s=0}^{n-1} \frac{a_{s,1}}{z^s} + O\left(\frac{1}{z^n}\right) \right\}$$

as $z \to \infty$ in the sector $|\mathrm{ph}\{(\lambda_2 - \lambda_1)z\}| \leqslant \pi$. By relabeling, we see that there is another analytic solution $w_{n,2}(z)$ such that

$$w_{n,2}(z) = e^{\lambda_2 z} z^{\mu_2} \left\{ \sum_{s=0}^{n-1} \frac{a_{s,2}}{z^s} + O\left(\frac{1}{z^n}\right) \right\}$$

as $z \to \infty$ in $|\mathrm{ph}\{(\lambda_1 - \lambda_2)z\}| \leqslant \pi$.

It remains to show that $w_{n,1}(z)$ and $w_{n,2}(z)$ are independent of n. If n_1 and n_2 are admissible values of n, then both $w_{n_1,1}(z)$ and $w_{n_2,1}(z)$ are recessive compared with $w_{n_1,2}(z)$ or $w_{n_2,2}(z)$ as $z \to \infty e^{-i\omega}$; hence their ratio is independent of z. That this ratio is unity again follows by letting $z \to \infty e^{-i\omega}$. Similarly for $w_{n,2}(z)$. This completes the proof of Theorem 2.1.

2.4 The regions of validity of the asymptotic expansions can be extended:

Theorem 2.2 *If δ is an arbitrary small positive constant, then the expansion* (2.04) *holds for the analytic continuation of $w_j(z)$ in the sector*

$$|\mathrm{ph}\{(\lambda_2 - \lambda_1)z\}| \leqslant \tfrac{3}{2}\pi - \delta \quad (j=1); \qquad |\mathrm{ph}\{(\lambda_1 - \lambda_2)z\}| \leqslant \tfrac{3}{2}\pi - \delta \quad (j=2). \tag{2.15}$$

Moreover, unless the expansion converges this sector of validity is maximal.

The extension can be achieved by modifying the analysis of §2.3. If, for example, $j = 1$ and

$$\pi \leqslant \mathrm{ph}(ze^{i\omega}) \leqslant \tfrac{3}{2}\pi - \delta, \tag{2.16}$$

then instead of the path mapped in Fig. 2.1 we use a straight line segment through $ze^{i\omega}$ parallel to the imaginary axis and crossing the real axis, together with the large circular arc and part of the real axis. An alternative proof, which also establishes the second part of the theorem, is as follows.

Let $w_1(z)$ and $w_2(z)$ be the solutions given by Theorem 2.1. Then $w_1(ze^{-2\pi i})$ is another solution of (2.02). This solution is dominant as $z \to \infty$ along the ray $\mathrm{ph}\, z = \pi - \omega$, and is therefore linearly independent of $w_2(z)$. Accordingly, there exist constants A and B such that

$$w_1(z) = Aw_1(ze^{-2\pi i}) + Bw_2(z). \tag{2.17}$$

Letting $z \to \infty e^{i(\pi - \omega)}$ we deduce that $A = e^{2\pi i \mu_1}$. The value of B cannot be determined this way, but since δ is positive $Bw_2(z)$ is uniformly exponentially small compared with $e^{2\pi i\mu_1}w_1(ze^{-2\pi i})$ for large z in the sector (2.16). Therefore in Poincaré's sense $Bw_2(z)$ does not contribute to the asymptotic expansion of the analytic continuation of $w_1(z)$. Accordingly, the expansion (2.04), with $j=1$, holds in (2.16). Similarly for the conjugate sector, and also for the second solution.

Next, if the constant B in (2.17) is nonzero, then it is evident that the region of validity of (2.04), with $j=1$, cannot be extended across $\mathrm{ph}(ze^{i\omega})=\frac{3}{2}\pi$. By applying Theorem 7.2 of Chapter 1 to the function $e^{-\lambda_1 z}z^{-\mu_1}w_1(z)$, we see from (2.17) that B vanishes if and only if (2.04) converges. Similarly for the ray $\mathrm{ph}(ze^{i\omega})=-\frac{3}{2}\pi$, and also for the second solution. The proof of Theorem 2.2 is complete.

Theorem 2.2 exemplifies the general rule that the monotonicity condition on the path of integration is necessary as well as sufficient. In other words, shadow zones (Chapter 6, §11.4) are genuine regions of exclusion.

Ex. 2.1 Show that for large z the equation $w''=\{(z+4)/z\}^{1/2}w$ has asymptotic solutions

$$e^{-z}\left(\frac{1}{z}-\frac{2}{z^2}+\frac{5}{z^3}-\frac{44}{3z^4}+\cdots\right),\qquad e^{z}\left(z+1-\frac{1}{2z}+\frac{2}{3z^2}+\cdots\right),$$

valid when $|\mathrm{ph}(\pm z)|\leqslant\frac{3}{2}\pi-\delta$ $(<\frac{3}{2}\pi)$, respectively.

Ex. 2.2 Show that the equation $w''+(z^{-4}\cos z)w=0$ has an asymptotic solution

$$w\sim(z+\tfrac{3}{32}z^3+\cdots)\cos(1/z)+(-\tfrac{1}{4}z^2+\tfrac{107}{1152}z^4+\cdots)\sin(1/z)$$

as $z\to 0$ in the sector $|\mathrm{ph}\, z|\leqslant\pi-\delta$ $(<\pi)$.

3 Equations Containing a Parameter

3.1 Suppose now that the coefficients in the given differential equation depend on a complex parameter u; thus

$$\frac{d^2w}{dz^2}+f(u,z)\frac{dw}{dz}+g(u,z)\,w=0. \tag{3.01}$$

Often it is important to know whether the solutions defined by Theorem 2.1 are holomorphic in u. This cannot be resolved by application of Theorem 3.2 of Chapter 5 because no ordinary point z_0 with the properties required by Condition (iv) is available.

Theorem 3.1 *Let u range over a fixed complex domain $\mathbf{U}$, and z range over a fixed annulus $\mathbf{A}$: $|z|>a$. Assume that for each u, $f(u,z)$ and $g(u,z)$ satisfy the conditions of Theorem 2.1, and*

(i) *The coefficients f_0 and g_0 in the series (2.01) are independent of u. Higher coefficients $f_s\equiv f_s(u)$ and $g_s\equiv g_s(u)$ are holomorphic functions of u.*
(ii) *If u is restricted to any compact domain $\mathbf{U}_c\subset\mathbf{U}$, then $|f_s(u)|\leqslant F_s^{(c)}$ and $|g_s(u)|\leqslant G_s^{(c)}$, where $F_s^{(c)}$ and $G_s^{(c)}$ are independent of u and the series $\sum F_s^{(c)}z^{-s}$ and $\sum G_s^{(c)}z^{-s}$ converge absolutely in $\mathbf{A}$.*
(iii) *$a_{0,1}$ and $a_{0,2}$ are holomorphic functions of u.*

Then at each point z of $\mathbf{A}$ each branch of $w_1(z)$, $w_2(z)$, and their first two partial z derivatives, is a holomorphic function of u.

Application of the M-test shows that $f(u, z)$ and $g(u, z)$ are continuous functions of both variables, and holomorphic functions of u for fixed z. To prove the theorem we retrace the steps of Theorem 2.1, bearing in mind that all quantities appearing in the proof may depend on u, with the exception of λ_1 and λ_2 (compare (1.09)).

From the given conditions and the definitions of §1.2, it is immediately seen that each of the quantities μ_1, μ_2, $a_{s,1}$, and $a_{s,2}$ is holomorphic in u. In consequence, the truncated series (2.05) is continuous in u and z, and holomorphic in u. The same is true of its partial derivatives $L_n'(z)$ and $L_n''(z)$, and therefore of $R_n(z)$. Furthermore, if $u \in \mathbf{U}_c$, then the quantity B_n in (2.08) is assignable independently of u. The only other quantities in the bound (2.13) which depend on u are M, β, and m_1. The definitions of M and β show that each may be replaced by an upper bound which is independent of u in $\mathbf{U}_c$. And since $|m_1|$ is bounded, it is easily seen that both the series (2.11) and its z-differentiated form converge uniformly in $\mathbf{U}_c$, for all $n \geqslant N_c$, an assignable constant.

Next, application of Theorem 1.1 of Chapter 2 to (2.12), with $s = 0$, shows that $h_1(z)$ is holomorphic in u. We also see, by uniform convergence, that $h_1(z)$ is continuous in u and z. Similarly, from the z-differentiated form of (2.12) we see that the same is true of $h_1'(z)$, and hence (by induction) of $h_s(z)$ and $h_s'(z)$, $s = 1, 2, \ldots$.

Summarizing so far, we have proved that if $u \in \mathbf{U}_c$, z lies in the intersection of $\mathbf{B}$ and the sector $|\mathrm{ph}(ze^{i\omega})| \leqslant \pi$, and $n \geqslant N_c$, then (a) each term in the series (2.11), and its z derivative, is holomorphic in u; (b) this series and its z-differentiated form converge uniformly with respect to u. In consequence, $\varepsilon_n(z)$, $\varepsilon_n'(z)$, $w_1(z)$, and $w_1'(z)$ are all holomorphic functions of u within $\mathbf{U}_c$. And since $w_1(z)$ is independent of n (§2.3), $w_1(z)$ and $w_1'(z)$ are holomorphic throughout $\mathbf{U}$. Holomorphicity of $w_1''(z)$ immediately follows from (3.01).

Since b can be arbitrarily close to a, Theorem 3.1 is established for $w_1(z)$ when z lies in $\mathbf{A}$ and the sector $|\mathrm{ph}\{(\lambda_2-\lambda_1)z\}| \leqslant \pi$. The extension to other branches follows from Theorem 3.2 of Chapter 5 by taking z_0 to be any finite point in the intersection of $\mathbf{A}$ and $|\mathrm{ph}\{(\lambda_2-\lambda_1)z\}| \leqslant \pi$. Similarly for the second solution. The proof is now complete.

3.2 The condition that f_0 and g_0 be independent of u is not essential, but without it the analysis becomes more difficult because ω and the z regions of validity of the asymptotic expansions (2.04) vary with u. In all applications in this book f_0 and g_0 are independent of u.

4 Hankel Functions; Stokes' Phenomenon

4.1 We apply the foregoing theory to Bessel's equation (Chapter 2, §9.2)

$$\frac{d^2w}{dz^2} + \frac{1}{z}\frac{dw}{dz} + \left(1 - \frac{\nu^2}{z^2}\right)w = 0. \tag{4.01}$$

The order ν may be real or complex. In the notation of §§1 and 2, we have $f_1 = 1$, $g_0 = 1$, $g_2 = -\nu^2$, all other coefficients being zero. From equations (1.09) and (1.10) we find that $\lambda_1 = i$, $\lambda_2 = -i$, and $\mu_1 = \mu_2 = -\frac{1}{2}$. With $a_{0,1} = a_{0,2} = 1$, the recurrence relation (1.11) gives $a_{s,1} = i^s A_s(\nu)$ and $a_{s,2} = (-i)^s A_s(\nu)$, where

$$A_s(\nu) = \frac{(4\nu^2-1^2)(4\nu^2-3^2)\cdots\{4\nu^2-(2s-1)^2\}}{s!\,8^s}. \tag{4.02}$$

On multiplying the solutions furnished by Theorems 2.1 and 2.2 by the normalizing factors $(2/\pi)^{1/2}\exp\{\mp(\frac{1}{2}\nu+\frac{1}{4})\pi i\}$, we see that equation (4.01) has unique solutions $H_\nu^{(1)}(z)$, $H_\nu^{(2)}(z)$, such that

$$H_\nu^{(1)}(z) \sim \left(\frac{2}{\pi z}\right)^{1/2} e^{i\zeta} \sum_{s=0}^{\infty} i^s \frac{A_s(\nu)}{z^s} \qquad (-\pi+\delta \leqslant \operatorname{ph} z \leqslant 2\pi-\delta), \tag{4.03}$$

$$H_\nu^{(2)}(z) \sim \left(\frac{2}{\pi z}\right)^{1/2} e^{-i\zeta} \sum_{s=0}^{\infty} (-i)^s \frac{A_s(\nu)}{z^s} \qquad (-2\pi+\delta \leqslant \operatorname{ph} z \leqslant \pi-\delta), \tag{4.04}$$

as $z \to \infty$, where δ is an arbitrary small positive constant,

$$\zeta = z - \tfrac{1}{2}\nu\pi - \tfrac{1}{4}\pi, \tag{4.05}$$

and the branch of $z^{1/2}$ is determined by

$$z^{1/2} = \exp(\tfrac{1}{2}\ln|z| + \tfrac{1}{2}i \operatorname{ph} z). \tag{4.06}$$

These solutions are called the *Hankel functions of order* ν, and (4.03) and (4.04) are sometimes called *Hankel's expansions*. Both $H_\nu^{(1)}(z)$ and $H_\nu^{(2)}(z)$ are analytic functions of z, their only possible singularities being the singularities of the defining differential equation, that is, 0 and ∞. Although the Hankel expansions hold only in certain sectors, the solutions themselves can be continued analytically to *any* value of $\operatorname{ph} z$ (Chapter 5, §3.1). *Principal branches* correspond to $-\pi < \operatorname{ph} z \leqslant \pi$.

As $z \to \infty$ in the sector $\delta \leqslant \operatorname{ph} z \leqslant \pi-\delta$, $H_\nu^{(1)}(z)$ is recessive and $H_\nu^{(2)}(z)$ is dominant; in $-\pi+\delta \leqslant \operatorname{ph} z \leqslant -\delta$ these roles are interchanged. Accordingly, the Hankel functions are linearly independent solutions and comprise a numerically satisfactory pair for large z in the sector $|\operatorname{ph} z| \leqslant \pi$ (but not elsewhere).

Next, *for fixed nonzero z each branch of $H_\nu^{(1)}(z)$, $H_\nu^{(2)}(z)$, $H_\nu^{(1)\prime}(z)$, and $H_\nu^{(2)\prime}(z)$ is an entire function of ν.* This follows immediately from Theorem 3.1.

Lastly, *$H_\nu^{(1)}(z)$ and $H_{\bar\nu}^{(2)}(\bar z)$ are complex conjugates*; this is because $\overline{H_{\bar\nu}^{(2)}(\bar z)}$ satisfies (4.01) and the same boundary conditions (4.03) as $H_\nu^{(1)}(z)$. This property enables formulas for one Hankel function to be transformed into corresponding formulas for the other function.

4.2 Since $H_\nu^{(1)}(z)$, $H_\nu^{(2)}(z)$, and $J_\nu(z)$ satisfy the same second-order differential equation there exists a connection formula

$$J_\nu(z) = AH_\nu^{(1)}(z) + BH_\nu^{(2)}(z).$$

By means of Laplace's method, it was shown in Chapter 4, §9 that

$$J_\nu(z) \sim \left(\frac{2}{\pi z}\right)^{1/2} \left\{\cos\zeta \sum_{s=0}^{\infty} (-)^s \frac{A_{2s}(\nu)}{z^{2s}} - \sin\zeta \sum_{s=0}^{\infty} (-)^s \frac{A_{2s+1}(\nu)}{z^{2s+1}}\right\} \tag{4.07}$$

as $z \to \infty$ in $|\mathrm{ph}\, z| \leqslant \pi - \delta\ (<\pi)$, where ζ and $A_s(\nu)$ are as in §4.1 above. By taking leading terms in (4.03), (4.04), and (4.07), and letting $z \to \infty e^{i\pi/2}$, we see that $B = \frac{1}{2}$. Similarly, by letting $z \to \infty e^{-i\pi/2}$ we have $A = \frac{1}{2}$. Thus

$$J_\nu(z) = \tfrac{1}{2}\{H_\nu^{(1)}(z) + H_\nu^{(2)}(z)\} \tag{4.08}$$

for all z, other than zero.

Next, since (4.01) is unaffected on changing the sign of ν, other solutions of this equation are $H_{-\nu}^{(1)}(z)$ and $H_{-\nu}^{(2)}(z)$. Like $H_\nu^{(1)}(z)$ the former is recessive at infinity in $\delta \leqslant \mathrm{ph}\, z \leqslant \pi - \delta$, hence its ratio to $H_\nu^{(1)}(z)$ must be independent of z. The actual ratio may be found by replacing ν with $-\nu$ in (4.03) and (4.05); thus

$$H_{-\nu}^{(1)}(z) = e^{\nu\pi i} H_\nu^{(1)}(z). \tag{4.09}$$

Similarly,

$$H_{-\nu}^{(2)}(z) = e^{-\nu\pi i} H_\nu^{(2)}(z). \tag{4.10}$$

From (4.08), (4.09), and (4.10), we derive

$$J_{-\nu}(z) = \tfrac{1}{2}\{e^{\nu\pi i} H_\nu^{(1)}(z) + e^{-\nu\pi i} H_\nu^{(2)}(z)\}. \tag{4.11}$$

Elimination of $H_\nu^{(2)}(z)$ and $H_\nu^{(1)}(z)$ in turn from (4.08) and (4.11) produces

$$H_\nu^{(1)}(z) = \frac{i\{e^{-\nu\pi i} J_\nu(z) - J_{-\nu}(z)\}}{\sin(\nu\pi)}, \qquad H_\nu^{(2)}(z) = -\frac{i\{e^{\nu\pi i} J_\nu(z) - J_{-\nu}(z)\}}{\sin(\nu\pi)}. \tag{4.12}$$

When ν is an integer or zero, each of these fractions may be replaced by its limiting value since the Hankel functions are continuous in ν.

Formulas for the analytic continuations $H_\nu^{(1)}(ze^{m\pi i})$ and $H_\nu^{(2)}(ze^{m\pi i})$, m being an arbitrary integer, are obtainable from (4.08), (4.11), (4.12), and the identities (Chapter 2, §9.3)

$$J_{\pm\nu}(ze^{m\pi i}) = e^{\pm m\nu\pi i} J_{\pm\nu}(z).$$

Thus

$$H_\nu^{(1)}(ze^{m\pi i}) = -[\sin\{(m-1)\nu\pi\}\, H_\nu^{(1)}(z) + e^{-\nu\pi i} \sin(m\nu\pi)\, H_\nu^{(2)}(z)]/\sin(\nu\pi), \tag{4.13}$$

$$H_\nu^{(2)}(ze^{m\pi i}) = [e^{\nu\pi i} \sin(m\nu\pi)\, H_\nu^{(1)}(z) + \sin\{(m+1)\nu\pi\}\, H_\nu^{(2)}(z)]/\sin(\nu\pi). \tag{4.14}$$

Again, limiting values are to be taken when ν is an integer or zero. These formulas confirm that $z = 0$ is a branch point of the Hankel functions for all values of ν.

4.3 Formulas (4.13) and (4.14) enable asymptotic expansions for the Hankel functions to be constructed for any phase range. For example, taking $m = 2$ in (4.13) we have

$$H_\nu^{(1)}(ze^{2\pi i}) = -H_\nu^{(1)}(z) - (1 + e^{-2\nu\pi i})\, H_\nu^{(2)}(z). \tag{4.15}$$

When $|\mathrm{ph}\, z| \leqslant \pi - \delta$ we may substitute on the right-hand side by means of (4.03) and (4.04). Then replacing z by $ze^{-2\pi i}$ we arrive at†

$$H_\nu^{(1)}(z) \sim \left(\frac{2}{\pi z}\right)^{1/2}\left\{e^{i\zeta}\sum_{s=0}^{\infty} i^s \frac{A_s(\nu)}{z^s} + (1+e^{-2\nu\pi i})e^{-i\zeta}\sum_{s=0}^{\infty}(-i)^s\frac{A_s(\nu)}{z^s}\right\}$$
$$(\pi+\delta \leqslant \operatorname{ph} z \leqslant 3\pi-\delta). \qquad (4.16)$$

It will be observed that (4.03) and (4.16) furnish apparently different representations of $H_\nu^{(1)}(z)$ in $\pi+\delta \leqslant \operatorname{ph} z \leqslant 2\pi-\delta$, the common region of validity. In this sector, however, $e^{-i\zeta}$ is exponentially small compared with $e^{i\zeta}$, hence the *whole* contribution of the second series in (4.16) is absorbable, in Poincaré's sense, in any of the error terms associated with the first series. Accordingly, there is no inconsistency.

Extensions to other phase ranges may be found in the same manner by taking appropriate values of m in (4.13). In every case we obtain a compound asymptotic expansion of the form (4.16) involving other multiples of the two series.

Stokes (1857) was the first to observe the *discontinuous* changes in the constants associated with a compound expansion when the phase range of the asymptotic variable changes in a continuous manner. The need for the discontinuities is called *Stokes' phenomenon*; it is by no means confined to solutions of Bessel's equation. A full understanding of the phenomenon requires a more complete error analysis; see §13.2 below.

4.4 Integral representations for the Hankel functions are obtainable as follows. In deriving the asymptotic expansion of $J_\nu(z)$ in Chapter 4, §9, we considered separately contributions from‡

$$\frac{1}{\pi i}\int_{-\infty}^{\infty+\pi i} e^{z\sinh t-\nu t}\,dt \qquad \text{and} \qquad -\frac{1}{\pi i}\int_{-\infty}^{\infty-\pi i} e^{z\sinh t-\nu t}\,dt. \qquad (4.17)$$

It is verifiable directly, by differentiation under the sign of integration, that these integrals satisfy (4.01) individually. And by examining their expansions for large z (Chapter 4, (9.07)), we perceive that *when* $|\operatorname{ph} z| < \frac{1}{2}\pi$, *the first of* (4.17) *equals* $H_\nu^{(1)}(z)$ *and the second equals* $H_\nu^{(2)}(z)$.

Corresponding representations for other phase ranges can be constructed by repeated deformations of the integration paths. For example, when $\delta \leqslant \operatorname{ph} z \leqslant \frac{1}{2}\pi-\delta$, we have

$$H_\nu^{(1)}(z) = \frac{1}{\pi i}\int_{-\infty+(\pi i/2)}^{\infty+(\pi i/2)} e^{z\sinh t-\nu t}\,dt. \qquad (4.18)$$

Analytic continuation then extends this result to $0 < \operatorname{ph} z < \pi$. The general formulas are easily seen to be

$$H_\nu^{(1)}(z) = \frac{1}{\pi i}\int_{-\infty+\alpha i}^{\infty+(\pi-\alpha)i} e^{z\sinh t-\nu t}\,dt, \qquad H_\nu^{(2)}(z) = -\frac{1}{\pi i}\int_{-\infty+\alpha i}^{\infty-(\pi+\alpha)i} e^{z\sinh t-\nu t}\,dt, \qquad (4.19)$$

† Again, the branch of $z^{1/2}$ is determined by $\operatorname{ph}(z^{1/2}) = \frac{1}{2}\operatorname{ph} z$.
‡ The sign of t has now been changed.

where α is arbitrary and $-\frac{1}{2}\pi+\alpha < \operatorname{ph} z < \frac{1}{2}\pi+\alpha$. These are *Sommerfeld's integrals.*

An immediate deduction from Sommerfeld's integrals is that $H_\nu^{(1)}(z)$ *and* $H_\nu^{(2)}(z)$ *satisfy the recurrence relations given for* $J_\nu(z)$ *in Chapter* 2, §9.5. It may be noted that anticipation of these recurrence relations underlay the choice of normalizing factors in (4.03) and (4.04).

4.5 Another type of contour integral for the Hankel functions is suggested by Poisson's integral (Chapter 2, Exercise 9.5):

$$J_\nu(z) = \frac{(\frac{1}{2}z)^\nu}{\pi^{1/2}\Gamma(\nu+\frac{1}{2})}\int_{-1}^{1}\cos(zt)(1-t^2)^{\nu-(1/2)}\,dt \qquad (\operatorname{Re}\nu > -\tfrac{1}{2}).$$

By differentiation under the sign of integration it is verifiable that Bessel's equation is satisfied by any contour integral of the form

$$z^\nu \int_{\mathscr{C}} e^{\pm izt}(t^2-1)^{\nu-(1/2)}\,dt,$$

provided that the branch of $(t^2-1)^{\nu-(1/2)}$ is continuous along the path $\mathscr{C}$ and the integrand returns to its initial value at the end of $\mathscr{C}$.

When $|\operatorname{ph} z| < \frac{1}{2}\pi$ and $\nu \neq \frac{1}{2}, \frac{3}{2}, \ldots$, an appropriate choice of paths and normalizing factors for the Hankel functions is given by

$$H_\nu^{(1)}(z) = \frac{\Gamma(\frac{1}{2}-\nu)(\frac{1}{2}z)^\nu}{\pi^{3/2}i}\int_{1+i\infty}^{(1+)} e^{izt}(t^2-1)^{\nu-(1/2)}\,dt, \tag{4.20}$$

and

$$H_\nu^{(2)}(z) = \frac{\Gamma(\frac{1}{2}-\nu)(\frac{1}{2}z)^\nu}{\pi^{3/2}i}\int_{1-i\infty}^{(1+)} e^{-izt}(t^2-1)^{\nu-(1/2)}\,dt. \tag{4.21}$$

Each path is a simple loop contour not enclosing $t=-1$, and $(t^2-1)^{\nu-(1/2)}$ takes its principal value at the intersection with the interval $(1,\infty)$. These representations are known as *Hankel's integrals.* To verify them, we note that they converge uniformly in the sector $|\operatorname{ph} z| \leqslant \frac{1}{2}\pi-\delta$ and that the integrands vanish at the endpoints. Therefore each integral satisfies Bessel's equation. And application of Watson's lemma for loop integrals (Chapter 4, §5.3) shows that the asymptotic forms of the right-hand sides of (4.20) and (4.21) agree with (4.03) and (4.04) in $|\operatorname{ph} z| \leqslant \frac{1}{2}\pi-\delta$.

Ex. 4.1 Verify the Wronskians

$$\mathscr{W}\{H_\nu^{(1)}(z), H_\nu^{(2)}(z)\} = -2\mathscr{W}\{J_\nu(z), H_\nu^{(1)}(z)\} = 2\mathscr{W}\{J_\nu(z), H_\nu^{(2)}(z)\} = -4i/(\pi z).$$

Ex. 4.2 Show that when ν is half an odd integer, positive or negative, the asymptotic expansions (4.03), (4.04), and (4.07) terminate and represent the left-hand sides exactly.

5 The Function $Y_\nu(z)$

5.1 We now have three standard solutions of Bessel's equation: $J_\nu(z)$, $H_\nu^{(1)}(z)$, and $H_\nu^{(2)}(z)$. Their characterizing properties are: (i) at the regular singularity at the origin $J_\nu(z)$ is recessive when $\operatorname{Re}\nu > 0$ or $\nu = 0$; (ii) at the irregular singularity at infinity $H_\nu^{(1)}(z)$ is recessive in the sector $\delta \leqslant \operatorname{ph} z \leqslant \pi-\delta$, and $H_\nu^{(2)}(z)$ is recessive in the conjugate sector. Accordingly, $J_\nu(z)$ and $H_\nu^{(1)}(z)$ comprise a numerically satisfactory pair of solutions throughout the *whole* of the sector $0 \leqslant \operatorname{ph} z \leqslant \pi$, provided that $\operatorname{Re} \nu \geqslant 0$. Similarly, $J_\nu(z)$ and $H_\nu^{(2)}(z)$ are satisfactory throughout $-\pi \leqslant \operatorname{ph} z \leqslant 0$.

In the special, but important, case of real variables $H_\nu^{(1)}(z)$ and $H_\nu^{(2)}(z)$ share the disadvantage of being complex functions. In consequence, yet another standard solution is needed. For this purpose $J_{-\nu}(z)$ is unsuitable because it is not linearly independent of $J_\nu(z)$ for all ν. The solution generally adopted is *Weber's function*, defined for all values of ν and z by†

$$Y_\nu(z) = \{H_\nu^{(1)}(z) - H_\nu^{(2)}(z)\}/(2i). \tag{5.01}$$

The *principal branch* is obtained by assigning principal values to $H_\nu^{(1)}(z)$ and $H_\nu^{(2)}(z)$. That $Y_\nu(z)$ is real when ν is real and z is positive follows from the last paragraph of §4.1. From (4.03) and (4.04) we derive the compound expansion

$$Y_\nu(z) \sim \left(\frac{2}{\pi z}\right)^{1/2} \left\{ \sin\zeta \sum_{s=0}^{\infty} (-)^s \frac{A_{2s}(\nu)}{z^{2s}} + \cos\zeta \sum_{s=0}^{\infty} (-)^s \frac{A_{2s+1}(\nu)}{z^{2s+1}} \right\} \tag{5.02}$$

as $z \to \infty$ in $|\operatorname{ph} z| \leqslant \pi - \delta$, whether or not ν be real. Comparison with (4.07) shows that, unlike $J_{-\nu}(z)$, the solution $Y_\nu(z)$ is linearly independent of $J_\nu(z)$ for all values of ν. If ν is real and z is large and positive, then asymptotically $Y_\nu(z)$ has the same amplitude of oscillation as $J_\nu(z)$ and is out of phase by $\frac{1}{2}\pi$; indeed, these properties motivate the actual choice of $Y_\nu(z)$ as a standard solution. Another welcome property is that $Y_\nu(z)$ *satisfies the same recurrence relations as* $J_\nu(z)$, $H_\nu^{(1)}(z)$, *and* $H_\nu^{(2)}(z)$.

The connections between the four standard solutions of Bessel's equation are conveniently memorizable as

$$H_\nu^{(1)}(z) = J_\nu(z) + iY_\nu(z), \qquad H_\nu^{(2)}(z) = J_\nu(z) - iY_\nu(z). \tag{5.03}$$

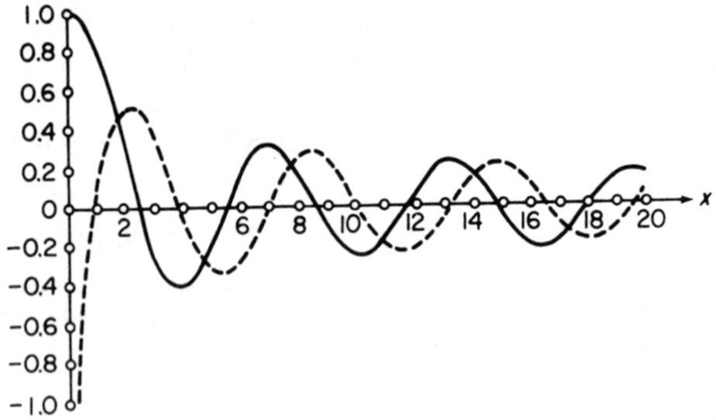

Fig. 5.1 $J_0(x)$ —— and $Y_0(x)$ - - - - -.

† Sometimes $Y_\nu(z)$ is denoted by $N_\nu(z)$. Often it is called the *Bessel function of the second kind*, $J_\nu(z)$ being of the *first kind*. In this terminology, the Hankel functions are *Bessel functions of the third kind*.

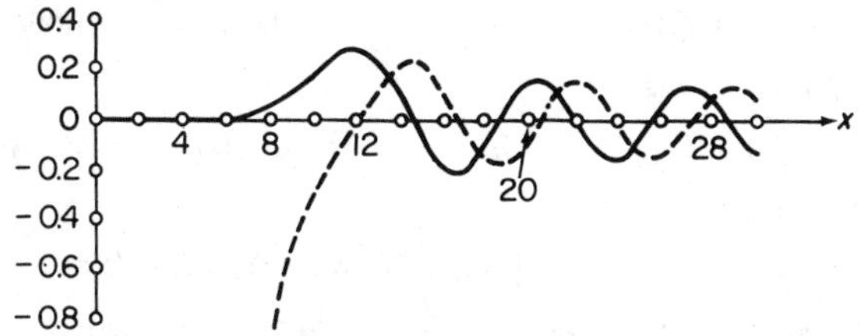

Fig. 5.2 $J_{10}(x)$ ——— and $Y_{10}(x)$ - - - - -.

But perhaps it needs stressing that $J_\nu(z)$ and $Y_\nu(z)$ form a numerically satisfactory pair only on the real axis, or in the neighborhood of $z = 0$ (when $\operatorname{Re}\nu \geqslant 0$). For large complex z, both solutions are dominant in all phase ranges.†

Graphs of $J_\nu(x)$ and $Y_\nu(x)$ for real variables are indicated in Figs. 5.1 and 5.2.

5.2 The expansion of $Y_\nu(z)$ in ascending powers of z is derivable from the power series for $J_{\pm\nu}(z)$ and the connection formula

$$Y_\nu(z) = \frac{J_\nu(z)\cos(\nu\pi) - J_{-\nu}(z)}{\sin(\nu\pi)}, \tag{5.04}$$

obtained from (4.12) and (5.01). Special interest attaches to the case in which ν is an integer, n, say, because both numerator and denominator vanish. Since $Y_\nu(z)$ is entire in ν, we find by L'Hôpital's rule‡

$$Y_n(z) = \frac{1}{\pi}\left[\frac{\partial J_\nu(z)}{\partial \nu}\right]_{\nu=n} + \frac{(-)^n}{\pi}\left[\frac{\partial J_\nu(z)}{\partial \nu}\right]_{\nu=-n} \qquad (n = 0, \pm 1, \pm 2, \ldots). \tag{5.05}$$

This relation immediately shows that $Y_{-n}(z) = (-)^n Y_n(z)$; hence in the following analysis we may suppose that $n \geqslant 0$.

From (9.09) of Chapter 2, we derive

$$\frac{\partial J_\nu(z)}{\partial \nu} = \left(\frac{1}{2}z\right)^\nu \sum_{s=0}^{\infty} \frac{(-)^s(\frac{1}{4}z^2)^s}{s!\,\Gamma(\nu+s+1)}\left\{\ln\left(\frac{1}{2}z\right) - \psi(\nu+s+1)\right\}, \tag{5.06}$$

where, as before, ψ denotes the logarithmic derivative of the Gamma function. Setting $\nu = \pm n$, and recalling that as z tends to a nonpositive integer m

$$1/\Gamma(z) \to 0, \qquad \psi(z)/\Gamma(z) \to (-)^{m+1}(-m)!,$$

we arrive at the desired expansion, given by

$$Y_n(z) = -\frac{(\frac{1}{2}z)^{-n}}{\pi}\sum_{s=0}^{n-1}\frac{(n-s-1)!}{s!}\left(\frac{1}{4}z^2\right)^s + \frac{2}{\pi}\ln\left(\frac{1}{2}z\right)J_n(z)$$
$$-\frac{(\frac{1}{2}z)^n}{\pi}\sum_{s=0}^{\infty}\{\psi(s+1) + \psi(n+s+1)\}\frac{(-)^s(\frac{1}{4}z^2)^s}{s!\,(n+s)!}. \tag{5.07}$$

This converges for all nonzero z.

† Verifiable with the aid of Exercise 5.3 below.

‡ This procedure is, in effect, Frobenius' method (Chapter 5, §5.3).

Ex. 5.1 Prove that $\mathscr{W}\{J_\nu(z), Y_\nu(z)\} = J_{\nu+1}(z)Y_\nu(z) - J_\nu(z)Y_{\nu+1}(z) = 2/(\pi z)$.

Ex. 5.2 Show that for any integer n

$$J_{-n-(1/2)}(z) = (-)^{n-1}Y_{n+(1/2)}(z), \qquad Y_{-n-(1/2)}(z) = (-)^n J_{n+(1/2)}(z).$$

Ex. 5.3 Show that for any integer m

$$Y_\nu(ze^{m\pi i}) = e^{-m\nu\pi i}Y_\nu(z) + 2i\sin(m\nu\pi)\cot(\nu\pi)J_\nu(z).$$

Ex. 5.4 From (5.05) and Chapter 2, Exercises 2.3 and 9.5, deduce that

$$Y_0(z) = 4\pi^{-2}\int_0^{\pi/2}\cos(z\cos\theta)\{\gamma+\ln(2z\sin^2\theta)\}\,d\theta \qquad (\gamma = \text{Euler's constant}).$$

Ex. 5.5 From Hankel's integrals (§4.5) derive the *Mehler–Sonine integrals*

$$J_\nu(x) = \frac{2(\frac{1}{2}x)^{-\nu}}{\pi^{1/2}\Gamma(\frac{1}{2}-\nu)}\int_1^\infty \frac{\sin(xt)\,dt}{(t^2-1)^{\nu+(1/2)}}, \qquad Y_\nu(x) = -\frac{2(\frac{1}{2}x)^{-\nu}}{\pi^{1/2}\Gamma(\frac{1}{2}-\nu)}\int_1^\infty \frac{\cos(xt)\,dt}{(t^2-1)^{\nu+(1/2)}},$$

when $|\mathrm{Re}\,\nu| < \frac{1}{2}$ and $x > 0$. Using the method of stationary phase confirm that the asymptotic forms of the right-hand sides match the leading terms in (4.07) and (5.02) when $\nu \in (-\frac{1}{2}, \frac{1}{2})$.

Ex. 5.6 Use induction to prove that when n is a positive integer or zero

$$\left[\frac{\partial J_\nu(z)}{\partial \nu}\right]_{\nu=n} = \frac{\pi}{2}Y_n(z) + \frac{n!}{2(\frac{1}{2}z)^n}\sum_{s=0}^{n-1}\frac{(\frac{1}{2}z)^s J_s(z)}{s!(n-s)},$$

$$\left[\frac{\partial Y_\nu(z)}{\partial \nu}\right]_{\nu=n} = -\frac{\pi}{2}J_n(z) + \frac{n!}{2(\frac{1}{2}z)^n}\sum_{s=0}^{n-1}\frac{(\frac{1}{2}z)^s Y_s(z)}{s!(n-s)}.$$

Ex. 5.7 By expanding $J_\nu(t)$ and integrating term by term, prove that when $\mathrm{Re}\,a > 0$ and $\mathrm{Re}(\mu+\nu) > 0$

$$\int_0^\infty e^{-at}t^{\mu-1}J_\nu(t)\,dt = \frac{\Gamma(\mu+\nu)}{2^\nu a^{\mu+\nu}}\mathbf{F}(\tfrac{1}{2}\mu+\tfrac{1}{2}\nu, \tfrac{1}{2}\mu+\tfrac{1}{2}\nu+\tfrac{1}{2}; \nu+1; -a^{-2}),$$

where $\mathbf{F}$ is the hypergeometric function of Chapter 5, and all functions have their principal values.

Ex. 5.8 By integrating by parts prove that when $\mathrm{Re}\,\nu > -1$

$$\lim_{a\to 0+}\int_0^\infty e^{-at}J_\nu(t)\,dt = \int_0^\infty J_\nu(t)\,dt.$$

By combining this result with the preceding exercise and equation (10.15) of Chapter 5, deduce that

$$\int_0^\infty J_\nu(t)\,dt = 1 \qquad (\mathrm{Re}\,\nu > -1),$$

and thence that

$$\int_0^\infty Y_\nu(t)\,dt = -\tan(\tfrac{1}{2}\pi\nu) \qquad (|\mathrm{Re}\,\nu| < 1).$$

6 Zeros of $J_\nu(z)$

6.1 In many applications of Bessel functions, including Chapter 12, properties of the zeros are essential. *We restrict the discussion in this section and §7 to real values of the order* ν.

Theorem 6.1 (i) *The* z *zeros of any solution of Bessel's equation are simple, with the possible exception of* $z = 0$.

(ii) *The z zeros of the derivative of any solution of Bessel's equation are simple, with the possible exception of* $z = 0$ *and* $\pm\nu$.

This theorem is a specialization of a general result for second-order differential equations. To prove it, suppose that $w(z_0) = w'(z_0) = 0$, z_0 being an ordinary point of the differential equation. Then from the proof of Theorem 1.1 of Chapter 5 it follows that $w(z) \equiv 0$. Alternatively if $w'(z_0) = w''(z_0) = 0$, then from Bessel's equation it follows that $w(z_0) = 0$, provided that $z_0 \neq \pm\nu$, and the proof concludes as before.

If z tends to infinity through positive real values, then the asymptotic expansion (4.07) shows that $J_\nu(z)$ changes sign infinitely often. Hence $J_\nu(z)$ and $J_\nu'(z)$ each have an infinity of positive real zeros. Also, since $J_\nu(ze^{m\pi i}) = e^{m\nu\pi i}J_\nu(z)$ when m is an integer, all branches of $J_\nu(z)$ and $J_\nu'(z)$ have an infinity of zeros on the positive and negative real axes.

When enumerated in ascending order of magnitude, the positive zeros of $J_\nu(z)$ are denoted by $j_{\nu,1}, j_{\nu,2}, \ldots$. Similarly, the sth positive zero of $J_\nu'(z)$ is denoted by $j_{\nu,s}'$.

6.2 Theorem 6.2† *The zeros of* $J_\nu(z)$ *are all real when* $\nu \geqslant -1$, *and the zeros of* $J_\nu'(z)$ *are all real when* $\nu \geqslant 0$.

In the first place, with the given conditions on ν no zero can be purely imaginary because all terms in the power series for $(\frac{1}{2}z)^{-\nu}J_\nu(z)$ and $(\frac{1}{2}z)^{1-\nu}J_\nu'(z)$ are positive or zero when $\operatorname{Re} z = 0$.

Next, consider the identity

$$(\alpha^2-\beta^2)\int_0^z tJ_\nu(\alpha t)J_\nu(\beta t)\,dt = z\left\{J_\nu(\alpha z)\frac{dJ_\nu(\beta z)}{dz} - J_\nu(\beta z)\frac{dJ_\nu(\alpha z)}{dz}\right\} \qquad (\nu \geqslant -1), \tag{6.01}$$

which is easily verifiable by differentiation and use of Bessel's equation. If α is a zero of either $J_\nu(z)$ or $J_\nu'(z)$, then by Schwarz's principle of symmetry the complex conjugate $\bar{\alpha}$ is also a zero. We may set $z = 1$ and $\beta = \bar{\alpha}$ in (6.01), and unless either $\operatorname{Re}\alpha = 0$ or $\operatorname{Im}\alpha = 0$ we deduce that

$$\int_0^1 tJ_\nu(\alpha t)J_\nu(\bar{\alpha}t)\,dt = 0.$$

This is a contradiction, however, because the integrand is positive. The theorem now follows.

When $-1 < \nu < 0$, the only modification of the result is that $J_\nu'(z)$ has a pair of purely imaginary zeros in addition to its real zeros; this is easily seen from (6.01) and the power series. When $\nu < -1$ and is nonintegral the method of proof fails because the integral in (6.01) diverges at its lower limit. As a matter of fact it can be shown that there *are* complex zeros in these circumstances.‡

† Lommel (1868, §19).
‡ Watson (1944, §15.27).

6.3 For the next theorem we require the following:

Lemma 6.1 *Corresponding to any positive number ε, a positive number δ can be assigned, independently of ν, such that in the x interval $(0,\delta]$, $J_\nu(x)$ has no zeros for all $\nu\in[-1+\varepsilon,\infty)$, and $J_\nu'(x)$ has no zeros for all $\nu\in[\varepsilon,\infty)$.*

When $\nu\geqslant -1+\varepsilon$ and $0<x\leqslant\delta$, we have from the power series

$$\left|\frac{\Gamma(\nu+1)\,J_\nu(x)}{(\frac{1}{2}x)^\nu}-1\right| = \left|\sum_{s=1}^{\infty}\frac{(-)^s(\frac{1}{4}x^2)^s}{(\nu+1)_s\,s!}\right| \leqslant \frac{\exp(\frac{1}{4}\delta^2)-1}{\varepsilon} < 1,$$

provided that $\delta^2<4\ln(1+\varepsilon)$. Thus $J_\nu(x)$ cannot vanish. Similarly for $J_\nu'(x)$.

Theorem 6.3 *For fixed s, $j_{\nu,s}$ is a differentiable function of ν in $(-1,\infty)$, and $j'_{\nu,s}$ is a differentiable function of ν in $(0,\infty)$.*

To establish the first result, let ε be an arbitrary positive number and a any point in $[-1+\varepsilon,\infty)$. From Theorem 6.1 $J_a'(j_{a,s})\neq 0$. Hence by the implicit function theorem there is a differentiable function $j(\nu)$ such that $j(a)=j_{a,s}$, and $J_\nu\{j(\nu)\}=0$ in an assignable ν neighborhood $\mathbf{N}(a)$, say. As ν varies continuously in $\mathbf{N}(a)$, the graph of $J_\nu(x)$ (Figs. 5.1 and 5.2) changes in a continuous manner. Lemma 6.1 shows that no new zero can enter the interval $0<x\leqslant j(\nu)$ from the left, nor can one of the existing $s-1$ zeros disappear at this end. Moreover, emergence or disappearance of a zero at any other point of the interval is precluded, because the graph shows that at the critical value of ν the zero would have to be a multiple zero, contrary to Theorem 6.1.

Thus $j_{\nu,s}=j(\nu)$ in $\mathbf{N}(a)$. Since a and ε are arbitrary, $j_{\nu,s}$ is continuous and differentiable throughout $(-1,\infty)$.

The proof for $j'_{\nu,s}$ is similar, except that first it is necessary to prove $j'_{\nu,s}$ cannot be a multiple zero of $J_\nu'(x)$ when $\nu>0$. The power series for $J_\nu(x)$ shows that both $J_\nu(x)$ and $xJ_\nu'(x)$ are positive and increasing when x is positive and sufficiently small. And Bessel's equation in the form

$$x\{xJ_\nu'(x)\}' = (\nu^2-x^2)\,J_\nu(x)$$

shows that in the interval $0<x<\nu$, the functions $\{xJ_\nu'(x)\}'$ and $J_\nu(x)$ either vanish together or not at all. Let x_ν be the smallest x for which this vanishing occurs, or if there is no vanishing let $x_\nu=\nu$. Then $\{xJ_\nu'(x)\}'$ is positive in $(0,x_\nu)$; hence $xJ_\nu'(x)$ and $J_\nu'(x)$ are positive in $(0,x_\nu]$. Therefore $J_\nu(x_\nu)>0$, implying that $x_\nu=\nu$ and thence that $J_\nu'(x)>0$ when $x\in(0,\nu]$. Consequently

$$j'_{\nu,1}>\nu \qquad (\nu>0).$$

Theorem 6.1 now shows that no $j'_{\nu,s}$ can be a multiple zero of $J_\nu'(x)$. This completes the proof of Theorem 6.3.

6.4 Theorem 6.4 *When ν is positive, $j_{\nu,s}$ is an increasing function of ν.*

Differentiation of the equation $J_\nu(j_{\nu,s})=0$ produces

$$J_\nu'(j_{\nu,s})\frac{dj_{\nu,s}}{d\nu}+\left[\frac{\partial J_\nu(x)}{\partial\nu}\right]_{x=j_{\nu,s}} = 0. \qquad (6.02)$$

To evaluate the second term we use the identity

$$\int \frac{J_\mu(x)\,J_\nu(x)}{x}\,dx = \frac{x\{J_\mu'(x)\,J_\nu(x) - J_\mu(x)\,J_\nu'(x)\}}{\mu^2-\nu^2} \qquad (\mu^2 \neq \nu^2),$$

which is verifiable by differentiation; compare (6.01). Letting $\mu \to \nu$ we obtain

$$\int \frac{J_\nu^2(x)}{x}\,dx = \frac{x}{2\nu}\left\{J_\nu(x)\frac{\partial J_\nu'(x)}{\partial \nu} - J_\nu'(x)\frac{\partial J_\nu(x)}{\partial \nu}\right\}.$$

Provided that $\nu > 0$, the integration limits can be set equal to 0 and $j_{\nu,s}$, yielding

$$\int_0^{j_{\nu,s}} \frac{J_\nu^2(x)}{x}\,dx = -\frac{j_{\nu,s}}{2\nu} J_\nu'(j_{\nu,s})\left[\frac{\partial J_\nu(x)}{\partial \nu}\right]_{x=j_{\nu,s}}.$$

Then by substitution in (6.02) we obtain

$$\frac{dj_{\nu,s}}{d\nu} = \frac{2\nu}{j_{\nu,s}\{J_\nu'(j_{\nu,s})\}^2}\int_0^{j_{\nu,s}} \frac{J_\nu^2(x)}{x}\,dx \qquad (\nu > 0),$$

from which the theorem immediately follows.

6.5 Asymptotic expansions for the large positive zeros of $J_\nu(z)$ can be found by reversion of (4.07). As a first approximation, we have

$$\cos(z - \tfrac{1}{2}\nu\pi - \tfrac{1}{4}\pi) + O(z^{-1}) = 0.$$

Hence, as in Chapter 1, §5.2,

$$z = s\pi + \tfrac{1}{2}\nu\pi - \tfrac{1}{4}\pi + O(s^{-1}),$$

where s is a large positive integer.

For higher terms, write $\alpha \equiv (s + \frac{1}{2}\nu - \frac{1}{4})\pi$. Then for large z

$$z - \alpha \sim -\tan^{-1}\left\{\sum_{s=0}^{\infty}(-)^s \frac{A_{2s+1}(\nu)}{z^{2s+1}} \Big/ \sum_{s=0}^{\infty}(-)^s \frac{A_{2s}(\nu)}{z^{2s}}\right\}$$

$$\sim -\frac{4\nu^2-1}{8z} - \frac{(4\nu^2-1)(4\nu^2-25)}{384z^3} - \cdots.$$

Successive resubstitutions produce *McMahon's expansion*†

$$z \sim \alpha - \frac{4\nu^2-1}{8\alpha} - \frac{(4\nu^2-1)(28\nu^2-31)}{384\alpha^3} - \cdots \qquad (s \to \infty). \tag{6.03}$$

Does this expansion actually represent the sth zero of $J_\nu(z)$ and not, for example, the $(s-1)$th zero? A general method for resolving questions of this kind is the *phase principle*‡:

† For additional terms see R. S. (1960). No explicit formula is available for the general term.

‡ Also called the *principle of the argument*. It is provable by applying the residue theorem to $f'(z)/f(z)$; see Levinson and Redheffer (1970, pp. 216–218). An application is made in §8.4.

Let $f(z)$ be holomorphic within a simply connected domain which contains a simple closed contour $\mathscr{C}$. Assume also that the zeros of $f(z)$ are counted according to their multiplicity and none are on $\mathscr{C}$. Then the number of zeros within $\mathscr{C}$ is $1/(2\pi)$ times the increase in any continuous branch of ph$\{f(z)\}$ *as z goes once round $\mathscr{C}$ in the positive sense.*

In the present instance, however, we may argue more simply, as follows. The expansion (6.03) is easily seen to be uniform with respect to ν in any compact interval. When $\nu = \frac{1}{2}$, the positive zeros of $J_\nu(z)$ are $\pi, 2\pi, 3\pi, \ldots$, exactly; compare Chapter 2, Exercise 9.3. *Hence* (6.03) *represents $j_{\nu,s}$ for this value of ν, and therefore, by continuity* (*Theorem* 6.3), *for all $\nu \in (-1, \infty)$.*

Ex. 6.1 Show that when ν is positive $j'_{\nu,1} < j_{\nu,1} < j'_{\nu,2} < j_{\nu,2} < j'_{\nu,3} < \cdots$.

Ex. 6.2 Using the method of §6.4, show that $dj'_{\nu,s}/d\nu > 0$ when $\nu > 0$.

Ex. 6.3 With the aid of Exercise 6.1 show that if ν is fixed and positive, and $\beta \equiv (s+\frac{1}{2}\nu-\frac{3}{4})\pi$, then

$$j'_{\nu,s} = \beta - \frac{4\nu^2+3}{8\beta} - \frac{112\nu^4+328\nu^2-9}{384\beta^3} + O\left(\frac{1}{s^5}\right).$$

Ex. 6.4 For any given positive integer s, let $\phi_s(\nu)$ be defined by $\phi_s(\nu) = j_{\nu,s}$ when $\nu > -1$, and by $\phi_s(\nu) = j_{\nu,s-k}$ when $-1-k < \nu \leqslant -k$ for each $k = 1, 2, \ldots, s-1$. Show that $\phi_s(\nu)$ is differentiable throughout $(-s, \infty)$.

7 Zeros of $Y_\nu(z)$ and Other Cylinder Functions

7.1 A function of the form

$$\mathscr{C}_\nu(x) = AJ_\nu(x) + BY_\nu(x), \tag{7.01}$$

in which A and B are independent of x (but may depend on ν) is called a *cylinder function of order* ν. The name stems from the importance of these functions in the solution of Laplace's equation in cylindrical coordinates.

Theorem 7.1 *The positive zeros of any two linearly independent, real, cylinder functions of the same order are interlaced.*

This is a special case of a general theorem concerning linear second-order differential equations. To prove Theorem 7.1, let one of the cylinder functions be (7.01) and the other be

$$\mathscr{D}_\nu(x) = CJ_\nu(x) + DY_\nu(x).$$

Using Exercise 5.1, we have

$$\mathscr{C}_\nu(x)\mathscr{D}'_\nu(x) - \mathscr{C}'_\nu(x)\mathscr{D}_\nu(x) = 2(AD-BC)/(\pi x).$$

Since $\mathscr{C}_\nu(x)$ and $\mathscr{D}_\nu(x)$ are independent, $AD-BC \neq 0$; compare Chapter 5, Theorem 1.2. At a positive zero of $\mathscr{C}_\nu(x)$, its derivative $\mathscr{C}'_\nu(x)$ is nonzero (Theorem 6.1); at consecutive zeros $\mathscr{C}'_\nu(x)$ has opposite signs, hence $\mathscr{D}_\nu(x)$ has opposite signs. Therefore an odd number of zeros of $\mathscr{D}_\nu(x)$ separates each consecutive pair of zeros of $\mathscr{C}_\nu(x)$. Similarly, an odd number of zeros of $\mathscr{C}_\nu(x)$ separates each consecutive pair of zeros of $\mathscr{D}_\nu(x)$. The theorem is now evident.

By taking $\mathscr{D}_\nu(x) = J_\nu(x)$, it is seen that all real cylinder functions have an infinity of positive zeros.

7.2 The sth positive zeros of $Y_\nu(x)$ and $Y_\nu'(x)$ are denoted by $y_{\nu,s}$ and $y'_{\nu,s}$, respectively.

Theorem 7.2 *When* $\nu > -\frac{1}{2}$,

$$y_{\nu,1} < j_{\nu,1} < y_{\nu,2} < j_{\nu,2} < \cdots. \tag{7.02}$$

Theorem 7.1 shows that there is exactly one zero of $Y_\nu(x)$ in each of the intervals $(j_{\nu,1}, j_{\nu,2}), (j_{\nu,2}, j_{\nu,3}), \ldots,$ and either one or none in $(0, j_{\nu,1})$. We have only to show that *one* is the correct alternative.

If $\nu > -1$, then $J_\nu(x)$ is positive as $x \to 0+$, implying that $J_\nu'(j_{\nu,1}) < 0$. Then by setting $x = j_{\nu,1}$ in the Wronskian relation of Exercise 5.1 we see that $Y_\nu(j_{\nu,1}) > 0$. Next, with the aid of equations (5.04) and (5.07) it is verifiable that as $x \to 0+$, $Y_\nu(x)$ is asymptotic to

$$-(1/\pi)\Gamma(\nu)(\tfrac{1}{2}x)^{-\nu}, \qquad (2/\pi)\ln x, \qquad \text{or} \qquad -(1/\pi)\cos(\nu\pi)\Gamma(-\nu)(\tfrac{1}{2}x)^{\nu}, \tag{7.03}$$

according as $\nu > 0$, $\nu = 0$, or $-\frac{1}{2} < \nu < 0$. In all three cases the sign is negative, which is opposite to the sign of $Y_\nu(x)$ at $j_{\nu,1}$. This completes the proof.

7.3 As in the proof of Theorem 6.3, each zero of $Y_\nu(x)$ is, locally, a differentiable function of ν. And Theorem 7.2 shows that the enumeration of the positive zeros of $Y_\nu(x)$ cannot change as ν varies continuously in $(-\frac{1}{2}, \infty)$. Accordingly, for fixed s, $y_{\nu,s}$ *is a differentiable function of* ν *throughout* $(-\frac{1}{2}, \infty)$.

No analog of Theorem 6.2 holds for $Y_\nu(z)$ and $Y_\nu'(z)$; this is evident from Exercise 7.5 below. The remaining theorem in §6, namely Theorem 6.4, does have an analog for the Y function. Indeed, $j_{\nu,s}$ is increasing in the ν interval $(-1, \infty)$ and $y_{\nu,s}$ is increasing in the ν interval $(-\frac{1}{2}, \infty)$. Available proofs† are somewhat recondite, however, and since we shall not use the results the proofs are omitted.

Ex. 7.1 Assume that the coefficients A and B in (7.01) are independent of ν. By considering the derivatives of $x^{-\nu}\mathscr{C}_\nu(x)$ and $x^{\nu+1}\mathscr{C}_{\nu+1}(x)$, show that the positive zeros of $\mathscr{C}_\nu(x)$ and $\mathscr{C}_{\nu+1}(x)$ are interlaced.

Ex. 7.2 In the notation of §7.1 show that the zeros of $\mathscr{C}_\nu'(x)$ and $\mathscr{D}_\nu'(x)$ which exceed $|\nu|$ are interlaced.

Ex. 7.3 Using the method of §6.2 prove that in the sector $|\mathrm{ph}\, z| < \frac{1}{2}\pi$ all zeros of $Y_0(z)$ and $Y_1(z)$ are real.

Ex. 7.4 Show that when $\nu > -\frac{1}{2}$ the asymptotic expansion of $y_{\nu,s}$ for large s is given by the right-hand side of (6.03) with $\alpha = (s + \frac{1}{2}\nu - \frac{3}{4})\pi$.

Ex. 7.5 Let n be zero or a positive integer. Using Exercise 5.3 show that $Y_n(z)$ has an infinite set of zeros in the sector $0 < \mathrm{ph}\, z < \pi$, and they lie on a curve having $\mathrm{ph}(z - \frac{1}{2}i \ln 3) = \pi$ as asymptote.

† Watson (1944, §15.6).

***Ex. 7.6** For fixed ν exceeding $-\frac{1}{2}$, let

$$\mathscr{C}_\nu(x,t) = J_\nu(x)\cos(\pi t) + Y_\nu(x)\sin(\pi t),$$

where t is a positive parameter. Show that the equation $\mathscr{C}_\nu(x,t) = 0$ is satisfied by $x = \rho(t)$, where $\rho(t)$ is an infinitely differentiable function with the properties

$$\rho(s) = j_{\nu,s}, \qquad \rho(s-\tfrac{1}{2}) = y_{\nu,s} \qquad (s = \text{any positive integer});$$
$$\rho(0+) = 0 \quad (\nu \geqslant 0), \qquad \rho(-\nu+) = 0 \quad (-\tfrac{1}{2} < \nu < 0).$$

Using primes to denote differentiations with respect to t, show also that

$$2\rho^2\rho'\rho''' - 3\rho^2\rho''^2 + (4\rho^2+1-4\nu^2)\rho'^4 - 4\pi^2\rho^2\rho'^2 = 0,$$

and

$$\left\{\left[\frac{\partial \mathscr{C}_\nu(x,t)}{\partial x}\right]_{x=\rho(t)}\right\}^2 = \frac{2}{\rho\rho'}.$$ [Olver, 1950.]

8 Modified Bessel Functions

8.1 In Chapter 2, §10, we constructed a solution $I_\nu(z)$ of the modified Bessel equation

$$\frac{d^2w}{dz^2} + \frac{1}{z}\frac{dw}{dz} - \left(1 + \frac{\nu^2}{z^2}\right)w = 0. \tag{8.01}$$

The distinguishing property of this solution is recession at the regular singularity $z = 0$ when $\operatorname{Re}\nu > 0$ or $\nu = 0$. Since Bessel's equation transforms into (8.01) on replacing z by iz, another solution is *Macdonald's function*

$$K_\nu(z) = \tfrac{1}{2}\pi i e^{\nu\pi i/2} H_\nu^{(1)}(ze^{\pi i/2}). \tag{8.02}$$

In this definition the right-hand side takes its principal value when $\operatorname{ph} z = 0$; for other values of $\operatorname{ph} z$ the branches of $K_\nu(z)$ are determined by continuity. The *principal branch* corresponds to $\operatorname{ph} z \in (-\pi, \pi]$.

Important properties of $K_\nu(z)$ are:

(i) *$K_\nu(z)$ is real when ν is real and z is positive (more precisely, when* $\operatorname{ph} z = 0$).
(ii) *$K_\nu(z)$ is recessive at infinity in the sector* $|\operatorname{ph} z| \leqslant \frac{1}{2}\pi - \delta$ $(< \frac{1}{2}\pi)$, *for all values of ν.*

Property (i) is immediately deducible from the following integral representation

$$K_\nu(z) = \int_0^\infty e^{-z\cosh t}\cosh(\nu t)\,dt \qquad (|\operatorname{ph} z| < \tfrac{1}{2}\pi), \tag{8.03}$$

which is derived from Sommerfeld's integral (4.18) by replacing z by iz and t by $t + \frac{1}{2}\pi i$. For (ii) we have, from (4.03),

$$K_\nu(z) \sim \left(\frac{\pi}{2z}\right)^{1/2} e^{-z} \sum_{s=0}^\infty \frac{A_s(\nu)}{z^s} \qquad (z \to \infty \text{ in } |\operatorname{ph} z| \leqslant \tfrac{3}{2}\pi - \delta). \tag{8.04}$$

Graphs of $I_\nu(x)$ and $K_\nu(x)$ for $\nu = 0$ and 10 are indicated in Fig. 8.1.

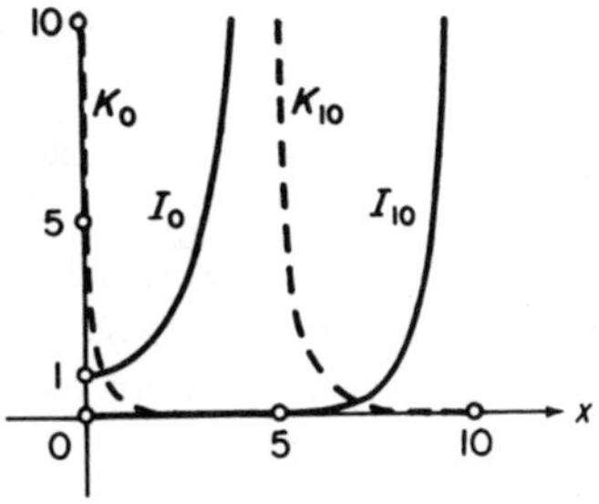

Fig. 8.1 Modified Bessel functions of orders 0 and 10.

8.2 Formulas for modified Bessel functions are readily derived from corresponding formulas for the unmodified functions. For example,

$$K_\nu(z) = \tfrac{1}{2}\pi\{I_{-\nu}(z) - I_\nu(z)\}/\sin(\nu\pi), \tag{8.05}$$

$$Y_\nu(ze^{\pi i/2}) = e^{(\nu+1)\pi i/2} I_\nu(z) - (2/\pi)\, e^{-\nu\pi i/2} K_\nu(z), \tag{8.06}$$

and

$$\mathscr{W}\{I_\nu(z), I_{-\nu}(z)\} = -2\sin(\nu\pi)/(\pi z), \qquad \mathscr{W}\{K_\nu(z), I_\nu(z)\} = 1/z. \tag{8.07}$$

In (8.05), the right-hand side is replaced by its limiting value when ν is an integer or zero. In (8.06) the branches take their principal values when $-\pi < \operatorname{ph} z \leqslant \frac{1}{2}\pi$. The Wronskians (8.07) show that $I_\nu(z)$ and $K_\nu(z)$ are linearly independent for all ν, but not $I_\nu(z)$ and $I_{-\nu}(z)$. And from properties stated in §8.1, it is seen that $I_\nu(z)$ and $K_\nu(z)$ comprise a numerically satisfactory pair *throughout* the sector $|\operatorname{ph} z| \leqslant \frac{1}{2}\pi$, provided that $\operatorname{Re} \nu \geqslant 0$.

The asymptotic expansion of $I_\nu(z)$ is available from (4.03), (4.04), and the connection formula

$$I_\nu(z) = \tfrac{1}{2} e^{-\nu\pi i/2}\{H_\nu^{(1)}(ze^{\pi i/2}) + H_\nu^{(2)}(ze^{\pi i/2})\}.$$

Neglecting exponentially small contributions,† we find that

$$I_\nu(z) \sim \frac{e^z}{(2\pi z)^{1/2}} \sum_{s=0}^{\infty} (-)^s \frac{A_s(\nu)}{z^s} \qquad (z \to \infty \text{ in } |\operatorname{ph} z| \leqslant \tfrac{1}{2}\pi - \delta). \tag{8.08}$$

8.3 The following result is required in a later chapter.

Theorem 8.1‡ (i) *If ν ($\geqslant 0$) is fixed, then throughout the x interval* $(0, \infty)$, *$I_\nu(x)$ is positive and increasing, and $K_\nu(x)$ is positive and decreasing.*

(ii) *If x (>0) is fixed, then throughout the ν interval* $(0, \infty)$, *$I_\nu(x)$ is decreasing, and $K_\nu(x)$ is increasing.*

Part (i) is easily deducible from the power series for $I_\nu(x)$ given in Chapter 2, (10.01), and the integral representation (8.03). The stated property of $K_\nu(x)$ in Part (ii) also follows immediately from (8.03).

† See also §13, especially Exercise 13.2.

‡ Part (ii) is a recent result; see Cochran (1967), A. L. Jones (1968), and Reudink (1968). The proof given is that of Cochran.

For the remaining property of $I_\nu(x)$, we find by differentiating the second of (8.07)

$$K_\nu'(x)\frac{\partial I_\nu(x)}{\partial\nu} - K_\nu(x)\frac{\partial I_\nu'(x)}{\partial\nu} = I_\nu'(x)\frac{\partial K_\nu(x)}{\partial\nu} - I_\nu(x)\frac{\partial K_\nu'(x)}{\partial\nu}. \tag{8.09}$$

Consider the right-hand side. The power series for $I_\nu(x)$ and $I_\nu'(x)$ show that both functions are positive when ν and x are positive. By differentiation of (8.03) it is seen that $\partial K_\nu(x)/\partial\nu$ is positive and $\partial K_\nu'(x)/\partial\nu$ is negative. Therefore the right-hand side of (8.09) is positive.

Now consider $\partial I_\nu(x)/\partial\nu$ for a fixed positive value of ν. As $x \to 0+$ we have

$$\frac{\partial I_\nu(x)}{\partial\nu} = \frac{(\frac{1}{2}x)^\nu}{\Gamma(\nu+1)}\left\{\ln\left(\frac{1}{2}x\right)+O(1)\right\}.$$

Hence $\partial I_\nu(x)/\partial\nu < 0$ for all sufficiently small x. As x increases continuously, either $\partial I_\nu(x)/\partial\nu$ stays negative, or we reach a value $x = x_\nu$, say, at which $\partial I_\nu(x)/\partial\nu$ vanishes. In the latter event it is easily seen graphically that the x derivative $\partial I_\nu'(x)/\partial\nu$ is nonnegative at x_ν. Hence the left-hand side of (8.09) is nonpositive at $x = x_\nu$, contradicting our findings concerning the right-hand side. Thus x_ν does not exist, that is, $\partial I_\nu(x)/\partial\nu < 0$ for all positive ν and x. The proof of Theorem 8.1 is complete.

8.4 Properties of the z zeros of $I_\nu(z)$ are derivable from those of $J_\nu(z)$ (§6) by rotation of the z plane through a right angle. For example, when $\nu \geqslant -1$, all zeros of $I_\nu(z)$ are purely imaginary and occur in conjugate pairs.

Theorem 8.2† *When ν is real, $K_\nu(z)$ has no zeros in the sector* $|\mathrm{ph}\, z| \leqslant \frac{1}{2}\pi$.

Equation (8.05) shows that

$$K_{-\nu}(z) = K_\nu(z) \qquad (\nu \text{ unrestricted}). \tag{8.10}$$

Also, by Schwarz's principle of symmetry

$$K_\nu(\bar{z}) = \overline{K_\nu(z)} \qquad (\nu \text{ real}). \tag{8.11}$$

Hence to establish the theorem it suffices to show that $K_\nu(z)$ is free from zeros in the sector $-\frac{1}{2}\pi \leqslant \mathrm{ph}\, z \leqslant 0$ when $\nu \geqslant 0$. We shall achieve this by applying the phase principle (§6.5) to the contour $ABCDA$ indicated in Fig. 8.2. In this diagram AB is the quarter-circle $|z| = r$, and CD is the quarter-circle $|z| = j_{\nu,s}$ (in the notation of §6). Ultimately we shall let $r \to 0$ and $s \to \infty$.

On AB, $|z|$ is small. From (8.06) and the complex version of (7.03) we derive

$$K_\nu(z) \sim \tfrac{1}{2}\Gamma(\nu)(\tfrac{1}{2}z)^{-\nu} \quad (\nu > 0); \qquad K_0(z) \sim -\ln z. \tag{8.12}$$

Hence for small r the change in the phase of $K_\nu(z)$ is given by

$$\underset{AB}{\Delta}\, \mathrm{ph}\{K_\nu(z)\} = \tfrac{1}{2}\nu\pi + o(1) \qquad (\nu \geqslant 0).$$

On BC, write $z = te^{-\pi i/2}$ so that t is positive. From (5.03) and (8.02) we have

$$K_\nu(te^{-\pi i/2}) = \tfrac{1}{2}\pi i e^{\nu\pi i/2}\{J_\nu(t) + iY_\nu(t)\}.$$

† Macdonald (1899).

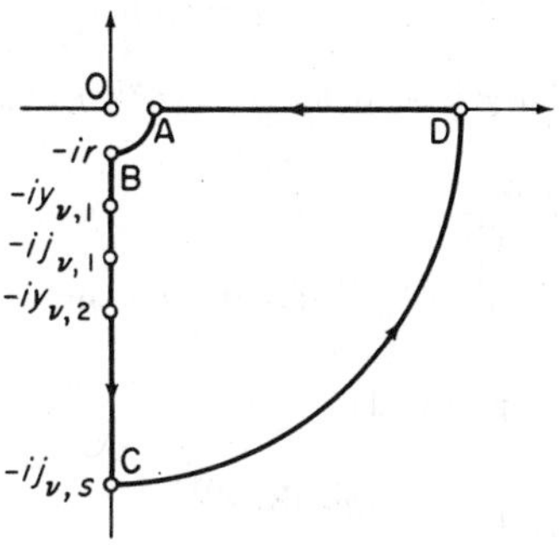

Fig. 8.2 z plane.

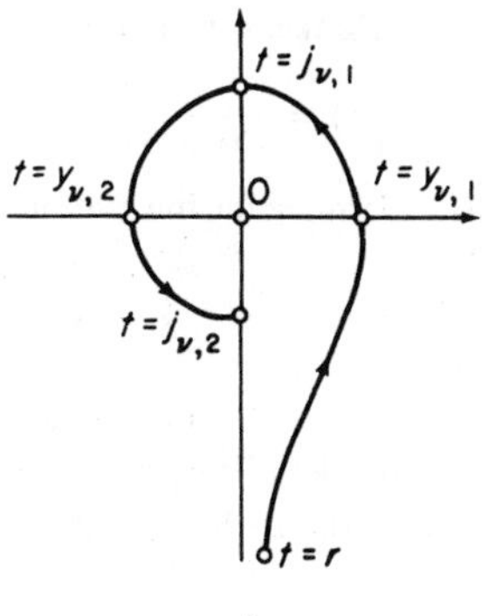

Fig. 8.3 $J_\nu(t)+iY_\nu(t)$.

At $t = r$, $Y_\nu(t)$ is negative and large in absolute value (compare (7.03)) and $J_\nu(t)$ is bounded and positive. Referring to properties of $J_\nu(t)$ and $Y_\nu(t)$ established in earlier sections, especially §7, we see that the map of $J_\nu(t)+iY_\nu(t)$ from $t = r$ to $t = j_{\nu,2}$ is essentially as indicated in Fig. 8.3, and its continuation from $t = j_{\nu,2}$ to $t = j_{\nu,s}$ consists of $\frac{1}{2}s-1$ positive circuits of the origin. Therefore for small r

$$\mathop{\Delta}_{BC} \operatorname{ph}\{K_\nu(z)\} = s\pi + o(1).$$

On CD, $|z|$ is large; from (8.04) we derive

$$\mathop{\Delta}_{CD} \operatorname{ph}\{K_\nu(z)\} = -\tfrac{1}{4}\pi - j_{\nu,s} + o(1).$$

Lastly, on DA there is no phase change since $K_\nu(z)$ is real and positive.

Collecting together the contributions, we arrive at

$$\mathop{\Delta}_{ABCDA} \operatorname{ph}\{K_\nu(z)\} = \tfrac{1}{2}\nu\pi + s\pi - \tfrac{1}{4}\pi - j_{\nu,s} + o(1). \tag{8.13}$$

Again, because s is large $j_{\nu,s}$ may be replaced by $(s+\frac{1}{2}\nu-\frac{1}{4})\pi$ with an error $o(1)$; compare §6.5. Then as $r \to 0$ and $s \to \infty$ the right-hand side of (8.13) vanishes, and the theorem follows.

Further information on the zeros of $K_\nu(z)$ is given in Exercise 8.7 below.

Ex. 8.1 Show that $e^{\nu\pi i}K_\nu(z)$ satisfies the same recurrence relations as $I_\nu(z)$, and also that

$$I_\nu(z)\,K_{\nu+1}(z) + I_{\nu+1}(z)\,K_\nu(z) = 1/z.$$

Ex. 8.2 Show that when m is any integer

$$K_\nu(ze^{m\pi i}) = e^{-m\nu\pi i}K_\nu(z) - \pi i \sin(m\nu\pi)\csc(\nu\pi)\,I_\nu(z).$$

Ex. 8.3 From (5.07) and (8.06) deduce that when n is a nonnegative integer

$$K_n(z) = \frac{1}{2}\left(\frac{1}{2}z\right)^{-n}\sum_{s=0}^{n-1}\frac{(n-s-1)!}{s!}\left(-\frac{1}{4}z^2\right)^s + (-)^{n+1}\ln\left(\frac{1}{2}z\right)I_n(z)$$
$$+(-)^n\frac{1}{2}\left(\frac{1}{2}z\right)^n\sum_{s=0}^{\infty}\{\psi(s+1)+\psi(n+s+1)\}\frac{(\frac{1}{4}z^2)^s}{s!\,(n+s)!}.$$

Ex. 8.4 From Hankel's integral (4.20) derive

$$K_\nu(z) = \frac{\pi^{1/2}(\frac{1}{2}z)^\nu}{\Gamma(\nu+\frac{1}{2})}\int_1^\infty e^{-zt}(t^2-1)^{\nu-(1/2)}\,dt \qquad (\operatorname{Re}\nu > -\tfrac{1}{2},\quad |\operatorname{ph} z| < \tfrac{1}{2}\pi).$$

Ex. 8.5 By deforming the integration path for Hankel's integral (4.20) for $H_{-\nu}^{(1)}(xze^{\pi i/2})$, establish *Basset's integral*

$$K_\nu(xz) = \frac{\Gamma(\nu+\frac{1}{2})(2z)^\nu}{\pi^{1/2}x^\nu}\int_0^\infty \frac{\cos(xt)\,dt}{(t^2+z^2)^{\nu+(1/2)}},$$

where $\operatorname{Re}\nu > -\frac{1}{2}$, $x > 0$, $|\operatorname{ph} z| < \frac{1}{2}\pi$, and the branch of $(t^2+z^2)^{\nu+(1/2)}$ is continuous and asymptotic to the principal value of $t^{2\nu+1}$ as $t \to +\infty$.

Ex. 8.6 Using Exercise 8.4 and the Beta-function integral prove that

$$\int_0^\infty t^{\mu-1}K_\nu(t)\,dt = 2^{\mu-2}\Gamma\left(\frac{\mu+\nu}{2}\right)\Gamma\left(\frac{\mu-\nu}{2}\right) \qquad (\operatorname{Re}\mu > |\operatorname{Re}\nu|).$$

Ex. 8.7 Let $\nu \geqslant 0$. By using the formula

$$K_\nu(te^{\pi i}) = e^{-\nu\pi i}\{K_\nu(t) - \pi i e^{\nu\pi i}I_\nu(t)\}$$

derived from Exercise 8.2, and applying the phase principle to the closed contour comprising (i) the circular arcs $z = Re^{i\theta}$ and $z = re^{i\theta}$ $(-\pi \leqslant \theta \leqslant \pi)$, R being large and r being small; (ii) the line segments $\operatorname{ph} z = \pm\pi$, $r \leqslant |z| \leqslant R$; prove that when $\nu - \frac{1}{2}$ is not an odd integer, the total number of zeros of $K_\nu(z)$ in the sector $|\operatorname{ph} z| \leqslant \pi$ is the even integer nearest to $\nu - \frac{1}{2}$.

Show also that if $\nu - \frac{1}{2}$ is an odd integer and each side of the cut $\operatorname{ph} z = \pm\pi$ is counted separately, then the total number of zeros is $\nu + \frac{1}{2}$. [Watson, 1944.]

9 Confluent Hypergeometric Equation

9.1 Bessel's equation may be regarded as a transformation of a special case of the *confluent hypergeometric equation*

$$\vartheta(\vartheta+c-1)\,w = z(\vartheta+a)\,w, \tag{9.01}$$

in which a and c are parameters and, as before, $\vartheta = z\,d/dz$. In turn, equation (9.01) is the special case $p = q = 1$ of the generalized hypergeometric equation (11.02) of Chapter 5.

On expansion, (9.01) becomes

$$z\frac{d^2w}{dz^2} + (c-z)\frac{dw}{dz} - aw = 0. \tag{9.02}$$

This differential equation has a regular singularity at the origin with exponents 0 and $1-c$, and an irregular singularity at infinity of rank 1. The name *confluent* originates in the following way. The hypergeometric function $F(a, b; c; z/b)$ satisfies the equation

$$z\left(1 - \frac{z}{b}\right)\frac{d^2w}{dz^2} + \left(c - z - \frac{a+1}{b}z\right)\frac{dw}{dz} - aw = 0.$$

This has regular singularities at 0, b, and ∞, and when $b \to \infty$ it reduces to (9.02). Thus there is a confluence of two of the regular singularities, producing an irregular singularity. Many properties of the solutions of (9.02) are easily derivable by this limiting procedure; see Exercises 9.2 and 9.4.

9.2 In the notation of Chapter 5, §11.1, the series solution of (9.02) which corresponds to the exponent 0 at $z=0$ is ${}_1F_1(a;c;z)$. Just as ${}_2F_1(a,b;c;z)$ is commonly denoted $F(a,b;c;z)$, so the simpler notations $M(a,c,z)$ or $\Phi(a,c;z)$ are often employed for ${}_1F_1(a;c;z)$. Using Pochhammer's symbol, we have

$$M(a,c,z) = \sum_{s=0}^{\infty} \frac{(a)_s}{(c)_s}\frac{z^s}{s!} \qquad (c \neq 0,-1,-2,\ldots). \tag{9.03}$$

This series converges for all finite z and defines an entire function, sometimes known as *Kummer's function.*

As in the case of the hypergeometric equation, fewer restrictions are needed when formulas are expressed in terms of the solution

$$\mathbf{M}(a,c,z) = \frac{1}{\Gamma(c)}M(a,c,z) = \sum_{s=0}^{\infty}\frac{(a)_s}{\Gamma(c+s)}\frac{z^s}{s!}. \tag{9.04}$$

By means of the Weierstrass M-test, it is easily verified that for fixed z, $\mathbf{M}(a,c,z)$ *is entire in a and entire in c*; compare Chapter 5, §9.2. In contrast, $M(a,c,z)$ is a meromorphic function of c, in general, with poles at $0,-1,-2,\ldots$.

For the second exponent at $z=0$ the corresponding solution is found to be

$$N(a,c,z) \equiv z^{1-c}M(1+a-c,\ 2-c,\ z) \qquad (c \neq 2,3,4,\ldots). \tag{9.05}$$

We shall also write

$$\mathbf{N}(a,c,z) = N(a,c,z)/\Gamma(2-c) = z^{1-c}\mathbf{M}(1+a-c,\ 2-c,\ z). \tag{9.06}$$

Chapter 5, (1.10) shows that the Wronskian of $\mathbf{M}(a,c,z)$ and $\mathbf{N}(a,c,z)$ is a constant multiple of $e^z z^{-c}$. Consideration of the limiting forms of these solutions and their derivatives at $z=0$ yields

$$\mathscr{W}\{\mathbf{M}(a,c,z),\ \mathbf{N}(a,c,z)\} = \pi^{-1}\sin(\pi c)e^z z^{-c}.$$

Thus $\mathbf{M}(a,c,z)$ and $\mathbf{N}(a,c,z)$ are linearly independent, except when c is an integer or zero.

9.3 An integral representation for $M(a,c,z)$ may be found by the method used for $\mathbf{F}(a,b;c;z)$ in Chapter 5, §9.4; thus

$$M(a,c,z) = \frac{\Gamma(c)}{\Gamma(a)\Gamma(c-a)}\int_0^1 t^{a-1}(1-t)^{c-a-1}e^{zt}\,dt \qquad (\operatorname{Re} c > \operatorname{Re} a > 0), \tag{9.07}$$

fractional powers taking their principal values.

By Pochhammer's method this integral can be transformed into a contour integral from which all restrictions on the parameters are removed by analytic continuation; see Exercise 9.4.

Ex. 9.1 Show that

$$J_\nu(z) = \frac{(\frac{1}{2}z)^\nu e^{-iz}}{\Gamma(\nu+1)}M(\nu+\tfrac{1}{2}, 2\nu+1, 2iz), \qquad I_\nu(z) = \frac{(\frac{1}{2}z)^\nu e^{-z}}{\Gamma(\nu+1)}M(\nu+\tfrac{1}{2}, 2\nu+1, 2z).$$

Ex. 9.2 Prove that for fixed a, c, and z, the hypergeometric series for $\mathbf{F}(a,b;c;z/b)$ converges uniformly with respect to $b \in [2|z|, \infty)$. Deduce that

$$\mathbf{M}(a,c,z) = \lim_{b\to\infty} \mathbf{F}(a,b;c;z/b),$$

and thence from Exercise 9.4 of Chapter 5 that

$$(c-a)\mathbf{M}(a-1,c,z) + (2a-c+z)\mathbf{M}(a,c,z) - a\mathbf{M}(a+1,c,z) = 0,$$
$$\mathbf{M}(a,c-1,z) + (1-c-z)\mathbf{M}(a,c,z) + (c-a)z\mathbf{M}(a,c+1,z) = 0.$$

Ex. 9.3 By transformation of the differential equation establish *Kummer's second transformation*†

$$M(a,2a,2z) = e^z {}_0F_1(a+\tfrac{1}{2}; \tfrac{1}{4}z^2) \qquad (2a \neq 0, -1, -2, \ldots).$$

Ex. 9.4 From Exercise 9.2 and Chapter 5, Exercise 9.5, deduce that

$$\mathbf{M}(a,c,z) = -\frac{\Gamma(1-a)\Gamma(1+a-c)}{4\pi^2 e^{c\pi i}} \int_\alpha^{(1+,0+,1-,0-)} t^{a-1}(1-t)^{c-a-1} e^{zt}\,dt,$$

where α is any point of $(0, 1)$, and the branches of t^{a-1} and $(1-t)^{c-a-1}$ are continuous on the path and take their principal values at the starting point.

10 Asymptotic Solutions of the Confluent Hypergeometric Equation

10.1 The theory of §2 is directly applicable to the irregular singularity of (9.02). Using Theorems 2.1 and 2.2, we find that there are unique solutions $U(a,c,z)$ and $V(a,c,z)$ with the properties

$$U(a,c,z) \sim z^{-a} \sum_{s=0}^{\infty} (-)^s \frac{(a)_s(1+a-c)_s}{s!\,z^s} \qquad (z\to\infty \text{ in } |\mathrm{ph}\, z| \leqslant \tfrac{3}{2}\pi - \delta), \quad (10.01)$$

and

$$V(a,c,z) \sim e^z(-z)^{a-c} \sum_{s=0}^{\infty} \frac{(c-a)_s(1-a)_s}{s!\,z^s} \qquad (z\to\infty \text{ in } |\mathrm{ph}(-z)| \leqslant \tfrac{3}{2}\pi - \delta), \quad (10.02)$$

where δ is an arbitrary small positive constant.‡ As $z \to \infty$ in the right half-plane $U(a,c,z)$ is recessive and $V(a,c,z)$ is dominant; in the left half-plane these roles are reversed. Thus the two solutions are linearly independent for all values of the parameters.

$U(a,c,z)$ and $V(a,c,z)$ are related in the following way. By a transformation of variables, which is suggested by comparing the right-hand sides of (10.01) and (10.02), we find that $e^z U(c-a, c, -z)$ satisfies equation (9.02). This solution is recessive as $z \to -\infty$, hence it is a constant multiple of $V(a,c,z)$. Inspection of leading terms shows that this multiple is unity; thus

$$V(a,c,z) = e^z U(c-a, c, -z). \quad (10.03)$$

† Kummer's first transformation is given in §10.2.

‡ Employing the notation of the generalized hypergeometric function, we may denote the right-hand side of (10.01) *formally* by $z^{-a}{}_2F_0(a, 1+a-c; -z^{-1})$; similarly for $V(a,c,z)$. Another notation often used for $U(a,c,z)$ is $\Psi(a,c;z)$.

An integral representation for $U(a, c, z)$ analogous to (9.07) is given by

$$U(a,c,z) = \frac{1}{\Gamma(a)} \int_0^\infty t^{a-1}(1+t)^{c-a-1} e^{-zt}\, dt \qquad (|\text{ph}\, z| < \tfrac{1}{2}\pi, \quad \text{Re}\, a > 0). \tag{10.04}$$

This can be verified by showing that the integral satisfies the confluent hypergeometric equation, and then comparing the asymptotic form yielded by Watson's lemma with (10.01).

10.2 The transformation leading to (10.03) also shows that $e^z \mathbf{M}(c-a, c, -z)$ satisfies (9.02). When $\text{Re}\, c > 1$ or $c = 1$, this solution is recessive at $z = 0$, and since it assumes the value $1/\Gamma(c)$ at this point, we deduce that

$$\mathbf{M}(a,c,z) = e^z \mathbf{M}(c-a, c, -z). \tag{10.05}$$

Analytic continuation removes all restrictions on the parameters in this result, which is known as *Kummer's transformation*. The transformation can also be established by multiplying the series (9.04) by the power series for e^{-z} and using Vandermonde's theorem.

Another transformation that leaves the confluent hypergeometric equation unchanged is to replace a and c simultaneously by $1+a-c$ and $2-c$, and then take a new dependent variable $z^{1-c}w$; this is inferable from (9.05). By comparing recessive solutions at $z = +\infty$, we deduce that

$$U(a,c,z) = z^{1-c} U(1+a-c, 2-c, z). \tag{10.06}$$

10.3 Let us seek the coefficients A and B in the connection formula

$$\mathbf{M}(a,c,z) = AU(a,c,z) + BV(a,c,z).$$

Because $U(a,c,z)$ and $V(a,c,z)$ are many-valued functions of z, the values of these coefficients depend on which branches we have in mind. To begin with, assume that $\text{ph}\, z \in [0, \pi]$ and $\text{ph}(-z) \in [-\pi, 0]$.

Temporarily restricting $\text{Re}\, c > \text{Re}\, a > 0$ and applying Laplace's method to (9.07), we find that

$$\mathbf{M}(a,c,z) \sim e^z z^{a-c}/\Gamma(a) \qquad (z \to \infty, \quad \text{ph}\, z = 0), \tag{10.07}$$

and

$$\mathbf{M}(a,c,z) \sim (-z)^{-a}/\Gamma(c-a) \qquad (z \to -\infty, \quad \text{ph}(-z) = 0). \tag{10.08}$$

Now as $z \to \infty$ with $\text{ph}\, z = 0$ and $\text{ph}(-z) = -\pi$, $U(a,c,z)$ is recessive and $V(a,c,z)$ is asymptotic to $e^z z^{a-c} e^{-(a-c)\pi i}$; compare (10.02). Comparison with (10.07) accordingly yields

$$B = e^{(a-c)\pi i}/\Gamma(a).$$

Similarly when $z \to -\infty$ with $\text{ph}\, z = \pi$ and $\text{ph}(-z) = 0$, $V(a,c,z)$ is recessive and $U(a,c,z)$ is asymptotic to $(-z)^{-a} e^{-a\pi i}$. Therefore

$$A = e^{a\pi i}/\Gamma(c-a).$$

Thus one version of the required connection formula is

$$\mathbf{M}(a,c,z) = \frac{e^{a\pi i}}{\Gamma(c-a)} U(a,c,z) + \frac{e^{(a-c)\pi i}}{\Gamma(a)} V(a,c,z), \tag{10.09}$$

the branches being determined by $\mathrm{ph}(-z) = -\pi$ when $\mathrm{ph}\, z = 0$, and by continuity elsewhere.

Formula (10.09) has been established with the restrictions $\mathrm{Re}\, c > \mathrm{Re}\, a > 0$. We have already noted in §9.2 that for fixed z, $\mathbf{M}(a, c, z)$ is entire in a and c. From Theorem 3.1 it follows that the same is true of $U(a, c, z)$ and $V(a, c, z)$, provided that z is nonzero. Hence by analytic continuation (10.09) *holds without restriction on the parameters.*

If we use instead the continuous branch of $V(a, c, z)$ determined by $\mathrm{ph}(-z) = \pi$ when $\mathrm{ph}\, z = 0$, then by symmetry the connection formula changes to

$$\mathbf{M}(a,c,z) = \frac{e^{-a\pi i}}{\Gamma(c-a)} U(a,c,z) + \frac{e^{(c-a)\pi i}}{\Gamma(a)} V(a,c,z). \tag{10.10}$$

The importance of (10.09) and (10.10) is that when combined with (10.01) and (10.02) they determine the asymptotic behavior of $\mathbf{M}(a, c, z)$ for large z throughout a wide range of $\mathrm{ph}\, z$, in fact, $|\mathrm{ph}\, z| \leqslant \frac{3}{2}\pi - \delta$. In various parts of this sector one of the functions $U(a, c, z)$ and $V(a, c, z)$ is exponentially small compared with the other, and therefore negligible in Poincaré's sense (although, as we shall see in §13, such neglect may incur loss of numerical accuracy). In other regions, notably the vicinities of $\mathrm{ph}\, z = \pm\frac{1}{2}\pi$, the contributions from $U(a, c, z)$ and $V(a, c, z)$ are equally significant.†

10.4 Another formula of importance connects U, $\mathbf{M}$, and $\mathbf{N}$:

$$U(a,c,z) = C\mathbf{M}(a,c,z) + D\mathbf{N}(a,c,z). \tag{10.11}$$

To find C and D we first let $z \to +\infty$. Then using (10.09) and (9.06), we have

$$\mathbf{M}(a,c,z) \sim e^z z^{a-c}/\Gamma(a), \qquad \mathbf{N}(a,c,z) \sim e^z z^{a-c}/\Gamma(1+a-c),$$

provided that neither a nor $1+a-c$ is a negative integer or zero. Since $U(a, c, z)$ is recessive in these circumstances, it follows that

$$C/\Gamma(a) = -D/\Gamma(1+a-c).$$

Next, assume that $\mathrm{Re}\, a > 0$ and $\mathrm{Re}\, c < 1$. Letting $z \to 0$ in (10.04), we obtain

$$U(a,c,0+) = \frac{1}{\Gamma(a)} \int_0^\infty t^{a-1}(1+t)^{c-a-1}\, dt = \frac{\Gamma(1-c)}{\Gamma(1+a-c)};$$

compare Exercise 1.3 of Chapter 2. On the right of (10.11), as $z \to 0$ the function $\mathbf{N}(a, c, z)$ vanishes and $\mathbf{M}(a, c, z)$ tends to $1/\Gamma(c)$. Hence

$$C = \frac{\Gamma(c)\Gamma(1-c)}{\Gamma(1+a-c)}, \qquad D = -\frac{\Gamma(c)\Gamma(1-c)}{\Gamma(a)},$$

† Much of the asymptotic information contained in (10.09) and (10.10) is also derivable from (9.07) by Laplace's method and the method of stationary phase. Contributions from both endpoints of the integration range need to be included; also, the restrictions $\mathrm{Re}\, c > \mathrm{Re}\, a > 0$ can be removed by the artifice of Chapter 4, §5.2.

and (10.11) becomes

$$U(a,c,z) = \frac{\pi}{\sin(\pi c)}\left\{\frac{\mathbf{M}(a,c,z)}{\Gamma(1+a-c)} - \frac{\mathbf{N}(a,c,z)}{\Gamma(a)}\right\}. \tag{10.12}$$

As in §10.3, analytic continuation removes all restrictions on the parameters.

In conjunction with (9.04) and (9.06) this formula describes the behavior of $U(a,c,z)$ near $z=0$. When c is an integer or zero, the right-hand side is replaced by its limiting value; see Exercise 10.6 below.

10.5 The Wronskian formula for U and $\mathbf{M}$ can be found by considering the limiting forms of these functions and their derivatives as $z\to\infty$ or as $z\to 0$. Either way yields

$$\mathscr{W}\{U(a,c,z),\, \mathbf{M}(a,c,z)\} = e^z z^{-c}/\Gamma(a).$$

Therefore unless a is a nonpositive integer these solutions are linearly independent. Their respective recessive properties at ∞ and 0 show that when $\operatorname{Re} c \geqslant 1$, $U(a,c,z)$ and $\mathbf{M}(a,c,z)$ comprise a numerically satisfactory pair of solutions *throughout* the sector $|\operatorname{ph} z| \leqslant \frac{1}{2}\pi$.

When $a = 0, -1, -2, \ldots$ and $\operatorname{Re} c > 1$ or $c = 1$, the recessive solution of (9.02) at the origin is also recessive at infinity in the sector $|\operatorname{ph} z| \leqslant \frac{1}{2}\pi - \delta$; it is in fact a polynomial in z of degree $-a$; see Exercise 10.3 below. As noted in Chapter 5, §7.3, combination of this solution with any linearly independent solution, for example $V(a,c,z)$, produces a numerically satisfactory pair throughout $|\operatorname{ph} z| \leqslant \frac{1}{2}\pi$.

Ex. 10.1 Show that

$$K_\nu(z) = \pi^{1/2}(2z)^\nu e^{-z} U(\nu+\tfrac{1}{2}, 2\nu+1, 2z).$$

Ex. 10.2 Show that the incomplete Gamma functions can be expressed

$$\gamma(\alpha,z) = (z^\alpha/\alpha)\, M(\alpha,\alpha+1,-z) \quad (\alpha \neq 0,-1,-2,\ldots); \qquad \Gamma(\alpha,z) = e^{-z}U(1-\alpha,1-\alpha,z).$$

Ex. 10.3 Show that the Laguerre and Hermite polynomials can be expressed

$$L_n^{(\alpha)}(x) = \frac{\Gamma(\alpha+1+n)}{n!}\mathbf{M}(-n,\alpha+1,x) = \frac{(-)^n}{n!}U(-n,\alpha+1,x),$$

$$H_n(x) = 2^n U(-\tfrac{1}{2}n,\tfrac{1}{2},x^2) = 2^n x U(\tfrac{1}{2}-\tfrac{1}{2}n,\tfrac{3}{2},x^2).$$

Ex. 10.4 Show that the parabolic cylinder function of Chapter 6, §6 is given by

$$U(a,z) = 2^{-(a/2)-(1/4)}\exp(-\tfrac{1}{4}z^2)\, U(\tfrac{1}{2}a+\tfrac{1}{4},\tfrac{1}{2},\tfrac{1}{2}z^2) = 2^{-(a/2)-(3/4)}\exp(-\tfrac{1}{4}z^2)\, zU(\tfrac{1}{2}a+\tfrac{3}{4},\tfrac{3}{2},\tfrac{1}{2}z^2).$$

Ex. 10.5 Verify that

$$U(a-1,c,z) + (c-2a-z)\,U(a,c,z) + a(1+a-c)\,U(a+1,c,z) = 0,$$

$$(c-a-1)\,U(a,c-1,z) + (1-c-z)\,U(a,c,z) + zU(a,c+1,z) = 0.$$

Ex. 10.6 Let m be a positive integer. By considering the limiting form of (10.12) show that

$$U(a,m,z) = \sum_{s=1}^{m-1}(-)^{s-1}\lambda_{m,-s}\frac{(s-1)!}{z^s} + \lambda_{m,0}M(a,m,z)\ln z + \sum_{s=0}^{\infty}\lambda_{m,s}\mu_{m,s}\frac{z^s}{s!},$$

where

$$\lambda_{m,s} = \frac{(-)^m\Gamma(a+s)}{\Gamma(a)\Gamma(1+a-m)(m+s-1)!}, \qquad \mu_{m,s} = \psi(a+s) - \psi(1+s) - \psi(m+s).$$

11 Whittaker Functions

11.1 The result of changing the dependent variable to eliminate the term in the first derivative from the confluent hypergeometric equation (9.02) can be expressed as

$$\frac{d^2W}{dz^2} = \left(\frac{1}{4} - \frac{k}{z} + \frac{m^2-\frac{1}{4}}{z^2}\right)W, \tag{11.01}$$

in which $k = \frac{1}{2}c-a$, $m = \frac{1}{2}c-\frac{1}{2}$, $W = e^{-z/2}z^{m+(1/2)}w$. This is *Whittaker's equation.* Standard solutions are

$$M_{k,m}(z) \equiv e^{-z/2}z^{m+(1/2)}M(m-k+\tfrac{1}{2},\ 2m+1,\ z), \tag{11.02}$$

$$W_{k,m}(z) \equiv e^{-z/2}z^{m+(1/2)}U(m-k+\tfrac{1}{2},\ 2m+1,\ z). \tag{11.03}$$

Each is a many-valued function of z. Principal branches correspond to the range $\operatorname{ph} z \in (-\pi, \pi]$.

All formulas in §§9 and 10 are reexpressible in terms of Whittaker's functions. In particular, the characterizing properties are

$$M_{k,m}(z) \sim z^{m+(1/2)} \qquad (z \to 0, \quad 2m \neq -1, -2, -3, \ldots), \tag{11.04}$$

and

$$W_{k,m}(z) \sim e^{-z/2}z^k \qquad (z \to \infty, \quad |\operatorname{ph} z| \leqslant \tfrac{3}{2}\pi-\delta). \tag{11.05}$$

11.2 An instructive application of the theory of Chapter 6 is to determine the behavior of $M_{k,m}(z)$ and $W_{k,m}(z)$ for large m. We sketch here the analysis in the case when the parameters and variable are real and positive.†

With the substitutions $x = z/m$, $l = k/m$, (11.01) becomes

$$d^2W/dx^2 = \{f(x)+g(x)\}\,W, \tag{11.06}$$

in which

$$f(x) = m^2\frac{x^2-4lx+4}{4x^2}, \qquad g(x) = -\frac{1}{4x^2}. \tag{11.07}$$

The zeros of $f(x)$ are located at $x = 2l \pm 2(l^2-1)^{1/2}$. If we restrict $l \in [0, \alpha]$, where α is a fixed number in the interval $[0, 1)$, then these zeros are complex, and $f(x)$ is positive throughout $(0, \infty)$. Accordingly, Theorem 2.1 of Chapter 6 is applicable to (11.06), with $a_1 = 0$ and $a_2 = \infty$. From §§4.2 and 4.3 of the same chapter—or directly—it is seen that the error-control function F constructed from (11.07) has a convergent variation at $x = \infty$ and 0, that is,

$$\mathcal{V}_{0,\infty}(F) \equiv \frac{1}{m}\int_0^\infty \left|\frac{(2x)^{1/2}}{(x^2-4lx+4)^{1/4}}\frac{d^2}{dx^2}\left\{\frac{(2x)^{1/2}}{(x^2-4lx+4)^{1/4}}\right\} + \frac{1}{2x(x^2-4lx+4)^{1/2}}\right| dx$$

is finite. Moreover, it is easily seen that

$$\mathcal{V}_{0,\infty}(F) = m^{-1}O(1), \tag{11.08}$$

† For similar results when k and z are purely imaginary see Chapter 11, §4.3.

uniformly with respect to $l \in [0, \alpha]$.

Again, the cited theorem asserts that solutions $w_1(x)$ and $w_2(x)$ of (11.06) exist such that

$$w_1(x) = f^{-1/4}(x) \exp\left\{\int f^{1/2}(x)\,dx\right\}\{1+\varepsilon_1(x)\}, \tag{11.09}$$

$$w_2(x) = f^{-1/4}(x) \exp\left\{-\int f^{1/2}(x)\,dx\right\}\{1+\varepsilon_2(x)\}, \tag{11.10}$$

where

$$|\varepsilon_1(x)| \leqslant \exp\{\tfrac{1}{2}\mathscr{V}_{0,x}(F)\} - 1, \qquad |\varepsilon_2(x)| \leqslant \exp\{\tfrac{1}{2}\mathscr{V}_{x,\infty}(F)\} - 1.$$

The first solution is recessive at $x = 0$; the second is recessive at $x = \infty$. Therefore

$$\frac{M_{k,m}(z)}{w_1(x)} = A(k,m), \qquad \frac{W_{k,m}(z)}{w_2(x)} = B(k,m),$$

where $A(k,m)$ and $B(k,m)$ are independent of x (or z).

The value of $A(k,m)$ can be found by letting $x \to 0$ and using (11.04), (11.09), and the fact that $\varepsilon_1(x) \to 0$. Similarly $B(k,m)$ is determinable by letting $x \to \infty$ and using (11.05), (11.10), and the limit $\varepsilon_2(x) \to 0$. These calculations depend on the following elementary identity

$$\int f^{1/2}(x)\,dx = \frac{1}{2}Z - k\ln(Z+z-2k) - m\ln\left(\frac{mZ-kz+2m^2}{z}\right),$$

in which

$$Z = (z^2 - 4kz + 4m^2)^{1/2}.$$

The final results are given by

$$M_{k,m}(z) = \frac{2^{k+2m+(1/2)} m^{2m+(1/2)} (m-k)^k z^{m+(1/2)} e^{Z/2}}{e^m Z^{1/2} (Z+z-2k)^k (mZ-kz+2m^2)^m}(1+\varepsilon_1),$$

and

$$W_{k,m}(z) = \frac{(Z+z-2k)^k (mZ-kz+2m^2)^m}{(m-k)^m (2e)^k z^{m-(1/2)} Z^{1/2} e^{Z/2}}(1+\varepsilon_2).$$

The error terms have the properties: (i) for fixed k and m, $\varepsilon_1 \to 0$ as $z \to 0$, and $\varepsilon_2 \to 0$ as $z \to \infty$; (ii) for large m, $\varepsilon_1 = O(m^{-1})$ and $\varepsilon_2 = O(m^{-1})$ uniformly with respect to $z \in (0, \infty)$ and $k \in [0, \alpha m]$, where α is any fixed number in $[0, 1)$.

It will be noted that Condition (ii) includes the case of fixed k. Some simplifications can then be made.

Ex. 11.1 From (10.12) derive

$$W_{k,m}(z) = \frac{\Gamma(-2m)}{\Gamma(\frac{1}{2}-m-k)} M_{k,m}(z) + \frac{\Gamma(2m)}{\Gamma(\frac{1}{2}+m-k)} M_{k,-m}(z),$$

the right-hand side being replaced by its limiting value when $2m$ is an integer or zero.

Ex. 11.2 Show that $M_{-k,m}(ze^{\pi i}) = ie^{m\pi i}M_{k,m}(z)$. Thence by using the preceding exercise prove that for any integer s

$$(-)^s W_{k,m}(ze^{2s\pi i}) = -\frac{e^{2k\pi i}\sin(2sm\pi)+\sin\{(2s-2)m\pi\}}{\sin(2m\pi)} W_{k,m}(z) - \frac{\sin(2sm\pi)}{\sin(2m\pi)} \frac{2\pi i e^{k\pi i}}{\Gamma(\frac{1}{2}+m-k)\Gamma(\frac{1}{2}-m-k)} W_{-k,m}(ze^{\pi i}).$$

*12 Error Bounds for the Asymptotic Solutions in the General Case

12.1 The method of proof used for Theorems 2.1 and 2.2 is insufficiently powerful to provide satisfactory bounds for the nth remainder term in the asymptotic expansion (2.04), particularly when n violates (2.14). To achieve realistic bounds, we construct an integral equation for the error term using complementary functions which approximate the required solutions more closely than $e^{\lambda_1 z}$ and $e^{\lambda_2 z}$. At the same time, we ease restrictions on $f(z)$ and $g(z)$:

The functions $f(z)$ and $g(z)$ are holomorphic in a domain containing the annular sector **S**: $\alpha \leqslant \operatorname{ph} z \leqslant \beta$, $|z| \geqslant a$, *and*

$$f(z) \sim \sum_{s=0}^{\infty} \frac{f_s}{z^s}, \qquad g(z) \sim \sum_{s=0}^{\infty} \frac{g_s}{z^s}, \qquad (z\to\infty \text{ in } \mathbf{S}), \tag{12.01}$$

where $f_0^2 \neq 4g_0$. The remainder terms associated with these expansions are denoted by

$$f(z) = \sum_{s=0}^{n-1} \frac{f_s}{z^s} + \frac{F_n(z)}{z^n}, \qquad g(z) = \sum_{s=0}^{n-1} \frac{g_s}{z^s} + \frac{G_n(z)}{z^n} \qquad (n=0,1,\ldots). \tag{12.02}$$

Thus for fixed n, $|F_n(z)|$ and $|G_n(z)|$ are bounded in **S**.

The steps which follow parallel the proof of Theorem 2.1, and we employ the same notation.

12.2 By using the identity

$$g(z)L_n(z) = e^{\lambda_1 z} z^{\mu_1} \sum_{s=0}^{n-1} \left\{ g_0 + \frac{g_1}{z} + \cdots + \frac{g_{n-s+1}}{z^{n-s+1}} + \frac{G_{n-s+2}(z)}{z^{n-s+2}} \right\} \frac{a_{s,1}}{z^s}$$

and a similar identity for $f(z)L_n'(z)$, we may verify that the residual term $R_n(z)$ of (2.06) is given by

$$R_n(z) = \frac{(f_0+2\lambda_1)na_{n,1}}{z^{n+1}} + \frac{\hat{R}_{n+1}(z)}{z^{n+2}}, \tag{12.03}$$

where

$$\hat{R}_{n+1}(z) = \sum_{s=0}^{n-1} a_{s,1}\{(\mu_1 - s)F_{n+1-s}(z) + \lambda_1 F_{n+2-s}(z) + G_{n+2-s}(z)\}, \tag{12.04}$$

and is bounded in **S**.

To construct a new integral equation equivalent to (2.07), we first seek a differential equation which approximates the given equation

$$w'' + f(z)\,w' + g(z)\,w = 0 \tag{12.05}$$

more closely than

$$w'' + f_0\,w' + g_0\,w = 0,$$

when $|z|$ is large. The most obvious choice is

$$w'' + \left(f_0 + \frac{f_1}{z}\right)w' + \left(g_0 + \frac{g_1}{z}\right)w = 0,$$

but this cannot be solved in terms of elementary functions, in general.

We apply the result of Exercise 1.2 of Chapter 6, determining the functions p and q in such a way that the expansions of the coefficients of dW/dz and W in powers of z^{-1} match the expansions of $f(z)$ and $g(z)$, respectively, as far as the terms in z^{-1}. Obviously the choice is not unique; for simplicity we take

$$q = f_0 + \frac{f_1}{z}, \qquad p = \frac{1}{2}\frac{dq}{dz} + \frac{1}{4}q^2 - g_0 - \frac{g_1}{z} + \frac{\text{constant}}{z^2},$$

choosing the constant in the second relation to make p a perfect square. Thus

$$p = \frac{1}{4}(f_0^2 - 4g_0)\left(1 + \frac{\rho}{z}\right)^2,$$

where

$$\rho = \frac{f_0 f_1 - 2g_1}{f_0^2 - 4g_0} = \frac{\mu_1 - \mu_2}{\lambda_1 - \lambda_2}; \tag{12.06}$$

compare (1.09) and (1.10). With these substitutions, the functions

$$W_1(z) = \left(1 + \frac{\rho}{z}\right)^{-1/2} e^{\lambda_1 z} z^{\mu_1}, \qquad W_2(z) = \left(1 + \frac{\rho}{z}\right)^{-1/2} e^{\lambda_2 z} z^{\mu_2}, \tag{12.07}$$

satisfy the differential equation

$$\frac{d^2W}{dz^2} + \left(f_0 + \frac{f_1}{z}\right)\frac{dW}{dz} + \left\{g_0 + \frac{g_1}{z} + \frac{\hat{g}_2}{z^2} + l(z)\right\}W = 0, \tag{12.08}$$

in which

$$\hat{g}_2 = \tfrac{1}{4}f_1(f_1 - 2) - \rho^2(\tfrac{1}{4}f_0^2 - g_0) = \mu_1\mu_2 + \tfrac{1}{2}(\mu_1 + \mu_2), \tag{12.09}$$

and

$$l(z) = -p^{1/4}(p^{-1/4})'' = \frac{\rho}{z^3}\left(1 + \frac{\rho}{4z}\right)\left(1 + \frac{\rho}{z}\right)^{-2}.$$

Clearly (12.08) has the desired matching with (12.05).

Using (12.02) with $n = 2$, we recast (2.07) in the form

$$\varepsilon_n''(z) + \left(f_0 + \frac{f_1}{z}\right)\varepsilon_n'(z) + \left\{g_0 + \frac{g_1}{z} + \frac{\hat{g}_2}{z^2} + l(z)\right\}\varepsilon_n(z)$$
$$= -e^{\lambda_1 z} z^{\mu_1} R_n(z) - \frac{\hat{G}_2(z)}{z^2}\varepsilon_n(z) - \frac{F_2(z)}{z^2}\varepsilon_n'(z), \tag{12.10}$$

where

$$\hat{G}_2(z) = G_2(z) - \hat{g}_2 - z^2 l(z). \tag{12.11}$$

Solution of (12.10) by variation of parameters yields the desired integral equation

$$\varepsilon_n(z) = \int_z^{z_1} \mathsf{K}(z,t)\left\{e^{\lambda_1 t} t^{\mu_1} R_n(t) + \frac{\hat{G}_2(t)}{t^2}\varepsilon_n(t) + \frac{F_2(t)}{t^2}\varepsilon_n'(t)\right\} dt, \tag{12.12}$$

in which

$$\mathsf{K}(z,t) = -\frac{W_1(z)\,W_2(t) - W_2(z)\,W_1(t)}{W_1(t)\,W_2'(t) - W_2(t)\,W_1'(t)} = \frac{W_1(z)\,W_2(t) - W_2(z)\,W_1(t)}{(\lambda_1 - \lambda_2)\,t^{\mu_1+\mu_2}e^{(\lambda_1+\lambda_2)t}}, \tag{12.13}$$

and z_1 is an arbitrary fixed point of $\mathbf{S}$, generally taken to be the point at infinity on the ray $\operatorname{ph} t = -\omega$; compare (2.12).

12.3 We solve equation (12.12) by application of Theorem 10.1 of Chapter 6. To bound the kernel we introduce the notations

$$\xi_1(z) = \lambda_1 z + \mu_1 \ln z, \qquad \xi_2(z) = \lambda_2 z + \mu_2 \ln z,$$

the branch of $\ln z$ being continuous on the path of integration, $\mathscr{P}_1$, say. Then from (12.07) and (12.13)

$$\mathsf{K}(z,t) = \left(1 + \frac{\rho}{z}\right)^{-1/2}\left(1 + \frac{\rho}{t}\right)^{-1/2}\frac{e^{\xi_1(z)-\xi_1(t)} - e^{\xi_2(z)-\xi_2(t)}}{\lambda_1 - \lambda_2}.$$

Let $\mathscr{P}_1$ be subject to the condition that $\operatorname{Re}\{\xi_2(t) - \xi_1(t)\}$ is nondecreasing as t passes from z to z_1. Then

$$|e^{\xi_2(z)-\xi_2(t)}| \leqslant |e^{\xi_1(z)-\xi_1(t)}|,$$

and therefore $|\mathsf{K}(z,t)| \leqslant P_0(z)\,Q(t)$, where

$$Q(t) = |e^{-\xi_1(t)}|, \qquad P_0(z) = 2c_1\left|\frac{e^{\xi_1(z)}}{\lambda_1 - \lambda_2}\right|, \qquad c_1 = \sup_{t\in\mathscr{P}_1}\left|1 + \frac{\rho}{t}\right|^{-1}. \tag{12.14}$$

Next,

$$\frac{W_j'(z)}{W_j(z)} = \lambda_j + \frac{\mu_j}{z} + \frac{\rho}{2z^2}\left(1 + \frac{\rho}{z}\right)^{-1} \qquad (j = 1, 2);$$

whence $|\partial\mathsf{K}(z,t)/\partial z| \leqslant P_1(z)\,Q(t)$, where $P_1(z) = c_2 P_0(z)$ and

$$c_2 = \frac{1}{2}\sup_{t\in\mathscr{P}_1}\left\{\left|\lambda_1 + \frac{\mu_1}{t} + \frac{\rho}{2t^2}\left(1 + \frac{\rho}{t}\right)^{-1}\right| + \left|\lambda_2 + \frac{\mu_2}{t} + \frac{\rho}{2t^2}\left(1 + \frac{\rho}{t}\right)^{-1}\right|\right\}. \tag{12.15}$$

Again, in the notation of the cited theorem,

$$\phi(t) = -R_n(t), \qquad J(t) = e^{\xi_1(t)}, \qquad \psi_0(t) = -t^{-2}\hat{G}_2(t), \qquad \psi_1(t) = -t^{-2}F_2(t).$$

Thus in Condition (vi) of Chapter 6, §10.2, we have

$$\Phi(z) = \int_z^{z_1} |R_n(t)\,dt|, \qquad \Psi_0(z) = \int_z^{z_1} \left|\frac{\hat{G}_2(t)}{t^2}\,dt\right|, \qquad \Psi_1(z) = \int_z^{z_1} \left|\frac{F_2(t)}{t^2}\,dt\right|,$$

and

$$\kappa = 1, \qquad \kappa_0 = 2c_1|\lambda_1-\lambda_2|^{-1}, \qquad \kappa_1 = 2c_1c_2|\lambda_1-\lambda_2|^{-1}.$$

Clearly,

$$\Psi_0(z) \leqslant c_3\mathcal{V}_{z,z_1}(t^{-1}), \qquad \Psi_1(z) \leqslant c_4\mathcal{V}_{z,z_1}(t^{-1}),$$

where

$$c_3 = \sup_{t\in\mathscr{P}_1} |\hat{G}_2(t)|, \qquad c_4 = \sup_{t\in\mathscr{P}_1} |F_2(t)|. \tag{12.16}$$

Also, referring to (12.03) and (12.04), and recalling that $f_0+2\lambda_1 = \lambda_1-\lambda_2$, we have

$$\Phi(z) \leqslant |\lambda_1-\lambda_2|\{\mathcal{V}_{z,z_1}(a_{n,1}t^{-n})+\mathcal{V}_{z,z_1}(r_{n+1,1}t^{-n-1})\},$$

where

$$r_{n+1,1} = \frac{\sup_{t\in\mathscr{P}_1}|\sum_{s=0}^{n-1}a_{s,1}\{(\mu_1-s)F_{n+1-s}(t)+\lambda_1F_{n+2-s}(t)+G_{n+2-s}(t)\}|}{(n+1)|\lambda_1-\lambda_2|}. \tag{12.17}$$

Applying the cited theorem we obtain a solution of the integral equation (12.12) along the chosen path $\mathscr{P}_1$, complete with bounds. That this solution also satisfies the differential equation (2.07) is established by analysis similar to §11.2 of Chapter 6.

12.4 Collecting together the foregoing results, we have:

Theorem 12.1 *With the italicized conditions of* §12.1, *equation* (12.05) *has, for each positive integer n, a solution*

$$w_{n,1}(z) = e^{\lambda_1 z}z^{\mu_1}\left(\sum_{s=0}^{n-1}\frac{a_{s,1}}{z^s}\right)+\varepsilon_{n,1}(z) \tag{12.18}$$

which depends on an arbitrary reference point z_1, *and is holomorphic in* **S**. *Here* λ_1, μ_1, *and* $a_{s,1}$ *are defined in* §1.2, *and the error term* $\varepsilon_{n,1}(z)$ *is bounded as follows. Let* $\mathbf{Z}_1(z_1)$ *be the z point set for which there exists a path* $\mathscr{P}_1$ *in* **S** *joining z with* z_1 *and having the properties:*

(i) $\mathscr{P}_1$ *consists of a finite chain of* R_2 *arcs.*
(ii) $\operatorname{Re}\{(\lambda_2-\lambda_1)t+(\mu_2-\mu_1)\ln t\}$ *is nondecreasing as t passes along* $\mathscr{P}_1$ *from z to* z_1.

Then in $\mathbf{Z}_1(z_1)$ *both* $|\varepsilon_{n,1}(z)|$ *and* $|\varepsilon'_{n,1}(z)|/c_2$ *are bounded by*

$$2c_1|e^{\lambda_1 z}z^{\mu_1}|\left\{\mathcal{V}_{\mathscr{P}_1}\left(\frac{a_{n,1}}{t^n}\right)+\mathcal{V}_{\mathscr{P}_1}\left(\frac{r_{n+1,1}}{t^{n+1}}\right)\right\}\exp\left\{\frac{2c_1(c_2c_4+c_3)}{|\lambda_1-\lambda_2|}\mathcal{V}_{\mathscr{P}_1}\left(\frac{1}{t}\right)\right\}, \tag{12.19}$$

where $r_{n+1,1}$ *is defined by* (12.17), c_1, c_2, *and* c_4 *are defined by* (12.14), (12.15), *and* (12.16), *with* ρ *given by* (12.06), *and*

$$c_3 = \sup_{t\in\mathscr{P}_1}\left|G_2(t) - \mu_1\mu_2 - \frac{1}{2}(\mu_1+\mu_2) - \frac{\rho}{t}\left(1+\frac{\rho}{4t}\right)\left(1+\frac{\rho}{t}\right)^{-2}\right|. \quad (12.20)$$

Remarks (a) Conditions (i) and (ii) on $\mathscr{P}_1$ are similar to conditions given in Chapter 6, §11.3. Again, in the context of the present theorem an admissible path may be referred to as a *progressive path*. If z_1 is the point at infinity on a progressive path $\mathscr{L}_1$ (and this will usually be the case), then $\mathscr{P}_1$ is required to coincide with $\mathscr{L}_1$ in a neighborhood of z_1.

(b) A similar result holds for a second solution of the differential equation: essentially, we interchange λ_1 with λ_2 and μ_1 with μ_2, replace $a_{s,1}$ by $a_{s,2}$, and introduce a new reference point z_2 and path $\mathscr{P}_2$.

(c) A similar theorem holds for real variables. In this case it suffices that $f(z)$ and $g(z)$ be continuous. Also, the bound (12.19) may be sharpened by replacing c_1 and c_2 simultaneously by $\frac{1}{2}c_1$ and $2c_2$, respectively.

12.5 In the next section Theorem 12.1 is applied to Hankel's expansions. In this case we have $\mu_1 = \mu_2$, which considerably simplifies the choice of progressive paths. A treatment of the more difficult case of Whittaker functions of large argument has been given by Olver (1965d).

Ex. 12.1 Prove that Condition (ii) on $\mathscr{P}_1$ is satisfied if $\cos\phi \geqslant |\rho/t|$, where $\phi - \mathrm{ph}(\lambda_2-\lambda_1)$ is the angle of slope of $\mathscr{P}_1$ at t.

Ex. 12.2 By means of the preceding exercise, show that Theorems 2.1 and 2.2 are a special case of Theorem 12.1.

Ex. 12.3 Using Exercise 12.1 show that the equation

$$\frac{d^2w}{dz^2} + \frac{z}{z-1}\frac{dw}{dz} + e^{-z}w = 0$$

has an analytic solution with the asymptotic expansion

$$\frac{e^{-z}}{z}\sum_{s=0}^{\infty}\left\{\sum_{j=0}^{s}\frac{(-1)^j}{j!}\right\}\frac{(-)^s s!}{z^s}$$

as $z\to\infty$ in the sector $|\mathrm{ph}\, z| \leqslant \frac{1}{2}\pi - \delta$ $(<\frac{1}{2}\pi)$.

*13 Error Bounds for Hankel's Expansions

13.1 Bounds for the errors in the truncated forms of (4.03) and (4.04) are obtainable from Theorem 12.1. With $n \geqslant 1$, write

$$H_\nu^{(1)}(z) = \left(\frac{2}{\pi z}\right)^{1/2} e^{i\zeta}\left\{\sum_{s=0}^{n-1} i^s\frac{A_s(\nu)}{z^s} + \eta_{n,1}(z)\right\}, \quad (13.01)$$

where, again, $\zeta = z - \frac{1}{2}\nu\pi - \frac{1}{4}\pi$. In the notation of §12,

$$F_s(z) = 0 \quad (s \geqslant 2), \qquad G_2(z) = -\nu^2, \qquad G_s(z) = 0 \quad (s \geqslant 3),$$

and

$$\rho = 0, \qquad c_1 = 1, \qquad c_3 = |\nu^2 - \tfrac{1}{4}|, \qquad c_4 = 0, \qquad r_{n+1,1} = 0 \quad (n \geqslant 0).$$

Taking $z_1 = i\infty$ and bearing in mind that $\eta_{n,1}(z) = e^{-iz} z^{1/2} \varepsilon_{n,1}(z)$, we derive from (12.19)

$$|\eta_{n,1}(z)| \leqslant 2|A_n(\nu)| \mathscr{V}_{z,i\infty}(t^{-n}) \exp\{|\nu^2 - \tfrac{1}{4}| \mathscr{V}_{z,i\infty}(t^{-1})\}, \tag{13.02}$$

the paths of variation being subject to the condition that $\operatorname{Im} t$ changes monotonically.

Bounds for the minimum variations are available from §13 of Chapter 6 on rotating the z plane through an angle $\frac{1}{2}\pi$. Thus

$$\mathscr{V}_{z,i\infty}(t^{-n}) \leqslant \begin{cases} |z|^{-n} & (0 \leqslant \operatorname{ph} z \leqslant \pi) \\ \chi(n)|z|^{-n} & (-\frac{1}{2}\pi \leqslant \operatorname{ph} z \leqslant 0 \quad \text{or} \quad \pi \leqslant \operatorname{ph} z \leqslant \frac{3}{2}\pi) \\ 2\chi(n)|\operatorname{Im} z|^{-n} & (-\pi < \operatorname{ph} z \leqslant -\frac{1}{2}\pi \quad \text{or} \quad \frac{3}{2}\pi \leqslant \operatorname{ph} z < 2\pi) \end{cases}, \tag{13.03}$$

where, again, $\chi(n) = \pi^{1/2}\Gamma(\frac{1}{2}n+1)/\Gamma(\frac{1}{2}n+\frac{1}{2})$.

When $|z| \gg |\nu^2 - \frac{1}{4}|$ and $0 \leqslant \operatorname{ph} z \leqslant \pi$, the ratio of the error bound (13.02) to the modulus of the first neglected term, $i^n A_n(\nu)/z^n$, is approximately 2. When $\frac{1}{2}\pi \leqslant |\operatorname{ph}(ze^{-\pi i/2})| \leqslant \pi$ this ratio is approximately $2\chi(n)$. Accordingly, (13.01) is very satisfactory for numerical computation in these phase ranges. But when $\pi \leqslant |\operatorname{ph}(ze^{-\pi i/2})| \leqslant \frac{3}{2}\pi - \delta$ we have

$$\mathscr{V}_{z,i\infty}(t^{-n}) \leqslant 2\chi(n)\csc^n\delta\, |z|^{-n}.$$

This bound grows sharply as $\delta \to 0$, warning us that if $\eta_{n,1}(z)$ is neglected, then (13.01) is inaccurate for numerical work near the boundaries $\operatorname{ph} z = -\pi$ and 2π.

For the second Hankel function, the corresponding results are

$$H_\nu^{(2)}(z) = \left(\frac{2}{\pi z}\right)^{1/2} e^{-i\zeta} \left\{ \sum_{s=0}^{n-1} (-i)^s \frac{A_s(\nu)}{z^s} + \eta_{n,2}(z) \right\}, \tag{13.04}$$

where

$$|\eta_{n,2}(z)| \leqslant 2|A_n(\nu)| \mathscr{V}_{z,-i\infty}(t^{-n}) \exp\{|\nu^2 - \tfrac{1}{4}| \mathscr{V}_{z,-i\infty}(t^{-1})\}. \tag{13.05}$$

The bounds (13.03) apply to $\mathscr{V}_{z,-i\infty}(t^{-n})$ in the conjugate sectors.

Error bounds for the corresponding expansions of $J_\nu(z)$ and $Y_\nu(z)$ are easily derivable from (13.02), (13.03), and (13.05) by means of the connection formulas (4.08) and (5.01).

13.2 Satisfactory asymptotic representations of $H_\nu^{(1)}(z)$ and $H_\nu^{(2)}(z)$ near the boundaries of the regions of validity of (4.03) and (4.04) can be constructed by means of the continuation formulas (4.13) and (4.14) in the manner of §4.3. Taking $m = 1$ and 2 in (4.13), we derive

$$H_\nu^{(1)}(ze^{2\pi i}) = -H_\nu^{(1)}(z) + 2\cos(\nu\pi) H_\nu^{(1)}(ze^{\pi i}).$$

When $\operatorname{ph} z \in (-\pi, \pi)$, both z and $ze^{\pi i}$ lie within the region of validity of (13.01) and (13.03). Substituting for $H_\nu^{(1)}(z)$ and $H_\nu^{(1)}(ze^{\pi i})$ and then replacing z by $ze^{-2\pi i}$, we obtain

$$H_\nu^{(1)}(z) = \left(\frac{2}{\pi z}\right)^{1/2} e^{i\zeta}\left\{\sum_{s=0}^{n-1} i^s \frac{A_s(\nu)}{z^s} + \eta_{n,1}(ze^{-2\pi i})\right\}$$
$$+ (1+e^{-2\nu\pi i})\left(\frac{2}{\pi z}\right)^{1/2} e^{-i\zeta}\left\{\sum_{s=0}^{n-1} (-i)^s \frac{A_s(\nu)}{z^s} + \eta_{n,1}(ze^{-\pi i})\right\}, \tag{13.06}$$

valid when $\pi < \operatorname{ph} z < 3\pi$. This is the full form of (4.16).

When $\pi < \operatorname{ph} z < 2\pi$, two different representations, (13.01) and (13.06), are available for $H_\nu^{(1)}(z)$; we noted in §4.3 that they are equivalent in the sense of Poincaré. In the sector $\frac{3}{2}\pi < \operatorname{ph} z < 2\pi$ the bounds (13.02) for the error terms in (13.06) depend on the first two rows of (13.03); the corresponding bound for (13.01) depends on the last row and is therefore larger. In a similar way the error bound for (13.01) is smaller than the combined bound for (13.06) when $\pi < \operatorname{ph} z < \frac{3}{2}\pi$.

We may view this result in the following way. Let $|z|$ have a given large value and n be fixed. As $\operatorname{ph} z$ increases continuously from $\frac{1}{2}\pi$, the right-hand side of (13.01), without the error term, gives a good approximation to $H_\nu^{(1)}(z)$ up to and including $\operatorname{ph} z = \frac{3}{2}\pi$. To achieve comparable numerical accuracy when $\frac{3}{2}\pi < \operatorname{ph} z < 2\pi$ it is necessary to add to this approximation the second series on the right of (13.06), *even though in this region $e^{-i\zeta}$ is exponentially small compared with $e^{i\zeta}$ and therefore negligible in Poincaré's sense.* As the value $\operatorname{ph} z = 2\pi$ is passed, $e^{-i\zeta}$ becomes large compared with $e^{i\zeta}$, causing the roles of the two series in (13.06) to interchange; the inclusion of the second is mandatory, and the first cannot be discarded without some loss of accuracy. Beyond $\operatorname{ph} z = \frac{5}{2}\pi$ the error bound for $\eta_{n,1}(ze^{-\pi i})$ in (13.06) becomes large, and to maintain accuracy a new multiple of the first series (obtainable from (4.13) with $m = 3$) has to be used. And so on. Thus it is possible to compute $H_\nu^{(1)}(z)$ for any value of $\operatorname{ph} z$ by one or two applications of (13.01), with $\operatorname{ph} z$ confined to the numerically acceptable range $[-\frac{1}{2}\pi, \frac{3}{2}\pi]$. Similarly for $H_\nu^{(2)}(z)$.

13.3 Another way of obtaining error bounds for the remainder terms in Hankel's expansions, which is particularly valuable when the variables are real, is to apply the methods of Chapters 3 and 4 to Hankel's integrals, as follows†:

Let us assume that $\nu > -\frac{1}{2}$ and $z > 0$. Then the path in (4.20) may be collapsed onto the two sides of the join of 1 and $1+i\infty$. Taking a new integration variable $\tau = (t-1)/i$ and using the reflection formula for the Gamma function, we arrive at

$$H_\nu^{(1)}(z) = \left(\frac{2}{\pi}\right)^{1/2} \frac{e^{i\zeta} z^\nu}{\Gamma(\nu+\frac{1}{2})} \int_0^\infty e^{-z\tau} \tau^{\nu-(1/2)} (1+\tfrac{1}{2}i\tau)^{\nu-(1/2)}\, d\tau, \tag{13.07}$$

the factors in the integrand having their principal values. For any positive integer n,

† The analysis in this subsection follows that of Watson (1944, §§7.2 and 7.3).

the form of Taylor's theorem obtained by repeated partial integrations shows that

$$(1+\tfrac{1}{2}i\tau)^{\nu-(1/2)} = \sum_{s=0}^{n-1}\binom{\nu-\frac{1}{2}}{s}\left(\frac{1}{2}i\tau\right)^s + \phi_n(\tau), \tag{13.08}$$

where

$$\phi_n(\tau) = \binom{\nu-\frac{1}{2}}{n}\left(\frac{1}{2}i\tau\right)^n n\int_0^1 (1-v)^{n-1}(1+\tfrac{1}{2}iv\tau)^{\nu-n-(1/2)}\,dv.$$

Substitution of the sum on the right-hand side of (13.08) into (13.07) produces the first n terms of (13.01). For the remainder term, assume that $n \geqslant \nu-\frac{1}{2}$. Then $|(1+\frac{1}{2}iv\tau)^{\nu-n-(1/2)}| \leqslant 1$. Hence

$$|\phi_n(\tau)| \leqslant \left|\binom{\nu-\frac{1}{2}}{n}\right|\left(\frac{1}{2}\tau\right)^n.$$

Substitution of this bound into (13.07) leads to the desired result: *If* $\nu > -\frac{1}{2}$ *and* z *is positive, then the* nth *remainder term in the expansion* (13.01) *is bounded in absolute value by the first neglected term, provided that* $n \geqslant \nu-\frac{1}{2}$. Similarly for $H_\nu^{(2)}(z)$.

Ex. 13.1 For positive x, let $\zeta = x-\frac{1}{2}\nu\pi-\frac{1}{4}\pi$ and $H_\nu^{(1)}(x) = \{2/(\pi x)\}^{1/2}e^{i\zeta}\{P(\nu,x)+iQ(\nu,x)\}$, so that

$J_\nu(x) = \{2/(\pi x)\}^{1/2}\{P(\nu,x)\cos\zeta - Q(\nu,x)\sin\zeta\}$, $Y_\nu(x) = \{2/(\pi x)\}^{1/2}\{P(\nu,x)\sin\zeta + Q(\nu,x)\cos\zeta\}$.

Show that if $\nu > -\frac{1}{2}$ and the asymptotic expansions

$$P(\nu,x) \sim \sum_{s=0}^{\infty}(-)^s\frac{A_{2s}(\nu)}{x^{2s}}, \qquad Q(\nu,x) \sim \sum_{s=0}^{\infty}(-)^s\frac{A_{2s+1}(\nu)}{x^{2s+1}} \qquad (x\to\infty),$$

are truncated at their nth terms, then the corresponding remainder term is bounded in absolute value by the first neglected term, provided that $n \geqslant \frac{1}{2}\nu-\frac{1}{4}$ in the case of $P(\nu,x)$, or $n \geqslant \frac{1}{2}\nu-\frac{3}{4}$ in the case of $Q(\nu,x)$.†

Ex. 13.2 Show that the modified Bessel functions are given by

$$K_\nu(z) = \left(\frac{\pi}{2z}\right)^{1/2}e^{-z}\left\{\sum_{s=0}^{n-1}\frac{A_s(\nu)}{z^s}+\gamma_n\right\},$$

$$I_\nu(z) = \frac{e^z}{(2\pi z)^{1/2}}\left\{\sum_{s=0}^{n-1}(-)^s\frac{A_s(\nu)}{z^s}+\delta_n\right\} - ie^{-\nu\pi i}\frac{e^{-z}}{(2\pi z)^{1/2}}\left\{\sum_{s=0}^{n-1}\frac{A_s(\nu)}{z^s}+\gamma_n\right\},$$

where $|\gamma_n|$ is bounded by

$$\begin{array}{ll} 2\exp\{|(\nu^2-\frac{1}{4})z^{-1}|\}\,|A_n(\nu)z^{-n}| & (|\mathrm{ph}\,z|\leqslant\frac{1}{2}\pi), \\ 2\chi(n)\exp\{\frac{1}{2}\pi|(\nu^2-\frac{1}{4})z^{-1}|\}\,|A_n(\nu)z^{-n}| & (\frac{1}{2}\pi\leqslant|\mathrm{ph}\,z|\leqslant\pi), \\ 4\chi(n)\exp\{\pi|(\nu^2-\frac{1}{4})(\mathrm{Re}\,z)^{-1}|\}\,|A_n(\nu)(\mathrm{Re}\,z)^{-n}| & (\pi\leqslant|\mathrm{ph}\,z|<\frac{3}{2}\pi), \end{array}$$

and $|\delta_n|$ is subject to the same bounds, except that the applicable sectors are respectively changed to

$$-\tfrac{3}{2}\pi \leqslant \mathrm{ph}\,z \leqslant -\tfrac{1}{2}\pi, \qquad -\tfrac{1}{2}\pi \leqslant \mathrm{ph}\,z \leqslant 0, \qquad 0 \leqslant \mathrm{ph}\,z < \tfrac{1}{2}\pi.$$ [Olver, 1964a.]

Ex. 13.3 Let γ_n be defined as in the preceding exercise. By means of Exercise 8.4 show that if ν is real, $z > 0$, and $n \geqslant |\nu|-\frac{1}{2}$, then $\gamma_n = \vartheta A_n(\nu)z^{-n}$, where $0 < \vartheta \leqslant 1$.

† With more intricate analysis it can be shown in each case that the remainder has the same sign as the first neglected term, provided that $\nu \geqslant 0$ (Watson, 1944, §7.32).

Ex. 13.4 With the notation of (13.01) prove that

$$\eta_{1,1}(z) = \frac{\frac{1}{4}-\nu^2}{2i}\int_z^{i\infty}\{1-e^{2i(t-z)}\}\{1+\eta_{1,1}(t)\}t^{-2}\,dt.$$

Then by use of Theorem 10.2 of Chapter 6, show that

$$|\eta_{1,1}(z)| \leqslant \exp\{|\nu^2-\tfrac{1}{4}|\mathscr{V}_{z,i\infty}(t^{-1})\}-1,$$

and hence that

$$|H_\nu^{(1)}(z)| \leqslant |2^{1/2}(\pi z)^{-1/2}e^{iz-(i\nu\pi/2)}|\exp\{|\nu^2-\tfrac{1}{4}|\mathscr{V}_{z,i\infty}(t^{-1})\},$$

where $\mathscr{V}_{z,i\infty}(t^{-1})$ is bounded by (13.03), with $n=1$ and $\chi(1)=\frac{1}{2}\pi$.

*14 Inhomogeneous Equations

14.1 Consider the differential equation

$$w''+f(z)w'+g(z)w = z^\alpha e^{\beta z}p(z), \tag{14.01}$$

in which α and β are real or complex constants, and $f(z)$, $g(z)$, and $p(z)$ are analytic functions of the complex variable z having convergent series expansions

$$f(z) = \sum_{s=0}^{\infty}\frac{f_s}{z^s}, \qquad g(z) = \sum_{s=0}^{\infty}\frac{g_s}{z^s}, \qquad p(z) = \sum_{s=0}^{\infty}\frac{p_s}{z^s}, \tag{14.02}$$

in the annulus **A**: $|z|>a$.† The general solution of (14.01) has the form

$$w(z) = Aw_1(z)+Bw_2(z)+W(z),$$

where A and B are arbitrary constants, $w_1(z)$ and $w_2(z)$ are independent solutions of the corresponding homogeneous differential equation, and $W(z)$ is a particular solution of (14.01). Asymptotic expansions for $w_1(z)$ and $w_2(z)$ have been derived earlier in this chapter; in this section we consider the construction of asymptotic approximations for $W(z)$.

We first observe that the substitution $w=e^{\beta z}v$ transforms (14.01) into

$$v''+\{f(z)+2\beta\}v'+\{g(z)+\beta f(z)+\beta^2\}v = z^\alpha p(z).$$

This is an equation of the same form as (14.01) but without an exponential factor in the inhomogeneous term. Accordingly, without loss of generality attention may be confined to the equation

$$w''+f(z)w'+g(z)w = z^\alpha p(z), \tag{14.03}$$

in which $f(z)$, $g(z)$, and $p(z)$ have the expansions (14.02).

Formal series solutions of (14.03) can be found by substituting

$$w = z^\alpha\sum_{s=0}^{\infty}\frac{a_s}{z^s} \tag{14.04}$$

† Actually, the analysis is easily extendible to cases in which the series (14.02) are merely asymptotic as $z\to\infty$ in a prescribed sector.

and equating coefficients. This yields

$$g_0 a_s + \sum_{j=1}^{s} \{g_j + f_{j-1}(\alpha - s + j)\} a_{s-j} + (\alpha - s + 2)(\alpha - s + 1) a_{s-2} = p_s, \tag{14.05}$$

for $s = 0, 1, \ldots$. Provided that $g_0 \neq 0$—and for simplicity in the text we assume this to be the case—equation (14.05) can be satisfied by recurrent determination of the a_s. In particular,

$$a_0 = g_0^{-1} p_0, \qquad a_1 = g_0^{-1} p_1 - g_0^{-2} p_0 (g_1 + \alpha f_0).$$

14.2 The structure of the recurrence relation (14.05) indicates that in general the series (14.04) diverges for all finite values of z; compare §2.1. To investigate the possible asymptotic nature of this expansion we first construct a differential equation for the nth partial sum. Following §12.1, let

$$f(z) = \sum_{s=0}^{n-1} \frac{f_s}{z^s} + \frac{F_n(z)}{z^n}, \qquad g(z) = \sum_{s=0}^{n-1} \frac{g_s}{z^s} + \frac{G_n(z)}{z^n}, \qquad p(z) = \sum_{s=0}^{n-1} \frac{p_s}{z^s} + \frac{P_n(z)}{z^n},$$

for $n = 0, 1, \ldots$, so that each of the functions $F_n(z)$, $G_n(z)$, and $P_n(z)$ is bounded in the closed annulus $\mathbf{B}$: $|z| \geqslant b$ for any b exceeding a. Denote

$$L_n(z) = z^{\alpha} \sum_{s=0}^{n-1} \frac{a_s}{z^s}, \tag{14.06}$$

and restrict $n \geqslant 1$. Then following §12.2 we find that

$$L_n''(z) + f(z) L_n'(z) + g(z) L_n(z) - z^{\alpha} p(z) = z^{\alpha} R_n(z),$$

where

$$R_n(z) = -\frac{g_0 a_n}{z^n} + \frac{\hat{R}_{n+1}(z)}{z^{n+1}},$$

and

$$\hat{R}_{n+1}(z) = (\alpha - n)(\alpha - n + 1) a_{n-1} - P_{n+1}(z) + \sum_{s=0}^{n-1} a_s \{(\alpha - s) F_{n-s}(z) + G_{n+1-s}(z)\}.$$

Accordingly,

$$|R_n(z)| \leqslant \frac{|g_0 a_n|}{|z|^n} + \frac{n r_{n+1}}{|z|^{n+1}}, \tag{14.07}$$

where

$$r_{n+1} = n^{-1} \sup_{z \in \mathbf{B}} |\hat{R}_{n+1}(z)|$$

and is finite.

Now suppose that

$$W_{n-1}(z) = L_n(z) + \varepsilon_n(z) \tag{14.08}$$

is a solution of (14.03). Then the error term satisfies the inhomogeneous equation

$$\varepsilon_n''(z) + f(z) \varepsilon_n'(z) + g(z) \varepsilon_n(z) = -z^{\alpha} R_n(z). \tag{14.09}$$

By variation of parameters we obtain

$$\varepsilon_n(z) = w_2(z)I_n^{(1)}(z) - w_1(z)I_n^{(2)}(z), \tag{14.10}$$

where $w_1(z)$ and $w_2(z)$ are the solutions defined by Theorem 2.1,

$$I_n^{(j)}(z) = \int_z^{\infty e^{-i\theta_j}} \frac{w_j(t)\,t^\alpha R_n(t)}{\mathscr{W}(t)}\,dt \qquad (j=1,2), \tag{14.11}$$

and

$$\mathscr{W}(t) = w_1(t)\,w_2'(t) - w_2(t)\,w_1'(t).$$

The direction θ_j of the upper limit in (14.11) is at our disposal, subject to the condition that the integral converges.

14.3 By hypothesis $f_0^2 \neq 4g_0$ and $g_0 \neq 0$. Hence the characteristic values λ_1 and λ_2 defined in §1.2 are unequal and nonzero. Using Abel's identity (Chapter 5, (1.10)) and inspecting the dominant terms in (2.04) and its differentiated form, we see that there exists a convergent expansion for $\mathscr{W}(t)$ of the form

$$\mathscr{W}(t) = (\lambda_2-\lambda_1)\,e^{(\lambda_1+\lambda_2)t}t^{\mu_1+\mu_2}\sum_{s=0}^{\infty}\frac{\omega_s}{t^s} \qquad (|t|>a),$$

with $\omega_0 = 1$. Therefore

$$\frac{w_1(t)}{\mathscr{W}(t)} \sim \frac{e^{-\lambda_2 t}t^{-\mu_2}}{\lambda_2-\lambda_1} \tag{14.12}$$

as $t\to\infty$ in the sector $\mathbf{S}_1$: $|\mathrm{ph}\{(\lambda_2-\lambda_1)z\}| \leqslant \frac{3}{2}\pi-\delta$, where δ is an arbitrary small positive constant; compare Theorem 2.2. Choosing $\theta_1 = \mathrm{ph}\,\lambda_2$ and restricting

$$-\tfrac{3}{2}\pi + \mathrm{ph}(\lambda_2-\lambda_1) + \delta \leqslant \mathrm{ph}\,\lambda_2 \leqslant \tfrac{3}{2}\pi + \mathrm{ph}(\lambda_2-\lambda_1) - \delta, \tag{14.13}$$

we see that the point $\infty e^{-i\theta_1}$ lies in $\mathbf{S}_1$ and the integral $I_n^{(1)}(z)$ converges. From (14.07) and (14.12) we derive

$$\left|\frac{w_1(t)\,t^\alpha R_n(t)}{\mathscr{W}(t)}\right| \leqslant K_1\left|e^{-\lambda_2 t}t^{\alpha-\mu_2}\right|\left\{\frac{|g_0 a_n|}{|t|^n} + \frac{nr_{n+1}}{|t|^{n+1}}\right\} \qquad (t\in\mathbf{S}_1\cap\mathbf{B}),$$

where K_1 is an assignable constant.

Now let $\mathbf{B}(\delta)$ be the annulus $|z| \geqslant b\csc\delta$, and denote by $\mathbf{T}_1$ the sector

$$-\tfrac{3}{2}\pi + \delta - \min\{\mathrm{ph}\,\lambda_2, \mathrm{ph}(\lambda_2-\lambda_1)\} \leqslant \mathrm{ph}\,z \leqslant \tfrac{3}{2}\pi - \delta - \max\{\mathrm{ph}\,\lambda_2, \mathrm{ph}(\lambda_2-\lambda_1)\},$$

so that $\mathbf{T}_1 \subset \mathbf{S}_1$ and $\infty e^{-i\theta_1} \in \mathbf{T}_1$. If $z \in \mathbf{T}_1 \cap \mathbf{B}(\delta)$, then a path can be found for $I_n^{(1)}(z)$ which lies in $\mathbf{T}_1 \cap \mathbf{B}$ and has the property $|e^{-\lambda_2 t}| \leqslant |e^{-\lambda_2 z}|$; see Fig. 14.1. Assuming that $n > \mathrm{Re}(\alpha-\mu_2)+1$, we deduce that

$$|I_n^{(1)}(z)| \leqslant K_1|e^{-\lambda_2 z}|\left\{\frac{|g_0 a_n|}{|n+\mu_2-\alpha-1|}\mathscr{V}_{z,\infty\exp(-i\,\mathrm{ph}\,\lambda_2)}(t^{-n-\mu_2+\alpha+1}) + \frac{nr_{n+1}}{|n+\mu_2-\alpha|}\mathscr{V}_{z,\infty\exp(-i\,\mathrm{ph}\,\lambda_2)}(t^{-n-\mu_2+\alpha})\right\}.$$

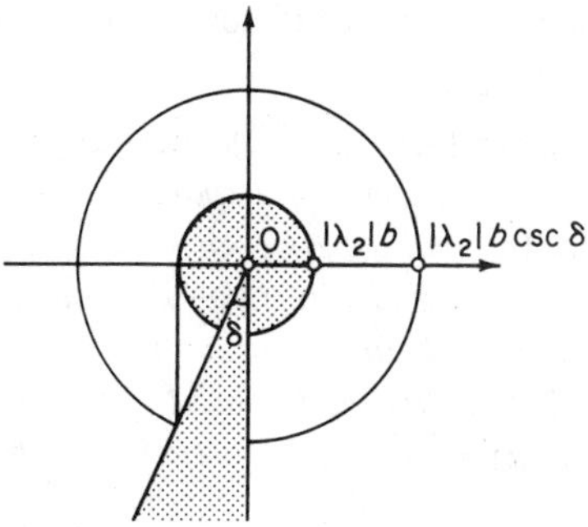

Fig. 14.1 $\lambda_2 t$ plane. $-\frac{1}{2}\pi \leqslant \mathrm{ph}(\lambda_2 t) \leqslant \frac{3}{2}\pi - \delta$.

By further supposing that the paths in the $\lambda_2 t$ plane are the same as those used in the t plane in §13 of Chapter 6—as we may—both variations appearing in the last inequality are $O(z^{-n-\mu_2+\alpha+1})$ as $z \to \infty$. Multiplication by $w_2(z)$ gives

$$w_2(z) I_n^{(1)}(z) = O(z^{\alpha-n+1})$$

as $z \to \infty$ in $\mathbf{S}_2 \cap \mathbf{T}_1$, where $\mathbf{S}_2$ is the sector $|\mathrm{ph}\{(\lambda_1 - \lambda_2)z\}| \leqslant \frac{3}{2}\pi - \delta$; compare (2.15).

Similarly, if $\theta_2 = \mathrm{ph}\,\lambda_1$ and

$$-\tfrac{3}{2}\pi + \mathrm{ph}(\lambda_1 - \lambda_2) + \delta \leqslant \mathrm{ph}\,\lambda_1 \leqslant \tfrac{3}{2}\pi + \mathrm{ph}(\lambda_1 - \lambda_2) - \delta, \tag{14.14}$$

then $I_n^{(2)}(z)$ converges; if also $n > \mathrm{Re}(\alpha - \mu_1) + 1$, then

$$w_1(z) I_n^{(2)}(z) = O(z^{\alpha-n+1})$$

as $z \to \infty$ in $\mathbf{S}_1 \cap \mathbf{T}_2$, where $\mathbf{T}_2$ is defined by

$$-\tfrac{3}{2}\pi + \delta - \min\{\mathrm{ph}\,\lambda_1, \mathrm{ph}(\lambda_1 - \lambda_2)\} \leqslant \mathrm{ph}\,z \leqslant \tfrac{3}{2}\pi - \delta - \max\{\mathrm{ph}\,\lambda_1, \mathrm{ph}(\lambda_1 - \lambda_2)\}.$$

Since $\mathbf{T}_1 \subset \mathbf{S}_1$ and $\mathbf{T}_2 \subset \mathbf{S}_2$, the common region of validity of these estimates is $\mathbf{T} \equiv \mathbf{T}_1 \cap \mathbf{T}_2$. Substitution in (14.10) produces

$$\varepsilon_n(z) = O(z^{\alpha-n+1}) \qquad (z \to \infty \text{ in } \mathbf{T}). \tag{14.15}$$

And on substituting (14.06) and (14.15) in (14.08) and absorbing the term $z^\alpha a_{n-1}/z^{n-1}$ in the error term $O(z^{\alpha-n+1})$, we see that there is a solution $W_{n-1}(z)$ of the given equation (14.03) such that

$$W_{n-1}(z) = z^\alpha\left\{\sum_{s=0}^{n-2} \frac{a_s}{z^s} + O\left(\frac{1}{z^{n-1}}\right)\right\} \qquad (z \to \infty \text{ in } \mathbf{T}). \tag{14.16}$$

The restrictions $n > \mathrm{Re}(\alpha - \mu_1) + 1$ and $n > \mathrm{Re}(\alpha - \mu_2) + 1$ introduced in the proof are unnecessary in this final result, for it is obvious from (14.16) that

$$W_{n-1}(z) = z^\alpha\left\{\sum_{s=0}^{m-2} \frac{a_s}{z^s} + O\left(\frac{1}{z^{m-1}}\right)\right\}$$

for any integer m in the range $1 \leqslant m \leqslant n$.

14.4 Collecting together the foregoing results and replacing n by $n+1$, we have the following:

Theorem 14.1 *Let $f(z)$, $g(z)$, and $p(z)$ be analytic functions of the complex variable z having convergent expansions of the form* (14.02) *for sufficiently large $|z|$, with $f_0^2 \neq 4g_0$ and $g_0 \neq 0$. Also let λ_1 and λ_2 be the zeros of the quadratic $\lambda^2+f_0\lambda+g_0$ with $\operatorname{ph}\lambda_1$, $\operatorname{ph}\lambda_2$, $\operatorname{ph}(\lambda_2-\lambda_1)$, and $\operatorname{ph}(\lambda_1-\lambda_2)$ chosen so that* (14.13) *and* (14.14) *are satisfied, δ being an arbitrary positive number. Then with coefficients a_s determined by* (14.05), *the differential equation* (14.03) *has a solution $W_n(z)$, depending on an arbitrary nonnegative integer n, such that*

$$W_n(z) = z^{\alpha}\left\{\sum_{s=0}^{n-1}\frac{a_s}{z^s} + O\left(\frac{1}{z^n}\right)\right\} \qquad (z\to\infty \text{ in } \mathbf{T}), \tag{14.17}$$

where $\mathbf{T}$ *is the sector*

$$-\tfrac{3}{2}\pi + \delta - \min\{\operatorname{ph}\lambda_1, \operatorname{ph}\lambda_2, \operatorname{ph}(\lambda_2-\lambda_1), \operatorname{ph}(\lambda_1-\lambda_2)\}$$
$$\leqslant \operatorname{ph} z \leqslant \tfrac{3}{2}\pi - \delta - \max\{\operatorname{ph}\lambda_1, \operatorname{ph}\lambda_2, \operatorname{ph}(\lambda_2-\lambda_1), \operatorname{ph}(\lambda_1-\lambda_2)\}. \tag{14.18}$$

In applying this theorem it should be observed that none of the phases of λ_1, λ_2, $\lambda_2-\lambda_1$, and $\lambda_1-\lambda_2$ need have its principal value. Moreover, by using different combinations which satisfy (14.13) and (14.14), we obtain differing sectors of validity $\mathbf{T}$. This does not provide a way of increasing the regions of validity, however: the proof shows that for given n, *distinct* solutions of the differential equation are associated with each region $\mathbf{T}$.

Suppose, for example, that $\lambda_1 > 0$ and $\lambda_2 < 0$. Then we may select $\operatorname{ph}(\lambda_2-\lambda_1) = \pi$ and $\operatorname{ph}(\lambda_1-\lambda_2) = 0$. Conditions (14.13) and (14.14) are satisfied with $\operatorname{ph}\lambda_1 = 0$ and $\operatorname{ph}\lambda_2 = \pi$, and the resulting $\mathbf{T}$ is $-\frac{3}{2}\pi+\delta \leqslant \operatorname{ph} z \leqslant \frac{1}{2}\pi-\delta$. Alternatively, if we select $\operatorname{ph}(\lambda_2-\lambda_1) = -\pi$ and $\operatorname{ph}(\lambda_1-\lambda_2) = 0$, then we would have $\operatorname{ph}\lambda_1 = 0$, $\operatorname{ph}\lambda_2 = -\pi$, and $\mathbf{T}$ becomes $-\frac{1}{2}\pi+\delta \leqslant \operatorname{ph} z \leqslant \frac{3}{2}\pi-\delta$. The solution having the property (14.17) in the phase range $[-\frac{3}{2}\pi+\delta, \frac{1}{2}\pi-\delta]$ is not the same as the solution having this property in $[-\frac{1}{2}\pi+\delta, \frac{3}{2}\pi-\delta]$.

Ex. 14.1 When $g_0 = 0$, show that (14.03) has a formal solution $z^{\alpha+1}\sum b_s z^{-s}$ in general. When may this construction fail?

Ex. 14.2 By transforming to the variable $\zeta = \frac{2}{3}z^{3/2}$ show that the equation $d^2w/dz^2 = zw - z^{-2}$ has solutions $w_j(z)$, $j = 0, \pm 1$, such that

$$w_j(z) \sim \sum_{s=0}^{\infty}\frac{s!\,3^s\cdot 4\cdot 7\cdot 10\cdots(3s+1)}{z^{3s+3}}$$

as $z\to\infty$ in the sector $|\operatorname{ph}(-ze^{2ij\pi/3})| \leqslant \frac{2}{3}\pi-\delta$ $(<\frac{2}{3}\pi)$.

*15 Struve's Equation

15.1 The following inhomogeneous form of Bessel's equation has solutions of physical and mathematical interest:

$$\frac{d^2w}{dz^2} + \frac{1}{z}\frac{dw}{dz} + \left(1 - \frac{\nu^2}{z^2}\right)w = \frac{(\frac{1}{2}z)^{\nu-1}}{\pi^{1/2}\Gamma(\nu+\frac{1}{2})}. \tag{15.01}$$

Using methods analogous to those of Chapter 5, §4, we readily verify that one solution is *Struve's function*

$$\mathbf{H}_\nu(z) = \left(\frac{1}{2}z\right)^{\nu+1} \sum_{s=0}^{\infty} \frac{(-)^s(\frac{1}{4}z^2)^s}{\Gamma(s+\frac{3}{2})\Gamma(\nu+s+\frac{3}{2})}. \tag{15.02}$$

This series converges for all finite z; indeed $z^{-\nu-1}\mathbf{H}_\nu(z)$ is entire in z. It is also readily established, by uniform convergence, that $\mathbf{H}_\nu(z)$ is entire in ν, provided that $z \neq 0$.

Another solution of (15.01) can be constructed by the theory of §14. Here

$$f(z) = \frac{1}{z}, \qquad g(z) = 1 - \frac{\nu^2}{z^2}, \qquad \alpha = \nu - 1, \qquad p(z) = \frac{1}{\pi^{1/2}2^{\nu-1}\Gamma(\nu+\frac{1}{2})}.$$

From (14.05) we derive $a_{2s+1} = 0$, and

$$a_{2s} = 2^{2s-\nu+1}\Gamma(s+\tfrac{1}{2})/\{\pi\Gamma(\nu-s+\tfrac{1}{2})\} \qquad (s = 0, 1, \ldots).$$

The characteristic values are $\lambda_1 = i$ and $\lambda_2 = -i$. With $\operatorname{ph}\lambda_1 = \operatorname{ph}(\lambda_1 - \lambda_2) = \frac{1}{2}\pi$ and $\operatorname{ph}\lambda_2 = \operatorname{ph}(\lambda_2 - \lambda_1) = -\frac{1}{2}\pi$, Conditions (14.13) and (14.14) are satisfied, and Theorem 14.1 shows that for any given positive integer n, there is a solution of (15.01) such that

$$W_{2n}(z) = z^{\nu-1}\left\{\sum_{s=0}^{n-1} \frac{a_{2s}}{z^{2s}} + O\left(\frac{1}{z^{2n}}\right)\right\}$$

as $z \to \infty$ in $|\operatorname{ph} z| \leqslant \pi - \delta\ (<\pi)$.

The solutions $W_{2n}(z)$ are all the same. To see this, write

$$W_{2n}(z) = W_2(z) + A_n H_\nu^{(1)}(z) + B_n H_\nu^{(2)}(z),$$

where A_n and B_n are independent of z. Letting $z \to \infty e^{\pm\pi i/2}$ and referring to Hankel's expansions (4.03) and (4.04), we see that $B_n = A_n = 0$. Thus equation (15.01) has a unique solution $\mathbf{K}_\nu(z)$, say, such that

$$\mathbf{K}_\nu(z) \sim z^{\nu-1} \sum_{s=0}^{\infty} \frac{a_{2s}}{z^{2s}} \qquad (z \to \infty \text{ in } |\operatorname{ph} z| \leqslant \pi - \delta). \tag{15.03}$$

15.2 To connect $\mathbf{H}_\nu(z)$ and $\mathbf{K}_\nu(z)$ we again proceed via an integral representation. Using the Beta-function integral and the duplication formula for the Gamma function, we have

$$\frac{1}{\Gamma(s+\frac{3}{2})\Gamma(\nu+s+\frac{3}{2})} = \frac{2^{2s+1}}{\pi^{1/2}(2s+1)!\,\Gamma(\nu+\frac{1}{2})} \int_0^1 \tau^s(1-\tau)^{\nu-(1/2)}\,d\tau. \tag{15.04}$$

Provided that $\operatorname{Re}\nu > -\frac{1}{2}$, (15.04) may be substituted in (15.02) and the order of integration and summation interchanged.† On taking a new integration variable

† Chapter 2, Theorem 8.1.

$t = \tau^{1/2}$, we arrive at

$$\mathbf{H}_\nu(z) = \frac{2(\frac{1}{2}z)^\nu}{\pi^{1/2}\Gamma(\nu+\frac{1}{2})}\int_0^1 \sin(zt)(1-t^2)^{\nu-(1/2)}\,dt \qquad (\operatorname{Re}\nu > -\tfrac{1}{2}).$$

To continue the analysis we need the asymptotic expansion of the last integral for large positive z. This could be found by the method of stationary phase,† but the need for this theory can be avoided by contour integration, as follows. We have

$$\mathbf{H}_\nu(z) = \frac{(\frac{1}{2}z)^\nu}{i\pi^{1/2}\Gamma(\nu+\frac{1}{2})}\{U_\nu(z) - V_\nu(z)\}, \tag{15.05}$$

where

$$U_\nu(z) = \int_0^1 e^{izt}(1-t^2)^{\nu-(1/2)}\,dt, \qquad V_\nu(z) = \int_0^1 e^{-izt}(1-t^2)^{\nu-(1/2)}\,dt.$$

Since $z > 0$, the integration path for $U_\nu(z)$ may be deformed into $\int_0^{i\infty} - \int_1^{1+i\infty}$. For the former we substitute $t = i\tau$. And because $\operatorname{Re}\nu > -\frac{1}{2}$, the second integral may be evaluated in terms of $H_\nu^{(1)}(z)$ by collapsing the loop path for Hankel's integral, as in §13.3. Thus

$$U_\nu(z) = i\int_0^\infty e^{-z\tau}(1+\tau^2)^{\nu-(1/2)}\,d\tau + \frac{\pi^{1/2}\Gamma(\nu+\frac{1}{2})}{2(\frac{1}{2}z)^\nu}H_\nu^{(1)}(z).$$

Similarly,

$$V_\nu(z) = -i\int_0^\infty e^{-z\tau}(1+\tau^2)^{\nu-(1/2)}\,d\tau + \frac{\pi^{1/2}\Gamma(\nu+\frac{1}{2})}{2(\frac{1}{2}z)^\nu}H_\nu^{(2)}(z).$$

Substitution in (15.05) and use of (5.01) yields

$$\mathbf{H}_\nu(z) - Y_\nu(z) = \frac{2(\frac{1}{2}z)^\nu}{\pi^{1/2}\Gamma(\nu+\frac{1}{2})}\int_0^\infty e^{-z\tau}(1+\tau^2)^{\nu-(1/2)}\,d\tau. \tag{15.06}$$

The restriction $\operatorname{Re}\nu > -\frac{1}{2}$ may now be removed by analytic continuation.

When Watson's lemma is applied to (15.06), the resulting asymptotic expansion is found to be identical with (15.03). In §15.1 we saw that the solution of (15.01) having this expansion is unique, hence

$$\mathbf{K}_\nu(z) = \mathbf{H}_\nu(z) - Y_\nu(z).$$

Again, analytic continuation extends this result from positive z to complex z, as long as branches are chosen in a continuous manner. This is the required connection formula. We have shown, incidentally, that the right-hand side of (15.06) furnishes an integral representation of $\mathbf{K}_\nu(z)$ when $|\operatorname{ph} z| < \frac{1}{2}\pi$.

15.3 The general solution of (15.01) may be expressed

$$w = \mathbf{H}_\nu(z) + AJ_\nu(z) + BY_\nu(z), \tag{15.07}$$

where A and B are arbitrary constants. Comparison of the power-series expansions

† Erdélyi (1955), Olver (1974).

of $\mathbf{H}_\nu(z)$, $J_\nu(z)$, and $Y_\nu(z)$ shows that this form of representation is numerically satisfactory for small or moderate values of $|z|$. But, with the possible exception of the real axis, (15.07) is unsatisfactory for large $|z|$ because all three functions $\mathbf{H}_\nu(z)$, $J_\nu(z)$, and $Y_\nu(z)$ have dominant asymptotic behavior.

For large z in $|\mathrm{ph}\, z| \leqslant \frac{1}{2}\pi$, a numerically satisfactory representation of the general solution is furnished by

$$w = \mathbf{K}_\nu(z) + AH_\nu^{(1)}(z) + BH_\nu^{(2)}(z),$$

where, again, A and B are arbitrary constants. In the upper part of this sector $H_\nu^{(1)}(z)$ is recessive, $H_\nu^{(2)}(z)$ is dominant, and $\mathbf{K}_\nu(z)$ has intermediate behavior. In the lower part the roles of $H_\nu^{(1)}(z)$ and $H_\nu^{(2)}(z)$ are interchanged.

In a similar way, the appropriate representation for large z in the sector $\frac{1}{2}\pi \leqslant \mathrm{ph}\, z \leqslant \frac{3}{2}\pi$ is given by

$$w = -e^{\nu\pi i}\mathbf{K}_\nu(ze^{-\pi i}) + AH_\nu^{(1)}(ze^{-\pi i}) + BH_\nu^{(2)}(ze^{-\pi i}),$$

it being easily verified, by transformation of variable, that the first term on the right-hand side *is* a solution of (15.01).†

Ex. 15.1 Prove that

$$\mathbf{H}_{\nu-1}(z) + \mathbf{H}_{\nu+1}(z) = \frac{2\nu}{z}\mathbf{H}_\nu(z) + \frac{(\frac{1}{2}z)^\nu}{\pi^{1/2}\Gamma(\nu+\frac{3}{2})}, \qquad \frac{d}{dz}\{z^\nu \mathbf{H}_\nu(z)\} = z^\nu \mathbf{H}_{\nu-1}(z),$$

$$\mathbf{H}_{\nu-1}(z) - \mathbf{H}_{\nu+1}(z) = 2\mathbf{H}_\nu'(z) - \frac{(\frac{1}{2}z)^\nu}{\pi^{1/2}\Gamma(\nu+\frac{3}{2})}, \qquad \frac{d}{dz}\{z^{-\nu}\mathbf{H}_\nu(z)\} = \frac{1}{\pi^{1/2}2^\nu\Gamma(\nu+\frac{3}{2})} - z^{-\nu}\mathbf{H}_{\nu+1}(z).$$

Ex. 15.2 If $\nu \neq -\frac{1}{2}$ and $\mathscr{C}_\nu(x)$ denotes a cylinder function, verify that

$$\int x^\nu \mathscr{C}_\nu(x)\, dx = \pi^{1/2}2^{\nu-1}\Gamma(\nu+\tfrac{1}{2})\, x\{\mathscr{C}_\nu(x)\mathbf{H}_\nu'(x) - \mathscr{C}_\nu'(x)\mathbf{H}_\nu(x)\}.$$

Ex. 15.3 By use of (15.06) show that if n is a nonnegative integer, $\mathbf{H}_{-n-(1/2)}(z) = (-)^n J_{n+(1/2)}(z)$. Show also that

$$\mathbf{H}_{1/2}(z) = 2^{1/2}(1-\cos z)/(\pi z)^{1/2}.$$

Ex. 15.4 Prove that $\mathbf{K}_\nu(ze^{-\pi i}) = 2i\cos(\nu\pi)H_\nu^{(1)}(z) - e^{-\nu\pi i}\mathbf{K}_\nu(z)$.

Ex. 15.5 By use of (15.06) show that if ν is real, z is positive, and $n \geqslant \nu - \frac{1}{2}$, then the nth remainder term in (15.03) is bounded in absolute value by the first neglected term and has the same sign.

Historical Notes and Additional References

The material on Bessel functions, confluent hypergeometric functions, and Struve's function, is classical, with greater emphasis than usual on derivation of properties directly from the defining differential equations. Principal reference sources include the books of Watson (1944), B.M.P. (1953a, b), Slater (1960), and N.B.S. (1964). The asymptotic theory and error analysis of irregular singularities is based on that of Olver (1964a, 1965d). Theorems 2.1 and 2.2 are due to Horn (1903); the present proofs are new. Theorems 3.1 and 14.1 appear to be new; a result related to the former has been given by Hsieh and Sibuya (1966).

† Or in Theorem 14.1, if we take $\mathrm{ph}\,\lambda_1 = \mathrm{ph}(\lambda_1-\lambda_2) = -\frac{3}{2}\pi$ and $\mathrm{ph}\,\lambda_2 = \mathrm{ph}(\lambda_2-\lambda_1) = -\frac{1}{2}\pi$, then the resulting solution is $-e^{\nu\pi i}\mathbf{K}_\nu(ze^{-\pi i})$.

§§1–2 The history of these series solutions has been sketched by Erdélyi (1956a, Chapter 3).

§§4–8 The major work on Bessel functions is still the treatise of Watson (1944). For a few further properties concerning zeros see R.S. (1960), and for extensive collections of definite and indefinite integrals see Luke (1962) and Oberhettinger (1972).

§5.1 The need to select numerically satisfactory pairs of solutions of Bessel's equation (and also Airy's equation) in the complex plane has not always been observed by table makers.

§6.5 Some error bounds for McMahon's expansion have been given by Hethcote (1970a, b).

§§9–11 Monographs on the confluent hypergeometric functions include those of Buchholz (1969), Tricomi (1954), and Slater (1960).

§11.2 These powerful approximations for Whittaker functions appear to be new, as do the related results in Chapter 11, §4.3. Other asymptotic approximations when m is large have been given by Kazarinoff (1955, 1957a). See also Chapter 10, Exercise 3.4.

§12 The extension of this error analysis to second-order equations having an irregular singularity of arbitrary finite rank is straightforward; details have been supplied by Olver and Stenger (1965). Much more difficult is the error analysis of a system of an arbitrary number of first-order differential equations having an irregular singularity of arbitrary rank; this has been treated by Stenger (1966a, b). See also the book by Wasow (1965, Chapters 4 and 5).

§15 (i) The notation $\mathbf{K}_\nu(z)$ is new; it has been introduced to emphasize the need to use numerically satisfactory solutions of Struve's equation.

(ii) Real zeros of $\mathbf{H}_\nu(z)$ have been studied by Steinig (1970).

8

SUMS AND SEQUENCES

1 The Euler–Maclaurin Formula and Bernoulli's Polynomials

1.1 If a and n are integers such that $a < n$, and $f(x)$ is a slowly varying function of x, then the sum

$$S = \tfrac{1}{2}f(a) + f(a+1) + f(a+2) + f(a+3) + \cdots + f(n-1) + \tfrac{1}{2}f(n)$$

is approximated by the integral

$$I = \int_a^n f(x)\,dx.$$

This observation forms the basis of a powerful method for determining the asymptotic behavior of S for large values of n.

The difference $S-I$ may be expressed in various ways. In the formula about to be derived the derivatives of $f(x)$ at a and n are employed. If j is an integer, then by partial integration

$$\int_j^{j+1} f(x)\,dx = [(x-j-\tfrac{1}{2})f(x)]_j^{j+1} - \int_j^{j+1} (x-j-\tfrac{1}{2})f'(x)\,dx.$$

Hence

$$\tfrac{1}{2}f(j) + \tfrac{1}{2}f(j+1) = \int_j^{j+1} f(x)\,dx + \int_j^{j+1} \varpi_1(x)f'(x)\,dx,$$

where†

$$\varpi_1(x) = x - [x] - \tfrac{1}{2}.$$

Summation from $j = a$ to $j = n-1$ produces

$$S = I + \int_a^n \varpi_1(x)f'(x)\,dx.$$

The function $\varpi_1(x)$ is sometimes called the *saw-tooth* function. It is periodic, with period 1, and has discontinuities at $x = 0, \pm 1, \pm 2, \ldots$; see Fig. 1.1.

† When x is negative, $[x]$ is to be interpreted as the integer in the interval $(x-1, x]$.

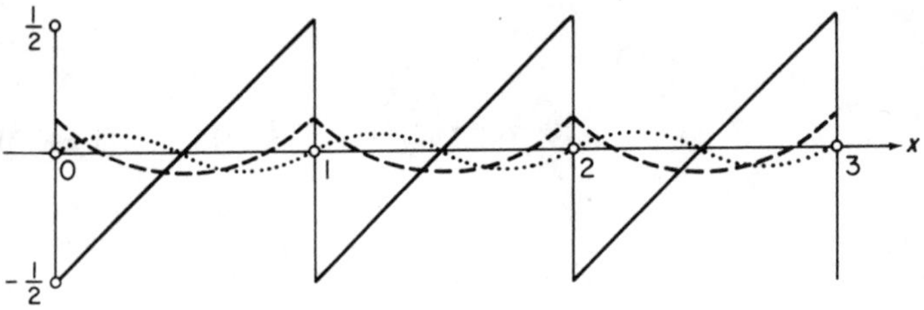

Fig. 1.1 ——— $\varpi_1(x)$; - - - - - $\varpi_2(x)$; ········ $\varpi_3(x)$. (Not to scale.)

To expand the integral of $\varpi_1(x)f'(x)$ we again integrate by parts:

$$\int_j^{j+1} \varpi_1(x)f'(x)\,dx = \varpi_2(j+1)f'(j+1) - \varpi_2(j)f'(j) - \int_j^{j+1} \varpi_2(x)f''(x)\,dx,$$

where

$$\varpi_2(x) = \int \varpi_1(x)\,dx, \tag{1.01}$$

and is continuous. Since

$$\int_j^{j+1} \varpi_1(x)\,dx = \int_0^1 (x-\tfrac{1}{2})\,dx = 0,$$

it follows that $\varpi_2(x)$ is periodic. Thus $\varpi_2(j+1) = \varpi_2(j) = \varpi_2(0)$. To enable the final expansion to be summed, we choose the arbitrary constant of integration in (1.01) in such a way that $\int_0^1 \varpi_2(x)\,dx = 0$. A further partial integration then gives

$$\int_j^{j+1} \varpi_1(x)f'(x)\,dx = \varpi_2(0)\{f'(j+1) - f'(j)\}$$
$$- \varpi_3(0)\{f''(j+1) - f''(j)\} + \int_j^{j+1} \varpi_3(x)f'''(x)\,dx,$$

where $\varpi_3(x) = \int \varpi_2(x)\,dx$.

Continuing the process we arrive at

$$\tfrac{1}{2}f(j) + \tfrac{1}{2}f(j+1) = \int_j^{j+1} f(x)\,dx + \sum_{s=1}^{m-1} (-)^{s+1}\varpi_{s+1}(0)\{f^{(s)}(j+1) - f^{(s)}(j)\}$$
$$+ (-)^{m+1}\int_j^{j+1} \varpi_m(x)f^{(m)}(x)\,dx,$$

where m is an arbitrary positive integer,

$$\varpi_s(x) = \int \varpi_{s-1}(x)\,dx, \tag{1.02}$$

and

$$\int_0^1 \varpi_s(x)\,dx = 0. \tag{1.03}$$

Summation gives the required result in the form

$$S = I + \sum_{s=1}^{m-1} (-)^{s+1} \varpi_{s+1}(0) \{f^{(s)}(n) - f^{(s)}(a)\} + (-)^{m+1} \int_a^n \varpi_m(x) f^{(m)}(x)\, dx. \tag{1.04}$$

1.2 The expansion (1.04) is known as the *Euler–Maclaurin formula*. To apply it we need properties of the periodic functions $\varpi_s(x)$. From the method of construction it is evident that $\varpi_s(x)$ can be represented as a polynomial of degree s in the interval $[0, 1)$. A generating function can be found as follows. Suppose that

$$\sum_{s=0}^{\infty} \varpi_s(x) t^s = G(x, t) \qquad (0 \leqslant x < 1),$$

where $\varpi_0(x)$ is defined to be 1. Since $\varpi_s'(x) = \varpi_{s-1}(x)$, we expect that

$$\partial G(x, t)/\partial x = tG(x, t),$$

and therefore $G(x, t)$ to have the form $g(t)e^{tx}$. The other necessary condition is given by (1.03) with $s \geqslant 1$. This implies that

$$\int_0^1 G(x, t)\, dx = 1,$$

and hence $g(t)(e^t - 1)/t = 1$. Thus we have the tentative result

$$\frac{te^{xt}}{e^t - 1} = \sum_{s=0}^{\infty} \varpi_s(x) t^s \qquad (0 \leqslant x < 1). \tag{1.05}$$

Now as a function of the complex variable t, the singularities of $te^{xt}/(e^t - 1)$ are located at $t = \pm 2\pi i, \pm 4\pi i, \ldots$. Therefore the series (1.05) converges at each point within the circle $|t| = 2\pi$; moreover with the aid of the M-test it can be seen that convergence is uniform with respect to x in the unit disk. Accordingly, the steps used in constructing the generating function are legitimate, and the coefficients in the expansion (1.05) satisfy (1.02) and (1.03).

1.3 It is customary to use a normalization slightly different from that given above. The *Bernoulli polynomials* $B_s(x)$ are defined by the expansion

$$\frac{te^{xt}}{e^t - 1} = \sum_{s=0}^{\infty} B_s(x) \frac{t^s}{s!} \qquad (|t| < 2\pi), \tag{1.06}$$

for all values of x. Therefore

$$s!\, \varpi_s(x) = B_s(x - [x]), \tag{1.07}$$

and from (1.02) and (1.03) we have the fundamental properties

$$B_s'(x) = sB_{s-1}(x), \qquad \int_0^1 B_s(x)\, dx = 0 \qquad (s \geqslant 1).$$

Further properties are derivable directly from the definition (1.06).

Of particular interest are the quantities $B_s(0)$; these are called the *Bernoulli numbers* and are denoted simply by B_s.† Thus

$$\frac{t}{e^t-1} = \sum_{s=0}^{\infty} B_s \frac{t^s}{s!}. \tag{1.08}$$

With the exception of B_1, all Bernoulli numbers of odd suffix vanish. This follows from the observation that

$$\frac{t}{e^t-1} + \frac{1}{2}t$$

is an even function of t.

Next, equating coefficients in the identity

$$e^{xt}\sum_{s=0}^{\infty} B_s \frac{t^s}{s!} = \sum_{s=0}^{\infty} B_s(x)\frac{t^s}{s!},$$

we see that

$$B_s(x) = \sum_{j=0}^{s} \binom{s}{j} B_{s-j}\, x^j. \tag{1.09}$$

In consequence, explicit expressions for individual Bernoulli polynomials can be constructed readily from the Bernoulli numbers.

The generating function (1.06) is unchanged when x and t are simultaneously replaced by $1-x$ and $-t$, respectively. Therefore

$$B_s(1-x) = (-)^s B_s(x). \tag{1.10}$$

Since $B_s = 0$ for $s = 3, 5, 7, \ldots$, it follows that

$$B_s(1) = B_s \qquad (s \geqslant 2).$$

Setting $x = 1$ in (1.09) we obtain the identity

$$B_{s-1} = -\frac{1}{s}\sum_{j=0}^{s-2} \binom{s}{j} B_j \qquad (s \geqslant 2), \tag{1.11}$$

which can be used to calculate the sequence $\{B_s\}$. The first few nonvanishing members are‡

$$B_0 = 1,\quad B_1 = -\tfrac{1}{2},\quad B_2 = \tfrac{1}{6},\quad B_4 = -\tfrac{1}{30},\quad B_6 = \tfrac{1}{42},\quad B_8 = -\tfrac{1}{30},\quad B_{10} = \tfrac{5}{66}.$$

From (1.09) the first seven Bernoulli polynomials are found to be

$$B_0(x) = 1, \qquad B_1(x) = x - \tfrac{1}{2}, \qquad B_2(x) = x^2 - x + \tfrac{1}{6},$$

$$B_3(x) = x^3 - \tfrac{3}{2}x^2 + \tfrac{1}{2}x = x(x-\tfrac{1}{2})(x-1), \qquad B_4(x) = x^4 - 2x^3 + x^2 - \tfrac{1}{30},$$

$$B_5(x) = x^5 - \tfrac{5}{2}x^4 + \tfrac{5}{3}x^3 - \tfrac{1}{6}x = x(x-\tfrac{1}{2})(x-1)(x^2-x-\tfrac{1}{3}),$$

$$B_6(x) = x^6 - 3x^5 + \tfrac{5}{2}x^4 - \tfrac{1}{2}x^2 + \tfrac{1}{42}.$$

† This notation, due to Nörlund (1924), is one of two in common use. In the other, and older, notation the present B_{2s} is denoted by $(-)^{s-1}B_s$.

‡ Listings up to B_{60} can be found, for example, in N.B.S. (1964, p. 810).

1.4 Theorem 1.1 *In the interval* $[0, 1]$

(i) *The only zeros of* $B_{2s}(x) - B_{2s}$, $s \geqslant 1$, *are* 0 *and* 1.
(ii) *The only zeros of* $B_{2s+1}(x)$, $s \geqslant 1$, *are* 0, $\frac{1}{2}$, *and* 1.
(iii) $|B_{2s}(x)| \leqslant |B_{2s}|$ *and* $|B_{2s}(x) - B_{2s}| \leqslant (2 - 2^{1-2s})|B_{2s}|$.

The results (i) and (ii) may be proved by induction. Suppose that the stated properties hold for a specified value of s—which is easily seen to be the case when $s = 1$. Equation (1.10) shows that $B_{2s+2}(x) - B_{2s+2}$ vanishes at 1, as well as at 0. Since the derivative of this function is $(2s+2)B_{2s+1}(x)$, its only stationary point within $(0, 1)$ is $x = \frac{1}{2}$. Therefore by Rolle's theorem $B_{2s+2}(x) - B_{2s+2}$ has no other zeros in $[0, 1]$.

Next, similar reasoning shows that $B_{2s+2}(x)$ has at most two distinct zeros in $(0, 1)$. Since $B_{2s+3}(x)$ has zeros at 0 and 1 and its derivative is $(2s+3)B_{2s+2}(x)$, it has at most one other zero in $[0, 1]$. Equation (1.10) shows that this zero is located at $x = \frac{1}{2}$. This establishes Parts (i) and (ii).

To prove (iii) we evaluate $B_{2s}(x)$ at its stationary point $x = \frac{1}{2}$. Setting $x = \frac{1}{2}$ in (1.06) we obtain

$$\frac{te^{t/2}}{e^t - 1} = \sum_{s=0}^{\infty} B_s\left(\frac{1}{2}\right)\frac{t^s}{s!}.$$

Addition to (1.08) produces

$$\frac{t}{e^{t/2} - 1} = \sum_{s=0}^{\infty} \left\{B_s\left(\frac{1}{2}\right) + B_s\right\}\frac{t^s}{s!}.$$

The function on the left can also be expanded by replacing t by $\frac{1}{2}t$ in (1.08). Comparison of the two series gives $B_s(\frac{1}{2}) + B_s = 2^{1-s}B_s$; whence

$$B_s(\tfrac{1}{2}) = -\left(1 - \frac{1}{2^{s-1}}\right)B_s. \tag{1.12}$$

Part (iii) now follows.

1.5 The final result we establish in this section is the expansion

$$\zeta(2s) \equiv \sum_{j=1}^{\infty} \frac{1}{j^{2s}} = (-)^{s-1}\frac{(2\pi)^{2s}B_{2s}}{2(2s)!} \qquad (s \geqslant 1). \tag{1.13}$$

To prove this, replace t by $2\pi it$ in (1.08). Then

$$\sum_{s=0}^{\infty} (-)^s \frac{(4\pi^2 t^2)^s}{(2s)!} B_{2s} = \frac{2\pi it}{e^{2\pi it} - 1} + \pi it = \pi t \cot(\pi t).$$

We also have the partial fraction expansion†

$$\pi t \cot(\pi t) = 1 + \sum_{j=1}^{\infty} \frac{2t^2}{t^2 - j^2} \qquad (t \text{ noninteger}).$$

† Knopp (1951, p. 207).

The desired result is obtained by expanding $2t^2/(t^2-j^2)$ in powers of t and comparing coefficients.

Equation (1.13) shows that the sign of B_{2s} is $(-)^{s-1}$ when $s \geqslant 1$. The expansion can also be used for the approximate computation of B_{2s} when s is large.

Ex. 1.1 If h is an arbitrary number in $[0,1]$, prove that

$$\int_j^{j+h} f(x)\,dx = \tfrac{1}{2}f(j+h) + (h-\tfrac{1}{2})f(j) - \int_j^{j+h} (x-j-h+\tfrac{1}{2})f'(x)\,dx,$$

$$\int_{j+h}^{j+1} f(x)\,dx = \tfrac{1}{2}f(j+h) - (h-\tfrac{1}{2})f(j+1) - \int_{j+h}^{j+1} (x-j-h-\tfrac{1}{2})f'(x)\,dx,$$

and

$$\int_j^{j+1} f(x)\,dx = f(j+h) - (h-\tfrac{1}{2})\{f(j+1)-f(j)\} - \int_j^{j+1} \varpi_1(x-h)f'(x)\,dx.$$

Thence establish the following generalization of (1.04)

$$\sum_{j=a}^{n-1} f(j+h) = \int_a^n f(x)\,dx + (h-\tfrac{1}{2})\{f(n)-f(a)\}$$

$$+\sum_{s=1}^{m-1} \varpi_{s+1}(h)\{f^{(s)}(n)-f^{(s)}(a)\} + (-)^{m+1}\int_a^n \varpi_m(x-h)f^{(m)}(x)\,dx.$$

Ex. 1.2 Show that

$$B_s(x+y) = \sum_{j=0}^{s} \binom{s}{j} B_j(x)\,y^{s-j}.$$

Ex. 1.3 Show that

$$\sum_{j=1}^{n} j^s = \frac{1}{s+1}\{B_{s+1}(n+1) - B_{s+1}\} \qquad (s = 1,2,3,\ldots).$$

Ex. 1.4 Prove the *multiplication theorem*

$$B_s(mx) = m^{s-1}\sum_{j=0}^{m-1} B_s\left(x+\frac{j}{m}\right),$$

in which m denotes an arbitrary positive integer.

Ex. 1.5 Let $\mathscr{S}_N$ be the square in the t plane with corners at $(\pm 1 \pm i)(2N+1)\pi$, N being a positive integer. By integrating t^{-s} times the left-hand side of (1.06) around $\mathscr{S}_N$ and letting $N\to\infty$, establish the Fourier expansions

$$B_{2s}(x) = \frac{2(-)^{s+1}(2s)!}{(2\pi)^{2s}}\sum_{j=1}^{\infty}\frac{\cos(2\pi jx)}{j^{2s}}, \qquad B_{2s+1}(x) = \frac{2(-)^{s+1}(2s+1)!}{(2\pi)^{2s+1}}\sum_{j=1}^{\infty}\frac{\sin(2\pi jx)}{j^{2s+1}},$$

when $s \geqslant 1$ and $0 \leqslant x \leqslant 1$. Show also that the second expansion holds when $s = 0$ and $0 < x < 1$.

2 Applications

2.1 The standardized form of the Euler–Maclaurin formula (1.04) is obtained on expressing the quantities $\varpi_{s+1}(0)$ and $\varpi_m(x)$ in terms of the Bernoulli numbers and polynomials by means of (1.07). Changing m to $2m$ and recalling that Bernoulli numbers of odd suffix exceeding unity are zero, we derive

$$\sum_{j=a}^{n} f(j) = \int_a^n f(x)\,dx + \frac{1}{2}f(a) + \frac{1}{2}f(n)$$
$$+ \sum_{s=1}^{m-1} \frac{B_{2s}}{(2s)!}\{f^{(2s-1)}(n) - f^{(2s-1)}(a)\} + R_m(n), \tag{2.01}$$

where a, m, and n are arbitrary integers such that $a < n$ and $m > 0$, and

$$R_m(n) = \frac{B_{2m}}{(2m)!}\{f^{(2m-1)}(n) - f^{(2m-1)}(a)\} - \int_a^n \frac{B_{2m}(x-[x])}{(2m)!} f^{(2m)}(x)\,dx. \tag{2.02}$$

The construction given in §1.1 shows that *a sufficient condition for the validity of this formula is that* $f^{(2m)}(x)$ *be absolutely integrable over* (a, n). Leading coefficients are given by

$$\frac{B_2}{2!} = \frac{1}{12}, \qquad \frac{B_4}{4!} = -\frac{1}{720}, \qquad \frac{B_6}{6!} = \frac{1}{30{,}240}, \qquad \frac{B_8}{8!} = -\frac{1}{1{,}209{,}600}.$$

For many purposes it is more convenient to express the remainder in the form

$$R_m(n) = \int_a^n \frac{B_{2m} - B_{2m}(x-[x])}{(2m)!} f^{(2m)}(x)\,dx. \tag{2.03}$$

From Theorem 1.1(iii) it follows that

$$|R_m(n)| \leqslant (2 - 2^{1-2m}) \frac{|B_{2m}|}{(2m)!} \mathscr{V}_{a,n}\{f^{(2m-1)}(x)\}. \tag{2.04}$$

In particular, if $f^{(2m)}(x)$ does not change sign in (a, n), then $R_m(n)$ is bounded in absolute value by $2 - 2^{1-2m}$ times the first neglected term in (2.01). It also has the same sign, because $|B_{2m}(x-[x])|$ is bounded by $|B_{2m}|$.

Next, except for integer x, the sign of $B_{2m} - B_{2m}(x-[x])$ is the same as the sign of B_{2m}, that is, $(-)^{m-1}$; compare (1.13). Referring to the error test of Chapter 3, §2.2, we obtain the sharper result that if $f^{(2m)}(x)$ and $f^{(2m+2)}(x)$ have the same constant sign in (a, n), then $R_m(n)$ is bounded in absolute value by the first neglected term in (2.01) (and has the same sign).

2.2 An application is provided by the asymptotic behavior of the sum

$$S(n) = \sum_{j=1}^{n} j \ln j \tag{2.05}$$

for large n. Here

$$f(x) = x \ln x, \qquad f'(x) = \ln x + 1, \qquad f^{(s)}(x) = (-)^s \frac{(s-2)!}{x^{s-1}} \qquad (s \geqslant 2). \tag{2.06}$$

Since

$$\int x \ln x \, dx = \tfrac{1}{2}x^2 \ln x - \tfrac{1}{4}x^2,$$

use of (2.01) and (2.02) with $a = 1$ and $m = 2$ yields

$$S(n) = \tfrac{1}{2}n^2 \ln n - \tfrac{1}{4}n^2 + \tfrac{1}{4} + \tfrac{1}{2}n \ln n + \tfrac{1}{12} \ln n + R_2(n), \tag{2.07}$$

where

$$R_2(n) = -\frac{1}{720}\left(1 - \frac{1}{n^2}\right) - \frac{1}{12}\int_1^n \frac{B_4(x-[x])}{x^3}\, dx.$$

Noting that $f^{(4)}(x)$ and $f^{(6)}(x)$ are positive throughout $[1, n]$, we see from the final result of §2.1 that $R_2(n)$ is negative and bounded by

$$|R_2(n)| \leqslant \frac{1}{720}\left(1 - \frac{1}{n^2}\right).$$

Smaller remainder terms can be found by taking higher values of m, for example,

$$|R_3(n)| \leqslant \frac{1}{5040}\left(1 - \frac{1}{n^4}\right).$$

However, for all m the asymptotic order of $R_m(n)$ as $n \to \infty$ is $O(1)$. To obtain an asymptotic expansion for $S(n)$ in descending powers of n, we express

$$R_2(n) = -\frac{1}{720}\left(1 - \frac{1}{n^2}\right) - \frac{1}{12}\int_1^\infty \frac{B_4(x-[x])}{x^3}\, dx + \frac{1}{12}\int_n^\infty \frac{B_4(x-[x])}{x^3}\, dx,$$

and integrate the last term repeatedly by parts, as in the construction of the Euler–Maclaurin formula. Substitution of the result in (2.07) produces

$$S(n) = \tfrac{1}{2}n^2 \ln n - \tfrac{1}{4}n^2 + \tfrac{1}{2}n \ln n + \tfrac{1}{12} \ln n + C - \sum_{s=2}^{m-1} \frac{B_{2s}}{2s(2s-1)(2s-2)n^{2s-2}} - \hat{R}_m(n), \tag{2.08}$$

where m is an arbitrary integer exceeding unity,

$$C = \frac{1}{4} - \frac{1}{720} - \frac{1}{12}\int_1^\infty \frac{B_4(x-[x])}{x^3}\, dx,$$

and

$$\hat{R}_m(n) = \int_n^\infty \frac{B_{2m} - B_{2m}(x-[x])}{2m(2m-1)x^{2m-1}}\, dx.$$

Again, the error test shows that

$$|\hat{R}_m(n)| \leqslant \frac{|B_{2m}|}{2m(2m-1)(2m-2)n^{2m-2}} \qquad (m \geqslant 2). \tag{2.09}$$

Thus (2.08) furnishes the desired expansion.

The remaining problem is to evaluate the constant C. An analytical method is given in §3.3 below. A numerical procedure is as follows. Suppose, for example, we set $m = 4$ and $n = 5$. Then (2.09) shows that

$$|\hat{R}_4(5)| \leqslant \frac{\frac{1}{30}}{8\cdot 7\cdot 6}\,\frac{1}{5^6} \doteqdot 0.6 \times 10^{-8}.$$

Direct numerical summation of (2.05) yields $S(5) = 18.27449\,823$. On subtracting the values of the known terms on the right-hand side of (2.08), we obtain

$$C = 0.24875\,449,$$

correct to eight decimal places. Taking these values of m and C in (2.08) and neglecting $\hat{R}_4(n)$, we have a representation of $S(n)$ for $n > 5$ which is correct to eight decimal places, that is, at least ten significant figures.

Greater accuracy in C is attainable by taking higher values of m and n. For example, twelve decimals are obtainable with $m = 5$ and $n = 10$; this is easily verified by evaluation of (2.09).

2.3 The method just employed to extend an asymptotic approximation yielded by the Euler–Maclaurin formula into an asymptotic expansion is quite typical. Suppose that successive derivatives of $f(x)$ form an asymptotic scale as $x \to \infty$, and $f^{(2M)}(x)$ is the lowest derivative of even order which is absolutely integrable over (a, ∞). Then we apply (2.01) with $m = M$, and rearrange the remainder term as

$$R_M(n) = \frac{B_{2M}}{(2M)!}\{f^{(2M-1)}(n) - f^{(2M-1)}(a)\} - \int_a^\infty \frac{B_{2M}(x-[x])}{(2M)!} f^{(2M)}(x)\,dx$$
$$+ \int_n^\infty \frac{B_{2M}(x-[x])}{(2M)!} f^{(2M)}(x)\,dx.$$

By integrating the last term repeatedly by parts, we obtain the required expansion in the form

$$\sum_{j=a}^{n} f(j) = \int_a^n f(x)\,dx + C + \frac{1}{2}f(n) + \sum_{s=1}^{m-1} \frac{B_{2s}}{(2s)!} f^{(2s-1)}(n)$$
$$- \int_n^\infty \frac{B_{2m} - B_{2m}(x-[x])}{(2m)!} f^{(2m)}(x)\,dx, \tag{2.10}$$

where m is an arbitrary integer not less than M, and C is independent of n.

The value of C may be computed from (2.10) to any desired accuracy by appropriate choice of m and n. When the infinite series $\sum f(j)$ converges, this computation is equivalent to finding the difference between the sum and $\int_a^\infty f(x)\,dx$, and is a commonly used process in numerical analysis.

2.4 As a second example consider

$$S(n) = \sum_{j=0}^{n-1} \left(1 - \frac{j^2}{n^2}\right)^{1/2}$$

for large n. A new feature is that individual terms are functions of n. Replacement of j by $n-j$ leads to the more convenient form

$$S(n) = \left(\frac{2}{n}\right)^{1/2} \sum_{j=1}^{n} f_n(j), \tag{2.11}$$

in which

$$f_n(x) = \left\{x\left(1 - \frac{x}{2n}\right)\right\}^{1/2}.$$

From (2.01) with $a = 1$ and $m = 1$, we derive

$$\sum_{j=1}^{n} f_n(j) = \int_1^n \left\{x\left(1 - \frac{x}{2n}\right)\right\}^{1/2} dx + \frac{1}{2}\left(1 - \frac{1}{2n}\right)^{1/2} + \frac{1}{2}\left(\frac{n}{2}\right)^{1/2} + R_1(n).$$

The range of the integral can be extended to $[0, n]$ with an error which is $O(1)$ for large n. Then by setting $x = n(1 - v^{1/2})$ it is evaluable by means of the Beta-function integral. Thus we arrive at

$$\sum_{j=1}^{n} f_n(j) = \frac{\pi}{2^{5/2}} n^{3/2} + \frac{n^{1/2}}{2^{3/2}} + O(1) + R_1(n). \tag{2.12}$$

To assess the asymptotic orders of magnitude of the derivatives of $f_n(x)$ uniformly for large n, we write $x = nt$, so that $t \in [1/n, 1]$. Then $f_n(x) = n^{1/2}\phi(t)$, where $\phi(t) = \{t(1 - \frac{1}{2}t)\}^{1/2}$. For each $s = 1, 2, \ldots$ it is easily seen that

$$\phi^{(s)}(t) = O(t^{-s+(1/2)}) \qquad (0 < t \leqslant 1).$$

Correspondingly,

$$f_n^{(s)}(x) = x^{-s+(1/2)}O(1) \tag{2.13}$$

as $n \to \infty$, uniformly for $x \in (0, n]$. Taking this result with $s = 2$ and substituting in (2.03), we deduce that $R_1(n) = O(1)$. Then from (2.11) and (2.12) we derive

$$S(n) = \tfrac{1}{4}\pi n + \tfrac{1}{2} + O(n^{-1/2}) \qquad (n \to \infty).$$

This is the first approximation.

Higher approximations cannot be obtained simply by taking a higher value for m, nor can the device of §2.3 be used because the derivatives $f_n^{(2m)}(x)$ are not integrable at $x = 2n$. Instead, recognizing that the difficulty stems from behavior of the derivatives for small x, we decompose $f_n(x)$ into $g_n(x) + x^{1/2}$, where

$$g_n(x) = x^{1/2}\left\{\left(1 - \frac{x}{2n}\right)^{1/2} - 1\right\}.$$

Corresponding to (2.13), we find that

$$g_n^{(s)}(x) = x^{-s+(3/2)}O(n^{-1}),$$

uniformly for large n. Then applying (2.01), with $m = 2$ and $f(x) = g_n(x)$, we have

$$\sum_{j=1}^{n} g_n(j) = \frac{\pi}{2^{5/2}} n^{3/2} - \frac{2}{3} n^{3/2} + \frac{n^{1/2}}{2}\left(\frac{1}{2^{1/2}} - 1\right) - \frac{1}{24 n^{1/2}} + O\left(\frac{1}{n}\right).$$

For the contribution of the term $x^{1/2}$ the method of §2.3 yields

$$\sum_{j=1}^{n} j^{1/2} = \frac{2}{3}n^{3/2} + \frac{1}{2}n^{1/2} + A + \frac{1}{24n^{1/2}} + O\left(\frac{1}{n^{5/2}}\right), \tag{2.14}$$

where A is a constant which can be evaluated numerically by taking a suitable value of n in this relation.† Addition of the last two results and substitution in (2.11) gives the next approximation to the required sum in the form

$$S(n) = \frac{\pi}{4}n + \frac{1}{2} + \frac{A\sqrt{2}}{n^{1/2}} + O\left(\frac{1}{n^{3/2}}\right).$$

Higher approximations may be found by an extension of the process.

Ex. 2.1 With C defined as in §2.2, prove that

$$1!\cdot 2!\cdot 3!\cdots n! \sim n^{(6n^2+12n+5)/12}\exp(-\tfrac{3}{4}n^2 - n + \tfrac{1}{12} - C)(2\pi)^{(n+1)/2} \qquad (n\to\infty).$$

Ex. 2.2 Verify that the value of A in (2.14) is -0.20789, to five decimal places.

Ex. 2.3 If α is fixed and $n\to\infty$, show that

$$\sum_{j=n}^{\infty}\frac{1}{j(j^2+\alpha^2)^{1/2}} \sim \frac{1}{n} + \frac{1}{2n^2} + \frac{1-\alpha^2}{6n^3} - \frac{\alpha^2}{4n^4} + \frac{9\alpha^4-20\alpha^2-4}{120n^5} + \frac{3\alpha^4}{16n^6} + \cdots.$$

Ex. 2.4 Show that

$$\sum_{j=2}^{n}\frac{1}{\ln j} \sim \frac{n}{\ln n} \qquad (n\to\infty),$$

with a relative error $O(1/\ln n)$, and also that this sum can be represented more accurately in the form

$$\mathrm{Ei}(\ln n) + \text{constant} + \frac{1}{2\ln n} - \frac{1}{12n(\ln n)^2} + O\left\{\frac{1}{n^3(\ln n)^2}\right\}.$$

Ex. 2.5 Prove that

$$\sum_{j=0}^{n}\frac{1}{j^2+n^2} = \frac{\pi}{4n} + \frac{3}{4n^2} + \phi(n),$$

where $|\phi(n)| \leqslant (3^{3/2}-2)/(32n^3)$. Show also that

$$\phi(n) = -\frac{1}{24n^3} + O\left(\frac{1}{n^7}\right) \qquad (n\to\infty).$$

3 Contour Integral for the Remainder Term

3.1 Throughout this section we again assume that a and n are integers such that $a < n$. We denote the strip $a \leqslant \operatorname{Re} t \leqslant n$ by **S**, and suppose that:

(i) $f(t)$ *is continuous throughout* **S** *and holomorphic in the interior of* **S**.
(ii) $f(t) = o(e^{2\pi|\operatorname{Im} t|})$ *as* $\operatorname{Im} t \to \pm\infty$ *in* **S**, *uniformly with respect to* $\operatorname{Re} t$.

† A numerical value is given in Exercise 2.2, and an analytical expression in Exercise 3.2.

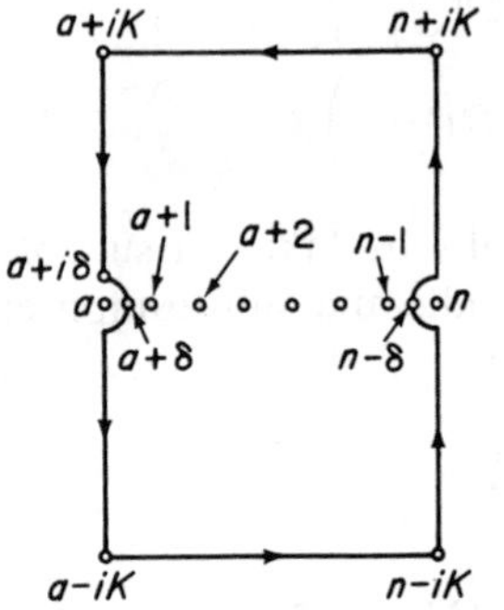

Fig. 3.1 t plane. Contour for Euler–Maclaurin formula.

With these assumptions the Euler–Maclaurin formula can be constructed in the following way. Let $\mathscr{C}$ be the closed contour depicted in Fig. 3.1, the radius δ of the semicircular indentations at a and n being less than $\frac{1}{2}$, and the distance K of the most remote points from the real axis being large. Since the residue of $\cot \pi t$ at any integer j is $1/\pi$, we have

$$\sum_{j=a+1}^{n-1} f(j) = (2i)^{-1} \int_{\mathscr{C}} \cot(\pi t) f(t)\, dt.$$

Denote by $\mathscr{C}_1$ and $\mathscr{C}_2$ the upper and lower parts of $\mathscr{C}$, respectively. Then by Cauchy's theorem

$$\sum_{j=a+1}^{n-1} f(j) - \int_{a+\delta}^{n-\delta} f(t)\, dt = (2i)^{-1} \int_{\mathscr{C}_1+\mathscr{C}_2} \cot(\pi t) f(t)\, dt + \tfrac{1}{2}\int_{\mathscr{C}_1} f(t)\, dt - \tfrac{1}{2}\int_{\mathscr{C}_2} f(t)\, dt$$

$$= \int_{\mathscr{C}_1} \frac{f(t)}{1-e^{-2\pi i t}}\, dt + \int_{\mathscr{C}_2} \frac{f(t)}{e^{2\pi i t}-1}\, dt.$$

Now let $K \to \infty$. In consequence of Condition (ii) the integrals along the parts of $\mathscr{C}_1$ and $\mathscr{C}_2$ parallel to the real axis both vanish. Next, let $\delta \to 0$. Since $f(t)$ is continuous, the integral of $f(t)/(1-e^{-2\pi i t})$ around the quarter circle linking $a + i\delta$ with $a+\delta$ tends to $-\frac{1}{4}f(a)$. Similarly for the other quarter circles. Thus we arrive at the *Abel–Plana formula*

$$\sum_{j=a}^{n} f(j) = \int_a^n f(t)\, dt + \frac{1}{2} f(a) + \frac{1}{2} f(n) + i \int_0^\infty \frac{f(a+iy) - f(n+iy) - f(a-iy) + f(n-iy)}{e^{2\pi y}-1}\, dy. \tag{3.01}$$

3.2 From now on for simplicity in exposition *we suppose that $f(t)$ is real on the intersection of* **S** *with the real axis.* The last formula may then be written

$$\sum_{j=a}^{n} f(j) = \int_a^n f(t)\, dt + \frac{1}{2} f(a) + \frac{1}{2} f(n) + 2\int_0^\infty \frac{\operatorname{Im}\{f(n+iy) - f(a+iy)\}}{e^{2\pi y}-1}\, dy. \tag{3.02}$$

We propose to substitute the Taylor expansions

$$\operatorname{Im}\{f(n+iy)\} = \sum_{s=1}^{m} (-)^{s-1} \frac{y^{2s-1}}{(2s-1)!} f^{(2s-1)}(n) + (-)^m \frac{y^{2m}}{(2m)!} \operatorname{Im}\{f^{(2m)}(n+i\vartheta_n y)\}, \tag{3.03}$$

$$\operatorname{Im}\{f(a+iy)\} = \sum_{s=1}^{m} (-)^{s-1} \frac{y^{2s-1}}{(2s-1)!} f^{(2s-1)}(a) + (-)^m \frac{y^{2m}}{(2m)!} \operatorname{Im}\{f^{(2m)}(a+i\vartheta_a y)\}, \tag{3.04}$$

in which m is an arbitrary positive integer, and ϑ_a and ϑ_n are certain numbers in the interval $(0, 1)$. These expansions are valid when $f^{(2m)}(t)$ is continuous on the boundaries of **S**.

Lemma 3.1

$$\int_0^\infty \frac{x^{2s-1}\,dx}{e^{2\pi x}-1} = (-)^{s-1} \frac{B_{2s}}{4s} \qquad (s \geqslant 1). \tag{3.05}$$

This identity is obtainable by combining equation (11.02) of Chapter 2 with equation (1.13) of the present chapter. It may also be verified directly by integrating the function $e^{-i\lambda x}/(e^{2\pi x}-1)$, in which $\lambda > 0$, around a rectangle with vertices at 0, K, $K+i$, and i, and indentations at 0 and i. Letting $K \to \infty$ and the indentations shrink to zero, we obtain a formula of Legendre, given by

$$\int_0^\infty \frac{\sin \lambda x}{e^{2\pi x}-1}\,dx = \frac{1}{2(e^\lambda-1)} - \frac{1}{2\lambda} + \frac{1}{4} \qquad (\lambda > 0).$$

Then (3.05) follows on repeatedly differentiating with respect to λ, and subsequently setting $\lambda = 0$; compare (1.08).

Returning to (3.02) and substituting by means of (3.03) and (3.04), we obtain the expansion (2.01) with a new formula for the remainder:

$$R_m(n) = \frac{B_{2m}}{(2m)!}\{f^{(2m-1)}(n) - f^{(2m-1)}(a)\}$$
$$+ 2\frac{(-)^m}{(2m)!}\int_0^\infty \operatorname{Im}\{f^{(2m)}(n+i\vartheta_n y) - f^{(2m)}(a+i\vartheta_a y)\}\frac{y^{2m}\,dy}{e^{2\pi y}-1}.$$

Here ϑ_a and $\vartheta_n \in (0, 1)$. The conditions for the validity of this formula are more restrictive than those associated with (2.02). The main interest in the present approach is in providing an analytical way of evaluating the constant term C in the asymptotic expansion of the sum. For this purpose it is more convenient to substitute in (3.02) by means of (3.03) but *not* by means of (3.04); thus

$$\sum_{j=a}^{n} f(j) = \int_a^n f(t)\,dt + \frac{1}{2}f(a) + \frac{1}{2}f(n) - 2\int_0^\infty \frac{\operatorname{Im}\{f(a+iy)\}}{e^{2\pi y}-1}\,dy$$
$$+ \sum_{s=1}^{m} \frac{B_{2s}}{(2s)!} f^{(2s-1)}(n) + 2\frac{(-)^m}{(2m)!}\int_0^\infty \operatorname{Im}\{f^{(2m)}(n+i\vartheta_n y)\}\frac{y^{2m}\,dy}{e^{2\pi y}-1}, \tag{3.06}$$

provided that the first infinite integral converges.† Since we are not using (3.04), the condition that $f^{(2m)}(t)$ be continuous on the line $\operatorname{Re} t = a$ is not needed here.

3.3 Consider again the example of §2.2. Here $f(x) = x \ln x$, hence Conditions (i) and (ii) of §3.1 are satisfied with $a = 0$ and n any positive integer. From (2.06) we see that when $m \geqslant 1$ the derivative $f^{(2m)}(n + i\vartheta_n y)$ is $O(n^{-2m+1})$ for large n, uniformly with respect to $y \in [0, \infty)$. Hence the last integral in (3.06) is $O(n^{-2m+1})$. Bearing in mind that $f(0) = 0$, we find by comparison with (2.08)

$$C = -2\int_0^\infty \frac{\operatorname{Im}\{f(iy)\}}{e^{2\pi y}-1}\,dy + \frac{B_2}{2!} = \frac{1}{12} - 2\int_0^\infty \frac{y \ln y}{e^{2\pi y}-1}\,dy.$$

Referring again to (11.02) of Chapter 2, we have

$$\int_0^\infty \frac{y^{\alpha-1}\,dy}{e^{2\pi y}-1} = \frac{\Gamma(\alpha)\zeta(\alpha)}{(2\pi)^\alpha} \qquad (\alpha > 1).$$

Differentiation with respect to α produces

$$\int_0^\infty \frac{y^{\alpha-1}\ln y}{e^{2\pi y}-1}\,dy = \frac{\Gamma'(\alpha)\zeta(\alpha)+\Gamma(\alpha)\zeta'(\alpha)-\ln(2\pi)\Gamma(\alpha)\zeta(\alpha)}{(2\pi)^\alpha}.$$

Setting $\alpha = 2$ and using (1.13) and the formula $\Gamma'(2) = \psi(2) = 1-\gamma$ obtained from Exercise 2.2 of Chapter 2, we obtain the desired result

$$C = \frac{\gamma+\ln(2\pi)}{12} - \frac{\zeta'(2)}{2\pi^2}.$$

Ex. 3.1 From the Abel–Plana formula derive *Jensen's formula*

$$\zeta(z) = \frac{1}{2} + \frac{1}{z-1} + 2\int_0^\infty \frac{\sin(z\tan^{-1}t)\,dt}{(1+t^2)^{z/2}(e^{2\pi t}-1)},$$

in which z has any real or complex value, other than 1, and all functions assume their principal values.

Ex. 3.2 Let α be any real constant, other than -1. Prove that

$$\sum_{j=1}^{n-1} j^\alpha - \zeta(-\alpha) \sim \frac{n^{\alpha+1}}{\alpha+1}\sum_{s=0}^{\infty}\binom{\alpha+1}{s}\frac{B_s}{n^s}$$

as $n \to \infty$. Show that if the expansion is truncated at the term $s = 2m-1$, where m is a positive integer, then the remainder is bounded in absolute value by the next term, and has the same sign, provided that $m \geqslant \frac{1}{2}(\alpha+1)$.

Verify also that the expansion agrees with Exercise 1.3 in the case when α is a positive integer.

Ex. 3.3 For the case $\alpha = -1$ of the preceding exercise, prove that

$$\sum_{j=1}^{n-1}\frac{1}{j} = \ln n + \gamma - \frac{1}{2n} - \sum_{s=1}^{m-1}\frac{B_{2s}}{2s}\frac{1}{n^{2s}} - \frac{B_{2m}}{2m}\frac{\vartheta_{m,n}}{n^{2m}},$$

where γ denotes Euler's constant, $m = 1, 2, 3, \ldots$, and $\vartheta_{m,n} \in (0, 1)$. Show also that

$$\gamma = \frac{1}{2} + 2\int_0^\infty \frac{y\,dy}{(1+y^2)(e^{2\pi y}-1)}.$$

† Convergence of the other infinite integral is then assured because the infinite integral in (3.02) converges with the conditions assumed in §3.1.

4 Stirling's Series for $\ln\Gamma(z)$

4.1 In preceding sections we have investigated the asymptotic behavior of finite sums as the number of terms tends to infinity. Another fruitful application of the Euler–Maclaurin formula is to infinite sums, or sequences, the terms of which depend on a real or complex asymptotic parameter z. An important example is furnished by Euler's limit formula for the Gamma function, derived in Chapter 2, §1.3. For the present purpose we take this in the form

$$\ln\Gamma(z) = \lim_{n\to\infty}\{S_{n-1}(1) - S_n(z) + z\ln n\} \qquad (|\mathrm{ph}\, z| < \pi),$$

where

$$S_n(z) = \ln z + \ln(z+1) + \ln(z+2) + \cdots + \ln(z+n).$$

The value of $\ln\Gamma(z)$ is not necessarily the principal one, but this is immaterial.

Applying (2.01) and (2.02) with $f(x) = \ln(x+z)$, $a = 0$, and $m = 1$, we obtain

$$S_n(z) = \left(n + z + \frac{1}{2}\right)\ln(n+z) - \left(z - \frac{1}{2}\right)\ln z - n + \frac{1}{12(n+z)} - \frac{1}{12z} + \int_0^n \frac{B_2(x-[x])\,dx}{2(x+z)^2},$$

and hence

$$S_{n-1}(1) - S_n(z) + z\ln n = \left(n + z + \frac{1}{2}\right)\ln\left(\frac{n}{n+z}\right) + \left(z - \frac{1}{2}\right)\ln z + \frac{11}{12} + \frac{1}{12n} - \frac{1}{12(n+z)} + \frac{1}{12z} + \int_0^{n-1}\frac{B_2(x-[x])\,dx}{2(x+1)^2} - \int_0^n \frac{B_2(x-[x])\,dx}{2(x+z)^2}.$$

If z is fixed and $n\to\infty$, then

$$\left(n + z + \frac{1}{2}\right)\ln\left(\frac{n}{n+z}\right) \to -z.$$

Therefore

$$\ln\Gamma(z) = \left(z - \frac{1}{2}\right)\ln z - z + C + \frac{1}{12z} - \int_0^\infty \frac{B_2(x-[x])\,dx}{2(x+z)^2}, \tag{4.01}$$

where

$$C = \frac{11}{12} + \int_0^\infty \frac{B_2(x-[x])\,dx}{2(x+1)^2}.$$

This constant can be evaluated by the methods of §3, but a simpler procedure here is to let $z \to \pm i\infty$ in (4.01) and substitute the results in the reflection formula

$$\Gamma(z)\Gamma(-z) = -\pi/\{z\sin(\pi z)\}. \tag{4.02}$$

This yields $C = \frac{1}{2}\ln(2\pi)$.

The required expansion is obtained from (4.01) by integrating the last term repeatedly by parts, as in the construction of the Euler–Maclaurin formula. This gives

$$\ln\Gamma(z) = \left(z-\frac{1}{2}\right)\ln z - z + \frac{1}{2}\ln(2\pi) + \sum_{s=1}^{m-1}\frac{B_{2s}}{2s(2s-1)z^{2s-1}} + R_m(z), \tag{4.03}$$

where m is an arbitrary positive integer, and

$$R_m(z) = \int_0^\infty \frac{B_{2m} - B_{2m}(x-[x])}{2m(x+z)^{2m}}\,dx.$$

To establish the asymptotic nature of this expansion for large $|z|$, write $\theta \equiv \operatorname{ph} z$ and assume that $|\theta| \leqslant \pi-\delta$, where δ is an arbitrary positive constant. Then using Theorem 1.1(iii) and the substitution $x = |z|\,\tau$, we derive

$$|R_m(z)| \leqslant \frac{|B_{2m}|}{m}\int_0^\infty \frac{dx}{|x+z|^{2m}} \leqslant \frac{|B_{2m}|}{m\,|z|^{2m-1}}\int_0^\infty \frac{d\tau}{(\tau^2-2\tau\cos\delta+1)^m} = O\left(\frac{1}{z^{2m-1}}\right).$$

4.2 Exponentiation of (4.03) gives

$$\Gamma(z) \sim e^{-z}z^z\left(\frac{2\pi}{z}\right)^{1/2}\left(1+\frac{1}{12z}+\frac{1}{288z^2}+\cdots\right) \tag{4.04}$$

as $z\to\infty$ in the sector $|\operatorname{ph} z| \leqslant \pi-\delta$. For positive z, this agrees with the result found in Chapter 3, §8.3 by Laplace's method. Comparing (4.03) and (4.04) we note that the former involves only alternate powers of $1/z$.

Another advantage of (4.03) is that realistic error bounds can be constructed in a simple way. Suppose first that z is real and positive. Then as in §2.1 the error test shows that

$$R_m(z) = \frac{B_{2m}}{2m(2m-1)}\frac{\vartheta_m}{z^{2m-1}} \qquad (z>0), \tag{4.05}$$

where ϑ_m is a number in the interval $(0,1)$.

Alternatively, let z be complex: $z = re^{i\theta}$. Then

$$|x+z|^2 = x^2+2xr\cos\theta+r^2 = (x+r)^2 - 4xr\sin^2\tfrac{1}{2}\theta \geqslant (x+r)^2\cos^2\tfrac{1}{2}\theta.$$

Hence

$$|R_m(z)| \leqslant \frac{1}{\cos^{2m}(\frac{1}{2}\theta)}\int_0^\infty \frac{|B_{2m}-B_{2m}(x-[x])|}{2m(x+r)^{2m}}\,dx = \sec^{2m}(\tfrac{1}{2}\theta)\,|R_m(r)|,$$

the last step being justified by the fact that $B_{2m}-B_{2m}(x-[x])$ does not change sign. Combination with (4.05) gives a result due to Stieltjes†

$$|R_m(z)| \leqslant \frac{|B_{2m}|\sec^{2m}(\frac{1}{2}\theta)}{2m(2m-1)|z|^{2m-1}} \qquad (|\theta|<\pi). \tag{4.06}$$

† Lindelöf (1905, §48). Another error bound can be derived from the formula given in Exercise 4.3 below; see Whittaker and Watson (1927, §12.33). See also Spira (1971).

As expected, this bound tends to infinity when θ approaches either of the extremes $\pm\pi$. However, the numerical use of (4.03) in the sectors $\frac{1}{2}\pi < |\theta| < \pi$ can be avoided—indeed, *should be avoided*—by use of (4.02).

Ex. 4.1 By use of (4.04) and Liouville's theorem establish *Gauss's multiplication formula*

$$\Gamma(nz) = (2\pi)^{(1-n)/2} n^{nz-(1/2)} \prod_{s=0}^{n-1} \Gamma\left(z+\frac{s}{n}\right),$$

in which n is a positive integer and $nz \neq 0, -1, -2, \dots$.

Ex. 4.2 From (4.03) derive the following generalization of Exercise 3.3:

$$\psi(z) = \ln z - \frac{1}{2z} - \sum_{s=1}^{m-1} \frac{B_{2s}}{2sz^{2s}} - U_m(z) \qquad (m \geqslant 1),$$

where

$$|U_m(z)| \leqslant \frac{|B_{2m}|\sec^{2m+1}(\frac{1}{2}\operatorname{ph} z)}{2m|z|^{2m}} \qquad (|\operatorname{ph} z| < \pi).$$

Ex. 4.3 From the Abel–Plana formula derive *Binet's formula*

$$\ln\Gamma(z) = \left(z-\frac{1}{2}\right)\ln z - z + \frac{1}{2}\ln(2\pi) + 2\int_0^\infty \frac{\tan^{-1}(t/z)}{e^{2\pi t}-1}\,dt \qquad (|\operatorname{ph} z| < \tfrac{1}{2}\pi),$$

in which the inverse tangent has its principal value.

Ex. 4.4 Let h be a constant such that $0 \leqslant h \leqslant 1$, and restrict $m \geqslant 2$ and $|\operatorname{ph} z| < \pi$. By using Exercise 1.1 obtain the following generalization of (4.03):

$$\ln\Gamma(z+h) = \left(z+h-\frac{1}{2}\right)\ln z - z + \frac{1}{2}\ln(2\pi) + \sum_{s=2}^{m} \frac{(-)^s B_s(h)}{s(s-1)z^{s-1}} - (m-1)!\int_0^\infty \frac{\varpi_m(x-h)}{(z+x)^m}\,dx.$$

*5 Summation by Parts

5.1 The Euler–Maclaurin formula is generally successful for sums whose terms $f(j)$ have a relatively slowly varying character, for in this event successive derivatives of $f(x)$ can be expected to diminish fairly rapidly. When this requirement is not fulfilled other methods are needed.

Cases in which the absolute values of individual terms change rapidly present little difficulty as a rule: the largest term approximates the whole sum. For example, if

$$S(n) = \sum_{j=1}^{n} j!,$$

then it is easily seen that

$$S(n) = n!\{1+O(n^{-1})\}.$$

Furthermore, inclusion of the second largest term produces

$$S(n) = n!\left\{1+\frac{1}{n}+O\left(\frac{1}{n^2}\right)\right\},$$

and so on. In effect, the sum is asymptotically represented by its own reversed partial sums. This kind of approximation may be regarded as a discrete analogue of Laplace's approximation for integrals of functions with sharp peaks (Chapter 3, §7).

Alternatively, suppose that individual terms change relatively slowly in magnitude, but have oscillating signs. Or, more generally, the terms are formed by multiplying a slowly varying function with a sine or cosine. In these cases the sum is the discrete analogue of a Fourier integral. Just as the method of integration by parts provides a valuable way of approximating the latter, so the method of *summation by parts* is useful for the former.

The formula for summation by parts is expressed by the identity

$$\sum_{j=1}^{n-1} u_j v_j = \sum_{j=1}^{n-1} U_j(v_j - v_{j+1}) + U_{n-1} v_n, \tag{5.01}$$

in which

$$U_j = u_1 + u_2 + \cdots + u_j. \tag{5.02}$$

In applications, the slowly varying factor is taken to be v_j; consequently $v_j - v_{j+1}$ is of smaller magnitude than v_j. The partial sum U_j of the oscillatory factors u_j is of the same order of magnitude as the u_j.

For example, when the terms alternate in sign, we have $u_j = (-1)^{j-1}$, $U_{2j+1} = 1$, and $U_{2j} = 0$. The method then consists of pairing the terms, and the rearranged sum can be handled by the Euler–Maclaurin formula.

5.2 As a nontrivial example consider

$$S(\alpha, \beta, n) = \sum_{j=1}^{n-1} e^{ij\beta} j^\alpha, \tag{5.03}$$

in which α and β are fixed real numbers. If β is zero or an integer multiple of 2π, then $e^{i\beta} = 1$ and the result of Exercise 3.2 applies. This case is excluded in what follows.

A first estimate of $S(\alpha, \beta, n)$ for large n is provided by

$$|S(\alpha, \beta, n)| \leqslant \sum_{j=1}^{n-1} j^\alpha = O(n^{\alpha+1}), \quad O(\ln n), \quad \text{or} \quad O(1), \tag{5.04}$$

according as $\alpha > -1$, $\alpha = -1$, or $\alpha < -1$; compare Exercises 3.2 and 3.3.

Next, with $u_j = e^{ij\beta}$ and $v_j = j^\alpha$, we have

$$U_j = \frac{e^{i\beta}}{e^{i\beta} - 1}(e^{ij\beta} - 1).$$

The transformation (5.01) gives

$$\begin{aligned} S(\alpha, \beta, n) &= \frac{e^{i\beta}}{e^{i\beta} - 1}\left[\sum_{j=1}^{n-1} (e^{ij\beta} - 1)\{j^\alpha - (j+1)^\alpha\} + \{e^{i(n-1)\beta} - 1\} n^\alpha\right] \\ &= \frac{e^{i\beta}}{e^{i\beta} - 1}\left[\sum_{j=1}^{n-1} e^{ij\beta}\{j^\alpha - (j+1)^\alpha\} + e^{i(n-1)\beta} n^\alpha - 1\right]. \end{aligned} \tag{5.05}$$

With the aid of (5.04) and the relation

$$j^{\alpha} - (j+1)^{\alpha} = -\alpha j^{\alpha-1} - \tfrac{1}{2}\alpha(\alpha-1)\,j^{\alpha-2} + O(j^{\alpha-3})$$

which holds for the set of all positive integers j, we derive

$$S(\alpha,\beta,n) = \frac{e^{i\beta}}{e^{i\beta}-1}\{-\alpha S(\alpha-1,\beta,n) + e^{i(n-1)\beta}n^{\alpha} + O(n^{\alpha-1}) + O(1)\}. \qquad (5.06)$$

Substituting for $S(\alpha-1,\beta,n)$ by means of (5.04), we see that

$$S(\alpha,\beta,n) = O(n^{\alpha}) + O(1).$$

Then replacing α in this relation by $\alpha-1$ and resubstituting in (5.06), we arrive at

$$S(\alpha,\beta,n) = \frac{e^{in\beta}}{e^{i\beta}-1}n^{\alpha} + O(n^{\alpha-1}) + O(1). \qquad (5.07)$$

5.3 Equation (5.07) furnishes a meaningful approximation to $S(\alpha,\beta,n)$, provided that $\alpha > 0$. To extend this approximation into an asymptotic expansion useful for any real value of α, we proceed as follows. Suppose first that $\alpha < -1$, and let $T(\alpha,\beta,n)$ denote the absolutely convergent series

$$T(\alpha,\beta,n) = \sum_{j=n}^{\infty} e^{ij\beta} j^{\alpha}.$$

Corresponding to (5.04) we have

$$T(\alpha,\beta,n) = O(n^{\alpha+1}) \qquad (n\to\infty). \qquad (5.08)$$

By partial summation, as in §5.2, we obtain

$$T(\alpha,\beta,n) = \frac{e^{i\beta}}{e^{i\beta}-1}\left[\sum_{j=n}^{\infty} e^{ij\beta}\{j^{\alpha}-(j+1)^{\alpha}\} - e^{i(n-1)\beta}n^{\alpha}\right],$$

and thence, by application of the Binomial theorem,

$$T(\alpha,\beta,n) = -\frac{e^{i\beta}}{e^{i\beta}-1}\sum_{p=1}^{q-1}\binom{\alpha}{p}T(\alpha-p,\beta,n) + O(n^{\alpha-q+1}) - \frac{e^{in\beta}}{e^{i\beta}-1}n^{\alpha}, \qquad (5.09)$$

where q is any positive integer.

Beginning with the estimate (5.08) and repeatedly resubstituting in the last relation, we see that for large n the function $T(\alpha,\beta,n)$ possesses an expansion of the form

$$T(\alpha,\beta,n) \sim -e^{in\beta}\sum_{s=0}^{\infty}\psi_s(\alpha,\beta)\,n^{\alpha-s}. \qquad (5.10)$$

The coefficients are given by

$$\psi_0(\alpha,\beta) = 1/(e^{i\beta}-1), \qquad (5.11)$$

and

$$\psi_s(\alpha,\beta) = -\frac{e^{i\beta}}{e^{i\beta}-1}\sum_{p=1}^{s}\binom{\alpha}{p}\psi_{s-p}(\alpha-p,\beta) \qquad (s \geqslant 1). \qquad (5.12)$$

In particular,

$$\psi_1(\alpha,\beta) = -\alpha\frac{e^{i\beta}}{(e^{i\beta}-1)^2}, \qquad \psi_2(\alpha,\beta) = \alpha(\alpha-1)\frac{e^{i\beta}(e^{i\beta}+1)}{2(e^{i\beta}-1)^3}. \tag{5.13}$$

The corresponding expansion of the original sum is

$$S(\alpha,\beta,n) \sim \phi(\alpha,\beta) + e^{in\beta}\sum_{s=0}^{\infty}\psi_s(\alpha,\beta)n^{\alpha-s}. \tag{5.14}$$

Here $\phi(\alpha,\beta)$ denotes $T(\alpha,\beta,1)$, that is,

$$\phi(\alpha,\beta) = \sum_{j=1}^{\infty} e^{ij\beta}j^{\alpha} \qquad (\alpha < -1). \tag{5.15}$$

5.4 Next, suppose that $-1 \leqslant \alpha < 0$. Again using the Binomial theorem, we rearrange (5.05) in the form

$$S(\alpha,\beta,n) = \frac{e^{i\beta}}{e^{i\beta}-1}\left[\sum_{j=1}^{\infty} e^{ij\beta}\{j^{\alpha}-(j+1)^{\alpha}\} + \sum_{p=1}^{q-1}\binom{\alpha}{p}T(\alpha-p,\beta,n) + O(n^{\alpha-q+1}) + e^{i(n-1)\beta}n^{\alpha} - 1\right], \tag{5.16}$$

where $q \geqslant 1$. Substituting by means of (5.10), we perceive that an expansion of the form (5.14) again holds with the $\psi_s(\alpha,\beta)$ satisfying (5.11) and (5.12), but in place of (5.15) we have

$$\phi(\alpha,\beta) = \frac{e^{i\beta}}{e^{i\beta}-1}\left[\sum_{j=1}^{\infty} e^{ij\beta}\{j^{\alpha}-(j+1)^{\alpha}\} - 1\right] \qquad (-1 \leqslant \alpha < 0). \tag{5.17}$$

More generally, assume that $m-2 \leqslant \alpha < m-1$, where m is any positive integer. Corresponding to (5.16) we use the decomposition

$$S(\alpha,\beta,n) = \frac{e^{i\beta}}{e^{i\beta}-1}\left[\sum_{j=1}^{\infty} e^{ij\beta}\left\{j^{\alpha}-(j+1)^{\alpha}+\sum_{p=1}^{m-1}\binom{\alpha}{p}j^{\alpha-p}\right\} - \sum_{p=1}^{m-1}\binom{\alpha}{p}S(\alpha-p,\beta,n) + \sum_{p=m}^{q-1}\binom{\alpha}{p}T(\alpha-p,\beta,n) + O(n^{\alpha-q+1}) + e^{i(n-1)\beta}n^{\alpha} - 1\right], \tag{5.18}$$

where q is an arbitrary integer such that $q \geqslant m$. Taking $m = 2, 3, \ldots$ in turn we see that an expansion of the form (5.14) again holds. By substituting (5.10) and (5.14) in (5.18) and comparing coefficients we reproduce (5.11) and (5.12), and also obtain

$$\phi(\alpha,\beta) = \frac{e^{i\beta}}{e^{i\beta}-1}\left[\sum_{j=1}^{\infty} e^{ij\beta}\left\{j^{\alpha}-(j+1)^{\alpha}+\sum_{p=1}^{m-1}\binom{\alpha}{p}j^{\alpha-p}\right\} - \sum_{p=1}^{m-1}\binom{\alpha}{p}\phi(\alpha-p,\beta) - 1\right]$$
$$(m-2 \leqslant \alpha < m-1). \tag{5.19}$$

5.5 Summarizing, we have shown that for all real values of α and β (other than $\beta = 0, \pm 2\pi, \pm 4\pi, \ldots$), and large positive values of n, the function $S(\alpha,\beta,n)$ defined by (5.03) has the asymptotic expansion (5.14). Formulas for the functions $\phi(\alpha,\beta)$ and $\psi_s(\alpha,\beta)$ are supplied by (5.11), (5.12), (5.13), (5.15), and (5.19).† It will be observed that the role of $\phi(\alpha,\beta)$ in (5.14) is analogous to that of $\zeta(-\alpha)$ in Exercise 3.2.

Ex. 5.1 If α is positive and fixed, show that

$$\frac{1}{n^\alpha}-\frac{1}{(n+1)^\alpha}+\frac{1}{(n+2)^\alpha}-\cdots=\frac{1}{2n^\alpha}\left\{1+O\left(\frac{1}{n}\right)\right\} \qquad (n\to\infty).$$

Ex. 5.2 For any given positive integer m, show that equation (5.19) holds for $\alpha\in(-\infty,m-1)$. Deduce that for fixed β, $\phi(\alpha,\beta)$ is continuous for all real α.

Ex. 5.3 By summing the right-hand side of (5.05) repeatedly by parts, prove the following generalization of (5.17):

$$\phi(\alpha,\beta)=\left(\frac{e^{i\beta}}{1-e^{i\beta}}\right)^m\sum_{j=1}^{\infty}e^{ij\beta}\Delta^m j^\alpha+\sum_{k=1}^{m}\left(\frac{e^{i\beta}}{1-e^{i\beta}}\right)^k[\Delta^{k-1}j^\alpha]_{j=1},$$

in which m is any positive integer such that $m\geqslant\alpha+1$, and Δ denotes the forward difference operator defined by

$$\Delta v_j=v_{j+1}-v_j, \qquad \Delta^2 v_j=\Delta v_{j+1}-\Delta v_j, \dots .$$

Ex. 5.4 From (5.15) derive

$$\phi(\alpha,\beta)=\frac{1}{\Gamma(-\alpha)}\int_0^\infty\frac{t^{-\alpha-1}\,dt}{e^{t-i\beta}-1} \qquad (\alpha<-1),$$

and hence

$$\phi(\alpha,\beta)=\frac{\Gamma(1+\alpha)}{2\pi i}\int_{-\infty}^{(0+)}\frac{t^{-\alpha-1}\,dt}{e^{-t-i\beta}-1} \qquad (\alpha\neq-1,-2,\dots),$$

provided that the contour in the second integral does not enclose any of the points $-\beta i+2l\pi i$, $l=0,\pm1,\pm2,\dots$. [Lerch, 1887.]

Ex. 5.5 If β is real and $e^{i\beta}\neq1$, prove that for large n

$$\sum_{j=1}^{n-1}e^{ij\beta}\ln j\sim\frac{e^{i\beta}}{e^{i\beta}-1}\left\{e^{i(n-1)\beta}\ln n+\mu(\beta)+e^{in\beta}\sum_{s=1}^{\infty}\frac{\lambda_s(\beta)}{n^s}\right\},$$

where

$$\mu(\beta)=\sum_{j=1}^{\infty}e^{ij\beta}\left\{\ln\left(\frac{j}{j+1}\right)+\frac{1}{j}\right\}-\phi(-1,\beta), \qquad \lambda_s(\beta)=\sum_{p=1}^{s}\frac{(-)^p}{p}\psi_{s-p}(-p,\beta).$$

6 Barnes' Integral for the Hypergeometric Function

6.1 The method used in §§3 and 4 is applicable to a substantial class of infinite sums

$$S(z)=\sum_{j=0}^{\infty}f(z,j),$$

in which the terms $f(z,j)$ depend on a real or complex parameter z, and $f(z,t)$ is

† An alternative formula for $\phi(\alpha,\beta)$ is provided by Exercise 5.3.

an analytic function of the complex variable t. The residue theorem enables the sum to be expressed in the form

$$S(z) = \frac{1}{2i} \int_{\mathscr{C}} \cot(\pi t)\, f(z,t)\, dt,$$

where $\mathscr{C}$ is a loop contour enclosing the points $t = 0, 1, 2, \ldots$, but not enclosing $-1, -2, -3, \ldots$ or the singularities of $f(z,t)$. The methods of the residue calculus and of Chapters 3 and 4 can then be brought to bear on the problem of approximating $S(z)$ asymptotically as z tends to infinity or some other distinguished point.

Use of the auxiliary factor $\cot(\pi t)$ is not essential; indeed, for certain types of series other factors may be more convenient. For example,

$$\sum_{j=0}^{\infty} (-)^j f(z,j) = \frac{1}{2i} \int_{\mathscr{C}} \csc(\pi t)\, f(z,t)\, dt,$$

and

$$\sum_{j=0}^{\infty} (-)^j \frac{f(z,j)}{j!} = -\frac{1}{2\pi i} \int_{\mathscr{C}} \Gamma(-t)\, f(z,t)\, dt.$$

In this section and §§7 and 8 the general procedure is illustrated by various examples.

6.2 When neither of the parameters a and b is zero or a negative integer the hypergeometric function $F(a,b;c;z)$ is given by

$$\frac{\Gamma(a)\Gamma(b)}{\Gamma(c)} F(a,b;c;z) = \sum_{j=0}^{\infty} \frac{\Gamma(a+j)\Gamma(b+j)}{\Gamma(c+j)} \frac{z^j}{j!} \qquad (|z|<1);$$

compare Chapter 5, §9.1. In accordance with the final equation of the preceding subsection, we consider the integral representation

$$I = \frac{1}{2\pi i} \int_{\mathscr{C}} \frac{\Gamma(a+t)\Gamma(b+t)\Gamma(-t)}{\Gamma(c+t)} (-z)^t\, dt. \tag{6.01}$$

The choice of endpoints of the path $\mathscr{C}$ is not crucial: we take them to be $\pm i\infty$. The important feature is that the poles of the integrand at $0, 1, 2, \ldots$ are on one side of $\mathscr{C}$ and the other poles $-a, -a-1, -a-2, \ldots$ and $-b, -b-1, -b-2, \ldots$ are on the opposite side; see the continuous curve in Fig. 6.1. Since, by hypothesis, neither a nor b is zero or a negative integer it is always possible to choose $\mathscr{C}$ in this way.

The question of convergence of I is resolvable by reference to Stirling's approximation (4.04). When α is a bounded real number and τ is a large real number this formula shows that

$$|\Gamma(\alpha+i\tau)| \sim (2\pi)^{1/2} |\tau|^{\alpha-(1/2)} e^{-\pi|\tau|/2}, \tag{6.02}$$

uniformly with respect to α. We also have

$$|(-z)^{\alpha+i\tau}| = |z|^{\alpha} \exp\{-\tau \operatorname{ph}(-z)\}. \tag{6.03}$$

Hence (6.01) converges absolutely, provided that $|\operatorname{ph}(-z)| < \pi$.

To apply the residue theorem, consider first the corresponding integrals around three sides of the rectangle with vertices at $\pm iT$, $N+\frac{1}{2}\pm iT$, where T is a positive

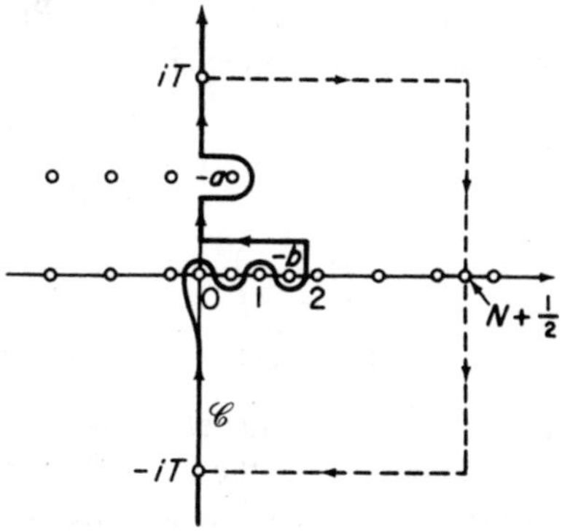

Fig. 6.1 Contour for Barnes' integral.

number exceeding $|\operatorname{Im} a|$ and $|\operatorname{Im} b|$, and N is a positive integer exceeding $-\operatorname{Re} a-\frac{1}{2}$ and $-\operatorname{Re} b-\frac{1}{2}$; see the broken lines in Fig. 6.1. From (6.02) and (6.03) it is seen that if N is kept fixed and $T \to \infty$, then the integrals along the horizontal sides both vanish. For the integral along the third side we write the integrand in the form

$$-\frac{\Gamma(a+t)\,\Gamma(b+t)}{\Gamma(c+t)\,\Gamma(1+t)}\,\frac{\pi}{\sin(\pi t)}(-z)^t.$$

From Chapter 4, §5, we have for large t in the right half-plane,

$$\frac{\Gamma(a+t)\,\Gamma(b+t)}{\Gamma(c+t)\,\Gamma(1+t)} \sim t^{a+b-c-1},$$

and is therefore bounded in absolute value for all N and all t on the path, provided that $\operatorname{Re} c \geqslant \operatorname{Re}(a+b)-1$. Next, if $t = N+\frac{1}{2}+i\tau$, then

$$|\csc(\pi t)| = |\operatorname{sech}(\pi\tau)| \leqslant 2e^{-\pi|\tau|}, \qquad |(-z)^t| = |z|^{N+(1/2)}\exp\{-\tau \operatorname{ph}(-z)\}.$$

By hypothesis $|z|<1$ and $|\operatorname{ph}(-z)|<\pi$; accordingly the line integral from $t = N+\frac{1}{2}+i\infty$ to $N+\frac{1}{2}-i\infty$ converges absolutely, and vanishes as $N \to \infty$.

From the residue theorem it follows that $-I$ is the sum of residues of the integrand at the poles $t = 0, 1, 2, \ldots$. Thus we arrive at *Barnes' integral*†

$$\frac{\Gamma(a)\,\Gamma(b)}{\Gamma(c)}F(a,b;c;z) = \frac{1}{2\pi i}\int_{-i\infty}^{i\infty}\frac{\Gamma(a+t)\,\Gamma(b+t)\,\Gamma(-t)}{\Gamma(c+t)}(-z)^t\,dt. \tag{6.04}$$

In establishing this result it has been assumed that $\operatorname{Re} c \geqslant \operatorname{Re}(a+b)-1$ and $|z|<1$. But for a fixed path the integral (6.04) converges and defines an entire function of c and an analytic function of z which is holomorphic in the sector $|\operatorname{ph}(-z)|<\pi$. Therefore by analytic continuation the only conditions needed for the validity of (6.04) are: (i) *neither a nor b is zero or a negative integer*; (ii) *the path separates the poles of $\Gamma(-t)$ from the poles of $\Gamma(a+t)\,\Gamma(b+t)$*; (iii) $|\operatorname{ph}(-z)|<\pi$.

6.3 The essence of the analysis just given is to evaluate the integral (6.01) by translating the integration path to the right. Suppose instead that the path is translated

† Barnes (1908).

across the poles of $\Gamma(a+t)\Gamma(b+t)$. Provided that $a-b$ is not an integer or zero each of these poles is simple. The residue of the integrand at $t=-a-j$ is

$$(-)^j \frac{\Gamma(b-a-j)\Gamma(a+j)}{j!\Gamma(c-a-j)}(-z)^{-a-j},$$

that is,

$$(-)^j \frac{\Gamma(b-a)\Gamma(1+a-b)}{\Gamma(c-a)\Gamma(1+a-c)} \frac{\Gamma(a+j)\Gamma(1+a-c+j)}{j!\Gamma(1+a-b+j)}(-z)^{-a-j}.$$

The corresponding residue at $t=-b-j$ is obtained by interchanging a and b.

Provided that $|z|>1$, the integral along the translated path vanishes as the path moves to infinity; this is provable by analysis similar to that of §6.2 and details are left as an exercise for the reader. The residue theorem then gives

$$\begin{aligned}\frac{F(a,b;c;z)}{\Gamma(c)} &= \frac{\Gamma(b-a)}{\Gamma(b)\Gamma(c-a)}(-z)^{-a}F(a,1+a-c;1+a-b;z^{-1})\\ &\quad+\frac{\Gamma(a-b)}{\Gamma(a)\Gamma(c-b)}(-z)^{-b}F(b,1+b-c;1+b-a;z^{-1}).\end{aligned} \tag{6.05}$$

Analytic continuation shows that the only restrictions needed now are (i) *$a-b$ is not an integer or zero*; (ii) $|\mathrm{ph}(-z)|<\pi$.

Formula (6.05) describes the behavior of the hypergeometric function for large $|z|$; it agrees with Chapter 5, (10.16).

Ex. 6.1 For the confluent hypergeometric functions prove that

$$\frac{\Gamma(a)}{\Gamma(c)}M(a,c,z) = \frac{1}{2\pi i}\int_{-i\infty}^{i\infty}\frac{\Gamma(a+t)\Gamma(-t)}{\Gamma(c+t)}(-z)^t\,dt$$

when $a \neq 0,-1,-2,\ldots$, and $|\mathrm{ph}(-z)|<\frac{1}{2}\pi$;

$$\Gamma(a)\Gamma(1+a-c)\,z^a U(a,c,z) = \frac{1}{2\pi i}\int_{-i\infty}^{i\infty}\Gamma(a+t)\Gamma(1+a-c+t)\Gamma(-t)\,z^{-t}\,dt$$

when $a\neq 0,-1,-2,\ldots$, $a-c\neq -1,-2,\ldots$, and $|\mathrm{ph}\,z|<\frac{3}{2}\pi$. In both integrals the path of integration separates the poles of $\Gamma(-t)$ from the other poles of the integrand.

7 Further Examples

7.1 Let us investigate the sum

$$S_\alpha(z) = \sum_{j=0}^{\infty}\frac{(-1)^j}{(z^2+j^2)^\alpha}$$

in the case when α is a real or complex constant such that $\mathrm{Re}\,\alpha>0$, z is a large complex parameter such that $|\mathrm{ph}\,z|<\frac{1}{2}\pi$ (ensuring that no term becomes infinite), and $(z^2+j^2)^\alpha$ has its principal value.†

† Series of this type occur in aerodynamic interference calculations (Olver, 1949).

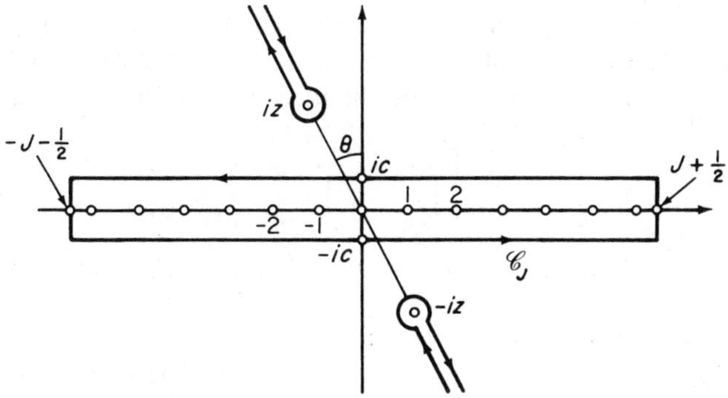

Fig. 7.1 t plane.

We first express $S_\alpha(z)$ in the form

$$S_\alpha(z) = \frac{1}{2z^{2\alpha}} + \frac{1}{2} \sum_{j=-\infty}^{\infty} (-)^j f(z, j),$$

where

$$f(z, t) = 1/(z^2+t^2)^\alpha.$$

As a function of $t, f(z, t)$ is analytic with branch points at $t = \pm iz$.

Consider the rectangular contour $\mathscr{C}_J$ depicted in Fig. 7.1. In this diagram J denotes an arbitrary positive integer and c an arbitrary positive number, subject to the restrictions

$$J > |\mathrm{Re}(iz)|, \qquad c < \mathrm{Im}(iz).$$

By the residue theorem

$$\sum_{j=-J}^{J} (-)^j f(z, j) = \frac{1}{2i} \int_{\mathscr{C}_J} \frac{\csc \pi t}{(z^2+t^2)^\alpha} dt. \tag{7.01}$$

On the right-hand vertical part of $\mathscr{C}_J$

$$|z^2+t^2| = |(t-iz)(t+iz)| \geqslant (J+\tfrac{1}{2})^2 - y^2,$$

where $z = x+iy$. Also

$$\mathrm{ph}(z^2+t^2) = \mathrm{ph}(t-iz) + \mathrm{ph}(t+iz) \in (-\tfrac{1}{2}\pi, \tfrac{1}{2}\pi).$$

Hence the contribution of this part of the path to (7.01) is bounded in absolute value by

$$\frac{e^{\pi|\mathrm{Im}\,\alpha|/2}}{2\{(J+\frac{1}{2})^2 - y^2\}^{\mathrm{Re}\,\alpha}} \int_{J+\frac{1}{2}-ic}^{J+\frac{1}{2}+ic} \left| \frac{dt}{\sin \pi t} \right|.$$

For fixed z this vanishes as $J \to \infty$, since $\mathrm{Re}\,\alpha > 0$ and the integral is independent of J. Similarly for the integral along the left-hand vertical part of $\mathscr{C}_J$. Accordingly,

$$S_\alpha(z) - \frac{1}{2z^{2\alpha}} = \frac{1}{4i} \int_{\mathscr{C}} \frac{\csc \pi t}{(z^2+t^2)^\alpha} dt, \tag{7.02}$$

where $\mathscr{C}$ comprises the doubly infinite straight lines $\mathrm{Im}\, t = \pm c$.

7.2 Critical points in the asymptotic evaluation of the integral (7.02) are the singularities $t = \pm iz$. We deform $\mathscr{C}$ into the two loop contours indicated in Fig. 7.1, this step being easily justified by considering the behavior of the integrand as $|t| \to \infty$. Thus

$$S_\alpha(z) - \frac{1}{2z^{2\alpha}} = \frac{1}{4i}\left\{\int_{\infty e^{i\theta}}^{(iz-)} + \int_{-\infty e^{i\theta}}^{(-iz-)}\right\} \frac{\csc \pi t}{(z^2+t^2)^\alpha}\, dt,$$

where $\theta = \operatorname{ph} z$. The branch of $(z^2+t^2)^\alpha$ takes its principal value on the real axis, and is determined by continuity elsewhere.

Suppose temporarily that $\operatorname{Re}\alpha < 1$. Then the loop contours may be collapsed onto the rays $\operatorname{ph} t = \pm\frac{1}{2}\pi + \theta$. Taking a new integration variable $\tau = \mp it/z$ we arrive at

$$S_\alpha(z) - \frac{1}{2z^{2\alpha}} = \frac{\sin(\pi\alpha)}{z^{2\alpha-1}} \int_1^\infty \frac{\operatorname{csch}(\pi z\tau)}{(\tau^2-1)^\alpha}\, d\tau. \tag{7.03}$$

Now

$$\operatorname{csch}(\pi z\tau) = 2e^{-\pi z\tau}(1-e^{-2\pi z\tau})^{-1} = 2\sum_{s=0}^{\infty} e^{-(2s+1)\pi z\tau}.$$

We may substitute this expansion in (7.03) and integrate term by term by means of Exercise 8.4 of Chapter 7. This yields

$$S_\alpha(z) = \frac{1}{2z^{2\alpha}} + \frac{2\pi^\alpha}{\Gamma(\alpha)\, z^{\alpha-(1/2)}} \sum_{s=0}^{\infty} \frac{K_{(1/2)-\alpha}\{(2s+1)\pi z\}}{(s+\frac{1}{2})^{(1/2)-\alpha}}, \tag{7.04}$$

the necessary inversion of the order of summation and integration being justifiable by reference to the dominated convergence theorem and the asymptotic form of the modified Bessel function K for large positive arguments (Chapter 7, §8.1). The temporary restriction $\operatorname{Re}\alpha < 1$ may now be lifted by appeal to analytic continuation.†

The series (7.04) converges rapidly for large z in the sector $|\operatorname{ph} z| < \frac{1}{2}\pi$; in consequence a few terms furnish a powerful approximation to $S_\alpha(z)$. In particular

$$S_\alpha(z) = \frac{1}{2z^{2\alpha}} + O\left(\frac{e^{-\pi z}}{z^\alpha}\right) \qquad (|\operatorname{ph} z| \leqslant \tfrac{1}{2}\pi - \delta < \tfrac{1}{2}\pi).$$

7.3 The second example to be treated in this section is the behavior of the series

$$F(\alpha, z) = \sum_{j=1}^{\infty} j^\alpha z^j \qquad (\alpha > 0),$$

as z approaches its circle of convergence.‡ Write

$$\beta = \operatorname{ph} z, \qquad \xi = \ln(1/|z|),$$

with β in the range $(-\pi, \pi]$. Then

$$F(\alpha, z) = \sum_{j=1}^{\infty} j^\alpha e^{ij\beta} e^{-j\xi};$$

† The restriction could have been avoided altogether by refraining from collapsing the loop contours of Fig. 7.1 and employing Hankel's integrals (Chapter 7, §4.5).

‡ Compare Chapter 1, Exercise 10.3. The method indicated is inapplicable to the present example because the series for $F(\alpha, 1)$ diverges when $\alpha \geqslant -1$.

compare (5.03). Our quest is the limiting form of $F(\alpha, z)$ as $\xi \to 0+$, β being kept fixed.

We have

$$F(\alpha, z) = \frac{1}{2i} \int_{\mathscr{C}} \cot(\pi t)\, t^{\alpha} e^{(i\beta - \xi)t}\, dt,$$

where $\mathscr{C}$ is a loop path enclosing the points $t = 1, 2, 3, \ldots$, but not $0, -1, -2, \ldots$. Following §3, we substitute

$$\frac{\cot \pi t}{2i} = -\frac{1}{2} - \frac{1}{e^{-2\pi i t} - 1} \qquad \text{(upper part of } \mathscr{C}\text{)}, \tag{7.05}$$

or

$$\frac{\cot \pi t}{2i} = \frac{1}{2} + \frac{1}{e^{2\pi i t} - 1} \qquad \text{(lower part of } \mathscr{C}\text{)}, \tag{7.06}$$

and deform the path to obtain

$$F(\alpha, z) = I_1 + I_2 + I_3,$$

where

$$I_1 = \int_0^{\infty} t^{\alpha} e^{(i\beta - \xi)t}\, dt, \qquad I_2 = \int_0^{i\infty} \frac{t^{\alpha} e^{(i\beta - \xi)t}}{e^{-2\pi i t} - 1}\, dt, \qquad I_3 = \int_0^{-i\infty} \frac{t^{\alpha} e^{(i\beta - \xi)t}}{e^{2\pi i t} - 1}\, dt.$$

The convergence of each integral is assured by the restrictions $\alpha > 0$ and $|\beta| \leqslant \pi$.

The first integral is explicitly evaluable:

$$I_1 = \Gamma(\alpha+1)(\xi - i\beta)^{-\alpha-1},$$

where $(\xi - i\beta)^{-\alpha-1}$ has its principal value. The second and third integrals are expansible in convergent power series in ξ; thus†

$$I_2 = \sum_{s=0}^{\infty} e^{(\alpha - s + 1)\pi i/2} \lambda_s(\alpha, \beta) \frac{\xi^s}{s!} \qquad (\xi < 2\pi + \beta), \tag{7.07}$$

$$I_3 = \sum_{s=0}^{\infty} e^{-(\alpha - s + 1)\pi i/2} \lambda_s(\alpha, -\beta) \frac{\xi^s}{s!} \qquad (\xi < 2\pi - \beta), \tag{7.08}$$

where

$$\lambda_s(\alpha, \beta) = e^{-(\alpha + s + 1)\pi i/2} \int_0^{i\infty} \frac{t^{\alpha + s} e^{i\beta t}}{e^{-2\pi i t} - 1}\, dt = \int_0^{\infty} \frac{\tau^{\alpha + s} e^{-\beta\tau}}{e^{2\pi\tau} - 1}\, d\tau.$$

Accordingly, if $\beta \neq 0$ then the limiting value of $F(\alpha, z)$ as $\xi \to 0$, that is, as $z \to e^{i\beta}$ along the ray $\operatorname{ph} z = \beta$, is given by

$$e^{\pm(\alpha+1)\pi i/2} \Gamma(\alpha+1) |\beta|^{-\alpha-1} + e^{(\alpha+1)\pi i/2} \lambda_0(\alpha, \beta) + e^{-(\alpha+1)\pi i/2} \lambda_0(\alpha, -\beta),$$

the upper or lower sign being taken according as $\beta \in (0, \pi]$ or $\beta \in (-\pi, 0)$.

† Again, necessary inversions of the order of summation and integration are justifiable by the dominated convergence theorem.

For the exceptional point $z = 1$, we have

$$F(\alpha, z) = \frac{\Gamma(\alpha+1)}{(-\ln z)^{\alpha+1}} + O(1) \qquad (z \to 1-).$$

And in this case the coefficients $\lambda_s(\alpha, \beta)$ can be expressed in terms of the Zeta function:

$$\lambda_s(\alpha, 0) = -\tfrac{1}{2}\csc(\tfrac{1}{2}\pi\alpha + \tfrac{1}{2}\pi s)\,\zeta(-\alpha - s);$$

compare Chapter 2, (11.02) and (11.05). Substitution in (7.07) and (7.08) yields

$$F(\alpha, z) = \frac{\Gamma(\alpha+1)}{(-\ln z)^{\alpha+1}} + \sum_{s=0}^{\infty} \zeta(-\alpha - s)\frac{(\ln z)^s}{s!}.$$

This expansion converges in the domain defined by $|\ln z| < 2\pi$, and furnishes the analytic continuation of $F(\alpha, z)$ in this region. In particular, $z = 1$ is seen to be a branch point or pole.

Ex. 7.1 Show that the function $F(\alpha, z)$ of §7.3 has the expansion

$$F(\alpha, z) = \Gamma(\alpha+1) \sum_{s=-\infty}^{\infty} \frac{1}{(2s\pi i - \ln z)^{\alpha+1}} \qquad (\alpha > 0, \quad 0 < z < 1).$$

By analytic continuation confirm that when $\operatorname{Re}\alpha > 0$ the only singularity of $F(\alpha, z)$ in the z plane is a branch point or pole at $z = 1$. [Evgrafov, 1961.]

Ex. 7.2 Prove that

$$\sum_{j=1}^{\infty} \frac{1}{j^2(z^2+j^2)^{1/2}} - \frac{\pi^2}{6z} + \frac{1}{z^2} - \frac{1}{4z^3} \sim -\frac{e^{-2\pi z}}{\pi^{1/2} z^{5/2}} \sum_{s=0}^{\infty} \frac{\Gamma(s+\frac{1}{2})\,a_s}{(2\pi z)^s}$$

as $z \to \infty$ in the sector $|\operatorname{ph} z| \leqslant \frac{1}{2}\pi - \delta$ $(< \frac{1}{2}\pi)$, where fractional powers take their principal values, $a_0 = 1$, and

$$a_s = \frac{1}{2^s}\binom{-\frac{1}{2}}{s} - \frac{2}{2^{s-1}}\binom{-\frac{1}{2}}{s-1} + \frac{3}{2^{s-2}}\binom{-\frac{1}{2}}{s-2} - \cdots + (-)^s(s+1) \qquad (s > 0).$$

Ex. 7.3 Assume that: (i) $f(t)$ is holomorphic within the union $\mathbf{T}(c)$ of the half-strip $|\operatorname{Im} t| \leqslant c$, $\operatorname{Re} t > 0$ and the disk $|t| \leqslant c$; (ii) $f(t)$ is real when $t \in (-c, \infty)$; (iii) $f(t) = O\{(\operatorname{Re} t)^{-1-\kappa}\}$ as $\operatorname{Re} t \to \infty$ in $\mathbf{T}(c)$, uniformly with respect to $\operatorname{Im} t$, where κ is a positive constant. By using the boundary of $\mathbf{T}(\delta)$ $(0 < \delta < c)$ as integration contour, expanding the factors $(e^{\mp 2\pi i t} - 1)^{-1}$ of (7.05) and (7.06) in series of powers of $e^{\pm 2\pi i t}$, and then letting $\delta \to 0$, establish *Poisson's summation formula*:†

$$\sum_{j=0}^{\infty} f(j) = \tfrac{1}{2}f(0) + \int_0^{\infty} f(t)\,dt + 2\sum_{s=1}^{\infty} \int_0^{\infty} f(t)\cos(2\pi s t)\,dt.$$

Ex. 7.4 By application of the preceding exercise show that

$$\sum_{j=0}^{\infty}{}' e^{-\alpha^2 j^2} = \frac{\pi^{1/2}}{2\alpha}\{1 + O(e^{-\pi^2/\alpha^2})\} \qquad (\alpha \to 0+),$$

where the prime indicates that the first term in the sum is to be halved.

† The assumed conditions are unnecessarily restrictive. By real-variable methods (Titchmarsh, 1948, §2.8; or Evgrafov, 1961, p. 138) it can be proved that sufficient conditions are (i) $f(t)$ is continuous and of bounded variation in $[0, \infty)$; (ii) $f(t) \to 0$ as $t \to \infty$; (iii) at least one of $\sum_{j=0}^{\infty} f(j)$ and $\int_0^{\infty} f(t)\,dt$ converges. Some interesting applications have been given by Lyness (1970, 1971a,c).

8 Asymptotic Expansions of Entire Functions

8.1 In this book we shall not attempt a systematic theory of the asymptotic behavior for large $|z|$ of functions defined by their Maclaurin series

$$F(z) = \sum_{j=0}^{\infty} a_j z^j \qquad (|z| < \infty).$$

The text is confined to a representative example amenable to the methods of the present chapter and previous chapters, given by

$$F_\rho(x) = \sum_{j=0}^{\infty} \left(\frac{x^j}{j!}\right)^\rho. \tag{8.01}$$

The problem is to find the asymptotic form of $F_\rho(x)$ for fixed positive ρ and large positive x. Special cases (which are useful later as checks) include $F_1(x) = e^x$ and $F_2(x) = I_0(2x)$, where I denotes the modified Bessel function.

8.2 Consider the contour $\mathscr{C}$ of Fig. 8.1, in which n denotes an arbitrary positive integer. By use of the residue theorem and equations (7.05) and (7.06), we derive

$$\begin{aligned}\sum_{j=0}^{n-1} \left(\frac{x^j}{j!}\right)^\rho &= \frac{1}{2i} \int_{\mathscr{C}} \left\{\frac{x^t}{\Gamma(t+1)}\right\}^\rho \cot(\pi t)\, dt \\ &= \int_{-1/2}^{n-(1/2)} \left\{\frac{x^t}{\Gamma(t+1)}\right\}^\rho dt - \int_{\mathscr{C}_1} \left\{\frac{x^t}{\Gamma(t+1)}\right\}^\rho \frac{dt}{e^{-2\pi i t}-1} \\ &\quad + \int_{\mathscr{C}_2} \left\{\frac{x^t}{\Gamma(t+1)}\right\}^\rho \frac{dt}{e^{2\pi i t}-1},\end{aligned}$$

where $\mathscr{C}_1$ and $\mathscr{C}_2$ are the upper and lower halves of $\mathscr{C}$, respectively.

With the aid of Stirling's approximation (4.04) we may verify that the integrals around the arc AB of $\mathscr{C}_1$ and the arc $\bar{B}A$ of $\mathscr{C}_2$ both vanish as $n \to \infty$, provided that $\rho \leqslant 4$. From here on we accept this restriction. Thus

$$F_\rho(x) = \int_{-1/2}^{\infty} \left\{\frac{x^t}{\Gamma(t+1)}\right\}^\rho dt + 2\,\mathrm{Re} \int_{-1/2}^{i\infty} \left\{\frac{x^t}{\Gamma(t+1)}\right\}^\rho \frac{dt}{e^{-2\pi i t}-1},$$

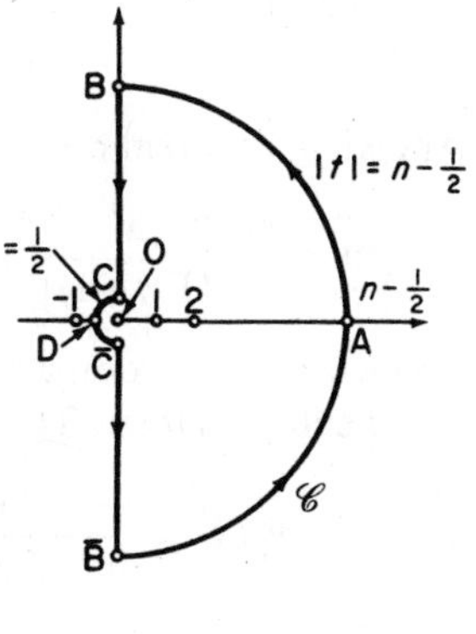

Fig. 8.1 t plane. Contour for $F_\rho(x)$.

where the path for the second integral consists of the quarter circle DC and the imaginary axis from C to $i\infty$. On this path—and also in the interval $[-\frac{1}{2}, 0]$—we have $|x^{t\rho}| \leqslant 1$ when $x \geqslant 1$. Therefore

$$F_\rho(x) = \int_0^\infty \left\{\frac{x^t}{\Gamma(t+1)}\right\}^\rho dt + O(1) \qquad (0 < \rho \leqslant 4). \tag{8.02}$$

The asymptotic behavior of the last integral is derivable by Laplace's method, as follows. Since

$$\frac{d}{dt}\left\{\frac{x^t}{\Gamma(t+1)}\right\} = \frac{x^t}{\Gamma(t+1)}\{\ln x - \psi(t+1)\}$$

and $\psi(t+1)$ is an increasing function, the integrand has a single peak, located at the root of the equation

$$\psi(t+1) = \ln x.$$

For large x this root is given by $t \sim x$; compare Exercise 4.2. With a view to replacing $\Gamma(t+1)$ by its Stirling approximant, we subdivide the integration range at λx, where λ is an arbitrary constant in the interval $(0, 1)$.

In the interval $0 \leqslant t \leqslant \lambda x$ the integrand in (8.02) attains its maximum at $t = \lambda x$. Therefore

$$\int_0^{\lambda x} \left\{\frac{x^t}{\Gamma(t+1)}\right\}^\rho dt \leqslant \left\{\frac{x^{\lambda x}}{\Gamma(\lambda x+1)}\right\}^\rho \lambda x = O\left\{\left(\frac{e}{\lambda}\right)^{\rho\lambda x} x^{1-(\rho/2)}\right\} \qquad (x \to \infty). \tag{8.03}$$

For the other interval, the standard procedure of Chapter 3 calls for a new variable $\tau = (t/x) - 1$. Then

$$\int_{\lambda x}^\infty \left\{\frac{x^t}{\Gamma(t+1)}\right\}^\rho dt = x\int_{-(1-\lambda)}^\infty \left\{\frac{x^{x(1+\tau)}}{\Gamma(x+x\tau+1)}\right\}^\rho d\tau$$

$$= \frac{x}{(2\pi x)^{\rho/2}}\left\{1 + O\left(\frac{1}{x}\right)\right\}\int_{-(1-\lambda)}^\infty \frac{e^{-\rho x\phi(\tau)}}{(1+\tau)^{\rho/2}}\, d\tau,$$

where

$$\phi(\tau) = (1+\tau)\{\ln(1+\tau) - 1\}. \tag{8.04}$$

For small τ,

$$\phi(\tau) = -1 + \tfrac{1}{2}\tau^2 - \tfrac{1}{6}\tau^3 + \cdots. \tag{8.05}$$

Applying Theorem 8.1 or Exercise 8.1 of Chapter 3, we obtain

$$\int_{\lambda x}^\infty \left\{\frac{x^t}{\Gamma(t+1)}\right\}^\rho dt = \frac{e^{\rho x}}{\rho^{1/2}(2\pi x)^{(\rho-1)/2}}\left\{1 + O\left(\frac{1}{x}\right)\right\} \qquad (x \to \infty). \tag{8.06}$$

Predictably, on comparing (8.03) with (8.06) we find that the latter estimate is exponentially larger. The required result is therefore given by

$$F_\rho(x) = \frac{e^{\rho x}}{\rho^{1/2}(2\pi x)^{(\rho-1)/2}}\left\{1 + O\left(\frac{1}{x}\right)\right\} \qquad (0 < \rho \leqslant 4, \quad x \to \infty). \tag{8.07}$$

It is easily verified that in the case $\rho = 2$ this approximation accords with the result of Chapter 7, §8.2.

Ex. 8.1 Show that in (8.07) the term $O(1/x)$ can be replaced by $\{(\rho^2-1)/(24\rho x)\} + O(1/x^2)$.

Ex. 8.2 Show that

$$\sum_{j=1}^{\infty} \frac{x^j}{j!} \ln j = e^x \left\{ \ln x - \frac{1}{2x} + O\left(\frac{1}{x^2}\right) \right\} \qquad (x \to +\infty).$$

[Pollak and Shepp, 1964.]

***Ex. 8.3** If the parameters a_j and c_j are fixed, none of the a_j being zero or a negative integer, show that

$$\left\{ \prod_{j=1}^{p} \Gamma(a_j) \Big/ \prod_{j=1}^{p} \Gamma(c_j) \right\} {}_pF_p(a_1, a_2, \ldots, a_p; c_1, c_2, \ldots, c_p; z) = z^k e^z \left\{ 1 + O\left(\frac{1}{z}\right) \right\}$$

as $z \to \infty$ in the sector $|\mathrm{ph}\, z| \leqslant \frac{1}{2}\pi - \delta$ $(< \frac{1}{2}\pi)$, where $k = \sum_{j=1}^{p} (a_j - c_j)$.

Also, if $\mathrm{Re}\, a_1$ is less than $\mathrm{Re}\, a_2, \mathrm{Re}\, a_3, \ldots, \mathrm{Re}\, a_p$, show that in the sector $|\mathrm{ph}(-z)| \leqslant \frac{1}{2}\pi - \delta$ the same function is approximated by

$$\left\{ \prod_{j=2}^{p} \Gamma(a_j - a_1) \Big/ \prod_{j=1}^{p} \Gamma(c_j - a_1) \right\} \Gamma(a_1)(-z)^{-a_1} + O(z^{-\alpha}) \qquad (z \to \infty),$$

where $\alpha = \min(1 + \mathrm{Re}\, a_1, \mathrm{Re}\, a_2, \mathrm{Re}\, a_3, \ldots, \mathrm{Re}\, a_p)$.

***Ex. 8.4** Define

$$G_\rho(x) = \sum_{j=0}^{\infty} (-)^j \left(\frac{x^j}{j!} \right)^\rho,$$

and let ρ be fixed and x be large and positive. If $\alpha = \pi/\rho$ and γ denotes Euler's constant, prove that

$$G_\rho(x) = \frac{2 \exp(x\rho \cos\alpha)}{\rho^{1/2} (2\pi x)^{(\rho-1)/2}} \{ \sin(\tfrac{1}{2}\alpha + x\rho \sin\alpha) + O(x^{-1}) \} \qquad (\rho \geqslant 2),$$

and

$$G_\rho(x) = \frac{1}{\Gamma(1-\rho)(\rho x \ln x)^\rho} \left[1 - \frac{\rho\gamma}{\ln x} + O\left\{ \frac{1}{(\ln x)^2} \right\} \right] \qquad (1 < \rho < 2).$$

9 Coefficients in a Power-Series Expansion; Method of Darboux

9.1 Let $f(t)$ be a given analytic function and

$$f(t) = \sum_{n=-\infty}^{\infty} a_n t^n$$

its Laurent expansion in an annulus $0 < |t| < r$. What is the asymptotic behavior of the sequence $\{a_n\}$ as $n \to \infty$ or $n \to -\infty$? More specially, what is the asymptotic behavior of the sequence of coefficients in a Maclaurin expansion?

Problems of this kind are immediately brought within the scope of earlier chapters by use of Cauchy's integral formula

$$a_n = \frac{1}{2\pi i} \int_{\mathscr{C}} \frac{f(t)}{t^{n+1}} \, dt, \tag{9.01}$$

in which $\mathscr{C}$ is a simple closed contour in the given annulus which encircles $t = 0$. A combination of the residue calculus and Laplace's method often suffices to determine the asymptotic form of the integral (9.01) for large $|n|$.

9.2 When $f(t)$ is not holomorphic throughout the region $|t| > 0$ an alternative approach, due to Darboux (1878), is available and frequently yields the required result with less labor. This method is as follows.

Let r be the distance from the origin of the nearest singularity of $f(t)$ and suppose that we can find a "comparison" function $g(t)$ having the following properties:

(i) *$g(t)$ is holomorphic in $0 < |t| < r$.*
(ii) *$f(t) - g(t)$ is continuous in $0 < |t| \leqslant r$.*
(iii) *The coefficients b_n in the Laurent expansion*

$$g(t) = \sum_{n=-\infty}^{\infty} b_n t^n \qquad (0 < |t| < r)$$

have known asymptotic behavior.

Then by allowing the contour in Cauchy's formula to expand we derive

$$a_n - b_n = \frac{1}{2\pi i} \int_{|t|=r} \frac{f(t) - g(t)}{t^{n+1}}\, dt = \frac{1}{2\pi r^n} \int_0^{2\pi} \{f(re^{i\theta}) - g(re^{i\theta})\}\, e^{-ni\theta}\, d\theta. \tag{9.02}$$

In consequence of the Riemann–Lebesgue lemma (Chapter 3, §4.2) the last integral vanishes as $n \to \infty$. Therefore

$$a_n = b_n + o(r^{-n}) \qquad (n \to \infty). \tag{9.03}$$

9.3 This simple result is refinable in two important ways. In the first place it is unnecessary for $f(t) - g(t)$ to be continuous on $|t| = r$; it is adequate that the integrals in (9.02) converge uniformly with respect to n. For example, Condition (ii) can be replaced by:

(iia) *On the circle $|t| = r$, the function $f(t) - g(t)$ has a finite number of singularities and at each singularity t_j, say,*

$$f(t) - g(t) = O\{(t - t_j)^{\sigma_j - 1}\} \qquad (t \to t_j),$$

where σ_j is an assignable positive constant.

The second refinement applies when $f(t) - g(t)$ is m times continuously differentiable on the circle $|t| = r$. In these circumstances the right-hand side of (9.02) may be integrated m times by parts to give the stronger result

$$a_n = b_n + o(r^{-n} n^{-m}) \qquad (n \to \infty), \tag{9.04}$$

also due to Darboux. And, again, this result is valid with the less restrictive condition

$$f^{(m)}(t) - g^{(m)}(t) = O\{(t - t_j)^{\sigma_j - 1}\} \tag{9.05}$$

in the neighborhood of each singularity t_j, provided that $\sigma_j > 0$.

10 Examples

10.1 The Legendre polynomials are generated by the function

$$f(t) \equiv \frac{1}{(1-2t\cos\alpha+t^2)^{1/2}} = \sum_{n=0}^{\infty} P_n(\cos\alpha)t^n \qquad (|t|<1);$$

compare Chapter 2, (7.20). The singularities of $f(t)$ are branch points at $t = e^{\pm i\alpha}$; we restrict $\alpha \in (0, \pi)$ to ensure that they are distinct.†

Let $(e^{i\alpha}-t)^{-1/2}$ denote the branch of this square root which is continuous in the t plane cut along the outward-drawn ray through $t = e^{i\alpha}$ and satisfies

$$(e^{i\alpha}-t)^{-1/2} \to e^{-i\alpha/2} \qquad (t \to 0);$$

see Fig. 10.1. Also, let $(e^{-i\alpha}-t)^{-1/2}$ denote the conjugate function. Then

$$f(t) = (e^{i\alpha}-t)^{-1/2}(e^{-i\alpha}-t)^{-1/2} \qquad (|t|<1).$$

Suppose that $t \to e^{\mp i\alpha}$ within the unit circle. Then from Fig. 10.2 we see that

$$(e^{i\alpha}-t)^{-1/2} \to e^{-\pi i/4}(2\sin\alpha)^{-1/2} \qquad (t \to e^{-i\alpha}),$$

$$(e^{-i\alpha}-t)^{-1/2} \to e^{\pi i/4}(2\sin\alpha)^{-1/2} \qquad (t \to e^{i\alpha}).$$

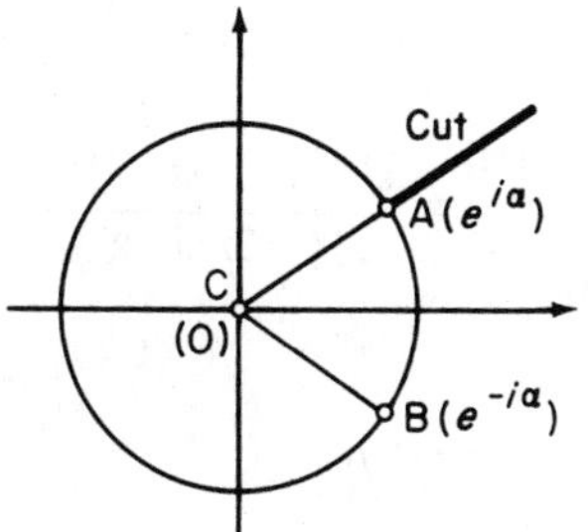

Fig. 10.1 t plane.

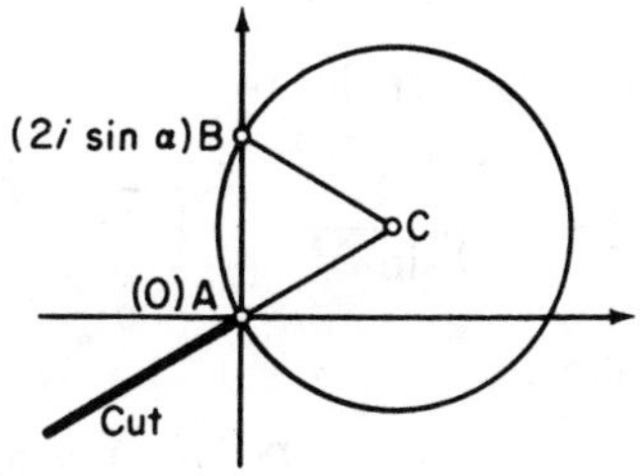

Fig. 10.2 $e^{i\alpha}-t$ plane.

Following the suggestions of §9.2, we take

$$g(t) = e^{-\pi i/4}(2\sin\alpha)^{-1/2}(e^{-i\alpha}-t)^{-1/2} + e^{\pi i/4}(2\sin\alpha)^{-1/2}(e^{i\alpha}-t)^{-1/2}.$$

Clearly $f(t)-g(t)$ is continuous within and on the unit circle, and the coefficient of t^n in the Maclaurin expansion of $g(t)$ is

$$b_n = \left(\frac{2}{\sin\alpha}\right)^{1/2}\binom{-\frac{1}{2}}{n}\cos\{(n+\tfrac{1}{2})\alpha + (n-\tfrac{1}{4})\pi\}.$$

Darboux's first formula (9.03) yields

$$P_n(\cos\alpha) = b_n + o(1) \qquad (n \to \infty).$$

† For $\alpha = 0$ or π we have $P_n(\pm 1) = (\pm 1)^n$.

However, since

$$\binom{-\frac{1}{2}}{n} = (-)^n \frac{\Gamma(n+\frac{1}{2})}{\pi^{1/2} n!} \sim \frac{(-1)^n}{(\pi n)^{1/2}} \qquad (n \to \infty),$$

and is therefore $o(1)$, this result is somewhat uninformative.

To improve the estimate for the error term, we observe that $f(t) - g(t)$ is const. $+O\{(t-e^{\mp i\alpha})^{1/2}\}$ as $t \to e^{\mp i\alpha}$. Condition (9.05) is satisfied with $m = 1$, and from (9.04) we derive

$$P_n(\cos\alpha) = b_n + o(n^{-1}) = \left(\frac{2}{\pi n \sin\alpha}\right)^{1/2} \cos(n\alpha + \tfrac{1}{2}\alpha - \tfrac{1}{4}\pi) + o(n^{-1}), \tag{10.01}$$

in agreement with Chapter 4, (8.01), apart from the error term.

10.2 To extend (10.01) into an asymptotic expansion we match $f(t)$ by a function $g_m(t)$ such that $f(t) - g_m(t)$ is m times continuously differentiable within and on the unit circle, m being an arbitrary integer. Since

$$(e^{i\alpha} - t)^{-1/2} = \frac{e^{-\pi i/4}}{(2\sin\alpha)^{1/2}} \sum_{s=0}^{\infty} \binom{-\frac{1}{2}}{s} \left(\frac{e^{-i\alpha} - t}{2i\sin\alpha}\right)^s \qquad (|e^{-i\alpha} - t| < 2\sin\alpha), \tag{10.02}$$

an appropriate choice is

$$g_m(t) = \frac{1}{(2\sin\alpha)^{1/2}} \sum_{s=0}^{m} \binom{-\frac{1}{2}}{s} \left\{ \frac{e^{-\pi i/4}(e^{-i\alpha} - t)^{s-(1/2)}}{(2i\sin\alpha)^s} + \frac{e^{\pi i/4}(e^{i\alpha} - t)^{s-(1/2)}}{(-2i\sin\alpha)^s} \right\}.$$

Accordingly,

$$P_n(\cos\alpha) = b_{n,m} + o(n^{-m}) \qquad (n \to \infty), \tag{10.03}$$

where $b_{n,m}$ is the coefficient of t^n in the Maclaurin expansion of $g_m(t)$.

An explicit expression for $b_{n,m}$ is given by

$$b_{n,m} = \left(\frac{2}{\sin\alpha}\right)^{1/2} \sum_{s=0}^{m} \binom{-\frac{1}{2}}{s} \binom{s-\frac{1}{2}}{n} \frac{\cos\alpha_{n,s}}{(2\sin\alpha)^s},$$

where $\alpha_{n,s} = (n-s+\frac{1}{2})\alpha + (n-\frac{1}{2}s-\frac{1}{4})\pi$. Moreover, since

$$b_{n,m+2} = b_{n,m} + O\left\{\binom{m+\frac{1}{2}}{n}\right\} + O\left\{\binom{m+\frac{3}{2}}{n}\right\} = b_{n,m} + O\left(\frac{1}{n^{m+(3/2)}}\right),$$

we see—by replacing m by $m+2$—that the error term $o(n^{-m})$ in (10.03) can be strengthened into $O(n^{-m-(3/2)})$. Thus in the notation of Chapter 1, §10.3

$$P_n(\cos\alpha) \sim \left(\frac{2}{\sin\alpha}\right)^{1/2} \sum_{s=0}^{\infty} \binom{-\frac{1}{2}}{s} \binom{s-\frac{1}{2}}{n} \frac{\cos\alpha_{n,s}}{(2\sin\alpha)^s}; \qquad \{n^{-s-(1/2)}\} \tag{10.04}$$

as $n \to \infty$.

A particular case of this result is that the error term $o(n^{-1})$ in (10.01) can be replaced by $O(n^{-3/2})$, agreeing with Chapter 4, (8.01).

10.3 Before leaving this example we draw attention to an interesting paradox.† It is easily verified that the series on the right-hand side of (10.04) converges when $2\sin\alpha > 1$, that is, when $\frac{1}{6}\pi < \alpha < \frac{5}{6}\pi$; it is natural to expect that the sum is $P_n(\cos\alpha)$ in these circumstances. But this is not the case. From (10.02) we derive

$$\frac{1}{(1-2t\cos\alpha+t^2)^{1/2}} = \frac{e^{-\pi i/4}}{(2\sin\alpha)^{1/2}}(e^{-i\alpha}-t)^{-1/2}\sum_{s=0}^{\infty}\binom{-\frac{1}{2}}{s}\frac{(e^{-i\alpha}-t)^{s-(1/2)}}{(2i\sin\alpha)^s},$$

this series converging uniformly when $|e^{-i\alpha}-t| \leqslant 2\sin\alpha-\delta$, where δ is an arbitrary small positive number. If $2\sin\alpha > 1$, then $t=0$ is includable in the region of uniform convergence. Differentiating n times, setting $t=0$, and equating real parts, we obtain

$$P_n(\cos\alpha) = \frac{1}{(2\sin\alpha)^{1/2}}\sum_{s=0}^{\infty}\binom{-\frac{1}{2}}{s}\binom{s-\frac{1}{2}}{n}\frac{\cos\alpha_{n,s}}{(2\sin\alpha)^s} \qquad (\tfrac{1}{6}\pi < \alpha < \tfrac{5}{6}\pi). \tag{10.05}$$

Comparison with (10.04) shows that the sum of the series (10.05) is represented as a generalized asymptotic expansion by *twice the same series* when n is large.

The explanation of the paradox is that the tail of (10.05), that is, the sum from $s=m$ (fixed) to $s=\infty$, is of a larger mathematical order than the first neglected term.‡ This is confirmable to some extent by considering higher terms. For example, apart from the oscillatory factor $\cos\alpha_{n,s}$ the contribution of the term for which $s=2n$ is

$$\frac{\Gamma(2n+\frac{1}{2})\Gamma(n+\frac{1}{4})\Gamma(n+\frac{3}{4})}{2\pi\Gamma(2n+1)\Gamma(n+\frac{1}{2})\Gamma(n+1)(\sin\alpha)^{2n+(1/2)}}.$$

For large n, this is asymptotic to $(\csc\alpha)^{2n+(1/2)}/(2^{3/2}\pi n)$ and therefore becomes infinitely large compared with the $(m+1)$th term, for any fixed value of m.

10.4 As a second example,§ consider the Maclaurin expansion

$$\phi(t) = \sum_{n=0}^{\infty} a_n t^{2n+1} \tag{10.06}$$

of the analytic function defined implicitly by the equation

$$\phi(t) - \sin\phi(t) = \tfrac{1}{6}t^3, \tag{10.07}$$

with the condition $\phi(t) \sim t$ as $t \to 0$. By reversion, the first four coefficients are found to be

$$a_0 = 1, \qquad a_1 = \tfrac{1}{60}, \qquad a_2 = \tfrac{1}{1400}, \qquad a_3 = \tfrac{1}{25200},$$

but an explicit fraction for a_n is unavailable for general n.

† Olver (1970b).
‡ This cannot happen with an ordinary Poincaré expansion; see Chapter 1, §3.3.
§ Watson (1944, §8.21).

From (9.01) we have

$$a_n = \frac{1}{2\pi i}\int_{\mathscr{C}} \frac{\phi(t)\, dt}{t^{2n+2}},$$

where $\mathscr{C}$ is a small closed contour encircling $t = 0$. Since the mapping between t and ϕ is conformal in the neighborhoods of the origins, transformation to the ϕ plane gives

$$a_n = \frac{1}{\pi i}\int_{\mathscr{S}} \frac{\phi(1-\cos\phi)\, d\phi}{\{6(\phi-\sin\phi)\}^{(2n+4)/3}} = \frac{1}{\pi i}\int_{\mathscr{S}} e^{-np(\phi)} q(\phi)\, d\phi, \tag{10.08}$$

where $\mathscr{S}$ is a small closed contour encircling $\phi = 0$, and

$$p(\phi) = \tfrac{2}{3}\ln\{6(\phi-\sin\phi)\}, \qquad q(\phi) = \phi(1-\cos\phi)\{6(\phi-\sin\phi)\}^{-4/3}.$$

The saddle points of the last integral are $\phi = \pm 2\pi, \pm 4\pi, \ldots$. To apply Laplace's approximation we deform $\mathscr{S}$ into straight lines parallel to the imaginary axis and passing through the saddle points at $\pm 2\pi$.† On these lines $\operatorname{Re}\{p(\phi)\}$ attains its minimum at the saddle points.

To record the expansions of $p(\phi)$ and $q(\phi)$ in the neighborhood of $\phi = 2\pi$, we write $\phi \equiv 2\pi + \psi$. Then

$$6(\phi - \sin\phi) = 12\pi + \psi^3 - \frac{\psi^5}{20} + \frac{\psi^7}{840} - \cdots,$$

$$p(\phi) = \frac{2}{3}\ln(12\pi) + \frac{\psi^3}{18\pi} - \frac{\psi^5}{360\pi} - \frac{\psi^6}{432\pi^2} + \frac{\psi^7}{15120\pi} + \cdots,$$

and

$$q(\phi) = \frac{1}{(12\pi)^{4/3}}\left\{\pi\psi^2 + \frac{\psi^3}{2} - \pi\frac{\psi^4}{12} - \frac{11\psi^5}{72} + \left(\frac{\pi}{360} - \frac{1}{18\pi}\right)\psi^6 + \cdots\right\}.$$

Use of Theorem 6.1 of Chapter 4 eventually leads to

$$\int_{2\pi}^{2\pi+i\infty} e^{-np(\phi)} q(\phi)\, d\phi \sim \frac{1}{(12\pi)^{(2n+4)/3}}\left\{\frac{\pi}{3}\left(\frac{18\pi}{n}\right) + e^{2\pi i/3}\frac{\Gamma(\frac{4}{3})}{6}\left(\frac{18\pi}{n}\right)^{4/3} - \frac{7}{1080}\left(\frac{18\pi}{n}\right)^2 - e^{2\pi i/3}\frac{\Gamma(\frac{7}{3})}{432\pi}\left(\frac{18\pi}{n}\right)^{7/3} + \cdots\right\} \qquad (n\to\infty).$$

For the corresponding integral from $2\pi - i\infty$ to 2π, we take the negative of the complex conjugate of this result. Furthermore, increasing the phase of ϕ by π increases that of $\phi - \sin\phi$ by 3π. Since n is an integer, it follows that the integrands in (10.08) are odd functions of ϕ. In consequence the integral down the line through $\phi = -2\pi$ equals that up the line through $\phi = 2\pi$. Combination of these results

† By application of the phase principle it is easily verified that $\phi - \sin\phi$ has no zeros in the strip $|\operatorname{Re}\phi| \leqslant 2\pi$, other than the triple zero at $\phi = 0$.

yields the desired expansion in the form

$$a_n = \frac{1}{(18)^{1/3}\Gamma(\frac{2}{3})n^{4/3}(12\pi)^{2n/3}}\left\{1 - \frac{1}{3n} + O\left(\frac{1}{n^{4/3}}\right)\right\}. \tag{10.09}$$

This result can also be obtained (and with less arithmetic) by the method of Darboux; see Exercise 10.2 below.

10.5 A third, and more difficult, example is the determination of the asymptotic behavior of the Maclaurin coefficients of $\exp(e^t)$. This is treated in Chapter 9, §3.

Ex. 10.1 Let a be a real or complex constant. By Darboux's method show that the $(n+1)$th Maclaurin coefficient of the function $2^{1-a}e^{at}/(2-e^t)$ is given by

$$a_n = \sum_{s=1-m}^{m-1} \frac{e^{2s\pi i a}}{(\ln 2 + 2s\pi i)^{n+1}} + O\left\{\frac{1}{|\ln 2 + 2m\pi i|^n}\right\} \qquad (n \to \infty),$$

where m is arbitrary. [Evgrafov, 1961.]

Ex. 10.2 Show by means of conformal mapping that the function $\phi(t)$ defined by (10.07) is holomorphic in the t plane cut along the real axis from $-\infty$ to $-(12\pi)^{1/3}$ and $(12\pi)^{1/3}$ to $+\infty$. Deduce that the coefficients a_n in the expansion (10.06) are given by

$$(12\pi)^{(2n+1)/3} a_n \sim 2\sum_{s=1}^{\infty} \binom{\frac{1}{3}s}{2n+1}\phi_s; \qquad \{n^{-(s/3)-1}\} \qquad (n \to \infty),$$

where $\phi_1, \phi_2, \ldots$ are the coefficients in the expansion of $\phi(t)$ in the neighborhoods of the branch points $t = \pm(12\pi)^{1/3}$:

$$\phi(t) = \pm 2\pi \mp \sum_{s=1}^{\infty} \phi_s \left\{1 \mp \frac{t}{(12\pi)^{1/3}}\right\}^{s/3}.$$

By evaluating ϕ_1, ϕ_2, and ϕ_4 verify that this result agrees with (10.09).

***Ex. 10.3** If $H_n(x)$ denotes the Hermite polynomial, then from Chapter 2, (7.09) derive

$$H_n(\nu^{1/2}x) = \frac{(-)^n n!}{2\pi i \nu^{n/2}} \int_{\mathscr{C}} \exp\{-\nu(2xt + t^2)\} \frac{dt}{t^{(\nu+1)/2}},$$

where $\nu = 2n+1$ and $\mathscr{C}$ is a simple closed contour encircling $t = 0$. Deduce that if n is large, and α and β are fixed numbers such that $0 < \alpha < \pi$ and $\beta > 0$, then†

(i) $H_n(\nu^{1/2}\cos\alpha) = (2\csc\alpha)^{1/2}\nu^{(\nu-1)/4}\exp(\frac{1}{4}\nu\cos 2\alpha)\{\sin(\frac{1}{2}\nu\alpha - \frac{1}{4}\nu\sin 2\alpha + \frac{1}{4}\pi) + O(n^{-1})\}$.

(ii) $H_n(\nu^{1/2}\cosh\beta) = (\frac{1}{2}\operatorname{csch}\beta)^{1/2}\nu^{(\nu-1)/4}\exp(\frac{1}{4}\nu e^{-2\beta} + \frac{1}{2}\nu\beta)\{1 + O(n^{-1})\}$.

(iii) $H_n(\nu^{1/2}) \sim \{2^{1/3}\pi^{1/2}\nu^{(3\nu-1)/12}e^{\nu/4}\}/\{3^{2/3}\Gamma(\frac{2}{3})\}$.

*11 Inverse Laplace Transforms; Haar's Method

11.1 The Fourier–Mellin inversion theorem for the Laplace transform can be stated in the following form:

Theorem 11.1‡ *Let $f(t)$ be a real or complex function of the real variable t with the following properties*:

† Higher terms in these approximations have been given by Watson (1918a) and Olver (1959).
‡ For a proof see, for example, Tolstov (1962, p. 188).

(i) $f(t) = 0 \quad (t < 0)$.
(ii) $f(t)$ *is continuous in* $(0, \infty)$ *save at a finite number of points.*
(iii) $\int_0^\infty e^{-\sigma t}|f(t)|\,dt < \infty$, *where* σ *is a constant.*

Then the function

$$F(p) = \int_0^\infty e^{-pt} f(t)\,dt \tag{11.01}$$

is holomorphic in the half-plane $\operatorname{Re} p > \sigma$, *and*

$$\frac{1}{2\pi i}\int_{c-i\infty}^{c+i\infty} e^{pt}F(p)\,dp = \tfrac{1}{2}f(t-) + \tfrac{1}{2}f(t+) \tag{11.02}$$

whenever $c \geqslant \sigma$ *and* $f(t)$ *has both a left-hand derivative and right-hand derivative at* t. *The last integral is to be interpreted as a Cauchy principal value*† *in cases where it does not converge separately at each limit.*

Formula (11.02) provides a way of estimating the asymptotic behavior of $f(t)$ for large t. Writing $p = c + iv$ and assuming, for simplicity, that $f(t)$ is differentiable for all sufficiently large t, we have

$$e^{-ct}f(t) = \frac{1}{2\pi}\int_{-\infty}^{\infty} e^{itv}F(c+iv)\,dv.$$

Suppose also that the last integral converges uniformly at each limit for all sufficiently large t. Then by the Riemann–Lebesgue lemma

$$f(t) = o(e^{ct}) \qquad (t \to +\infty). \tag{11.03}$$

Next, suppose that in addition to the foregoing conditions each of the integrals

$$\int_{-\infty}^{\infty} e^{itv}F^{(s)}(c+iv)\,dv \qquad (s = 1, 2, \ldots, m)$$

converges uniformly for all sufficiently large t. Then $F(p)$, $F'(p), \ldots, F^{(m-1)}(p)$ all vanish as $p \to c \pm i\infty$,‡ and by repeated integrations by parts we obtain the improved estimate

$$f(t) = o(t^{-m}e^{ct}) \qquad (t \to \infty). \tag{11.04}$$

11.2 The further the contour in (11.02) can be translated to the left, the smaller the estimates (11.03) and (11.04) become. The method of Haar (1926) for approximating $f(t)$ is similar to Darboux's procedure for approximating coefficients in Laurent expansions (§9) and provides another illustration of the general principle of extraction of singular parts.

Let r be the abscissa of the rightmost singularity (or singularities) of $F(p)$, and assume that the contour in (11.02) can be deformed into a path $\mathscr{C}(r)$, comprising the line $\operatorname{Re} p = r$ with indentations passing to the right of the singularities. Suppose that a function $G(p)$ can be found having singularities at the same points on

† That is, the limit of $\int_{c-i\omega}^{c+i\omega}$ as $\omega \to \infty$.
‡ Compare Chapter 3, §5.2.

$\operatorname{Re} p = r$ and the properties:

(i) *$G(p)$ is holomorphic when* $\operatorname{Re} p > r$.

(ii) *$F^{(m)}(p) - G^{(m)}(p)$ is continuous when* $\operatorname{Re} p \geqslant r$, *for some nonnegative integer m.*

(iii) *Each integral* $\int_{r-i\infty}^{r+i\infty} e^{t(p-r)}\{F^{(s)}(p) - G^{(s)}(p)\}\,dp$, $s = 0, 1, \ldots, m$, *converges uniformly for all sufficiently large t.*

(iv) *The integral*

$$g(t) = \frac{1}{2\pi i}\int_{\mathscr{C}(r)} e^{tp} G(p)\,dp$$

has known asymptotic behavior for large t.

We then have

$$f(t) - g(t) = \frac{1}{2\pi i}\int_{\mathscr{C}(r)} e^{tp}\{F(p) - G(p)\}\,dp.$$

In consequence of Condition (ii), $\mathscr{C}(r)$ may now be replaced by the line $\operatorname{Re} p = r$. From the analysis of §11.1 we have the required result

$$f(t) = g(t) + o(t^{-m} e^{rt}) \qquad (t \to \infty).$$

Usually the most difficult requirement to satisfy in applications is the convergence of the integrals in Conditions (iii) and (iv). One way of overcoming this stumbling block—and thereby greatly increasing the power of the method—is to bend the contour of integration into the left half-plane. The factor e^{tp} then plays a strong role in the convergence.

We illustrate the method by two examples. The first is from Haar's paper (1926).† The second has been studied independently by several writers.‡

11.3 From Chapter 2, Exercise 9.8, we have, by analytic continuation,

$$\int_0^\infty e^{-pt} t^\nu J_\nu(t)\,dt = \frac{2^\nu \Gamma(\nu + \frac{1}{2})}{\pi^{1/2}(p^2+1)^{\nu+(1/2)}} \qquad (\operatorname{Re} p > 0, \quad \nu > -\tfrac{1}{2}).$$

Application of Theorem 11.1, with σ taken to be any positive number, yields

$$t^\nu J_\nu(t) = \pi^{-1/2} 2^\nu \Gamma(\nu + \tfrac{1}{2}) f(t) \qquad (t > 0), \tag{11.05}$$

where

$$f(t) = \frac{1}{2\pi i}\int_{c-i\infty}^{c+i\infty} \frac{e^{tp}\,dp}{(p^2+1)^{\nu+(1/2)}} \qquad (\nu > -\tfrac{1}{2}, \quad c > 0), \tag{11.06}$$

$(p^2+1)^{\nu+(1/2)}$ taking its principal value.

To find the asymptotic expansion of $f(t)$ for large t, we observe that the integrand in (11.06) has singularities at $p = \pm i$, and express $f(t)$ as the sum of two loop integrals

$$\frac{1}{2\pi i}\left\{\int_{-\infty+i}^{(i+)} + \int_{-\infty-i}^{(-i+)}\right\}. \tag{11.07}$$

† See also Doetsch (1955, pp. 168–170).

‡ Hull and Froese (1955), Wyman and Wong (1969), and Dorning, Nicolaenko, and Thurber (1969). See also Riekstiņš (1973).

In the first of these we substitute the expansion

$$\frac{1}{(p^2+1)^{\nu+(1/2)}} = \sum_{s=0}^{n-1} \frac{1}{2^{\nu+(1/2)}e^{(2\nu+1)\pi i/4}} \binom{-\nu-\frac{1}{2}}{s} \frac{(p-i)^{s-\nu-(1/2)}}{(2i)^s}$$
$$+ O\{(p-i)^{n-\nu-(1/2)}\}, \tag{11.08}$$

in which n is an arbitrary integer and the O term is uniform with respect to p on the contour. Using Hankel's loop integral for the reciprocal of the Gamma function we arrive at

$$\frac{1}{2\pi i}\int_{-\infty+i}^{(i+)} \frac{e^{tp}\,dp}{(p^2+1)^{\nu+(1/2)}}$$
$$= \frac{1}{2^{\nu+(1/2)}e^{(2\nu+1)\pi i/4}} \sum_{s=0}^{n-1} \binom{-\nu-\frac{1}{2}}{s} \frac{1}{(2i)^s} \frac{e^{it}}{\Gamma(\nu+\frac{1}{2}-s)\,t^{s-\nu+(1/2)}} + \varepsilon_n(t),$$

where

$$\varepsilon_n(t) = \frac{1}{2\pi i}\int_{-\infty+i}^{(i+)} e^{tp} O\{(p-i)^{n-\nu-(1/2)}\}\,dp.$$

Provided that $n > \nu-\frac{1}{2}$ the contour in the last integral may be collapsed onto the cut through $p = i$ parallel to the negative real axis. Thence we see that

$$\varepsilon_n(t) = O(1/t^{n-\nu+(1/2)}) \qquad (t \to \infty).$$

Similar analysis applies to the second of the integrals (11.07). Substituting in (11.05) and (11.06) by means of these results, we obtain the required expansion

$$J_\nu(t) = \left(\frac{2}{\pi t}\right)^{1/2} \sum_{s=0}^{n-1} \binom{-\nu-\frac{1}{2}}{s} \frac{\Gamma(\nu+\frac{1}{2})}{\Gamma(\nu+\frac{1}{2}-s)} \frac{\cos\{t-(\frac{1}{2}s+\frac{1}{2}\nu+\frac{1}{4})\pi\}}{(2t)^s} + O\left(\frac{1}{t^{n+(1/2)}}\right)$$

as $t \to +\infty$. The restriction $n > \nu-\frac{1}{2}$ is now easily removed, and the only one remaining is $\nu > -\frac{1}{2}$. It is readily verified that this expansion agrees with (9.09) of Chapter 4.

From the standpoint of Haar, the role of $F(p)$ is played here by $(p^2+1)^{-\nu-(1/2)}$ and that of $G(p)$ by

$$\sum_{s=0}^{n-1} \frac{1}{2^{\nu+(1/2)}} \binom{-\nu-\frac{1}{2}}{s} \left\{\frac{(p-i)^{s-\nu-(1/2)}}{e^{(2\nu+1)\pi i/4}(2i)^s} + \frac{(p+i)^{s-\nu-(1/2)}}{e^{-(2\nu+1)\pi i/4}(-2i)^s}\right\}.$$

Conditions (i) and (ii) of §11.2 are fulfilled with $r = 0$, provided that $n \geqslant \nu+\frac{1}{2}$ and $m \leqslant [n-\nu-\frac{1}{2}]$. Condition (iii) is not necessarily satisfied, but we overcome this difficulty, and at the same time ensure the convergence of the integral in Condition (iv), by deforming the integration path into the two loops.

11.4 The second example is furnished by the function

$$\Psi(\alpha, t) = \int_\alpha^\infty \frac{t^{v-1}}{\Gamma(v)}\,dv \qquad (\alpha \geqslant 0, \quad t > 0). \tag{11.09}$$

In Exercise 8.7 of Chapter 3 it is stated that

$$\Psi(\alpha, t) \sim e^t \qquad (\alpha \text{ fixed}, \quad t \to \infty). \tag{11.10}$$

This is verifiable by means of Laplace's method and Stirling's formula. We now propose to extend (11.10) into an asymptotic expansion.

Provided that $\alpha > 0$, the Laplace transform of $\Psi(\alpha, t)$ is obtainable by integrating the identity

$$\int_0^\infty \frac{e^{-pt} t^{v-1}}{\Gamma(v)} dt = \frac{1}{p^v} \qquad (\operatorname{Re} p > 0, \quad v > 0)$$

with respect to v from α to ∞. This gives

$$\int_0^\infty e^{-pt} \Psi(\alpha, t)\, dt = \frac{1}{p^\alpha \ln p} \qquad (\operatorname{Re} p > 1), \tag{11.11}$$

the inversion of the order of the integrations being justifiable by uniform convergence. Uniform convergence also enables (11.11) to be extended to $\alpha = 0$, because $\Psi(\alpha, t) \leqslant \Psi(0, t)$ and (in consequence of Exercises 8.6 and 8.7 of Chapter 3) the integral

$$\int_0^\infty e^{-pt} \Psi(0, t)\, dt$$

converges at both limits when $p > 1$.

Application of Theorem 11.1 yields

$$\Psi(\alpha, t) = \frac{1}{2\pi i} \int_{c-i\infty}^{c+i\infty} \frac{e^{tp}\, dp}{p^\alpha \ln p} \qquad (c > 1).$$

The singularities of the integrand consist of a simple pole of residue e^t at $p = 1$, and a branch point at $p = 0$. The stationary points satisfy

$$t - \frac{\alpha}{p} - \frac{1}{p \ln p} = 0.$$

For large t this equation has a root of the form

$$p = \frac{\alpha}{t} + O\left(\frac{1}{t \ln t}\right),$$

which tends to zero as $t \to \infty$. To enable the integration path to pass through this point it has to be translated across the pole. Bending the path to the left and allowing for the residue at the pole, we derive

$$\Psi(\alpha, t) = e^t + F(t), \tag{11.12}$$

where

$$F(t) = \frac{1}{2\pi i} \int_{-\infty}^{(0+)} \frac{e^{tp}\, dp}{p^\alpha \ln p}.$$

The standard procedure of Chapter 4 now calls for the substitution $p = q/t$. This yields

$$F(t) = -\frac{t^{\alpha-1}}{2\pi i}\int_{-\infty}^{(0+)} \frac{e^q\,dq}{q^\alpha(\ln t - \ln q)}$$

$$= -\frac{t^{\alpha-1}}{2\pi i}\int_{-\infty}^{(0+)} \left\{\sum_{s=0}^{n-1} \frac{(\ln q)^s}{(\ln t)^{s+1}} + \frac{(\ln q)^n}{(\ln t)^n(\ln t - \ln q)}\right\}\frac{e^q}{q^\alpha}\,dq,$$

where n is an arbitrary positive integer.

Parametric differentiation of Hankel's loop integral for $\mathrm{Rg}(\alpha) \equiv 1/\Gamma(\alpha)$ produces

$$\frac{1}{2\pi i}\int_{-\infty}^{(0+)} \frac{(\ln q)^s e^q}{q^\alpha}\,dq = (-)^s\,\mathrm{Rg}^{(s)}(\alpha) \qquad (s = 0, 1, 2, \ldots).$$

Also, it is easily seen that

$$(\ln t - \ln q)^{-1} = O(1) \qquad (t \to \infty),$$

uniformly with respect to q on the contour. Hence

$$F(t) = -t^{\alpha-1}\left[\sum_{s=0}^{n-1} (-)^s \frac{\mathrm{Rg}^{(s)}(\alpha)}{(\ln t)^{s+1}} + O\left\{\frac{1}{(\ln t)^n}\right\}\right].$$

Clearly the upper summation limit may be changed to $n-2$ without affecting the error term. Substitution in (11.12) then gives the desired result in the form

$$\Psi(\alpha, t) \sim e^t - t^{\alpha-1}\sum_{s=0}^{\infty} (-)^s \frac{\mathrm{Rg}^{(s)}(\alpha)}{(\ln t)^{s+1}} \qquad (t \to \infty). \tag{11.13}$$

Ex. 11.1 Prove that

$$\frac{1}{2\pi i}\int_{c-i\infty}^{c+i\infty} \frac{e^{tp}\ln p}{p^2+1}\,dp \sim \frac{1}{2}\pi\cos t - \sum_{s=0}^{\infty} (-)^s \frac{(2s)!}{t^{2s+1}} \qquad (t \to \infty),$$

provided that $c > 0$ and $\ln p$ has its principal value.

Confirm the result by showing that the left-hand side equals $\cos t\,\mathrm{Si}(t) - \sin t\,\mathrm{Ci}(t)$ and using Exercise 1.3 of Chapter 3. [Hull and Froese, 1955.]

Ex. 11.2† Let

$$f(t) = \frac{1}{2\pi i}\int_{c-i\infty}^{c+i\infty} e^{tp}p^p\,dp,$$

where $c > 0$ and p^p has its principal value. Show that

(i) If $t \to -\infty$ and $T \equiv e^{-t-1}$, then

$$f(t) = e^{-T}\left(\frac{T}{2\pi}\right)^{1/2}\left\{1 + \frac{1}{24T} + O\left(\frac{1}{T^2}\right)\right\}.$$

(ii) If $t \to +\infty$ and γ denotes Euler's constant, then

$$f(t) = \frac{1}{t^2} + \frac{2\ln t + 2\gamma - 3}{t^3} + O\left\{\frac{(\ln t)^2}{t^4}\right\}.$$

† Börsch-Supan (1961). The integral arises in the calculation of energy loss of fast particles due to ionization.

Historical Notes and Additional References

Most of the textual material in this chapter is classical. Considerable use has been made of the works of Lindelöf (1905), Ford (1916), Hardy (1949), Oberhettinger (1953), de Bruijn (1961), and Evgrafov (1961).

§§1–3 For further properties of the Bernoulli polynomials and related polynomials see, for example, Milne-Thomson (1933) and N.B.S. (1964). For notes on the discovery of the Euler–Maclaurin formula see Barnes (1905) or Whittaker and Watson (1927, p. 127). For recent extensions and applications see Duncan (1957), Navot (1961), Lyness and Ninham (1967), and Chakravarti (1970).

§5 The method of summation by parts has seldom been exploited in the manner given here.

§6.1 An analogous procedure based on the Mellin transform has been given by Macfarlane (1949).

§6.2 Integrals of Barnes' type have been discussed extensively by Braaksma (1963) and Luke (1969a).

§8 For extensive treatments of the problem of determining the asymptotic behavior of entire functions from their Maclaurin series see Barnes (1906), Watson (1913), Ford (1936), Wright (1948), Evgrafov (1961), Riekstiņa (1968), and references cited in these works.

§§9–10 For recent work on the asymptotic expansion of Laurent coefficients and the method of Darboux, see Wyman (1959) and Fields (1968).

§11 Further development and applications of Haar's method have been made by Reudink (1965) and Fields and Ismail (1974). The latter reference includes derivations of error bounds.

For an account of analogous methods based on the Parseval formula for Mellin transforms instead of the Fourier–Mellin inversion theorem, see Bleistein and Handelsman (1974).

9

INTEGRALS: FURTHER METHODS

1 Logarithmic Singularities

1.1 In Chapter 3, §§7 and 8 we discussed the asymptotic expansion of integrals of the form

$$\int_a^b e^{-xp(t)}q(t)\,dt,$$

in which x is large and positive, $p(t)$ attains its minimum at $x = a$, and as $t \to a$ from the right $p(t)$ and $q(t)$ have asymptotic expansions in ascending fractional powers of $t-a$. A similar approach can be used for integrals of the form

$$\int_a^b e^{-xp(t)}q(t)\ln(t-a)\,dt.$$

We first select any convenient point k of (a, b) such that $p'(t) > 0$ when $t \in (a, k]$. Again, the integral over the range (k, b) is exponentially small compared with the contribution from (a, k), and may be ignored in the construction of the asymptotic expansion.

For the range (a, k), the integration variable $v \equiv p(t)-p(a)$ is used. As before, suppose that the asymptotic behavior of $p(t)$ and $q(t)$ at $t = a$ is expressed by

$$p(t) \sim p(a) + \sum_{s=0}^{\infty} p_s(t-a)^{s+\mu}, \qquad p'(t) \sim \sum_{s=0}^{\infty} (s+\mu)\,p_s(t-a)^{s+\mu-1},$$

$$q(t) \sim \sum_{s=0}^{\infty} q_s(t-a)^{s+\lambda-1},$$

where p_0, μ, and λ are positive, and q_0 is nonzero. Then

$$\ln(t-a) = \frac{1}{\mu}\ln v + g(v), \qquad \frac{q(t)}{p'(t)} = f(v),$$

where $g(v)$ and $f(v)$ have asymptotic expansions in ascending fractional powers of v which are free from logarithms. Accordingly,

$$\int_a^k e^{-xp(t)}q(t)\ln(t-a)\,dt = \frac{e^{-xp(a)}}{\mu}\int_0^{\kappa} e^{-xv}f(v)\ln v\,dv + e^{-xp(a)}\int_0^{\kappa} e^{-xv}f(v)g(v)\,dv,$$

where $\kappa = p(k)-p(a)$. For large x the asymptotic expansion of the last integral can be found by the theory of Chapter 3. For the integral containing $\ln v$ Laplace's approach suggests we substitute for $f(v)$ by means of its expansion in powers of v, replace the upper limit κ by infinity, and integrate term by term. The last step is easily carried out by use of the identity

$$\int_0^\infty e^{-xv}v^{\alpha-1}\ln v\,dv = \frac{\Gamma'(\alpha)-\Gamma(\alpha)\ln x}{x^\alpha} \qquad (\alpha>0,\quad x>0), \tag{1.01}$$

obtained by α-wise differentiation of Euler's integral

$$\int_0^\infty e^{-xv}v^{\alpha-1}\,dv = \frac{\Gamma(\alpha)}{x^\alpha} \qquad (\alpha>0,\quad x>0). \tag{1.02}$$

1.2 The following theorem places the foregoing approach on a sound basis.

Theorem 1.1† *Let $f(v)$ be sectionally continuous in* $(0,\infty)$ *and*

$$f(v) \sim \sum_{s=0}^{\infty} a_s v^{(s+\lambda-\mu)/\mu} \qquad (v\to 0+), \tag{1.03}$$

where λ and μ are positive constants. Then as $x\to\infty$

$$\int_0^\infty e^{-xv}f(v)\ln v\,dv \sim \sum_{s=0}^{\infty}\Gamma'\left(\frac{s+\lambda}{\mu}\right)\frac{a_s}{x^{(s+\lambda)/\mu}} - \ln x\sum_{s=0}^{\infty}\Gamma\left(\frac{s+\lambda}{\mu}\right)\frac{a_s}{x^{(s+\lambda)/\mu}}, \tag{1.04}$$

provided that this integral converges at its upper limit for all sufficiently large x.

We note that the convergence of the given integral at its lower limit is assured by (1.03). We also observe that by admitting sectionally continuous behavior in $f(v)$ (as in earlier chapters) we automatically treat integrals of the same form over any finite interval $(0,\kappa)$.

The proof of Theorem 1.1 parallels that of Theorem 3.1 of Chapter 3. Let n be an arbitrary nonnegative integer, and define $\phi_n(v)$ by

$$f(v) = \sum_{s=0}^{n-1} a_s v^{(s+\lambda-\mu)/\mu} + \phi_n(v) \qquad (v>0). \tag{1.05}$$

Then in consequence of (1.01)

$$\int_0^\infty e^{-xv}f(v)\ln v\,dv = \sum_{s=0}^{n-1}\frac{a_s}{x^{(s+\lambda)/\mu}}\left\{\Gamma'\left(\frac{s+\lambda}{\mu}\right)-\Gamma\left(\frac{s+\lambda}{\mu}\right)\ln x\right\} + \int_0^\infty e^{-xv}\phi_n(v)\ln v\,dv. \tag{1.06}$$

To estimate the remainder term, subdivide the integration range at $v=1$, and write

$$K_n = \sup_{v\in(0,1]} |\phi_n(v)\,v^{-(n+\lambda-\mu)/\mu}|.$$

† Doetsch (1955, p. 50).

Then K_n is finite, and

$$\left|\int_0^1 e^{-xv}\phi_n(v)\ln v\,dv\right| \leqslant K_n\int_0^1 e^{-xv}v^{(n+\lambda-\mu)/\mu}\ln\left(\frac{1}{v}\right)dv$$

$$= \frac{K_n}{x^{(n+\lambda)/\mu}}\int_0^x e^{-\tau}\tau^{(n+\lambda-\mu)/\mu}(\ln x-\ln\tau)\,d\tau$$

$$\leqslant \frac{K_n}{x^{(n+\lambda)/\mu}}\int_0^\infty e^{-\tau}\tau^{(n+\lambda-\mu)/\mu}(|\ln x|+|\ln\tau|)\,d\tau,$$

which is plainly $O(x^{-(n+\lambda)/\mu}\ln x)$ as $x\to\infty$.

For the tail, assume that X is a value of x for which the integrals in (1.06) converge. Then for $x > X$ partial integration yields

$$\int_1^\infty e^{-xv}\phi_n(v)\ln v\,dv = (x-X)\int_1^\infty e^{-(x-X)v}\left\{\int_1^v e^{-Xw}\phi_n(w)\ln w\,dw\right\}dv$$

$$= O(e^{-(x-X)});$$

compare Chapter 3, (3.07) and (3.08).

The expansion (1.04) is obtained by combining these results.

1.3 The essential message of Theorem 1.1 is that differentiation of the expansion

$$\int_0^\infty e^{-xv}f(v)\,dv \sim \sum_{s=0}^\infty \Gamma\left(\frac{s+\lambda}{\mu}\right)\frac{a_s}{x^{(s+\lambda)/\mu}}$$

with respect to λ (or μ) is legitimate. Using the same method and conditions we can verify without difficulty that the process of differentiation is repeatable; thus

$$\int_0^\infty e^{-xv}f(v)(\ln v)^m\,dv \sim \sum_{j=0}^m \binom{m}{j}(-\ln x)^j\sum_{s=0}^\infty \Gamma^{(m-j)}\left(\frac{s+\lambda}{\mu}\right)\frac{a_s}{x^{(s+\lambda)/\mu}}, \tag{1.07}$$

where m is any positive integer.

The next question is: can a similar result be constructed with m replaced by any positive real number β? To avoid complex-valued integrands the integration range is now standardized as $(0, 1)$, rather than $(0, \infty)$.

Theorem 1.2 *Let $f(v)$ be sectionally continuous in* $(0, 1]$ *and satisfy* (1.03). *Then for any positive constant* β

$$\int_0^1 e^{-xv}f(v)\left(\ln\frac{1}{v}\right)^\beta dv = \sum_{s=0}^{n-1} a_s L\left(\frac{s+\lambda}{\mu}, \beta, x\right) + O\left\{L\left(\frac{n+\lambda}{\mu}, \beta, x\right)\right\} \qquad (x\to\infty), \tag{1.08}$$

where n is an arbitrary integer, and

$$L(\alpha, \beta, x) = \int_0^1 e^{-xv}v^{\alpha-1}\left(\ln\frac{1}{v}\right)^\beta dv. \tag{1.09}$$

This result is provable in a similar way, but the proof is omitted as the result is of limited usefulness. This is because individual functions L can be evaluated in closed form only for special values of the parameters α and β, and are scarcely less recondite than the original integral. An asymptotic expansion for $L(\alpha, \beta, x)$ for large x can be constructed in descending powers of $\ln x$ by the methods of Chapter 3, however, and this result is stated in Exercise 1.4 below. Then by substituting in the first term of (1.08) we arrive at

$$\int_0^1 e^{-xv} f(v)\left(\ln\frac{1}{v}\right)^{\beta} dv \sim \frac{a_0 (\ln x)^{\beta}}{x^{\lambda/\mu}} \sum_{s=0}^{\infty} (-)^s \binom{\beta}{s} \frac{\Gamma^{(s)}(\lambda/\mu)}{(\ln x)^s}, \tag{1.10}$$

contributions from higher terms, and also from the error term, being absorbed in the error terms associated with (1.10). The expansion (1.10) is simpler than (1.08), but much less powerful.

Ex. 1.1 When α and β are fixed and positive and x is large and positive, show that

$$\int_0^\infty \exp(-t - xt^{\alpha})\, t^{\beta-1} \ln t \, dt \sim \frac{1}{\alpha^2} \sum_{s=0}^{\infty} \Gamma'\left(\frac{s+\beta}{\alpha}\right) \frac{(-1)^s}{s!\, x^{(s+\beta)/\alpha}} - \frac{\ln x}{\alpha^2} \sum_{s=0}^{\infty} \Gamma\left(\frac{s+\beta}{\alpha}\right) \frac{(-1)^s}{s!\, x^{(s+\beta)/\alpha}}.$$

Ex. 1.2 By using Theorem 1.1 and the integral representation for Macdonald's function given in Chapter 7, Exercise 8.4, show that for fixed ν and large positive z

$$\frac{\partial}{\partial \nu} K_\nu(z) \sim \left(\frac{\pi}{2z}\right)^{1/2} \frac{\nu e^{-z}}{z} \sum_{s=0}^{\infty} \frac{\alpha_s(\nu)}{(8z)^s},$$

where

$$\alpha_s(\nu) = \frac{(4\nu^2 - 1^2)(4\nu^2 - 3^2)\cdots\{4\nu^2 - (2s+1)^2\}}{(s+1)!} \left\{\frac{1}{4\nu^2 - 1^2} + \frac{1}{4\nu^2 - 3^2} + \cdots + \frac{1}{4\nu^2 - (2s+1)^2}\right\}.$$

Ex. 1.3 Show that the expansion given in the preceding exercise also holds throughout the sector $|\mathrm{ph}\, z| \leq \frac{3}{2}\pi - \delta$ $(< \frac{3}{2}\pi)$.

Ex. 1.4 Show that for fixed positive values of α and β the integral (1.09) has the expansion

$$L(\alpha, \beta, x) \sim \frac{(\ln x)^{\beta}}{x^{\alpha}} \sum_{s=0}^{\infty} (-)^s \binom{\beta}{s} \frac{\Gamma^{(s)}(\alpha)}{(\ln x)^s} \qquad (x \to \infty).$$

[Erdélyi, 1961.]

Ex. 1.5 Let $f(v)$ satisfy the conditions of Theorem 1.1. By subdivision of the integration range at $v = x^{-1/2}$ and use of Exercise 1.4, prove that

$$\int_0^1 e^{-xv} f(v) \ln\left(\ln\frac{1}{v}\right) dv \sim \frac{a_0}{x^{\lambda/\mu}} \left\{\Gamma\left(\frac{\lambda}{\mu}\right) \ln(\ln x) - \sum_{s=1}^{\infty} \frac{\Gamma^{(s)}(\lambda/\mu)}{s (\ln x)^s}\right\} \qquad (x \to \infty).$$

2 Generalizations of Laplace's Method

2.1 The motivating idea of Laplace can be applied to integrals in which the large parameter x enters in a more general way than in the integrals of Chapters 3 and 4. As a start, consider

$$I(x) = \int_a^b \exp\{-xp(t) + x^{\alpha} r(t)\}\, q(t)\, dt, \tag{2.01}$$

in which $p(t)$ and $q(t)$ satisfy the conditions of Chapter 3, §7.2, α (<1) is a constant, and $r(t)$ is independent of x. What kind of behavior in $r(t)$ can be tolerated without disturbing the result of Theorem 7.1 of Chapter 3?

To obtain an answer to this question, assume that

$$r(t) = O\{(t-a)^{\nu}\} \qquad (t \to a+),$$

where ν is a nonnegative constant. With the previous transformation of variables, it is seen that the contribution to $I(x)$ from the range (a,k) is

$$e^{-xp(a)} \int_0^{\kappa} e^{-xv} \exp\{x^{\alpha} g(v)\} f(v)\, dv,$$

where $g(v) = r(t)$. For small v we have

$$f(v) \sim Qv^{(\lambda/\mu)-1}/(\mu P^{\lambda/\mu}), \qquad g(v) = O(v^{\nu/\mu}),$$

and

$$\exp\{x^{\alpha} g(v)\} = 1 + x^{\alpha} O(v^{\nu/\mu}) + \frac{x^{2\alpha}}{2!} O(v^{2\nu/\mu}) + \cdots.$$

Retention of just the first term in this series produces the previous approximation to $I(x)$ for large x, namely

$$\frac{Q}{\mu}\Gamma\left(\frac{\lambda}{\mu}\right)\frac{e^{-xp(a)}}{(Px)^{\lambda/\mu}}. \tag{2.02}$$

The corresponding contribution from the second term is (2.02) multiplied by $x^{\alpha}O(x^{-\nu/\mu})$. From the next term we obtain (2.02) times $x^{2\alpha}O(x^{-2\nu/\mu})$, and so on. In consequence, the original approximation should continue to apply, provided that $\mu\alpha < \nu$.

A similar extension suggests itself for $q(t)$. If this function depends on x in such a way that

$$q(t) = Q(t-a)^{\lambda-1} + x^{\beta} O\{(t-a)^{\lambda_1 - 1}\} \qquad (t \to a+),$$

where β and λ_1 are constants, then the relative effect of the second term upon the original approximation is $x^{\beta} O\{x^{-(\lambda_1-\lambda)/\mu}\}$, and should therefore be negligible when $\mu\beta < \lambda_1 - \lambda$.

We are now ready to formulate a precise theorem. To simplify the final result, we concentrate on a range $(a,k]$ in which $p'(t)$ is positive, and require k to be finite. In consequence, a separate investigation of the tail may be needed in applications, as in Chapter 3. For notational simplicity, we shall also suppose that $a = 0$.

2.2 Theorem 2.1 *Let k and X be fixed positive numbers, and*

$$I(x) = \int_0^{k} e^{-xp(t)+r(x,t)} q(x,t)\, dt. \tag{2.03}$$

Assume that

(i) *$p'(t)$ is continuous and positive in $(0,k]$, and as $t \to 0+$*

$$p(t) = p(0) + Pt^{\mu} + O(t^{\mu_1}), \qquad p'(t) = \mu P t^{\mu-1} + O(t^{\mu_1 - 1}),$$

where $P > 0$ and $\mu_1 > \mu > 0$.

(ii) *For each $x \in [X, \infty)$, the real or complex functions $r(x,t)$ and $q(x,t)$ are continuous in $0 < t \leqslant k$. Moreover,*

$$|r(x,t)| \leqslant Rx^{\alpha}t^{\nu}, \qquad |q(x,t) - Qt^{\lambda-1}| \leqslant Q_1 x^{\beta} t^{\lambda_1 - 1},$$

where R, α, ν, Q, λ, Q_1, β, and λ_1 are independent of x and t, and†

$$\nu \geqslant 0, \qquad \lambda > 0, \qquad \lambda_1 > 0, \qquad \alpha < \min(1, \nu/\mu), \qquad \beta < (\lambda_1 - \lambda)/\mu.$$

Then

$$I(x) = \frac{Q}{\mu}\Gamma\left(\frac{\lambda}{\mu}\right)\frac{e^{-xp(0)}}{(Px)^{\lambda/\mu}}\left\{1 + O\left(\frac{1}{x^{\varpi_1/\mu}}\right)\right\} \qquad (x \to \infty), \tag{2.04}$$

where

$$\varpi_1 = \min(\mu_1 - \mu, \nu - \mu\alpha, \lambda_1 - \lambda - \mu\beta). \tag{2.05}$$

The proof of this result is similar to that of Theorem 7.1 of Chapter 3. We express

$$e^{xp(0)}I(x) = \frac{Q}{\mu P^{\lambda/\mu}}\left\{\int_0^\infty e^{-xv}v^{(\lambda/\mu)-1}\,dv - \varepsilon_1(x) + \varepsilon_3(x)\right\} + \varepsilon_2(x), \tag{2.06}$$

where

$$\varepsilon_1(x) = \int_\kappa^\infty e^{-xv}v^{(\lambda/\mu)-1}\,dv, \qquad \varepsilon_2(x) = \int_0^\kappa e^{-xv+r(x,t)}\left\{\frac{q(x,t)}{p'(t)} - \frac{Qv^{(\lambda/\mu)-1}}{\mu P^{\lambda/\mu}}\right\}dv, \tag{2.07}$$

$$\varepsilon_3(x) = \int_0^\kappa e^{-xv}v^{(\lambda/\mu)-1}(e^{r(x,t)} - 1)\,dv. \tag{2.08}$$

In estimating these error terms we use the symbol A generically to denote a positive number which is independent of x and v, and O to denote an order term which is uniform with respect to $x \in [X, \infty)$ and $v \in (0, \kappa]$. From Chapter 3, (7.11) we have

$$\varepsilon_1(x) = O(x^{-1}e^{-\kappa x}). \tag{2.09}$$

Next,

$$t = (v/P)^{1/\mu}\{1 + O(v^{(\mu_1-\mu)/\mu})\} \qquad (v \to 0+). \tag{2.10}$$

Hence, from Condition (ii) we derive

$$|r(x,t)| \leqslant Ax^\alpha v^{\nu/\mu}. \tag{2.11}$$

Write $\alpha_0 = \max(\alpha, 0)$. Then

$$|e^{r(x,t)}| \leqslant \exp(Ax^{\alpha_0}v^{\nu/\mu}) \leqslant \exp(Ax^{\alpha_0}v^{\alpha_0}), \tag{2.12}$$

since $\alpha_0 \leqslant \nu/\mu$ and $v \leqslant \kappa$. Using the given conditions and the relation (2.10), we see that

$$\begin{aligned}\frac{q(x,t)}{p'(t)} &= \frac{Qt^{\lambda-1} + x^\beta O(t^{\lambda_1-1})}{\mu P t^{\mu-1} + O(t^{\mu_1-1})} \\ &= \frac{Qt^{\lambda-\mu}}{\mu P} + O(t^{\lambda+\mu_1-2\mu}) + x^\beta O(t^{\lambda_1-\mu}) \\ &= \frac{Qv^{(\lambda/\mu)-1}}{\mu P^{\lambda/\mu}} + O(v^{(\lambda+\mu_1-2\mu)/\mu}) + x^\beta O(v^{(\lambda_1-\mu)/\mu}).\end{aligned}$$

† None of the quantities α, β, or $\lambda_1 - \lambda$ is required to be positive.

Substitution of this result and (2.12) in the second of (2.07) produces

$$|\varepsilon_2(x)| \leqslant \int_0^\kappa \exp\{-xv+A(xv)^{\alpha_0}\}\{O(v^{(\lambda+\mu_1-2\mu)/\mu})+x^\beta O(v^{(\lambda_1-\mu)/\mu})\}\,dv$$

$$= O\left(\frac{1}{x^{(\lambda+\mu_1-\mu)/\mu}}\right)\int_0^\infty \exp(-\tau+A\tau^{\alpha_0})\,\tau^{(\lambda+\mu_1-2\mu)/\mu}\,d\tau$$

$$+\,O\left(\frac{x^\beta}{x^{\lambda_1/\mu}}\right)\int_0^\infty \exp(-\tau+A\tau^{\alpha_0})\,\tau^{(\lambda_1-\mu)/\mu}\,d\tau. \tag{2.13}$$

Because $0 \leqslant \alpha_0 < 1$, $\mu_1 > \mu$, $\lambda > 0$, and $\lambda_1 > 0$, the last two integrals converge. Accordingly,

$$\varepsilon_2(x) = O(x^{-(\lambda+\varpi_1)/\mu}), \tag{2.14}$$

where ϖ_1 is defined by (2.05).

For $\varepsilon_3(x)$ we employ the elementary inequality $|e^y-1| \leqslant |y|\,e^{|y|}$. Combination with (2.11) produces

$$|e^{r(x,t)}-1| \leqslant Ax^\alpha v^{\nu/\mu}\exp(Ax^\alpha v^{\nu/\mu}) \leqslant Ax^\alpha v^{\nu/\mu}\exp(Ax^{\alpha_0}v^{\alpha_0});$$

compare (2.12). Substitution of this result in (2.08) gives

$$|\varepsilon_3(x)| \leqslant Ax^\alpha \int_0^\kappa \exp\{-xv+A(xv)^{\alpha_0}\}\,v^{(\lambda+\nu-\mu)/\mu}\,dv$$

$$= O\left(\frac{x^\alpha}{x^{(\lambda+\nu)/\mu}}\right) = O\left(\frac{1}{x^{(\lambda+\varpi_1)/\mu}}\right); \tag{2.15}$$

compare (2.13) and (2.14).

The proof of Theorem 2.1 is completed by combining (2.06), (2.09), (2.14), and (2.15).

Ex. 2.1 Assume that a and n are positive numbers. Starting from the integral representation

$$K_n(a) = \tfrac{1}{2}\int_{-\infty}^{\infty} e^{nw-a\cosh w}\,dw$$

substitute $w = t+\sinh^{-1}(n/a)$, and from Theorem 2.1 deduce that

$$K_n(a) = \left(\frac{\pi}{2n}\right)^{1/2}\left(\frac{2n}{ea}\right)^n\left\{1+O\left(\frac{1}{n^{1/2}}\right)\right\} \qquad (a \text{ fixed}, \quad n\to\infty).$$

Confirm this result by means of Chapter 7, (8.05).

Ex. 2.2 Show that

$$\int_0^\infty \sin(xt)\exp\left\{-x^{3/2}\sinh\left(\frac{t^{1/2}}{x^{1/2}}\right)\right\}dt = \frac{12}{x^3}+O\left(\frac{1}{x^5}\right) \qquad (x\to\infty).$$

Ex. 2.3 Show that

$$\int_0^\infty \exp\{2xt^{1/2}-(1+t^2)^{1/2}\}\,dt = 2\pi^{1/2}x\exp(x^2)\{1+O(x^{-1})\} \qquad (x\to\infty).$$

Ex. 2.4 If $r(x,t)$ is real and nonpositive and all conditions of Theorem 2.1 are satisfied except that the inequality $\alpha < \min(1,\nu/\mu)$ is relaxed to $\alpha < \nu/\mu$, show that (2.04) again applies.

Ex. 2.5 With the aid of the preceding exercise prove that

$$\int_0^\infty \exp\{-\sinh(x+t)\}\,dt = \operatorname{sech} x\, e^{-\sinh x}\{1+O(e^{-x})\} \qquad (x\to\infty).$$

Ex. 2.6 Assume that $\chi(t)$ is continuously differentiable in $[0,\infty)$ and $\chi'(t) = O(t^{-\delta})$ as $t\to\infty$, where $\delta > \frac{1}{2}$. Show that

$$\int_0^\infty e^{-t+\chi(t)} t^{x-1}\,dt \sim \Gamma(x)\, e^{\chi(x)} \qquad (x\to\infty).$$

[Berg, 1958.]

*3 Example from Combinatoric Theory

3.1 To illustrate the use of Theorem 2.1 for contour integrals, consider the problem† of determining the asymptotic form, for large n, of the Maclaurin coefficients defined by

$$\exp(e^z) = \sum_{n=0}^{\infty} a_n z^n.$$

From Cauchy's integral formula, we have

$$a_n = \frac{1}{2\pi i}\int_{\mathscr{C}} \frac{\exp(e^w)}{w^{n+1}}\,dw,$$

where $\mathscr{C}$ is a simple closed contour encircling $w = 0$. One of the stationary points of $\exp(e^w)/w^n$ is $w = \ln N$, where N is the positive root of the equation

$$N \ln N = n. \tag{3.01}$$

By taking logarithms and applying Theorem 5.1 of Chapter 1, we see that

$$N \sim n/\ln n \qquad (n\to\infty). \tag{3.02}$$

Following the methods of Chapter 4, we try to deform $\mathscr{C}$ in such a way that the absolute value of the integrand attains its maximum value on the contour at $w = \ln N$. This condition is fulfilled by the contour indicated in Fig. 3.1, in which the radius R of the circular arc exceeds $\ln N$. This is because $|w|^{-n-1}$ obviously attains its maximum at $w = \ln N$, and

$$|\exp(e^w)| \leqslant \exp|e^w| = \exp(e^{\operatorname{Re} w}) \leqslant e^N, \tag{3.03}$$

equality being attained when $w = \ln N$. Letting $R\to\infty$ we see that $\mathscr{C}$ may be further deformed into the doubly infinite straight line $\operatorname{Re} w = \ln N$ and hence, by taking complex conjugates, that

$$a_n = \operatorname{Im}\left\{\frac{1}{\pi}\int_{\ln N}^{\ln N + i\infty} \frac{\exp(e^w)}{w^{n+1}}\,dw\right\}.$$

† de Bruijn (1961, §6.2). The problem arises in connection with the number of class partitions of a finite set. More extensive work on similar problems may be found in the papers by Rubin (1967) and Harris and Schoenfeld (1968).

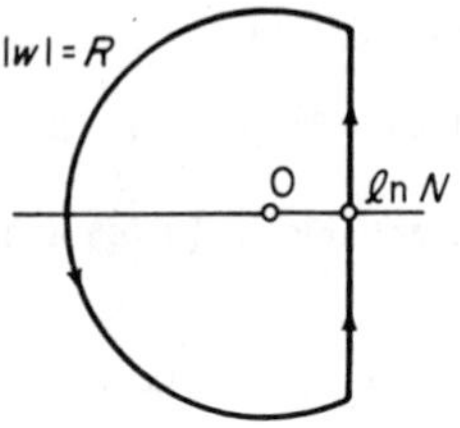

Fig. 3.1 w plane.

3.2 Consider the contribution to the last integral from the segment defined by

$$w = \ln N + it \qquad (0 \leqslant t \leqslant k),$$

k being any positive constant. We have

$$\begin{aligned}\exp(e^w)/w^n &= \exp\{Ne^{it} - n\ln(\ln N + it)\} \\ &= \exp\left[N\left\{1 + it + \frac{(it)^2}{2!} + \cdots\right\} - n\left\{\ln(\ln N) + \frac{it}{\ln N} - \frac{(it)^2}{2(\ln N)^2} + \cdots\right\}\right] \\ &= e^N(\ln N)^{-n}\exp\{-Mt^2 + t^3 O(N)\}\end{aligned}$$

for large N, uniformly with respect to $t \in [0, k]$. Here

$$M = \frac{1}{2}N + \frac{n}{2(\ln N)^2} = \frac{1}{2}N\left(1 + \frac{1}{\ln N}\right); \qquad (3.04)$$

compare (3.01). Thus $M \sim \frac{1}{2}N$ as $N \to \infty$. We also have, again for large N,

$$\frac{1}{w} = \frac{1}{\ln N + it} = \frac{1}{\ln N}\left\{1 + O\left(\frac{t}{\ln N}\right)\right\} = \frac{1}{\ln N}\{1 + tO(1)\},$$

uniformly with respect to t. Therefore

$$\frac{\exp(e^w)}{w^{n+1}} = \frac{e^N}{(\ln N)^{n+1}}\{1 + tO(1)\}\exp\{-Mt^2 + t^3 O(M)\} \qquad (n \to \infty).$$

In the notation of Theorem 2.1, $x = M$, $p(t) = t^2$, $P = 1$, $\mu = 2$, $\mu_1 = \infty$, $\alpha = 1$, $\nu = 3$, $Q = 1$, $\lambda = 1$, $\beta = 0$, $\lambda_1 = 2$, and $\varpi_1 = 1$. Hence†

$$\begin{aligned}\frac{1}{\pi}\int_{\ln N}^{\ln N + ik}\frac{\exp(e^w)}{w^{n+1}}\,dw &= \frac{ie^N}{\pi(\ln N)^{n+1}}\frac{\pi^{1/2}}{2M^{1/2}}\left\{1 + O\left(\frac{1}{M^{1/2}}\right)\right\} \\ &= \frac{ie^N}{\{2\pi(n+N)\}^{1/2}}\left(\frac{N}{n}\right)^{n+(1/2)}\left\{1 + O\left(\frac{1}{N^{1/2}}\right)\right\}; \qquad (3.05)\end{aligned}$$

compare (3.01) and (3.04).

† The reader will notice that the condition $\alpha < 1$ of the theorem is violated. Provided that k is sufficiently small this is permissible. This is because in (2.12) we have

$$x^{\alpha_0}v^{\nu/\mu} = x^{\alpha_0}v^{\alpha_0}v^{(\nu/\mu)-\alpha_0} \leqslant \kappa^{(\nu/\mu)-\alpha_0}x^{\alpha_0}v^{\alpha_0}.$$

In the present instance $(\nu/\mu) - \alpha_0 = \frac{1}{2}$; hence by taking κ to be sufficiently small the second generic constant A appearing in (2.12) can be made less than unity. In consequence, although $\alpha_0 = 1$, the integrals in (2.13) still converge at their upper limits and (2.14) applies. Similarly for (2.15).

For the contribution from the remaining part of the contour, we set

$$w = \ln N + i(k+t) \qquad (0 \leqslant t < \infty).$$

Again using the inequality (3.03), we obtain

$$\left|\frac{1}{\pi}\int_{\ln N+ik}^{\ln N+i\infty} \frac{\exp(e^w)}{w^{n+1}}\, dw\right| \leqslant \frac{e^N}{\pi}\int_0^\infty \frac{dt}{\{(\ln N)^2+k^2+t^2\}^{(n+1)/2}}$$

$$\leqslant \frac{Ae^N}{\{(\ln N)^2+k^2\}^{n/2}} = \frac{Ae^N N^n}{(n^2+k^2N^2)^{n/2}},$$

where A is independent of N.

By taking logarithms it is readily verified that the last bound can be absorbed in the error term of (3.05). Accordingly, the final result has the form

$$a_n = \frac{e^N}{\{2\pi(n+N)\}^{1/2}}\left(\frac{N}{n}\right)^{n+(1/2)}\left\{1+O\left(\frac{1}{N^{1/2}}\right)\right\} \qquad (n\to\infty), \qquad (3.06)$$

where N is the positive root of equation (3.01).†

3.3 It should be observed that if the asymptotic expression $n/\ln n$ is substituted for N throughout (3.06), then the asymptotic character of (3.06) is destroyed. This is also true if N is replaced by higher approximations obtained by resubstitutions in (3.01) in the manner of Chapter 1, §5.

4 Generalizations of Laplace's Method (continued)

4.1 Let us return to the exploratory analysis of §2.1 and consider what happens when

$$r(t) \sim R(t-a)^\nu \qquad (t\to a+),$$

but the condition $\nu > \mu\alpha$ is violated. If $\nu < \mu\alpha$, then we can regard x^α instead of x as the large parameter, and interchange the roles of $p(t)$ and $r(t)$; compare Exercise 2.4. It remains to consider the case $\nu = \mu\alpha$.

The term $x^\alpha r(t)$ can no longer be disregarded in constructing an approximation for $I(x)$. Approximating this term by $x^\alpha R(t-a)^\nu$ and repeating the analysis, we expect the approximation to $I(x)$ to be modified by the replacement of

$$\int_0^\infty e^{-xv} v^{(\lambda/\mu)-1}\, dv, \qquad \text{that is,}\quad \Gamma(\lambda/\mu)x^{-\lambda/\mu},$$

by

$$\int_0^\infty \exp(-xv+yx^\alpha v^\alpha)\, v^{(\lambda/\mu)-1}\, dv, \qquad \text{that is,}\quad x^{-\lambda/\mu}\int_0^\infty \exp(-\tau+y\tau^\alpha)\,\tau^{(\lambda/\mu)-1}\, d\tau.$$

Here $y = RP^{-\nu/\mu}$, and P again denotes the limiting value of $(t-a)^{-\mu}\{p(t)-p(a)\}$

† An extension of the analysis to higher terms shows that the error term $O(N^{-1/2})$ in (3.06) can be strengthened to $O(N^{-1})$.

as $t \to a$. In general the last integral cannot be evaluated explicitly, but because it is independent of x this form of approximation is acceptable for many purposes.

The integral

$$\mathrm{Fi}(\alpha, \beta; y) = \int_0^\infty \exp(-\tau + y\tau^\alpha)\,\tau^{\beta-1}\,d\tau \qquad (0 \leqslant \operatorname{Re}\alpha < 1, \quad \operatorname{Re}\beta > 0) \tag{4.01}$$

is called *Faxén's integral.*† With the parenthetic conditions it converges at both limits and defines an entire function of y. Special cases are

$$\mathrm{Fi}(\alpha, \beta; 0) = \Gamma(\beta), \qquad \mathrm{Fi}(0, \beta; y) = e^y\Gamma(\beta). \tag{4.02}$$

Commonly needed pairs of values of the parameters are $\alpha = \beta = \frac{1}{2}$ and $\alpha = \beta = \frac{1}{3}$. Substituting $\tau = t^2$, we obtain

$$\mathrm{Fi}(\tfrac{1}{2}, \tfrac{1}{2}; y) = 2\exp(\tfrac{1}{4}y^2)\int_0^\infty \exp\{-(t - \tfrac{1}{2}y)^2\}\,dt = \pi^{1/2}\exp(\tfrac{1}{4}y^2)\{1 + \operatorname{erf}(\tfrac{1}{2}y)\}; \tag{4.03}$$

compare Chapter 2, §4.1. Also,

$$\mathrm{Fi}(\tfrac{1}{2}, \tfrac{1}{2}; y) + \mathrm{Fi}(\tfrac{1}{2}, \tfrac{1}{2}; -y) = 2\pi^{1/2}\exp(\tfrac{1}{4}y^2). \tag{4.04}$$

When $\alpha = \beta = \frac{1}{3}$ we have

$$\mathrm{Fi}(\tfrac{1}{3}, \tfrac{1}{3}; y) = 3^{2/3}\pi\,\mathrm{Hi}(3^{-1/3}y), \tag{4.05}$$

where $\mathrm{Hi}(x)$ is *Scorer's function*, defined by

$$\mathrm{Hi}(x) = \frac{1}{\pi}\int_0^\infty \exp(-\tfrac{1}{3}t^3 + xt)\,dt, \tag{4.06}$$

and discussed further in Chapter 11, §12. With the aid of Chapter 2, (8.02) it may be verified that

$$e^{-\pi i/6}\,\mathrm{Fi}(\tfrac{1}{3}, \tfrac{1}{3}; ye^{\pi i/3}) + e^{\pi i/6}\,\mathrm{Fi}(\tfrac{1}{3}, \tfrac{1}{3}; ye^{-\pi i/3}) = 3^{2/3}2\pi\,\mathrm{Ai}(-3^{-1/3}y). \tag{4.07}$$

The asymptotic forms of $\mathrm{Fi}(\alpha, \beta; y)$ when $y \to +\infty$ or $y \to -\infty$ are stated in Chapter 3, Exercise 7.3.

4.2 Theorem 4.1 *Let*

$$I(x) = \int_0^k \exp\{-xp(t) + x^{\nu/\mu}r(t)\}\,q(t)\,dt, \tag{4.08}$$

where k is fixed and positive, and assume that:

(i) *In the interval $(0, k]$, $p'(t)$ is continuous and positive, and the real or complex functions $q(t)$ and $r(t)$ are continuous.*

(ii) *As $t \to 0+$*

$$p(t) = p(0) + Pt^\mu + O(t^{\mu_1}), \qquad p'(t) = \mu P t^{\mu-1} + O(t^{\mu_1 - 1}),$$
$$q(t) = Qt^{\lambda-1} + O(t^{\lambda_1 - 1}), \qquad r(t) = Rt^\nu + O(t^{\nu_1}),$$

† Faxén (1921). This notation was introduced by Bleistein, Handelsman, and Lew (1972).

where

$$P > 0, \qquad \mu_1 > \mu > 0, \qquad \lambda_1 > \lambda > 0, \qquad \mu > \nu \geqslant 0, \qquad \nu_1 > \nu.$$

Then

$$I(x) = \frac{Q}{\mu} \operatorname{Fi}\left(\frac{\nu}{\mu}, \frac{\lambda}{\mu}; \frac{R}{P^{\nu/\mu}}\right) \frac{e^{-xp(0)}}{(Px)^{\lambda/\mu}} \left\{1 + O\left(\frac{1}{x^{\varpi/\mu}}\right)\right\} \qquad (x \to \infty), \tag{4.09}$$

where

$$\varpi = \min(\lambda_1 - \lambda, \mu_1 - \mu, \nu_1 - \nu). \tag{4.10}$$

The proof of this result is similar to that of Theorem 2.1. With the notations

$$y = R/P^{\nu/\mu}, \qquad v = p(t) - p(0), \qquad \kappa = p(k) - p(0),$$

$$g(v) = r(t) - yv^{\nu/\mu}, \qquad f(v) = q(t)/p'(t),$$

the principal identities are

$$e^{xp(0)} I(x) = \frac{Q}{\mu P^{\lambda/\mu}} \left[\int_0^\infty \exp\{-xv + y(xv)^{\nu/\mu}\}\, v^{(\lambda/\mu)-1}\, dv - \varepsilon_1(x) + \varepsilon_3(x)\right] + \varepsilon_2(x),$$

$$\varepsilon_1(x) = \int_\kappa^\infty \exp\{-xv + y(xv)^{\nu/\mu}\}\, v^{(\lambda/\mu)-1}\, dv,$$

$$\varepsilon_2(x) = \int_0^\kappa \exp\{-xv + y(xv)^{\nu/\mu} + x^{\nu/\mu} g(v)\} \left\{f(v) - \frac{Qv^{(\lambda/\mu)-1}}{\mu P^{\lambda/\mu}}\right\} dv,$$

$$\varepsilon_3(x) = \int_0^\kappa \exp\{-xv + y(xv)^{\nu/\mu}\} [\exp\{x^{\nu/\mu} g(v)\} - 1]\, v^{(\lambda/\mu)-1}\, dv.$$

And with $v \in (0, \kappa]$ and A denoting a generic constant, required inequalities are

$$\left| f(v) - \frac{Qv^{(\lambda/\mu)-1}}{\mu P^{\lambda/\mu}} \right| \leqslant Av^{(\lambda+\varpi-\mu)/\mu}, \qquad |g(v)| \leqslant Av^{(\nu+\varpi)/\mu} \leqslant Av^{\nu/\mu},$$

$$|\exp\{x^{\nu/\mu} g(v)\} - 1| \leqslant \exp\{Ax^{\nu/\mu} v^{(\nu+\varpi)/\mu}\} - 1 \leqslant Ax^{\nu/\mu} v^{(\nu+\varpi)/\mu} \exp\{A(xv)^{\nu/\mu}\}.$$

Other details of the proof are left to the reader.

4.3 Theorems 2.1 and 4.1 can be combined into a more general theorem, which is stated in the exercise which follows and is provable in the same manner.

Ex. 4.1 Let

$$I(x) = \int_0^k \exp\{-xp(t) + x^{\sigma/\mu} s(t) + r(x,t)\}\, q(x,t)\, dt,$$

where $p(t)$, $r(x,t)$, and $q(x,t)$ satisfy the conditions of Theorem 2.1, and $s(t)$ is a real or complex function of t which is continuous in $[0,k]$, and has the property

$$s(t) = St^\sigma + O(t^{\sigma_1}) \qquad (t \to 0+),$$

σ and σ_1 being constants such that $\mu > \sigma \geqslant 0$ and $\sigma_1 > \sigma$. Prove that

$$I(x) = \frac{Q}{\mu} \operatorname{Fi}\left(\frac{\sigma}{\mu}, \frac{\lambda}{\mu}; \frac{S}{P^{\sigma/\mu}}\right) \frac{e^{-xp(0)}}{(Px)^{\lambda/\mu}} \left\{1 + O\left(\frac{1}{x^{\varpi_2/\mu}}\right)\right\} \qquad (x \to \infty),$$

where $\varpi_2 = \min(\mu_1 - \mu, \nu - \mu\alpha, \lambda_1 - \lambda - \mu\beta, \sigma_1 - \sigma)$.

5 Examples

5.1 The theory of §4 is illustrated in the present section by two examples. First, we consider the following integral for the parabolic cylinder function derived in Chapter 6, Exercise 6.1:

$$U\left(n+\frac{1}{2}, a\right) = \frac{\exp(-\frac{1}{4}a^2)}{\Gamma(n+1)} \int_0^\infty \exp\left(-at-\frac{1}{2}t^2\right) t^n \, dt \qquad (n > -1).$$

We seek the asymptotic form of $U(n+\frac{1}{2}, a)$ for large n and fixed real values of the argument a.†

The integrand attains its maximum value at $t = \xi$, where $a\xi + \xi^2 - n = 0$, giving

$$\xi = -\tfrac{1}{2}a + (\tfrac{1}{4}a^2 + n)^{1/2}.$$

Since $\xi \sim n^{1/2}$ as $n \to \infty$, we substitute $t = n^{1/2}(1+\tau)$. This yields

$$U\left(n+\frac{1}{2}, a\right) = \exp\left(-an^{1/2} - \frac{1}{2}n - \frac{1}{4}a^2\right) \frac{n^{(n+1)/2}}{\Gamma(n+1)} \int_{-1}^\infty \exp\{-np(\tau) - a\tau n^{1/2}\} \, d\tau,$$

where

$$p(\tau) = \tau + \tfrac{1}{2}\tau^2 - \ln(1+\tau) = \tau^2 - \tfrac{1}{3}\tau^3 + \cdots \qquad (\tau \to 0).$$

Application of Theorem 4.1, with $x = n$, $r(\tau) = -a\tau$, and $q(\tau) = 1$, yields

$$\int_0^k \exp\{-np(\tau) - a\tau n^{1/2}\} \, d\tau = \frac{\mathrm{Fi}(\frac{1}{2}, \frac{1}{2}; -a)}{2n^{1/2}} \left\{1 + O\left(\frac{1}{n^{1/2}}\right)\right\},$$

for any fixed value of the positive number k. Similarly

$$\int_{-k}^0 \exp\{-np(\tau) - a\tau n^{1/2}\} \, d\tau = \frac{\mathrm{Fi}(\frac{1}{2}, \frac{1}{2}; a)}{2n^{1/2}} \left\{1 + O\left(\frac{1}{n^{1/2}}\right)\right\},$$

when $k \in (0, 1)$. The contributions of the tails $\int_k^\infty$ and $\int_{-1}^{-k}$ are exponentially small; hence by addition and use of (4.04) we obtain the required result

$$U\left(n+\frac{1}{2}, a\right) = \exp\left(-an^{1/2} - \frac{1}{2}n\right) \frac{\pi^{1/2} n^{n/2}}{\Gamma(n+1)} \left\{1 + O\left(\frac{1}{n^{1/2}}\right)\right\}$$
$$= \frac{\exp(-an^{1/2} + \frac{1}{2}n)}{2^{1/2} n^{(n+1)/2}} \left\{1 + O\left(\frac{1}{n^{1/2}}\right)\right\}.$$

The reader will easily verify that this result is a special case of Chapter 6, (6.06).

5.2 As a second example,‡ consider Schläfli's integral

$$J_\rho(z) = \frac{1}{2\pi i} \int_{\infty - \pi i}^{\infty + \pi i} e^{z \sinh t - \rho t} \, dt \qquad (|\mathrm{ph}\, z| < \tfrac{1}{2}\pi),$$

† This example is due to Copson (1965, §20).
‡ Olver (1952). This reference includes additional terms in the final result.

(Chapter 2, (9.13)) in the case

$$z = \rho + a\rho^{1/3},$$

where a is a real or complex constant and ρ is large and positive.

The dominant term in the exponent is $\rho(\sinh t - t)$, the t derivative of which vanishes at $t = 0, \pm 2\pi i, \pm 4\pi i, \ldots$. We choose a path through the origin along which

$$v \equiv t - \sinh t$$

is real and nonnegative. The existence of this path is easily demonstrated by considering the v map of the half-strip $|\mathrm{Im}\, t| \leqslant \pi$, $\mathrm{Re}\, t \geqslant 0$; see Fig. 5.1.

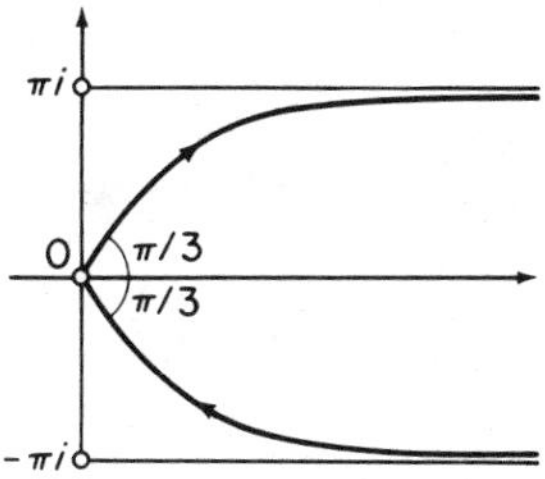

Fig. 5.1 t plane. Integration path for $J_\rho(\rho + a\rho^{1/3})$.

For small t,

$$v = -\tfrac{1}{6}t^3 - \tfrac{1}{120}t^5 - \tfrac{1}{5040}t^7 - \cdots.$$

On the upper branch of the path reversion gives

$$t = e^{\pi i/3}(6v)^{1/3} + O(v) \qquad (v \to 0),$$

fractional powers having their principal values. We express the integral along this branch $J_\rho^{(1)}(\rho + a\rho^{1/3})$, say, in the form

$$J_\rho^{(1)}(\rho + a\rho^{1/3}) = \frac{1}{2\pi i}\int_0^\infty \exp\{-\rho v + \rho^{1/3} r(v)\}\, q(v)\, dv,$$

where

$$r(v) = a \sinh t, \qquad q(v) = dt/dv.$$

For small v

$$r(v) = ae^{\pi i/3}(6v)^{1/3} + O(v), \qquad q(v) = 2e^{\pi i/3}(6v)^{-2/3} + O(1).$$

Application of Theorem 4.1 produces†

$$2\pi i J_\rho^{(1)}(\rho + a\rho^{1/3}) = \frac{2e^{\pi i/3}}{6^{2/3}\rho^{1/3}}\, \mathrm{Fi}(\tfrac{1}{3}, \tfrac{1}{3}; 6^{1/3}e^{\pi i/3}a) + O\left(\frac{1}{\rho}\right).$$

For the corresponding integral along the lower branch we make appropriate changes

† Theorem 4.1 applies only to the integral over a finite range $\int_0^k$. The negligibility of the tail $\int_k^\infty$ is easily demonstrated by Laplace's method.

to the sign of i in this result:

$$2\pi i J_\rho^{(2)}(\rho+a\rho^{1/3}) = -\frac{2e^{-\pi i/3}}{6^{2/3}\rho^{1/3}}\,\mathrm{Fi}(\tfrac{1}{3},\tfrac{1}{3};6^{1/3}e^{-\pi i/3}a) + O\left(\frac{1}{\rho}\right).$$

Addition and use of (4.07) gives the desired approximation

$$J_\rho(\rho+a\rho^{1/3}) = (2/\rho)^{1/3}\,\mathrm{Ai}(-2^{1/3}a) + O(\rho^{-1}). \qquad (5.01)$$

Ex. 5.1 Let $\mathbf{A}_\nu(x)$ be defined as in Chapter 3, Exercise 7.2. Show that for fixed l and large positive ν

$$\mathbf{A}_{-\nu}(\nu+l\nu^{1/3}) = \pi 2^{1/3}\nu^{-1/3}\,\mathrm{Hi}(-2^{1/3}l)\{1+O(\nu^{-2/3})\}.$$

Ex. 5.2 Show that for large n the coefficient of z^n in the Maclaurin expansion of $\exp(z+\frac{1}{2}z^2)$ is asymptotic to $\frac{1}{2}\pi^{-1/2}n^{-(n+1)/2}\exp(\frac{1}{2}n+n^{1/2}-\frac{1}{4})$. [Chowla, Herstein, and Moore, 1951.]

6 More General Kernels

6.1 The central result in the Laplace method for obtaining asymptotic expansions may be stated as follows. If $f(v)$ can be expanded in ascending powers of v in the form

$$f(v) \sim \sum_{s=0}^{\infty} a_s v^{(s+\lambda-\mu)/\mu} \qquad (v \to 0+), \qquad (6.01)$$

then

$$\int_0^\infty e^{-xv} f(v)\,dv \sim \sum_{s=0}^{\infty} \Gamma\left(\frac{s+\lambda}{\mu}\right)\frac{a_s}{x^{(s+\lambda)/\mu}} \qquad (x \to +\infty), \qquad (6.02)$$

provided that $\lambda > 0$, $\mu > 0$, and the integral converges.

The theory of §4 shows that similar induction of series takes place with other types of integral. Let us now consider the general form

$$I(x) = \int_0^\infty g(xv)\,f(v)\,dv. \qquad (6.03)$$

Substituting the expansion (6.01) and integrating formally term by term, we obtain the series

$$\sum_{s=0}^{\infty} G\left(\frac{s+\lambda}{\mu}\right)\frac{a_s}{x^{(s+\lambda)/\mu}}, \qquad (6.04)$$

in which $G(\alpha)$ denotes the *Mellin transform* of $g(t)$, defined by

$$G(\alpha) = \int_0^\infty g(t)\,t^{\alpha-1}\,dt.$$

In what circumstances does (6.04) furnish an asymptotic expansion of the integral (6.03) for large x? This question is answerable by investigating successive remainder terms

$$\varepsilon_n(x) = \int_0^\infty g(xv)\,\phi_n(v)\,dv \qquad (n = 0, 1, \ldots),$$

in which

$$\phi_n(v) = f(v) - \sum_{s=0}^{n-1} a_s v^{(s+\lambda-\mu)/\mu} \tag{6.05}$$

and is $O(v^{(n+\lambda-\mu)/\mu})$ as $v \to 0+$. The needed result is $\varepsilon_n(x) = O(x^{-(n+\lambda)/\mu})$; we proceed to establish its validity with conditions which are satisfied in many applications.

6.2 Lemma 6.1 *Let $g(v)$ and $\phi(v)$ be real or complex functions of v which are sectionally continuous in the interval* $(0, \infty)$ *and have the properties*

$$|g(v)| \leqslant A \exp(-av^\kappa) \qquad (0 \leqslant v < \infty), \tag{6.06}$$

$$\phi(v) = O(v^{\alpha-1}) \quad (v \to 0), \qquad \phi(v) = O(e^{\sigma v^\kappa}) \quad (v \to \infty), \tag{6.07}$$

where A, a, α, σ, and κ are positive constants. Then

$$\int_0^\infty g(xv)\,\phi(v)\,dv = O\left(\frac{1}{x^\alpha}\right) \qquad (x \to \infty). \tag{6.08}$$

This is easily proved. The conditions (6.07) show that

$$|\phi(v)| \leqslant Bv^{\alpha-1}\frac{1+v}{1+v^\alpha}e^{\sigma v^\kappa} \qquad (0 < v < \infty),$$

where B is an assignable constant. Therefore when $x > (\sigma/a)^{1/\kappa}$, we have

$$\left|\int_0^\infty g(xv)\,\phi(v)\,dv\right| \leqslant AB\int_0^\infty v^{\alpha-1}\frac{1+v}{1+v^\alpha}\exp(\sigma v^\kappa - ax^\kappa v^\kappa)\,dv.$$

Theorem 7.1 of Chapter 3 shows that the last integral is $O(x^{-\alpha})$.

6.3 As an example, consider

$$I(x) = \int_0^\infty \mathrm{Ai}(xv)\,f(v)\,dv,$$

where x is positive and $f(v)$ is a sectionally continuous function in $(0, \infty)$ having the expansion (6.01). From Chapter 4, §4.1

$$\mathrm{Ai}(t) \sim \tfrac{1}{2}\pi^{-1/2}t^{-1/4}\exp(-\tfrac{2}{3}t^{3/2}) \qquad (t \to \infty).$$

Hence the given integral converges for large x if we postulate that for an assignable σ

$$f(v) = O\{\exp(\sigma v^{3/2})\} \qquad (v \to \infty).$$

With $g(v) = \mathrm{Ai}(v)$ and $\phi(v)$ set equal to the right-hand side of (6.05), the conditions of Lemma 6.1 are satisfied. Therefore

$$\int_0^\infty \mathrm{Ai}(xv)\,\phi_n(v)\,dv = O(x^{-(n+\lambda)/\mu}),$$

and thence we obtain the desired result

$$\int_0^\infty \mathrm{Ai}(xv)\,f(v)\,dv \sim \sum_{s=0}^\infty G\left(\frac{s+\lambda}{\mu}\right)\frac{a_s}{x^{(s+\lambda)/\mu}} \qquad (x \to \infty), \tag{6.09}$$

where

$$G(\alpha) = \int_0^\infty \mathrm{Ai}(t)\, t^{\alpha-1}\, dt \qquad (\mathrm{Re}\,\alpha > 0).$$

The required transforms can be evaluated as follows. From Chapter 2, (8.02) we derive

$$G(\alpha) = \frac{1}{2\pi i}\int_0^\infty t^{\alpha-1}\, dt \int_{\infty e^{-\pi i/3}}^{\infty e^{\pi i/3}} \exp(\tfrac{1}{3}v^3 - tv)\, dv.$$

Inversion of the order of integration is readily justified, and gives

$$G(\alpha) = \frac{\Gamma(\alpha)}{2\pi i}\int_{\infty e^{-\pi i/3}}^{\infty e^{\pi i/3}} \frac{\exp(\frac{1}{3}v^3)}{v^\alpha}\, dv = \frac{\Gamma(\alpha)}{2\pi i}\int_{-\infty}^{(0+)} \frac{e^w}{(3w)^{(\alpha+2)/3}}\, dw = \frac{\Gamma(\alpha)}{3^{(\alpha+2)/3}\,\Gamma(\frac{1}{3}\alpha+\frac{2}{3})}. \tag{6.10}$$

A special case of (6.09) is as follows. Let $f(v)$ be sectionally continuous in $[0, \infty)$ and

$$f(v) = f(0) + O(v^\delta) \quad (v \to 0), \qquad f(v) = O\{\exp(\sigma v^{3/2})\} \quad (v \to \infty),$$

where δ and σ are positive constants. Then†

$$\int_0^\infty \mathrm{Ai}(xv)\, f(v)\, dv = \frac{f(0)}{3x} + O\left(\frac{1}{x^{1+\delta}}\right) \qquad (x \to \infty). \tag{6.11}$$

6.4 As a second example, consider

$$I(x) = \int_0^\infty K_0(xv)\, f(v)\, dv,$$

where, again, $f(v)$ is sectionally continuous in $(0, \infty)$ and has the expansion (6.01). For large argument, the modified Bessel function is given by

$$K_0(t) \sim \{\pi/(2t)\}^{1/2} e^{-t} \qquad (t \to \infty);$$

hence convergence of the given integral for large x is assured if

$$f(v) = O(e^{\sigma v}) \qquad (v \to \infty),$$

for an assignable σ. Without loss of generality we may assume that $\sigma > 0$.

An extra complication in the present case stems from the logarithmic singularity in the kernel:

$$K_0(t) = -\ln t + O(1) \qquad (t \to 0+);$$

see Chapter 7, Exercise 8.3. Set

$$g(t) = K_0(t) + e^{-t}\ln\left(\frac{t}{t+1}\right),$$

† Muldoon (1970) has investigated the relation (6.11) with weaker restrictions on $f(v)$.

so that $g(t)$ is bounded at the origin and $O(t^{-1/2}e^{-t})$ at infinity. Then for each nonnegative integer n

$$I(x) = \sum_{s=0}^{n-1} \frac{a_s}{x^{(s+\lambda)/\mu}} \int_0^\infty K_0(t)\, t^{(s+\lambda-\mu)/\mu}\, dt + \varepsilon_n(x) - \eta_n(x), \tag{6.12}$$

where

$$\varepsilon_n(x) = \int_0^\infty g(xv)\, \phi_n(v)\, dv, \qquad \eta_n(x) = \int_0^\infty e^{-xv} \ln\left(\frac{xv}{xv+1}\right) \phi_n(v)\, dv,$$

$\phi_n(v)$ again being defined by (6.05). Thus $\phi_n(v)$ is $O(v^{(n+\lambda-\mu)/\mu})$ near the origin and $O(e^{\sigma v})$ at infinity. Application of Lemma 6.1 immediately yields

$$\varepsilon_n(x) = O(x^{-(n+\lambda)/\mu}) \qquad (x \to \infty).$$

Next, if n is large enough to ensure that $n+\lambda-\mu \geqslant 0$, then

$$|\phi_n(v)| \leqslant A_n v^{(n+\lambda-\mu)/\mu} e^{\sigma v} \qquad (0 \leqslant v < \infty),$$

for an assignable A_n. Hence

$$|\eta_n(x)| \leqslant \frac{A_n}{x^{(n+\lambda)/\mu}} \int_0^\infty e^{-t} \ln\left(\frac{t+1}{t}\right) t^{(n+\lambda-\mu)/\mu} e^{\sigma t/x}\, dt = O\left(\frac{1}{x^{(n+\lambda)/\mu}}\right).$$

Thus (6.12) represents an asymptotic expansion for large x. Evaluating the Mellin transform with the aid of Chapter 7, Exercise 8.6, we have the desired result

$$\int_0^\infty K_0(xv)\, f(v)\, dv \sim \frac{1}{4} \sum_{s=0}^\infty \left\{\Gamma\left(\frac{s+\lambda}{2\mu}\right)\right\}^2 \frac{a_s}{(\frac{1}{2}x)^{(s+\lambda)/\mu}} \qquad (x \to \infty).$$

Ex. 6.1 Prove that

$$\int_0^1 (1-v)^{1/2} \exp(-x^2 v^2) \sin(xv)\, dv \sim \sum_{s=0}^\infty (-)^s \binom{\frac{1}{2}}{s} \frac{A_s}{x^{s+1}} \qquad (x \to \infty),$$

where $A_0 = F(\frac{1}{2})$, F denoting Dawson's integral, and

$$A_s = \int_0^\infty t^s \exp(-t^2) \sin t\, dt.$$

Ex. 6.2 Let $f(v)$ be sectionally continuous in $(0, \infty)$, and

$$f(v) = v^{\alpha-1} \ln v\{1+O(v^\delta)\} \quad (v \to 0), \qquad f(v) = O\{\exp(\sigma v^{3/2})\} \quad (v \to \infty),$$

where α, δ, and σ are positive constants. Prove that

$$\int_0^\infty \mathrm{Ai}(xv)\, f(v)\, dv = \frac{G(\alpha)}{x^\alpha} \left\{\psi(\alpha) - \frac{1}{3}\psi\left(\frac{1}{3}\alpha + \frac{2}{3}\right) - \frac{1}{3}\ln 3 - \ln x + O\left(\frac{\ln x}{x^\delta}\right)\right\}$$

as $x \to +\infty$, where $G(\alpha)$ is given by (6.10) and ψ denotes Γ'/Γ.

Ex. 6.3 Show that for large positive x

$$\int_0^\infty \exp(-xt + x^{1/2} \sin t) \sin(xt) \Gamma(t)\, t^{-t}\, dt = \frac{\pi}{4} + O\left(\frac{1}{x^{1/2}}\right).$$

7 Nicholson's Integral for $J_\nu^2(z) + Y_\nu^2(z)$

7.1 In this section we illustrate the theory of §6 by an interesting integral which is of importance in a later chapter. The integral can be found by multiplying together Sommerfeld's contour integrals for the Hankel functions (Chapter 7, §4.4) in the following way. Suppose, temporarily, that z is a positive real variable. Then

$$H_\nu^{(1)}(z) = \frac{1}{\pi i}\int_{-\infty}^{\infty+\pi i} e^{z\sinh t-\nu t}\,dt, \qquad H_\nu^{(2)}(z) = -\frac{1}{\pi i}\int_{-\infty}^{\infty-\pi i} e^{z\sinh t-\nu t}\,dt.$$

Ignoring all questions of convergence for the moment, we deform the integration paths into the lines $t = \frac{1}{2}\pi i+\tau_1$ $(-\infty<\tau_1<\infty)$ and $t = -\frac{1}{2}\pi i+\tau_2$ $(-\infty<\tau_2<\infty)$, respectively. Then

$$\begin{aligned} &J_\nu^2(z)+Y_\nu^2(z)\\ &\quad= H_\nu^{(1)}(z)\,H_\nu^{(2)}(z) = \frac{1}{\pi^2}\int_{-\infty}^{\infty} e^{iz\cosh\tau_1-\nu\tau_1}\,d\tau_1\int_{-\infty}^{\infty} e^{-iz\cosh\tau_2-\nu\tau_2}\,d\tau_2\\ &\quad= \frac{1}{\pi^2}\int_{-\infty}^{\infty}\int_{-\infty}^{\infty}\exp\left\{2iz\sinh\left(\frac{\tau_1-\tau_2}{2}\right)\sinh\left(\frac{\tau_1+\tau_2}{2}\right)-\nu(\tau_1+\tau_2)\right\}d\tau_1\,d\tau_2. \end{aligned}$$

Changing to new integration variables $v = \frac{1}{2}(\tau_1-\tau_2)$ and $w = \frac{1}{2}(\tau_1+\tau_2)$, we obtain

$$\frac{2}{\pi^2}\int_{-\infty}^{\infty} e^{-2\nu w}\,dw\int_{-\infty}^{\infty} e^{2iz\sinh v\sinh w}\,dv,$$

that is,

$$\frac{2}{\pi^2}\int_0^{\infty} e^{-2\nu w}\,dw\int_{-\infty}^{\infty} e^{2iz\sinh v\sinh w}\,dv + \frac{2}{\pi^2}\int_0^{\infty} e^{2\nu w}\,dw\int_{-\infty}^{\infty} e^{-2iz\sinh v\sinh w}\,dv.$$

Finally, we displace the paths for the inner integrals into the lines $\operatorname{Im} v = \frac{1}{2}\pi$ and $\operatorname{Im} v = -\frac{1}{2}\pi$, respectively, and then refer to equation (8.03) of Chapter 7. Thus we arrive at *Nicholson's integral*:

$$N_\nu(z) \equiv \frac{8}{\pi^2}\int_0^{\infty} K_0(2z\sinh w)\cosh(2\nu w)\,dw. \tag{7.01}$$

Although the foregoing derivation can be placed on a sound footing, the analysis is difficult.† Instead, following Wilkins (1948) we shall verify the suggested result by quite a different method.

7.2 Theorem 7.1 *The integral* (7.01) *converges when* $|\operatorname{ph} z| < \frac{1}{2}\pi$ *and equals* $J_\nu^2(z)+Y_\nu^2(z)$.

From relations (8.04) and (8.12) of Chapter 7 it is seen that for positive real values of w

$$K_0(2z\sinh w) \sim \pi^{1/2}(4z\sinh w)^{-1/2}e^{-2z\sinh w} \qquad (w\to\infty),$$

† Given by Watson (1944, §13.73).

and

$$K_0(2z\sinh w) \sim -\ln w \qquad (w \to 0),$$

uniformly with respect to z in any compact domain $\mathbf{Z}$ within $|\mathrm{ph}\, z| < \frac{1}{2}\pi$. Hence the integral (7.01) converges uniformly in $\mathbf{Z}$ and defines a holomorphic function in the sector.

From Chapter 5, Exercise 11.1 we record the following differential equation

$$\frac{d^3W}{dz^3} + \frac{3}{z}\frac{d^2W}{dz^2} + \left(4 + \frac{1-4\nu^2}{z^2}\right)\frac{dW}{dz} + \frac{4}{z}W = 0 \tag{7.02}$$

for the square of any solution of Bessel's equation. Differentiation of (7.01) under the sign of integration yields

$$N_\nu'(z) = \frac{16}{\pi^2}\int_0^\infty K_0'(2z\sinh w)\cosh(2\nu w)\sinh w\, dw,$$

and

$$N_\nu''(z) = \frac{32}{\pi^2}\int_0^\infty K_0''(2z\sinh w)\cosh(2\nu w)\sinh^2 w\, dw.$$

Then with the aid of the modified Bessel equation, we find that

$$N_\nu''(z) = 4L_\nu(z) - 4N_\nu(z) - z^{-1}N_\nu'(z), \tag{7.03}$$

where

$$L_\nu(z) = \frac{8}{\pi^2}\int_0^\infty K_0(2z\sinh w)\cosh(2\nu w)\cosh^2 w\, dw.$$

Similarly,

$$L_\nu'(z) = z^{-1}N_\nu(z) - 2z^{-1}L_\nu(z) + \nu^2 z^{-2}N_\nu'(z). \tag{7.04}$$

Elimination of $L_\nu(z)$ from (7.03) and (7.04) confirms that $N_\nu(z)$ satisfies (7.02). Accordingly, $N_\nu(z)$ can be expressed as a linear combination of any three independent solutions; thus

$$N_\nu(z) = A\{J_\nu^2(z)+Y_\nu^2(z)\} + B\{H_\nu^{(1)}(z)\}^2 + C\{H_\nu^{(2)}(z)\}^2. \tag{7.05}$$

The constants A, B, and C can be found by letting $z \to +\infty$, as follows. From the theory of §6.4 with $x = 2z$ and $v = \sinh w$, it is easily seen that

$$N_\nu(z) \sim 2/(\pi z).$$

Substituting in (7.05) by means of this result and the asymptotic forms of $J_\nu(z)$, $Y_\nu(z)$, and the Hankel functions given in Chapter 7, we obtain

$$1 = A + \exp\{2i(z-\tfrac{1}{2}\nu\pi-\tfrac{1}{4}\pi)\}\,B + \exp\{-2i(z-\tfrac{1}{2}\nu\pi-\tfrac{1}{4}\pi)\}\,C + o(1).$$

Hence $B = C = 0$, $A = 1$, and the theorem is proved.

7.3 An immediate observation from Nicholson's integral is that $J_\nu^2(z)+Y_\nu^2(z)$ is an even function of ν. Next, since $K_0(t)$ is a decreasing function of t (Chapter 7,

§8.3) another deduction is that *for any fixed real value of* ν *the function* $J_\nu^2(x)+Y_\nu^2(x)$ *is decreasing in the interval* $0 < x < \infty$.

In Chapter 11 we are interested in the growth of $x\{J_\nu^2(x)+Y_\nu^2(x)\}$ in the same interval. Differentiating under the sign of integration and integrating by parts, we readily verify that

$$\frac{d}{dx}[x\{J_\nu^2(x)+Y_\nu^2(x)\}] = \frac{8}{\pi^2}\int_0^\infty K_0(2x\sinh w)\tanh w\cosh(2\nu w)$$
$$\times\{\tanh w - 2\nu\tanh(2\nu w)\}\,dw.$$

For positive values of w, the factor $\tanh w - 2\nu\tanh(2\nu w)$ is positive or negative according as $|2\nu| <$ or >1. Hence $x\{J_\nu^2(x)+Y_\nu^2(x)\}$ *is an increasing function of* x *when* $-\frac{1}{2} < \nu < \frac{1}{2}$, *and a decreasing function of* x *when* $\nu > \frac{1}{2}$ *or* $\nu < -\frac{1}{2}$.

Ex. 7.1 Let $A_s(\nu)$ be defined as in Chapter 7, (4.02). With the aid of Nicholson's integral and the expansion (9.04) of Chapter 4, deduce that

$$J_\nu^2(z) + Y_\nu^2(z) \sim \frac{2}{\pi z}\sum_{s=0}^\infty 1\cdot3\cdot5\cdots(2s-1)\frac{A_s(\nu)}{z^{2s}}$$

as $z\to\infty$ in the sector $|\mathrm{ph}\,z| \leqslant \pi-\delta$ $(<\pi)$. [Watson, 1944.]

8 Oscillatory Kernels

8.1 Results analogous to those of earlier sections can also be found for the corresponding integrals obtained on replacing the real parameter x by ix.† The proofs are more difficult, however, in the same way that the method of stationary phase is more difficult to justify than Laplace's method. We confine the text to an example which can be treated by the comparatively simple method of integration by parts.

Consider

$$I(x) = \int_0^\infty \mathrm{Ai}(-xv)\,f(v)\,dv,$$

in which $x > 0$ and $f(v)$ is infinitely differentiable in $[0,\infty)$. Repeated partial integrations produce

$$I(x) = \left[-\frac{A_1(xv)}{x}f(v) - \frac{A_2(xv)}{x^2}f'(v) - \cdots - \frac{A_n(xv)}{x^n}f^{(n-1)}(v)\right]_0^\infty$$
$$+\frac{1}{x^n}\int_0^\infty A_n(xv)\,f^{(n)}(v)\,dv, \tag{8.01}$$

where

$$A_1(v) = \int_v^\infty \mathrm{Ai}(-t)\,dt, \qquad A_s(v) = \int_v^\infty A_{s-1}(t)\,dt \qquad (s \geqslant 2).$$

† See, for example, Ludwig (1967).

When v is large and positive, the asymptotic expansion of $\mathrm{Ai}(-v)$ has the form

$$\mathrm{Ai}(-v) \sim \cos(\tfrac{2}{3}v^{3/2}-\tfrac{1}{4}\pi)\sum_{s=0}^{\infty}(-)^s\frac{\alpha_{2s}}{v^{3s+(1/4)}}+\sin(\tfrac{2}{3}v^{3/2}-\tfrac{1}{4}\pi)\sum_{s=0}^{\infty}(-)^s\frac{\alpha_{2s+1}}{v^{3s+(7/4)}},$$

in which the α_s are constants; see Chapter 4, §4.2. Integration by parts shows that

$$A_1(v) \sim \sin(\tfrac{2}{3}v^{3/2}-\tfrac{1}{4}\pi)\sum_{s=0}^{\infty}(-)^s\frac{\beta_{2s}}{v^{3s+(3/4)}}-\cos(\tfrac{2}{3}v^{3/2}-\tfrac{1}{4}\pi)\sum_{s=0}^{\infty}(-)^s\frac{\beta_{2s+1}}{v^{3s+(9/4)}},$$

where the β_s are constants. In particular, $A_1(v) = O(v^{-3/4})$. Repetition of the process yields

$$A_s(v) = O(v^{-(s/2)-(1/4)}). \tag{8.02}$$

Suppose that $f(v)$ satisfies

$$f^{(s)}(v) = O(v^{(2s-3-\delta_s)/4}) \qquad (v \to +\infty),$$

for each $s = 1, 2, \ldots$, where the δ_s are positive constants. Taking $s = 1$ and integrating, we obtain

$$f(v) = O(v^{(3-\delta_1)/4}).$$

Then from (8.01) and (8.02) we derive the desired result

$$I(x) \sim \sum_{s=0}^{\infty}\frac{A_{s+1}(0)\,f^{(s)}(0)}{x^{s+1}} \qquad (x \to +\infty).$$

8.2 The necessary values of the coefficients $A_s(0)$ may be found with the aid of the formula

$$\mathrm{Ai}(-t) = 2\,\mathrm{Re}\{e^{\pi i/3}\,\mathrm{Ai}(te^{\pi i/3})\} \qquad (t \text{ real});$$

compare Chapter 2, (8.06). Now

$$\int_0^\infty e^{\pi i/3}\,\mathrm{Ai}(te^{\pi i/3})\,dt = \int_0^{\infty e^{\pi i/3}} \mathrm{Ai}(t)\,dt = \int_0^\infty \mathrm{Ai}(t)\,dt,$$

the rotation of the path of integration being justifiable by use of the relation

$$\mathrm{Ai}(t) = O\{t^{-1/4}\exp(-\tfrac{2}{3}t^{3/2})\} \qquad (|\mathrm{ph}\,t| \leqslant \tfrac{1}{3}\pi), \tag{8.03}$$

(Chapter 4, §4.1), and Jordan's inequality. Equation (6.10), with $\alpha = 1$, shows that the value of the last integral is $\tfrac{1}{3}$, and hence that $A_1(0) = \tfrac{2}{3}$.

In a similar way

$$\int_0^\infty dv\int_v^\infty e^{\pi i/3}\,\mathrm{Ai}(te^{\pi i/3})\,dt = \int_0^\infty dv\int_{ve^{\pi i/3}}^\infty \mathrm{Ai}(t)\,dt = e^{-\pi i/3}\int_0^{\infty e^{\pi i/3}} dv\int_v^\infty \mathrm{Ai}(t)\,dt.$$

Partial integration of (8.03) produces

$$\int_v^\infty \mathrm{Ai}(t)\,dt = O\{v^{-3/4}\exp(-\tfrac{2}{3}v^{3/2})\} \qquad (|\mathrm{ph}\,v| \leqslant \tfrac{1}{3}\pi).$$

Accordingly, in the last repeated integral the path in the v plane may be rotated to give

$$\int_0^\infty dv \int_v^\infty e^{\pi i/3} \operatorname{Ai}(te^{\pi i/3})\, dt = e^{-\pi i/3} \int_0^\infty dv \int_v^\infty \operatorname{Ai}(t)\, dt = e^{-\pi i/3} \int_0^\infty v \operatorname{Ai}(v)\, dv.$$

Referring again to (6.10), this time with $\alpha = 2$, we derive

$$A_2(0) = \{3^{4/3}\Gamma(\tfrac{4}{3})\}^{-1} = \{3^{1/3}\Gamma(\tfrac{1}{3})\}^{-1}.$$

Repetition of the procedure yields the desired formula

$$A_s(0) = 2\operatorname{Re}\left\{\frac{e^{-(s-1)\pi i/3} G(s)}{(s-1)!}\right\} = \frac{2\cos\{\frac{1}{3}(s-1)\pi\}}{3^{(s+2)/3}\Gamma(\frac{1}{3}s+\frac{2}{3})}.$$

Ex. 8.1 With the aid of Chapter 7, Exercise 5.8, show that

$$\int_0^\infty J_0(xt) f(t)\, dt = \frac{f(0)}{x} + O\left(\frac{1}{x^{3/2}}\right) \qquad (x \to \infty),$$

provided that $f(t)$ is continuously differentiable in $[0, \infty)$ and

$$f'(t) = O(t^{-(1+\delta)/2}) \qquad (t \to \infty),$$

where δ is a positive constant.

Ex. 8.2† Show that

$$\int_0^\infty \frac{J_0^2(t)}{t+x}\, dt = \frac{\ln x}{\pi x} + \frac{c}{x} + O\left(\frac{\ln x}{x^2}\right) \qquad (x \to \infty), \qquad c = \int_0^\infty \left\{J_0^2(t) - \frac{1}{\pi(1+t)}\right\} dt.$$

9 Bleistein's Method

9.1 Consider an integral of the form

$$I(\alpha, x) = \int_0^k e^{-xp(\alpha,t)} q(\alpha, t)\, t^{\lambda-1}\, dt, \tag{9.01}$$

in which k and λ are positive constants (k possibly being infinite), α is a variable parameter in the interval $[0, k)$, and x is a large positive parameter. Let us suppose that $\partial^2 p(\alpha, t)/\partial t^2$ and $q(\alpha, t)$ are continuous functions of α and t, and also that for given α the minimum value of $p(\alpha, t)$ in $[0, k)$ is attained at $t = \alpha$, at which point $\partial p(\alpha, t)/\partial t$ vanishes but both $\partial^2 p(\alpha, t)/\partial t^2$ and $q(\alpha, t)$ are nonzero. By Laplace's method the asymptotic form of $I(\alpha, x)$ for large x may be verified to be

$$I(\alpha, x) \sim e^{-xp(\alpha,\alpha)} q(\alpha, \alpha)\, \alpha^{\lambda-1} \left\{\frac{x}{2\pi}\left[\frac{\partial^2 p(\alpha, t)}{\partial t^2}\right]_{t=\alpha}\right\}^{-1/2} \tag{9.02}$$

if $\alpha \neq 0$, or

$$I(\alpha, x) \sim \tfrac{1}{2} e^{-xp(0,0)} q(0, 0)\, \Gamma(\tfrac{1}{2}\lambda) \left\{\frac{x}{2}\left[\frac{\partial^2 p(0, t)}{\partial t^2}\right]_{t=0}\right\}^{-\lambda/2} \tag{9.03}$$

if $\alpha = 0$; compare Chapter 3, §7.

† Generalizations of this result have been given by Luke (1968).

When $\lambda \neq 1$, it is obvious that the right-hand side of (9.02) does not reduce to the right-hand side of (9.03) as $\alpha \to 0$. When $\lambda = 1$, the limiting value of the right-hand side of (9.02) is finite and nonzero, but differs by a factor of 2 from (9.03).† This abrupt change in the form of the approximation at $\alpha = 0$ indicates that the accuracy of (9.02) deteriorates as α diminishes. To put this another way, the approximation (9.02) cannot be uniformly valid for arbitrarily small values of α.

9.2 We now outline a method, due to Bleistein (1966), which yields a generalized asymptotic expansion for $I(\alpha, x)$ which *is* uniformly valid in a closed α interval containing $\alpha = 0$. The underlying idea is to transform (9.01) into an integral which closely resembles the simplest kind of integral maintaining the essential features:

$$J(\alpha, x) = \int_0^\infty \exp\{-x(\tfrac{1}{2}w^2 - aw)\}\, w^{\lambda-1}\, dw,$$

where $a = a(\alpha)$ depends on α and vanishes at $\alpha = 0$. In terms of Faxén's integral (§4)

$$J(\alpha, x) = \tfrac{1}{2}(\tfrac{1}{2}x)^{-\lambda/2}\, \mathrm{Fi}\{\tfrac{1}{2}, \tfrac{1}{2}\lambda; a(2x)^{1/2}\}.$$

The appropriate transformation of integration variables is supplied by

$$p(\alpha, t) = \tfrac{1}{2}w^2 - aw + b, \tag{9.04}$$

the values of a and b being chosen in such a way that the points $t = 0$ and $t = \alpha$ correspond respectively to $w = 0$ and $w = a$. This gives

$$b = p(\alpha, 0), \qquad a = \{2p(\alpha, 0) - 2p(\alpha, \alpha)\}^{1/2}.$$

Thus

$$w = \{2p(\alpha, 0) - 2p(\alpha, \alpha)\}^{1/2} \pm \{2p(\alpha, t) - 2p(\alpha, \alpha)\}^{1/2}, \tag{9.05}$$

the upper or lower sign being taken according as $t > \alpha$ or $t < \alpha$. Equation (9.05) determines a one-to-one relationship between t and w. Moreover, since

$$\frac{dw}{dt} = \pm \frac{1}{\{2p(\alpha, t) - 2p(\alpha, \alpha)\}^{1/2}} \frac{\partial p(\alpha, t)}{\partial t},$$

the relationship is free from singularity at $t = \alpha$.

Making the transformation (9.04), we obtain

$$I(\alpha, x) = e^{-xp(\alpha, 0)} \int_0^\kappa \exp\{-x(\tfrac{1}{2}w^2 - aw)\}\, f(\alpha, w)\, w^{\lambda-1}\, dw,$$

where

$$f(\alpha, w) = q(\alpha, t) \left(\frac{t}{w}\right)^{\lambda-1} \frac{dt}{dw},$$

and $\kappa \equiv \kappa(\alpha)$ is the value of w at $t = k$. We now expand $f(\alpha, w)$ in a Taylor series *centered at the peak value* $w = a$ of the exponential factor in the integrand. This series has the form

$$f(\alpha, w) = \sum_{s=0}^{\infty} \phi_s(\alpha)(w - a)^s, \tag{9.06}$$

† The reader is advised to understand from Laplace's method why this is so.

in which the coefficients $\phi_s(\alpha)$ are continuous at $\alpha = 0$. The required expansion is then obtained in a manner analogous to Laplace's method: the limit κ is replaced by infinity and termwise integration performed in a formal manner. Thus

$$I(\alpha, x) = \frac{e^{-xp(\alpha,0)}}{x^{\lambda/2}} \sum_{s=0}^{\infty} \phi_s(\alpha) \frac{F_s(a\sqrt{x})}{x^{s/2}}, \tag{9.07}$$

where

$$F_s(y) = \int_0^{\infty} \exp(-\tfrac{1}{2}\tau^2 + y\tau)(\tau - y)^s \tau^{\lambda-1}\, d\tau. \tag{9.08}$$

9.3 By applying Laplace's method to (9.08)† it is found that for large positive values of y

$$F_s(y) \sim 2^{(s+1)/2}\Gamma(\tfrac{1}{2}s+\tfrac{1}{2})\, e^{y^2/2} y^{\lambda-1} \quad (s \text{ even}); \qquad F_s(y) = O(e^{y^2/2} y^{\lambda-2}) \quad (s \text{ odd}). \tag{9.09}$$

Accordingly, for fixed nonzero α and large x successive terms of the series (9.07), taken in pairs, are of diminishing asymptotic order. And this is also true when $\alpha = a = 0$, since in this event the factors $F_s(a\sqrt{x})$ are independent of x. However, the method of derivation gives no indication whether the series is an asymptotic expansion (in the generalized sense) for large x, and the equality sign in (9.07) has only formal significance. In the next section we treat a specific example and prove that the resulting series does have the claimed uniform asymptotic property.

Ex. 9.1 Prove that $F_1(y) = F_0'(y) - yF_0(y)$, $F_2(y) = -yF_1(y) + \lambda F_0(y)$, and

$$F_s(y) = -yF_{s-1}(y) + (s+\lambda-2)F_{s-2}(y) + (s-2)\, yF_{s-3}(y) \qquad (s \geqslant 3).$$

Ex. 9.2 With the aid of (9.09) confirm that when α is fixed the leading term of (9.07) reduces to (9.02) or (9.03) according as $\alpha > 0$ or $\alpha = 0$.

Ex. 9.3‡ Let $p(\alpha, t)$, $q(\alpha, t)$, and a sufficient number of their partial derivatives be continuous for all real values of α and t. Assume that: (i) $p(\alpha, t) \to \infty$ as $t \to \pm\infty$; (ii) as a function of t the minimum value of $p(\alpha, t)$ is attained at $t = \alpha$, at which point $\partial^2 p(\alpha, t)/\partial t^2$ and $q(\alpha, t)$ are nonzero. With the definitions of §9.2 prove that

$$\int_{-\infty}^{\infty} e^{-xp(\alpha,t)} q(\alpha,t) \frac{dt}{t} = 2\pi^{1/2} f(\alpha, 0)\, e^{-xp(\alpha,\alpha)} F[x^{1/2}\{p(\alpha,0) - p(\alpha,\alpha)\}^{1/2}]$$
$$+ e^{-xp(\alpha,0)} \int_{-\infty}^{\infty} \exp\{-x(\tfrac{1}{2}w^2 - aw)\} \frac{f(\alpha,w) - f(\alpha,0)}{w}\, dw,$$

where F denotes Dawson's integral, $f(\alpha, w) = q(\alpha, t)(w/t)(dt/dw)$, and the left-hand integral has its Cauchy principal value.

10 Example

10.1 Let

$$I(\alpha, x) = \int_0^{\pi/2} e^{x(\cos\theta + \theta\sin\alpha)}\, d\theta, \tag{10.01}$$

† Set $\tau = y(1+t)$ and apply Theorem 8.1 of Chapter 3.

‡ Communicated by G. F. Miller. In effect the result shows how the case $\lambda = 0$ of (9.01) may be transformed into the case $\lambda = 1$.

where $0 \leqslant \alpha < \frac{1}{2}\pi$ and x is a large positive parameter. In the notation of §9.1 we have

$$p(\alpha,\theta) = -\cos\theta - \theta\sin\alpha, \qquad \partial p(\alpha,\theta)/\partial\theta = \sin\theta - \sin\alpha.$$

The only minimum of $p(\alpha,\theta)$ in the range of integration is at $\theta = \alpha$. By Laplace's method it is readily verified that for fixed α and large x the asymptotic expansion of $I(\alpha, x)$ begins

$$I(\alpha,x) \sim e^{x(\cos\alpha+\alpha\sin\alpha)}\left(\frac{2\pi}{x\cos\alpha}\right)^{1/2}\left\{1 + \frac{5-2\cos^2\alpha}{24x\cos^3\alpha} + \cdots\right\} \tag{10.02}$$

if $\alpha \neq 0$, or

$$I(\alpha,x) \sim e^{x}\left(\frac{\pi}{2x}\right)^{1/2}\left\{1 + \frac{1^2}{1!(8x)} + \frac{1^2\cdot 3^2}{2!(8x)^2} + \cdots\right\} \tag{10.03}$$

if $\alpha = 0$.

We apply the method of §9 to derive an asymptotic expansion for $I(\alpha, x)$ which is uniformly valid for $\alpha \in [0, \alpha_0]$, where α_0 is any fixed number in the interval $0 < \alpha_0 < \frac{1}{2}\pi$. Since

$$p(\alpha,0) = -1, \qquad p(\alpha,\alpha) = -\cos\alpha - \alpha\sin\alpha, \qquad p(\alpha,\tfrac{1}{2}\pi) = -\tfrac{1}{2}\pi\sin\alpha,$$

we have, in the notation of §9.2,

$$a = 2^{1/2}(\cos\alpha + \alpha\sin\alpha - 1)^{1/2}, \tag{10.04}$$

$$\cos\theta + \theta\sin\alpha = 1 + aw - \tfrac{1}{2}w^2, \tag{10.05}$$

and

$$w = a \pm 2^{1/2}\{\cos\alpha + (\alpha-\theta)\sin\alpha - \cos\theta\}^{1/2} \qquad (\theta \gtrless \alpha).$$

The transformed integral is given by

$$I(\alpha,x) = e^{x}\int_0^{\kappa} \exp\{-x(\tfrac{1}{2}w^2 - aw)\}\frac{d\theta}{dw}\,dw, \tag{10.06}$$

where

$$\frac{d\theta}{dw} = \frac{w-a}{\sin\theta - \sin\alpha}, \tag{10.07}$$

and

$$\kappa = a + 2^{1/2}\{\cos\alpha + (\alpha - \tfrac{1}{2}\pi)\sin\alpha\}^{1/2}. \tag{10.08}$$

Let the Taylor-series expansion of θ at $w = a$ be denoted by

$$\theta = \alpha + \sum_{s=0}^{\infty}\frac{\phi_s(\alpha)}{s+1}(w-a)^{s+1}, \tag{10.09}$$

so that

$$\frac{d\theta}{dw} = \sum_{s=0}^{\infty}\phi_s(\alpha)(w-a)^s; \tag{10.10}$$

compare (9.06). Leading coefficients are easily calculated by substituting in (10.07) and equating coefficients. The first three are found to be

$$\phi_0(\alpha) = \frac{1}{(\cos\alpha)^{1/2}}, \qquad \phi_1(\alpha) = \frac{\sin\alpha}{3\cos^2\alpha}, \qquad \phi_2(\alpha) = \frac{5-2\cos^2\alpha}{24(\cos\alpha)^{7/2}}. \quad (10.11)$$

The next step is to replace the upper limit in (10.06) by infinity, substitute for $d\theta/dw$ by means of (10.10), and integrate formally term by term. Making the substitution $w = a+(2/x)^{1/2}\tau$ and referring to (10.04), we see that

$$\int_0^\infty \exp\left\{-x\left(\frac{1}{2}w^2 - aw\right)\right\}(w-a)^s\,dw$$

$$= \exp\left(\frac{1}{2}a^2 x\right)\left(\frac{2}{x}\right)^{(s+1)/2}\int_{-a(x/2)^{1/2}}^{\infty} \exp(-\tau^2)\tau^s\,d\tau$$

$$= \frac{1}{2}e^{x(\cos\alpha+\alpha\sin\alpha-1)}\left(\frac{2}{x}\right)^{(s+1)/2} X_s(\alpha,x), \quad (10.12)$$

where

$$X_s(\alpha,x) = \Gamma(\tfrac{1}{2}s+\tfrac{1}{2}) + (-)^s\gamma\{\tfrac{1}{2}s+\tfrac{1}{2},\, x(\cos\alpha+\alpha\sin\alpha-1)\}, \quad (10.13)$$

γ denoting the incomplete Gamma function (Chapter 2, §5.1). Therefore the desired expansion is given formally by

$$I(\alpha,x) = \frac{e^{x(\cos\alpha+\alpha\sin\alpha)}}{(2x)^{1/2}}\sum_{s=0}^{\infty}\phi_s(\alpha)X_s(\alpha,x)\left(\frac{2}{x}\right)^{s/2}. \quad (10.14)$$

10.2 It is easily seen from the definition (10.13) that the function $X_s(\alpha,x)$ satisfies

$$\Gamma(\tfrac{1}{2}s+\tfrac{1}{2}) \leqslant X_s(\alpha,x) < 2\Gamma(\tfrac{1}{2}s+\tfrac{1}{2}) \quad (s \text{ even}); \qquad 0 < X_s(\alpha,x) \leqslant \Gamma(\tfrac{1}{2}s+\tfrac{1}{2}) \quad (s \text{ odd}).$$

This suggests that the series (10.14) may be a generalized asymptotic expansion with scale $\{e^{x(\cos\alpha+\alpha\sin\alpha)}x^{-(s+1)/2}\}$. Before embarking on the proof we confirm that we are on the right track by examining the reduced form of (10.14) in cases when α is fixed.

When $\alpha = 0$ we have from (10.11) and (10.13)

$$\phi_0(0) = 1, \qquad \phi_1(0) = 0, \qquad \phi_2(0) = \tfrac{1}{8}, \qquad X_s(0,x) = \Gamma(\tfrac{1}{2}s+\tfrac{1}{2}).$$

Accordingly, the first three terms of (10.14) reduce to

$$\frac{e^x}{(2x)^{1/2}}\left\{\Gamma(\tfrac{1}{2}) + \frac{\Gamma(\tfrac{3}{2})}{8}\frac{2}{x}\right\},$$

in agreement with (10.03).

Alternatively, if $\alpha \neq 0$ then for large x the function $\gamma\{\tfrac{1}{2}s+\tfrac{1}{2}, x(\cos\alpha+\alpha\sin\alpha-1)\}$ differs from $\Gamma(\tfrac{1}{2}s+\tfrac{1}{2})$ by an exponentially small quantity. Neglecting this difference, we readily perceive that the first three terms of (10.14) accord with (10.02).

10.3 To establish the asymptotic nature of (10.14) we first prove that for sufficiently small $|w-a|$ the series (10.10) converges uniformly with respect to α and w. This is most easily achieved by complex-variable theory. From Cauchy's formula for the derivatives of an analytic function we have

$$\phi_s(\alpha) = \frac{1}{s!}\left[\frac{d^{s+1}\theta}{dw^{s+1}}\right]_{w=a} = \frac{1}{2\pi i}\int_{\mathscr{W}} \frac{d\theta}{dw}\frac{dw}{(w-a)^{s+1}} = \frac{1}{2\pi i}\int_{\mathscr{C}} \frac{d\theta}{(w-a)^{s+1}}, \tag{10.15}$$

where $\mathscr{W}$ is a simple closed contour encircling $w = a$ but not enclosing any singularity of $d\theta/dw$, and $\mathscr{C}$ is the map of $\mathscr{W}$ in the θ plane.

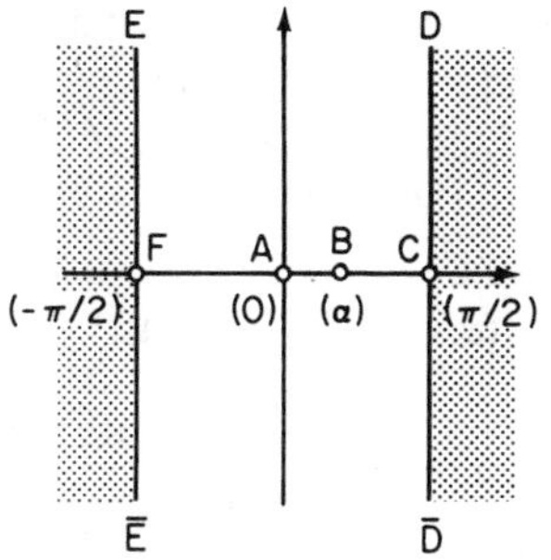

Fig. 10.1 θ plane.

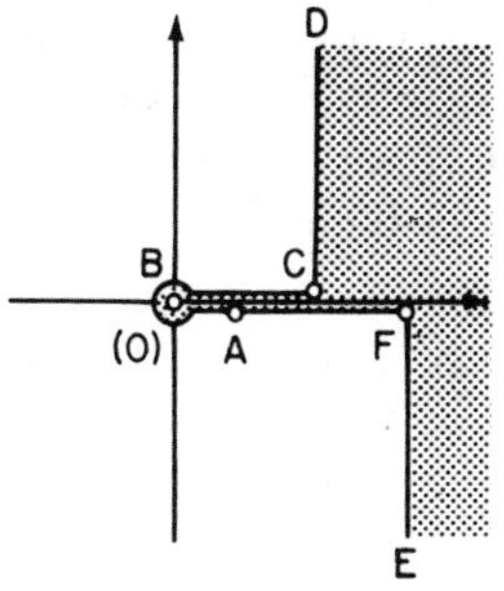

Fig. 10.2 v plane.

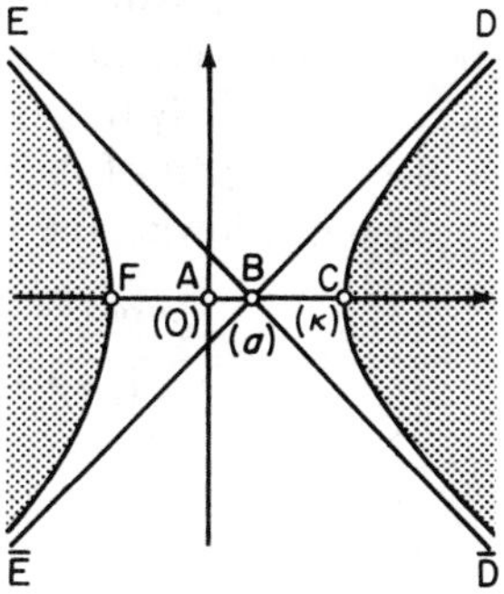

Fig. 10.3 w plane.

The mapping between the planes of θ and w is easily constructed by intermediate passage through the plane of the variable v, defined by

$$\cos\alpha + \alpha\sin\alpha - \cos\theta - \theta\sin\alpha = v = \tfrac{1}{2}(w-a)^2.$$

It suffices to consider the strip $|\mathrm{Re}\,\theta| \leqslant \frac{1}{2}\pi$. Corresponding points are indicated in Figs. 10.1–10.3. Only one sheet of the v map is depicted: the map of the lower half of the θ strip is obtainable by reflection in the real v axis. The affixes of the points C and F in the v plane are $\cos\alpha + (\alpha \mp \frac{1}{2}\pi)\sin\alpha$; both are real and positive. The boundary curves $DC\bar{D}$ and $EF\bar{E}$ in the w plane are rectangular hyperbolas with asymptotes $DB\bar{E}$ and $EB\bar{D}$.

Within the boundaries of the regions $ABCDEFA$ the $\theta \leftrightarrow v$ transformation is one-to-one, as is the $v \leftrightarrow w$ transformation. Hence the relation between θ and w is one-to-one. This result and its conjugate form show that θ and w are analytic functions of each other throughout the whole of the unshaded domains of Figs. 10.1 and 10.3, except possibly at $w = a$ and $\theta = \alpha$. That these points *are* unexceptional follows from the expansion (10.09) and the fact that $\phi_0(\alpha) \neq 0$.

We now specify the integration path $\mathscr{C}$ in (10.15) to be a fixed simple closed contour lying within the strip $|\mathrm{Re}\,\theta| < \frac{1}{2}\pi$ and containing the interval $[0, \alpha_0]$ in its interior, α_0 being the given fixed number in the interval $(0, \frac{1}{2}\pi)$. By continuity, as θ passes around $\mathscr{C}$ and α ranges over the interval $[0, \alpha_0]$ the function $|w-a|$ attains a minimum value d, say. Moreover, d is necessarily positive because $\mathscr{C}$ does not pass through the point $\theta = \alpha$. From (10.15) we immediately derive the key inequality

$$|\phi_s(\alpha)| \leqslant l d^{-s-1}, \tag{10.16}$$

where $2\pi l$ is the length of $\mathscr{C}$.

10.4 We follow our usual procedure and define a remainder coefficient $\Phi_n(\alpha, w)$ by the equation

$$\frac{d\theta}{dw} = \sum_{s=0}^{n-1} \phi_s(\alpha)(w-a)^s + \Phi_n(\alpha, w)(w-a)^n, \tag{10.17}$$

in which n is an arbitrary nonnegative integer. Majorizing the series

$$\Phi_n(\alpha, w) = \phi_n(\alpha) + \phi_{n+1}(\alpha)(w-a) + \phi_{n+2}(\alpha)(w-a)^2 + \cdots$$

by use of (10.16), we see that $|\Phi_n(\alpha, w)|$ is bounded when $\alpha \in [0, \alpha_0]$ and $|w-a| \leqslant \frac{1}{2}d$. And since $d\theta/dw$, $\phi_s(\alpha)$, and a are all continuous functions of α and w, it follows directly from (10.17) that $|\Phi_n(\alpha, w)|$ is also bounded when $\alpha \in [0, \alpha_0]$ and w lies in the part of the interval $[0, \kappa]$ exterior to the circle $|w-a| = \frac{1}{2}d$.

Again, following the Laplace method we write (10.06) in the form

$$I(\alpha, x) = \frac{e^{x(\cos\alpha + \alpha\sin\alpha)}}{(2x)^{1/2}} \left\{ \sum_{s=0}^{n-1} \phi_s(\alpha) X_s(\alpha, x) \left(\frac{2}{x}\right)^{s/2} - \varepsilon_{n,1}(\alpha, x) \right\} + \varepsilon_{n,2}(\alpha, x), \tag{10.18}$$

where

$$\varepsilon_{n,1}(\alpha, x) = \sum_{s=0}^{n-1} \phi_s(\alpha) \Gamma\{\tfrac{1}{2}s + \tfrac{1}{2}, \tfrac{1}{2}(\kappa - a)^2 x\} \left(\frac{2}{x}\right)^{s/2},$$

and

$$\varepsilon_{n,2}(\alpha, x) = e^x \int_0^\kappa \exp\{-x(\tfrac{1}{2}w^2 - aw)\} \Phi_n(\alpha, w)(w-a)^n \, dw.$$

Here Γ denotes the complementary incomplete Gamma function. From (10.08) we see that $\frac{1}{2}(\kappa - a)^2 \geqslant \kappa_0$, where

$$\kappa_0 \equiv \cos\alpha_0 + (\alpha_0 - \tfrac{1}{2}\pi)\sin\alpha_0$$

and is positive. Accordingly,

$$\varepsilon_{n,1}(\alpha, x) = O(x^{-1/2} e^{-\kappa_0 x}) \qquad (x \to \infty) \tag{10.19}$$

uniformly with respect to α; compare Chapter 3, (1.04). Next, if A_n denotes the maximum value of $|\Phi_n(\alpha, w)|$ when $\alpha \in [0, \alpha_0]$ and $w \in [0, \kappa]$, we have

$$|\varepsilon_{n,2}(\alpha, x)| \leqslant A_n e^x \int_0^\infty \exp\{-x(\tfrac{1}{2}w^2 - aw)\} |w-a|^n \, dw.$$

By means of the previous substitution $w = a + (2/x)^{1/2}\tau$ it is found that

$$|\varepsilon_{n,2}(\alpha, x)| \leqslant A_n e^{x(\cos\alpha + \alpha\sin\alpha)} \left(\frac{2}{x}\right)^{(n+1)/2} \Gamma\left(\frac{1}{2}n + \frac{1}{2}\right). \tag{10.20}$$

Combining (10.18), (10.19), and (10.20), and remembering that κ_0 and A_n are independent of α and x we have, for any nonnegative integer n,

$$I(\alpha, x) = \frac{e^{x(\cos\alpha + \alpha\sin\alpha)}}{(2x)^{1/2}} \left\{ \sum_{s=0}^{n-1} \phi_s(\alpha) X_s(\alpha, x) \left(\frac{2}{x}\right)^{s/2} + O\left(\frac{1}{x^{n/2}}\right) \right\}$$

as $x \to \infty$, uniformly with respect to $\alpha \in [0, \alpha_0]$. Here $X_s(\alpha, x)$ is defined by (10.13) and the coefficients $\phi_s(\alpha)$ are calculable by the procedure of §10.1. This is the required result.

Ex. 10.1 Prove that

$$\int_0^\infty \frac{\exp\{-x(t-\alpha)^2(t+1)\}}{(t+t^2)^{1/2}}\,dt \sim \frac{\exp(-x\alpha^2)}{2(1+\alpha)x^{1/4}}\,\mathrm{Fi}\left(\frac{1}{2}, \frac{1}{4}; 2\alpha\sqrt{x}\right)$$

as $x \to \infty$, uniformly with respect to $\alpha \in [0, \alpha_0]$, where α_0 is any constant in the interval $(0, 2)$.

11 The Method of Chester, Friedman, and Ursell

11.1 Let

$$I(\alpha, x) = \int_{\mathscr{P}} e^{-xp(\alpha,t)} q(\alpha, t)\,dt \tag{11.01}$$

be a contour integral in which x is a large parameter, and $p(\alpha, t)$ and $q(\alpha, t)$ are analytic functions of the complex variable t and continuous functions of the parameter α. Suppose that in the region of integration there are two simple saddle points, that is, zeros of $\partial p(\alpha, t)/\partial t$, which coincide for a certain value $\hat{\alpha}$, say, of α. The problem of obtaining an asymptotic approximation to $I(\alpha, x)$ which is uniformly valid for α in a neighborhood of $\hat{\alpha}$ is similar to the problem of §9. In the present case we employ a cubic transformation of variables, given by

$$p(\alpha, t) = \tfrac{1}{3}w^3 + aw^2 + bw + c; \tag{11.02}$$

compare (9.04). The stationary points of the right-hand side are the roots $w_1(\alpha)$ and $w_2(\alpha)$, say, of the equation

$$w^2 + 2aw + b = 0.$$

The values of the coefficients $a = a(\alpha)$ and $b = b(\alpha)$ are chosen in such a way that in (11.02) the values $w = w_1(\alpha)$ and $w = w_2(\alpha)$ correspond to the zeros of $\partial p(\alpha, t)/\partial t$. The other coefficient c is not crucial and is prescribed in any convenient way.

The given integral transforms into

$$I(\alpha, x) = e^{-xc} \int_{\mathscr{Q}} \exp\{-x(\tfrac{1}{3}w^3 + aw^2 + bw)\}\, f(\alpha, w)\,dw,$$

where $\mathscr{Q}$ is the w map of the original path $\mathscr{P}$, and

$$f(\alpha, w) = q(\alpha, t)\frac{dt}{dw} = q(\alpha, t)\frac{w^2 + 2aw + b}{\partial p(\alpha, t)/\partial t}.$$

Owing to the choice of a and b, the function $f(\alpha, w)$ is analytic at $w_1(\alpha)$ and $w_2(\alpha)$ when $\alpha \neq \hat{\alpha}$, and at the confluence of these points when $\alpha = \hat{\alpha}$. For large x, $I(\alpha, x)$ is approximated by the corresponding integral with $f(\alpha, w)$ replaced by a constant, that is, by an Airy or Scorer function, depending on the path $\mathscr{Q}$. In essence, this is the method of Chester, Friedman, and Ursell (1957). In the next two sections we illustrate its use and indicate its scope by means of an example and exercises. In

Chapter 11 we derive the same uniform asymptotic approximations in another way, applicable when $I(\alpha, x)$ satisfies a differential equation of appropriate type.

12 Anger Functions of Large Order

12.1 In Chapter 3, Exercise 7.2, we stated the following asymptotic approximations for the function†

$$\mathbf{A}_{-\nu}(a\nu) = \int_0^\infty e^{-\nu(a\sinh t - t)}\,dt \tag{12.01}$$

in three cases when a is real and fixed, and ν tends to infinity through positive real values:

$$\mathbf{A}_{-\nu}(a\nu) \sim \frac{1}{(a-1)\nu} \qquad (a > 1), \tag{12.02}$$

$$\mathbf{A}_{-\nu}(a\nu) \sim \left(\frac{2}{9}\right)^{1/3} \frac{\Gamma(\frac{1}{3})}{\nu^{1/3}} \qquad (a = 1), \tag{12.03}$$

$$\mathbf{A}_{-\nu}(a\nu) \sim \left(\frac{2\pi}{\nu}\right)^{1/2} \left(\frac{1+(1-a^2)^{1/2}}{a}\right)^{\nu} \frac{\exp\{-\nu(1-a^2)^{1/2}\}}{(1-a^2)^{1/4}} \qquad (0 < a < 1). \tag{12.04}$$

These results are obtainable by the method of Laplace: when $a > 1$ the interval of integration contains no saddle points; when $a = 1$ there is a double saddle point at the endpoint $t = 0$; when $0 < a < 1$ there is a simple saddle point interior to the interval of integration.

The differences in form of the three approximations indicate that (12.02) and (12.04) become useless as a approaches 1. To bridge the gaps another form of approximation is available, given by

$$\mathbf{A}_{-\nu}(\nu + l\nu^{1/3}) = \pi\,\mathrm{Hi}(-2^{1/3}l)(2/\nu)^{1/3} + O(\nu^{-1}), \tag{12.05}$$

Hi denoting Scorer's function; see Exercise 5.1 of the present chapter. This approximation reduces to (12.03) when the constant l is zero, and is a considerable improvement on (12.02) or (12.04) when a is close to 1.

In this section we apply the method of Chester, Friedman, and Ursell to approximate $\mathbf{A}_{-\nu}(a\nu)$ for large ν by a *single asymptotic expansion which is uniformly valid in any compact a interval contained in* $(0, 1]$ *and embraces* (12.03), (12.04), *and* (12.05) *as particular cases.*‡ An analogous expansion for the a interval $[1, \infty)$ is stated as an exercise at the end of the section.

12.2 Let us write $a = \operatorname{sech}\alpha$, where $\alpha \geqslant 0$. Then

$$\mathbf{A}_{-\nu}(\nu\operatorname{sech}\alpha) = \int_0^\infty e^{-\nu p(\alpha, t)}\,dt, \tag{12.06}$$

† The symbol a used here is not to be confused with the a of §11.1.

‡ The approximations (12.02) and (12.04) also are uniform with respect to a, but not in intervals which include $a = 1$.

where

$$p(\alpha, t) = \operatorname{sech}\alpha \sinh t - t.$$

The saddle points are located at $t = \pm\alpha$, $\pm\alpha \pm 2\pi i, \ldots$. Of these both α and $-\alpha$ are relevant to the problem; the former because it lies in the range of integration, and the latter because it coalesces with the former as $\alpha \to 0$. Following §11.1 we make the transformation

$$\operatorname{sech}\alpha \sinh t - t = \tfrac{1}{3}w^3 + \gamma w^2 - \zeta w, \tag{12.07}$$

where (to avoid confusion) a has been replaced by γ, and b by $-\zeta$. Also, we have taken $c = 0$ to make the endpoint $t = 0$ correspond to $w = 0$.

The stationary points of the right-hand side of (12.07) are $-\gamma \pm (\gamma^2 + \zeta)^{1/2}$. By setting $\gamma = 0$ and

$$\tfrac{2}{3}\zeta^{3/2} = \alpha - \tanh\alpha, \tag{12.08}$$

we arrange that these points correspond to $t = \pm\alpha$. Equation (12.08) determines $\zeta = \zeta(\alpha)$ as a nonnegative increasing function of α. On introduction of an intermediate variable v, the transformation (12.07) becomes

$$\operatorname{sech}\alpha \sinh t - t = v = \tfrac{1}{3}w^3 - \zeta w. \tag{12.09}$$

Since

$$dv/dt = \operatorname{sech}\alpha \cosh t - 1, \qquad dv/dw = w^2 - \zeta,$$

it follows that as t increases from 0 to α, or w increases from 0 to $\zeta^{1/2}$, v decreases monotonically from 0 to $-\frac{2}{3}\zeta^{3/2}$. Moreover, as t increases from α to ∞, or w increases from $\zeta^{1/2}$ to ∞, v increases monotonically from $-\frac{2}{3}\zeta^{3/2}$ to ∞. Thus t and w are definable, in a one-to-one manner, as increasing functions of each other in the interval $[0, \infty)$.

With the transformation (12.09) the integral (12.06) becomes

$$\mathbf{A}_{-\nu}(\nu \operatorname{sech}\alpha) = \int_0^\infty \exp\{-\nu(\tfrac{1}{3}w^3 - \zeta w)\}\frac{dt}{dw}\,dw. \tag{12.10}$$

The peak value of the exponential factor in the range of integration occurs at $w = \zeta^{1/2}$. Expanding v in Taylor series at this point, we derive from (12.08) and (12.09)

$$v = -\tfrac{2}{3}\zeta^{3/2} + \zeta^{1/2}(w - \zeta^{1/2})^2 + \tfrac{1}{3}(w - \zeta^{1/2})^3,$$

and

$$v = -\tfrac{2}{3}\zeta^{3/2} + \tfrac{1}{2}(t-\alpha)^2 \tanh\alpha + \tfrac{1}{6}(t-\alpha)^3 + \cdots.$$

Whether or not α is zero, reversion and substitution show that t can be expanded as a series of powers of $w - \zeta^{1/2}$, which we denote in the form

$$t = \alpha + \sum_{s=0}^{\infty} \frac{q_s(\alpha)}{s+1}(w - \zeta^{1/2})^{s+1}. \tag{12.11}$$

Leading coefficients are found to be

$$q_0(\alpha) = \left(\frac{4\zeta}{\tanh^2\alpha}\right)^{1/4}, \qquad q_1(\alpha) = \frac{2-\{q_0(\alpha)\}^3}{3q_0(\alpha)\tanh\alpha}, \tag{12.12}$$

if $\alpha \neq 0$, or the limiting values of these expressions if $\alpha = 0$.

Substituting the differentiated form of (12.11) in (12.10) and integrating formally term by term, we obtain the required expansion in the form

$$\mathbf{A}_{-\nu}(\nu \operatorname{sech}\alpha) = \sum_{s=0}^{\infty} q_s(\alpha) \frac{\pi \operatorname{Qi}_s(\nu^{2/3}\zeta)}{\nu^{(s+1)/3}}, \tag{12.13}$$

where

$$\operatorname{Qi}_s(y) = \frac{1}{\pi}\int_0^\infty \exp(-\tfrac{1}{3}t^3 + yt)(t-y^{1/2})^s\,dt. \tag{12.14}$$

It can be seen from the following approximations for large y and fixed s (obtained by Laplace's method) that successive terms in (12.13) form a diminishing asymptotic sequence for large ν:

$$\pi \operatorname{Qi}_s(y) = \Gamma(\tfrac{1}{2}s+\tfrac{1}{2})\, y^{-(s+1)/4} \exp(\tfrac{2}{3}y^{3/2})\{1+O(y^{-3/2})\} \qquad (s \text{ even}), \tag{12.15}$$

and

$$\pi \operatorname{Qi}_s(y) = -\tfrac{1}{3}\Gamma(\tfrac{1}{2}s+2)\, y^{-(s+4)/4} \exp(\tfrac{2}{3}y^{3/2})\{1+O(y^{-3/2})\} \qquad (s \text{ odd}). \tag{12.16}$$

12.3 We now establish the uniform asymptotic nature of the formal expansion (12.13). In the present subsection we restrict α to a fixed compact interval $[0, \alpha_0]$, where α_0 is arbitrary.

The integral corresponding to (10.15) is given by

$$q_s(\alpha) = \frac{1}{2\pi i}\int_{\mathscr{T}} \frac{dt}{(w-\zeta^{1/2})^{s+1}}, \tag{12.17}$$

where the path $\mathscr{T}$ is a simple closed contour whose map in the w plane encloses $w = \zeta^{1/2}$ but no singularity of dt/dw. The mapping between the planes of t and w is constructed by passage through the plane of the intermediate variable v introduced in (12.09). Corresponding points of the transformations are indicated in Figs. 12.1–12.3. The parametric equation of the boundary curve ED in Fig. 12.3 is given by

$$\tfrac{1}{3}w^3 - \zeta w = -i\pi - \tau \qquad (0 \leqslant \tau < \infty).$$

When $\alpha = 0$ the points O, A, and B coincide in each map. Whether or not α is zero, it is seen that within the boundaries of the regions $OABCDEO$ the $t \leftrightarrow v$ transformation is one-to-one, as is the $v \leftrightarrow w$ transformation. Hence the $t \leftrightarrow w$ transformation is one-to-one.

The mapping of the rest of the strip $|\operatorname{Im} t| \leqslant \pi$ is deducible from Figs. 12.1 and 12.3 by reflection in the real and imaginary axes. Since $q_0(\alpha) \neq 0$ the expansion (12.11)

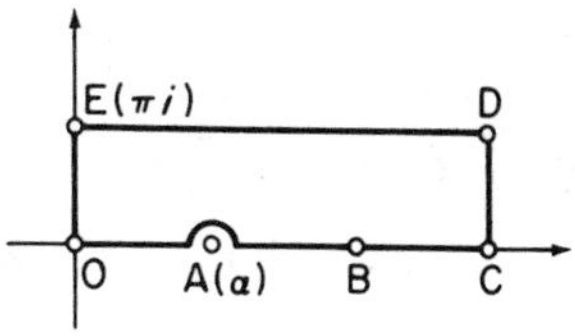

Fig. 12.1 t plane.

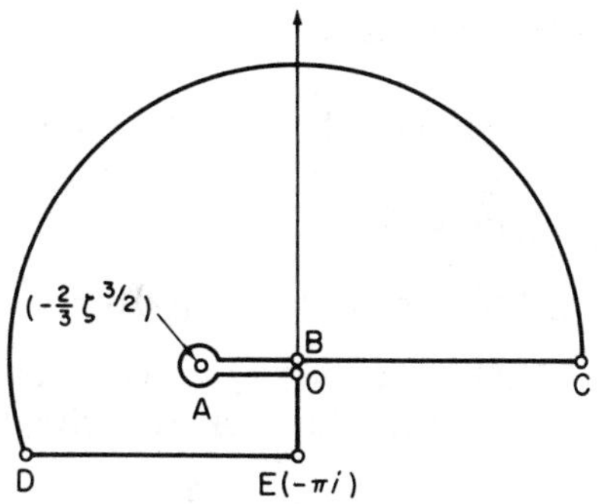

Fig. 12.2 v plane.

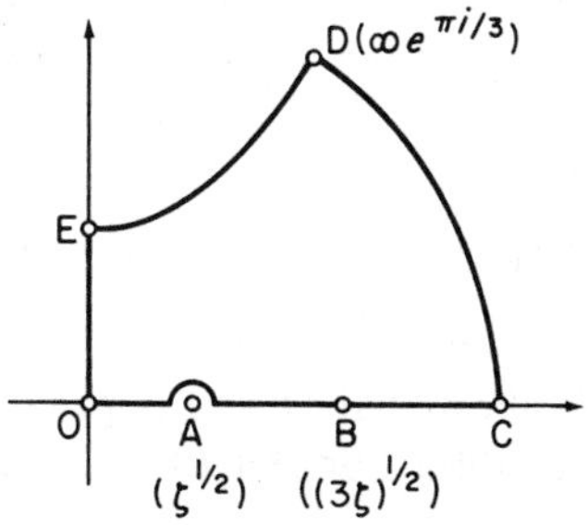

Fig. 12.3 w plane.

shows that t and w are analytic functions of each other in the neighborhoods of the points $w = \pm\zeta^{1/2}$ and $t = \pm\alpha$ if $\alpha \neq 0$, or $w = 0$ and $t = 0$ if $\alpha = 0$.

In formula (12.17) we take $\mathscr{T}$ to be the boundary of the rectangle **T** defined by $|\operatorname{Im} t| \leqslant \pi$ and $|\operatorname{Re} t| \leqslant T$, where T is a fixed positive number exceeding α_0, the upper endpoint of the α interval. Let d be the minimum value of $|w - \zeta^{1/2}|$ as t passes around **T** and α ranges over the interval $[0, \alpha_0]$. By continuity, d must be positive; compare §10.3. Then from (12.17) we derive

$$|q_s(\alpha)| \leqslant \frac{2(T+\pi)}{\pi d^{s+1}} \qquad (0 \leqslant \alpha \leqslant \alpha_0). \tag{12.18}$$

It is also evident from (12.17) that each coefficient $q_s(\alpha)$ is continuous in this closed interval.

We now define $Q_n(\alpha, w)$ by the equation

$$\frac{dt}{dw} = \sum_{s=0}^{n-1} q_s(\alpha)(w - \zeta^{1/2})^s + Q_n(\alpha, w)(w - \zeta^{1/2})^n, \tag{12.19}$$

where n is an arbitrary integer. Then from (12.10)

$$\mathbf{A}_{-\nu}(\nu \operatorname{sech} \alpha) = \sum_{s=0}^{n-1} q_s(\alpha) \frac{\pi \operatorname{Qi}_s(\nu^{2/3}\zeta)}{\nu^{(s+1)/3}} + \varepsilon_n(\alpha, \nu), \tag{12.20}$$

where the error term is given by

$$\varepsilon_n(\alpha, \nu) = \int_0^\infty \exp(-\tfrac{1}{3}\nu w^3 + \nu\zeta w)\, Q_n(\alpha, w)\,(w-\zeta^{1/2})^n\, dw. \tag{12.21}$$

Majorizing the series

$$Q_n(\alpha, w) = q_n(\alpha) + q_{n+1}(\alpha)(w-\zeta^{1/2}) + q_{n+2}(\alpha)(w-\zeta^{1/2})^2 + \cdots \tag{12.22}$$

by use of (12.18), we see that $|Q_n(\alpha, w)|$ is bounded when $|w-\zeta^{1/2}| \leqslant \frac{1}{2}d$. For positive real values of w outside this disk we derive, from (12.09),

$$\frac{dt}{dw} \sim \frac{3}{w} \qquad (w \to \infty),$$

uniformly with respect to $\alpha \in [0, \alpha_0]$. Substituting in (12.19) by means of this result and (12.18), we see that $|Q_n(\alpha, w)|$ is bounded outside the disk. Thus $|Q_n(\alpha, w)|$ is bounded when $\alpha \in [0, \alpha_0]$ and $w \in [0, \infty)$. If n is even, then $(w-\zeta^{1/2})^n$ is of constant sign in (12.21), and we obtain

$$\varepsilon_n(\alpha, \nu) = O\{\nu^{-(n+1)/3}\,\mathrm{Qi}_n(\nu^{2/3}\zeta)\} \qquad (\nu \to \infty), \tag{12.23}$$

uniformly with respect to $\alpha \in [0, \alpha_0]$. This is the desired result. When n is odd, equations (12.14) and (12.16) show that $\mathrm{Qi}_n(y)$ is positive at $y = 0$ and negative as $y \to \infty$. Thus $\mathrm{Qi}_n(y)$ vanishes at least once in $(0, \infty)$, implying that (12.23) cannot be extended to odd values of n.

12.4 We have attained our primary objective: an approximation has been constructed for $\mathbf{A}_{-\nu}(\nu\,\mathrm{sech}\,\alpha)$ for large ν which is satisfactory throughout a closed α interval which includes $\alpha = 0$. *All* of the earlier asymptotic approximations for $\mathbf{A}_{-\nu}(a\nu)$ for $0 < a \leqslant 1$ can be regarded as special cases of the expansion (12.20). For example, when $\alpha = \zeta = 0$ (12.20) reduces to a Poincaré expansion in descending powers of $\nu^{1/3}$,† the first term of which agrees with (12.03). Next, if ζ is positive and fixed, then for large ν the functions $\mathrm{Qi}_s(\nu^{2/3}\zeta)$ can be replaced by their asymptotic expansions for large argument; compare (12.15) and (12.16). This yields an expansion for $\mathbf{A}_{-\nu}(\nu\,\mathrm{sech}\,\alpha)$ which descends in powers of ν and agrees with (12.04). Similarly in the case of (12.05) with $l \leqslant 0$.

As in Bleistein's method, this unifying property is elegant from a theoretical standpoint. There are also obvious practical advantages in being able to construct a computing routine based on a single formula and applicable to a wide range of variables. But the price of generality is that the evaluation of individual terms in (12.20) is somewhat complicated. Special tables or computing subroutines are needed for the functions $\mathrm{Qi}_s(y)$,‡ and although the first few coefficients $q_s(\alpha)$ can be found by the procedure of §12.2, or, equivalently, by repeatedly differentiating (12.09) and setting $t = \alpha$ and $w = \zeta^{1/2}$, an expression for general s is unavailable.§

† Actually, descending odd powers of $\nu^{1/3}$.

‡ See also Exercise 12.2.

§ An easier way of deriving the leading $q_s(\alpha)$ in the present example is via differential-equation theory (Chapter 11). This method may be inapplicable in other cases, however.

Ex. 12.1 By using an expansion of the form

$$\frac{dt}{dw} = \sum_{s=0}^{\infty} p_s(\alpha)(w^2-\zeta)^s$$

in place of the differentiated form of (12.11), prove that

$$\mathrm{A}_{-\nu}(\nu \operatorname{sech} \alpha) = \sum_{s=0}^{n-1} p_s(\alpha) \frac{\pi \operatorname{Pi}_s(\nu^{2/3}\zeta)}{\nu^{(2s+1)/3}} + \theta_n(\alpha, \nu),$$

where

$$\operatorname{Pi}_s(y) = \frac{1}{\pi}\int_0^{\infty} \exp(-\tfrac{1}{3}t^3 + yt)(t^2-y)^s \, dt,$$

and $\theta_n(\alpha, \nu) = O\{\nu^{-(2n+1)/3} \operatorname{Pi}_n(\nu^{2/3}\zeta)\}$ as $\nu \to \infty$, uniformly with respect to $\alpha \in [0, \alpha_0]$, provided that n is even and the positive constant α_0 is sufficiently small.

Show also that for fixed s and large positive y

$$\pi \operatorname{Pi}_s(y) = 2^s \Gamma(\tfrac{1}{2}s+\tfrac{1}{2})\, y^{(s-1)/4} \exp(\tfrac{2}{3}y^{3/2})\{1+O(y^{-3/2})\} \qquad (s = 0, 2, 4, \ldots),$$

$$\pi \operatorname{Pi}_s(y) = \tfrac{1}{3}2^s(s-1)\Gamma(\tfrac{1}{2}s+1)\, y^{(s-4)/4} \exp(\tfrac{2}{3}y^{3/2})\{1+O(y^{-3/2})\} \qquad (s = 3, 5, 7, \ldots).$$

Ex. 12.2† In the notation of (12.14) and Exercise 12.1, show that

$$\operatorname{Qi}_0(y) = \operatorname{Hi}(y), \qquad \operatorname{Qi}_1(y) = \operatorname{Hi}'(y) - y^{1/2}\operatorname{Hi}(y), \qquad \operatorname{Qi}_2(y) = \pi^{-1} - 2y^{1/2}\operatorname{Qi}_1(y),$$

$$\operatorname{Qi}_s(y) = (-)^s \pi^{-1} y^{(s-2)/2} - 2y^{1/2}\operatorname{Qi}_{s-1}(y) + (s-2)\operatorname{Qi}_{s-3}(y) \qquad (s \geqslant 3),$$

and

$$\operatorname{Pi}_0(y) = \operatorname{Hi}(y), \qquad \operatorname{Pi}_1(y) = \pi^{-1}, \qquad \operatorname{Pi}_2(y) = -\pi^{-1}y + 2\operatorname{Hi}'(y), \qquad \operatorname{Pi}_3(y) = \pi^{-1}y^2 + 4\operatorname{Hi}(y),$$

$$\operatorname{Pi}_s(y) = \pi^{-1}(-y)^{s-1} + 2(s-1)\{(2s-5)\operatorname{Pi}_{s-3}(y) + 2(s-3)\, y \operatorname{Pi}_{s-4}(y)\} \qquad (s \geqslant 4).$$

***Ex. 12.3** Let $\beta \in [0, \frac{1}{2}\pi)$ and ζ be the real root of the equation $\frac{2}{3}\zeta^{3/2} = \tan\beta - \beta$. Prove that for any nonnegative integer n and any fixed positive acute angle β_0,

$$\mathrm{A}_{-\nu}(\nu \sec\beta) = \sum_{s=0}^{n-1} h_{2s}(\beta) \frac{\pi \operatorname{Hi}^{(2s)}(-\nu^{2/3}\zeta)}{\nu^{(2s+1)/3}} + \eta_n(\beta, \nu),$$

where $h_0(\beta) = \zeta/(\sec\beta - 1)$, higher coefficients $h_{2s}(\beta)$ are continuous functions of β, and

$$\eta_n(\beta, \nu) = O\{\nu^{-(2n+1)/3} \operatorname{Hi}^{(2n)}(-\nu^{2/3}\zeta)\}$$

as $\nu \to \infty$, uniformly with respect to $\beta \in [0, \beta_0]$.

***Ex. 12.4** For nonnegative α and large positive ν, let

$$I(\alpha, \nu) = \int_0^{\infty} \exp[\nu\{\alpha t - \tfrac{1}{2}t^2 - \alpha \tan^{-1}t + \tfrac{1}{2}\ln(1+t^2)\}]\, dt.$$

Show that if $\alpha \neq 0$, then the integrand has a double saddle point at $t = 0$ and a single saddle point at $t = \alpha$; alternatively if $\alpha = 0$, then the integrand has a triple saddle point at $t = 0$.

Show also that

$$I(\alpha, \nu) = \frac{\zeta(1+\alpha^2)^{1/2}}{2^{3/2}\alpha\nu^{1/4}} \operatorname{Fi}\left(\frac{3}{4}, \frac{1}{4}; \frac{2^{3/2}\nu^{1/4}\zeta}{3}\right)\left\{1 + \frac{O(1)}{\nu^{1/4} + \nu\alpha^3}\right\}$$

uniformly for $\alpha \in [0, \alpha_0]$ and $\nu \in [\nu_0, \infty)$, where α_0 and ν_0 are arbitrary fixed positive numbers, and $\zeta = \{6\alpha^2 - 12\alpha \tan^{-1}\alpha + 6\ln(1+\alpha^2)\}^{1/4}$.

† This exercise shows that the sequences $\operatorname{Pi}_s(y)$ and $\operatorname{Qi}_s(y)$, $s = 0, 1, 2, \ldots$, can be computed by recurrence. When y is large, however, serious numerical cancellation occurs in alternate recurrence steps for the $\operatorname{Qi}_s(y)$ but not for the $\operatorname{Pi}_s(y)$: this is discernible from the asymptotic forms given in §12.2 and Exercise 12.1.

*13 Extension of the Region of Validity

13.1 The uniform estimate (12.23) for the error term in (12.20) has been established for the compact interval $[0, \alpha_0]$, where α_0 is an arbitrary positive number. For the remaining part of the interval $[0, \infty)$, the method of Laplace is available; it yields a simpler form of expansion, the leading term of which is given by (12.04). It is of considerable interest, however, both in the present example and related problems, to show how the interval of uniform validity of (12.20) and (12.23) may be extended to $0 \leqslant \alpha < \infty$: in terms of the original variable a of $\mathbf{A}_{-\nu}(a\nu)$ this corresponds to $0 < a \leqslant 1$.

An indirect approach, due to Ursell (1965), is to prove that the Laplace expansion is uniformly valid for $\alpha \in [\alpha_0, \infty)$. A matching process then shows that the same is true of (12.20) and (12.23). In the present section we solve the problem in a direct manner by extending the mappings of §12.3.

13.2 Throughout the present subsection we suppose that $\alpha \geqslant \alpha_0$. As integration contour $\mathscr{T}$ in (12.17) we employ the circle

$$t = \alpha + \rho e^{i\theta} \qquad (0 \leqslant \theta \leqslant 2\pi),$$

where ρ is independent of α. Provided that $\rho < \min(\pi, \alpha_0)$, this circle lies within the half-strip $\operatorname{Re} t > 0$, $|\operatorname{Im} t| < \pi$. The map of $\mathscr{T}$ in the v plane has the equation

$$v = \operatorname{sech}\alpha \sinh(\alpha + \rho e^{i\theta}) - \alpha - \rho e^{i\theta}.$$

Hence from (12.08)

$$v + \tfrac{2}{3}\zeta^{3/2} = \tfrac{1}{2}\rho^2 e^{2i\theta}\tanh\alpha + O(\rho^3),$$

uniformly with respect to α and θ. Accordingly, for sufficiently small ρ the v map of $\mathscr{T}$ lies outside the circle

$$|v + \tfrac{2}{3}\zeta^{3/2}| = \tfrac{1}{4}\rho^2 \tanh\alpha.$$

Now consider the circle $\mathscr{C}$ in the w plane defined by

$$w = \zeta^{1/2} + \sigma\zeta^{-1/4}e^{i\phi} \qquad (0 \leqslant \phi \leqslant 2\pi),$$

where σ is independent of α, and $\sigma^2 < \tfrac{3}{2}\alpha_0 - \tfrac{3}{2}\tanh\alpha_0$ (ensuring that $\mathscr{C}$ lies in the right half-plane). The v map of $\mathscr{C}$ is given by

$$v = -\tfrac{2}{3}\zeta^{3/2} + \sigma^2 e^{2i\phi} + \tfrac{1}{3}\sigma^3\zeta^{-3/4}e^{3i\phi}.$$

Since $\alpha \geqslant \alpha_0$, $\zeta^{-3/4}$ is bounded. Hence for all sufficiently small σ (independently of α) the v map of $\mathscr{C}$ lies in the disk $|v + \tfrac{2}{3}\zeta^{3/2}| \leqslant 2\sigma^2$.

If we take $\sigma = 2^{-3/2}\rho(\tanh\alpha_0)^{1/2} = \rho_0$, say, then $\tfrac{1}{4}\rho^2\tanh\alpha \geqslant 2\sigma^2$ and the v map of $\mathscr{T}$ contains the v map of $\mathscr{C}$. This implies

$$|w - \zeta^{1/2}| \geqslant \rho_0\zeta^{-1/4} \qquad (t \in \mathscr{T}).$$

Substitution in (12.17) yields the key result

$$|q_s(\alpha)| \leqslant \rho(\zeta^{1/4}/\rho_0)^{s+1} \qquad (\alpha \geqslant \alpha_0), \tag{13.01}$$

where ρ and ρ_0 are positive numbers, assignable independently of α and s. Thus†

$$q_s(\alpha) = O(\zeta^{(s+1)/4}) \qquad (\alpha \to \infty). \tag{13.02}$$

We now establish that with the definition (12.19)

$$Q_n(\alpha, w) = \zeta^{(n+1)/4}O(1) \qquad (n \geqslant 2), \tag{13.03}$$

uniformly for $\alpha \in [\alpha_0, \infty)$ and $w \in [0, \infty)$.

Suppose first that $|w-\zeta^{1/2}| < r\zeta^{-1/4}$, where r is a small positive number chosen independently of ζ. Provided that $r < \rho_0$, the relation (13.03) follows on majorizing the series (12.22) by means of (13.01).

Alternatively, suppose that

$$|w-\zeta^{1/2}| \geqslant r\zeta^{-1/4}; \tag{13.04}$$

then either

$$0 \leqslant w \leqslant \zeta^{1/2} - r\zeta^{-1/4} \qquad \text{or} \qquad \zeta^{1/2} + r\zeta^{-1/4} \leqslant w < \infty. \tag{13.05}$$

To treat these intervals we use the identity

$$Q_n(\alpha, w) = \left\{\frac{2\zeta^{1/2}}{w-\zeta^{1/2}} + 1\right\} \frac{1}{(w-\zeta^{1/2})^{n-2}} \frac{1}{\operatorname{sech}\alpha \cosh t - 1} - \sum_{s=0}^{n-1} \frac{q_s(\alpha)}{(w-\zeta^{1/2})^{n-s}}, \tag{13.06}$$

obtained from (12.19) and (12.09). For each value of α, the factor $(\operatorname{sech}\alpha \cosh t - 1)^{-1}$ attains its maximum modulus in the second of the intervals (13.05) at the endpoint $\zeta^{1/2} + r\zeta^{-1/4}$. The corresponding value of v is given by

$$v = -\tfrac{2}{3}\zeta^{3/2} + r^2 + \tfrac{1}{3}r^3\zeta^{-3/4} = -\tfrac{2}{3}\zeta^{3/2} + r^2 + O(r^3), \tag{13.07}$$

the O term being uniform with respect to α. Denoting the corresponding value of t by $\alpha + \tau$, we have

$$v = \operatorname{sech}\alpha \sinh(\alpha+\tau) - \alpha - \tau = -\tfrac{2}{3}\zeta^{3/2} + \tfrac{1}{2}\tau^2 \tanh\alpha + O(\tau^3), \tag{13.08}$$

as $\tau \to 0$, uniformly with respect to α. Comparison of (13.07) with (13.08) shows that

$$\tau = (\tfrac{1}{2}\tanh\alpha)^{-1/2} r + O(r^2).$$

In consequence,

$$\operatorname{sech}\alpha \cosh t - 1 = (2\tanh\alpha)^{1/2} r + O(r^2), \tag{13.09}$$

again uniformly with respect to α. For all sufficiently small values of r, the right-hand side of (13.09) exceeds $(\tanh\alpha_0)^{1/2} r$. Hence $(\operatorname{sech}\alpha \cosh t - 1)^{-1}$ is bounded in the interval $\zeta^{1/2} + r\zeta^{-1/4} \leqslant w < \infty$. In a similar way we can show that this function

† In the cases $s = 0$ and 1, it is easily verified that (13.02) accords with the asymptotic forms of the expressions (12.12).

is bounded in absolute value in the first of the intervals (13.05). The proof of (13.03) is completed by applying these results, together with (13.01) and (13.04), to the identity (13.06).

13.3 On bounding the integral (12.21) by use of (13.03) we see that

$$\varepsilon_n(\alpha, \nu) = \zeta^{(n+1)/4}\nu^{-(n+1)/3}\,\mathrm{Qi}_n(\nu^{2/3}\zeta)\,O(1) \qquad (n = 2, 4, 6, \ldots),$$

uniformly with respect to $\alpha \in [\alpha_0, \infty)$ and $\nu \in (0, \infty)$. Combination with (12.23) gives the desired extension

$$\varepsilon_n(\alpha, \nu) = (1+\zeta)^{(n+1)/4}\nu^{-(n+1)/3}\,\mathrm{Qi}_n(\nu^{2/3}\zeta)\,O(1) \qquad (n = 2, 4, 6, \ldots), \tag{13.10}$$

uniformly with respect to $\alpha \in [0, \infty)$ and $\nu \in (0, \infty)$.

By use of (12.15) the error term may be recast in the form

$$\varepsilon_n(\alpha, \nu) = \left(\frac{1+\zeta}{1+\nu^{2/3}\zeta}\right)^{(n+1)/4} \frac{\exp(\frac{2}{3}\nu\zeta^{3/2})}{\nu^{(n+1)/3}}\,O(1) \qquad (n = 2, 4, 6, \ldots). \tag{13.11}$$

Similar results for odd n, and also $n = 0$, can be found by use of (12.16) and the identity

$$\varepsilon_n(\alpha, \nu) = q_n(\alpha)\frac{\pi\,\mathrm{Qi}_n(\nu^{2/3}\zeta)}{\nu^{(n+1)/3}} + \varepsilon_{n+1}(\alpha, \nu).$$

***Ex. 13.1** Show that the error term in Exercise 12.3 is given by

$$\eta_n(\beta, \nu) = (1+\zeta^{1/2})^{-2n-1}\nu^{-(2n+1)/3}\,\mathrm{Hi}^{(2n)}(-\nu^{2/3}\zeta)\,O(1),$$

uniformly with respect to $\beta \in [0, \frac{1}{2}\pi)$ and $\nu \in (0, \infty)$.

***Ex. 13.2** Define ζ as in (12.08), and the coefficients $p_s(\alpha)$ as in Exercise 12.1. Starting from Schläfli's integral (Chapter 2, (9.13)), prove that

$$J_\nu(\nu\,\mathrm{sech}\,\alpha) = \sum_{s=0}^{n-1} p_s(\alpha)\frac{\mathrm{Ri}_s(\nu^{2/3}\zeta)}{\nu^{(2s+1)/3}} + \left(\frac{1+\nu^{2/3}\zeta}{1+\zeta}\right)^{(n-1)/4}\frac{\exp(-\frac{2}{3}\nu\zeta^{3/2})}{\nu^{(2n+1)/3}}\,O(1),$$

where n is an arbitrary positive integer,

$$\mathrm{Ri}_s(y) = \frac{1}{2\pi i}\int_{\infty e^{-\pi i/3}}^{\infty e^{\pi i/3}} \exp(\tfrac{1}{3}t^3 - yt)(t^2-y)^s\,dt,$$

and the O term is uniform with respect to $\alpha \in [0, \infty)$ and $\nu \in (0, \infty)$.

***Ex. 13.3** Show that the expansion given in the preceding exercise can be rearranged as

$$J_\nu(\nu\,\mathrm{sech}\,\alpha) = \frac{\mathrm{Ai}(\nu^{2/3}\zeta)}{\nu^{1/3}}\sum_{s=0}^{n}\frac{a_s(\alpha)}{\nu^{2s}} + \frac{\mathrm{Ai}'(\nu^{2/3}\zeta)}{\nu^{5/3}}\sum_{s=0}^{n-1}\frac{b_s(\alpha)}{\nu^{2s}} + \left(\frac{1+\nu^{2/3}\zeta}{1+\zeta}\right)^{1/4}\frac{\exp(-\frac{2}{3}\nu\zeta^{3/2})}{\nu^{2n+(5/3)}}\,O(1),$$

where the O term is uniform with respect to $\alpha \in [0, \infty)$ and ν bounded away from zero, and

$a_0(\alpha) = p_0(\alpha), \quad b_0(\alpha) = 2p_2(\alpha), \quad a_1(\alpha) = 4p_3(\alpha) + 12\zeta p_4(\alpha), \quad b_1(\alpha) = 80p_5(\alpha) + 120\zeta p_6(\alpha)$

Show also that for every s

$$a_s(\alpha) = \sum_{j=0}^{s}\lambda_{s,j}\zeta^{s-j}p_{4s-j}(\alpha), \qquad b_s(\alpha) = \sum_{j=0}^{s}\mu_{s,j}\zeta^{s-j}p_{4s-j+2}(\alpha),$$

where the coefficients $\lambda_{s,j}$ and $\mu_{s,j}$ are integers or zero.

Historical Notes and Additional References

Generalizations of Laplace's method for single integrals have been published in numerous papers. For extensive bibliographies see Berg (1968), Riekstiņš and Cīrulis (1970), D. S. Jones (1972), and Bleistein and Handelsman (1974). Papers giving extensions to multiple integrals include those by Hsu (1951) and Riedel (1965).

§1 The method of stationary phase can be extended in a similar way to integrals of the form

$$\int_a^b e^{ixp(t)} q(t) \ln(t-a)\, dt.$$

Results have been supplied by Erdélyi (1956b), with a correction by McKenna (1967).

§6 Extensions of this theory will be found in the paper by Handelsman and Lew (1971) and references cited therein. For other approaches, see Stenger (1970a, 1972).

§8 More general results have been given by Handelsman and Bleistein (1973) and Soni and Soni (1973).

§§9–13 The paper of Chester, Friedman, and Ursell (1957) was an important innovation in the asymptotic approximation of definite integrals. Bleistein's method is based on the transformation of variables introduced by these earlier authors. Several modifications to the methods have been introduced in the present account. For other developments see Fedoryuk (1964), Bleistein (1967), Rice (1968), Erdélyi (1970), Bleistein and Handelsman (1974), and references cited in these works.

For other treatments of the case in which an integrand has a pole or other singularity close to a saddle point see Clemmow (1950), van der Waerden (1951), Oberhettinger (1959), and Cīrulis (1968).

10

DIFFERENTIAL EQUATIONS WITH A PARAMETER: EXPANSIONS IN ELEMENTARY FUNCTIONS

1 Classification and Preliminary Transformations

1.1 In this chapter and the two which follow, we seek approximate solutions of differential equations of the form

$$d^2w/dz^2 = \{u^2 f(z) + g(z)\} w, \tag{1.01}$$

in which u is a large parameter, real or complex, and the independent variable z ranges over a real interval $\mathbf{I}(u)$ or a complex domain $\mathbf{D}(u)$, neither of which need be bounded. In developing the theory we shall have in mind equations in which $f(z)$ and $g(z)$ are independent of u. But because the error bounds obtained for the approximate solutions do not relate to this condition, the theory provides meaningful results for many equations in which $f(z)$ and $g(z)$ depend on u.

As in Chapter 6, zeros of $f(z)$ are called turning points or transition points of the differential equation. We sometimes extend the name *transition point* to include singularities of $f(z)$ or $g(z)$. The forms of the approximate solutions of (1.01) depend on the number and nature of the transition points. In the case in which $\mathbf{I}(u)$ or $\mathbf{D}(u)$ is free from transition points, which we shall refer to henceforth as *Case I*, we saw in Chapter 6 that the LG functions

$$f^{-1/4}(z) \exp\left\{\pm u \int f^{1/2}(z)\, dz\right\}$$

furnish asymptotic solutions of (1.01) with uniform relative error $O(u^{-1})$ as $|u| \to \infty$.

The purpose of the present chapter is to reconsider Case I and construct asymptotic *expansions* in descending powers of u for solutions of the differential equation. The leading term in each expansion is one of the LG functions.

Chapters 11 and 12 are concerned with cases in which $\mathbf{I}(u)$ or $\mathbf{D}(u)$ contains exactly one transition point z_0, say. If z_0 is a pole of $f(z)$ of order $m \geqslant 2$, and $g(z)$ is analytic at z_0 or has a pole of order less than $\frac{1}{2}m+1$, then the LG approximation can be used (Chapter 6, §5.3). We shall soon see that this is also true of the higher approximations for Case I. Other cases to be considered include *Case II* in which z_0 is a simple zero of $f(z)$ and an analytic point of $g(z)$, and *Case III* in which z_0 is a simple pole of $f(z)$ and $(z-z_0)^2 g(z)$ is analytic. We assume that $f(z)$ and $g(z)$ are analytic functions of z except possibly at the transition point, with some easing of this requirement when the variables are real.

1.2 Basically the same approach is made in all cases. First, the Liouville transformation (Chapter 6, §1) is applied. This introduces new variables W and ξ, related by

$$W = \dot{z}^{-1/2} w,$$

the dot denoting differentiation with respect to ξ. Equation (1.01) becomes

$$d^2W/d\xi^2 = \{u^2\dot{z}^2 f(z) + \psi(\xi)\}\, W, \tag{1.02}$$

where

$$\psi(\xi) = \dot{z}^2 g(z) - \tfrac{1}{2}\{z, \xi\} = \dot{z}^2 g(z) + \dot{z}^{1/2} \frac{d^2}{d\xi^2}(\dot{z}^{-1/2}). \tag{1.03}$$

The transformation is now fixed by specifying the relation between z and ξ in such a manner that (i) ξ and z are analytic functions of each other at the transition point (if any); (ii) the approximating differential equation obtained by neglecting all or part of $\psi(\xi)$ in (1.02) has solutions which are functions of a single variable. The choices are as follows:

Case I. $\dot{z}^2 f(z) = 1$, giving $\xi = \int f^{1/2}(z)\, dz$.

Case II. $\dot{z}^2 f(z) = \xi$, giving $\frac{2}{3}\xi^{3/2} = \int_{z_0}^{z} f^{1/2}(t)\, dt$.

Case III. $\dot{z}^2 f(z) = 1/\xi$, giving $2\xi^{1/2} = \int_{z_0}^{z} f^{1/2}(t)\, dt$.

Then equation (1.02) reduces to

$$d^2W/d\xi^2 = \{u^2\xi^m + \psi(\xi)\}\, W, \tag{1.04}$$

with $m = 0$ (Case I), $m = 1$ (Case II), or $m = -1$ (Case III).

The function $\xi = \xi(z)$ is obviously free from singularity in Case I. In Cases II and III we have $\xi \sim a(z - z_0)$, as $z \to z_0$, where a is a nonzero constant; again there is no singularity. Thus for real variables there is a one-to-one continuous relationship between z and ξ, and when z is complex the domain $\mathbf{D}(u)$ is mapped conformally on a certain domain $\mathbf{\Delta}(u)$ in the ξ plane. The function $\psi(\xi)$ defined by (1.03) is holomorphic in $\mathbf{\Delta}(u)$ in Cases I and II. In Case III $\psi(\xi)$ may have a single or double pole at $\xi = 0$, corresponding to the same kind of pole of $g(z)$ at $z = z_0$.

In Cases I and II first approximations to solutions of equation (1.04) are obtained by neglecting $\psi(\xi)$. In Case I this is the LG approximating procedure used in Chapter 6. In Case II, considered in Chapter 11, this leads to approximations in terms of Airy functions. In Case III, considered in Chapter 12, the basic approximating equation is

$$\frac{d^2W}{d\xi^2} = \left(\frac{u^2}{\xi} + \frac{c}{\xi^2}\right) W,$$

where c is the value of $\xi^2\psi(\xi)$ at $\xi = 0$. The solutions are expressible in terms of modified Bessel functions of order $\pm(1+4c)^{1/2}$ and argument $2u\xi^{1/2}$.

Ex. 1.1 If z_0 is a zero of $f(z)$ of multiplicity m and $g(z)$ is analytic at z_0, show that equation (1.01) may be transformed into the form (1.04) with $\psi(\xi)$ analytic at $\xi = 0$.

2 Case I: Formal Series Solutions

2.1 We now consider the construction of asymptotic expansions of solutions of (1.01) in a domain free from transition points. In developing the theory it is convenient to work in terms of the transformed variables W and ξ of §1.2. (We hasten to add, however, that the evaluation of the conformal transformation from the z plane to the ξ plane is often a major and unavoidable task in applications.) Thus we consider the differential equation

$$d^2W/d\xi^2 = \{u^2 + \psi(\xi)\}\, W, \tag{2.01}$$

in which u is a parameter ranging over an unbounded real interval or complex domain, and $\psi(\xi)$ is holomorphic in a bounded or unbounded complex domain $\mathbf{\Delta}$. Both $\psi(\xi)$ and $\mathbf{\Delta}$ may depend on u, except where indicated otherwise. In terms of the original variables of equation (1.01) we have

$$\xi = \int f^{1/2}(z)\, dz, \qquad w = f^{-1/4}(z)\, W, \tag{2.02}$$

and

$$\psi(\xi) = \frac{g(z)}{f(z)} - \frac{1}{f^{3/4}(z)} \frac{d^2}{dz^2}\left\{\frac{1}{f^{1/4}(z)}\right\}; \tag{2.03}$$

compare Chapter 6, (1.07).

2.2 From one point of view uniform asymptotic expansions of solutions of (2.01) are available from the proofs of the main theorems in Chapter 6. For example, if ξ ranges over an interval (α_1, α_2) and u is positive, then on introducing u into the proof of Theorem 2.1 the Liouville–Neumann expansion becomes

$$W(u, \xi) = e^{u\xi}\left[1 + \sum_{s=0}^{\infty} \{h_{s+1}(u, \xi) - h_s(u, \xi)\}\right]. \tag{2.04}$$

From (2.13) of the same chapter we derive

$$|h_s(u, \xi) - h_{s-1}(u, \xi)| \leqslant \frac{1}{s!\,(2u)^s}\left\{\int_{\alpha_1}^{\alpha_2} |\psi(v)|\, dv\right\}^s.$$

Therefore when $\psi(\xi)$, α_1, and α_2 are independent of u, the sum in (2.04) may be regarded as a generalized asymptotic expansion with scale $\{u^{-s-1}\}$ as $u \to \infty$. Unfortunately, the functions $h_s(u, \xi)$ are generally as difficult to evaluate as the wanted solutions of (2.01), rendering (2.04) useless for numerical (and most analytical) purposes.

2.3 What is needed is an expansion for W in which individual terms have the variables u and ξ separated. We consider the following form

$$W(u, \xi) = e^{u\xi} \sum_{s=0}^{\infty} \frac{A_s(\xi)}{u^s}. \tag{2.05}$$

Differentiation with respect to ξ produces

$$\frac{dW}{d\xi} = ue^{u\xi}\sum_{s=0}^{\infty}\frac{B_s(\xi)}{u^s},$$

where

$$B_s(\xi) = A_s(\xi) + A'_{s-1}(\xi). \tag{2.06}$$

Again,

$$\frac{d^2W}{d\xi^2} = u^2e^{u\xi}\sum_{s=0}^{\infty}\frac{B_s(\xi)+B'_{s-1}(\xi)}{u^s} = u^2e^{u\xi}\sum_{s=0}^{\infty}\frac{A_s(\xi)+2A'_{s-1}(\xi)+A''_{s-2}(\xi)}{u^s}.$$

Substituting in (2.01) and equating coefficients of $e^{u\xi}/u^s$, we see that this differential equation is satisfied formally if

$$2A'_{s+1}(\xi) = -A''_s(\xi) + \psi(\xi)A_s(\xi). \tag{2.07}$$

Taking $s = -1$ we see that $A_0(\xi)$ = constant, which we may take to be unity without loss of generality. Then higher coefficients are determined recursively by

$$A_{s+1}(\xi) = -\tfrac{1}{2}A'_s(\xi) + \tfrac{1}{2}\int\psi(\xi)A_s(\xi)\,d\xi \qquad (s = 0, 1, \ldots), \tag{2.08}$$

the constants of integration being arbitrary. Clearly each coefficient $A_s(\xi)$ is holomorphic in Δ, and independent of u when $\psi(\xi)$ is independent of u. In terms of the original coefficients of (1.01), (2.08) may be expressed

$$A_{s+1} = -\tfrac{1}{2}f^{-1/2}(z)\frac{dA_s}{dz} + \int\Lambda(z)A_s\,dz, \tag{2.09}$$

where

$$\Lambda(z) = \{16f^2(z)g(z)+4f(z)f''(z)-5f'^2(z)\}/\{32f^{5/2}(z)\}; \tag{2.10}$$

compare again Chapter 6, (1.07).

A second formal solution of (2.01) is obtainable from (2.05) simply by replacing u by $-u$:

$$e^{-u\xi}\sum_{s=0}^{\infty}(-)^s\frac{A_s(\xi)}{u^s}. \tag{2.11}$$

In general, the expansions (2.05) and (2.11) diverge, and the most that can be established is that in certain regions they provide uniform asymptotic expansions of solutions of the differential equation. Divergence is illustrated by the example $\psi(\xi) = (\xi+1)^{-2}$. With the lower limit of integration in (2.08) taken as ∞, for simplicity, it is verifiable by induction that

$$A_s(\xi) = \frac{1}{(2\xi+2)^s}\prod_{j=1}^{s}\left(j-1-\frac{1}{j}\right).$$

Therefore for any given value of ξ, $A_{s+1}(\xi)/A_s(\xi)\to\infty$ as $s\to\infty$, showing that (2.05) and (2.11) diverge for all values of u.

Ex. 2.1 Show that equation (2.01) has formal solutions

$$W = \exp\left\{\pm u\xi + \sum_{s=1}^{\infty} (\pm)^s \frac{E_s(\xi)}{u^s}\right\},$$

where $E_1(\xi) = \frac{1}{2}\int \psi(\xi)\, d\xi$, $E_2(\xi) = -\frac{1}{4}\psi(\xi)$, and

$$E_{s+1}(\xi) = -\tfrac{1}{2}E_s'(\xi) - \tfrac{1}{2}\sum_{j=1}^{s-1}\int E_j'(\xi)\, E_{s-j}'(\xi)\, d\xi \qquad (s \geqslant 1).$$

Show also that

$$E_s(\xi) = A_s(\xi) - s^{-1}\sum_{j=1}^{s-1} jA_{s-j}(\xi)\, E_j(\xi).$$

Ex. 2.2 Show that an expansion equivalent to that of Exercise 2.1 can be generated formally from equation (2.01) by repeated applications of the Liouville transformation.

[Moriguchi, 1959.]

3 Error Bounds for the Formal Solutions

3.1 Theorem 3.1 *With the conditions of* §2.1, *the differential equation*

$$d^2W/d\xi^2 = \{u^2 + \psi(\xi)\}\, W \tag{3.01}$$

has, for each value of u and each positive integer n, solutions $W_{n,j}(u, \xi)$, $j = 1, 2$, *which are holomorphic in* Δ, *depend on arbitrary reference points* $\alpha_j \equiv \alpha_j(u)$, *and are given by*

$$W_{n,1}(u, \xi) = e^{u\xi}\sum_{s=0}^{n-1}\frac{A_s(\xi)}{u^s} + \varepsilon_{n,1}(u, \xi), \tag{3.02}$$

$$W_{n,2}(u, \xi) = e^{-u\xi}\sum_{s=0}^{n-1}(-)^s\frac{A_s(\xi)}{u^s} + \varepsilon_{n,2}(u, \xi). \tag{3.03}$$

Here the coefficients $A_s(\xi)$ *are defined recursively by* (2.08), *with* $A_0(\xi) = 1$, *and*

$$|\varepsilon_{n,j}(u, \xi)|,\ \left|\frac{\partial\varepsilon_{n,j}(u, \xi)}{u\,\partial\xi}\right| \leqslant 2\left|e^{(-)^{j-1}u\xi}\right| \exp\left\{\frac{2\mathscr{V}_{\alpha_j,\xi}(A_1)}{|u|}\right\}\frac{\mathscr{V}_{\alpha_j,\xi}(A_n)}{|u|^n} \tag{3.04}$$

when $\xi \in \Xi_j(u, \alpha_j)$ *(defined below).*

This theorem is proved in the next subsection. The region of validity $\Xi_j(u, \alpha_j)$ comprises the ξ point set for which there exists a path $\mathscr{Q}_j$ linking ξ with α_j in Δ and having the properties:

(i) *$\mathscr{Q}_j$ consists of a finite chain of R_2 arcs.*

(ii) *As v passes along $\mathscr{Q}_j$ from α_j to ξ, the real part of uv is nondecreasing if $j = 1$ or nonincreasing if $j = 2$.*

The variations in (3.04) are evaluated along $\mathscr{Q}_j$. The point α_j may be at infinity on an infinite R_2 arc $\mathscr{L}_j \equiv \mathscr{L}_j(u)$, provided that (a) $\mathscr{Q}_j$ coincides with $\mathscr{L}_j$ in a neighborhood of α_j, (b) the variations converge.

Again, in the context of the present theorem we shall say that a path fulfilling Conditions (i) and (ii) is *progressive*, or more precisely, *$(u\xi)$-progressive*. It should be observed that a path may be progressive for some values of $\operatorname{ph} u$, but not for others. In consequence, the regions $\Xi_j(u, \alpha_j)$ vary with $\operatorname{ph} u$. They may also vary with $|u|$ if α_j and Δ depend on u. As in Chapter 6, §11.4, the regions $\Xi_j(u, \alpha_j)$ are not confined to a single Riemann sheet.

3.2 The proof of Theorem 3.1 parallels proofs of similar theorems in Chapters 6 and 7. By differentiation of (3.02) and use of (2.07), we find that

$$\frac{\partial^2 \varepsilon_{n,1}}{\partial \xi^2} - \{u^2 + \psi(\xi)\} \varepsilon_{n,1} = \frac{2e^{u\xi} A_n'(\xi)}{u^{n-1}}. \tag{3.05}$$

Transferring the term $\psi(\xi)\varepsilon_{n,1}$ to the right-hand side and using the method of variation of parameters, we obtain

$$\varepsilon_{n,1}(u, \xi) = \int_{\alpha_1}^{\xi} \mathsf{K}(\xi, v) \left\{ \frac{2e^{uv} A_n'(v)}{u^n} + \frac{\psi(v)\, \varepsilon_{n,1}(u, v)}{u} \right\} dv, \tag{3.06}$$

where

$$\mathsf{K}(\xi, v) = \tfrac{1}{2}\{e^{u(\xi - v)} - e^{u(v - \xi)}\}.$$

The integration path is taken to be $\mathscr{Q}_1$. In consequence of Condition (ii), the following bound holds for the kernel and its ξ derivatives:

$$|\mathsf{K}(\xi, v)|,\ |u^{-1}\,\partial \mathsf{K}(\xi, v)/\partial\xi|,\ |u^{-2}\,\partial^2 \mathsf{K}(\xi, v)/\partial\xi^2| \leqslant |e^{u(\xi - v)}|.$$

Applying Theorem 10.1 of Chapter 6 to (3.06) with $J(v) = e^{uv}$, and observing from (2.07) that

$$\int_{\alpha_1}^{\xi} |\psi(v)\, dv| = 2\mathscr{V}_{\alpha_1, \xi}(A_1), \tag{3.07}$$

we obtain (3.04) in the case $j = 1$. That this solution of the integral equation satisfies (3.05) in the complex plane is establishable as in Chapter 6, §11.2. The proof for $j = 2$ is similar.

Ex. 3.1 Let u be positive, and $\psi(\xi)$ and its first $n-1$ derivatives be continuous real or complex functions in a given finite or infinite interval (α_1, α_2). Show that equation (3.01) has solutions (3.02) and (3.03) which are $n+1$ times continuously differentiable in (α_1, α_2) and satisfy

$$|\varepsilon_{n,j}(u, \xi)|,\ \tfrac{1}{2}u^{-1}|\partial\varepsilon_{n,j}(u, \xi)/\partial\xi| \leqslant u^{-n} \exp\{(-)^{j-1} u\xi\} \mathscr{V}_{\alpha_j, \xi}(A_n) \exp\{u^{-1} \mathscr{V}_{\alpha_j, \xi}(A_1)\}.$$

Ex. 3.2 Show that if ξ is real and $u > 0$, then the equation

$$\frac{d^2 W}{d\xi^2} + 2iu \frac{dW}{d\xi} - e^{i\xi} W = 0$$

has a solution such that

$$\left| W - 1 + \frac{e^{i\xi}}{2u} - \frac{2e^{i\xi} + e^{2i\xi}}{8u^2} + \frac{6e^{i\xi} + 9e^{2i\xi} + e^{3i\xi}}{48u^3} \right| \leqslant \frac{67|\xi|}{48u^4} \exp\left(\frac{|\xi|}{u}\right).$$

Ex. 3.3 If δ is an arbitrary constant in $(0, \tfrac{1}{4}\pi)$, show that the equation

$$d^2 w/dz^2 = \{u^2 + \exp(-z^2)\} z^2 w$$

has a solution of the form

$$\frac{e^{-uz^2/2}}{z^{1/2}}\left[1+\left(-\frac{3}{16z^2}+\frac{e^{-z^2}}{4}\right)\frac{1}{u}+\left\{\frac{105}{512z^4}-\left(\frac{1}{4}+\frac{3}{64z^2}\right)e^{-z^2}+\frac{e^{-2z^2}}{32}\right\}\frac{1}{u^2}+O\left(\frac{1}{u^3}\right)\right],$$

as $u\to\infty$ in the sector $|\mathrm{ph}\, u|\leqslant\pi-4\delta$, uniformly with respect to z in the annular sector $|z|\geqslant\delta$, $|\mathrm{ph}\, z|\leqslant\frac{1}{4}\pi-\delta$.

Ex. 3.4 Show that for real or complex values of z, k, and m, the Whittaker function $M_{k,m}(z)$ (Chapter 7, §11.1) is given by

$$M_{k,m}(z)=z^{m+(1/2)}\left\{\sum_{s=0}^{n-1}\frac{A_s(z)}{m^s}+\varepsilon_n(k,m,z)\right\}\qquad(n=1,2,\ldots),$$

provided that $\mathrm{Re}\, m>0$, where $A_0(z)=1$,

$$A_{s+1}(z)=-\tfrac{1}{2}zA_s'(z)+\tfrac{1}{8}\int_0^z(t-4k)\,A_s(t)\,dt\qquad(s\geqslant 0),$$

and

$$|\varepsilon_n(k,m,z)|\leqslant 2\exp\left\{\frac{2\mathscr{V}_{0,z}(A_1)}{|m|}\right\}\frac{\mathscr{V}_{0,z}(A_n)}{|m|^n},$$

the paths of variation being straight lines.†

Ex. 3.5 Show that when $u>0$, the equation

$$\frac{d^2w}{dz^2}=\left\{u^2+\frac{1}{u(z^2+1)}\right\}w$$

has solutions $w(u,\pm z)$, such that for large u

$$w(u,z)=e^{uz}\left[1+\frac{\tan^{-1}z}{2u^2}-\frac{1}{4u^3}\frac{1}{z^2+1}+\frac{1}{8u^4}\left\{(\tan^{-1}z)^2-\frac{2z}{(z^2+1)^2}\right\}+O\left(\frac{1}{u^5}\right)\right],$$

uniformly throughout the z plane, except for the two half-strips $|\mathrm{Re}\, z|\leqslant\delta$, $|\mathrm{Im}\, z|\geqslant 1-\delta$. Here δ is an arbitrary small positive constant, and the branch of $\tan^{-1}z$ is continuous.

Can this region of validity be extended?

Ex. 3.6‡ Let k be a constant in $[0,1)$, and u a large positive parameter. If the equation

$$\frac{d^2w}{dx^2}+\frac{u^2}{1-k^2\sin^2 x}w=0$$

has solutions of period π, show that

$$u=\frac{s\pi}{K(k^2)}+\frac{(1-\frac{1}{2}k^2)K(k^2)-E(k^2)}{4\pi s}+O\left(\frac{1}{s^2}\right),$$

where s is an integer, and $K(k^2)$, $E(k^2)$ are the elliptic integrals defined in Chapter 5, Exercise 9.2.

4 Behavior of the Coefficients at a Singularity

4.1 In §5.3 of Chapter 6 we showed that the LG functions provide meaningful approximations to solutions of (1.01) in the neighborhoods of poles of $f(z)$ and

† This exercise furnishes an asymptotic approximation to $M_{k,m}(z)$ for large $|m|$ which is uniformly valid in bounded z regions. In contrast, the approximation of Chapter 7, §11.2 is uniform throughout an infinite z interval.

‡ Communicated by W. H. Reid. The problem arises in the theory of gravity waves in water.

$g(z)$, subject to certain restrictions on the orders of the poles. The same restrictions apply for higher approximations:

Theorem 4.1 *If, at a given finite point z_0, either*

(i) *$f(z)$ has a pole of order $m > 2$, and $g(z)$ is analytic or has a pole of order less than $\frac{1}{2}m+1$,*

or

(ii) *$f(z)$ has a double pole and $(z-z_0)^2 g(z) \to -\frac{1}{4}$ as $z \to z_0$,*

then each coefficient A_s defined by (2.09) *and* (2.10) *has a convergent variation at $z = z_0$.*

Without loss of generality it may be supposed that $z_0 = 0$. We are given

$$f(z) \sim f_0 z^{-m}, \qquad g(z) = O(z^{-p}) \qquad (z \to 0),$$

where $f_0 \neq 0$, $m \geqslant 2$, and $p \geqslant 0$. Furthermore, because these asymptotic approximations are the initial terms in convergent power-series expansions they are differentiable any number of times.

In Case (i) equation (2.10) yields

$$\Lambda(z) = O(z^{q-1})$$

as $z \to 0$, where $q = \min(\frac{1}{2}m-1, \frac{1}{2}m-p+1)$ and is positive. Recalling that A_0 is a constant, we derive from (2.09)

$$A_s = \text{constant} + O(z^q)$$

in the case $s = 1$. Hence by repeated substitutions in (2.09) we see that the same result holds for general s. Since the O terms are differentiable it follows that $\mathcal{V}(A_s)$ is finite at $z = 0$.

In Case (ii) it is verifiable from (2.10) that $\Lambda(z)$ is analytic at $z = 0$, and hence from (2.09) that each coefficient A_s is analytic in z at this point. The theorem now follows.

4.2 An illustration of Case (ii) is provided by Exercise 3.4. As in Chapter 6, any case in which $f(z)$ has a double pole at z_0 and $g(z)$ has, at worst, a double pole at the same point, is transformable into Case (ii) by replacing u^2 by $u^2 - f_0^{-1}(\frac{1}{4}+g_0)$, where f_0 and g_0 are the limiting values at z_0 of $(z-z_0)^2 f(z)$ and $(z-z_0)^2 g(z)$, respectively.

Ex. 4.1 Suppose that in the neighborhood of $z = \infty$, $f(z)$ and $g(z)$ can be expanded in convergent series

$$f(z) = z^m \sum_{s=0}^{\infty} f_s z^{-s}, \qquad g(z) = z^p \sum_{s=0}^{\infty} g_s z^{-s},$$

where $f_0 \neq 0$, $g_0 \neq 0$, and m and p are integers or zero. Verify that each $\mathcal{V}(A_s)$ converges at $z = \infty$ when $m > -2$ and $p < \frac{1}{2}m - 1$, or when $m = p = -2$ and $g_0 = -\frac{1}{4}$.

*5 Behavior of the Coefficients at a Singularity (continued)

5.1 The variations of the coefficients A_s converge at singularities belonging to a broader class than that specified by Theorem 4.1 and Exercise 4.1. In extending these results it is more convenient to work in terms of the variable ξ, and we no longer suppose that $\psi(\xi)$ is independent of u. The singular point is located at $\xi = \infty$ in all cases, and the main condition adopted is given by:

(i) $|\psi(\xi)| \leqslant k/\{1+|\xi|^{1+\rho}\}$ *when* $\xi \in \mathbf{\Delta}$, *where* k *and* ρ *are positive numbers independent of* ξ *and* u.

Next, to assure uniform convergence at $\xi = \infty$ with respect to u, we suppose that the integration constants in (2.08) are chosen in such a way that:

(ii) *At a prescribed point* α *of* $\mathbf{\Delta}$, $|A_s(\alpha)| \leqslant \hat{k}_s$, $s = 0, 1, \ldots$, *where* $\hat{k}_s$ *is independent of* u.

The reference point α may depend on u, and it may also be the point at infinity on a given path $\mathscr{L}$.† Of course, Condition (ii) is automatically fulfilled in the case when $\mathbf{\Delta}$, $\psi(\xi)$, and the $A_s(\xi)$ *are* independent of u.

Lastly, to derive a bound for $A_{s+1}(\xi)$ from a bound for the preceding coefficient $A_s(\xi)$ we restrict the final result to a subdomain $\mathbf{\Gamma}$, say, of $\mathbf{\Delta}$. This is because the recurrence relation (2.08) entails differentiation at each step; compare Ritt's theorem in Chapter 1, §4.3. To be precise, $\mathbf{\Gamma}$ is subject to the conditions:

(iii) *The distance between each point of* $\mathbf{\Gamma}$ *and each boundary point of* $\mathbf{\Delta}$ *has a positive lower bound* d, *say, which is independent of* u.

(iv) *For each point* ξ *of* $\mathbf{\Gamma}$ *a path can be found lying wholly in* $\mathbf{\Gamma}$, *such that* $\int_\alpha^\xi (1+|v|^{1+\rho})^{-1}\,|dv| \leqslant l$, *where* ρ *and* α *are the quantities introduced in Conditions* (i) *and* (ii), *and* l *is independent of* ξ *and* u.

Condition (iv) is fulfilled, for example, when the path is a straight line or a bounded set of straight-line segments. It will be noticed, incidentally, that the condition requires $\alpha \in \mathbf{\Gamma}$.

5.2 Theorem 5.1 *If Conditions* (i)–(iv) *are satisfied for all values of* u *under consideration, then*

$$|A_s(\xi)| \leqslant k_s, \qquad |A_s'(\xi)| \leqslant l_s/(1+|\xi|^{1+\rho}) \qquad (\xi \in \mathbf{\Gamma}), \tag{5.01}$$

where k_s *and* l_s *are independent of* ξ *and* u.

Convergence of $\mathscr{V}(A_s)$ at $\xi = \infty$ is, of course, a simple consequence of the second inequality and Condition (iv).

The theorem is provable by induction. Let $\delta \in (0, d)$, and $\mathbf{\Gamma}(\delta)$ be the union of $\mathbf{\Gamma}$ and the set of points which lie within a distance δ of any boundary point of $\mathbf{\Gamma}$. Clearly $\mathbf{\Gamma}(\delta) \subset \mathbf{\Delta}$. If $\xi \in \mathbf{\Gamma}(\delta)$, then there exists a path which links α with ξ, lies in $\mathbf{\Gamma}(\delta)$, and satisfies

$$\int_\alpha^\xi \frac{|dv|}{1+|v|^{1+\rho}} < l + d. \tag{5.02}$$

For if $\xi \in \mathbf{\Gamma}$, then (5.02) follows at once from Condition (iv). If ξ is any other point of $\mathbf{\Gamma}$, then there exists a boundary point ξ_B of $\mathbf{\Gamma}$ such that $|\xi - \xi_B| < \delta\ (<d)$. We

† All paths are assumed to be finite chains of R_2 arcs.

take the path in (5.02) to be the known path linking α and ξ_{B}, plus the join of ξ_{B} and ξ.

Assume that the second of (5.01) holds when $\xi \in \mathbf{\Gamma}(\delta_s)$ for some $\delta_s \in (0, d)$. We have

$$A_s(\xi) = A_s(\alpha) + \int_{\alpha}^{\xi} A_s'(v)\, dv.$$

Substitution by means of (5.02) and Condition (ii) produces

$$|A_s(\xi)| \leqslant \hat{k}_s + l_s(l+d) \qquad (\xi \in \mathbf{\Gamma}(\delta_s)).$$

This agrees with the first of (5.01) if we take $k_s = \hat{k}_s + l_s(l+d)$.

Next let δ_{s+1} be an arbitrary number in $(0, \delta_s)$, so that $\mathbf{\Gamma}(\delta_{s+1}) \subset \mathbf{\Gamma}(\delta_s)$. From Cauchy's integral formula we have

$$A_s''(\xi) = \frac{1}{2\pi i} \int_{|v-\xi| = \delta_s - \delta_{s+1}} \frac{A_s'(v)}{(v-\xi)^2}\, dv. \tag{5.03}$$

If $\xi \in \mathbf{\Gamma}(\delta_{s+1})$, then $v \in \mathbf{\Gamma}(\delta_s)$.† Also, we have

$$\frac{1+|\xi|^{1+\rho}}{1+|v|^{1+\rho}} \leqslant \frac{1+(|v|+d)^{1+\rho}}{1+|v|^{1+\rho}} < c, \tag{5.04}$$

where c is an assignable number depending only on d and ρ. Substituting the second of (5.01) in (5.03), and using (5.04), we derive

$$|A_s''(\xi)| \leqslant \frac{l_s}{2\pi} \int_{|v-\xi| = \delta_s - \delta_{s+1}} \frac{|dv|}{|v-\xi|^2(1+|v|^{1+\rho})}$$

$$< \frac{cl_s}{(\delta_s - \delta_{s+1})(1+|\xi|^{1+\rho})} \qquad (\xi \in \mathbf{\Gamma}(\delta_{s+1})).$$

From this result, (2.07), Condition (i), and the first of (5.01), we obtain

$$|A_{s+1}'(\xi)| < l_{s+1}/(1+|\xi|^{1+\rho}) \qquad (\xi \in \mathbf{\Gamma}(\delta_{s+1})),$$

where $l_{s+1} = \frac{1}{2}cl_s(\delta_s - \delta_{s+1})^{-1} + \frac{1}{2}kk_s$. The truth of Theorem 5.1 is now evident.

Ex. 5.1 In a given sector $\mathbf{S}$ let $\psi(\xi)$ be holomorphic and $\psi(\xi) = O(\xi^{-1-\rho})$ as $\xi \to \infty$, where ρ is a positive constant. Starting with $A_0(\xi) = 1$, define higher coefficients by (2.08) with the condition $A_{s+1}(\xi) \to 0$ as $\xi \to \infty$ in $\mathbf{S}$. Show that for large ξ in any proper subsector of $\mathbf{S}$ having the same vertex, $A_s(\xi) = O(\xi^{-s\rho_1})$ and $A_s'(\xi) = O(\xi^{-s\rho_1 - 1})$, where $\rho_1 = \min(\rho, 1)$.

*6 Asymptotic Properties with Respect to the Parameter

6.1 The solutions furnished by Theorem 3.1 depend on the positive integer n. It is natural to enquire whether, in general, the differential equation (3.01) has solutions

† This is obviously true if $\xi \in \mathbf{\Gamma}$. If ξ is any other point of $\mathbf{\Gamma}(\delta_{s+1})$, then there is a point ξ_{B} on the boundary of $\mathbf{\Gamma}$ such that $|\xi - \xi_{\mathrm{B}}| < \delta_{s+1}$. Hence $|v - \xi_{\mathrm{B}}| = |(v-\xi) + (\xi - \xi_{\mathrm{B}})| < \delta_s$.

$W_1(u,\xi)$ and $W_2(u,\xi)$ with the properties

$$W_1(u,\xi) \sim e^{u\xi}\sum_{s=0}^{\infty}\frac{A_s(\xi)}{u^s}, \qquad \frac{1}{u}\frac{\partial W_1(u,\xi)}{\partial \xi} \sim e^{u\xi}\sum_{s=0}^{\infty}\frac{B_s(\xi)}{u^s}, \tag{6.01}$$

$$W_2(u,\xi) \sim e^{-u\xi}\sum_{s=0}^{\infty}(-)^s\frac{A_s(\xi)}{u^s}, \qquad \frac{1}{u}\frac{\partial W_2(u,\xi)}{\partial \xi} \sim -e^{-u\xi}\sum_{s=0}^{\infty}(-)^s\frac{B_s(\xi)}{u^s}, \tag{6.02}$$

as $u \to \infty$, uniformly with respect to $\xi \in \Xi_1(u,\alpha_1)$ for (6.01) or $\xi \in \Xi_2(u,\alpha_2)$ for (6.02).† That is, in a sense, can we take $n = \infty$ in the result of Theorem 3.1? The question is answerable in the affirmative if Conditions (i)–(iv) of §5.1 are adopted with $\alpha = \alpha_1$ for (6.01) or $\alpha = \alpha_2$ for (6.02), with the added provisos that the path in (iv) must be progressive, and when α_j is at infinity $(-)^j \operatorname{Re}(u\xi) \to \infty$ as $\xi \to \alpha_j$ along $\mathscr{L}_j$.

Let $A_s(\alpha)$ be the value of $A_s(\xi)$ at α, or, when α is at infinity, the limiting value of $A_s(\xi)$ as $\xi \to \alpha$ along $\mathscr{L}$, the existence of this limit being a corollary of the second of (5.01). Similarly, let $B_s(\alpha)$ be the value, or limiting value, of the coefficient defined by (2.06). Define $K(u)$ and $L(u)$ to be functions with the expansions‡

$$K(u) \sim \sum_{s=0}^{\infty}\frac{A_s(\alpha)}{u^s}, \qquad L(u) \sim \sum_{s=0}^{\infty}\frac{B_s(\alpha)}{u^s} \qquad (u \to \infty). \tag{6.03}$$

Then $W_1(u,\xi)$ is defined to be the solution of (2.01) satisfying the conditions

$$W_1(u,\alpha) = e^{u\alpha}K(u), \qquad \partial W_1(u,\alpha)/\partial\alpha = ue^{u\alpha}L(u), \tag{6.04}$$

when α is finite, or

$$W_1(u,\xi) \sim e^{u\xi}K(u) \qquad (\xi \to \alpha \text{ along } \mathscr{L}_1), \tag{6.05}$$

when α is at infinity.

6.2 Let ξ_1 be any point of the domain $\mathbf{\Gamma}$ (§5.1), and $W_{n,1}(u,\xi)$ and $W_{n,2}(u,\xi)$ be the solutions given by Theorem 3.1 with $\alpha_1 = \alpha$, $\alpha_2 = \xi_1$, and arbitrary choice of n. Since, by hypothesis, ξ_1 can be joined to α by a progressive path lying in $\mathbf{\Gamma}$, and therefore also in $\mathbf{\Delta}$, we have $\alpha \in \Xi_2(u,\xi_1)$. To verify that $W_1(u,\xi)$ has the properties (6.01), we express

$$W_1(u,\xi) = \lambda_{n,1}W_{n,1}(u,\xi) + \lambda_{n,2}W_{n,2}(u,\xi), \tag{6.06}$$

$$\frac{\partial W_1(u,\xi)}{\partial\xi} = \lambda_{n,1}\frac{\partial W_{n,1}(u,\xi)}{\partial\xi} + \lambda_{n,2}\frac{\partial W_{n,2}(u,\xi)}{\partial\xi}, \tag{6.07}$$

and seek the values of the coefficients $\lambda_{n,1}$ and $\lambda_{n,2}$.

† When the coefficients $A_s(\xi)$ and $B_s(\xi)$ depend on u, the expansions (6.01) are to be interpreted as generalized asymptotic expansions with respect to the scale $\{e^{u\xi}u^{-s}\}$. Similarly for (6.02).

‡ When $A_s(\alpha)$ and $B_s(\alpha)$ are independent of u, the existence of $K(u)$ and $L(u)$ follows directly from Chapter 1, §9.1. In other cases the expansions (6.03) are to be interpreted as generalized asymptotic expansions with scale $\{u^{-s}\}$, and the existence of $K(u)$ and $L(u)$ follows from (5.01) and analysis similar to Chapter 1, §9.1.

When α is finite we set $\xi = \alpha$ in (3.02), (3.03), and their ξ differentiated forms. Then using (3.04), (5.01), and (6.04), we obtain

$$K(u) = \lambda_{n,1}\left\{\sum_{s=0}^{n-1}\frac{A_s(\alpha)}{u^s} + O\left(\frac{1}{u^n}\right)\right\} + \lambda_{n,2}\,e^{-2u\alpha}\left\{\sum_{s=0}^{n-1}(-)^s\frac{A_s(\alpha)}{u^s} + O\left(\frac{1}{u^n}\right)\right\},$$

and

$$L(u) = \lambda_{n,1}\left\{\sum_{s=0}^{n-1}\frac{B_s(\alpha)}{u^s} + O\left(\frac{1}{u^n}\right)\right\} - \lambda_{n,2}\,e^{-2u\alpha}\left\{\sum_{s=0}^{n-1}(-)^s\frac{B_s(\alpha)}{u^s} + O\left(\frac{1}{u^n}\right)\right\}.$$

These equations are solvable for $\lambda_{n,1}$ and $\lambda_{n,2}$. With the aid of (6.03), it is easily seen that

$$\lambda_{n,1} = 1 + O(u^{-n}), \qquad \lambda_{n,2} = e^{2u\alpha}O(u^{-n}). \tag{6.08}$$

We now set $\xi = \xi_1$ in (6.06) and substitute by means of (3.02), (3.03), and (6.08). The monotonicity condition shows that $|e^{2u\alpha - u\xi_1}| \leqslant |e^{u\xi_1}|$, hence the whole contribution from the second solution on the right-hand side of (6.06) may be absorbed in the uniform error term associated with the first solution. Thus we derive

$$W_1(u, \xi_1) = e^{u\xi_1}\left\{\sum_{s=0}^{n-1}\frac{A_s(\xi_1)}{u^s} + O\left(\frac{1}{u^n}\right)\right\}. \tag{6.09}$$

Since ξ_1 and n are arbitrary, we have established that the first of (6.01) holds uniformly with respect to $\xi \in \mathbf{\Gamma} \cap \Xi_1(u, \alpha)$. Similarly for the second of (6.01).

Alternatively, if α is at infinity then by letting $\xi \to \alpha$ in (6.06) and the results of Theorem 3.1, and using (6.05) and the assumption that $\operatorname{Re}(u\xi) \to -\infty$, we find that $\lambda_{n,2} = 0$. This limiting process also shows that $\lambda_{n,1} = 1 + O(u^{-n})$; compare (6.03). Equation (6.09) follows immediately, and the same conclusion is reached.

We observe, incidentally, that because there is an infinity of functions $K(u)$ and $L(u)$ satisfying (6.03), the solution $W_1(u, \xi)$ is not unique.

Similar deductions may be made concerning the existence and nonuniqueness of a solution $W_2(u, \xi)$ with the properties (6.02).

Ex. 6.1 With the conditions of §§2.1, 5.1, and 6.1 let $\psi(\xi) = \psi(u, \xi)$ depend on u in such a way that for each value of u under consideration

$$\left|\psi(u,\xi) - \sum_{s=0}^{n-1}\frac{\psi_s(\xi)}{u^s}\right| < \frac{1}{1+|\xi|^{1+\rho}}\frac{c_n}{u^n} \qquad (\xi \in \mathbf{\Delta}, \quad n = 0, 1, 2, \ldots),$$

where (i) the functions $\psi_s(\xi)$ are independent of u and holomorphic in $\mathbf{\Delta}$, (ii) c_n and ρ are positive numbers independent of both u and ξ. From the results of §6 deduce that for large $|u|$ there are solutions $W_1(u, \xi)$ and $W_2(u, \xi)$ of (3.01) with the uniform asymptotic expansions (6.01) and (6.02) when ξ lies in $\mathbf{\Gamma} \cap \Xi_1(u, \alpha_1)$ for (6.01) or $\mathbf{\Gamma} \cap \Xi_2(u, \alpha_2)$ for (6.02), where the coefficients $A_s(\xi)$ are now defined by $A_0(\xi) = 1$ and

$$A_{s+1}(\xi) = -\tfrac{1}{2}A_s'(\xi) + \tfrac{1}{2}\int\{\psi_0(\xi)A_s(\xi) + \psi_1(\xi)A_{s-1}(\xi) + \cdots + \psi_s(\xi)A_0(\xi)\}\,d\xi \qquad (s \geqslant 0).$$

[Olver, 1958.]

7 Modified Bessel Functions of Large Order

7.1 For large values of the order ν and fixed values of the argument z, the asymptotic behavior of any Bessel function is immediately available from its series expansion in ascending powers of z. For example, from the definition

$$I_\nu(z) = (\tfrac{1}{2}z)^\nu \sum_{s=0}^{\infty} \frac{(\tfrac{1}{4}z^2)^s}{s!\,\Gamma(\nu+s+1)},$$

it is seen that

$$I_\nu(z) = \frac{(\tfrac{1}{2}z)^\nu}{\Gamma(\nu+1)}\left\{1 + O\left(\frac{1}{\nu}\right)\right\} \tag{7.01}$$

as $\nu \to \infty$ in the sector $|\mathrm{ph}\,\nu| \leqslant \pi-\delta$, where δ is an arbitrary small positive number. Not only is this result valid for fixed z: it holds uniformly in any bounded z domain. Clearly, however, the approximation is numerically meaningful only when $|\nu| \gg \frac{1}{4}|z|^2$.

Greater physical and mathematical interest attaches to the problem of representing Bessel functions for large $|\nu|$ in a uniform manner in *unbounded* z domains. This problem may be attacked by applying Laplace's method to integral representations for the Bessel functions derived in Chapters 2 and 7, or by applying the theory of the present chapter. As a general rule, when both an integral representation and an ordinary differential equation are available, the integral provides the easiest way of finding the first approximation, but higher approximations and error bounds are obtained more readily by use of the differential equation. An application of Laplace's method to the present problem was made in Chapter 4, §9.4. We turn now to the differential equation approach, concentrating on the modified Bessel functions in the present chapter and on the unmodified functions in Chapter 11.

7.2 On removing the term in the first derivative by changing the dependent variable in the modified form of Bessel's equation (Chapter 2, (10.03)), we obtain

$$\frac{d^2w}{dz^2} = \left(\frac{\nu^2-\frac{1}{4}}{z^2} + 1\right)w,$$

where $w = z^{1/2}I_\nu(z)$, $z^{1/2}K_\nu(z)$, or any linear combination of these functions. With ν playing the role of the parameter u, the transformations (2.02) and (2.03) could be applied with

$$f(z) = \frac{1}{z^2}, \qquad g(z) = -\frac{1}{4z^2} + 1.$$

Theorem 4.1(ii) then shows that in the resulting asymptotic series the variations of the coefficients converge at the singularity $z = 0$. Unfortunately, as suggested by Exercise 4.1 and verifiable by means of (2.09) and (2.10), these variations diverge at the other singularity $z = \infty$. To overcome this obstacle we replace z by νz; thus

$$\frac{d^2w}{dz^2} = \left\{\nu^2\frac{1+z^2}{z^2} - \frac{1}{4z^2}\right\}w, \qquad w = z^{1/2}I_\nu(\nu z) \quad\text{or}\quad z^{1/2}K_\nu(\nu z). \tag{7.02}$$

With

$$f(z) = \frac{1+z^2}{z^2}, \qquad g(z) = -\frac{1}{4z^2}, \tag{7.03}$$

Theorem 4.1(ii) and Exercise 4.1 show that the variations of the new coefficients converge at both $z = 0$ and $z = \infty$.

The only regions in the z plane that cannot now be treated successfully by means of Theorem 3.1 are the neighborhoods of the turning points $z = \pm i$ (or, in terms of the original argument of the modified Bessel functions, $z = \pm i\nu$). Consideration of these neighborhoods is deferred until the next chapter.

For the remainder of the present section *we adopt the simplifying restrictions that ν is real and positive, and* $|\mathrm{ph}\, z| < \frac{1}{2}\pi$. Extensions to complex ν and other z regions (excluding the turning points) are considered in §8.

7.3 Applying the transformations (2.02) and (2.03), with $f(z)$ and $g(z)$ given by (7.03), we find that

$$\xi = \int \frac{(1+z^2)^{1/2}}{z}\, dz, \qquad w = \left(\frac{z^2}{1+z^2}\right)^{1/4} W, \tag{7.04}$$

and

$$d^2 W/d\xi^2 = \{\nu^2 + \psi(\xi)\}\, W, \tag{7.05}$$

where

$$\psi(\xi) = \tfrac{1}{4} z^2 (4 - z^2)/(1+z^2)^3. \tag{7.06}$$

On integration, the first of (7.04) yields

$$\xi = (1+z^2)^{1/2} + \ln \frac{z}{1+(1+z^2)^{1/2}}, \tag{7.07}$$

it being convenient to take the arbitrary constant of integration to be zero.

The domain $\mathbf{D}$ is the sector $|\mathrm{ph}\, z| < \frac{1}{2}\pi$, and its mapping on the ξ plane is quickly determined by the following considerations:

(i) When z is real and positive, ξ is real; $z = 0, \infty$ corresponding to $\xi = -\infty, +\infty$, respectively.

(ii) For small $|z|$, $\xi = \ln(\frac{1}{2}z) + 1 + o(1)$; for large $|z|$, $\xi = z + o(1)$.

(iii) If $z = iy$ $(0 < y < 1)$, then $i\, d\xi/dz$ is real and positive.

(iv) As $z \to i$, $\xi - \frac{1}{2}\pi i \equiv \int_i^z t^{-1}(1+t^2)^{1/2}\, dt$ is asymptotic to $\frac{1}{3} 2^{3/2} e^{-\pi i/4} (z-i)^{3/2}$.

(v) If $z = iy$ $(y > 1)$, then $d\xi/dz$ is real and positive.

Thus $\mathbf{D}$ is transformed into the domain $\mathbf{\Delta}$ comprising the union of the half-plane $\mathrm{Re}\, \xi > 0$ and the strip $|\mathrm{Im}\, \xi| < \frac{1}{2}\pi$. Corresponding points of the transformation are indicated in Figs. 7.1 and 7.2. Since z does not take the values $\pm i$ when $\xi \in \mathbf{\Delta}$, it follows from (7.06) that $\psi(\xi)$ is holomorphic in $\mathbf{\Delta}$.

With a view to achieving maximal regions of validity in the final expansions, we select α_1 and α_2 to be convenient points in $\mathbf{\Delta}$ at which $\mathrm{Re}(\nu\xi)$ has its "least" and "greatest" values, respectively. Thus α_1 is at infinity on the negative real axis and α_2 is at infinity on the positive real axis. From Conditions (i) and (ii) of §3.1 it then follows that $\Xi_1(\nu, \alpha_1)$ and $\Xi_2(\nu, \alpha_2)$ each coincide with $\mathbf{\Delta}$.

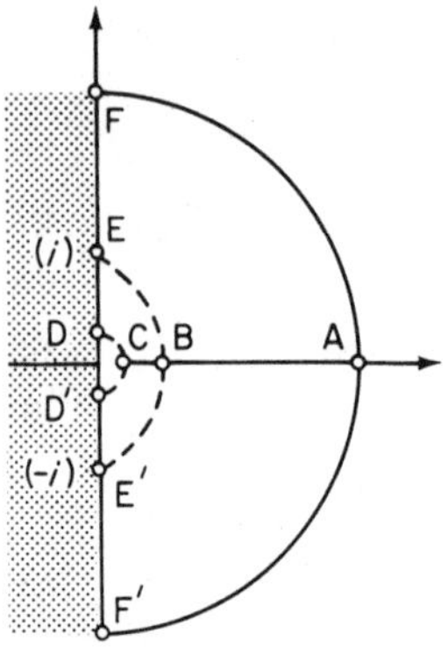

Fig. 7.1 z plane. Domain **D**.

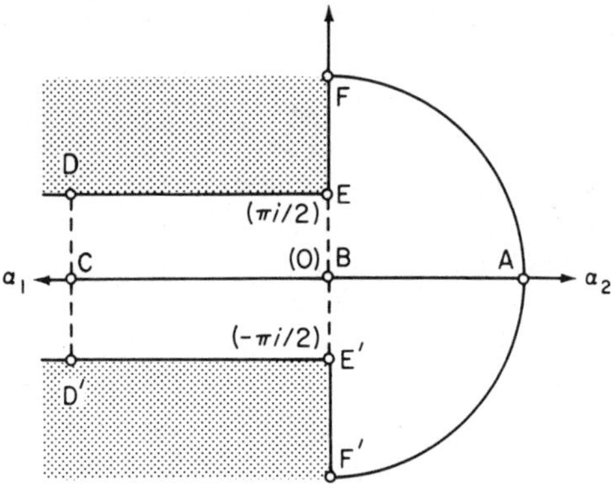

Fig. 7.2 ξ plane. Domain **Δ**.

Solutions of (7.05) are given by (3.02) and (3.03) with $u = \nu$ and the error terms bounded by (3.04) in **Δ**. In terms of z, the coefficients A_s are defined recursively by $A_0 = 1$ and

$$A_{s+1} = -\frac{1}{2}\frac{z}{(1+z^2)^{1/2}}\frac{dA_s}{dz} + \frac{1}{8}\int \frac{z(4-z^2)}{(1+z^2)^{5/2}} A_s\,dz \qquad (s \geqslant 0); \tag{7.08}$$

compare (2.09) and (2.10). In particular,

$$A_1 = \frac{1}{8(1+z^2)^{1/2}} - \frac{5}{24(1+z^2)^{3/2}} + \text{constant}.$$

The form of this result suggests that higher coefficients may be expressed conveniently in terms of the variable

$$p = (1+z^2)^{-1/2}. \tag{7.09}$$

And for later convenience we now change the symbol A to U; thus $A_s = U_s(p)$. On specifying the integration constants in (7.08) by the condition $A_{s+1} \to 0$ as $z \to \infty$, so that $U_{s+1}(0) = 0$, we find that in terms of p

$$U_{s+1}(p) = \tfrac{1}{2}p^2(1-p^2)\,U_s'(p) + \tfrac{1}{8}\int_0^p (1-5q^2)\,U_s(q)\,dq. \tag{7.10}$$

From this relation it is seen that $U_s(p)$ is a polynomial in p of degree $3s$. The first four polynomials are found to be†

$$\begin{gathered} U_0 = 1, \qquad U_1 = (3p-5p^3)/24, \qquad U_2 = (81p^2-462p^4+385p^6)/1152, \\ U_3 = (30375p^3-369603p^5+765765p^7-425425p^9)/414720. \end{gathered} \tag{7.11}$$

Returning to the original variables of (7.02), we write

$$\eta_{n,j}(\nu,z) = \exp\{(-)^j\nu\xi\}\,\varepsilon_{n,j}(\nu,\xi),$$

† Expressions for the next three have been given by B.A. (1952).

and

$$w_{n,1}(\nu,z) \equiv \frac{z^{1/2}}{(1+z^2)^{1/4}} W_{n,1}(\nu,\xi) = \frac{z^{1/2}e^{\nu\xi}}{(1+z^2)^{1/4}}\left\{\sum_{s=0}^{n-1}\frac{U_s(p)}{\nu^s}+\eta_{n,1}(\nu,z)\right\}, \tag{7.12}$$

$$w_{n,2}(\nu,z) \equiv \frac{z^{1/2}}{(1+z^2)^{1/4}} W_{n,2}(\nu,\xi) = \frac{z^{1/2}e^{-\nu\xi}}{(1+z^2)^{1/4}}\left\{\sum_{s=0}^{n-1}(-)^s\frac{U_s(p)}{\nu^s}+\eta_{n,2}(\nu,z)\right\}. \tag{7.13}$$

The new error terms are bounded by

$$|\eta_{n,1}(\nu,z)| \leqslant 2\exp\left\{\frac{2\mathscr{V}_{1,p}(U_1)}{\nu}\right\}\frac{\mathscr{V}_{1,p}(U_n)}{\nu^n}, \tag{7.14}$$

$$|\eta_{n,2}(\nu,z)| \leqslant 2\exp\left\{\frac{2\mathscr{V}_{0,p}(U_1)}{\nu}\right\}\frac{\mathscr{V}_{0,p}(U_n)}{\nu^n}, \tag{7.15}$$

the variations being taken along ξ-progressive paths.

7.4 To identify $z^{1/2}I_\nu(\nu z)$ and $z^{1/2}K_\nu(\nu z)$ in terms of $w_{n,1}(\nu,z)$ and $w_{n,2}(\nu,z)$, we note that for fixed ν, $w_{n,1}(\nu,z)$ is recessive as $\xi\to-\infty$, that is, as $z\to 0$, and $w_{n,2}(\nu,z)$ is recessive as $\xi\to+\infty$, that is, as $z\to+\infty$. Accordingly, $z^{1/2}I_\nu(\nu z)$ is a multiple† of $w_{n,1}(\nu,z)$ and $z^{1/2}K_\nu(\nu z)$ is a multiple of $w_{n,2}(\nu,z)$. A convenient way of determining these multiples is to compare the limiting forms of the solutions as $z\to+\infty$. From Chapter 7, (8.04) and (8.08), we have

$$z^{1/2}I_\nu(\nu z) \sim \frac{e^{\nu z}}{(2\pi\nu)^{1/2}}, \qquad z^{1/2}K_\nu(\nu z) \sim \left(\frac{\pi}{2\nu}\right)^{1/2} e^{-\nu z},$$

whereas in (7.12) and (7.13) $\xi = z+O(z^{-1})$, $e^{\nu\xi}\sim e^{\nu z}$, $U_s(p)\to 0$ $(s\geqslant 1)$, and $\eta_{n,2}(\nu,z)\to 0$. Thus we derive

$$I_\nu(\nu z) = \frac{1}{1+\eta_{n,1}(\nu,\infty)}\frac{e^{\nu\xi}}{(2\pi\nu)^{1/2}(1+z^2)^{1/4}}\left\{\sum_{s=0}^{n-1}\frac{U_s(p)}{\nu^s}+\eta_{n,1}(\nu,z)\right\}, \tag{7.16}$$

$$K_\nu(\nu z) = \left(\frac{\pi}{2\nu}\right)^{1/2}\frac{e^{-\nu\xi}}{(1+z^2)^{1/4}}\left\{\sum_{s=0}^{n-1}(-)^s\frac{U_s(p)}{\nu^s}+\eta_{n,2}(\nu,z)\right\}. \tag{7.17}$$

These are our main results. They are valid when $\nu>0$, $|\mathrm{ph}\, z|<\frac{1}{2}\pi$, and n is any positive integer. The function ξ is given by (7.07), and the coefficients $U_s(p)$ by (7.09), (7.10), and (7.11), all branches taking their principal values when $z\in(0,\infty)$ and being continuous elsewhere. The error terms $\eta_{n,1}(\nu,z)$ and $\eta_{n,2}(\nu,z)$ are bounded by (7.14) and (7.15); $\eta_{n,1}(\nu,\infty)$ is the limiting value‡ of $\eta_{n,1}(\nu,z)$ as $z\to+\infty$ and is bounded by (7.14) with $z=\infty$, that is, $p=0$.

7.5 Asymptotic properties of the expansions (7.16) and (7.17) for large ν are derivable from the error bounds. In the ξ plane (Fig. 7.2) the variational paths can be constructed by traveling from ξ parallel to the imaginary axis until the real axis

† By "multiple" we mean here a number which is independent of z, but may depend on ν and n.
‡ The existence of this limit follows from the evaluation of $z^{1/2}I_\nu(\nu z)/w_{n,1}(\nu,z)$ at $z=+\infty$.

is reached, then proceeding along the real axis to α_1 or α_2. Let δ be an arbitrary constant in $(0, \frac{1}{2}\pi)$ and $\mathbf{\Delta}(\delta)$ the part of the domain $\mathbf{\Delta}$ remaining after removal of all points at a distance less than δ from its boundaries. When $\xi \in \mathbf{\Delta}(\delta)$ the shortest distance from the variational paths to the singularities at $\xi = \pm\pi i/2$ is δ, from which it follows that the variations of each coefficient $U_s(p)$ are bounded.

Thus we arrive at

$$I_\nu(\nu z) \sim \frac{e^{\nu\xi}}{(2\pi\nu)^{1/2}(1+z^2)^{1/4}} \sum_{s=0}^{\infty} \frac{U_s(p)}{\nu^s}, \tag{7.18}$$

$$K_\nu(\nu z) \sim \left(\frac{\pi}{2\nu}\right)^{1/2} \frac{e^{-\nu\xi}}{(1+z^2)^{1/4}} \sum_{s=0}^{\infty} (-)^s \frac{U_s(p)}{\nu^s}, \tag{7.19}$$

as $\nu \to \infty$, uniformly in the z map $\mathbf{D}(\delta)$, say, of $\mathbf{\Delta}(\delta)$. It is easily verified that if δ' is any given small positive number, then $\mathbf{D}(\delta)$ includes the sector $|\mathrm{ph}\, z| \leqslant \frac{1}{2}\pi - \delta'$ for all sufficiently small δ.

Ex. 7.1 By letting $z \to 0$ in (7.16) show that $U_s(1) = \gamma_s$, where $\gamma_0, \gamma_1, \gamma_2, \ldots$ are the coefficients in the expansion

$$\frac{(2\pi)^{1/2}\nu^{\nu-(1/2)}}{e^\nu \Gamma(\nu)} \sim \gamma_0 + \frac{\gamma_1}{\nu} + \frac{\gamma_2}{\nu^2} + \cdots \qquad (\nu \to \infty).$$

Thence derive

$$1 + \eta_{n,1}(\nu, \infty) = \frac{e^\nu \Gamma(\nu)}{(2\pi)^{1/2}\nu^{\nu-(1/2)}} \left(\sum_{s=0}^{n-1} \frac{\gamma_s}{\nu^s}\right).$$

Ex. 7.2 With the same conditions as for (7.16) and (7.17), prove that

$$I_\nu'(\nu z) = \frac{(1+z^2)^{1/4}}{1+\eta_{n,1}(\nu,\infty)} \frac{e^{\nu\xi}}{(2\pi\nu)^{1/2} z} \left[\sum_{s=0}^{n-1} \frac{V_s}{\nu^s} + \frac{V_n - U_n}{\nu^n} + \kappa_{n,1}(\nu, z) - \frac{z^2 p^3}{2\nu} \eta_{n,1}(\nu, z)\right],$$

$$K_\nu'(\nu z) = -\left(\frac{\pi}{2\nu}\right)^{1/2} \frac{(1+z^2)^{1/4} e^{-\nu\xi}}{z} \left[\sum_{s=0}^{n-1} (-)^s \frac{V_s}{\nu^s} + (-)^n \frac{V_n - U_n}{\nu^n} - \kappa_{n,2}(\nu, z) + \frac{z^2 p^3}{2\nu} \eta_{n,2}(\nu, z)\right],$$

where $V_0 = 1$, $V_s = U_s - \frac{1}{2}p(1-p^2)U_{s-1} - p^2(1-p^2)(dU_{s-1}/dp)$ when $s \geqslant 1$, and $|\kappa_{n,1}(\nu, z)|$ and $|\kappa_{n,2}(\nu, z)|$ are bounded by the right-hand sides of (7.14) and (7.15), respectively.

Show also that

$$U_{2s}V_0 - U_{2s-1}V_1 + U_{2s-2}V_2 - \cdots + U_0 V_{2s} = 0 \qquad (s \geqslant 1).$$

Ex. 7.3 Verify (as far as the first few terms) that if $z \to \infty$ in the sector $|\mathrm{ph}\, z| \leqslant \frac{1}{2}\pi - \delta$ $(< \frac{1}{2}\pi)$, ν being fixed, then the expansion (7.17) agrees with (8.04) of Chapter 7.

*8 Extensions of the Regions of Validity for the Expansions of the Modified Bessel Functions

8.1† All possible extensions of the regions of validity of (7.16) and (7.17), or more precisely, (7.14) and (7.15), can be found by extending the ξ domain $\mathbf{\Delta}$ of Fig. 7.2 as far as the monotonicity condition will allow.

† In this subsection and the next we continue to suppose that ν is real and positive.

Consider first the expansion of $I_\nu(\nu z)$. Here the reference point is $\alpha_1 = -\infty$, and the monotonicity condition precludes extension across the boundaries EF and $E'F'$, but not across DE and $D'E'$. The z maps of the shaded regions can be constructed by use of the relation

$$\xi(ze^{\pi i}) = \xi(z) + \pi i$$

obtained from (7.07) by making a small half-circuit about $z = 0$. The first stage in the continuation of the mapping across DE is indicated in Figs. 8.1 and 8.2. The former is the reflection of Fig. 7.1 in the imaginary z axis; the latter is Fig. 7.2 translated through πi. There is a new singularity of $\psi(\xi)$ at the point E_1 of affix $3\pi i/2$. The next continuation—across D_1E_1—is achieved by another reflection in the imaginary z axis, and a further translation πi in the ξ plane. And so on.

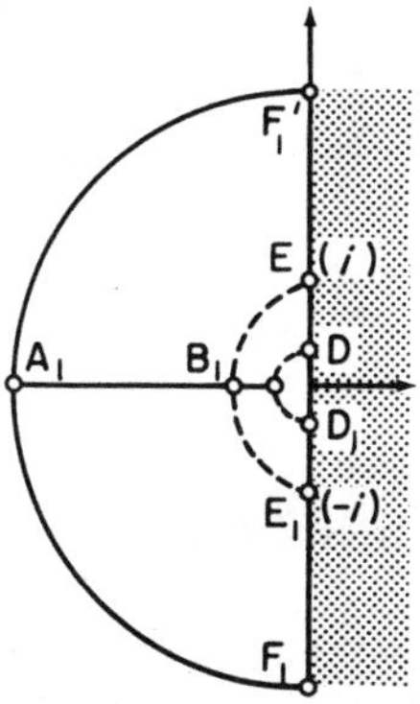

Fig. 8.1 z plane.

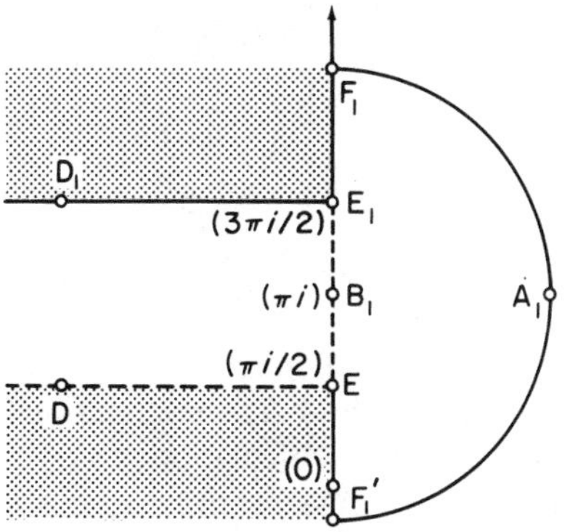

Fig. 8.2 ξ plane.

Points in the unshaded region of Fig. 8.2 can be linked to α_1 by a ξ-progressive path lying in the union of this region and the unshaded region of Fig. 7.2. In consequence, this region is admissible. The corresponding z region is $\frac{1}{2}\pi \leqslant \operatorname{ph} z \leqslant \frac{3}{2}\pi$, with the exclusion of the boundaries EF_1' and E_1F_1 (Fig. 8.1). Continuing the process, we conclude that (7.14) *and* (7.16) *are valid in the z plane cut along the imaginary axis from i to $i\infty$ and $-i$ to $-i\infty$: any branch of* ph z *is allowable, but points on the cuts must be excluded.*

This result is mainly of theoretical interest. In practice the use of (7.16) can be confined to the original region of validity ($|\operatorname{ph} z| < \frac{1}{2}\pi$ plus the points on the imaginary axis between i and $-i$); elsewhere, the continuation formula

$$I_\nu(\nu z e^{m\pi i}) = e^{\nu m\pi i} I_\nu(\nu z) \qquad (m = \text{an integer}) \tag{8.01}$$

is employed. Indeed, except for an inessential change in the variational paths for the error term, the extension just obtained can be regarded as a reformulation of (8.01).

We hasten to add that besides the actual cuts, points *near* the cuts should be avoided in applications because of the largeness of the error terms there. Instead, results given in Exercise 8.2 below or (in the proximity of $z = \pm i$) Chapter 11, §10, should be employed.

8.2 In the case of $K_\nu(\nu z)$ the reference point α_2 is at $+\infty$, and the maximal regions of validity are quite different because continuation is feasible across *all* the boundaries of Fig. 7.2.

On crossing DE and referring to Figs. 8.1 and 8.2, we see that we cannot reenter the right half of the ξ plane without violating the monotonicity condition. Hence EB_1E_1 is a true boundary. Similarly, on crossing D_1E_1 we must again keep to the left of the imaginary ξ axis. Continuing the process, we see that (7.15) *and* (7.17) *are valid with unrestricted values of* ph z *when* z *lies within the eye-shaped domain* **K** *bounded by the curve* EBE' *of* Fig. 7.1 *and the image of this curve in the imaginary axis.*

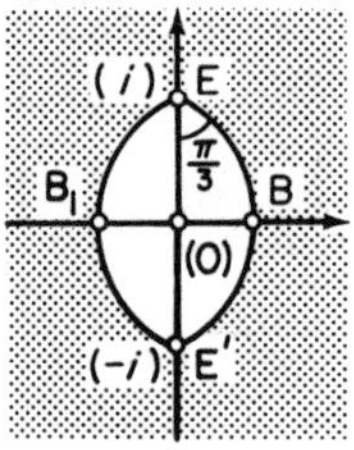

Fig. 8.3 z plane. Domain **K**.

The domain **K** is depicted in Fig. 8.3. At the vertices $z = \pm i$, the angle between the boundaries is $2\pi/3$. The intersections of these boundaries with the real axis are found by numerical calculation† to be $z = \pm 0.66274\ldots$.

Next, the allowable extension across EF is indicated by the unshaded regions in the mapping shown in Figs. 8.4 and 8.5. In the z plane $E\hat{B}\hat{E}$ is a boundary of **K**. A similar extension across $E'F'$ is allowable, and combining these results we see that (7.15) *and* (7.17) *are also valid outside the closure of* **K**, *provided that* $|\text{ph}\, z| < \frac{3}{2}\pi$.

No further extensions can be made. In practice use of (7.17) would be restricted to the sector $|\text{ph}\, z| \leqslant \pi$, other regions being covered by means of the continuation

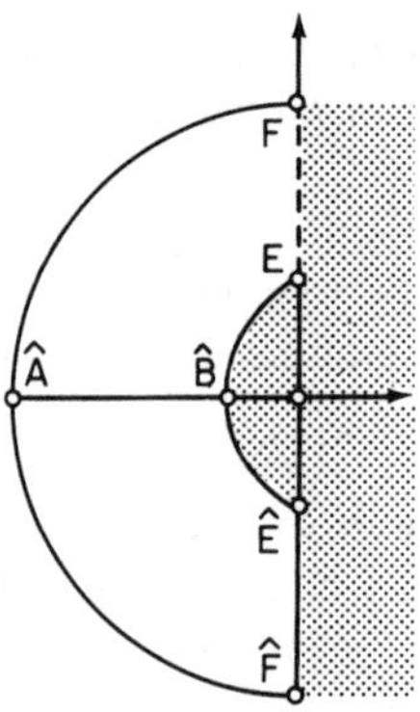

Fig. 8.4 z plane. Extension for $K_\nu(\nu z)$.

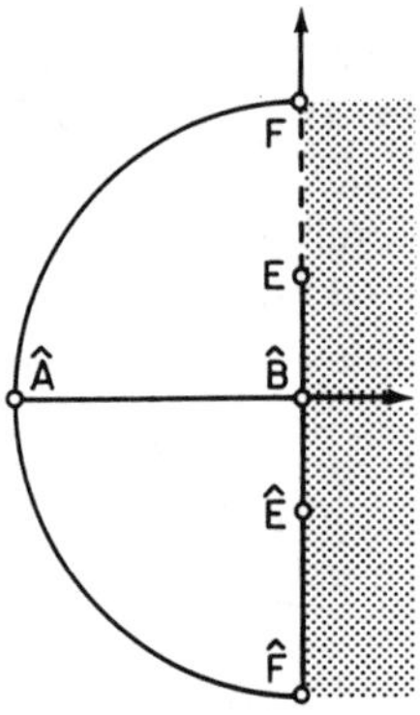

Fig. 8.5 ξ plane.

† Based on Exercise 8.1 below.

formula

$$\sin(\nu\pi)\,K_\nu(\nu z e^{m\pi i}) = \sin(m\nu\pi)\,K_\nu(\nu z e^{\pi i}) - \sin\{(m-1)\nu\pi\}\,K_\nu(\nu z) \qquad (m = \text{an integer}) \tag{8.02}$$

obtainable from Chapter 7, (4.13) and (8.02). Again, numerical applications near the left-hand boundary of **K** (both inside and outside **K**) should be avoided: more accurate representations for these regions are supplied in Chapter 11, §10.

8.3 We turn now to complex values of ν. In the text we shall suppose that $|\mathrm{ph}\,\nu| < \frac{1}{2}\pi$ in the interest of brevity, but certain extensions can be made beyond these limits, as indicated in Exercise 8.3 below. The expansions (7.16) and (7.17) are unchanged in form; modifications apply only to the regions of validity, stemming from the requirement that $\nu\xi$ be progressive on the linking paths.

In the case of $I_\nu(\nu z)$, the reference point α_1 is chosen to be the point at infinity in the direction of strongest recession, given by $\mathrm{ph}(-\xi) = -\omega$, where $\omega \equiv \mathrm{ph}\,\nu$. The resulting modifications to Fig. 7.2 are indicated in Fig. 8.7. The region of validity includes the unshaded domain, and continuations may be made across HE and $H'E'$, but not EG and $E'G'$. Proceeding as in §8.1, we see that the net effect is to change the cuts in the final z regions of validity from segments of the imaginary axis to the maps of EG and $E'G'$, that is,

$$\mathrm{ph}(\xi - \tfrac{1}{2}\pi i) = \tfrac{1}{2}\pi - \omega, \qquad \mathrm{ph}(\xi + \tfrac{1}{2}\pi i) = -\tfrac{1}{2}\pi - \omega.$$

In the z plane these cuts begin at an angle $-2\omega/3$ with the imaginary axis and have the lines $\mathrm{ph}(z \mp \frac{1}{2}\pi i) = \pm\frac{1}{2}\pi - \omega$ as asymptotes; see Fig. 8.6. Except on these cuts the relations (7.14) and (7.16) apply with any value of $\mathrm{ph}\,z$.†

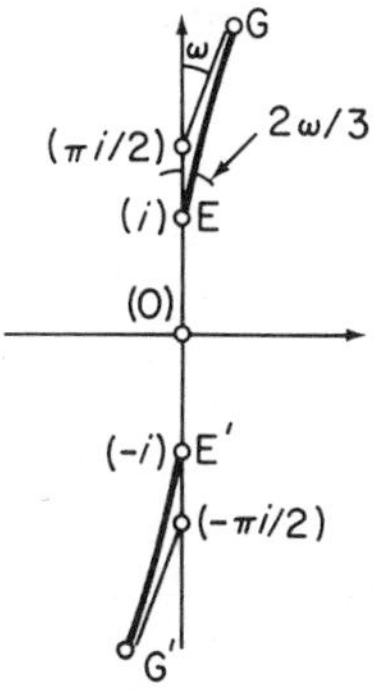

Fig. 8.6 z plane. Cuts for $I_\nu(\nu z)$.

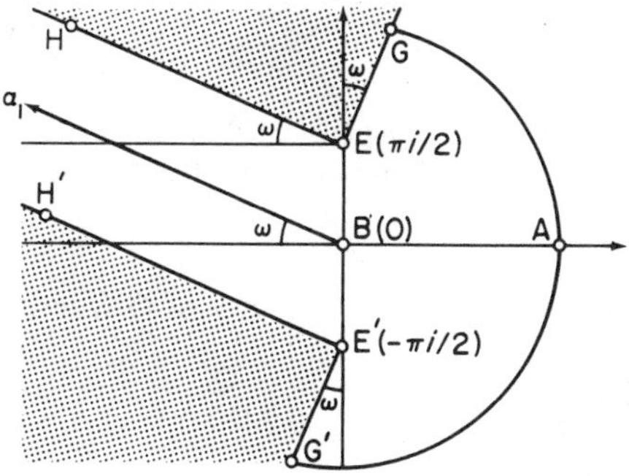

Fig. 8.7 ξ plane. $\omega = \mathrm{ph}\,\nu$.

In the expansion of $K_\nu(\nu z)$ the appropriate choice of α_2 is $\infty e^{-i\omega}$. Clearly the unshaded region of Fig. 7.2 is still admissible, which means that (7.15) and (7.17) are valid for $|\mathrm{ph}\,z| < \frac{1}{2}\pi$. Continuation across all the boundaries is again allowable, but we shall not attempt to specify the maximal z regions in detail because of their

† There is an error in the specification of these regions in the paper by Olver (1954b, §3). The author is indebted to J. A. Cochran for pointing this out.

complexity. Again, the use of the expansion near the true boundaries of its region of validity should be avoided by use of the continuation formula (8.02), where this is possible, or by use of expansions of the kind developed in Chapter 11.

Ex. 8.1 From (7.07) show that if $t = \operatorname{csch}^{-1} z$, then $\xi = \coth t - t$. By taking the real part of t as parameter τ, show that a parametric equation for the boundaries of **K** is given by

$$z = \pm(\tau^2 - \tau \tanh \tau)^{1/2} \pm i(\tau \coth \tau - \tau^2)^{1/2} \qquad (0 \leqslant \tau \leqslant \tau_0),$$

where τ_0 is the positive root of $\coth \tau = \tau$.

Ex. 8.2† Verify the connection formula

$$\pi i J_\nu(z) = e^{-\nu\pi i/2} K_\nu(ze^{-\pi i/2}) - e^{\nu\pi i/2} K_\nu(ze^{\pi i/2}).$$

Thence show that when $\nu > 0$ and $z > 1$,

$$J_\nu(\nu z) = \left(\frac{2}{\pi\nu}\right)^{1/2} \frac{1}{(z^2-1)^{1/4}} \left[\cos\{\nu\hat{\xi}(z) - \tfrac{1}{4}\pi\} \sum_{s=0}^{n-1} (-)^s \frac{\hat{U}_{2s}(\hat{p})}{\nu^{2s}} \right.$$

$$\left. + \sin\{\nu\hat{\xi}(z) - \tfrac{1}{4}\pi\} \sum_{s=0}^{n-1} (-)^s \frac{\hat{U}_{2s+1}(\hat{p})}{\nu^{2s+1}} + \theta_{2n}(\nu, z)\right],$$

where $\hat{\xi}(z) = (z^2-1)^{1/2} - \sec^{-1} z$, $\hat{p} = (z^2-1)^{-1/2}$, $\hat{U}_s(\hat{p}) = i^s U_s(-i\hat{p})$, and

$$|\theta_{2n}(\nu, z)| \leqslant 2 \exp\left\{\frac{2\mathscr{V}_{0,\hat{p}}(\hat{U}_1)}{\nu}\right\} \frac{\mathscr{V}_{0,\hat{p}}(\hat{U}_{2n})}{\nu^{2n}}.$$

Ex. 8.3‡ By replacing ν by $i\nu$ and z by $-iz$ in equation (7.02) and then applying Theorem 3.1, show that when ν is real and positive and $|\mathrm{ph}(z-1)| < \pi$

$$K_{i\nu}(\nu z) = \left(\frac{\pi}{2\nu}\right)^{1/2} \frac{\exp\{-\frac{1}{2}\nu\pi - \nu\hat{\xi}(z)\}}{(z^2-1)^{1/4}} \left\{\sum_{s=0}^{n-1} (-)^s \frac{\hat{U}_s(\hat{p})}{\nu^s} + \phi_n(\nu, z)\right\},$$

where $\hat{\xi}(z)$, $\hat{p}$, and $\hat{U}_s(\hat{p})$ are defined as in Exercise 8.2, and

$$|\phi_n(\nu, z)| \leqslant 2 \exp\left\{\frac{2\mathscr{V}_{0,\hat{p}}(\hat{U}_1)}{\nu}\right\} \frac{\mathscr{V}_{0,\hat{p}}(\hat{U}_n)}{\nu^n},$$

the variations being evaluated along $\hat{\xi}$-progressive paths.

Are the lines $\mathrm{ph}(z-1) = \pm\pi$ true boundaries? [Balogh, 1967.]

*9 More General Forms of Differential Equation

9.1 The methods of this chapter can be extended to differential equations of the type

$$d^2w/dz^2 = u^{2l} f(u, z) w, \tag{9.01}$$

in which l is a positive integer and $f(u, z)$ is an analytic function of z having a uniform asymptotic expansion of the form

$$f(u, z) \sim f_0(z) + \frac{f_1(z)}{u} + \frac{f_2(z)}{u^2} + \cdots$$

† Compare Chapter 4, (9.11). Except for the error bounds the result is again due to Debye (1909).

‡ Except for the present error bound this result can also be found by Laplace's method; compare Chapter 4, Exercise 9.3.

as $|u| \to \infty$. For simplicity in exposition, we confine attention to the case $l = 1$, but similar analyses may be constructed for higher values of l.

As before, the parameter u is supposed to range over an unbounded real interval or complex domain, and we also assume that $f(u, z)$ and the coefficients $f_s(z)$ are holomorphic functions of z in a complex domain $\mathbf{D}(u)$, bounded or otherwise. Following §1.2 we find that the transformations

$$\xi = \int f_0^{1/2}(z)\,dz, \qquad w = f_0^{-1/4}(z)\,W,$$

throw (9.01), with $l = 1$, into the form

$$d^2W/d\xi^2 = \{u^2 + u\phi(\xi) + \psi(u, \xi)\}\,W, \tag{9.02}$$

where $\phi(\xi) = f_1(z)/f_0(z)$ and

$$\psi(u, \xi) \sim \sum_{s=0}^{\infty} \frac{\psi_s(\xi)}{u^s}. \tag{9.03}$$

In the last expansion

$$\psi_0(\xi) = \frac{f_2(z)}{f_0(z)} - \frac{1}{f_0^{3/4}(z)}\frac{d^2}{dz^2}\left\{\frac{1}{f_0^{1/4}(z)}\right\}, \qquad \psi_s(\xi) = \frac{f_{s+2}(z)}{f_0(z)} \qquad (s \geqslant 1).$$

Provided that $f_0(z)$ does not vanish in $\mathbf{D}(u)$, that is, turning points are again excluded, the functions $\phi(\xi)$, $\psi(u, \xi)$, and $\psi_s(\xi)$ are holomorphic in the corresponding ξ domain $\Delta(u)$, say.

9.2 Use of the LG approximation suggests that (9.02) may have a series solution of the type

$$W = e^{u\xi} X(\xi) \sum_{s=0}^{\infty} \frac{A_s(\xi)}{u^s},$$

with $A_0(\xi) = 1$. Differentiating this expansion twice with respect to ξ, substituting in (9.02), and formally equating coefficients, we obtain

$$2X' = \phi X, \tag{9.04}$$

and

$$2XA_{s+1}' + X''A_s + 2X'A_s' + XA_s'' = X(\psi_0 A_s + \psi_1 A_{s-1} + \cdots + \psi_s A_0). \tag{9.05}$$

Integration gives

$$X = \exp\left(\tfrac{1}{2}\int \phi\,d\xi\right), \tag{9.06}$$

and

$$A_{s+1} = -\tfrac{1}{2}\phi A_s - \tfrac{1}{2}A_s' + \tfrac{1}{2}\int \{\psi_0 A_s + \psi_1 A_{s-1} + \cdots + \psi_s A_0 + (\tfrac{1}{2}\phi' - \tfrac{1}{4}\phi^2)A_s\}\,d\xi$$
$$(s = 0, 1, \ldots). \tag{9.07}$$

Clearly X and each coefficient A_s is a holomorphic function of ξ in $\mathbf{\Delta}(u)$.

Let the nth remainder term in the expansion (9.03) be expressed in the form

$$\psi(u,\xi) = \sum_{s=0}^{n-1} \frac{\psi_s(\xi)}{u^s} + \frac{\Psi_n(u,\xi)}{u^n} \qquad (n = 0, 1, \ldots), \tag{9.08}$$

so that $\Psi_n(u,\xi)$ is holomorphic and bounded in $\mathbf{\Delta}(u)$. By use of (9.04), (9.05), and (9.08), we may verify that if

$$W_n(u,\xi) \equiv e^{u\xi} X(\xi) \sum_{s=0}^{n-1} \frac{A_s(\xi)}{u^s} + \varepsilon_n(u,\xi) \tag{9.09}$$

satisfies (9.02), then $\varepsilon_n(u,\xi)$ satisfies the inhomogeneous equation

$$\varepsilon_n''(u,\xi) - \{u^2 + u\phi(\xi) + \psi(u,\xi)\}\, \varepsilon_n(u,\xi) = X(\xi)\, e^{u\xi} \left\{ \frac{2A_n'(\xi)}{u^{n-1}} + \frac{R_n'(u,\xi)}{u^n} \right\}, \tag{9.10}$$

in which primes denote partial differentiations with respect to ξ, and

$$R_n(u,\xi) = \sum_{s=0}^{n-1} \int A_s(\xi)\, \Psi_{n-s}(u,\xi)\, d\xi. \tag{9.11}$$

9.3 In order to solve (9.10) by previous methods, we need a homogeneous differential equation with known solutions which approximates the left-hand side of (9.10) as far as the term $u\phi(\xi)\,\varepsilon_n(u,\xi)$ when $|u|$ is large. Such an equation can be constructed by referring again to Exercise 1.2 of Chapter 6. On setting $q = 0$ and $p = (u + \tfrac{1}{2}\phi)^2$, we find that the functions

$$Z_1(u,\xi) = \frac{e^{u\xi} X(\xi)}{\{u + \tfrac{1}{2}\phi(\xi)\}^{1/2}}, \qquad Z_2(u,\xi) = \frac{e^{-u\xi}}{X(\xi)\{u + \tfrac{1}{2}\phi(\xi)\}^{1/2}},$$

satisfy

$$d^2Z/d\xi^2 = \{u^2 + u\phi(\xi) + \zeta(u,\xi)\}\, Z, \tag{9.12}$$

with

$$\zeta(u,\xi) = \frac{1}{4}\phi^2 - \frac{\phi''}{4u + 2\phi} + \frac{3\phi'^2}{(4u + 2\phi)^2}.$$

The Wronskian of Z_1 and Z_2 may be calculated from the relations

$$\frac{Z_1'}{Z_1} = u + \frac{1}{2}\phi - \frac{\phi'}{4u + 2\phi}, \qquad \frac{Z_2'}{Z_2} = -u - \frac{1}{2}\phi - \frac{\phi'}{4u + 2\phi}. \tag{9.13}$$

Thus $\mathscr{W}(Z_1, Z_2) = -2$.

Using Z_1 and Z_2 as complementary functions and solving (9.10) by variation of parameters, we obtain

$$\varepsilon_n(u,\xi) = \int_\alpha^\xi K(\xi,v) \left[\{\psi(u,v) - \zeta(u,v)\}\, \varepsilon_n(u,v) + X(v)\, e^{uv} \left\{ \frac{2A_n'(v)}{u^{n-1}} + \frac{R_n'(u,v)}{u^n} \right\} \right] dv, \tag{9.14}$$

where

$$K(\xi, v) = \tfrac{1}{2}\{Z_1(u,\xi)Z_2(u,v) - Z_2(u,\xi)Z_1(u,v)\},$$

and $\alpha \equiv \alpha(u)$ is an arbitrary point of $\mathbf{\Delta}(u)$; compare (3.06).

As usual, a monotonicity condition is imposed on the integration path. Denoting this path by $\mathscr{Q}$, we stipulate that $\mathrm{Re}\{uv + \frac{1}{2}\int \phi(v)\,dv\}$ be nondecreasing as v passes along $\mathscr{Q}$ from α to ξ. In consequence,

$$|e^{uv}X(v)| \leqslant |e^{u\xi}X(\xi)|.$$

Use of this inequality and the relations (9.13) produces the following bounds for the kernel and its derivative:

$$|K(\xi,v)| \leqslant \frac{1}{\lambda}\left|\frac{e^{u\xi}X(\xi)}{e^{uv}X(v)}\right|, \qquad \left|\frac{\partial K(\xi,v)}{\partial \xi}\right| \leqslant \frac{\mu}{\lambda}\left|\frac{e^{u\xi}X(\xi)}{e^{uv}X(v)}\right|,$$

in which

$$\lambda \equiv \lambda(u,\xi) = \inf_{v\in\mathscr{Q}} |u + \tfrac{1}{2}\phi(v)|, \tag{9.15}$$

and

$$\mu \equiv \mu(u,\xi) = \lambda^{1/2}\left\{\left|u + \frac{1}{2}\phi(\xi)\right|^{1/2} + \frac{|\phi'(\xi)|}{4|u+\frac{1}{2}\phi(\xi)|^{3/2}}\right\}, \tag{9.16}$$

provided that $\lambda \neq 0$.

9.4 Applying Theorem 10.1 of Chapter 6 to equation (9.14) in the manner of §3.2, and collecting together the results, we arrive at the following:

Theorem 9.1 *Suppose that u is a real or complex parameter, $\phi(\xi)$ is holomorphic in a complex domain $\mathbf{\Delta}(u)$, and $\psi(u,\xi)$ can be expanded in the form* (9.08) *in which the $\psi_s(\xi)$ and $\Psi_n(u,\xi)$ are holomorphic in $\mathbf{\Delta}(u)$. Then for each nonzero value of u and each positive integer n, the differential equation* (9.02) *has a solution $W_n(u,\xi)$ which* (i) *depends on an arbitrary reference point $\alpha \equiv \alpha(u)$ of $\mathbf{\Delta}(u)$,* (ii) *is holomorphic in $\mathbf{\Delta}(u)$,* (iii) *has the form* (9.09) *with $|\varepsilon_n(u,\xi)|$ and $\mu^{-1}|\partial\varepsilon_n(u,\xi)/\partial\xi|$ both bounded by*

$$\frac{|e^{u\xi}X(\xi)|}{\lambda}\left\{\frac{2\mathscr{V}_{\mathscr{Q}}(A_n)}{|u|^{n-1}} + \frac{\mathscr{V}_{\mathscr{Q}}(R_n)}{|u|^n}\right\}\exp\left\{\frac{\mathscr{V}_{\mathscr{Q}}(H)}{\lambda}\right\}, \tag{9.17}$$

when $\xi \in \mathbf{\Xi}(u,\alpha)$ (defined below).

In this theorem the functions $X(\xi)$, $A_s(\xi)$, $R_n(u,\xi)$, λ, and μ are defined by (9.06), (9.07), (9.11), (9.15), and (9.16). The function $H \equiv H(u,\xi)$ is defined by

$$H(u,\xi) = \int\left[\psi(u,\xi) - \frac{\phi^2(\xi)}{4} + \frac{\phi''(\xi)}{4u+2\phi(\xi)} - \frac{3\phi'^2(\xi)}{\{4u+2\phi(\xi)\}^2}\right]d\xi. \tag{9.18}$$

The variational path $\mathscr{Q}$ must lie in $\mathbf{\Delta}(u)$ and comprise a finite chain of R_2 arcs linking α with ξ such that when v passes along $\mathscr{Q}$ from α to ξ, $\mathrm{Re}\{uv + \frac{1}{2}\int\phi(v)\,dv\}$ is nondecreasing. The region of validity $\mathbf{\Xi}(u,\alpha)$ comprises the ξ point set for which such a path can be found. The point α can be at infinity on an infinite R_2 arc $\mathscr{L}(u)$, provided that $\mathscr{Q}$ coincides with $\mathscr{L}(u)$ in the neighborhood of α and the variations in (9.17) converge.

For given values of u, n, and α, Theorem 9.1 provides one solution of the differential equation. An independent, and numerically satisfactory, second solution can be constructed by applying the same theorem to (9.02) with u, $\phi(\xi)$, and the $\psi_s(\xi)$ and $\Psi_n(u,\xi)$ of odd suffix all replaced by their negatives, and making another choice of α.

Asymptotic properties of the solutions for large $|u|$ can be established in a general way by analysis similar to that of §§5 and 6 or directly from the error bounds in most applications. Roughly speaking, in a wide range of circumstances we have $\lambda \sim |u|$, $\mu \sim |u|$, $\mathscr{V}_{\mathscr{Q}}(R_n) = O(1)$, $\mathscr{V}_{\mathscr{Q}}(H) = O(1)$, and the right-hand side of (9.17) is asymptotic to $2|e^{u\xi}X(\xi)|\mathscr{V}_{\mathscr{Q}}(A_n)|u|^{-n}$.

Ex. 9.1 Let $\phi(\xi)$ and $\psi_s(\xi)$, $s = 0, 1, \ldots$, be holomorphic in a sector **S**, and

$$\phi(\xi) = O(\xi^{-(1/2)-\sigma}), \qquad \psi_s(\xi) = O(\xi^{-1-\rho-s\rho}),$$

as $\xi \to \infty$ in **S**, where σ and ρ are positive constants. If the lower limit of integration in equation (9.07) is taken to be at infinity in **S**, show that $A_s(\xi) = O(\xi^{-s\rho_1})$ as $\xi \to \infty$ in any proper subsector of **S** having the same vertex, where ρ_1 is the least of ρ, 2σ, $\frac{1}{2}+\sigma$, and 1.

Ex. 9.2 Let κ be a real or complex constant, δ a constant in the interval $(0, \frac{3}{4}\pi)$, and u a large positive parameter. By applying Theorem 9.1 show that the equation

$$d^2w/dz^2 = \{u^2z^2 - 2u\kappa - \tfrac{1}{4}z^{-2} - (\kappa+1)^2 z^{-1}\}\, w$$

has a solution with the property

$$w \sim \exp(-\tfrac{1}{2}uz^2)\, z^{\kappa-(1/2)} \sum_{s=0}^{\infty} \frac{A_s(z)}{u^s} \qquad (u \to \infty),$$

uniformly in the annular sector $|\operatorname{ph} z| \leqslant \frac{3}{4}\pi - \delta$, $|z| \geqslant \delta$, where $A_0(z) = 1$, and

$$A_{s+1} = \frac{\kappa}{z^2} A_s + \frac{1}{2z}\frac{dA_s}{dz} + \frac{(\kappa+1)^2}{2} \int \left(\frac{1}{z^3} + \frac{1}{z^2}\right) A_s\, dz.$$

Ex. 9.3 Show that in any fixed compact z domain the equation $d^2w/dz^2 = u(u^2 + u\cos z)^{1/2}\, w$ has solutions $w(u, \pm z)$ such that

$$w(u,z) \sim \exp(uz + \tfrac{1}{4}\sin z) \sum_{s=0}^{\infty} \frac{A_s(z)}{u^s}$$

as $|u| \to \infty$, uniformly with respect to z, where $A_0(z) = 1$, and

$$A_{s+1} = -\frac{1}{4}\cos z\, A_s - \frac{1}{2}\frac{dA_s}{dz} - \int \left(\frac{\sin z}{8} + \frac{3\cos^2 z}{32}\right) A_s\, dz + \frac{1}{2}\sum_{j=1}^{s} \binom{\frac{1}{2}}{j+2} \int \cos^{j+2} z\, A_{s-j}\, dz.$$

*10 Inhomogeneous Equations

10.1 Consider the differential equation

$$d^2W/d\xi^2 - \{u^2 + \psi(\xi)\}\, W = \varpi(\xi), \tag{10.01}$$

where, again, the parameter u ranges over an unbounded real interval or complex domain, and $\psi(\xi)$ and $\varpi(\xi)$ are holomorphic in a complex domain $\mathbf{\Delta}$. For simplicity, we suppose that $\psi(\xi)$ and $\varpi(\xi)$ are independent of u, but this restriction could be eased somewhat without major complications; in particular, a factor $\mu(u)$ in the inhomogeneous term is immediately removable by taking $W/\mu(u)$ as new dependent variable.†

The general solution of (10.01) has the form

$$W(u,\xi) = G(u,\xi) + A(u)W_1(u,\xi) + B(u)W_2(u,\xi),$$

where $G(u,\xi)$ is a particular solution, $A(u)$ and $B(u)$ are arbitrary functions of u, and $W_1(u,\xi)$ and $W_2(u,\xi)$ are independent solutions of the corresponding homogeneous differential equation. Asymptotic approximations for $W_1(u,\xi)$ and $W_2(u,\xi)$ have been investigated earlier in this chapter. We consider now the construction of approximations for $G(u,\xi)$.

A first attempt is to neglect terms on the left-hand side of (10.01) not containing u explicitly. This yields

$$G(u,\xi) \risingdotseq -\varpi(\xi)/u^2.$$

Then treating the neglected terms as corrections and iterating formally, we are led to consider as a possible solution a series of the form

$$G(u,\xi) = \frac{1}{u^2}\sum_{s=0}^{\infty}\frac{G_s(\xi)}{u^{2s}}. \tag{10.02}$$

Assuming the $G_s(\xi)$ to be independent of u, and substituting this expansion for W in (10.01) and equating coefficients, we find that $G_0(\xi) = -\varpi(\xi)$ (as we expect), and

$$G_{s+1}(\xi) = G_s''(\xi) - \psi(\xi)G_s(\xi) \qquad (s \geqslant 0). \tag{10.03}$$

This determines higher coefficients recursively; each is holomorphic in $\mathbf{\Delta}$. As in the case of (2.05), however, the differentiations at each step may be expected to induce divergence of (10.02) for all finite values of u.

10.2 Let $L_n(u,\xi)$ denote the partial sum

$$L_n(u,\xi) = \frac{1}{u^2}\sum_{s=0}^{n-1}\frac{G_s(\xi)}{u^{2s}}, \tag{10.04}$$

in which n is an arbitrary nonnegative integer. With the aid of (10.03) it may be verified that

$$\frac{d^2L_n}{d\xi^2} - \{u^2+\psi(\xi)\}L_n = \varpi(\xi) + \frac{G_n(\xi)}{u^{2n}}.$$

Subtraction of this result from (10.01) gives

$$\frac{d^2}{d\xi^2}(W-L_n) - \{u^2+\psi(\xi)\}(W-L_n) = -\frac{G_n(\xi)}{u^{2n}}.$$

Hence by variation of parameters we see that (10.01) has a solution

$$W(u,\xi) = L_n(u,\xi) + \varepsilon_n(u,\xi),$$

† A less trivial extension which can be accommodated straightforwardly is replacement of the right-hand side of (10.01) by a series of the form $\sum u^{-2s}\varpi_s(\xi)$.

where

$$\varepsilon_n(u,\xi) = \frac{W_2(u,\xi)}{u^{2n}\mathscr{W}(u)}\int_\xi^{\alpha_1} W_1(u,v)\,G_n(v)\,dv - \frac{W_1(u,\xi)}{u^{2n}\mathscr{W}(u)}\int_\xi^{\alpha_2} W_2(u,v)\,G_n(v)\,dv. \tag{10.05}$$

Here $\mathscr{W}(u)$ denotes the Wronskian of $W_1(u,\xi)$ and $W_2(u,\xi)$, and α_1 and α_2 are arbitrarily chosen reference points in $\mathbf{\Delta}$.

The complementary functions $W_1(u,\xi)$ and $W_2(u,\xi)$ can be taken to be respectively the solutions $W_{1,1}(u,\xi)$ and $W_{1,2}(u,\xi)$ given by Theorem 3.1. Then

$$W_1 = e^{u\xi}\{1+O(u^{-1})\}, \qquad W_2 = e^{-u\xi}\{1+O(u^{-1})\},$$

$$dW_1/d\xi = ue^{u\xi}\{1+O(u^{-1})\}, \qquad dW_2/d\xi = -ue^{-u\xi}\{1+O(u^{-1})\},$$

uniformly with respect to ξ. Hence

$$\mathscr{W}(u) = -2u\{1+O(u^{-1})\}.$$

Let the integration paths in (10.05) be $(u\xi)$-progressive, or more precisely, $\mathrm{Re}(uv)$ be nonincreasing for the first integral and nondecreasing for the second integral. Then by use of the estimates just obtained, we see that

$$|\varepsilon_n(u,\xi)| \leqslant \frac{1+O(u^{-1})}{2|u|^{2n+1}}\left\{\int_\xi^{\alpha_1}|G_n(v)\,dv| + \int_\xi^{\alpha_2}|G_n(v)\,dv|\right\} = O\left(\frac{1}{u^{2n+1}}\right), \tag{10.06}$$

provided that these integrals are finite. The result then holds in the intersection of the regions $\Xi_1(u,\alpha_1)$ and $\Xi_2(u,\alpha_2)$, defined in §3.1.†

Ex. 10.1 For positive real values of the parameter u, show that the equation

$$\frac{d^2w}{dz^2} = \left\{u^2 + \frac{1}{u(z^2+1)}\right\}w - u^2$$

has a solution of the form

$$1 - \frac{1}{z^2+1}\frac{1}{u^3} + \frac{2(1-3z^2)}{(z^2+1)^3u^5} + \frac{1}{(z^2+1)^2u^6} + O\left(\frac{1}{u^7}\right)$$

as $u\to\infty$, uniformly in the z region of validity given in Exercise 3.5.

*11 Example: An Inhomogeneous Form of the Modified Bessel Equation

11.1 Consider the equation

$$z^2\frac{d^2w}{dz^2} + z\frac{dw}{dz} - (z^2+\nu^2)w = z^m, \tag{11.01}$$

† A more elaborate way of estimating $|\varepsilon_n(u,\xi)|$ is to retain the factors $|e^{uv}|$ and $|e^{-uv}|$ under the signs of integration and apply Laplace's method (Chapter 4, §6). Advantages of this modification include obviating convergence of the integrals in (10.06) and strengthening the final estimate to $O(u^{-2n-2})$.

in which m is a fixed nonnegative integer. As in the case of the homogeneous equation (§7) we begin by replacing z with νz. Then

$$\frac{d^2w}{dz^2} + \frac{1}{z}\frac{dw}{dz} - \nu^2\left(\frac{1+z^2}{z^2}\right)w = \nu^m z^{m-2}. \tag{11.02}$$

Application of the theory of §10 calls for new variables

$$\xi = \int z^{-1}(1+z^2)^{1/2}\,dz, \qquad W = (1+z^2)^{1/4}w;$$

compare (7.04). The desired expansion can be obtained more easily, however, by applying the idea expressed in the third paragraph of §10.1 directly to equation (11.02). Thus we consider the formal expansion

$$w = \nu^{m-2}\sum_{s=0}^{\infty}\frac{G_s(z)}{\nu^{2s}},$$

in which the coefficients $G_s(z)$ are independent of ν.

Substitution of the last expansion in (11.02) yields

$$G_0(z) = -z^m/(1+z^2), \tag{11.03}$$

and

$$G_{s+1}(z) = \frac{z^2}{1+z^2}G_s''(z) + \frac{z}{1+z^2}G_s'(z) \qquad (s = 0, 1, \ldots). \tag{11.04}$$

In particular,

$$G_1(z) = -\frac{m^2z^m + 2(m^2-2m-2)z^{m+2} + (m-2)^2z^{m+4}}{(1+z^2)^4}.$$

Inspection of (11.04) shows that $G_s(z)$ is a rational function of z with poles at $z = \pm i$. Also,

$$G_s(z) = O(z^{m-2s-2}) \qquad (z \to \infty), \tag{11.05}$$

and $G_s(0) = 0$ except when $s = m = 0$.

Following §10.2, we define

$$L_n(\nu, z) = \nu^{m-2}\sum_{s=0}^{n-1}\frac{G_s(z)}{\nu^{2s}},$$

where n is an arbitrary nonnegative integer. Then

$$\frac{d^2}{dz^2}(w-L_n) + \frac{1}{z}\frac{d}{dz}(w-L_n) - \nu^2\left(\frac{1+z^2}{z^2}\right)(w-L_n) = -\frac{1+z^2}{z^2}\frac{G_n(z)}{\nu^{2n-m}}. \tag{11.06}$$

Using the Wronskian relation (Chapter 7, §8.2)

$$K_\nu(\nu z)I_\nu'(\nu z) - I_\nu(\nu z)K_\nu'(\nu z) = 1/(\nu z)$$

and solving (11.06) by variation of parameters, we obtain a particular solution $L_n(\nu, z) + \varepsilon_n(\nu, z)$ of (11.02), where

$$\varepsilon_n(\nu, z) = \frac{K_\nu(\nu z)}{\nu^{2n-m}} \int_0^z \frac{1+t^2}{t} G_n(t) I_\nu(\nu t)\, dt + \frac{I_\nu(\nu z)}{\nu^{2n-m}} \int_z^\infty \frac{1+t^2}{t} G_n(t) K_\nu(\nu t)\, dt. \tag{11.07}$$

11.2 Let us establish the asymptotic nature of the approximate solution $L_n(\nu, z)$ in the case when ν is positive and $|\mathrm{ph}\, z| \leqslant \frac{1}{2}\pi - \delta \ (< \frac{1}{2}\pi)$. In these circumstances we have from §7.5,

$$I_\nu(\nu z) = \frac{e^{\nu\xi(z)}}{(2\pi\nu)^{1/2}(1+z^2)^{1/4}}\{1 + O(\nu^{-1})\},$$

$$K_\nu(\nu z) = \left(\frac{\pi}{2\nu}\right)^{1/2} \frac{e^{-\nu\xi(z)}}{(1+z^2)^{1/4}}\{1 + O(\nu^{-1})\},$$

as $\nu \to \infty$, uniformly with respect to z, where $\xi(z)$ is defined by (7.07). As in §7.5, the integration paths in (11.07) can be chosen so that $|e^{\nu\xi(t)}|$ is nondecreasing, and we derive

$$|\varepsilon_n(\nu, z)| \leqslant \frac{1 + O(\nu^{-1})}{2\nu^{2n-m+1}|1+z^2|^{1/4}} \left(\int_0^z + \int_z^\infty\right) \left| \frac{(1+t^2)^{3/4}}{t} G_n(t)\, dt \right|.$$

Provided that $n > 0$ and $n \geqslant \frac{1}{2}m$ these integrals converge at $t = 0$ and $t = \infty$, and as in §7.5 the paths can be chosen so that the integrals are bounded with respect to z. Therefore $\varepsilon_n(\nu, z)$ is uniformly $O(\nu^{-2n+m-1})$.

Thus we have shown that for positive ν, and positive n not less than $\frac{1}{2}m$, equation (11.02) has a holomorphic solution

$$G_n(\nu, z) = \nu^{m-2} \sum_{s=0}^{n-1} \frac{G_s(z)}{\nu^{2s}} + O(\nu^{-2n+m-1})$$

as $\nu \to \infty$, uniformly in the sector $|\mathrm{ph}\, z| \leqslant \frac{1}{2}\pi - \delta$, where the coefficients $G_s(z)$ are defined by (11.03) and (11.04). It should be noticed that $G_n(\nu, z)$ is a numerically satisfactory companion to the complementary functions $I_\nu(\nu z)$ and $K_\nu(\nu z)$ both for small z and large z.

Clearly the analysis is extendible to complex values of ν, provided that z is suitably restricted. Furthermore, error bounds are derivable by the same method.†

Ex. 11.1 With the conditions of §11.2 prove that $G_n(\nu, z)$ is independent of n.

Ex. 11.2 Using the result of Chapter 6, §6.2, show that if x is real and u is positive, then the equation

$$d^2 w/dx^2 = u^2(x^2+1)w - \cos x$$

has a solution with the property

$$w(u, x) \sim \frac{1}{u^2} \sum_{s=0}^{\infty} \frac{1}{u^{2s}} \left(\frac{1}{x^2+1} \frac{d^2}{dx^2}\right)^s \frac{\cos x}{x^2+1}$$

as $u \to \infty$, uniformly with respect to $x \in (-\infty, \infty)$.

† Again, these bounds can be improved by applying Laplace's method to (11.07).

Historical Notes and Additional References

This chapter is based on papers by Olver (1954a, b; 1961). Much of the original material has been considerably expanded and improved.

The earliest work on differential equations establishing asymptotic expansions of solutions in descending powers of a parameter appears to be that of Horn (1899). He considered the case of a second-order equation with bounded real values of the independent variable z, and large complex values of the parameter u. Since then the asymptotic theory of linear differential equations in regions free from transition points has been extended systematically to equations of higher order; contributors include Schlesinger (1907), Birkhoff (1908), Turrittin (1936), Hukuhara (1937), Kiyek (1963), and Wasow (1965). Birkhoff and Turrittin consider the case of a single equation of arbitrary order n; the other writers study a set of n simultaneous first-order equations (much the commoner case in physical applications). Like Horn, earlier writers restrict z to be real and bounded. Some of the most general results are those of Kiyek. He constructs asymptotic expansions for large $|u|$ of solutions of matrix differential equations of the form

$$d\mathbf{w}/dz = u^l \mathbf{f}(u, z)\mathbf{w},$$

in which u and z are unbounded complex variables and l is a positive integer. The principal assumption is that the coefficient matrix $\mathbf{f}(u, z)$ has a uniform asymptotic expansion of the form

$$\mathbf{f}(u, z) \sim \sum_{s=0}^{\infty} \frac{\mathbf{f}_s(z)}{u^s},$$

in which the matrices $\mathbf{f}_s(z)$ are holomorphic in z and the latent roots of $\mathbf{f}_0(z)$ are distinct for all z.

§3 For similar results for a pair of simultaneous first-order differential equations see Stenger (1970b).

§9 Asymptotic solutions of equation (9.01) for general values of the positive integer l have been given by Erdélyi (1956a, §§4.2 and 4.3) and Thorne (1960). The methods of these writers are less adaptable to the determination of error bounds than the natural extension of the analysis we have given for the case $l = 1$.

11

DIFFERENTIAL EQUATIONS WITH A PARAMETER: TURNING POINTS

1 Airy Functions of Real Argument

1.1 In the approximate solution of second-order linear differential equations having a simple turning point, the solutions of the simplest equation of this kind, given by

$$d^2w/dx^2 = xw, \tag{1.01}$$

play the role of the exponential functions in Chapters 6 and 10.

One solution of (1.01) is Airy's integral $\mathrm{Ai}(x)$ introduced in Chapter 2, §8. The following properties of this function were established in Chapters 2, 4, and 7, or are easily derivable from results given in these chapters.

$$\mathrm{Ai}(x) = \frac{1}{\pi}\int_0^\infty \cos(\tfrac{1}{3}t^3 + xt)\,dt, \tag{1.02}$$

$$\mathrm{Ai}(0) = 1/\{3^{2/3}\Gamma(\tfrac{2}{3})\}, \qquad \mathrm{Ai}'(0) = -1/\{3^{1/3}\Gamma(\tfrac{1}{3})\}, \tag{1.03}$$

$$\mathrm{Ai}(x) = \pi^{-1}(\tfrac{1}{3}x)^{1/2}K_{1/3}(\xi), \qquad \mathrm{Ai}'(x) = -\pi^{-1}3^{-1/2}xK_{2/3}(\xi), \tag{1.04}$$

$$\mathrm{Ai}(-x) = \tfrac{1}{3}x^{1/2}\{J_{1/3}(\xi)+J_{-1/3}(\xi)\} = (\tfrac{1}{3}x)^{1/2}\,\mathrm{Re}\{e^{\pi i/6}H_{1/3}^{(1)}(\xi)\}, \tag{1.05}$$

$$\mathrm{Ai}'(-x) = \tfrac{1}{3}x\{J_{2/3}(\xi)-J_{-2/3}(\xi)\} = 3^{-1/2}x\,\mathrm{Re}\{e^{-\pi i/6}H_{2/3}^{(1)}(\xi)\}. \tag{1.06}$$

In (1.04), (1.05), and (1.06), x is positive and $\xi = \tfrac{2}{3}x^{3/2}$.

For large positive x, with the same definition of ξ,

$$\mathrm{Ai}(x) \sim \frac{e^{-\xi}}{2\pi^{1/2}x^{1/4}}\sum_{s=0}^{\infty}(-)^s\frac{u_s}{\xi^s}, \qquad \mathrm{Ai}'(x) \sim -\frac{x^{1/4}e^{-\xi}}{2\pi^{1/2}}\sum_{s=0}^{\infty}(-)^s\frac{v_s}{\xi^s}, \tag{1.07}$$

$$\mathrm{Ai}(-x) \sim \frac{1}{\pi^{1/2}x^{1/4}}\left\{\cos(\xi-\tfrac{1}{4}\pi)\sum_{s=0}^{\infty}(-)^s\frac{u_{2s}}{\xi^{2s}} + \sin(\xi-\tfrac{1}{4}\pi)\sum_{s=0}^{\infty}(-)^s\frac{u_{2s+1}}{\xi^{2s+1}}\right\}, \tag{1.08}$$

$$\mathrm{Ai}'(-x) \sim \frac{x^{1/4}}{\pi^{1/2}}\left\{\sin(\xi-\tfrac{1}{4}\pi)\sum_{s=0}^{\infty}(-)^s\frac{v_{2s}}{\xi^{2s}} - \cos(\xi-\tfrac{1}{4}\pi)\sum_{s=0}^{\infty}(-)^s\frac{v_{2s+1}}{\xi^{2s+1}}\right\}, \tag{1.09}$$

where $u_0 = v_0 = 1$ and

$$u_s = \frac{(2s+1)(2s+3)(2s+5)\cdots(6s-1)}{(216)^s s!}, \qquad v_s = -\frac{6s+1}{6s-1}u_s \qquad (s \geqslant 1).$$

1.2 The characteristic feature of $\mathrm{Ai}(x)$ in relation to the differential equation (1.01) is recession as $x \to +\infty$. The criteria for defining the standard solution of the second kind, $\mathrm{Bi}(x)$, are based on asymptotic behavior for large negative x: $\mathrm{Bi}(x)$ has the same amplitude of oscillation as $\mathrm{Ai}(x)$ and differs in phase by $\frac{1}{2}\pi$. Corresponding properties, with the same definitions of ξ, u_s, and v_s are as follows:

$$\mathrm{Bi}(x) = \frac{1}{\pi}\int_0^\infty \{\exp(-\tfrac{1}{3}t^3+xt) + \sin(\tfrac{1}{3}t^3+xt)\}\,dt, \tag{1.10}$$

$$\mathrm{Bi}(0) = 1/\{3^{1/6}\Gamma(\tfrac{2}{3})\}, \qquad \mathrm{Bi}'(0) = 3^{1/6}/\Gamma(\tfrac{1}{3}), \tag{1.11}$$

$$\mathrm{Bi}(x) = (\tfrac{1}{3}x)^{1/2}\{I_{1/3}(\xi)+I_{-1/3}(\xi)\} = (\tfrac{1}{3}x)^{1/2}\,\mathrm{Re}\{e^{\pi i/6}H_{1/3}^{(1)}(-i\xi)\}, \tag{1.12}$$

$$\mathrm{Bi}'(x) = 3^{-1/2}x\{I_{2/3}(\xi)+I_{-2/3}(\xi)\} = 3^{-1/2}x\,\mathrm{Re}\{e^{\pi i/3}H_{2/3}^{(1)}(-i\xi)\}, \tag{1.13}$$

$$\mathrm{Bi}(-x) = (\tfrac{1}{3}x)^{1/2}\{J_{-1/3}(\xi)-J_{1/3}(\xi)\} = -(\tfrac{1}{3}x)^{1/2}\,\mathrm{Im}\{e^{\pi i/6}H_{1/3}^{(1)}(\xi)\}, \tag{1.14}$$

$$\mathrm{Bi}'(-x) = 3^{-1/2}x\{J_{-2/3}(\xi)+J_{2/3}(\xi)\} = -3^{-1/2}x\,\mathrm{Im}\{e^{-\pi i/6}H_{2/3}^{(1)}(\xi)\}, \tag{1.15}$$

$$\mathrm{Bi}(x) \sim \frac{e^\xi}{\pi^{1/2}x^{1/4}}\sum_{s=0}^\infty \frac{u_s}{\xi^s}, \qquad \mathrm{Bi}'(x) \sim \frac{x^{1/4}e^\xi}{\pi^{1/2}}\sum_{s=0}^\infty \frac{v_s}{\xi^s}, \tag{1.16}$$

$$\mathrm{Bi}(-x) \sim \frac{1}{\pi^{1/2}x^{1/4}}\left\{-\sin(\xi-\tfrac{1}{4}\pi)\sum_{s=0}^\infty (-)^s\frac{u_{2s}}{\xi^{2s}} + \cos(\xi-\tfrac{1}{4}\pi)\sum_{s=0}^\infty (-)^s\frac{u_{2s+1}}{\xi^{2s+1}}\right\}, \tag{1.17}$$

$$\mathrm{Bi}'(-x) \sim \frac{x^{1/4}}{\pi^{1/2}}\left\{\cos(\xi-\tfrac{1}{4}\pi)\sum_{s=0}^\infty (-)^s\frac{v_{2s}}{\xi^{2s}} + \sin(\xi-\tfrac{1}{4}\pi)\sum_{s=0}^\infty (-)^s\frac{v_{2s+1}}{\xi^{2s+1}}\right\}. \tag{1.18}$$

Also,

$$\mathscr{W}\{\mathrm{Ai}(x), \mathrm{Bi}(x)\} = 1/\pi. \tag{1.19}$$

Verification of (1.10) to (1.19) is left as an exercise for the reader.

1.3 Graphs of $\mathrm{Ai}(x)$ and $\mathrm{Bi}(x)$ are given in Fig. 1.1. When x is positive, $\mathrm{Ai}(x)$, $\mathrm{Bi}(x)$, and $\mathrm{Bi}'(x)$ are positive and $\mathrm{Ai}'(x)$ is negative; moreover, all four functions are monotonic. In the case of $\mathrm{Ai}(x)$ and its derivative this result follows from

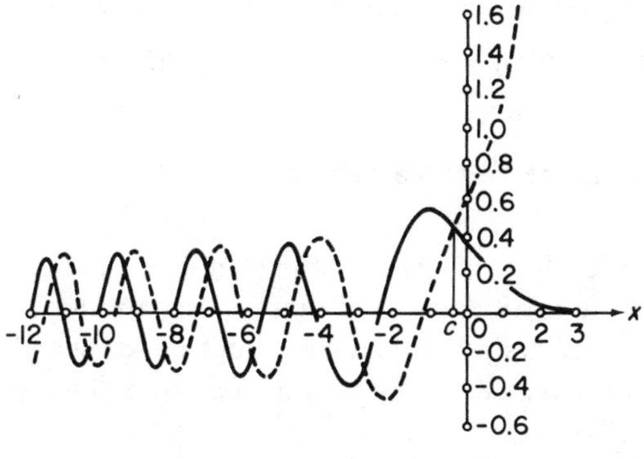

Fig. 1.1 $\mathrm{Ai}(x)$ ——— and $\mathrm{Bi}(x)$ - - - - -.

(1.01), (1.04), and the fact that $K_{1/3}(\xi)$ and $K_{2/3}(\xi)$ are both positive (Chapter 7, §8.3). For $\mathrm{Bi}(x)$ and its derivative we refer to the Maclaurin expansion of $\mathrm{Bi}(x)$; by repeated differentiations of (1.01) and use of (1.11) it is seen that all terms in the expansion are positive.

***1.4** Error bounds for the asymptotic expansions given in §§1.1 and 1.2 are available from results proved in earlier chapters.

For (1.07) we see from Chapter 4, §4.1 and Chapter 7, Exercise 13.3, that the nth error term is less than the first neglected term in absolute value, and has the same sign, provided that $n \geqslant 0$ for $\mathrm{Ai}(x)$, and $n \geqslant 1$ for $\mathrm{Ai}'(x)$. In particular,

$$\mathrm{Ai}(x) \leqslant \frac{e^{-\xi}}{2\pi^{1/2}x^{1/4}}, \qquad |\mathrm{Ai}'(x)| \leqslant \frac{x^{1/4}e^{-\xi}}{2\pi^{1/2}}\left(1+\frac{7}{72\xi}\right) \qquad (x>0).$$

For (1.16) we use (1.12) and (1.13), and refer to Chapter 7, §13.1. The nth error term is bounded in magnitude by the first neglected term, multiplied by $2\chi(n)\exp\{5\pi/(72\xi)\}$ in the case of $\mathrm{Bi}(x)$, or $2\chi(n)\exp\{7\pi/(72\xi)\}$ in the case of $\mathrm{Bi}'(x)$. In both cases $n \geqslant 1$ and $\chi(n)$ denotes $\pi^{1/2}\Gamma(\frac{1}{2}n+1)/\Gamma(\frac{1}{2}n+\frac{1}{2})$.

Next, for (1.08), (1.09), (1.17), and (1.18) we have

$$\begin{aligned}
\mathrm{Ai}(-x) &= \pi^{-1/2}x^{-1/4}\{\cos(\xi-\tfrac{1}{4}\pi)\,P(\tfrac{1}{3},\xi) - \sin(\xi-\tfrac{1}{4}\pi)\,Q(\tfrac{1}{3},\xi)\},\\
\mathrm{Ai}'(-x) &= \pi^{-1/2}x^{1/4}\{\sin(\xi-\tfrac{1}{4}\pi)\,P(\tfrac{2}{3},\xi) + \cos(\xi-\tfrac{1}{4}\pi)\,Q(\tfrac{2}{3},\xi)\},\\
\mathrm{Bi}(-x) &= -\pi^{-1/2}x^{-1/4}\{\sin(\xi-\tfrac{1}{4}\pi)\,P(\tfrac{1}{3},\xi) + \cos(\xi-\tfrac{1}{4}\pi)\,Q(\tfrac{1}{3},\xi)\},\\
\mathrm{Bi}'(-x) &= \pi^{-1/2}x^{1/4}\{\cos(\xi-\tfrac{1}{4}\pi)\,P(\tfrac{2}{3},\xi) - \sin(\xi-\tfrac{1}{4}\pi)\,Q(\tfrac{2}{3},\xi)\},
\end{aligned}$$

where $P(\nu,\xi)$ and $Q(\nu,\xi)$ are defined in Chapter 7, Exercise 13.1. Thus if the expansions of $P(\frac{1}{3},\xi)$, $Q(\frac{1}{3},\xi)$, $P(\frac{2}{3},\xi)$, and $Q(\frac{2}{3},\xi)$ are truncated at their nth terms, then the error terms are bounded in absolute value by the first neglected terms, provided that $n \geqslant 0$, 0, 1, and 0, respectively. In brief, the error is less than the next term provided that the following term is of opposite sign.

Ex. 1.1 Show that $u_{2s}v_0 - u_{2s-1}v_1 + u_{2s-2}v_2 - \cdots + u_0 v_{2s} = 0$ when $s>0$.

Ex. 1.2 Verify that

$$\int \mathrm{Ai}^2(x)\,dx = x\,\mathrm{Ai}^2(x) - \mathrm{Ai}'^2(x), \qquad \int \mathrm{Ai}(x)\,\mathrm{Bi}(x)\,dx = x\,\mathrm{Ai}(x)\,\mathrm{Bi}(x) - \mathrm{Ai}'(x)\,\mathrm{Bi}'(x),$$

and also that in the interval $(0,\infty)$, $\sup\{x^{1/2}\mathrm{Ai}(x)/|\mathrm{Ai}'(x)|\} = 1$.

2 Auxiliary Functions for Real Variables

2.1 The properties given in §1 show that the character of the Airy functions for positive arguments is quite different from that for negative arguments. In order to have a convenient way of assessing the magnitudes of $\mathrm{Ai}(x)$ and $\mathrm{Bi}(x)$ for all real values of x we introduce a *modulus function* $M(x)$, *phase function* $\theta(x)$, and *weight function* $E(x)$, related by

$$E(x)\,\mathrm{Ai}(x) = M(x)\sin\theta(x), \qquad E^{-1}(x)\,\mathrm{Bi}(x) = M(x)\cos\theta(x), \qquad (2.01)$$

where $E^{-1}(x) = 1/E(x)$. Once $E(x)$ has been prescribed the other two functions are determined by

$$M(x) = \{E^2(x)\,\mathrm{Ai}^2(x) + E^{-2}(x)\,\mathrm{Bi}^2(x)\}^{1/2},$$
$$\theta(x) = \tan^{-1}\{E^2(x)\,\mathrm{Ai}(x)/\mathrm{Bi}(x)\}. \qquad (2.02)$$

For negative x, $\mathrm{Ai}(x)$ and $\mathrm{Bi}(x)$ are of the same order of magnitude, and the natural choice of $E(x)$ is unity. For positive x, however, $\mathrm{Bi}(x)$ is exponentially large compared with $\mathrm{Ai}(x)$; unless $E^2(x)$ is correspondingly large the contributions to $M(x)$ from $E(x)\,\mathrm{Ai}(x)$ and $E^{-1}(x)\,\mathrm{Bi}(x)$ would be quite disparate.

2.2 Our actual choice of $E(x)$ is as follows. Let $x = c$ be the negative root of the equation

$$\mathrm{Ai}(x) = \mathrm{Bi}(x)$$

of smallest absolute value. Then

$$E(x) = \{\mathrm{Bi}(x)/\mathrm{Ai}(x)\}^{1/2} \quad (c \leqslant x < \infty), \qquad E(x) = 1 \quad (-\infty < x \leqslant c). \qquad (2.03)$$

Using the Wronskian (1.19), we derive

$$\frac{d}{dx}\frac{\mathrm{Bi}(x)}{\mathrm{Ai}(x)} = \frac{1}{\pi\,\mathrm{Ai}^2(x)}.$$

Accordingly, as x decreases from ∞ to 0, $\mathrm{Bi}(x)/\mathrm{Ai}(x)$ decreases strictly from ∞ to $3^{1/2}$; compare (1.03) and (1.11). The decrease continues as x enters negative values; hence a zero of $\mathrm{Bi}(x)$ is reached ahead of the first zero of $\mathrm{Ai}(x)$.† But before this occurs the point $x = c$ is passed and $E(x)$ is no longer defined as $\{\mathrm{Bi}(x)/\mathrm{Ai}(x)\}^{1/2}$. Accordingly, *$E(x)$ is a monotonic function of x which is never less than unity.*

The location of c in relation to the zeros of $\mathrm{Ai}(x)$ and $\mathrm{Bi}(x)$ is indicated in Fig. 1.1. Numerical calculation yields

$$c = -0.36605,$$

correct to five decimal places.

With the definition (2.03), equations (2.02) yield

$$M(x) = \{2\,\mathrm{Ai}(x)\,\mathrm{Bi}(x)\}^{1/2}, \qquad \theta(x) = \tfrac{1}{4}\pi \qquad (x \geqslant c), \qquad (2.04)$$

$$M(x) = \{\mathrm{Ai}^2(x) + \mathrm{Bi}^2(x)\}^{1/2}, \qquad \theta(x) = \tan^{-1}\{\mathrm{Ai}(x)/\mathrm{Bi}(x)\} \qquad (x \leqslant c), \qquad (2.05)$$

the branch of the inverse tangent being continuous and equal to $\frac{1}{4}\pi$ at $x = c$.

From the results of §1 we obtain the asymptotic forms

$$E(x) \sim 2^{1/2}\exp(\tfrac{2}{3}x^{3/2}), \qquad M(x) \sim \pi^{-1/2}x^{-1/4} \qquad (x \to +\infty), \qquad (2.06)$$

and‡

$$M(x) \sim \pi^{-1/2}(-x)^{-1/4} \qquad (x \to -\infty). \qquad (2.07)$$

† That $\mathrm{Bi}(x)$ and $\mathrm{Ai}(x)$ have no common zeros is seen from their Wronskian relation.

‡ The formula for $\theta(x)$ corresponding to (2.07) is stated in Exercise 2.2 below.

Modulus and phase functions are also needed for the derivatives of the Airy functions. They are defined by

$$E(x)\,\mathrm{Ai}'(x) = N(x)\sin\omega(x), \qquad E^{-1}(x)\,\mathrm{Bi}'(x) = N(x)\cos\omega(x), \tag{2.08}$$

with the same definition (2.03) for $E(x)$. Thus

$$N(x) = \left\{\frac{\mathrm{Ai}'^2(x)\,\mathrm{Bi}^2(x)+\mathrm{Bi}'^2(x)\,\mathrm{Ai}^2(x)}{\mathrm{Ai}(x)\,\mathrm{Bi}(x)}\right\}^{1/2}, \qquad \omega(x) = \tan^{-1}\left\{\frac{\mathrm{Ai}'(x)\,\mathrm{Bi}(x)}{\mathrm{Bi}'(x)\,\mathrm{Ai}(x)}\right\}$$

$$(x \geqslant c), \tag{2.09}$$

$$N(x) = \{\mathrm{Ai}'^2(x)+\mathrm{Bi}'^2(x)\}^{1/2}, \qquad \omega(x) = \tan^{-1}\{\mathrm{Ai}'(x)/\mathrm{Bi}'(x)\}$$

$$(x \leqslant c). \tag{2.10}$$

Again, the branches of the inverse tangents are chosen in a continuous manner, and fixed by the condition

$$\omega(x) \to -\tfrac{1}{4}\pi \qquad (x \to +\infty). \tag{2.11}$$

For large $|x|$,

$$N(x) \sim \pi^{-1/2}|x|^{1/4} \qquad (x \to \pm\infty). \tag{2.12}$$

2.3 Representative numerical values of the auxiliary functions are given in Table 2.1, correct to two decimal places. Notice should be taken of the slowly varying nature of the modulus functions for all x, and the phase functions for $x \geqslant c$.

Table 2.1

x	$E(x)$	$M(x)$	$\theta(x)$	$N(x)$	$\omega(x)$
$-\infty$	1	0.00^a	∞	∞^a	∞
-10	1	0.32	21.86	1.00	20.30
-5	1	0.38	8.23	0.84	6.68
-4	1	0.40	6.11	0.80	4.57
-3	1	0.43	4.23	0.75	2.71
-2	1	0.47	2.64	0.68	1.15
-1	1	0.55	1.38	0.59	-0.02
-0.8	1	0.57	1.17	0.57	-0.19
-0.6	1	0.59	0.98	0.55	-0.33
-0.4	1	0.63	0.81	0.54	-0.43
$-0.36\ldots$ c	1	0.63	0.79	0.53	-0.45
-0.2	1.14	0.65	0.79	0.49	-0.61
0.0	1.32	0.66	0.79	0.48	-0.79
0.2	1.52	0.65	0.79	0.49	-0.90
0.4	1.77	0.64	0.79	0.51	-0.97
0.6	2.08	0.62	0.79	0.53	-1.00
0.8	2.48	0.60	0.79	0.55	-1.00

Table 2.1 *(continued)*

x	$E(x)$	$M(x)$	$\theta(x)$	$N(x)$	$\omega(x)$
1	2.99	0.57	0.79	0.57	−0.99
2	9.72	0.48	0.79	0.67	−0.89
3	4.61×10	0.43	0.79	0.74	−0.84
4	2.97×10^2	0.40	0.79	0.80	−0.82
5	2.46×10^3	0.38	0.79	0.84	−0.81
10	2.03×10^9	0.32	0.79	1.00	−0.79
∞	∞	0.00[a]	0.79	∞[a]	−0.79

[a] For $|x| > 10$, the asymptotic forms (2.06), (2.07), and (2.12) for $M(x)$ and $N(x)$ are numerically correct to at least 0.1%.

2.4 Various constants associated with the auxiliary functions occur in later sections, including

$$\lambda = \sup_{(-\infty,\infty)} \{\pi |x|^{1/2} M^2(x)\}, \tag{2.13}$$

$$\mu_1 = \sup_{(-\infty,\infty)} \{\pi E(x)\, M(x)\, |x^{1/2}\,\mathrm{Ai}(x)|\},$$
$$\mu_2 = \sup_{(-\infty,\infty)} \{\pi E^{-1}(x)\, M(x)\, |x^{1/2}\,\mathrm{Bi}(x)|\}. \tag{2.14}$$

Examination of the asymptotic forms of $\mathrm{Ai}(x)$, $\mathrm{Bi}(x)$, $E(x)$, and $M(x)$ for large $|x|$ shows that each supremum is finite. Numerical calculations† yield the values

$$\lambda = 1.04\ldots, \qquad \mu_1 = 1, \qquad \mu_2 = 1, \tag{2.15}$$

the suprema being approached at $x = 1.33\ldots$, $-\infty$, and $-\infty$, respectively.

Ex. 2.1 Show that

$$|x|^{1/2} M^2(x) = \tfrac{1}{2}\xi\{J_{1/3}^2(\xi) + Y_{1/3}^2(\xi)\} \qquad (x \leqslant c),$$

where $\xi = \tfrac{2}{3}|x|^{3/2}$. Thence with the aid of Chapter 9, §7.3, show that if the x range in (2.13) and (2.14) is curtailed from $(-\infty, \infty)$ to $(-\infty, c]$, then the corresponding suprema are all 1.

Ex. 2.2 Show that when $x \leqslant c$,

$$\theta(x) = \tfrac{2}{3}\pi + \tan^{-1}\{Y_{1/3}(\xi)/J_{1/3}(\xi)\}, \qquad \omega(x) = \tfrac{1}{3}\pi + \tan^{-1}\{Y_{2/3}(\xi)/J_{2/3}(\xi)\},$$

where $\xi = \tfrac{2}{3}|x|^{3/2}$. Hence with the aid of Chapter 7, §§6.5 and 7.2 verify that for large negative x

$$\theta(x) = \xi + \tfrac{1}{4}\pi + O(\xi^{-1}), \qquad \omega(x) = \xi - \tfrac{1}{4}\pi + O(\xi^{-1}).$$

3 The First Approximation

3.1 By analogy with the theory of the LG approximation given in Chapter 6, the first form of turning-point problem we consider is given by the differential equation

$$d^2w/dx^2 = \{u^2 f(x) + g(x)\}\, w \tag{3.01}$$

† Eased by Exercise 2.1 below.

in which u is a positive parameter, x ranges over a real interval (a, b), finite or infinite, and $f(x)$ is real and has just one zero x_0, say, in (a, b), with $f'(x_0) \neq 0$. Generally the parameter u is to be thought of as being large, but this is inessential in the error analysis, nor need $f(x)$ and $g(x)$ be independent of u. Roughly speaking, the resulting approximations are meaningful when $|g(x)|$ is small compared with $|u^2 f(x)/(x-x_0)|$ in the neighborhood of x_0, and $|g(x)|$ is small compared with $|u^2 f(x)|$ elsewhere.

It should be noticed that decomposition of a given differential equation into the form (3.01) is not unique. For example, $f(x)$ and $g(x)$ may be replaced by $f(x)+ku^{-2}\,\mathrm{Re}\{g(x)\}$ and $(1-k)\,\mathrm{Re}\{g(x)\}+i\,\mathrm{Im}\{g(x)\}$, respectively, where k is any constant. This perturbs x_0; in consequence, *a turning point is not an invariant of a differential equation in the manner of a singularity.*

Following Chapter 10, §1.2 we take a new independent variable, which for convenience we shall denote by ζ instead of ξ, and a new dependent variable W, given by

$$\zeta\left(\frac{d\zeta}{dx}\right)^2 = f(x), \qquad w = \left(\frac{d\zeta}{dx}\right)^{-1/2} W. \tag{3.02}$$

Without loss of generality $f(x)$ is assumed to have the same sign as $x-x_0$. Adoption of the conditions $\zeta(x_0)=0$, $\zeta'(x_0)>0$, and integration of the first of (3.02) yields

$$\tfrac{2}{3}\zeta^{3/2} = \int_{x_0}^{x} f^{1/2}(t)\,dt \quad (x \geqslant x_0), \qquad \tfrac{2}{3}(-\zeta)^{3/2} = \int_{x}^{x_0} \{-f(t)\}^{1/2}\,dt \quad (x \leqslant x_0). \tag{3.03}$$

These relations determine a one-to-one correspondence between x and ζ. Setting

$$\hat{f} \equiv \hat{f}(x) = \left(\frac{d\zeta}{dx}\right)^2 = \frac{f(x)}{\zeta},$$

we find that equation (3.01) transforms into

$$d^2W/d\zeta^2 = \{u^2\zeta + \psi(\zeta)\}\,W, \tag{3.04}$$

in which

$$\psi(\zeta) = \frac{1}{\hat{f}^{1/4}}\frac{d^2(\hat{f}^{1/4})}{d\zeta^2} + \frac{g}{\hat{f}} = -\frac{1}{\hat{f}^{3/4}}\frac{d^2}{dx^2}\left(\frac{1}{\hat{f}^{1/4}}\right) + \frac{g}{\hat{f}}, \tag{3.05}$$

or, equivalently,

$$\psi(\zeta) = \frac{5}{16\zeta^2} + \{4f(x)f''(x) - 5f'^2(x)\}\frac{\zeta}{16f^3(x)} + \frac{\zeta g(x)}{f(x)}. \tag{3.06}$$

Equation (3.04) is the standard form of differential equation for Case II (Chapter 10, §1.2).

The essence of the approximating procedure is to disregard the term $\psi(\zeta)$ in (3.04), enabling this equation to be solved in terms of Airy functions. As a preliminary step towards assessment of the errors in representing the solutions of (3.04) by $\mathrm{Ai}(u^{2/3}\zeta)$ and $\mathrm{Bi}(u^{2/3}\zeta)$, we introduce restrictions on $f(x)$ and $g(x)$ to guarantee continuity of $\psi(\zeta)$.

3.2 Lemma 3.1 *In a given interval* (a, b), *let* $f(x)/(x-x_0)$ *be a positive, twice continuously differentiable function and* $g(x)$ *a continuous real or complex function. Then* $\zeta(x)/(x-x_0)$ *is positive and twice continuously differentiable in* (a, b), *and* $\psi(\zeta)$ *is continuous in the corresponding* ζ *interval.*

To establish this result, consider first $x > x_0$ and set

$$p(x) = \left\{\frac{f(x)}{x-x_0}\right\}^{1/2}, \qquad q(x) = \frac{1}{(x-x_0)^{3/2}} \int_{x_0}^{x} (t-x_0)^{1/2} p(t)\, dt.$$

Then $p(x)$ is positive and twice continuously differentiable. By use of the mean-value theorem for integrals, we see that $q(x)$ is right-continuous at x_0, and takes the (nonzero) value $\frac{2}{3}p(x_0)$ there.

Next, integration by parts yields

$$q'(x) = \frac{1}{(x-x_0)^{5/2}} \int_{x_0}^{x} (t-x_0)^{3/2} p'(t)\, dt \qquad (x > x_0).$$

Therefore $q'(x)$ is continuous in (x_0, b). A further application of the mean-value theorem shows that $q'(x) \to \frac{2}{5}p'(x_0)$ as $x \to x_0$. The existence and continuity of the right derivative $q'(x)$ at $x = x_0$ now follow from the mean-value theorem for derivatives. In a similar way we may show that $q''(x)$ shares these properties. Since

$$\zeta(x)/(x-x_0) = \{\tfrac{3}{2}q(x)\}^{2/3} \qquad (x > x_0),$$

it follows that $\zeta(x)/(x-x_0)$ is nonvanishing and twice continuously differentiable on the right at x_0. In an exactly similar manner we can show that the same properties hold on the left at x_0, and moreover that the corresponding left and right derivatives are equal.

Finally, continuity of $\psi(\zeta)$ follows from (3.05) and the relation

$$\hat{f}(x) = \{p(x)\}^2 \{\tfrac{3}{2}q(x)\}^{-2/3}.$$

3.3 Before stating our main theorem for real variables we introduce an error-control function:

$$H(x) \equiv -\int_0^{\zeta} |v|^{-1/2} \psi(v)\, dv. \tag{3.07}$$

From (3.02) and (3.06) we have, equivalently,

$$H(x) = \int_{x_0}^{x} \left\{ \frac{1}{|f|^{1/4}} \frac{d^2}{dx^2}\left(\frac{1}{|f|^{1/4}}\right) - \frac{g}{|f|^{1/2}} - \frac{5|f|^{1/2}}{16|\zeta|^3} \right\} dx. \tag{3.08}$$

Theorem 3.1 *With the conditions of Lemma* 3.1 *and the definitions of* §§2 *and* 3.1, *equation* (3.01) *has twice continuously differentiable solutions* $w_1(u, x)$ *and* $w_2(u, x)$, *such that*

$$\begin{aligned} w_1(u,x) &= \hat{f}^{-1/4}(x)\{\mathrm{Bi}(u^{2/3}\zeta) + \varepsilon_1(u,x)\}, \\ w_2(u,x) &= \hat{f}^{-1/4}(x)\{\mathrm{Ai}(u^{2/3}\zeta) + \varepsilon_2(u,x)\}, \end{aligned} \tag{3.09}$$

where

$$\frac{|\varepsilon_1(u,x)|}{M(u^{2/3}\zeta)}, \ \frac{|\partial\varepsilon_1(u,x)/\partial x|}{u^{2/3}\hat{f}^{1/2}(x) N(u^{2/3}\zeta)} \leqslant \frac{E(u^{2/3}\zeta)}{\lambda}\left[\exp\left\{\frac{\lambda \mathcal{V}_{a,x}(H)}{u}\right\} - 1\right], \tag{3.10}$$

and

$$\frac{|\varepsilon_2(u,x)|}{M(u^{2/3}\zeta)}, \frac{|\partial\varepsilon_2(u,x)/\partial x|}{u^{2/3}\hat{f}^{1/2}(x)\,N(u^{2/3}\zeta)} \leqslant \frac{E^{-1}(u^{2/3}\zeta)}{\lambda}\left[\exp\left\{\frac{\lambda\mathscr{V}_{x,b}(H)}{u}\right\}-1\right]. \tag{3.11}$$

The proof of this result is similar to that for the LG approximation. Writing $\eta_j(\zeta) = \varepsilon_j(u,x)$ and applying the transformation of §3.1, we obtain the following differential equation for the error term in the case $j = 1$:

$$\eta_1''(\zeta) - u^2\zeta\eta_1(\zeta) = \psi(\zeta)\{\eta_1(\zeta) + \mathrm{Bi}(u^{2/3}\zeta)\}.$$

Variation of parameters and use of the Wronskian (1.19) produces

$$\eta_1(\zeta) = \pi u^{-2/3}\int_\alpha^\zeta \mathsf{K}(\zeta,v)\psi(v)\{\eta_1(v) + \mathrm{Bi}(u^{2/3}v)\}\,dv \tag{3.12}$$

in which α is the value of ζ corresponding to $x = a$, and

$$\mathsf{K}(\zeta,v) = \mathrm{Bi}(u^{2/3}\zeta)\,\mathrm{Ai}(u^{2/3}v) - \mathrm{Ai}(u^{2/3}\zeta)\,\mathrm{Bi}(u^{2/3}v). \tag{3.13}$$

Bounds for the kernel are expressible in terms of the auxiliary functions. We have

$$\begin{aligned}|\mathsf{K}(\zeta,v)| &= M(u^{2/3}\zeta)\,M(u^{2/3}v)\,|\,E(u^{2/3}\zeta)\,E^{-1}(u^{2/3}v)\cos\theta(u^{2/3}\zeta)\sin\theta(u^{2/3}v)\\ &\quad - E(u^{2/3}v)\,E^{-1}(u^{2/3}\zeta)\sin\theta(u^{2/3}\zeta)\cos\theta(u^{2/3}v)\,|\\ &\leqslant E(u^{2/3}\zeta)\,E^{-1}(u^{2/3}v)\,M(u^{2/3}\zeta)\,M(u^{2/3}v) \qquad (v \leqslant \zeta),\end{aligned} \tag{3.14}$$

since $E(u^{2/3}v)/E(u^{2/3}\zeta) \leqslant E(u^{2/3}\zeta)/E(u^{2/3}v)$. Similarly,

$$|\partial\mathsf{K}(\zeta,v)/\partial\zeta| \leqslant u^{2/3}E(u^{2/3}\zeta)\,E^{-1}(u^{2/3}v)\,N(u^{2/3}\zeta)\,M(u^{2/3}v) \qquad (v \leqslant \zeta). \tag{3.15}$$

Also $\partial^2\mathsf{K}(\zeta,v)/\partial\zeta^2 = u^2\zeta\mathsf{K}(\zeta,v)$. Equation (3.12) is solved by applying Theorem 10.2 of Chapter 6 with $\phi(v) = \psi_0(v) = |v|^{-1/2}\psi(v)$, $\psi_1(v) = 0$, $J(v) = \mathrm{Bi}(u^{2/3}v)$, and referring to (2.13), (2.14), (2.15), and (3.07). This yields the required bounds (3.10).

Similarly for the second solution $w_2(u,x)$.

Ex. 3.1 If the continuity conditions on $f''(x)$ and $g(x)$ are relaxed to sectional continuity, show that Theorem 3.1 holds, except that the second derivatives of the solutions are discontinuous. May x_0 be a point of discontinuity of $f''(x)$ or $g(x)$?

Ex. 3.2 If κ is a real or complex constant and $x \in (-\infty,\infty)$, show that the differential equation $w'' = \{x + \kappa\exp(-x^2)\}w$ has solutions of the form $\mathrm{Bi}(x) + \varepsilon_1(x)$ and $\mathrm{Ai}(x) + \varepsilon_2(x)$, where

$$E^{-1}(x)|\varepsilon_1(x)|,\ E(x)|\varepsilon_2(x)| \leqslant \lambda^{-1}M(x)[\exp\{\Gamma(\tfrac{1}{4})\lambda|\kappa|\} - 1].$$

Ex. 3.3 Show that when $x \in [0,\infty)$ the equation $w'' = 25x(x+1)^2 w$ has an approximate solution $(3x+5)^{1/6}(x+1)^{-1/2}\mathrm{Ai}\{x(3x+5)^{2/3}\}$, with a relative error not exceeding

$$(2^{1/2}/\lambda)[\exp\{\lambda\mathscr{V}_{x,\infty}(H)\} - 1],$$

where

$$H = \frac{7\tan^{-1}x^{1/2}}{80} + \frac{(21x^2 + 70x + 61)x^{1/2}}{80(x+1)^2(3x+5)}.$$

4 Asymptotic Properties of the Approximation; Whittaker Functions with *m* Large

4.1 Since $|f|^{1/2}\,dx = |\zeta|^{1/2}\,d\zeta$, the relation between the error-control function $H(x)$ of §3.3 and the corresponding function $F(x)$ associated with the LG approximation (Chapter 6, §2) is given by

$$H(x) = F(x) \pm \frac{5}{24|\zeta|^{3/2}}, \tag{4.01}$$

the upper or lower sign being taken according as $x > x_0$ or $x < x_0$. And since $\mathscr{V}(\zeta^{-3/2}) < \infty$ when $\zeta \to \pm\infty$, it follows that $\mathscr{V}(H)$ and $\mathscr{V}(F)$ either converge together or diverge together. Accordingly, when x approaches a singularity of the differential equation the approximations supplied by Theorem 3.1 are asymptotic representations of actual solutions in the same circumstances that the LG functions have this property (Chapter 6, §§3 and 4). Moreover, if the Airy functions are replaced by their asymptotic forms for large $\pm\zeta$, then the approximations reduce to the LG functions; compare Exercise 4.1 below.

4.2 Now consider varying values of the parameter u. If $\mathscr{V}_{a,b}(H)$ is independent of u, or, more generally, a bounded function of u, then from (3.10) we have

$$\varepsilon_1(u, x) = E(u^{2/3}\zeta)\, M(u^{2/3}\zeta)\, O(u^{-1}) \qquad (u \to \infty),$$

uniformly with respect to $x \in (a, b)$. The function $E(u^{2/3}\zeta)\, M(u^{2/3}\zeta)$ is of the same order of magnitude as $\mathrm{Bi}(u^{2/3}\zeta)$ except, of course, in the neighborhoods of the zeros of $\mathrm{Bi}(u^{2/3}\zeta)$. Hence $\hat{f}^{-1/4}(x)\,\mathrm{Bi}(u^{2/3}\zeta)$ provides an asymptotic representation of $w_1(u, x)$ for large u, which is satisfactory uniformly for all x in the closure of (a, b). Similarly for the approximation $\hat{f}^{-1/4}(x)\,\mathrm{Ai}(u^{2/3}\zeta)$.

Theorem 3.1 may also furnish uniform asymptotic representations when a parameter enters the differential equation in other ways. A simple illustration is provided by Exercise 3.2 on letting the parameter κ tend to zero.

***4.3** As an interesting application, consider the equation

$$\frac{d^2W}{dz^2} = \left(-\frac{1}{4} + \frac{k}{z} + \frac{m^2 - \frac{1}{4}}{z^2}\right) W \tag{4.02}$$

obtained from the Whittaker equation (11.01) of Chapter 7 on replacing k by ik and z by iz. We indicate briefly how to obtain asymptotic solutions for positive real values of k and z, and large positive m. Making the substitutions $x = -z/m$† and $l = k/m$, we obtain

$$d^2W/dx^2 = \{m^2 f(x) + g(x)\}\, W,$$

where

$$f(x) = \frac{4 - 4lx - x^2}{4x^2}, \qquad g(x) = -\frac{1}{4x^2};$$

† The minus sign in the definition of x is introduced to make the sign of $f(x)$ the same as that of $x - x_0$ in the subsequent analysis.

compare Chapter 7, §11.2. The x range is $(-\infty, 0)$ and contains a turning point at $x = x_0$, given by

$$x_0 = -2l - 2(l^2+1)^{1/2}.$$

Application of Theorem 3.1 produces solutions of the form

$$W_1 = \left(\frac{\zeta x^2}{4-4lx-x^2}\right)^{1/4} \{\mathrm{Bi}(m^{2/3}\zeta)+\varepsilon_1\}, \quad W_2 = \left(\frac{\zeta x^2}{4-4lx-x^2}\right)^{1/4} \{\mathrm{Ai}(m^{2/3}\zeta)+\varepsilon_2\}, \tag{4.03}$$

where ζ is defined by

$$\frac{2}{3}\zeta^{3/2} = \int_{x_0}^{x} \left(\frac{4-4lt-t^2}{4t^2}\right)^{1/2} dt \quad \text{or} \quad \frac{2}{3}(-\zeta)^{3/2} = \int_{x}^{x_0} \left(\frac{t^2+4lt-4}{4t^2}\right)^{1/2} dt, \tag{4.04}$$

according as $x \in [x_0, 0)$ or $x \in (-\infty, x_0]$. The error terms are bounded by (3.10) and (3.11), with $u = m$, $a = -\infty$, $b = 0$, and H given by (3.08).

Referring to (4.01), and using analysis similar to that of Chapter 7, §11.2, we see that $\mathscr{V}(H)$ converges at $x = -\infty$ and $x = 0$ (as well, of course, at $x = x_0$). Moreover, $\mathscr{V}_{0,\infty}(H)$ is bounded when l ranges over any bounded interval $[0, \alpha]$, say.

Restoring the original variables and writing

$$Z = |4m^2+4kz-z^2|^{1/2},$$

we find that the required solutions become

$$W_1 = z^{1/2}Z^{-1/2}|\zeta|^{1/4}\{\mathrm{Bi}(m^{2/3}\zeta)+\varepsilon_1\}, \quad W_2 = z^{1/2}Z^{-1/2}|\zeta|^{1/4}\{\mathrm{Ai}(m^{2/3}\zeta)+\varepsilon_2\}. \tag{4.05}$$

Evaluation of the integrals (4.04) yields

$$\frac{2}{3}\zeta^{3/2} = -\frac{Z}{2m} + \frac{k}{m}\cos^{-1}\left\{\frac{z-2k}{2(k^2+m^2)^{1/2}}\right\} + \ln\left\{\frac{mZ+kz+2m^2}{z(k^2+m^2)^{1/2}}\right\} \quad (0 < z \leqslant z_0),$$

$$\frac{2}{3}(-\zeta)^{3/2} = \frac{Z}{2m} - \frac{k}{m}\ln\left\{\frac{Z+z-2k}{2(k^2+m^2)^{1/2}}\right\} - \cos^{-1}\left\{\frac{kz+2m^2}{z(k^2+m^2)^{1/2}}\right\} \quad (z_0 \leqslant z < \infty),$$

where $z_0 = 2k+2(k^2+m^2)^{1/2}$. The error terms in (4.05) have the properties: (i) for fixed k and m, $\varepsilon_1 \to 0$ as $z \to \infty$ and $\varepsilon_2 \to 0$ as $z \to 0$, (ii) for large m, $\varepsilon_1 = E(m^{2/3}\zeta)\,M(m^{2/3}\zeta)\,O(m^{-1})$ and $\varepsilon_2 = E^{-1}(m^{2/3}\zeta)\,M(m^{2/3}\zeta)\,O(m^{-1})$, uniformly with respect to $z \in (0, \infty)$ and $k \in [0, \alpha m]$, where α is any positive constant.

Identifications of W_1 and W_2 in terms of standard solutions of (4.02) are stated in Exercise 4.5 below.

Ex. 4.1 In the notation of Theorem 3.1, suppose that $\mathscr{V}_{a,b}(H) < \infty$ and $\int_{x_0}^{x} f^{1/2}(t)\,dt \to \infty$ as $x \to b-$. Show that

$$w_2(u,x) \sim \frac{1}{2\pi^{1/2}u^{1/6}f^{1/4}(x)} \exp\left\{-u\int_{x_0}^{x} f^{1/2}(t)\,dt\right\} \quad (x \to b-).$$

If, also, $\mathscr{V}_{a,b}(H) < u\lambda^{-1}\ln(1+2^{-1/2}\lambda)$, then by expressing $w_1(u,x)$ as a linear combination of the solutions with the properties (3.02) and (3.03) of Chapter 6 show that $\varepsilon_1(u,x)/\mathrm{Bi}(u^{2/3}\zeta)$

tends to a constant as $x \to b-$. Hence prove

$$w_1(u,x) \sim \frac{1+\sigma}{\pi^{1/2}u^{1/6}f^{1/4}(x)} \exp\left\{u\int_{x_0}^{x} f^{1/2}(t)\,dt\right\} \qquad (x \to b-),$$

where σ is a constant which is bounded in absolute value by $2^{1/2}\lambda^{-1}[\exp\{u^{-1}\lambda\mathscr{V}_{a,b}(H)\}-1]$.

Ex. 4.2 If $H_m(x)$ denotes the Hermite polynomial (Chapter 2, §7) and $\nu \equiv (2m+1)^{1/2}$, verify that $\exp(-x^2/2)H_m(x)$ is a solution of the differential equation

$$d^2w/dx^2 = (x^2-\nu^2)w.$$

Thence show that when $x > -1$

$$H_m(\nu x) = (2\pi)^{1/2}e^{-\nu^2/4}\nu^{(3\nu^2-1)/6}e^{\nu^2x^2/2}\left(\frac{\zeta}{x^2-1}\right)^{1/4}\{\mathrm{Ai}(\nu^{4/3}\zeta)+\varepsilon(x)\},$$

where ζ is given by

$$\{\tfrac{3}{4}x(x^2-1)^{1/2}-\tfrac{3}{4}\cosh^{-1}x\}^{2/3} \quad (x \geqslant 1) \qquad \text{or} \qquad -\{\tfrac{3}{4}\cos^{-1}x-\tfrac{3}{4}x(1-x^2)^{1/2}\}^{2/3} \quad (-1 < x \leqslant 1),$$

and

$$|\varepsilon(x)| \leqslant \lambda^{-1}E^{-1}(\nu^{4/3}\zeta)M(\nu^{4/3}\zeta)[\exp\{\lambda\nu^{-2}\mathscr{V}_{x,\infty}(H)\}-1]; \qquad H(x) \equiv \frac{5}{24|\zeta|^{3/2}}+\frac{x^3-6x}{12|x^2-1|^{3/2}}.$$

Ex. 4.3 Show that the derivative of the function $H(x)$ defined in the preceding exercise does not vanish when $x \in (-1,\infty)$. Deduce that

$$|\varepsilon(x)| \leqslant 1.36\{\exp(0.09\nu^{-2})-1\}\,\mathrm{Ai}(\nu^{4/3}\zeta) \qquad (x \geqslant 1),$$
$$|\varepsilon(x)| \leqslant 0.97\{\exp(0.28\nu^{-2})-1\}E^{-1}(\nu^{4/3}\zeta)M(\nu^{4/3}\zeta) \qquad (0 \leqslant x \leqslant 1).$$

Ex. 4.4 Show that the equation

$$\frac{d^2w}{dx^2} = \left(u^2x+\frac{ux}{x+1}\right)w$$

has solutions of the form

$$\mathrm{Bi}(\zeta)+E(\zeta)M(\zeta)O(u^{-1}), \qquad \mathrm{Ai}(\zeta)+E^{-1}(\zeta)M(\zeta)O(u^{-1}),$$

as $u \to +\infty$, uniformly in the interval $-1+\delta \leqslant x < \infty$, where δ is an arbitrary positive constant, and

$$\zeta = x[u+\tfrac{3}{2}x^{-3/2}\{x^{1/2}-\tan^{-1}(x^{1/2})\}]^{2/3} \qquad (x \geqslant 0),$$
$$\zeta = x[u+\tfrac{3}{2}(-x)^{-3/2}\{\tanh^{-1}(-x)^{1/2}-(-x)^{1/2}\}]^{2/3} \quad (x \leqslant 0).$$

Ex. 4.5 Prove that, in the notation of §4.3,

$$W_1 = \frac{e^{\pi k/2}}{\pi^{1/2}m^{1/6}}\mathrm{Re}\{e^{i\delta}W_{ik,m}(iz)\}, \qquad \delta \equiv k-\frac{1}{4}\pi+m\tan^{-1}\left(\frac{m}{k}\right)-\frac{1}{2}k\ln(k^2+m^2);$$

$$W_2 = \frac{(k^2+m^2)^{m/2}}{\pi^{1/2}2^{2m+(3/2)}m^{2m+(2/3)}}\exp\left\{m-\frac{1}{2}\pi k-k\tan^{-1}\left(\frac{k}{m}\right)-\frac{1}{2}m\pi i-\frac{1}{4}\pi i\right\}M_{ik,m}(iz).$$

*5 Real Zeros of the Airy Functions

5.1 In §1.3 it was seen that none of the functions $\mathrm{Ai}(x)$, $\mathrm{Bi}(x)$, $\mathrm{Ai}'(x)$, and $\mathrm{Bi}'(x)$ vanishes when x is positive. On the other hand, the asymptotic expansions of §§1.1 and 1.2 show that each function has an infinite sequence of negative zeros. These sequences are usually denoted by $\{a_n\}$, $\{b_n\}$, $\{a'_n\}$, and $\{b'_n\}$, respectively, each being arranged in ascending order of absolute value for $n = 1, 2, 3, \ldots$.

Lemma 5.1 *As x increases from $-\infty$ to c (defined in §2.2), the functions $1/\{|x|^{1/4}M(x)\}$, $N(x)/|x|^{1/4}$, $M'(x)$, $M(x)$, and $M(x)N(x)$ are increasing, and the functions $\theta(x)$, $|\theta'(x)|$, $\omega(x)$, and $|\omega'(x)|$ are decreasing.*

The proof of the given property of $1/\{|x|^{1/4}M(x)\}$ is indicated in Exercise 2.1. In a similar way, for the second function we have

$$|x|^{-1/2}N^2(x) = \tfrac{1}{2}\xi\{J^2_{2/3}(\xi)+Y^2_{2/3}(\xi)\} \qquad (x \leqslant c),$$

and again §7.3 of Chapter 9 shows that this is a decreasing function of ξ.

Next, since $x \leqslant c$ we have from (2.02) and (2.03)

$$\tan\theta(x) = \mathrm{Ai}(x)/\mathrm{Bi}(x).$$

Differentiation and use of the Wronskian relation yields

$$\theta'(x) = -1/\{\pi M^2(x)\}. \tag{5.01}$$

By substituting $w = Me^{i\theta}$ in equation (1.01), separating real and imaginary parts and eliminating θ' by means of (5.01), we derive the following nonlinear differential equation for M:

$$M''(x) = xM(x) + 1/\{\pi^2 M^3(x)\}.$$

Since $\pi|x|^{1/2}M^2(x) < 1$ (Exercise 2.1), it follows that $M''(x) > 0$. Combination of this result with (2.07) and the corresponding result

$$M'(x) \sim \tfrac{1}{4}\pi^{-1/2}(-x)^{-5/4} \qquad (x \to -\infty),$$

obtained from §§1.1 and 1.2, establishes that $M'(x)$ and $M(x)$ are increasing.

For the product $M(x)N(x)$, it is easily verified that

$$d\{M^2(x)N^2(x)\}/dx = 2M(x)M'(x)\{N^2(x)+xM^2(x)\}. \tag{5.02}$$

The infimum of $|x|^{-1/2}N^2(x)$ and supremum of $|x|^{1/2}M^2(x)$ are both approached as $x \to -\infty$, and are equal (compare (2.07) and (2.12)). Therefore the right-hand side of (5.02) is positive; that is, $M(x)N(x)$ is increasing.

The stated properties of the phase function $\theta(x)$ follow immediately from results just obtained and (5.01). For $\omega(x)$, the corresponding equation is

$$\omega'(x) = x/\{\pi N^2(x)\}. \tag{5.03}$$

Since x is negative, $\omega(x)$ is decreasing. Also, $|x|^{1/2}/N^2(x)$ is decreasing; *a fortiori* $|\omega'(x)|$ is decreasing. This completes the proof of the lemma.

5.2 From Lemma 5.1 and properties of $\theta(x)$ and $\omega(x)$ given in §2, we see that at the zeros of the Airy functions and their derivatives,

$$\theta(a_n) = n\pi, \qquad \theta(b_n) = (n-\tfrac{1}{2})\pi, \qquad \omega(a'_n) = (n-1)\pi, \qquad \omega(b'_n) = (n-\tfrac{1}{2})\pi, \tag{5.04}$$

and thence that

$$c > b_1 > a_1 > b_2 > a_2 > \cdots, \qquad c > a'_1 > b'_1 > a'_2 > b'_2 > \cdots.$$

By reversion of the asymptotic expansions given in §§1.1 and 1.2, the following asymptotic expansions are obtained for large n:

$$\begin{aligned} a_n &= -T\{\tfrac{3}{8}\pi(4n-1)\}, & \mathrm{Ai}'(a_n) &= (-)^{n-1}V\{\tfrac{3}{8}\pi(4n-1)\}, \\ b_n &= -T\{\tfrac{3}{8}\pi(4n-3)\}, & \mathrm{Bi}'(b_n) &= (-)^{n-1}V\{\tfrac{3}{8}\pi(4n-3)\}, \\ a_n' &= -U\{\tfrac{3}{8}\pi(4n-3)\}, & \mathrm{Ai}(a_n') &= (-)^{n-1}W\{\tfrac{3}{8}\pi(4n-3)\}, \\ b_n' &= -U\{\tfrac{3}{8}\pi(4n-1)\}, & \mathrm{Bi}(b_n') &= (-)^{n}W\{\tfrac{3}{8}\pi(4n-1)\}, \end{aligned} \tag{5.05}$$

where†

$$\begin{aligned} T(t) &\sim t^{2/3}(1+\tfrac{5}{48}t^{-2}-\tfrac{5}{36}t^{-4}+\cdots), \\ U(t) &\sim t^{2/3}(1-\tfrac{7}{48}t^{-2}+\tfrac{35}{288}t^{-4}-\cdots), \\ V(t) &\sim \pi^{-1/2}t^{1/6}(1+\tfrac{5}{48}t^{-2}-\tfrac{1525}{4608}t^{-4}+\cdots), \\ W(t) &\sim \pi^{-1/2}t^{-1/6}(1-\tfrac{7}{96}t^{-2}+\tfrac{1673}{6144}t^{-4}-\cdots). \end{aligned}$$

The process of reversion does not, in itself, establish that these expansions pertain to the nth zeros and not, for example, the $(n+1)$th zeros. This point is resolvable by use of (5.04) and Exercise 2.2, or by application of the phase principle in the manner of Exercise 8.4 below.

Ex. 5.1 By taking $m = \frac{1}{3}$ in Exercise 8.2 of Chapter 6, show that for $n = 1, 2, \ldots$, $\mathrm{Ai}(x)$ has exactly one zero of the form $-\{\frac{3}{8}\pi(4n-1)+\alpha_n\}^{2/3}$, and for $n = 2, 3, \ldots$, $\mathrm{Bi}(x)$ has exactly one zero of the form $-\{\frac{3}{8}\pi(4n-3)+\beta_n\}^{2/3}$, where

$$|\alpha_n| \leqslant \frac{3\pi}{4}\left[\exp\left\{\frac{5}{9\pi(4n-3)}\right\}-1\right], \qquad |\beta_n| \leqslant \frac{3\pi}{4}\left[\exp\left\{\frac{5}{9\pi(4n-5)}\right\}-1\right].$$

*6 Zeros of the First Approximation

6.1 Let $X = X(\zeta)$ be the function inverse to $\zeta = \zeta(x)$, defined by (3.03). If the error terms $\varepsilon_1(u,x)$ and $\varepsilon_2(u,x)$ in (3.09) are negligibly small, then approximations to the zeros of $w_1(u,x)$ and $w_2(u,x)$ are $X(u^{-2/3}b_n)$ and $X(u^{-2/3}a_n)$, respectively. Let us make a rigorous assessment of the errors in these approximations, beginning with the first solution. Throughout the analysis it is supposed that u is large enough to satisfy the inequality

$$\lambda^{-1}[\exp\{\lambda\mathscr{V}_{a,b}(H)u^{-1}\}-1] < \tfrac{1}{2}. \tag{6.01}$$

We write

$$\rho_1(u,\zeta) = \varepsilon_1(u,x)/\{E(u^{2/3}\zeta)\,M(u^{2/3}\zeta)\}, \qquad \sigma_1(u,\zeta) = \lambda^{-1}[\exp\{\lambda\mathscr{V}_{a,x}(H)u^{-1}\}-1], \tag{6.02}$$

so that $|\rho_1(u,\zeta)| \leqslant \sigma_1(u,\zeta)$. Moreover, in consequence of (6.01) both $|\rho_1(u,\zeta)|$ and $\sigma_1(u,\zeta)$ are less than $\frac{1}{2}$. Substituting in (3.09) for $\mathrm{Bi}(u^{2/3}\zeta)$ by means of (2.01), we see

† Additional terms have been given by B.A. (1946b) and Olver (1954b).

that at a zero of $w_1(u, x)$

$$\cos\theta(u^{2/3}\zeta) = -\rho_1(u,\zeta). \tag{6.03}$$

Since the left-hand side equals $2^{-1/2}$ when $\zeta \geqslant u^{-2/3}c$ there can be no zeros in this range.

Next, when $\zeta < u^{-2/3}c$ we have

$$\theta(u^{2/3}\zeta) - (n-\tfrac{1}{2})\pi + (-)^n \sin^{-1}\rho_1(u,\zeta) = 0, \tag{6.04}$$

where n is an arbitrary positive integer and the inverse sine has its principal value. From the bound

$$|\sin^{-1}\rho_1(u,\zeta)| < \sin^{-1}\tfrac{1}{2} = \tfrac{1}{6}\pi$$

and equations (5.04), it is seen that when $n \geqslant 2$ the left-hand side of equation (6.04) is positive when $\zeta = u^{-2/3}a_n$ and negative when $\zeta = u^{-2/3}a_{n-1}$. This statement also holds for $n = 1$ if we define $a_0 = c$; compare (2.04). Accordingly, there is a zero in the interval

$$u^{-2/3}a_n < \zeta < u^{-2/3}a_{n-1} \qquad (n = 1, 2, \ldots). \tag{6.05}$$

This zero may be delimited in a shorter interval, as follows. Applying the mean-value theorem, we have

$$\theta(u^{2/3}\zeta) = \theta(b_n) + (u^{2/3}\zeta - b_n)\,\theta'(\tau),$$

where $\tau \in (a_n, a_{n-1})$. Combination of this result with (5.04) and (6.04) yields

$$u^{2/3}\zeta - b_n = (-)^{n-1}\sin^{-1}\{\rho_1(u,\zeta)\}/\theta'(\tau). \tag{6.06}$$

Hence

$$|u^{2/3}\zeta - b_n| \leqslant \sin^{-1}\{\sigma_1(u,\zeta)\}/|\theta'(\tau)| < \tfrac{1}{3}\pi\sigma_1(u,\zeta)/|\theta'(\tau)|,$$

since $\sigma_1(u,\zeta) < \frac{1}{2}$. In view of the fact that $\sigma_1(u,\zeta)$ is a nondecreasing function of ζ and $|\theta'(\tau)|$ is a decreasing function of τ (Lemma 5.1), we have

$$|u^{2/3}\zeta - b_n| < \beta_n,$$

where

$$\beta_n = \tfrac{1}{3}\pi\sigma_1(u, u^{-2/3}a_{n-1})/|\theta'(a_{n-1})| = \tfrac{1}{3}\pi^2 M^2(a_{n-1})\,\sigma_1(u, u^{-2/3}a_{n-1}); \tag{6.07}$$

compare (5.01). This is the first desired result.

6.2 That there is not *more* than one zero in the interval (6.05) can be established by proving that

$$\{1-\rho_1^2(u,\zeta)\}^{-1/2}\,|\partial\rho_1(u,\zeta)/\partial\zeta| < u^{2/3}\,|\theta'(u^{2/3}\zeta)|; \tag{6.08}$$

compare the differentiated form of (6.04). Since $u^{2/3}\zeta < c$ we have

$$\frac{\partial\rho_1(u,\zeta)}{\partial\zeta} = \frac{\partial}{\partial\zeta}\left\{\frac{\varepsilon_1(u,x)}{M(u^{2/3}\zeta)}\right\} = \frac{1}{\hat{f}^{1/2}(x)\,M(u^{2/3}\zeta)}\,\frac{\partial\varepsilon_1(u,x)}{\partial x} - \frac{M'(u^{2/3}\zeta)}{M^2(u^{2/3}\zeta)}\,u^{2/3}\varepsilon_1(u,x).$$

With the aid of (3.10) and (6.02) we derive

$$\frac{1}{\{1-\rho_1^2(u,\zeta)\}^{1/2}}\left|\frac{\partial\rho_1(u,\zeta)}{\partial\zeta}\right| \leqslant \frac{u^{2/3}\sigma_1(u,\zeta)}{\{1-\sigma_1^2(u,\zeta)\}^{1/2}}\,\frac{N(u^{2/3}\zeta)+M'(u^{2/3}\zeta)}{M(u^{2/3}\zeta)}.$$

From this inequality and (5.01) it is seen that (6.08) is certainly satisfied if

$$\frac{\sigma_1(u,\zeta)}{\{1-\sigma_1^2(u,\zeta)\}^{1/2}} < \frac{1}{\pi M(u^{2/3}\zeta)\{N(u^{2/3}\zeta)+M'(u^{2/3}\zeta)\}}. \tag{6.09}$$

Since $\sigma_1(u,\zeta) < \frac{1}{2}$ the left-hand side is bounded by $1/\sqrt{3} = 0.577\ldots$. Lemma 5.1 shows that the right-hand side is a decreasing function of $u^{2/3}\zeta$. At $u^{2/3}\zeta = c$ its value is

$$1/[\pi \operatorname{Ai}(c)\{\operatorname{Ai}'(c)+\operatorname{Bi}'(c)+\sqrt{2\operatorname{Ai}'^2(c)+2\operatorname{Bi}'^2(c)}\}].$$

Numerical computation yields $0.708\ldots$; hence the required results (6.09) and (6.08) follow.†

6.3 The results of §§6.1 and 6.2 may be summarized as follows. Define $\sigma_1(u,\zeta)$ by (6.02) and β_n by (6.07), with the convention that $a_0 = c$. If (6.01) holds, then the solution $w_1(u,x)$ has no zeros in the interval $X(u^{-2/3}c) \leqslant x < b$ and exactly one zero in the interval $X(u^{-2/3}a_n) < x < X(u^{-2/3}a_{n-1})$ for each positive integer n. This zero also lies in the interval $X\{u^{-2/3}(b_n-\beta_n)\} < x < X\{u^{-2/3}(b_n+\beta_n)\}$.

In a similar way, we may prove that with the condition (6.01) the solution $w_2(u,x)$ has no zeros in the interval $X(u^{-2/3}b_1) \leqslant x < b$ and exactly one zero in the interval $X(u^{-2/3}b_{n+1}) < x < X(u^{-2/3}b_n)$ for each positive integer n. Furthermore, this zero lies in the interval $X\{u^{-2/3}(a_n-\alpha_n)\} < x < X\{u^{-2/3}(a_n+\alpha_n)\}$, where

$$\alpha_n = \tfrac{1}{3}\pi^2 M^2(b_n)\,\sigma_2(u, u^{-2/3}b_{n+1}), \qquad \sigma_2(u,\zeta) = \lambda^{-1}[\exp\{\lambda \mathscr{V}_{x,b}(H)u^{-1}\}-1].$$

6.4 Asymptotic properties of the approximations $X(u^{-2/3}b_n)$ and $X(u^{-2/3}a_n)$ are deducible from the error bounds just established. If u is fixed and the ζ interval extends to $-\infty$, then from (6.07), (6.02), (5.05), and (2.07), we see that as $n\to\infty$,

$$\beta_n = M^2(a_{n-1})\,o(1) = o(n^{-1/3}).$$

Therefore the nth zero of $w_1(u,x)$ to the left of the turning point x_0 has the form $X\{u^{-2/3}b_n + o(n^{-1/3})\}$ as $n\to\infty$. Similarly the nth zero of $w_2(u,x)$ to the left of x_0 has the form $X\{u^{-2/3}a_n + o(n^{-1/3})\}$. These results are included in simpler ones available from the LG approximation (Chapter 6, §8.1).

Next, if the variation of the error-control function H over (a,b) is a bounded function of u, then for large u

$$\beta_n = M^2(a_{n-1})\,O(u^{-1}), \qquad \alpha_n = M^2(b_n)\,O(u^{-1}).$$

Thus, for example, the nth zero of $w_1(u,x)$ has the form $X\{u^{-2/3}b_n + O(u^{-5/3})\}$. By use of the mean-value theorem this can be reexpressed as

$$X(u^{-2/3}b_n) + O(u^{-5/3}), \tag{6.10}$$

or on expansion as

$$x_0 + \frac{b_n}{\{f'(x_0)\}^{1/3}}\frac{1}{u^{2/3}} - \frac{b_n^2}{10}\frac{f''(x_0)}{\{f'(x_0)\}^{5/3}}\frac{1}{u^{4/3}} + O\left(\frac{1}{u^{5/3}}\right), \tag{6.11}$$

† An alternative way of completing the analysis, which avoids the need for numerical computation, is to extend the definitions (2.05) and (2.10) temporarily from $x \in (-\infty, c]$ to $x \in (-\infty, 0]$. All steps in the proof of Lemma 5.1 hold in the extended range. Evaluating the right-hand side of (6.09) at $\zeta = 0$ by means of (1.03) and (1.11) we obtain $3^{-1/2}$; hence (6.09) is established.

since

$$X'(0) = \frac{1}{\{f'(x_0)\}^{1/3}}, \qquad X''(0) = -\frac{f''(x_0)}{5\{f'(x_0)\}^{5/3}}.$$

The approximation (6.11) is more explicit than (6.10), but because it depends on Taylor's theorem applied at $\zeta = 0$, the implied constant in the error term grows with n. On the other hand, even if the ζ interval extends to $-\infty$, the error term in (6.10) is uniform for all n, provided that $X'(\zeta)$, that is, $\hat{f}^{-1/2}(x)$, is bounded as $\zeta \to -\infty$. Indeed, owing to the factor $M^2(a_{n-1})$ in the estimate for β_n, the error term in (6.10) can be improved in these circumstances to $n^{-1/3}O(u^{-5/3})$.

Ex. 6.1 Show that at the nth zero of $w_1(u,x)$ to the left of x_0

$$|u^{-2/3}\hat{f}^{-1/4}(x)(\partial w_1/\partial x) - \mathrm{Bi}'(b_n)| < N(a_n)\{\sigma_1(u, u^{-2/3}a_{n-1}) + |a_n|\beta_n^2\},$$

and at the nth zero of $w_2(u,x)$ to the left of x_0

$$|u^{-2/3}\hat{f}^{-1/4}(x)(\partial w_2/\partial x) - \mathrm{Ai}'(a_n)| < N(b_{n+1})\{\sigma_2(u, u^{-2/3}b_{n+1}) + |b_{n+1}|\alpha_n^2\}.$$

Ex. 6.2 If (6.01) is satisfied and n is a positive integer, show that the function $\hat{f}^{1/4}(x)\,w_1(u,x)$ has at least one stationary value in the interval $X(u^{-2/3}a'_{n+1}) < x < X(u^{-2/3}a'_n)$, and that this value also lies in the interval $X\{u^{-2/3}(b'_n - \beta'_n)\} < x < X\{u^{-2/3}(b'_n + \beta'_n)\}$, where

$$|a'_n|\beta'_n = \tfrac{1}{3}\pi^2 N^2(a'_n)\,\sigma_1(u, u^{-2/3}a'_n).$$

Ex. 6.3 Using the notation and results of Exercises 4.2 and 4.3, show that the nth zero of the Hermite polynomial $H_m(t)$ to the left of $t = 1$ is given by

$$\nu X(\nu^{-4/3}a_n) + \vartheta_n \frac{\delta_n}{\nu}\left\{\frac{-a_n + \delta_n}{1 - X^2(\nu^{-4/3}a_n - \nu^{-4/3}\delta_n)}\right\}^{1/2},$$

provided that $X(\nu^{-4/3}b_{n+1}) \geqslant 0$, where $x = X(\zeta)$ is the inverse function to $\zeta(x)$,

$$\delta_n = 3.2M^2(b_n)\{\exp(0.28\nu^{-2}) - 1\},$$

and ϑ_n is some number in the interval $(-1, 1)$.

Ex. 6.4 Show that $w = x^{1/2}J_\nu(\nu x)$ satisfies

$$\frac{d^2 w}{dx^2} = \left(\nu^2 \frac{1-x^2}{x^2} - \frac{1}{4x^2}\right) w$$

and thence that (in the notation of Chapter 7, §6.1)

$$j_{\nu,n} = \nu - a_n(\tfrac{1}{2}\nu)^{1/3} + \tfrac{3}{20}a_n^2(\tfrac{1}{2}\nu)^{-1/3} + O(\nu^{-2/3})$$

as $\nu \to +\infty$, n being fixed.

Show also that

$$j_{\nu,n} = \nu X(\nu^{-2/3}a_n) + n^{-1/3}O(\nu^{-2/3}) + O(\nu^{-1})$$

as $\nu \to \infty$, uniformly with respect to all n, where $X(\zeta)$ is defined implicitly by

$$\tfrac{2}{3}(-\zeta)^{3/2} = (X^2 - 1)^{1/2} - \sec^{-1} X.$$

7 Higher Approximations

7.1 The approximations (3.09) may be regarded as leading terms in asymptotic expansions. To obtain higher terms, we again apply the transformations (3.02) to obtain

$$d^2W/d\zeta^2 = \{u^2\zeta + \psi(\zeta)\}\,W, \tag{7.01}$$

where $\psi(\zeta)$ is given by (3.05) or (3.06). As before, we suppose that the parameter u is positive and the variable ζ ranges over a real interval (α, β) which contains $\zeta = 0$ and may be infinite. We also assume, in the text, that the real or complex function $\psi(\zeta)$ is infinitely differentiable.

Analogy with Case I (Chapter 10, §2.3) suggests we try for a solution of the form

$$W = \mathrm{Ai}(u^{2/3}\zeta) \sum_{s=0}^{\infty} \frac{A_s(\zeta)}{u^s}, \tag{7.02}$$

where the coefficients $A_s(\zeta)$ are independent of u. Since $\mathrm{Ai}''(u^{2/3}\zeta) = u^{2/3}\zeta\,\mathrm{Ai}(u^{2/3}\zeta)$, formal differentiations produce

$$\frac{dW}{d\zeta} = \mathrm{Ai}(u^{2/3}\zeta) \sum_{s=0}^{\infty} \frac{A_s'(\zeta)}{u^s} + u^{2/3}\,\mathrm{Ai}'(u^{2/3}\zeta) \sum_{s=0}^{\infty} \frac{A_s(\zeta)}{u^s}, \tag{7.03}$$

and

$$\frac{d^2W}{d\zeta^2} = \mathrm{Ai}(u^{2/3}\zeta) \sum_{s=-2}^{\infty} \frac{A_s''(\zeta) + \zeta A_{s+2}(\zeta)}{u^s} + u^{2/3}\,\mathrm{Ai}'(u^{2/3}\zeta) \sum_{s=0}^{\infty} \frac{2A_s'(\zeta)}{u^s}, \tag{7.04}$$

coefficients with negative suffix being interpreted as zero. Substituting in (7.01) and comparing like powers of u, we derive

$$A_s''(\zeta) = \psi(\zeta)\,A_s(\zeta), \qquad A_s'(\zeta) = 0.$$

Only when $\psi(\zeta)$ vanishes identically are these equations satisfiable.

7.2 Inspection of (7.02) to (7.04) suggests that we try instead

$$W = \mathrm{Ai}(u^{2/3}\zeta) \sum_{s=0}^{\infty} \frac{A_s(\zeta)}{u^{2s}} + \frac{\mathrm{Ai}'(u^{2/3}\zeta)}{u^{4/3}} \sum_{s=0}^{\infty} \frac{B_s(\zeta)}{u^{2s}}, \tag{7.05}$$

where $A_s(\zeta)$ and $B_s(\zeta)$ are independent of u. This time we obtain

$$\frac{dW}{d\zeta} = \mathrm{Ai}(u^{2/3}\zeta) \sum_{s=0}^{\infty} \frac{C_s(\zeta)}{u^{2s}} + u^{2/3}\,\mathrm{Ai}'(u^{2/3}\zeta) \sum_{s=0}^{\infty} \frac{D_s(\zeta)}{u^{2s}},$$

where

$$C_s(\zeta) = A_s'(\zeta) + \zeta B_s(\zeta), \qquad D_s(\zeta) = A_s(\zeta) + B_{s-1}'(\zeta).$$

Differentiating again and comparing like powers of u after substituting in (7.01), we find that

$$A_s''(\zeta) - \psi(\zeta)\,A_s(\zeta) + 2\zeta B_s'(\zeta) + B_s(\zeta) = 0, \tag{7.06}$$

and

$$2A_{s+1}'(\zeta) + B_s''(\zeta) - \psi(\zeta)\,B_s(\zeta) = 0. \tag{7.07}$$

Integration of these two equations produces

$$\begin{aligned} B_s(\zeta) &= \frac{1}{2\zeta^{1/2}} \int_0^{\zeta} \{\psi(v)\,A_s(v) - A_s''(v)\} \frac{dv}{v^{1/2}} \qquad (\zeta > 0), \\ B_s(\zeta) &= \frac{1}{2(-\zeta)^{1/2}} \int_{\zeta}^{0} \{\psi(v)\,A_s(v) - A_s''(v)\} \frac{dv}{(-v)^{1/2}} \qquad (\zeta < 0), \end{aligned} \tag{7.08}$$

and

$$A_{s+1}(\zeta) = -\tfrac{1}{2}B_s'(\zeta) + \tfrac{1}{2}\int \psi(\zeta)\, B_s(\zeta)\, d\zeta. \tag{7.09}$$

Alternate application of (7.08) and (7.09) determines $B_s(\zeta)$ and $A_s(\zeta)$ recursively, apart from arbitrary constants of integration associated with (7.09). In particular, $A_0(\zeta) =$ constant, which we take to be unity without loss of generality.†

That $A_s(\zeta)$ and $B_s(\zeta)$ are infinitely differentiable is a consequence of the following result, the proof of which is similar to that of Lemma 3.1 and is omitted.

Lemma 7.1 *In a given interval* (α, β) *containing the origin, assume that* $h(\zeta)$ *and its first* n *derivatives are continuous real or complex functions, and let* $I(\zeta)$ *be defined by*

$$\frac{1}{\zeta^{1/2}}\int_0^\zeta h(v)\frac{dv}{v^{1/2}}, \qquad 2h(0), \qquad or \qquad \frac{1}{(-\zeta)^{1/2}}\int_\zeta^0 h(v)\frac{dv}{(-v)^{1/2}},$$

according as $\zeta > 0$, $\zeta = 0$, *or* $\zeta < 0$. *Then* $I(\zeta)$ *is* n *times continuously differentiable in* (α, β). *Also,* $I^{(n+1)}(\zeta)$ *is continuous except possibly at* $\zeta = 0$.

7.3 The following theorem supplies a bound for the difference between a partial sum of the formal series (7.05) and an actual solution of the differential equation, together with the corresponding result when Bi replaces Ai.

Theorem 7.1 *With the conditions of the first paragraph of* §7.1 *and the definitions of* §2 *equation* (7.01) *has, for each value of* u *and each nonnegative integer* n, *a pair of infinitely differentiable solutions* $W_{2n+1,1}(u,\zeta)$ *and* $W_{2n+1,2}(u,\zeta)$, *given by*

$$W_{2n+1,1}(u,\zeta) = \mathrm{Bi}(u^{2/3}\zeta)\sum_{s=0}^{n}\frac{A_s(\zeta)}{u^{2s}} + \frac{\mathrm{Bi}'(u^{2/3}\zeta)}{u^{4/3}}\sum_{s=0}^{n-1}\frac{B_s(\zeta)}{u^{2s}} + \varepsilon_{2n+1,1}(u,\zeta), \tag{7.10}$$

$$W_{2n+1,2}(u,\zeta) = \mathrm{Ai}(u^{2/3}\zeta)\sum_{s=0}^{n}\frac{A_s(\zeta)}{u^{2s}} + \frac{\mathrm{Ai}'(u^{2/3}\zeta)}{u^{4/3}}\sum_{s=0}^{n-1}\frac{B_s(\zeta)}{u^{2s}} + \varepsilon_{2n+1,2}(u,\zeta), \tag{7.11}$$

where

$$\frac{|\varepsilon_{2n+1,1}(u,\zeta)|}{M(u^{2/3}\zeta)}, \quad \frac{|\partial\varepsilon_{2n+1,1}(u,\zeta)/\partial\zeta|}{u^{2/3}N(u^{2/3}\zeta)} \leqslant 2E(u^{2/3}\zeta)\exp\left\{\frac{2\lambda\mathcal{V}_{\alpha,\zeta}(|\zeta|^{1/2}B_0)}{u}\right\}\frac{\mathcal{V}_{\alpha,\zeta}(|\zeta|^{1/2}B_n)}{u^{2n+1}}, \tag{7.12}$$

$$\frac{|\varepsilon_{2n+1,2}(u,\zeta)|}{M(u^{2/3}\zeta)}, \quad \frac{|\partial\varepsilon_{2n+1,2}(u,\zeta)/\partial\zeta|}{u^{2/3}N(u^{2/3}\zeta)} \leqslant 2E^{-1}(u^{2/3}\zeta)\exp\left\{\frac{2\lambda\mathcal{V}_{\zeta,\beta}(|\zeta|^{1/2}B_0)}{u}\right\}\frac{\mathcal{V}_{\zeta,\beta}(|\zeta|^{1/2}B_n)}{u^{2n+1}}. \tag{7.13}$$

† Another way of obtaining the coefficients (which is useful in applications) is indicated in Exercise 7.4 below.

This theorem may be proved in a similar manner to Theorem 3.1. In the case of the first solution, for example, the integral equation for the error term corresponding to (3.12) is found to be

$$\varepsilon_{2n+1,1}(u,\zeta) = \frac{\pi}{u^{2/3}}\int_{\alpha}^{\zeta} \mathsf{K}(\zeta,v)\{\psi(v)\,\varepsilon_{2n+1,1}(u,v) - R_{2n+1}(u,v)\,\mathrm{Bi}(u^{2/3}v)\}\,dv, \tag{7.14}$$

where $\mathsf{K}(\zeta,v)$ is defined by (3.13), and

$$R_{2n+1}(u,\zeta) = -\frac{2\zeta B_n'(\zeta)+B_n(\zeta)}{u^{2n}} = \mp\frac{2|\zeta|^{1/2}}{u^{2n}}\frac{d\{|\zeta|^{1/2}B_n(\zeta)\}}{d\zeta}, \tag{7.15}$$

the upper or lower sign being taken according as $\zeta \gtrless 0$. In solving (7.14) by means of Theorem 10.1 of Chapter 6, we use the relation

$$\int |v^{-1/2}\psi(v)|\,dv = 2\mathscr{V}(|\zeta|^{1/2}B_0),$$

obtained from (7.08) with $s=0$.

A similar theorem can be constructed for the case in which the formal series (7.05) is truncated at an even number of terms.

7.4 Asymptotic properties of the expansions (7.10) and (7.11) follow directly from the error bounds (7.12) and (7.13). Thus if $\psi(\zeta)$ is independent of u and the functions $|\zeta|^{1/2}B_0(\zeta)$ and $|\zeta|^{1/2}B_n(\zeta)$ are of bounded variation over the interval (α,β),† then for large u

$$\varepsilon_{2n+1,1}(u,\zeta) = E(u^{2/3}\zeta)\,M(u^{2/3}\zeta)\,O(u^{-2n-1}),$$
$$\varepsilon_{2n+1,2}(u,\zeta) = E^{-1}(u^{2/3}\zeta)\,M(u^{2/3}\zeta)\,O(u^{-2n-1}),$$

uniformly with respect to ζ.

To express (7.10), for example, in the form of a compound asymptotic expansion, write $\tau \equiv u^{2/3}\zeta$ for brevity and decompose the factor $E(u^{2/3}\zeta)\,M(u^{2/3}\zeta)$ in the error term according to the identity

$$E(\tau)\,M(\tau) = \mathfrak{A}(\tau)\,\mathrm{Bi}(\tau) + \mathfrak{B}(\tau)\,\mathrm{Bi}'(\tau),$$

where

$$\mathfrak{A}(\tau) = \frac{|\tau|\,E(\tau)\,M(\tau)\,\mathrm{Bi}(\tau)}{|\tau|\,\mathrm{Bi}^2(\tau)+\mathrm{Bi}'^2(\tau)}, \qquad \mathfrak{B}(\tau) = \frac{E(\tau)\,M(\tau)\,\mathrm{Bi}'(\tau)}{|\tau|\,\mathrm{Bi}^2(\tau)+\mathrm{Bi}'^2(\tau)}.$$

It is readily verified that both $|\mathfrak{A}(\tau)|$ and $|\mathfrak{B}(\tau)|$ are bounded when $\tau\in(-\infty,\infty)$. Accordingly, if the coefficients $A_s(\zeta)$ and $B_s(\zeta)$ appearing in (7.10) are bounded in (α,β),‡ then this expansion is a uniform compound asymptotic expansion of the solution $W_{2n+1,1}(u,\zeta)$ to $2n+1$ terms. Similarly for $W_{2n+1,2}(u,\zeta)$.

† Compare Exercise 7.1 below.

‡ Again, sufficient conditions that the coefficients be bounded in an infinite interval are stated in Exercise 7.1.

The existence of a solution which is independent of n and has the *infinite* series (7.05) as a compound asymptotic expansion may be demonstrated by the method of Chapter 10, §6.

Ex. 7.1 Assume that for each nonnegative integer s

$$\psi^{(s)}(\zeta) = O(|\zeta|^{-s-(1/2)-\sigma}),$$

as $\zeta \to \pm\infty$, where σ is a positive constant. By means of induction show that $A_s(\zeta)$, $D_s(\zeta)$, and $\mathcal{V}(|\zeta|^{1/2} B_s)$ all tend to finite limits as $\zeta \to \pm\infty$.

Show also that

$$A_s^{(r)}(\zeta) = O(\zeta^{-r-\sigma_1}), \quad B_s^{(r)}(\zeta) = O(\zeta^{-r-(1/2)}), \quad C_s^{(r)}(\zeta) = O(\zeta^{-r+(1/2)}), \quad D_s^{(r)}(\zeta) = O(\zeta^{-r-\sigma_1}),$$

where $\sigma_1 = \min(\sigma, \frac{3}{2})$, provided that $r \geqslant 1$ in the case of $A_s^{(r)}(\zeta)$ and $D_s^{(r)}(\zeta)$, and $r \geqslant 0$ in the case of $B_s^{(r)}(\zeta)$ and $C_s^{(r)}(\zeta)$.

Ex. 7.2 With the notation of Exercise 4.2, show that

$$H_m(\nu x) = (2\pi)^{1/2} e^{-\nu^2/4} \nu^{(3\nu^2-1)/6} e^{\nu^2 x^2/2} \left(\frac{\zeta}{x^2-1}\right)^{1/4} \left(1 - \sum_{s=0}^{n-1} \frac{\beta_s}{\nu^{4s+2}}\right)^{-1} W_{2n+1,2},$$

where

(i) $W_{2n+1,2}$ is given by (7.11) and (7.13) with $u = 2m+1$, $\beta = \infty$;
(ii) the coefficients $A_s(\zeta)$ and $B_s(\zeta)$ are defined by (7.08) and (7.09), with

$$\psi(\zeta) = \frac{5}{16\zeta^2} - \frac{3x^2+2}{4(x^2-1)^3}\zeta, \qquad A_{s+1}(\infty) = 0 \qquad (s \geqslant 0);$$

(iii) $\beta_s = \lim\{\zeta^{1/2} B_s(\zeta)\}$ as $\zeta \to +\infty$.

***Ex. 7.3**† Let $W_{k,m}(x)$ denote the Whittaker function (Chapter 7, §11.1). Show that when k and x are positive and m is unrestricted

$$W_{k,m}(4kx) = \frac{2^{4/3}\pi^{1/2}k^{k+(1/6)}}{\phi_n(k,m)\, e^k} \left(\frac{x\zeta}{x-1}\right)^{1/4}$$
$$\times \left[\mathrm{Ai}\{(4k)^{2/3}\zeta\} \sum_{s=0}^{n} \frac{A_s(\zeta)}{(4k)^{2s}} + \frac{\mathrm{Ai}'\{(4k)^{2/3}\zeta\}}{(4k)^{4/3}} \sum_{s=0}^{n-1} \frac{B_s(\zeta)}{(4k)^{2s}} + \varepsilon_{2n+1,2}(4k,\zeta)\right],$$

provided that $\phi_n(k,m) \neq 0$, where

(i)
$$\begin{cases} \frac{2}{3}\zeta^{3/2} = (x^2-x)^{1/2} - \ln\{x^{1/2} + (x-1)^{1/2}\} & (x \geqslant 1), \\ \frac{2}{3}(-\zeta)^{3/2} = \cos^{-1}(x^{1/2}) - (x-x^2)^{1/2} & (0 < x \leqslant 1); \end{cases}$$

(ii) $A_s(\zeta)$ and $B_s(\zeta)$ are defined by (7.08) and (7.09), with

$$\psi(\zeta) = \frac{(4m^2-1)\zeta}{x(x-1)} + \frac{(3-8x)\zeta}{4x(x-1)^3} + \frac{5}{16\zeta^2};$$

(iii) $$\phi_n(k,m) = \sum_{s=0}^{n} \frac{A_s(\infty)}{(4k)^{2s}} - \sum_{s=0}^{n-1} \frac{1}{(4k)^{2s+1}} \left[\lim_{\zeta\to\infty} \{\zeta^{1/2} B_s(\zeta)\}\right];$$

(iv) $\varepsilon_{2n+1,2}(4k,\zeta)$ satisfies (7.13) with $u = 4k$, $\beta = \infty$.

Ex. 7.4 By application of the theory of Chapter 10 to equation (7.01) in the interval $\delta \leqslant \zeta < \infty$, δ being an arbitrary positive constant, a uniform asymptotic expansion can be constructed for

† This result furnishes asymptotic expansions (complete with error bounds) for $W_{k,m}(x)$ when k is large, the expansions being uniform with respect to $x \in [\delta, \infty)$ and bounded $|m|$, δ being any positive constant. For similar expansions when the variables are complex see Skovgaard (1966), and Erdélyi and Swanson (1957).

the solution $W_{2n+1,2}(u,\zeta)$ of Theorem 7.1 in the form

$$W_{2n+1,2}(u,\zeta) \sim \frac{e^{-u\xi}}{2\pi^{1/2}u^{1/6}\zeta^{1/4}} \sum_{s=0}^{\infty} (-)^s \frac{\mathsf{A}_s(\xi)}{u^s} \qquad (u\to\infty),$$

where $\xi = \frac{2}{3}\zeta^{3/2}$ and the coefficients $\mathsf{A}_s(\xi)$ are independent of u. If u_j and v_j are defined as in §1.1, then by expanding the Airy functions in (7.11) and using induction and Exercise 1.1, prove that

$$A_s(\zeta) = \sum_{j=0}^{2s} (-)^j v_j \xi^{-j} \mathsf{A}_{2s-j}(\xi), \qquad \zeta^{1/2}B_s(\zeta) = \sum_{j=0}^{2s+1} (-)^j u_j \xi^{-j} \mathsf{A}_{2s-j+1}(\xi). \quad \text{[Olver, 1954b.]}$$

Ex. 7.5 Suppose that not all derivatives of $\psi(\zeta)$ are continuous. Show that (i) if $\psi^{(2N-1)}(\zeta)$ is continuous in (α,β) and $\psi^{(3N-1)}(\zeta)$ is continuous in a neighborhood of $\zeta = 0$, then $A_N'(\zeta)$ is continuous in (α,β); (ii) if $\psi^{(2N)}(\zeta)$ is continuous in (α,β) and $\psi^{(3N)}(\zeta)$ is continuous in a neighborhood of $\zeta = 0$, then $2\zeta B_N'(\zeta)+B_N(\zeta)$ is continuous in (α,β).

Show also that with the conditions (ii) equation (7.01) has twice continuously differentiable solutions with the properties (7.10) to (7.13), provided that $n \leqslant N$.

8 Airy Functions of Complex Argument

8.1 For complex values of z, solutions of the differential equation

$$d^2w/dz^2 = zw \tag{8.01}$$

include $\mathrm{Ai}(z)$, $\mathrm{Ai}(ze^{\pm 2\pi i/3})$, $\mathrm{Bi}(z)$, and $\mathrm{Bi}(ze^{\pm 2\pi i/3})$. For convenience, we sometimes employ the notations

$$\mathrm{Ai}_0(z) = \mathrm{Ai}(z), \qquad \mathrm{Ai}_1(z) = \mathrm{Ai}(ze^{-2\pi i/3}), \qquad \mathrm{Ai}_{-1}(z) = \mathrm{Ai}(ze^{2\pi i/3}). \tag{8.02}$$

We also denote the sectors $|\mathrm{ph}\, z| \leqslant \frac{1}{3}\pi$, $\frac{1}{3}\pi \leqslant \mathrm{ph}\, z \leqslant \pi$, and $-\pi \leqslant \mathrm{ph}\, z \leqslant -\frac{1}{3}\pi$ by $\mathbf{S}_0$, $\mathbf{S}_1$, and $\mathbf{S}_{-1}$, respectively; see Fig. 8.1. And in considering the $\mathrm{Ai}_j(z)$ and $\mathbf{S}_j$, and also auxiliary functions introduced in §8.3 below, we enumerate the suffix j modulo 3; thus

$$\mathbf{S}_2 = \mathbf{S}_{-1}, \qquad \mathrm{Ai}_j(ze^{2\pi i/3}) = \mathrm{Ai}_{j-1}(z).$$

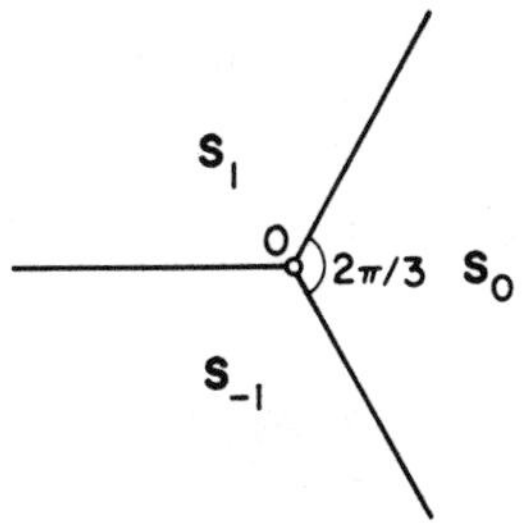

Fig. 8.1 Sectors $\mathbf{S}_j$.

From Chapter 4, §4 and Chapter 1, §8.3, we know that *the asymptotic expansions* (1.07), (1.08), *and* (1.09) *are valid with x replaced by z and ξ denoting the principal value of $\frac{2}{3}z^{3/2}$, provided that $|\mathrm{ph}\, z| \leqslant \pi-\delta$ in the case of* (1.07), *and $|\mathrm{ph}\, z| \leqslant \frac{2}{3}\pi-\delta$ in the case of* (1.08) *and* (1.09). Accordingly, *at infinity* $\mathrm{Ai}(z)$ *is recessive within* $\mathbf{S}_0$, *and dominant within* $\mathbf{S}_1$ *and* $\mathbf{S}_{-1}$. From this it follows that $\mathrm{Ai}_j(z)$ is recessive within

$\mathbf{S}_j$ and dominant within $\mathbf{S}_{j-1}$ and $\mathbf{S}_{j+1}$. Hence $\mathrm{Ai}_j(z)$ and $\mathrm{Ai}_{j+1}(z)$ comprise a numerically satisfactory pair of solutions of (8.01) in $\mathbf{S}_j \cup \mathbf{S}_{j+1}$, but not within $\mathbf{S}_{j-1}$.

In the case of $\mathrm{Bi}(z)$ *we may replace* x *by* z *and* ξ *by the principal value of* $\frac{2}{3}z^{3/2}$ *in* (1.16), (1.17), *and* (1.18), *provided that* $|\mathrm{ph}\, z| \leqslant \frac{1}{3}\pi - \delta$ *in the case of* (1.16), *and* $|\mathrm{ph}\, z| \leqslant \frac{2}{3}\pi - \delta$ *in the case of* (1.17) *and* (1.18). This assertion may be verified by using the analytic continuations of formulas (1.12) to (1.15) and referring to Hankel's expansions (Chapter 7, §4.1). Thus $\mathrm{Bi}(z)$ is dominant at infinity within all three sectors $\mathbf{S}_j$; in consequence, $\mathrm{Ai}(z)$ and $\mathrm{Bi}(z)$ comprise a numerically satisfactory pair of solutions of (8.01) only in $\mathbf{S}_0$ or on the negative real axis.

Connection formulas for the various solutions are easily derived from (1.03) and (1.11). For example, we have

$$\mathrm{Ai}(z) + e^{-2\pi i/3}\,\mathrm{Ai}_1(z) + e^{2\pi i/3}\,\mathrm{Ai}_{-1}(z) = 0, \tag{8.03}$$

(as in Chapter 2, (8.06)), and

$$\mathrm{Ai}(ze^{\pm 2\pi i/3}) = \tfrac{1}{2}e^{\pm \pi i/3}\{\mathrm{Ai}(z) \mp i\,\mathrm{Bi}(z)\}. \tag{8.04}$$

Error bounds for the asymptotic expansions of the Airy functions and their derivatives for complex arguments can be found from the error bounds for Hankel's expansions given in Chapter 7, §13.

8.2 The distribution of the zeros of the Airy functions in the complex plane may be investigated by Lommel's method.† Using (8.01), we derive

$$\frac{d}{dz}\{b\,\mathrm{Ai}(az)\,\mathrm{Ai}'(bz) - a\,\mathrm{Ai}(bz)\,\mathrm{Ai}'(az)\} = (b^3 - a^3)\, z\,\mathrm{Ai}(az)\,\mathrm{Ai}(bz),$$

where a and b are any constants. Hence if $b^3 \neq a^3$, then

$$\int_0^1 t\,\mathrm{Ai}(at)\,\mathrm{Ai}(bt)\,dt = \frac{b\,\mathrm{Ai}(a)\,\mathrm{Ai}'(b) - a\,\mathrm{Ai}(b)\,\mathrm{Ai}'(a)}{b^3 - a^3} - \frac{b-a}{b^3 - a^3}\,\mathrm{Ai}(0)\,\mathrm{Ai}'(0).$$

Similarly,

$$\int_0^1 \mathrm{Ai}'(at)\,\mathrm{Ai}'(bt)\,dt = \frac{a^2\,\mathrm{Ai}(a)\,\mathrm{Ai}'(b) - b^2\,\mathrm{Ai}(b)\,\mathrm{Ai}'(a)}{a^3 - b^3} - \frac{a^2 - b^2}{a^3 - b^3}\,\mathrm{Ai}(0)\,\mathrm{Ai}'(0).$$

Suppose that $a = re^{i\theta}$ is a nonreal zero of $\mathrm{Ai}(z)$ or $\mathrm{Ai}'(z)$. Then $b = re^{-i\theta}$ is also a zero, and we obtain

$$\int_0^1 t\,\mathrm{Ai}(at)\,\mathrm{Ai}(bt)\,dt = -\frac{1}{r^2}\frac{\sin\theta}{\sin 3\theta}\,\mathrm{Ai}(0)\,\mathrm{Ai}'(0), \tag{8.05}$$

and

$$\int_0^1 \mathrm{Ai}'(at)\,\mathrm{Ai}'(bt)\,dt = -\frac{1}{r}\frac{\sin 2\theta}{\sin 3\theta}\,\mathrm{Ai}(0)\,\mathrm{Ai}'(0). \tag{8.06}$$

The integrals on the left-hand sides of these two equations are positive, and

† Chapter 7, §6.2.

$\mathrm{Ai}(0)\,\mathrm{Ai}'(0)$ is negative. To avoid a contradiction $\sin\theta/\sin 3\theta$ and $\sin 2\theta/\sin 3\theta$ must both be positive and finite. In consequence $a \in \mathbf{S}_0$. From (1.04) and Chapter 7, Theorem 8.2, however, it is seen that there can be no zeros in $\mathbf{S}_0$. Therefore *the zeros of* $\mathrm{Ai}(z)$ *and* $\mathrm{Ai}'(z)$ *are all real and negative.*

Equations (8.05) and (8.06) also hold with Ai replaced throughout by Bi. In this case, however, $\mathrm{Bi}(0)\,\mathrm{Bi}'(0)$ is positive; this implies that nonreal zeros are confined to the sectors $\frac{1}{3}\pi < |\mathrm{ph}\, z| < \frac{1}{2}\pi$. By constructing the asymptotic forms

$$\mathrm{Bi}(ze^{\pm\pi i/3}) = (2/\pi)^{1/2} e^{\pm\pi i/6} z^{-1/4} \{\cos(\xi - \tfrac{1}{4}\pi \mp \tfrac{1}{2} i \ln 2) + e^{|\mathrm{Im}\,\xi|} O(\xi^{-1})\},$$

$$\mathrm{Bi}'(ze^{\pm\pi i/3}) = (2/\pi)^{1/2} e^{\mp\pi i/6} z^{1/4} \{\cos(\xi + \tfrac{1}{4}\pi \mp \tfrac{1}{2} i \ln 2) + e^{|\mathrm{Im}\,\xi|} O(\xi^{-1})\},$$

when $|\mathrm{ph}\, z| \leqslant \frac{2}{3}\pi - \delta$, where $\xi = \frac{2}{3} z^{3/2}$, the reader will have no difficulty in perceiving that $\mathrm{Bi}(z)$ and $\mathrm{Bi}'(z)$ each have an infinity of zeros in these sectors.†

8.3 Suitable auxiliary functions analogous to those of §2 can be constructed by introducing the weight functions

$$E_j(z) = |\exp(\tfrac{2}{3} z^{3/2})| \qquad (j = 0, \pm 1),$$

the branch of the fractional power being chosen in such a way that $E_j(z) \geqslant 1$ in $\mathbf{S}_j$ and $E_j(z) \leqslant 1$ in $\mathbf{S}_{j-1} \cup \mathbf{S}_{j+1}$. Thus $z^{3/2}$ has its principal value in the case $j = 0$. Moreover,

$$E_j(ze^{2\pi i/3}) = E_{j-1}(z), \qquad E_0(\bar{z}) = E_0(z), \qquad E_1(\bar{z}) = E_{-1}(z). \tag{8.07}$$

Next, we define

$$E_{j+1}(z)\,|\mathrm{Ai}_{j+1}(z)| = M_j(z) \sin\theta_j(z), \qquad E_{j-1}(z)\,|\mathrm{Ai}_{j-1}(z)| = M_j(z) \cos\theta_j(z), \tag{8.08}$$

$$E_{j+1}(z)\,|\mathrm{Ai}'_{j+1}(z)| = N_j(z) \sin\omega_j(z), \qquad E_{j-1}(z)\,|\mathrm{Ai}'_{j-1}(z)| = N_j(z) \cos\omega_j(z), \tag{8.09}$$

$M_j(z)$ and $N_j(z)$ being real and positive, and $\theta_j(z)$ and $\omega_j(z)$ being real. Thus

$$M_j(z) = \{E_{j+1}^2(z)\,|\mathrm{Ai}_{j+1}^2(z)| + E_{j-1}^2(z)\,|\mathrm{Ai}_{j-1}^2(z)|\}^{1/2},$$

$$\theta_j(z) = \tan^{-1}\{E_{j+1}(z)\,|\mathrm{Ai}_{j+1}(z)| / E_{j-1}(z)\,|\mathrm{Ai}_{j-1}(z)|\},$$

the inverse tangent having its principal value. Again,

$$M_j(ze^{2\pi i/3}) = M_{j-1}(z), \qquad M_0(\bar{z}) = M_0(z), \qquad M_1(\bar{z}) = M_{-1}(z). \tag{8.10}$$

Similarly for $N_j(z)$ and $\omega_j(z)$.

The role of the constants λ and μ_s of §2.4 is played by

$$\upsilon_1 = \sup\{\pi |z|^{1/2} M_j^2(z)\}, \tag{8.11}$$

and

$$\upsilon_2 = \sup\{\pi E_{j-1}(z) M_j(z) |z^{1/2} \mathrm{Ai}_{j-1}(z)|\} = \sup\{\pi E_{j+1}(z) M_j(z) |z^{1/2} \mathrm{Ai}_{j+1}(z)|\}, \tag{8.12}$$

† See also Exercise 8.4 below.

the suprema being taken for $z \in \mathbf{S}_{j-1} \cup \mathbf{S}_{j+1}$. That the suprema are finite is seen by considering asymptotic behavior for large $|z|$. That the suprema are independent of j is seen by replacing z by $ze^{2\pi i/3}$; the suffix j on all functions and sectors is then decreased by unity. Lastly, equality of the suprema in (8.12) is verifiable by taking $j = \pm 1$ and using the reflection properties given in (8.07) and (8.10).

Ex. 8.1 Verify that

$$\mathscr{W}\{\mathrm{Ai}(z), \mathrm{Ai}_{\pm 1}(z)\} = \frac{e^{\pm \pi i/6}}{2\pi}, \qquad \mathscr{W}\{\mathrm{Ai}_1(z), \mathrm{Ai}_{-1}(z)\} = \frac{1}{2\pi i}.$$

Ex. 8.2 Show that $M_j(z) \sim (2\pi)^{-1/2}|z|^{-1/4}$ as $z \to \infty$ in any closed sector, with vertex at the origin, not containing the boundaries of $\mathbf{S}_j$.

Show also that $(1+|z|^{1/4})\,M_j(z)$ and its reciprocal are bounded for all z.

Ex. 8.3 Show that $\upsilon_1 \geqslant \upsilon_2$, $\upsilon_1 \geqslant \frac{5}{4}$, and $\upsilon_2 \geqslant (\frac{5}{4})^{1/2}$.

Ex. 8.4 Let δ be an arbitrary positive constant less than $\frac{1}{3}\pi$. Also, let $\mathscr{S}$ denote the contour depicted in Fig 8.2, the equation of the curve BCD being $\mathrm{Im}(z^{3/2}) = R^{3/2}\cos\frac{3}{2}\delta$. Show that for all sufficiently large positive integers n, $\mathscr{S}$ contains n zeros of $\mathrm{Bi}(z)$ if $R = \{(\frac{3}{2}n + \frac{3}{8})\pi \sec\frac{3}{2}\delta\}^{2/3}$.

With the aid of this result prove that in the sector $\frac{1}{3}\pi < \mathrm{ph}\, z < \frac{1}{2}\pi$ the nth zero of $\mathrm{Bi}(z)$, enumerated in ascending order of magnitude, has the asymptotic expansion

$$e^{\pi i/3}T\{\tfrac{3}{8}\pi(4n-1) + \tfrac{3}{4}i\ln 2\},$$

where $T(t)$ is defined in §5.2. [Olver, 1954b.]

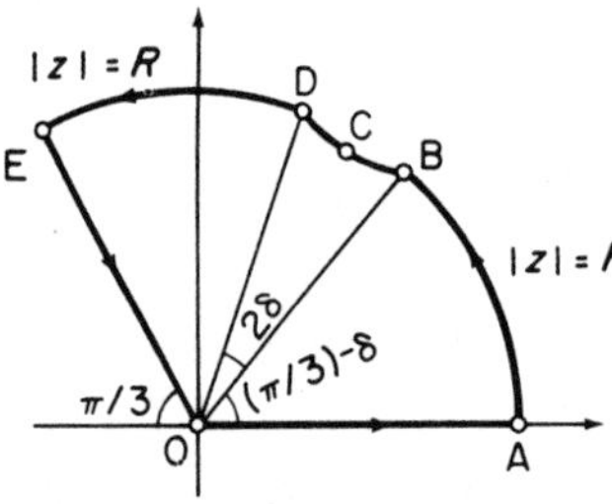

Fig. 8.2 Contour $\mathscr{S}$.

9 Asymptotic Approximations for Complex Variables

9.1 Consider now the differential equation (7.01) in the case when u is a large real or complex parameter, and $\psi(\zeta)$ is holomorphic in a bounded or unbounded domain $\boldsymbol{\Delta}$. To describe the asymptotic nature of the formal solution (7.05) in these circumstances, we introduce the following definitions, similar to those of Chapter 10, §3.1.

Let $\omega = \mathrm{ph}\, u$, and denote by $\alpha_j \equiv \alpha_j(u)$, $j = 0, \pm 1$, an arbitrary reference point interior to the intersection of $\boldsymbol{\Delta}$ and the sector $e^{-2i\omega/3}\mathbf{S}_j$.† Define $\mathbf{Z}_j(u, \alpha_j)$ to be the set of points ζ for which there exists a path $\mathscr{Q}_j$ linking ζ with α_j in $\boldsymbol{\Delta}$ and having the properties:

(i) *$\mathscr{Q}_j$ consists of a finite chain of R_2 arcs.*

† That is, $\mathbf{S}_j$ rotated through an angle $-2\omega/3$.

(ii) *As v passes along $\mathcal{Q}_j$ from ζ to α_j the real part of $uv^{3/2}$ is nondecreasing, the branch of this function being continuous and chosen so that* $\mathrm{Re}(uv^{3/2}) > 0$ *in the neighborhood of* α_j.

The point α_j may be at infinity on an infinite R_2 arc $\mathcal{L}_j \equiv \mathcal{L}_j(u)$, say, provided that $\mathcal{Q}_j$ coincides with $\mathcal{L}_j$ in a neighborhood of α_j.

The significance of Condition (ii) is indicated in the accompanying diagrams. Figure 9.1 illustrates the family of level curves of the principal branch of the function $\exp(\frac{2}{3}\zeta^{3/2})$. Figures 9.2, 9.3, and 9.4 show a typical domain $\mathbf{\Delta}$ with corresponding

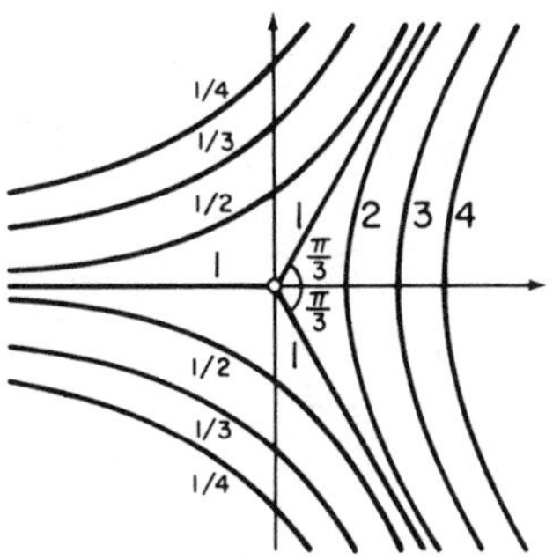

Fig. 9.1 Level curves of $\exp(\frac{2}{3}\zeta^{3/2})$. Numbers indicate values of $|\exp(\frac{2}{3}\zeta^{3/2})|$.

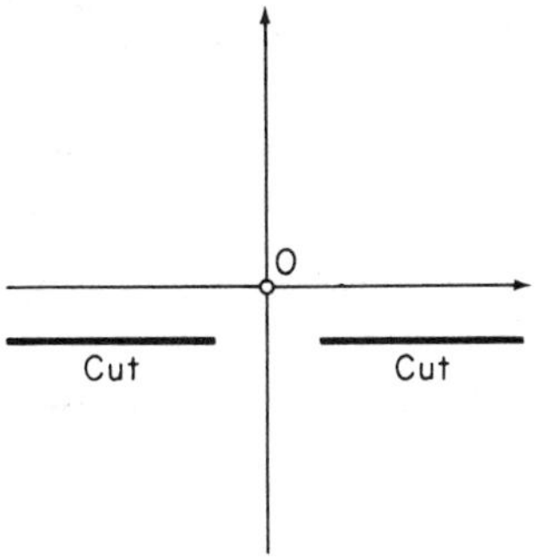

Fig. 9.2 Domain $\mathbf{\Delta}$.

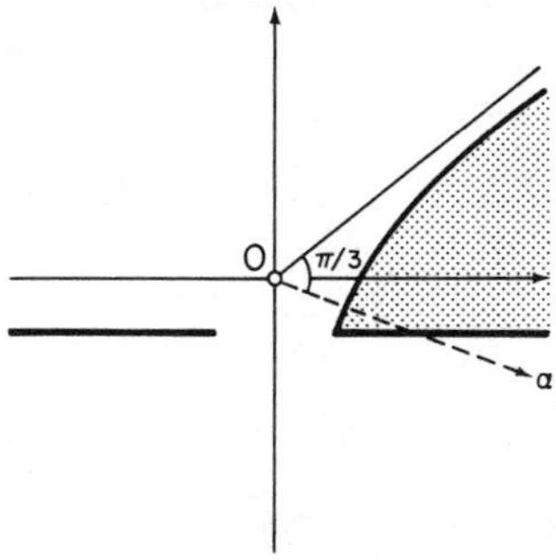

Fig. 9.3 $\mathbf{Z}_0(u, \alpha_0)$. $\alpha_0 = \infty e^{-2i\omega/3}$.

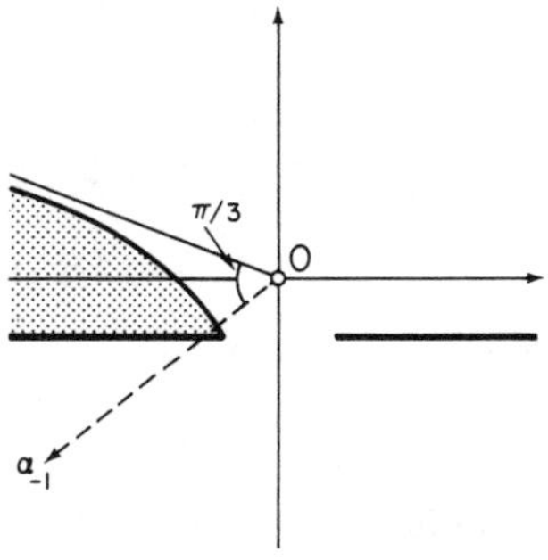

Fig. 9.4 $\mathbf{Z}_{-1}(u, \alpha_{-1})$. $\alpha_{-1} = \infty e^{-2i(\pi+\omega)/3}$.

domains $\mathbf{Z}_0(u, \infty e^{-2i\omega/3})$ and $\mathbf{Z}_{-1}(u, \infty e^{-2i(\pi+\omega)/3})$, ω being a positive acute angle. The shaded regions in Figs. 9.3 and 9.4 are excluded shadow zones (Chapter 6, §11.4); their curved boundaries belong to the family of curves of Fig. 9.1 rotated through an angle $-2\omega/3$. Often shadow zones can be eliminated by arranging that cuts in the original domain $\mathbf{\Delta}$ lie along appropriate level curves.

9.2 Theorem 9.1 *With the definitions of* §§8.3 *and* 9.1, *the differential equation*

$$d^2W/d\zeta^2 = \{u^2\zeta + \psi(\zeta)\} W, \tag{9.01}$$

has, for each value of u and each nonnegative integer n, solutions $W_{2n+1,j}(u,\zeta)$, $j=0,\pm 1$, *which are holomorphic in* $\mathbf{\Delta}$, *depend on* α_j, *and are given by*

$$W_{2n+1,j}(u,\zeta) = \mathrm{Ai}_j(u^{2/3}\zeta)\sum_{s=0}^{n}\frac{A_s(\zeta)}{u^{2s}} + \frac{\mathrm{Ai}_j'(u^{2/3}\zeta)}{u^{4/3}}\sum_{s=0}^{n-1}\frac{B_s(\zeta)}{u^{2s}} + \varepsilon_{2n+1,j}(u,\zeta). \quad (9.02)$$

Here the coefficients $A_s(\zeta)$ *and* $B_s(\zeta)$ *are the analytic continuations of the functions defined by* (7.08) *and* (7.09), *and the error term is subject to the bounds*

$$\frac{|\varepsilon_{2n+1,j}(u,\zeta)|}{M_{j\pm1}(u^{2/3}\zeta)},\ \frac{|\partial\varepsilon_{2n+1,j}(u,\zeta)/\partial\zeta|}{|u|^{2/3}N_{j\pm1}(u^{2/3}\zeta)} \leqslant \frac{4\upsilon_2}{E_j(u^{2/3}\zeta)}\exp\left\{\frac{4\upsilon_1}{|u|}\mathscr{V}_{\mathscr{Q}_j}(\zeta^{1/2}B_0)\right\}\frac{\mathscr{V}_{\mathscr{Q}_j}(\zeta^{1/2}B_n)}{|u|^{2n+1}}, \quad (9.03)$$

when $\zeta \in \mathbf{Z}_j(u,\alpha_j)$. *In* (9.03) *the suffix on M and N is* $j+1$ *when* $u^{2/3}\zeta \in \mathbf{S}_{j-1}\cup\mathbf{S}_j$ *and* $j-1$ *when* $u^{2/3}\zeta\in\mathbf{S}_j\cup\mathbf{S}_{j+1}$.†

In the typical case $j=0$ and $u^{2/3}\zeta\in\mathbf{S}_0\cup\mathbf{S}_1$, an outline of the proof of this theorem is as follows. We first observe that $A_s(\zeta)$ and $B_s(\zeta)$ are analytic everywhere that $\psi(\zeta)$ is analytic, including the origin.‡ The error term satisfies the integral equation

$$\varepsilon_{2n+1,0}(u,\zeta) = \frac{2\pi e^{-\pi i/6}}{u^{2/3}}\int_{\zeta}^{\alpha_1}\mathsf{K}(\zeta,v)\{\psi(v)\,\varepsilon_{2n+1,0}(u,v) - R_{2n+1}(u,v)\,\mathrm{Ai}(u^{2/3}v)\}\,dv,$$

in which $R_{2n+1}(u,\zeta)$ is given by the first of (7.15), and

$$\mathsf{K}(\zeta,v) = \mathrm{Ai}(u^{2/3}\zeta)\,\mathrm{Ai}_1(u^{2/3}v) - \mathrm{Ai}_1(u^{2/3}\zeta)\,\mathrm{Ai}(u^{2/3}v).$$

The monotonicity condition on the path of integration implies that

$$E_0(u^{2/3}v) \geqslant E_0(u^{2/3}\zeta),$$

and also that $u^{2/3}v\in\mathbf{S}_0\cup\mathbf{S}_1$. Therefore

$$E_1(u^{2/3}v) = 1/E_0(u^{2/3}v), \qquad E_1(u^{2/3}\zeta) = 1/E_0(u^{2/3}\zeta).$$

In consequence, we derive

$$|\mathsf{K}(\zeta,v)| \leqslant \{E_0(u^{2/3}v)/E_0(u^{2/3}\zeta)\}\,M_{-1}(u^{2/3}\zeta)\,M_{-1}(u^{2/3}v),$$

with a similar bound for $\partial\mathsf{K}/\partial\zeta$; compare (3.14) and (3.15). Application of Theorem 10.1 of Chapter 6 in the manner of §3.3 leads to the stated bounds on $\varepsilon_{2n+1,0}(u,\zeta)$ and its ζ derivative.

***9.3** When the domain $\mathbf{\Delta}$ is unbounded, Theorem 9.1 yields meaningful approximations for large ζ whenever the variations of the functions $\zeta^{1/2}B_s(\zeta)$ converge at infinity. In the case of real variables sufficient conditions are given by Exercise 7.1. For complex ζ, an investigation on the lines of Chapter 10, §5 leads to the following theorem§:

† Thus two bounds are available for the error term and its derivative when $u^{2/3}\zeta\in\mathbf{S}_j$.

‡ Analyticity at $\zeta=0$ is verifiable by considering Maclaurin expansions.

§ Olver (1958, §11). Although the result is interesting it is not required in most applications of Theorem 9.1. The proof is therefore omitted.

Assume that for all values of u under consideration:

(i) $|\psi(\zeta)| \leqslant k/(1+|\zeta|^{(1/2)+\rho})$ *when* $\zeta \in \Delta$, *where* k *and* ρ *are positive numbers independent of* ζ *and* u.

(ii) *At a prescribed point* α *of* Δ, $|A_s(\alpha)| \leqslant \hat{k}_s$, $s = 0, 1, \ldots,$ *where* $\hat{k}_s$ *is independent of* u.

(iii) *A subdomain* Γ *of* Δ *exists such that the distance between each boundary point* ζ_B *of* Γ *and any boundary point of* Δ *is not less than* $d|\zeta_B|^{-1/2}$, *where* $d\,(>0)$ *is assignable independently of* ζ_B *and* u.†

(iv) Γ *contains* α *and the disk* $|\zeta| \leqslant d$.

(v) *For each point* ζ *of* Γ, *a path can be found in* Γ *along which*

$$\int_\alpha^\zeta \frac{|dv|}{1+|v|^{1+\rho_1}} \leqslant l,$$

where $\rho_1 = \min(\rho, \frac{3}{2})$, *and* l *is independent of* ζ *and* u.

Theorem 9.2 *With the conditions of this subsection, there exist finite numbers* k_s, *independent of* ζ *and* u, *such that when* $\zeta \in \Gamma$

$$|A_s(\zeta)| \leqslant k_s, \qquad |A_s'(\zeta)| \leqslant \frac{k_s}{1+|\zeta|^{1+\rho_1}}, \qquad |A_s''(\zeta)| \leqslant \frac{k_s}{1+|\zeta|^{(1/2)+\rho_1}},$$

$$|B_s(\zeta)| \leqslant \frac{k_s}{1+|\zeta|^{1/2}}, \qquad |B_s'(\zeta)| \leqslant \frac{k_s}{1+|\zeta|^{3/2}}, \qquad |B_s''(\zeta)| \leqslant \frac{k_s}{1+|\zeta|^{1+\rho_1}}.$$

Since from (7.06)

$$\{\zeta^{1/2}B_s(\zeta)\}' = \tfrac{1}{2}\zeta^{-1/2}\{\psi(\zeta)A_s(\zeta) - A_s''(\zeta)\},$$

a corollary of this theorem is that $\mathscr{V}_{\alpha,\zeta}(\zeta^{1/2}B_s)$ is bounded in Γ for each s, provided that the path satisfies Condition (v).

Ex. 9.1 Show that Condition (v) of §9.3 is fulfilled when the path of integration comprises a bounded number of arcs of level curves of the function $\exp(\frac{2}{3}u\zeta^{3/2})$.

10 Bessel Functions of Large Order

10.1 In Chapter 10, §§7 and 8 we constructed asymptotic expansions of the modified Bessel functions $I_\nu(\nu z)$ and $K_\nu(\nu z)$ for large ν in terms of elementary functions. The z regions of validity included the singularities 0 and ∞, but not the turning points $\pm i$. In this section we establish asymptotic expansions that hold at one of the turning points as well as at the singularities. It is more convenient now to work with the unmodified Bessel functions; this means in effect that the z plane of Chapter 10 is rotated through an angle $\frac{1}{2}\pi$.

† This condition implies that in the $\frac{2}{3}\zeta^{3/2}$ plane there is a positive lower bound on the distance between the boundaries of the maps of Γ and Δ.

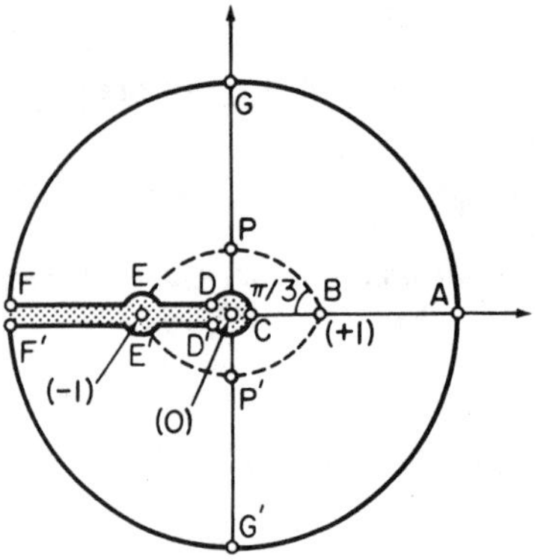

Fig. 10.1 z plane.

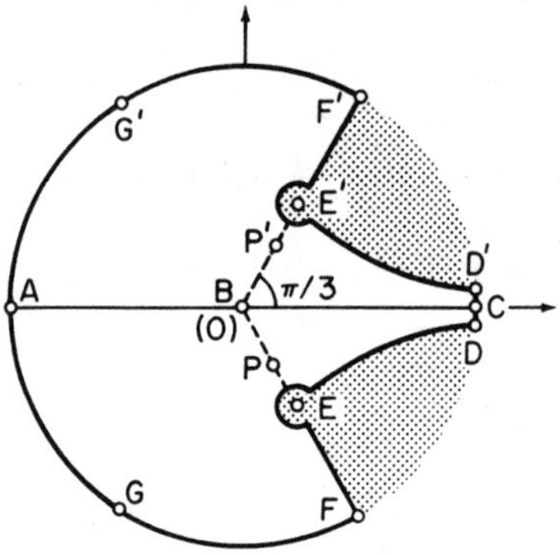

Fig. 10.2 ζ plane.

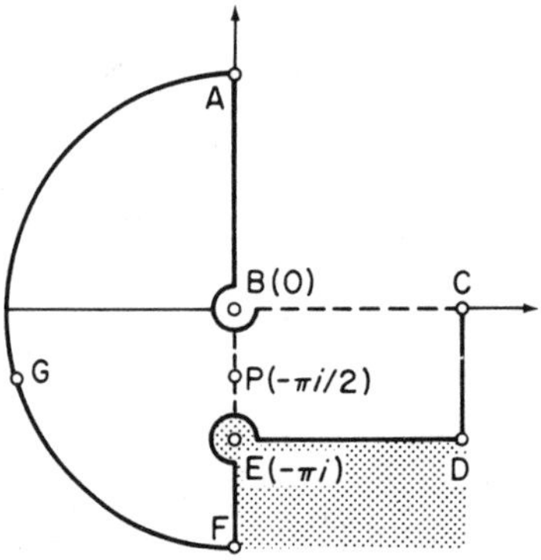

Fig. 10.3 ξ plane (i).

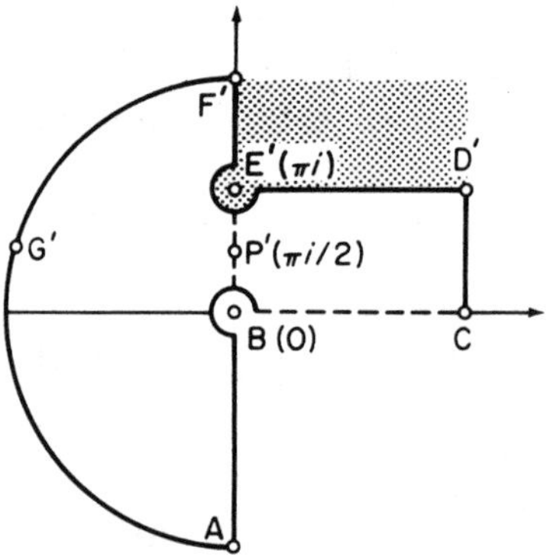

Fig. 10.4 ξ plane (ii).

Corresponding to the differential equation (7.02) of Chapter 10, we start with the equation

$$\frac{d^2w}{dz^2} = \left\{v^2\frac{1-z^2}{z^2} - \frac{1}{4z^2}\right\} w \tag{10.01}$$

satisfied by $z^{1/2}\mathscr{C}_v(vz)$, where $\mathscr{C}_v$ denotes any cylinder function of order v. Equation (10.01) has turning points at $z = \pm 1$. With v playing the role of the parameter u, application of the transformations of §3.1 yields†

$$\left(\frac{d\zeta}{dz}\right)^2 = \frac{1-z^2}{\zeta z^2} = \hat{f}, \qquad \frac{2}{3}\zeta^{3/2} = -\int_1^z \frac{(1-t^2)^{1/2}}{t}\,dt, \tag{10.02}$$

and

$$d^2W/d\zeta^2 = \{v^2\zeta + \psi(\zeta)\}\,W, \tag{10.03}$$

† The external negative sign in the second of (10.02) is introduced to make ζ real when $z \in (0, \infty)$.

where $W = \hat{f}^{1/4} w$ and

$$\psi(\zeta) = \hat{f}^{-1/4} \frac{d^2}{d\zeta^2}(\hat{f}^{1/4}) - \frac{1}{4z^2\hat{f}} = \frac{5}{16\zeta^2} + \frac{\zeta z^2(z^2+4)}{4(z^2-1)^3}; \tag{10.04}$$

compare (3.06).

Evaluation of the integral in (10.02) gives

$$\frac{2}{3}\zeta^{3/2} = \ln\frac{1+(1-z^2)^{1/2}}{z} - (1-z^2)^{1/2}, \tag{10.05}$$

it being supposed that the branches take their principal values when $z \in (0,1)$ and $\zeta \in (0,\infty)$, and are continuous elsewhere. The mapping of the z plane on the ζ plane is facilitated by the introduction of the intermediate variable $\xi = \frac{2}{3}\zeta^{3/2}$; the z to ξ mapping is closely related to the mapping of Figs. 7.1 and 7.2 of Chapter 10. Corresponding points of the present transformation are indicated in Figs. 10.1–10.4.

In the ζ plane the points E and E', both of which correspond to the turning point $z = -1$, have affixes $(\frac{3}{2}\pi)^{2/3} e^{\mp\pi i/3}$. The curves ED and $E'D'$ have parametric equations

$$\zeta = (\tfrac{3}{2})^{2/3}(\tau \mp i\pi)^{2/3} \qquad (0 \leqslant \tau < \infty).$$

In the z plane the broken curves BPE and $BP'E'$ correspond to the line segments $\operatorname{ph}\zeta = \mp\frac{1}{3}\pi$, $|\zeta| \leqslant (\frac{3}{2}\pi)^{2/3}$. They are the same as the broken curve EBE' of Chapter 10, Fig. 7.1, rotated through angles $\pm\frac{1}{2}\pi$. The eye-shaped domain bounded by BPE and $BP'E'$ will be denoted by **J**.

10.2 Let us apply Theorem 9.1 to equation (10.03). To avoid devoting too much space in the text to this problem, *we adopt the restrictions* $\nu > 0$ *and* $|\operatorname{ph} z| < \pi$. Extensions to complex ν and other z regions may be made in a manner similar to that used for the modified Bessel functions in Chapter 10, §8.

Define $\mathbf{\Delta}$ to be the ζ domain corresponding to $|\operatorname{ph} z| < \pi$, indicated in Fig. 10.2. Since $\zeta(z)$ is holomorphic within $|\operatorname{ph} z| < \pi$ and $d\zeta/dz$ does not vanish in this sector, the inverse function $z = z(\zeta)$ is holomorphic within $\mathbf{\Delta}$. From (10.02) and (10.04) we see that the same is true of $\hat{f}$ and $\psi(\zeta)$. It may, in fact, be verified that the Maclaurin expansions of $z(\zeta)$ and $\psi(\zeta)$ begin as follows:

$$z(\zeta) = 1 - 2^{-1/3}\zeta + \tfrac{3}{10}2^{-2/3}\zeta^2 + \tfrac{1}{700}\zeta^3 + \cdots, \qquad \psi(\zeta) = \tfrac{1}{70}2^{1/3} + \cdots.$$

Both $z(\zeta)$ and $\psi(\zeta)$ have singularities at the points E and E'; this is a consequence of the vanishing of $d\zeta/dz$ at $z = -1$.

The coefficients $A_s(\zeta)$ and $B_s(\zeta)$ are defined recursively by $A_0(\zeta) = 1$,

$$B_s(\zeta) = \frac{1}{2\zeta^{1/2}} \int_0^\zeta \{\psi(v)A_s(v) - A_s''(v)\} \frac{dv}{v^{1/2}}, \tag{10.06}$$

and

$$A_{s+1}(\zeta) = -\tfrac{1}{2}B_s'(\zeta) + \tfrac{1}{2}\int \psi(\zeta)B_s(\zeta)\,d\zeta + \text{constant}. \tag{10.07}$$

The branches in (10.06) assume their principal values when $\zeta \in (0, \infty)$, and are continuous elsewhere. Each coefficient is holomorphic throughout $\mathbf{\Delta}$, including $\zeta = 0$.

Consider what happens as $\zeta \to \infty$. First, we suppose that $|\mathrm{ph}(-\zeta)| \leqslant \frac{2}{3}\pi$. From Figs. 10.1 and 10.2 it is seen that $|z| \to \infty$. More precisely, from the analytic continuation of (10.05) we derive

$$z = \tfrac{2}{3}(-\zeta)^{3/2} + \tfrac{1}{2}\pi + O(\zeta^{-3/2}), \tag{10.08}$$

and thence, from (10.04),

$$\psi(\zeta) \sim -1/(4\zeta^2). \tag{10.09}$$

Starting from $A_0(\zeta) = 1$ and (10.09), and referring to Ritt's theorem (Chapter 1, §4.3), we readily establish the following relations by induction:

$$A_s(\zeta) = k_s + O(\zeta^{-3/2}), \qquad B_s(\zeta) = l_s(-\zeta)^{-1/2} + O(\zeta^{-2}), \tag{10.10}$$

as $\zeta \to \infty$ in $|\mathrm{ph}(-\zeta)| \leqslant \frac{2}{3}\pi - \delta$. Here δ is an arbitrary positive constant, and k_s and l_s are independent of ζ. And by the same method it is provable that $\mathscr{V}\{\zeta^{1/2}B_s(\zeta)\}$ converges in these circumstances.†

Next, let $\zeta \to \infty$ in the region between the boundary curves ED and $E'D'$ of Fig. 10.2. Then we have

$$z \sim 2\exp(-\tfrac{2}{3}\zeta^{3/2} - 1), \qquad \psi(\zeta) \sim 5/(16\zeta^2). \tag{10.11}$$

In this case Ritt's theorem is inapplicable because the region under consideration cannot be included in a sector. However, it is easily seen from (10.02), (10.04), and (10.05) that the asymptotic relations (10.11) are differentiable any number of times; hence by induction we derive from (10.06) and (10.07)

$$A_s(\zeta) = \hat{k}_s + O(\zeta^{-3/2}), \qquad B_s(\zeta) = \hat{l}_s\zeta^{-1/2} + O(\zeta^{-2}), \tag{10.12}$$

where $\hat{k}_s$ and $\hat{l}_s$ are independent of ζ.‡ And $\mathscr{V}\{\zeta^{1/2}B_s(\zeta)\}$ again converges.

To facilitate identification of standard solutions, and also to attain maximal regions of validity, we take the reference points required by Theorem 9.1 to be the points at infinity in the directions of strongest recession of the $\mathrm{Ai}_j(\nu^{2/3}\zeta)$; thus $\alpha_0 = +\infty$, $\alpha_1 = \infty e^{2\pi i/3}$, and $\alpha_{-1} = \infty e^{-2\pi i/3}$. This choice is permissible, because we have just shown that the variations of the functions $\zeta^{1/2}B_s(\zeta)$ converge at infinity. Since the condition on progressive paths is that $\mathrm{Re}\,\xi$ be monotonic, we see from Figs. 10.3 and 10.4 that there are no shadow zones; that is, $\mathbf{Z}_j(\nu, \alpha_j) = \mathbf{\Delta}$ for $j = 0, 1$, and -1. Accordingly, throughout $\mathbf{\Delta}$ there exist solutions of (10.03) of the form (9.02) with $u = \nu$ and the error terms bounded by (9.03).

10.3 We now identify the standard Bessel functions in terms of the asymptotic solutions. From (10.08) and (10.11) it is seen that if $\zeta \to \alpha_0$, α_1, or α_{-1}, then $z \to 0$, $\frac{1}{2}\pi - i\infty$, or $\frac{1}{2}\pi + i\infty$, respectively. From Chapter 7, solutions of (10.01) which are

† These results should be compared with Exercise 7.1 and Theorem 9.2.

‡ Compare Exercise 7.1. Because the present region is separated from $|\mathrm{ph}(-\zeta)| \leqslant \frac{2}{3}\pi - \delta$ there is no reason to suppose that $k_s = \hat{k}_s$ and $l_s = \hat{l}_s$.

recessive in these circumstances are $z^{1/2}J_\nu(\nu z)$, $z^{1/2}H_\nu^{(2)}(\nu z)$, $z^{1/2}H_\nu^{(1)}(\nu z)$. These functions are therefore multiples of $W_{2n+1,0}(\nu,\zeta)$, $W_{2n+1,1}(\nu,\zeta)$, $W_{2n+1,-1}(\nu,\zeta)$, respectively.

In particular,

$$\Lambda_n J_\nu(\nu z) = \left(\frac{\zeta}{1-z^2}\right)^{1/4} W_{2n+1,0}(\nu,\zeta), \tag{10.13}$$

where Λ_n is independent of z. To determine Λ_n, we fix the constants of integration in (10.07) by the condition

$$A_{s+1}(-\infty) = 0 \qquad (s = 0, 1, \ldots), \tag{10.14}$$

and compare the two sides of (10.13) as $\zeta \to -\infty$, that is, as $z \to +\infty$, ν and n being fixed. From Chapter 4, (9.09) we have

$$J_\nu(\nu z) = \{2/(\pi\nu z)\}^{1/2}\{\cos(\nu z - \tfrac{1}{2}\nu\pi - \tfrac{1}{4}\pi) + O(z^{-1})\}. \tag{10.15}$$

And from (1.08), (1.09), (10.08), and (10.10) we find that the right-hand side of (10.13) takes the form

$$\frac{1}{\nu^{1/6}(\pi z)^{1/2}}\left\{\cos(\nu z - \tfrac{1}{2}\nu\pi - \tfrac{1}{4}\pi) + \sin(\nu z - \tfrac{1}{2}\nu\pi - \tfrac{1}{4}\pi)\sum_{s=0}^{n-1}\frac{l_s}{\nu^{2s+1}} + O\left(\frac{1}{z}\right)\right\}$$
$$+\left(\frac{-\zeta}{z^2-1}\right)^{1/4}\varepsilon_{2n+1,0}(\nu,\zeta). \tag{10.16}$$

By letting z tend to ∞ through the sequence of values for which $\nu z - \frac{1}{2}\nu\pi - \frac{1}{4}\pi$ is an even multiple of π, we deduce that

$$\Lambda_n = 2^{-1/2}\nu^{1/3}(1+\delta_{2n+1}),$$

where

$$\delta_{2n+1} = \pi^{1/2}\nu^{1/6}\lim\{(-\zeta)^{1/4}\varepsilon_{2n+1,0}(\nu,\zeta)\}, \tag{10.17}$$

the existence of this limit being a consequence of the known existence of Λ_n.

10.4 Collecting together the results obtained so far, we have

$$J_\nu(\nu z) = \frac{1}{1+\delta_{2n+1}}\frac{1}{\nu^{1/3}}\left(\frac{4\zeta}{1-z^2}\right)^{1/4}\left\{\mathrm{Ai}(\nu^{2/3}\zeta)\sum_{s=0}^{n}\frac{A_s(\zeta)}{\nu^{2s}}\right.$$
$$\left.+\frac{\mathrm{Ai}'(\nu^{2/3}\zeta)}{\nu^{4/3}}\sum_{s=0}^{n-1}\frac{B_s(\zeta)}{\nu^{2s}} + \varepsilon_{2n+1,0}(\nu,\zeta)\right\} \tag{10.18}$$

valid when $\nu > 0$ and $|\mathrm{ph}\, z| < \pi$, where ζ is defined by (10.05), $A_s(\zeta)$ and $B_s(\zeta)$ are defined by (10.06), (10.07), and (10.14), and

$$|\varepsilon_{2n+1,0}(\nu,\zeta)| \leqslant 4\upsilon_2\frac{M_j(\nu^{2/3}\zeta)}{E_0(\nu^{2/3}\zeta)}\exp\left\{\frac{4\upsilon_1}{\nu}\mathscr{V}_{\zeta,\infty}(\zeta^{1/2}B_0)\right\}\frac{\mathscr{V}_{\zeta,\infty}(\zeta^{1/2}B_n)}{\nu^{2n+1}}, \tag{10.19}$$

the paths of variation being chosen so that $\mathrm{Re}(\zeta^{3/2})$ is monotonic. When z lies within the closure of the eye-shaped domain $\mathbf{J}$ bounded by the broken curves of Fig. 10.1, the suffix j on M may be either 1 or -1. Outside $\mathbf{J}$, $j = 1$ if $0 \leqslant \mathrm{ph}\, z < \pi$, or $j = -1$ if $-\pi < \mathrm{ph}\, z \leqslant 0$.

A bound for the quantity δ_{2n+1} can be found from (9.03). Since, however, (10.17) involves only real values of ζ, a better bound is available from Theorem 7.1. Thus we derive

$$|\delta_{2n+1}| \leqslant 2\nu^{-2n-1} e^{\nu_0/\nu} \mathscr{V}_{-\infty,\infty}(|\zeta|^{1/2} B_n), \tag{10.20}$$

where $\nu_0 = 2\lambda \mathscr{V}_{-\infty,\infty}(|\zeta|^{1/2} B_0)$ and λ is defined by (2.13). Use of Theorem 7.1 also yields a sharper bound for the error term $\varepsilon_{2n+1,0}(\nu,\zeta)$, again when z is real and positive:

$$|\varepsilon_{2n+1,0}(\nu,\zeta)| \leqslant \frac{2M(\nu^{2/3}\zeta)}{E(\nu^{2/3}\zeta)} \exp\left\{\frac{2\lambda}{\nu} \mathscr{V}_{\zeta,\infty}(|\zeta|^{1/2} B_0)\right\} \frac{\mathscr{V}_{\zeta,\infty}(|\zeta|^{1/2} B_n)}{\nu^{2n+1}}, \tag{10.21}$$

where E and M are defined in §2.

An interesting and important property of the coefficients $B_s(\zeta)$ is obtained from (10.15) and (10.16) on letting z tend to ∞ through the sequence of values for which $\nu z - \frac{1}{2}\nu\pi + \frac{1}{4}\pi$ is an even multiple of π. In this event the cosine terms vanish, and we obtain

$$\sum_{s=0}^{n-1} \frac{l_s}{\nu^{2s+1}} = \pi^{1/2} \nu^{1/6} \lim \{(-\zeta)^{1/4} \varepsilon_{2n+1,0}(\nu,\zeta)\}. \tag{10.22}$$

For large ν, the content of the braces is $O(\nu^{-2n-1})$, uniformly with respect to ζ. Hence $l_0 = l_1 = \cdots = l_{n-1} = 0$. Since n is arbitrary, this means that

$$\lim_{\zeta \to -\infty} \{(-\zeta)^{1/2} B_s(\zeta)\} = 0 \qquad (s = 0, 1, \ldots). \tag{10.23}$$

10.5 The corresponding results for the Hankel functions can be found by letting $\zeta \to \infty e^{\mp 2\pi i/3}$. For $s \geqslant 0$, the limiting values of $A_{s+1}(\zeta)$ and $(-\zeta)^{1/2} B_s(\zeta)$ are all zero; this follows from the equations $k_{s+1} = l_s = 0$ and Cauchy's theorem. Comparing the asymptotic forms of the Hankel functions (Chapter 7, §4.1) with those of the $\{\zeta/(1-z^2)\}^{1/4} W_{2n+1,j}(\nu,\zeta)$, we arrive at

$$H_\nu^{(1)}(\nu z) = \frac{2e^{-\pi i/3}}{\nu^{1/3}} \left(\frac{4\zeta}{1-z^2}\right)^{1/4} W_{2n+1,-1}(\nu,\zeta), \tag{10.24}$$

$$H_\nu^{(2)}(\nu z) = \frac{2e^{\pi i/3}}{\nu^{1/3}} \left(\frac{4\zeta}{1-z^2}\right)^{1/4} W_{2n+1,1}(\nu,\zeta). \tag{10.25}$$

The need for *separate* identification of $J_\nu(\nu z)$, $H_\nu^{(1)}(\nu z)$, and $H_\nu^{(2)}(\nu z)$ should be noticed. If, for example, we attempted to deduce (10.25) from (10.18) and (10.24) by means of the connection formula

$$H_\nu^{(2)}(\nu z) = 2J_\nu(\nu z) - H_\nu^{(1)}(\nu z),$$

then we would fail because everywhere in the z plane at least one of the error terms associated with the right-hand side is of dominant character, *including* the region in which $H_\nu^{(2)}(\nu z)$ is recessive. On the other hand, the corresponding expansion for $Y_\nu(\nu z)$ can be deduced by use of connection formulas since $Y_\nu(\nu z)$ is dominant everywhere.

10.6 The problem treated in this section is also amenable to the method of Chester, Friedman, and Ursell; see Chapter 9, especially Exercises 13.2 and 13.3. Again, the differential-equation approach provides the easiest way of calculating higher terms, error bounds, and regions of validity in the complex plane.

Ex. 10.1 Establish the compound expansions

$$J_\nu(\nu z) \sim \frac{1}{\nu^{1/3}}\left(\frac{4\zeta}{1-z^2}\right)^{1/4}\left\{\mathrm{Ai}(\nu^{2/3}\zeta)\sum_{s=0}^{\infty}\frac{A_s(\zeta)}{\nu^{2s}} + \frac{\mathrm{Ai}'(\nu^{2/3}\zeta)}{\nu^{4/3}}\sum_{s=0}^{\infty}\frac{B_s(\zeta)}{\nu^{2s}}\right\},$$

$$Y_\nu(\nu z) \sim -\frac{1}{\nu^{1/3}}\left(\frac{4\zeta}{1-z^2}\right)^{1/4}\left\{\mathrm{Bi}(\nu^{2/3}\zeta)\sum_{s=0}^{\infty}\frac{A_s(\zeta)}{\nu^{2s}} + \frac{\mathrm{Bi}'(\nu^{2/3}\zeta)}{\nu^{4/3}}\sum_{s=0}^{\infty}\frac{B_s(\zeta)}{\nu^{2s}}\right\},$$

as $\nu \to \infty$, uniformly with respect to z in the sector $|\mathrm{ph}\, z| \leqslant \pi - \delta$ $(< \pi)$.

Ex. 10.2 Show how to deduce the approximation (5.01) of Chapter 9 from the preceding exercise.

Ex. 10.3 By use of Exercise 7.4 show that

$$A_s(\zeta) = \sum_{j=0}^{2s} (\tfrac{3}{2})^j v_j \zeta^{-3j/2} U_{2s-j}(p), \qquad \zeta^{1/2} B_s(\zeta) = -\sum_{j=0}^{2s+1} (\tfrac{3}{2})^j u_j \zeta^{-3j/2} U_{2s-j+1}(p),$$

where u_s and v_s are defined in §1.1, $U_s(p)$ is defined in Chapter 10, §7.3, $p = (1-z^2)^{-1/2}$, and the branches of the fractional powers take their principal values when $z \in (0, 1)$ and are continuous elsewhere.

Show also that, in the notation of (10.12) and Exercise 7.1 of Chapter 10, $\hat{k}_s = \gamma_{2s}$ and $\hat{l}_s = -\gamma_{2s+1}$.

Ex. 10.4 Show that with proper choices of branch of $\zeta(z)$ and progressive path, relations (10.18) and (10.19) hold for $\nu > 0$ and *any* value of ph z, provided that z does not lie in either of the intervals $(-\infty, -1]$ and $[1, \infty)$.

What are the corresponding extensions of (10.24) and (10.25)?

***Ex. 10.5** Let (i) $f(t)$ and $f'(t)$ be sectionally continuous in $(0, \infty)$; (ii) $f(t) = O(t^{c_1})$ as $t \to 0$; (iii) $f(t)$ and $f'(t) = O(t^{-(1/2)-c_2})$ as $t \to \infty$. Here c_1 and c_2 are real constants, c_2 being positive. With the aid of Chapter 9, §§6.3 and 8, prove that

$$\lim_{\nu \to +\infty}\left\{\nu \int_0^\infty f(t) J_\nu(\nu t)\, dt\right\} = \tfrac{1}{3}f(1-) + \tfrac{2}{3}f(1+).$$

***Ex. 10.6** Show that when $\nu > 0$ and $|\mathrm{ph}\, z| < \pi$,

$$K_{i\nu}(\nu z) = \frac{\pi e^{-\nu\pi/2}}{\nu^{1/3}}\left(\frac{4\zeta}{1-z^2}\right)^{1/4}\left\{\mathrm{Ai}(-\nu^{2/3}\zeta)\sum_{s=0}^{n}(-)^s\frac{A_s(\zeta)}{\nu^{2s}}\right.$$

$$\left. + \frac{\mathrm{Ai}'(-\nu^{2/3}\zeta)}{\nu^{4/3}}\sum_{s=0}^{n-1}(-)^s\frac{B_s(\zeta)}{\nu^{2s}} + \varepsilon_{2n+1}(\nu, \zeta)\right\},$$

where ζ, $A_s(\zeta)$, and $B_s(\zeta)$ are defined as in §§10.2 and 10.3, and

$$|\varepsilon_{2n+1}(\nu, \zeta)| \leqslant 4\upsilon_2 \frac{M_j(-\nu^{2/3}\zeta)}{E_0(-\nu^{2/3}\zeta)} \exp\left\{\frac{4\upsilon_1}{\nu}\mathscr{V}_{-\infty,\zeta}(\zeta^{1/2}B_0)\right\}\frac{\mathscr{V}_{-\infty,\zeta}(\zeta^{1/2}B_n)}{\nu^{2n+1}},$$

the paths of variation being chosen so that $\mathrm{Im}(\zeta^{3/2})$ is monotonic, and the suffix j on M being -1 if $0 \leqslant \mathrm{ph}\, z < \pi$ and 1 if $-\pi < \mathrm{ph}\, z \leqslant 0$. [Balogh, 1967.]

***Ex. 10.7** As in Chapter 7, §6, let $j_{\nu,n}$ denote the nth positive zero of $J_\nu(z)$. Show that as $\nu \to +\infty$,

$$j_{\nu,n} \sim \nu \sum_{s=0}^{\infty} \frac{l_s(\nu^{-2/3}a_n)}{\nu^{2s}}, \qquad J_\nu'(j_{\nu,n}) \sim -\frac{\mathrm{Ai}'(a_n)}{\nu^{2/3}} \sum_{s=0}^{\infty} \frac{L_s(\nu^{-2/3}a_n)}{\nu^{2s}},$$

uniformly with respect to n, where $l_s(\zeta)$ and $L_s(\zeta)$ are holomorphic throughout $\mathbf{\Delta}$. Obtain explicit expressions for l_0, l_1, and L_0.

***Ex. 10.8** By deforming the contour $\mathscr{C}$ in the integral (9.03) of Chapter 2 into the best possible circle centered at the origin, prove *Kapteyn's inequality*

$$|J_n(nz)| \leqslant |z^n \exp\{n(1-z^2)^{1/2}\}\{1+(1-z^2)^{1/2}\}^{-n}|,$$

where n is a nonnegative integer, z is unrestricted, and $(1-z^2)^{1/2}$ has its principal value.

With the aid of Exercise 8.1 of Chapter 10, deduce that $|J_n(nz)| \leqslant 1$ when $z \in \mathbf{J}$, defined in §10.1.

*11 More General Form of Differential Equation

11.1 In Chapter 10, §9 we considered equations of the form

$$d^2w/dz^2 = u^2 f(u,z) w, \tag{11.01}$$

in which $f(u,z)$ is an analytic function of z having a uniform asymptotic expansion

$$f(u,z) \sim f_0(z) + \frac{f_1(z)}{u} + \frac{f_2(z)}{u^2} + \cdots$$

as $|u| \to \infty$. In the case when the coefficients $f_s(z)$ are holomorphic in a complex domain $\mathbf{D}(u)$ and $f_0(z)$ vanishes at just one point z_0 (say) of $\mathbf{D}(u)$, we have an extended form of turning-point problem. We indicate briefly how to construct uniform asymptotic solutions for this more general problem.

Following §3.1, we make the transformations

$$\zeta = \left\{\frac{3}{2}\int_{z_0}^{z} f_0^{1/2}(t)\,dt\right\}^{2/3}, \qquad W = \left(\frac{d\zeta}{dz}\right)^{1/2} w = \hat{f}^{1/4}(z) w,$$

where $\hat{f}(z) = f_0(z)/\zeta$. Then (11.01) becomes

$$d^2W/d\zeta^2 = \{u^2\zeta + u\phi(\zeta) + \psi(u,\zeta)\} W, \tag{11.02}$$

with

$$\phi(\zeta) = f_1(z)/\hat{f}(z), \tag{11.03}$$

and

$$\psi(u,\zeta) \sim \sum_{s=0}^{\infty} \frac{\psi_s(\zeta)}{u^s}, \tag{11.04}$$

the coefficients in the last expansion being given by

$$\psi_0(\zeta) = \frac{f_2(z)}{\hat{f}(z)} - \frac{1}{\hat{f}^{3/4}(z)} \frac{d^2}{dz^2}\left\{\frac{1}{\hat{f}^{1/4}(z)}\right\}, \qquad \psi_s(\zeta) = \frac{f_{s+2}(z)}{\hat{f}(z)} \qquad (s \geqslant 1).$$

Since $\zeta(z)$, $f_s(z)$, and $1/\hat{f}(z)$ are all holomorphic in $\mathbf{D}(u)$, the functions $\phi(\zeta)$ and $\psi_s(\zeta)$ are holomorphic in the corresponding ζ domain $\mathbf{\Delta}(u)$, say.

11.2 Can the differential equation (11.02) be satisfied formally by a series of the previous form (7.05)? By direct trial we find that the answer is negative, unless $\phi(\zeta)$ and the $\psi_s(\zeta)$ of odd suffix all vanish. Nor can (11.02) be satisfied by the more general expansion

$$W = \mathrm{Ai}(u^{2/3}\zeta)\sum_{s=0}^{\infty}\frac{A_s(\zeta)}{u^s} + \frac{\mathrm{Ai}'(u^{2/3}\zeta)}{u^{4/3}}\sum_{s=0}^{\infty}\frac{B_s(\zeta)}{u^s},$$

unless $\phi(\zeta)$ vanishes.

To discover a suitable series solution we first use the method of §3.1 to arrive at a tentative approximation. We introduce a variable $\hat{\zeta}$ by means of the equation

$$\hat{\zeta}(d\hat{\zeta}/d\zeta)^2 = u^2\zeta + u\phi(\zeta) + \hat{\psi}(u,\zeta), \tag{11.05}$$

in which $\hat{\psi}(u,\zeta)$ is any function which is bounded in $\mathbf{\Delta}(u)$ when $|u|\to\infty$. Provided that $\phi(0)=0$—and temporarily we assume this to be the case—we may choose

$$\hat{\psi}(u,\zeta) = \phi^2(\zeta)/(4\zeta),$$

thereby making the right-hand side of (11.05) a perfect square. On taking the square root and integrating, we derive

$$\frac{2}{3}\hat{\zeta}^{3/2} = \frac{2}{3}u\zeta^{3/2} + \frac{1}{2}\int_0^{\zeta}\frac{\phi(v)}{v^{1/2}}\,dv,$$

giving

$$\hat{\zeta} = u^{2/3}\zeta\left\{1 + \frac{3}{4u\zeta^{3/2}}\int_0^{\zeta}\frac{\phi(v)}{v^{1/2}}\,dv\right\}^{2/3} = u^{2/3}\zeta + \frac{\Phi(\zeta)}{u^{1/3}} + O\left(\frac{1}{u^{4/3}}\right),$$

where

$$\Phi(\zeta) = \frac{1}{2\zeta^{1/2}}\int_0^{\zeta}\frac{\phi(v)}{v^{1/2}}\,dv. \tag{11.06}$$

The resulting approximate solution of (11.02) is $(d\hat{\zeta}/d\zeta)^{-1/2}\mathrm{Ai}(\hat{\zeta})$. Expanding by Taylor's theorem and discarding a factor $u^{-1/3}$, we find that this approximation becomes

$$\{1+O(u^{-1})\}\,\mathrm{Ai}\{u^{2/3}\zeta+u^{-1/3}\Phi(\zeta)\} + O(u^{-4/3})\,\mathrm{Ai}'\{u^{2/3}\zeta+u^{-1/3}\Phi(\zeta)\}.$$

11.3 The last expression suggests that (11.02) may be satisfiable formally by a series of the form

$$W = \mathrm{Ai}\left(u^{2/3}\zeta + \frac{\Phi}{u^{1/3}}\right)\sum_{s=0}^{\infty}\frac{A_s(\zeta)}{u^s} + \frac{1}{u^{4/3}}\mathrm{Ai}'\left(u^{2/3}\zeta + \frac{\Phi}{u^{1/3}}\right)\sum_{s=0}^{\infty}\frac{B_s(\zeta)}{u^s}, \tag{11.07}$$

where $\Phi = \Phi(\zeta)$ is defined by (11.06). Since $\Phi(\zeta)$ is analytic at $\zeta=0$ whether or not $\phi(0)$ is zero, we drop the restriction $\phi(0)=0$ in what follows.

We have

$$\frac{d}{d\zeta}\operatorname{Ai}\left(u^{2/3}\zeta+\frac{\Phi}{u^{1/3}}\right)=u^{2/3}\left(1+\frac{\Phi'}{u}\right)\operatorname{Ai}'\left(u^{2/3}\zeta+\frac{\Phi}{u^{1/3}}\right),$$

and

$$\frac{d}{d\zeta}\operatorname{Ai}'\left(u^{2/3}\zeta+\frac{\Phi}{u^{1/3}}\right)=u^{4/3}\left(\zeta+\frac{\Phi}{u}\right)\left(1+\frac{\Phi'}{u}\right)\operatorname{Ai}\left(u^{2/3}\zeta+\frac{\Phi}{u^{1/3}}\right),$$

primes denoting differentiations with respect to ζ, except those on Ai. Consequently, differentiation of (11.07) produces

$$\frac{dW}{d\zeta}=\operatorname{Ai}\left(u^{2/3}\zeta+\frac{\Phi}{u^{1/3}}\right)\sum_{s=0}^{\infty}\frac{C_s(\zeta)}{u^s}+u^{2/3}\operatorname{Ai}'\left(u^{2/3}\zeta+\frac{\Phi}{u^{1/3}}\right)\sum_{s=0}^{\infty}\frac{D_s(\zeta)}{u^s},$$

where

$$C_s=A_s'+\zeta B_s+(\zeta\Phi'+\Phi)B_{s-1}+\Phi\Phi'B_{s-2},\qquad D_s=A_s+\Phi'A_{s-1}+B_{s-2}'.$$

We differentiate again, substitute the result in (11.02), and equate coefficients of like powers of u. Using (11.04), (11.07), and the relation

$$2\zeta\Phi'+\Phi=\phi$$

(obtained from (11.06)), we find on reduction that

$$\begin{aligned}B_s+2\zeta B_s'={}&\psi_0A_s+\psi_1A_{s-1}+\cdots+\psi_sA_0-(2\Phi\Phi'+\zeta\Phi'^2)A_s-A_s''-\Phi\Phi'^2A_{s-1}\\&-(2\Phi'+\zeta\Phi'')B_{s-1}-2(\zeta\Phi'+\Phi)B_{s-1}'-(\Phi\Phi''+\Phi'^2)B_{s-2}-2\Phi\Phi'B_{s-2}',\end{aligned}$$

and

$$\begin{aligned}2A_{s+1}'={}&-\Phi''A_s-2\Phi'A_s'+\psi_0B_{s-1}+\psi_1B_{s-2}+\cdots+\psi_{s-1}B_0\\&-(2\Phi\Phi'+\zeta\Phi'^2)B_{s-1}-B_{s-1}''-\Phi\Phi'^2B_{s-2}.\end{aligned}$$

The last two equations correspond to (7.06) and (7.07), and in the previous manner their integrated forms determine the $B_s(\zeta)$ and $A_s(\zeta)$ recursively, apart from arbitrary constants of integration. Each coefficient is holomorphic throughout $\mathbf{\Delta}(u)$, including the turning point $\zeta=0$. In particular, $A_0(\zeta)=$ constant, which we may take to be unity; then

$$B_0(\zeta)=\frac{1}{2\zeta^{1/2}}\int_0^{\zeta}\{\psi_0(v)-2\Phi(v)\Phi'(v)-v\Phi'^2(v)\}\frac{dv}{v^{1/2}}.$$

11.4 The investigation may be continued on the lines of §§7.3 and 9.2 with a view to establishing the asymptotic nature of the formal expansion (11.07); compare Theorem 9.1 of Chapter 10. We shall not pursue the details except in one respect. A preliminary requirement in the construction of an integral equation for the error term is a differential equation, with known solutions, which approximates (11.02) as far as the term $u\phi(\zeta)W$ when $|u|$ is large. This can be found by applying the Liouville transformation

$$z=u^{2/3}\zeta+u^{-1/3}\Phi,\qquad w=(dz/d\zeta)^{1/2}W,$$

to the Airy equation $d^2w/dz^2 = zw$. The result is given by

$$\frac{d^2W}{d\zeta^2} = \left\{u^2\zeta + u\phi + \zeta\Phi'^2 + 2\Phi\Phi' + \frac{\Phi\Phi'^2}{u} + \frac{3\Phi''^2 - 2\Phi'\Phi''' - 2u\Phi'''}{4(u+\Phi')^2}\right\} W, \tag{11.08}$$

with solutions $(u+\Phi')^{-1/2}\,\mathrm{Ai}(u^{2/3}\zeta + u^{-1/3}\Phi)$ and $(u+\Phi')^{-1/2}\,\mathrm{Bi}(u^{2/3}\zeta + u^{-1/3}\Phi)$. Clearly (11.08) has the desired properties; compare also Chapter 10, (9.12).

*12 Inhomogeneous Equations

12.1 In §10 of Chapter 10 we considered asymptotic solutions of certain inhomogeneous differential equations belonging to Case I. Analogous turning-point problems are posed by equations of the form

$$d^2W/d\zeta^2 - \{u^2\zeta + \psi(\zeta)\}\,W = \varpi(\zeta), \tag{12.01}$$

in which u is a large real or complex parameter, and the functions $\psi(\zeta)$ and $\varpi(\zeta)$ are independent of u and holomorphic in a complex domain $\mathbf{\Delta}$ containing $\zeta = 0$.

The simplest representative equation of this type is given by

$$d^2W/d\zeta^2 - u^2\zeta W = 1. \tag{12.02}$$

In this section we show how to construct formal series solutions of (12.01) in terms of solutions of (12.02). We first observe that a simple change of variables transforms (12.02) into the parameter-free form

$$\mathrm{Wi}''(z) - z\,\mathrm{Wi}(z) = 1, \tag{12.03}$$

with $W = u^{-4/3}\,\mathrm{Wi}(u^{2/3}\zeta)$. By analogy with §7.2, we try for a solution of (12.01) in the form

$$W = \frac{\mathrm{Wi}(u^{2/3}\zeta)}{u^{4/3}}\sum_{s=0}^{\infty}\frac{A_s(\zeta)}{u^{2s}} + \frac{\mathrm{Wi}'(u^{2/3}\zeta)}{u^{8/3}}\sum_{s=0}^{\infty}\frac{B_s(\zeta)}{u^{2s}} + \frac{1}{u^2}\sum_{s=0}^{\infty}\frac{G_s(\zeta)}{u^{2s}}, \tag{12.04}$$

with $A_s(\zeta)$, $B_s(\zeta)$, and $G_s(\zeta)$ independent of u. Differentiation yields

$$\frac{dW}{d\zeta} = \frac{\mathrm{Wi}(u^{2/3}\zeta)}{u^{4/3}}\sum_{s=0}^{\infty}\frac{C_s(\zeta)}{u^{2s}} + \frac{\mathrm{Wi}'(u^{2/3}\zeta)}{u^{2/3}}\sum_{s=0}^{\infty}\frac{D_s(\zeta)}{u^{2s}} + \frac{1}{u^2}\sum_{s=0}^{\infty}\frac{H_s(\zeta)}{u^{2s}},$$

where

$$C_s(\zeta) = A_s'(\zeta) + \zeta B_s(\zeta), \qquad D_s(\zeta) = A_s(\zeta) + B_{s-1}'(\zeta), \qquad H_s(\zeta) = G_s'(\zeta) + B_s(\zeta).$$

Differentiating again and comparing coefficients, we find that (12.01) is satisfied if

$$A_s''(\zeta) + 2\zeta B_s'(\zeta) + B_s(\zeta) = \psi(\zeta)\,A_s(\zeta), \qquad 2A_{s+1}'(\zeta) + B_s''(\zeta) = \psi(\zeta)\,B_s(\zeta),$$

$$A_0(\zeta) = \zeta G_0(\zeta) + \varpi(\zeta),$$

and

$$A_s(\zeta) + 2B_{s-1}'(\zeta) + G_{s-1}''(\zeta) = \zeta G_s(\zeta) + \psi(\zeta)\,G_{s-1}(\zeta) \qquad (s \geqslant 1).$$

Hence

$$B_s(\zeta) = \frac{1}{2\zeta^{1/2}} \int_0^\zeta \{\psi(v) A_s(v) - A_s''(v)\} \frac{dv}{v^{1/2}},$$
$$A_{s+1}(\zeta) = -\frac{1}{2} B_s'(\zeta) + \frac{1}{2} \int \psi(\zeta) B_s(\zeta)\, d\zeta, \tag{12.05}$$

$$G_0(\zeta) = \frac{A_0(\zeta) - \varpi(\zeta)}{\zeta},$$
$$G_s(\zeta) = \frac{A_s(\zeta) + 2B_{s-1}'(\zeta) + G_{s-1}''(\zeta) - \psi(\zeta) G_{s-1}(\zeta)}{\zeta} \qquad (s \geqslant 1). \tag{12.06}$$

Equations (12.05) agree with (7.08) and (7.09). When combined with (12.06) they determine successively $A_s(\zeta)$, $G_s(\zeta)$, $B_s(\zeta)$ for $s = 0, 1, \ldots$. Each function is holomorphic in $\mathbf{\Delta}$, including the turning point $\zeta = 0$, *provided that* the integration constants in the second of (12.05) are chosen in such a way that the numerators of the fractions in (12.06) vanish at $\zeta = 0$. This gives, for example,

$$A_0(\zeta) = A_0(0) = \varpi(0),$$

and

$$A_1(0) = -2B_0'(0) - G_0''(0) + \psi(0) G_0(0) = -\tfrac{2}{3}\psi'(0)\varpi(0) + \tfrac{1}{3}\varpi'''(0) - \psi(0)\varpi'(0).$$

12.2 The asymptotic nature of the formal expansion (12.04) for large $|u|$ can be established in the manner of Chapter 10, §10. We shall not elaborate on the details, except to discuss the more important properties of solutions of the comparison equation (12.03). For historical reasons we renormalize this equation in the form

$$w''(z) - zw(z) = 1/\pi. \tag{12.07}$$

The solutions are obviously related by $\mathrm{Wi}(z) = \pi w(z)$. Referring to the Wronskian relation (1.19) and using the method of variation of parameters, we obtain

$$w(z) = \mathrm{Bi}(z) \int \mathrm{Ai}(z)\, dz - \mathrm{Ai}(z) \int \mathrm{Bi}(z)\, dz.$$

For real values of $z = x$, say, standard solutions are $-\mathrm{Gi}(x)$ and $\mathrm{Hi}(x)$ where $\mathrm{Gi}(x)$ and $\mathrm{Hi}(x)$ are *Scorer's functions*, given by

$$\mathrm{Gi}(x) = \mathrm{Bi}(x) \int_x^\infty \mathrm{Ai}(t)\, dt + \mathrm{Ai}(x) \int_0^x \mathrm{Bi}(t)\, dt, \tag{12.08}$$

and

$$\mathrm{Hi}(x) = \mathrm{Bi}(x) \int_{-\infty}^x \mathrm{Ai}(t)\, dt - \mathrm{Ai}(x) \int_{-\infty}^x \mathrm{Bi}(t)\, dt. \tag{12.09}$$

The general solution of (12.07), with $z = x$, can be expressed in either of the forms

$$w(x) = A\,\mathrm{Ai}(x) + B\,\mathrm{Bi}(x) - \mathrm{Gi}(x), \qquad w(x) = C\,\mathrm{Ai}(x) + D\,\mathrm{Bi}(x) + \mathrm{Hi}(x), \tag{12.10}$$

where A, B, C, and D are arbitrary constants.

The connection between Gi(x) and Hi(x) can be found by evaluating these functions and their derivatives at $x = 0$. From Chapter 9, §§6.3 and 8, we have

$$\int_0^\infty \mathrm{Ai}(t)\,dt = \tfrac{1}{3}, \qquad \int_{-\infty}^0 \mathrm{Ai}(t)\,dt = \tfrac{2}{3}.$$

Similarly, from (1.14) and Chapter 7, Exercise 5.8, we see that

$$\int_{-\infty}^0 \mathrm{Bi}(t)\,dt = 3^{-1/2}\int_0^\infty \{J_{-1/3}(\xi)-J_{1/3}(\xi)\}\,d\xi = 0.$$

Substituting in (12.08) and (12.09) and their differentiated forms by means of these relations and (1.11), we find that

$$\mathrm{Gi}(0) = \tfrac{1}{2}\,\mathrm{Hi}(0) = \tfrac{1}{3}\,\mathrm{Bi}(0) = 1/\{3^{7/6}\Gamma(\tfrac{2}{3})\},$$
$$\mathrm{Gi}'(0) = \tfrac{1}{2}\,\mathrm{Hi}'(0) = \tfrac{1}{3}\,\mathrm{Bi}'(0) = 1/\{3^{5/6}\Gamma(\tfrac{1}{3})\},$$

and thence that

$$\mathrm{Gi}(x) + \mathrm{Hi}(x) = \mathrm{Bi}(x). \tag{12.11}$$

An integral representation for Hi(x) analogous to Airy's integral is furnished by

$$\mathrm{Hi}(x) = \frac{1}{\pi}\int_0^\infty \exp(-\tfrac{1}{3}t^3+xt)\,dt. \tag{12.12}$$

This is easily verified by showing that the right-hand side satisfies the differential equation (12.07) and assumes the correct starting values at $x = 0$. From the last two equations and (1.10), we have

$$\mathrm{Gi}(x) = \frac{1}{\pi}\int_0^\infty \sin(\tfrac{1}{3}t^3+xt)\,dt. \tag{12.13}$$

Equations (12.12) and (12.13) were used as definitions by Scorer (1950). The former was also used as a definition in Chapter 9, §4.1.

12.3 For large negative x the asymptotic expansion of Hi(x) is found by applying Watson's lemma to the integral (12.12); thus

$$\mathrm{Hi}(x) \sim -\frac{1}{\pi x}\left\{1 + \frac{1}{x^3}\sum_{s=0}^\infty \frac{(3s+2)!}{s!\,(3x^3)^s}\right\} \qquad (x \to -\infty). \tag{12.14}$$

The corresponding expansion for Gi(x) may be written down from this result and (1.17) by means of (12.11). Clearly, for sufficiently large $-x$, Gi(x) behaves numerically like Bi(x). It is therefore an unsatisfactory companion to the complementary functions Ai(x) and Bi(x) for the purpose of representing the general solution in the form (12.10). For $x \in (-\infty, 0]$, the *only* numerically satisfactory particular solution is, in fact, Hi(x).

For large positive x, the asymptotic form of Gi(x) can be found from the integral (12.13) by the method of integration by parts. Two such integrations yield

$$\mathrm{Gi}(x) = \frac{1}{\pi x} + \varepsilon(x), \qquad \varepsilon(x) \equiv \frac{2}{\pi}\int_0^\infty \sin(\tfrac{1}{3}t^3+xt)\frac{d}{dt}\left\{\frac{t}{(t^2+x)^3}\right\}dt.$$

Clearly

$$|\varepsilon(x)| \leqslant (2/\pi)\mathscr{V}_{t=0,\infty}\{t(t^2+x)^{-3}\}.$$

Substituting $t = x^{1/2}v$ we see that $\varepsilon(x) = O(x^{-5/2})$. Higher terms can be calculated in the same way, but the theory of Chapter 7, §14 shows immediately that, apart from a change of sign, they are the same as in (12.14):

$$\mathrm{Gi}(x) \sim \frac{1}{\pi x}\left\{1 + \frac{1}{x^3}\sum_{s=0}^{\infty}\frac{(3s+2)!}{s!\,(3x^3)^s}\right\} \qquad (x \to +\infty). \tag{12.15}$$

Thus the asymptotic behavior of $\mathrm{Gi}(x)$ is intermediate to that of $\mathrm{Ai}(x)$ and $\mathrm{Bi}(x)$. Accordingly, $-\mathrm{Gi}(x)$ is a numerically satisfactory companion to the complementary functions $\mathrm{Ai}(x)$ and $\mathrm{Bi}(x)$ when $x \in [0, \infty)$.

12.4 In the complex plane, the asymptotic expansion (12.14)—with x replaced by z—holds for the analytic continuation $\mathrm{Hi}(z)$ of $\mathrm{Hi}(x)$, provided that $|\mathrm{ph}(-z)| \leqslant \frac{2}{3}\pi - \delta$, δ being an arbitrary positive constant. This can be seen by applying Theorem 3.3 of Chapter 4 to the integral (12.12), taking $\alpha_1 = -\frac{1}{6}\pi$ and $\alpha_2 = \frac{1}{6}\pi$; alternatively, the theory of Chapter 7, §14 can be used. In consequence of this result, $\mathrm{Ai}_1(z)$, $\mathrm{Ai}_{-1}(z)$, and $\mathrm{Hi}(z)$ comprise a numerically satisfactory set of solutions of (12.07) in $\mathbf{S}_1 \cup \mathbf{S}_{-1}$.

For the sectors $\mathbf{S}_0 \cup \mathbf{S}_1$ and $\mathbf{S}_0 \cup \mathbf{S}_{-1}$ appropriate particular solutions are $e^{2\pi i/3}\,\mathrm{Hi}(ze^{2\pi i/3})$ and $e^{-2\pi i/3}\,\mathrm{Hi}(ze^{-2\pi i/3})$, respectively. Within these sectors, the asymptotic expansions of these functions for large $|z|$ are given by the right-hand side of (12.14) with $x = z$.

Ex. 12.1 Show that

$$\int_0^z \mathrm{Ai}(t)\,dt = \tfrac{1}{3} + \pi\{\mathrm{Ai}'(z)\,\mathrm{Gi}(z) - \mathrm{Ai}(z)\,\mathrm{Gi}'(z)\} = -\tfrac{2}{3} + \pi\{\mathrm{Ai}(z)\,\mathrm{Hi}'(z) - \mathrm{Ai}'(z)\,\mathrm{Hi}(z)\},$$

and

$$\int_0^z \mathrm{Bi}(t)\,dt = \pi\{\mathrm{Bi}'(z)\,\mathrm{Gi}(z) - \mathrm{Bi}(z)\,\mathrm{Gi}'(z)\} = \pi\{\mathrm{Bi}(z)\,\mathrm{Hi}'(z) - \mathrm{Bi}'(z)\,\mathrm{Hi}(z)\}.$$

Ex. 12.2 Show that for large $|z|$, $-\mathrm{Gi}(z)$ is a numerically satisfactory solution of (12.07) in the sector $\mathbf{S}_0$, but not elsewhere.

***Ex. 12.3** Consider the inhomogeneous form of Bessel's equation given by

$$z^2\frac{d^2w}{dz^2} + z\frac{dw}{dz} + (z^2 - \nu^2)\,w = z^m,$$

in which m is a fixed nonnegative integer. On replacing z by νz and w by $z^{-1/2}w$, this equation becomes

$$\frac{d^2w}{dz^2} = \left\{\nu^2\frac{1-z^2}{z^2} - \frac{1}{4z^2}\right\}w + \nu^m z^{m-(3/2)}.$$

Let (i) ζ, $f(z)$, $\psi(\zeta)$, $A_s(\zeta)$, and $B_s(\zeta)$ be defined by (10.02), (10.04), (10.06), and (10.07); (ii) $G_s(\zeta)$ be defined by (12.06), with $\varpi(\zeta) = z^{m-(3/2)}f^{-3/4}(z)$; (iii) the values of $A_s(0)$ be chosen to make the numerators of the fractions in (12.06) vanish when $\zeta = 0$. Show that for each nonnegative integer n the last equation has a solution

$$\frac{\nu^m}{f^{1/4}(z)}\left\{\frac{\pi\,\mathrm{Hi}(\nu^{2/3}\zeta)}{\nu^{4/3}}\sum_{s=0}^{n}\frac{A_s(\zeta)}{\nu^{2s}} + \frac{\pi\,\mathrm{Hi}'(\nu^{2/3}\zeta)}{\nu^{8/3}}\sum_{s=0}^{n-1}\frac{B_s(\zeta)}{\nu^{2s}} + \frac{1}{\nu^2}\sum_{s=0}^{n}\frac{G_s(\zeta)}{\nu^{2s}} + \varepsilon_{2n+1}(\nu, z)\right\},$$

in which the error term satisfies

$$\frac{d^2\varepsilon_{2n+1}}{d\zeta^2} - \{v^2\zeta + \psi(\zeta)\}\varepsilon_{2n+1} = \frac{\pi\,\mathrm{Hi}(v^{2/3}\zeta)}{v^{2n+(4/3)}}\{B_n(\zeta)+2\zeta B_n'(\zeta)\} + \frac{\psi(\zeta)\,G_n(\zeta)-G_n''(\zeta)}{v^{2n+2}}.$$

By variation of parameters, show also that when $n \geqslant \frac{1}{2}(m-1)$ there exists a solution with the properties (i) $\varepsilon_{2n+1}(v,z)\to 0$ as $z\to+\infty$, v being fixed and positive; (ii)

$$\varepsilon_{2n+1}(v,z) = E(v^{2/3}\zeta)\,M(v^{2/3}\zeta)\,O(v^{-2n-(7/3)})$$

as $v\to\infty$, uniformly with respect to $z\in(0,\infty)$, where E and M are defined in §2.

***Ex. 12.4** Show that the equation

$$\eta(d^2w/dx^2) + x(1+x)^2 w = (1+x)^{3/2},$$

in which η is a small positive parameter, has a solution of the form

$$w = \frac{(1+\frac{3}{5}x)^{1/6}}{(1+x)^{1/2}}\left[\frac{\pi}{\eta^{1/3}}\,\mathrm{Hi}\left\{-\frac{x}{\eta^{1/3}}\left(1+\frac{3}{5}x\right)^{2/3}\right\} + \frac{(1+\frac{3}{5}x)^{1/2}-1}{x(1+\frac{3}{5}x)^{2/3}} + \varepsilon(\eta,x)\right],$$

where $\varepsilon(\eta,x)\to 0$ as $x\to\infty$, and $\varepsilon(\eta,x)=O(\eta^{1/3})$ as $\eta\to 0$, uniformly for $x\in[0,\infty)$.

Historical Notes and Additional References

The earliest successful investigations of uniform asymptotic solutions of turning-point problems were those of Langer (1931, 1932, 1949). Previously only local approximations in the neighborhood of the turning point were available. In the first two references Langer supplied the necessary modification of Liouville's transformation (§3.1) and derived the leading term of the asymptotic solution together with the corresponding Liouville–Neumann series. In the 1949 paper he showed how to construct formal series solutions (§7.2), and established their asymptotic nature for bounded real values of the independent variable ζ, and complex values of the large parameter u. Extensions to unrestricted ζ were given by Cherry (1950a) and Olver (1954a, 1958), and explicit bounds for the error terms were derived by Olver (1963, 1964b).

For differential equations of order higher than two, researches on turning-point problems have been surveyed by Wasow (1965, §31.3), McHugh (1971), and Nayfeh (1973, p. 360). Notable papers include those of Langer (1962), Kiyek (1963), Turrittin (1964), Hanson (1966), and Wasow (1966).

§1 Analytic properties and numerical tabulations of the integral

$$\int_0^\infty \cos\left\{\frac{1}{2}\pi(t^3-mt)\right\}dt \equiv 2\left(\frac{\pi^2}{12}\right)^{1/3}\mathrm{Ai}\left\{-\left(\frac{\pi^2}{12}\right)^{1/3}m\right\}$$

were introduced by Airy (1838, 1849). The modern notations $\mathrm{Ai}(x)$ and $\mathrm{Bi}(x)$ are due to Jeffreys (1942, 1953) and J. C. P. Miller (B.A., 1946b). In an earlier paper (Jeffreys, 1928) $\mathrm{Ai}(x)$ was denoted by $\mathrm{Ai}(-x)$. The graphs in Fig. 1.1 and the entries in Table 2.1 are based on the B.A. tables.

§2 The choice of weight functions and auxiliary functions is not unique; another was employed by Olver (1963, 1964b). The present functions—which are almost the same as those used by Erdélyi (1960)—are more elegant because they assign equal weighting to $\mathrm{Ai}(x)$ and $\mathrm{Bi}(x)$ for positive x. In consequence, slightly simpler and sharper error bounds emerge.

§4.3 Equation (4.02) is of importance in the scattering theory of particles in a Coulomb field; see, for example, Mott and Massey (1965, p. 60). The present results appear to be the first to cover the interval $0<z<\infty$ by means of a single uniform approximation.

§5 Sharper versions of the result stated in Exercise 5.1 have been given by Hethcote (1970a,b).

§7.2 An alternative way of synthesizing the formal expansion (7.05), based on the Mellin transform and Nörlund summation operator, has been given by Jorna (1964a). In subsequent papers (for example, 1964b and 1965) Jorna applies his method to various special functions. Mostly the results obtained are of a formal character, and regions of validity in the complex plane are not investigated.

§10 Extensions to the region $|\mathrm{ph}\, \nu| < \frac{1}{2}\pi$ have been given by Olver (1954b). In unpublished work J. A. Cochran has noted that the asymptotic expansion (10.24) for $H_\nu^{(1)}(\nu z)$ is valid when $-\frac{1}{2}\pi < \mathrm{ph}\, \nu < \frac{3}{2}\pi$ and $|\mathrm{ph}\, z| < \pi$, with a similar extension for $H_\nu^{(2)}(\nu z)$. For asymptotic approximations to zeros of the Bessel and Hankel functions of large order see Olver (1954b), R.S. (1960), Cochran (1965), Boyer (1969), and Hethcote (1970b).

Generalizations of the result given in Exercise 10.5 have been published by Lorch and Szego (1955) and Muldoon (1970).

§12.1 This procedure for constructing a formal series solution of equation (12.01) is due to Clark (1963). Clark considered the following form of equation

$$\frac{d^2 W}{d\zeta^2} - \left\{u^2\zeta + u \sum_{s=0}^{\infty} \frac{\psi_s(\zeta)}{u^s}\right\} W = u^2 \sum_{s=0}^{\infty} \frac{\varpi_s(\zeta)}{u^s},$$

in which $\psi_s(\zeta)$ and $\varpi_s(\zeta)$ are independent of u, and established the asymptotic nature of the formal solution for bounded real ζ and large complex u. See also Nayfeh (1973, pp. 354–359).

For further properties and lists of numerical tables of $\mathrm{Gi}(x)$, $\mathrm{Hi}(x)$, and the derivatives and integrals of these functions, see Luke (1962, Chapter 6) and Nosova and Tumarkin (1965). Exercise 12.4 is based on the last reference.

12

DIFFERENTIAL EQUATIONS WITH A PARAMETER: SIMPLE POLES AND OTHER TRANSITION POINTS

1 Bessel Functions and Modified Bessel Functions of Real Order and Argument

1.1 In the theory of differential equations belonging to Case III, the comparison functions are Bessel functions and modified Bessel functions. For reference, we collect here properties of these functions which will be needed frequently. These properties have been established in Chapters 2 and 7 or are readily derivable from results given in these chapters. *Until* §8 *we suppose that the symbols* v *and* x *denote variables restricted by* $v \geqslant 0$ *and* $x > 0$.

The modified Bessel functions $I_v(x)$ and $K_v(x)$ are continuous positive functions of v and x. For fixed v and increasing x, $I_v(x)$ is increasing and $K_v(x)$ is decreasing. On the other hand, for fixed x and increasing v, $I_v(x)$ is decreasing and $K_v(x)$ is increasing.

If v is fixed and $x \to 0$, then

$$I_v(x) \sim (\tfrac{1}{2}x)^v/\Gamma(v+1), \tag{1.01}$$

and

$$K_0(x) \sim \ln\left(\frac{1}{x}\right), \qquad K_v(x) \sim \frac{\Gamma(v)}{2(\frac{1}{2}x)^v} \qquad (v > 0). \tag{1.02}$$

As $x \to \infty$

$$I_v(x) \sim \frac{e^x}{(2\pi x)^{1/2}}, \qquad K_v(x) \sim \left(\frac{\pi}{2x}\right)^{1/2} e^{-x}, \tag{1.03}$$

each of these relations being uniformly valid with respect to bounded v.

Next,

$$\mathscr{W}\{K_v(x), I_v(x)\} = I_v(x)K_{v+1}(x) + I_{v+1}(x)K_v(x) = 1/x, \tag{1.04}$$

$$\left.\begin{aligned} \mathscr{Z}_{v-1}(x) - \mathscr{Z}_{v+1}(x) &= (2v/x)\mathscr{Z}_v(x), & \mathscr{Z}_v'(x) &= \mathscr{Z}_{v-1}(x) - (v/x)\mathscr{Z}_v(x),\\ \mathscr{Z}_{v-1}(x) + \mathscr{Z}_{v+1}(x) &= 2\mathscr{Z}_v'(x), & \mathscr{Z}_v'(x) &= \mathscr{Z}_{v+1}(x) + (v/x)\mathscr{Z}_v(x).\end{aligned}\right\} \tag{1.05}$$

In (1.05) $\mathscr{Z}_v = I_v$ or $e^{v\pi i}K_v$.

Functions of negative order are given by

$$I_{-\nu}(x) = I_\nu(x) + (2/\pi)\sin(\nu\pi)K_\nu(x), \qquad K_{-\nu}(x) = K_\nu(x). \tag{1.06}$$

Relations (1.03), (1.04), and (1.05) hold with ν replaced throughout by $-\nu$. Relation (1.01) holds with this change provided that $\nu \neq 1, 2, 3, \ldots$.

1.2 The unmodified Bessel functions, also, are continuous in ν and x. If ν is fixed and $x \to 0$, then

$$J_\nu(x) \sim (\tfrac{1}{2}x)^\nu/\Gamma(\nu+1), \tag{1.07}$$

$$Y_0(x) \sim -\frac{2}{\pi}\ln\left(\frac{1}{x}\right), \qquad Y_\nu(x) \sim -\frac{\Gamma(\nu)}{\pi(\frac{1}{2}x)^\nu} \qquad (\nu > 0). \tag{1.08}$$

As $x \to \infty$,

$$\begin{aligned} J_\nu(x) &= \left(\frac{2}{\pi x}\right)^{1/2}\left\{\cos(x-\tfrac{1}{2}\nu\pi-\tfrac{1}{4}\pi) + O\left(\frac{1}{x}\right)\right\}, \\ Y_\nu(x) &= \left(\frac{2}{\pi x}\right)^{1/2}\left\{\sin(x-\tfrac{1}{2}\nu\pi-\tfrac{1}{4}\pi) + O\left(\frac{1}{x}\right)\right\}, \end{aligned} \tag{1.09}$$

uniformly with respect to bounded ν.

Next,

$$\mathscr{W}\{J_\nu(x), Y_\nu(x)\} = J_{\nu+1}(x)Y_\nu(x) - J_\nu(x)Y_{\nu+1}(x) = 2/(\pi x), \tag{1.10}$$

$$\begin{aligned} &\mathscr{C}_{\nu-1}(x) + \mathscr{C}_{\nu+1}(x) = (2\nu/x)\mathscr{C}_\nu(x), && \mathscr{C}_\nu'(x) = \mathscr{C}_{\nu-1}(x) - (\nu/x)\mathscr{C}_\nu(x), \\ &\mathscr{C}_{\nu-1}(x) - \mathscr{C}_{\nu+1}(x) = 2\mathscr{C}_\nu'(x), && \mathscr{C}_\nu'(x) = -\mathscr{C}_{\nu+1}(x) + (\nu/x)\mathscr{C}_\nu(x), \end{aligned} \tag{1.11}$$

where $\mathscr{C}_\nu = J_\nu$ or Y_ν.

Functions of negative order are given by

$$J_{-\nu}(x) = \cos(\nu\pi)J_\nu(x) - \sin(\nu\pi)Y_\nu(x), \qquad Y_{-\nu}(x) = \sin(\nu\pi)J_\nu(x) + \cos(\nu\pi)Y_\nu(x). \tag{1.12}$$

Relations (1.09), (1.10), and (1.11) hold with ν replaced throughout by $-\nu$. Relation (1.07) holds with this change if $\nu \neq 1, 2, 3, \ldots$.

The positive zeros $j_{\nu,s}$ of $J_\nu(x)$ and $y_{\nu,s}$ of $Y_\nu(x)$ interlace according to the inequalities

$$y_{\nu,1} < j_{\nu,1} < y_{\nu,2} < j_{\nu,2} < \cdots. \tag{1.13}$$

For fixed ν and large s,

$$j_{\nu,s} = (s+\tfrac{1}{2}\nu-\tfrac{1}{4})\pi + O(s^{-1}), \qquad y_{\nu,s} = (s+\tfrac{1}{2}\nu-\tfrac{3}{4})\pi + O(s^{-1}). \tag{1.14}$$

1.3 As in the case of the Airy functions, auxiliary modulus and phase functions $M_\nu(x)$ and $\theta_\nu(x)$, say, are needed to make a combined assessment of the magnitudes of $J_\nu(x)$ and $Y_\nu(x)$. Let $x = X_\nu$ denote the smallest positive root of the equation

$$J_\nu(x) + Y_\nu(x) = 0.$$

From (1.07), (1.08), and (1.13) it is seen that $J_\nu(x)+Y_\nu(x)$ is negative as $x \to 0+$, and positive at $x = y_{\nu,1}$. Hence

$$0 < X_\nu < y_{\nu,1}. \tag{1.15}$$

That X_ν is a continuous function of ν is provable by the methods of Chapter 7, §7.

The weight function we employ for real variables is defined by

$$E_\nu(x) = \{-Y_\nu(x)/J_\nu(x)\}^{1/2} \quad (0 < x \leqslant X_\nu), \qquad E_\nu(x) = 1 \quad (x \geqslant X_\nu). \tag{1.16}$$

From (1.10) we derive

$$\frac{d}{dx}\{E_\nu^2(x)\} = -\frac{2}{\pi x J_\nu^2(x)} \qquad (0 < x < X_\nu).$$

Accordingly, $E_\nu(x)$ is a continuous, positive, nonincreasing function of x. It is also continuous in ν.

We set

$$J_\nu(x) = E_\nu^{-1}(x)\, M_\nu(x) \cos\theta_\nu(x), \qquad Y_\nu(x) = E_\nu(x)\, M_\nu(x) \sin\theta_\nu(x), \tag{1.17}$$

giving

$$M_\nu(x) = \{2\,|Y_\nu(x)|\, J_\nu(x)\}^{1/2}, \qquad \theta_\nu(x) = -\tfrac{1}{4}\pi \qquad (0 < x \leqslant X_\nu), \tag{1.18}$$

or

$$M_\nu(x) = \{J_\nu^2(x)+Y_\nu^2(x)\}^{1/2}, \qquad \theta_\nu(x) = \tan^{-1}\{Y_\nu(x)/J_\nu(x)\} \qquad (x \geqslant X_\nu), \tag{1.19}$$

the branch of the inverse tangent being chosen to make $\theta_\nu(x)$ continuous. Differentiation and use of (1.10) yields

$$\theta_\nu'(x) = \frac{2}{\pi x M_\nu^2(x)} \qquad (x > X_\nu). \tag{1.20}$$

Hence $\theta_\nu(x)$ is a nondecreasing function of x. From (1.13) and (1.15) it follows that

$$\theta_\nu(y_{\nu,s}) = (s-1)\,\pi, \qquad \theta_\nu(j_{\nu,s}) = (s-\tfrac{1}{2})\,\pi. \tag{1.21}$$

The asymptotic forms of $E_\nu(x)$ and $M_\nu(x)$ as $x \to 0$ are found to be

$$E_0(x) \sim \left\{\frac{2}{\pi}\ln\left(\frac{1}{x}\right)\right\}^{1/2}, \qquad E_\nu(x) \sim \left(\frac{\nu}{\pi}\right)^{1/2} \frac{\Gamma(\nu)}{(\frac{1}{2}x)^\nu} \qquad (\nu > 0), \tag{1.22}$$

and

$$M_0(x) \sim 2\left\{\frac{1}{\pi}\ln\left(\frac{1}{x}\right)\right\}^{1/2}, \qquad M_\nu(x) \to \left(\frac{2}{\pi\nu}\right)^{1/2} \qquad (\nu > 0). \tag{1.23}$$

As $x \to \infty$, we have

$$M_\nu(x) \sim \left(\frac{2}{\pi x}\right)^{1/2}, \qquad \theta_\nu(x) = x - \tfrac{1}{2}\nu\pi - \tfrac{1}{4}\pi + O(x^{-1}). \tag{1.24}$$

To prove the last result we observe from (1.09) and (1.17) that

$$\theta_\nu(x) = x - \tfrac{1}{2}\nu\pi - \tfrac{1}{4}\pi + 2m\pi + O(x^{-1}),$$

where m is an integer.† Since $\theta_\nu(x)$ is continuous in ν, the value of m is independent of ν. That $m = 0$ follows by setting $\nu = \tfrac{1}{2}$, for then $J_{1/2}(x) = 2^{1/2}(\pi x)^{-1/2}\sin x$, $Y_{1/2}(x) = -2^{1/2}(\pi x)^{-1/2}\cos x$, $X_{1/2} = \tfrac{1}{4}\pi$, and $\theta_{1/2}(x) = x - \tfrac{1}{2}\pi$, exactly.

Numerical values of X_ν, computed from tables of Bessel functions by inverse interpolation, are as follows‡:

ν	0	$\tfrac{1}{2}$	1	$1\tfrac{1}{2}$	2	$2\tfrac{1}{2}$	3	10
X_ν	0.230	0.785	1.329	1.864	2.393	2.918	3.441	10.639

Ex. 1.1 Show that when ν is large $X_\nu = \nu - c(\tfrac{1}{2}\nu)^{1/3} + O(\nu^{-1/3})$, where $c = -0.36605\ldots$ is defined in Chapter 11, §2.2.

2 Case III: Formal Series Solutions

2.1 Our primary concern in this chapter is the solution of equations of the form

$$d^2w/dx^2 = \{u^2 f(x) + g(x)\}\, w, \tag{2.01}$$

in which u is a large parameter, and at $x = x_0$, say, $f(x)$ has a simple pole and $(x-x_0)^2 g(x)$ is analytic. It is supposed that there are no other transition points in the x region under consideration.

Following Chapter 10, §1.2 we take a new independent variable, which for convenience we shall denote by $\tfrac{1}{4}\zeta$ instead of ξ, and a new dependent variable W, related by

$$\frac{1}{\zeta}\left(\frac{d\zeta}{dx}\right)^2 = 4f(x), \qquad w = \left(\frac{d\zeta}{dx}\right)^{-1/2} W. \tag{2.02}$$

Since the variables are real we assume, without loss of generality, that $f(x)$ has the same sign as $x - x_0$. Integration then yields

$$\zeta^{1/2} = \int_{x_0}^{x} f^{1/2}(t)\, dt \quad (x \geqslant x_0), \qquad (-\zeta)^{1/2} = \int_{x}^{x_0} \{-f(t)\}^{1/2}\, dt \quad (x \leqslant x_0). \tag{2.03}$$

These equations determine a continuous one-to-one correspondence between the variables x and ζ.

The transformed differential equation is given by

$$\frac{d^2W}{d\zeta^2} = \left\{\frac{u^2}{4\zeta} + \hat{\psi}(\zeta)\right\} W, \tag{2.04}$$

† Compare Chapter 1, Exercise 5.4.

‡ The values suggest, correctly, that X_ν is an increasing function of ν. A theorem that includes this result has been proved by Watson (1944, §15.6).

where

$$\hat{\psi}(\zeta) = \frac{g(x)}{\hat{f}(x)} + \frac{1}{\hat{f}^{1/4}(x)} \frac{d^2\hat{f}^{1/4}(x)}{d\zeta^2}, \qquad \hat{f}(x) = \left(\frac{d\zeta}{dx}\right)^2 = 4\zeta f(x).$$

If $g(x)$ has a simple or double pole at $x = x_0$, then $\hat{\psi}(\zeta)$ has the same kind of singularity at $\zeta = 0$. We denote the value of $\zeta^2\hat{\psi}(\zeta)$ at $\zeta = 0$ by $\frac{1}{4}(\nu^2-1)$, and rearrange (2.04) in the form

$$\frac{d^2W}{d\zeta^2} = \left\{\frac{u^2}{4\zeta} + \frac{\nu^2-1}{4\zeta^2} + \frac{\psi(\zeta)}{\zeta}\right\} W, \tag{2.05}$$

in which $\psi(\zeta) \equiv \zeta\hat{\psi}(\zeta) - \frac{1}{4}(\nu^2-1)\zeta^{-1}$ and is analytic at $\zeta = 0$. In terms of the original variables, $\frac{1}{4}(\nu^2-1)$ is the value of $(x-x_0)^2 g(x)$ at $x = x_0$, and†

$$\psi(\zeta) = \frac{1-4\nu^2}{16\zeta} + \frac{g(x)}{4f(x)} + \frac{4f(x)f''(x) - 5f'^2(x)}{64f^3(x)}. \tag{2.06}$$

Equation (2.05) *will be taken as the standard form of differential equation for Case III.* At first, we suppose that $\nu \geqslant 0$, $u > 0$, and the range of ζ is a real interval (α, β) which contains $\zeta = 0$ and may be unbounded. Later, we consider the more general case in which u and ζ are complex. For simplicity, it will be supposed that the real or complex function $\psi(\zeta)$ and all its derivatives are continuous in (α, β), but as in Cases I and II this requirement can be eased without undue complication. It is not necessary that $\psi(\zeta)$ be independent of u and ν: if this function depends on these variables, then we assume that the dependence is continuous.‡

The differential equation (2.05) has a regular singularity at $\zeta = 0$. Therefore a solution which is real on one side of this point becomes complex, as a rule, when continued to the other side; compare Chapter 5, §§4 and 5. In constructing real solutions we consider separately the intervals $[0, \beta)$ and $(\alpha, 0]$, beginning with the former.

2.2 If the term $\psi(\zeta)/\zeta$ is neglected, then (2.05) reduces to

$$\frac{d^2W}{d\zeta^2} = \left(\frac{u^2}{4\zeta} + \frac{\nu^2-1}{4\zeta^2}\right) W \tag{2.07}$$

with independent (and numerically satisfactory) solutions $\zeta^{1/2} I_\nu(u\zeta^{1/2})$ and $\zeta^{1/2} K_\nu(u\zeta^{1/2})$. Analogy with §7.2 of Chapter 11 suggests that (2.05) may be satisfiable by an expansion of the form

$$W = \zeta^{1/2}\mathscr{Z}_\nu(u\zeta^{1/2}) \sum_{s=0}^{\infty} \frac{A_s(\zeta)}{u^{2s}} + \frac{1}{u^2}\frac{d}{d\zeta}\{\zeta^{1/2}\mathscr{Z}_\nu(u\zeta^{1/2})\} \sum_{s=0}^{\infty} \frac{B_s(\zeta)}{u^{2s}},$$

where $\mathscr{Z}_\nu$ again denotes I_ν or $e^{\nu\pi i}K_\nu$. From (1.05) we obtain

$$\begin{aligned}(d/d\zeta)\{\zeta^{1/2}\mathscr{Z}_\nu(u\zeta^{1/2})\} &= \tfrac{1}{2}\zeta^{-1/2}\mathscr{Z}_\nu(u\zeta^{1/2}) + \tfrac{1}{2}u\mathscr{Z}_\nu'(u\zeta^{1/2}) \\ &= \tfrac{1}{2}(1+\nu)\zeta^{-1/2}\mathscr{Z}_\nu(u\zeta^{1/2}) + \tfrac{1}{2}u\mathscr{Z}_{\nu+1}(u\zeta^{1/2}).\end{aligned} \tag{2.08}$$

† $\psi(\zeta)$ differs from the function $\psi(\xi)$ of Chapter 10, §1.2.
‡ An example in which $\psi(\zeta)$ depends on u is given by Exercise 5.3 below.

Bearing in mind that the function $\mathscr{Z}_{\nu+1}$ is better known than $\mathscr{Z}'_\nu$ from both the analytical and tabular standpoints, we try instead a solution of the form

$$W = \zeta^{1/2}\mathscr{Z}_\nu(u\zeta^{1/2})\sum_{s=0}^{\infty}\frac{A_s(\zeta)}{u^{2s}} + \frac{\zeta}{u}\mathscr{Z}_{\nu+1}(u\zeta^{1/2})\sum_{s=0}^{\infty}\frac{B_s(\zeta)}{u^{2s}}. \tag{2.09}$$

Formal differentiation of (2.09) and use of (2.08) and the companion relation

$$\frac{d}{d\zeta}\left\{\frac{\zeta}{u}\mathscr{Z}_{\nu+1}(u\zeta^{1/2})\right\} = \frac{1}{2}\zeta^{1/2}\mathscr{Z}_\nu(u\zeta^{1/2}) + \frac{1-\nu}{2u}\mathscr{Z}_{\nu+1}(u\zeta^{1/2}),$$

yields

$$\frac{dW}{d\zeta} = \zeta^{-1/2}\mathscr{Z}_\nu(u\zeta^{1/2})\sum_{s=0}^{\infty}\frac{C_s(\zeta)}{u^{2s}} + u\mathscr{Z}_{\nu+1}(u\zeta^{1/2})\sum_{s=0}^{\infty}\frac{D_s(\zeta)}{u^{2s}},$$

where

$$C_s(\zeta) = \tfrac{1}{2}(1+\nu)A_s(\zeta) + \zeta A'_s(\zeta) + \tfrac{1}{2}\zeta B_s(\zeta),$$
$$D_s(\zeta) = \tfrac{1}{2}A_s(\zeta) + \tfrac{1}{2}(1-\nu)B_{s-1}(\zeta) + \zeta B'_{s-1}(\zeta).$$

Differentiating again and comparing like powers of u, we find that (2.05) is formally satisfied if

$$\zeta A''_s(\zeta) + (\nu+1)A'_s(\zeta) - \psi(\zeta)A_s(\zeta) + \zeta B'_s(\zeta) + \tfrac{1}{2}B_s(\zeta) = 0, \tag{2.10}$$

and

$$A'_{s+1}(\zeta) + \zeta B''_s(\zeta) + (1-\nu)B'_s(\zeta) - \psi(\zeta)B_s(\zeta) = 0. \tag{2.11}$$

Integration of these two equations yields

$$B_s(\zeta) = -A'_s(\zeta) + \frac{1}{\zeta^{1/2}}\int_0^\zeta\{\psi(v)A_s(v) - (\nu+\tfrac{1}{2})A'_s(v)\}\frac{dv}{v^{1/2}}, \tag{2.12}$$

$$A_{s+1}(\zeta) = \nu B_s(\zeta) - \zeta B'_s(\zeta) + \int\psi(\zeta)B_s(\zeta)\,d\zeta. \tag{2.13}$$

Relations (2.12) and (2.13) determine successively $A_0(\zeta)$, $B_0(\zeta)$, $A_1(\zeta)$, $B_1(\zeta)$, ..., apart from arbitrary constants of integration associated with (2.13).† In particular, $A_0(\zeta) = \text{constant}$, which we take to be unity without loss of generality. The continuity and repeated differentiability of each $A_s(\zeta)$ and $B_s(\zeta)$ in the interval $[0, \beta)$ is an immediate consequence of Lemma 7.1 of Chapter 11. We also observe that the functions $A_s(\zeta)$, $B_s(\zeta)$, and their ζ derivatives are continuous functions of ν, provided that the same is true of each integration constant associated with (2.13).

3 Error Bounds: Positive ζ

3.1 In order to bound the error terms associated with the partial sums of the formal series (2.09), we introduce the quantity

$$\lambda_1(\nu) = \sup_{x\in(0,\infty)}\{2xI_\nu(x)K_\nu(x)\}, \tag{3.01}$$

† An easier way of evaluating higher coefficients in practice is indicated in Exercise 5.2 below.

analogous to the constant λ of Chapter 11, §2.4. Relations (1.01), (1.02), and (1.03) show that $\lambda_1(\nu)$ is finite and not less than unity. Regarded as a function of ν it is continuous. Numerical computations yield, for example, $\lambda_1(0) = 1.07$ and $\lambda_1(1) = 1.00$, to two decimal places, the values of x corresponding to these suprema being 1.08 and ∞, respectively.

3.2 Theorem 3.1 *With the conditions of* §2.1, *equation* (2.05) *has, for each value of* u *and each nonnegative integer* n, *solutions* $W_{2n+1,1}(u,\zeta)$ *and* $W_{2n+1,2}(u,\zeta)$ *which are repeatedly differentiable in the* ζ *interval* $(0,\beta)$, *and are given by*

$$W_{2n+1,1}(u,\zeta) = \zeta^{1/2} I_\nu(u\zeta^{1/2}) \sum_{s=0}^{n} \frac{A_s(\zeta)}{u^{2s}} + \frac{\zeta}{u} I_{\nu+1}(u\zeta^{1/2}) \sum_{s=0}^{n-1} \frac{B_s(\zeta)}{u^{2s}} + \varepsilon_{2n+1,1}(u,\zeta), \tag{3.02}$$

$$W_{2n+1,2}(u,\zeta) = \zeta^{1/2} K_\nu(u\zeta^{1/2}) \sum_{s=0}^{n} \frac{A_s(\zeta)}{u^{2s}} - \frac{\zeta}{u} K_{\nu+1}(u\zeta^{1/2}) \sum_{s=0}^{n-1} \frac{B_s(\zeta)}{u^{2s}} + \varepsilon_{2n+1,2}(u,\zeta), \tag{3.03}$$

where

$$|\varepsilon_{2n+1,1}(u,\zeta)| \leqslant \lambda_1(\nu)\zeta^{1/2} I_\nu(u\zeta^{1/2}) \exp\left\{\frac{\lambda_1(\nu)}{u} \mathcal{V}_{0,\zeta}(\zeta^{1/2}B_0)\right\} \frac{\mathcal{V}_{0,\zeta}(\zeta^{1/2}B_n)}{u^{2n+1}}, \tag{3.04}$$

$$|\varepsilon_{2n+1,2}(u,\zeta)| \leqslant \lambda_1(\nu)\zeta^{1/2} K_\nu(u\zeta^{1/2}) \exp\left\{\frac{\lambda_1(\nu)}{u} \mathcal{V}_{\zeta,\beta}(\zeta^{1/2}B_0)\right\} \frac{\mathcal{V}_{\zeta,\beta}(\zeta^{1/2}B_n)}{u^{2n+1}}, \tag{3.05}$$

$$\left| \frac{\partial \varepsilon_{2n+1,1}(u,\zeta)}{\partial \zeta} - \frac{\nu+1}{2\zeta} \varepsilon_{2n+1,1}(u,\zeta) \right| \leqslant \lambda_1(\nu) \frac{K_{\nu+1}(u\zeta^{1/2})}{K_\nu(u\zeta^{1/2})} I_\nu(u\zeta^{1/2}) \exp\left\{\frac{\lambda_1(\nu)}{u} \mathcal{V}_{0,\zeta}(\zeta^{1/2}B_0)\right\} \frac{\mathcal{V}_{0,\zeta}(\zeta^{1/2}B_n)}{u^{2n}}, \tag{3.06}$$

$$\left| \frac{\partial \varepsilon_{2n+1,2}(u,\zeta)}{\partial \zeta} - \frac{\nu+1}{2\zeta} \varepsilon_{2n+1,2}(u,\zeta) \right| \leqslant \lambda_1(\nu) K_{\nu+1}(u\zeta^{1/2}) \exp\left\{\frac{\lambda_1(\nu)}{u} \mathcal{V}_{\zeta,\beta}(\zeta^{1/2}B_0)\right\} \frac{\mathcal{V}_{\zeta,\beta}(\zeta^{1/2}B_n)}{u^{2n}}. \tag{3.07}$$

The steps in proving this theorem by our standard method are as follows. For the first solution, it is found that

$$\varepsilon''_{2n+1,1}(u,\zeta) - \left(\frac{u^2}{4\zeta} + \frac{\nu^2-1}{4\zeta^2}\right)\varepsilon_{2n+1,1}(u,\zeta) = \frac{\psi(\zeta)}{\zeta}\varepsilon_{2n+1,1}(u,\zeta) + I_\nu(u\zeta^{1/2})\frac{\{\zeta^{1/2}B_n(\zeta)\}'}{u^{2n}},$$

primes denoting differentiations with respect to ζ. The complementary functions associated with the left-hand side of this equation are $\zeta^{1/2}I_\nu(u\zeta^{1/2})$ and $\zeta^{1/2}K_\nu(u\zeta^{1/2})$, with Wronskian $-\frac{1}{2}$; compare (1.04). Therefore

$$\varepsilon_{2n+1,1}(u,\zeta) = 2\int_0^\zeta (\zeta v)^{1/2}\{I_\nu(u\zeta^{1/2})K_\nu(uv^{1/2}) - K_\nu(u\zeta^{1/2})I_\nu(uv^{1/2})\}$$
$$\times\left[\frac{\psi(v)}{v}\varepsilon_{2n+1,1}(u,v) + \frac{I_\nu(uv^{1/2})}{u^{2n}}\frac{d}{dv}\{v^{1/2}B_n(v)\}\right]dv. \qquad (3.08)$$

Direct solution of the last equation by means of Theorem 10.1 of Chapter 6 leads to (3.04) but not (3.06). To obtain both results simultaneously we recast (3.08) in the form

$$\hat{\varepsilon}_{2n+1,1}(u,\zeta) = \int_0^\zeta \mathsf{K}(\zeta,v)\left[\frac{\psi(v)}{v^{1/2}}\hat{\varepsilon}_{2n+1,1}(u,v) + \frac{I_\nu(uv^{1/2})}{u^{2n}v^{\nu/2}}\frac{d}{dv}\{v^{1/2}B_n(v)\}\right]dv, \qquad (3.09)$$

in which $\hat{\varepsilon}_{2n+1,1}(u,\zeta) = \zeta^{-(\nu+1)/2}\varepsilon_{2n+1,1}(u,\zeta)$, and

$$\mathsf{K}(\zeta,v) = 2\zeta^{-\nu/2}v^{(\nu+1)/2}\{I_\nu(u\zeta^{1/2})K_\nu(uv^{1/2}) - K_\nu(u\zeta^{1/2})I_\nu(uv^{1/2})\}. \qquad (3.10)$$

As v increases, $I_\nu(uv^{1/2})$ increases and $K_\nu(uv^{1/2})$ decreases. Accordingly,

$$0 \leqslant \mathsf{K}(\zeta,v) < 2\zeta^{-\nu/2}v^{(\nu+1)/2}I_\nu(u\zeta^{1/2})K_\nu(uv^{1/2}).$$

Next,

$$\partial\mathsf{K}(\zeta,v)/\partial\zeta = u(v/\zeta)^{(\nu+1)/2}\{I_{\nu+1}(u\zeta^{1/2})K_\nu(uv^{1/2}) + K_{\nu+1}(u\zeta^{1/2})I_\nu(uv^{1/2})\}, \qquad (3.11)$$

and by use of the properties stated in the second paragraph of §1.1, we derive

$$\frac{\partial\mathsf{K}(\zeta,v)}{\partial\zeta} \leqslant 2u\left(\frac{v}{\zeta}\right)^{(\nu+1)/2}K_{\nu+1}(u\zeta^{1/2})I_\nu(u\zeta^{1/2})\frac{K_\nu(uv^{1/2})}{K_\nu(u\zeta^{1/2})}.$$

The desired bounds (3.04) and (3.06) are now obtainable from (3.09) by applying Theorem 10.1 of Chapter 6, and referring to the definition (3.01) and the identity

$$\int v^{-1/2}|\psi(v)|\,dv = \mathscr{V}\{v^{1/2}B_0(v)\}.$$

The proof of (3.05) and (3.07) is similar.

In the case $\beta = \infty$, Theorem 3.1 provides a meaningful result for $W_{2n+1,2}(u,\zeta)$ only when the variations of $\zeta^{1/2}B_0(\zeta)$ and $\zeta^{1/2}B_n(\zeta)$ converge at infinity. Sufficient conditions for this purpose are stated in Exercise 4.2 below.

Ex. 3.1 By substituting in (3.02) and (3.03) by means of (1.05), show that

$$W_{2n+1,1}(u,\zeta) = \zeta^{1/2}I_\nu(u\zeta^{1/2})\sum_{s=0}^{n}\frac{\hat{A}_s(\zeta)}{u^{2s}} + \frac{\zeta}{u}I_{\nu-1}(u\zeta^{1/2})\sum_{s=0}^{n-1}\frac{\hat{B}_s(\zeta)}{u^{2s}} + \varepsilon_{2n+1,1}(u,\zeta),$$

$$W_{2n+1,2}(u,\zeta) = \zeta^{1/2}K_\nu(u\zeta^{1/2})\sum_{s=0}^{n}\frac{\hat{A}_s(\zeta)}{u^{2s}} - \frac{\zeta}{u}K_{\nu-1}(u\zeta^{1/2})\sum_{s=0}^{n-1}\frac{\hat{B}_s(\zeta)}{u^{2s}} + \varepsilon_{2n+1,2}(u,\zeta),$$

where $\hat{A}_s(\zeta)$ and $\hat{B}_s(\zeta)$ satisfy the same recurrence relations as $A_s(\zeta)$ and $B_s(\zeta)$, except that ν is replaced by $-\nu$.

4 Error Bounds: Negative ζ

4.1 When ζ is negative, a numerically satisfactory pair of solutions of the comparison equation (2.07) is $|\zeta|^{1/2}J_\nu(u|\zeta|^{1/2})$ and $|\zeta|^{1/2}Y_\nu(u|\zeta|^{1/2})$. By replacing ζ in (2.09) by $|\zeta|e^{\pi i}$, or by direct substitution, we may verify that equation (2.05) is formally satisfied by the expansion

$$W = |\zeta|^{1/2}\mathscr{C}_\nu(u|\zeta|^{1/2})\sum_{s=0}^{\infty}\frac{A_s(\zeta)}{u^{2s}} - \frac{|\zeta|}{u}\mathscr{C}_{\nu+1}(u|\zeta|^{1/2})\sum_{s=0}^{\infty}\frac{B_s(\zeta)}{u^{2s}}, \tag{4.01}$$

in which $A_s(\zeta)$ and $B_s(\zeta)$ denote the analytic continuations across $\zeta = 0$ of the coefficients introduced in §2.2. Equations (2.10) and (2.11) again hold; their integrated forms are given by (2.13) and

$$B_s(\zeta) = -A_s'(\zeta) + \frac{1}{|\zeta|^{1/2}}\int_\zeta^0 \{\psi(v)A_s(v) - (\nu+\tfrac{1}{2})A_s'(v)\}\frac{dv}{|v|^{1/2}}. \tag{4.02}$$

Corresponding to $\lambda_1(\nu)$, we define

$$\lambda_2(\nu) = \sup\{\pi x M_\nu^2(x)\}, \qquad \lambda_3(\nu) = \sup\{\pi x|J_\nu(x)|E_\nu(x)M_\nu(x)\},$$
$$\lambda_4(\nu) = \sup\{\pi x|Y_\nu(x)|E_\nu^{-1}(x)M_\nu(x)\}, \tag{4.03}$$

each supremum being taken over the x interval $(0, \infty)$. Here $E_\nu(x)$ and $M_\nu(x)$ are the auxiliary functions introduced in §1.3. The results of §1 show that each quantity $\lambda_j(\nu)$ is finite. Moreover, each is continuous in ν, and

$$\lambda_3(\nu) \leqslant \lambda_2(\nu), \qquad \lambda_4(\nu) \leqslant \lambda_2(\nu), \qquad \lambda_2(\nu) \geqslant 2.$$

Numerical values for $\nu = 0$ and 1, together with the values of x corresponding to the suprema, are as follows:

$$\lambda_2(0) = 2.00 \ (x=\infty), \qquad \lambda_3(0) = 2.00 \ (x=\infty), \qquad \lambda_4(0) = 2.00 \ (x=\infty),$$

$$\lambda_2(1) = 2.34 \ (x=1.26), \qquad \lambda_3(1) = 2.14 \ (x=2.13), \qquad \lambda_4(1) = 2.05 \ (x=3.82).$$

We shall also need the following result.

Lemma 4.1

$$E_{\nu+1}(x) \geqslant E_\nu(x) \qquad (\nu \geqslant 0, \quad 0 < x < \infty). \tag{4.04}$$

To prove this inequality we have from (1.13), (1.15), and Theorem 6.4 of Chapter 7

$$X_\nu < y_{\nu,1} < j_{\nu,1} < j_{\nu+1,1}.$$

Accordingly, if $x \in (0, X_\nu]$, then $Y_\nu(x) < 0$ and $J_{\nu+1}(x) > 0$. Now suppose that $x \in (0, \hat{X}_\nu]$, where $\hat{X}_\nu \equiv \min(X_\nu, X_{\nu+1})$. From (1.16) and (1.10)

$$\frac{E_{\nu+1}^2(x)}{E_\nu^2(x)} = \frac{J_\nu(x)Y_{\nu+1}(x)}{J_{\nu+1}(x)Y_\nu(x)} = 1 - \frac{2}{\pi x J_{\nu+1}(x)Y_\nu(x)} > 1.$$

The first conclusion to be drawn from this inequality is that (4.04) holds in $(0, \hat{X}_\nu]$. A second conclusion is that $X_{\nu+1} > X_\nu$; compare (1.16).† Thus $\hat{X}_\nu = X_\nu$. And if $x > X_\nu$, then $E_\nu(x) = 1$ and $E_{\nu+1}(x) \geqslant 1$. Accordingly, (4.04) also holds in (X_ν, ∞) and the lemma is proved.

4.2 Theorem 4.1 *With the conditions of* §2.1, *equation* (2.05) *has, for each value of* u *and each nonnegative integer* n, *solutions* $W_{2n+1,3}(u,\zeta)$ *and* $W_{2n+1,4}(u,\zeta)$ *which are repeatedly differentiable in the* ζ *interval* $(\alpha, 0)$, *and are given by*

$$W_{2n+1,3}(u,\zeta) = |\zeta|^{1/2} J_\nu(u|\zeta|^{1/2}) \sum_{s=0}^{n} \frac{A_s(\zeta)}{u^{2s}} - \frac{|\zeta|}{u} J_{\nu+1}(u|\zeta|^{1/2}) \sum_{s=0}^{n-1} \frac{B_s(\zeta)}{u^{2s}} + \varepsilon_{2n+1,3}(u,\zeta), \tag{4.05}$$

$$W_{2n+1,4}(u,\zeta) = |\zeta|^{1/2} Y_\nu(u|\zeta|^{1/2}) \sum_{s=0}^{n} \frac{A_s(\zeta)}{u^{2s}} - \frac{|\zeta|}{u} Y_{\nu+1}(u|\zeta|^{1/2}) \sum_{s=0}^{n-1} \frac{B_s(\zeta)}{u^{2s}} + \varepsilon_{2n+1,4}(u,\zeta), \tag{4.06}$$

where

$$|\varepsilon_{2n+1,3}(u,\zeta)| \leqslant \lambda_3(\nu) |\zeta|^{1/2} E_\nu^{-1}(u|\zeta|^{1/2}) M_\nu(u|\zeta|^{1/2}) \times \exp\left\{\frac{\lambda_2(\nu)}{u} \mathscr{V}_{\zeta,0}(|\zeta|^{1/2} B_0)\right\} \frac{\mathscr{V}_{\zeta,0}(|\zeta|^{1/2} B_n)}{u^{2n+1}}, \tag{4.07}$$

$$|\varepsilon_{2n+1,4}(u,\zeta)| \leqslant \lambda_4(\nu) |\zeta|^{1/2} E_\nu(u|\zeta|^{1/2}) M_\nu(u|\zeta|^{1/2}) \times \exp\left\{\frac{\lambda_2(\nu)}{u} \mathscr{V}_{\alpha,\zeta}(|\zeta|^{1/2} B_0)\right\} \frac{\mathscr{V}_{\alpha,\zeta}(|\zeta|^{1/2} B_n)}{u^{2n+1}}, \tag{4.08}$$

$$\left|\frac{\partial \varepsilon_{2n+1,3}(u,\zeta)}{\partial \zeta} - \frac{\nu+1}{2\zeta} \varepsilon_{2n+1,3}(u,\zeta)\right| \leqslant \frac{1}{2} \lambda_3(\nu) \frac{E_{\nu+1}(u|\zeta|^{1/2})}{E_\nu^2(u|\zeta|^{1/2})} M_{\nu+1}(u|\zeta|^{1/2}) \times \exp\left\{\frac{\lambda_2(\nu)}{u} \mathscr{V}_{\zeta,0}(|\zeta|^{1/2} B_0)\right\} \frac{\mathscr{V}_{\zeta,0}(|\zeta|^{1/2} B_n)}{u^{2n}}, \tag{4.09}$$

$$\left|\frac{\partial \varepsilon_{2n+1,4}(u,\zeta)}{\partial \zeta} - \frac{\nu+1}{2\zeta} \varepsilon_{2n+1,4}(u,\zeta)\right| \leqslant \frac{1}{2} \lambda_4(\nu) E_{\nu+1}(u|\zeta|^{1/2}) M_{\nu+1}(u|\zeta|^{1/2}) \times \exp\left\{\frac{\lambda_2(\nu)}{u} \mathscr{V}_{\alpha,\zeta}(|\zeta|^{1/2} B_0)\right\} \frac{\mathscr{V}_{\alpha,\zeta}(|\zeta|^{1/2} B_n)}{u^{2n}}. \tag{4.10}$$

† This also follows from the result of Watson mentioned in the footnote on p. 438.

The proof is similar to that of Theorem 3.1. Writing

$$\hat{\varepsilon}_{2n+1,3}(u,-\zeta) = (-\zeta)^{-(\nu+1)/2}\varepsilon_{2n+1,3}(u,\zeta),$$

and then replacing ζ by $-\zeta$, we find that

$$\hat{\varepsilon}_{2n+1,3}(u,\zeta) = \int_0^\zeta \mathsf{K}(\zeta,v)\left[\frac{\psi(-v)}{v^{1/2}}\hat{\varepsilon}_{2n+1,3}(u,v) + \frac{J_\nu(uv^{1/2})}{u^{2n}v^{\nu/2}}\frac{d}{dv}\{v^{1/2}B_n(-v)\}\right]dv, \tag{4.11}$$

ζ now being positive, and

$$\mathsf{K}(\zeta,v) = \pi\zeta^{-\nu/2}v^{(\nu+1)/2}\{J_\nu(u\zeta^{1/2})Y_\nu(uv^{1/2}) - Y_\nu(u\zeta^{1/2})J_\nu(uv^{1/2})\}.$$

Using (1.17) and the fact that $E_\nu(uv^{1/2})$ is a nonincreasing function of v, we derive

$$|\mathsf{K}(\zeta,v)| \leqslant \pi\zeta^{-\nu/2}v^{(\nu+1)/2}E_\nu^{-1}(u\zeta^{1/2})E_\nu(uv^{1/2})M_\nu(u\zeta^{1/2})M_\nu(uv^{1/2}) \qquad (0<v\leqslant\zeta); \tag{4.12}$$

compare Chapter 11, (3.14). The derivative of the kernel is given by

$$\begin{aligned}\partial\mathsf{K}(\zeta,v)/\partial\zeta &= -\tfrac{1}{2}\pi u(v/\zeta)^{(\nu+1)/2}\{J_{\nu+1}(u\zeta^{1/2})Y_\nu(uv^{1/2}) - Y_{\nu+1}(u\zeta^{1/2})J_\nu(uv^{1/2})\}\\ &= \tfrac{1}{2}\pi u(v/\zeta)^{(\nu+1)/2}E_{\nu+1}(u\zeta^{1/2})E_\nu^{-2}(u\zeta^{1/2})M_{\nu+1}(u\zeta^{1/2})\\ &\quad\times E_\nu(uv^{1/2})M_\nu(uv^{1/2})T_\nu(\zeta,v),\end{aligned} \tag{4.13}$$

where

$$\begin{aligned}T_\nu(\zeta,v) = &-E_{\nu+1}^{-2}(u\zeta^{1/2})E_\nu^2(u\zeta^{1/2})\cos\theta_{\nu+1}(u\zeta^{1/2})\sin\theta_\nu(uv^{1/2})\\ &+E_\nu^2(u\zeta^{1/2})E_\nu^{-2}(uv^{1/2})\sin\theta_{\nu+1}(u\zeta^{1/2})\cos\theta_\nu(uv^{1/2}).\end{aligned}$$

Use of Lemma 4.1 and the fact that $E_\nu^2(u\zeta^{1/2})E_\nu^{-2}(uv^{1/2}) \leqslant 1$ shows that

$$|T_\nu(\zeta,v)| \leqslant 1. \tag{4.14}$$

The inequalities (4.07) and (4.09) may now be obtained by applying Theorem 10.1 of Chapter 6 to (4.11), and subsequently replacing ζ by $-\zeta$. The proof of (4.08) and (4.10) is similar.

Ex. 4.1 Show that $\lambda_1(\nu)\to 1$ as $\nu\to\infty$. Also, by use of Exercise 1.1 of the present chapter and Exercise 10.1 of Chapter 11, show that when $\nu\to\infty$,

$$X_\nu M_\nu^2(X_\nu) \sim 2^{5/3}\nu^{1/3}\,\mathrm{Ai}^2(c),$$

and thence that $\lambda_2(\nu)$ is unbounded in these circumstances.

Ex. 4.2 Assume that for each nonnegative integer s

$$\psi^{(s)}(\zeta) = O(|\zeta|^{-s-(1/2)-\sigma})$$

as $\zeta\to\pm\infty$, where σ is a positive constant. Prove that when $\zeta\to\pm\infty$, $A_s(\zeta)$, $D_s(\zeta)$, and $\mathscr{V}(|\zeta|^{1/2}B_s)$ all tend to finite limits, and also that

$$A_s^{(r)}(\zeta) = O(|\zeta|^{-r-\sigma_1}), \qquad D_s^{(r)}(\zeta) = O(|\zeta|^{-r-\sigma_1}) \qquad (r\geqslant 1),$$

$$B_s^{(r)}(\zeta) = O(|\zeta|^{-r-(1/2)}), \qquad C_s^{(r)}(\zeta) = O(|\zeta|^{-r+(1/2)}) \qquad (r\geqslant 0),$$

where $\sigma_1 = \min(\sigma,\tfrac{1}{2})$.

Ex. 4.3 If κ is a real or complex constant and $l_j \equiv \frac{1}{4}\Gamma(\frac{1}{4})\lambda_j(1)$, show that the equation

$$d^2w/dx^2 = \{x^{-1} + \kappa \exp(-x^2)\}\, w$$

has solutions of the form

$$x^{1/2}\{I_1(2x^{1/2}) + \varepsilon_1(x)\}, \qquad x^{1/2}\{K_1(2x^{1/2}) + \varepsilon_2(x)\},$$

$$|x|^{1/2}\{J_1(2|x|^{1/2}) + \varepsilon_3(x)\}, \qquad |x|^{1/2}\{Y_1(2|x|^{1/2}) + \varepsilon_4(x)\},$$

where

$$|\varepsilon_1(x)| \leqslant l_1|\kappa| \exp(l_1|\kappa|)\, I_1(2x^{1/2}), \qquad |\varepsilon_2(x)| \leqslant l_1|\kappa| \exp(l_1|\kappa|)\, K_1(2x^{1/2}) \qquad (x > 0),$$

$$E_1(2|x|^{1/2})|\varepsilon_3(x)|/l_3, \;\; E_1^{-1}(2|x|^{1/2})|\varepsilon_4(x)|/l_4 \leqslant |\kappa| \exp(l_2|\kappa|)\, M_1(2|x|^{1/2}) \qquad (x < 0).$$

Ex. 4.4† In the notation of Theorem 4.1 apply Theorem 10.2 of Chapter 6 to show that the error terms $|\varepsilon_{1,3}(u,\zeta)|$ and $|\varepsilon_{1,4}(u,\zeta)|$ are respectively bounded by

$$\frac{\lambda_3(\nu)}{\lambda_2(\nu)} |\zeta|^{1/2} E_\nu^{-1}(u|\zeta|^{1/2})\, M_\nu(u|\zeta|^{1/2}) \left[\exp\left\{\frac{\lambda_2(\nu)}{u} \mathscr{V}_{\zeta,0}(|\zeta|^{1/2} B_0)\right\} - 1\right],$$

$$\frac{\lambda_4(\nu)}{\lambda_2(\nu)} |\zeta|^{1/2} E_\nu(u|\zeta|^{1/2})\, M_\nu(u|\zeta|^{1/2}) \left[\exp\left\{\frac{\lambda_2(\nu)}{u} \mathscr{V}_{\alpha,\zeta}(|\zeta|^{1/2} B_0)\right\} - 1\right].$$

Ex. 4.5 Let $M_{k,m}(x)$ denote the Whittaker function (Chapter 7, §11), and $k > 0$, $m \geqslant 0$, $x \geqslant 0$. Show that for each nonnegative integer n

$$M_{k,m}(x) = \frac{\Gamma(2m+1)}{k^m} \left\{ x^{1/2} J_{2m}(2k^{1/2}x^{1/2}) \sum_{s=0}^{n} \frac{a_s(m,x)}{k^s} + \frac{x}{k^{1/2}} J_{2m+1}(2k^{1/2}x^{1/2}) \sum_{s=0}^{n-1} \frac{b_s(m,x)}{k^s} + \varepsilon_{2n+1}(k,m,x) \right\},$$

where $a_s(m,x)$ and $b_s(m,x)$ are polynomials in x of degrees $3s$ and $3s+1$, respectively, and $|\varepsilon_{2n+1}(k,m,x)|$ is bounded by

$$\lambda_3(2m)\, x^{1/2} \frac{M_{2m}(2k^{1/2}x^{1/2})}{E_{2m}(2k^{1/2}x^{1/2})} \exp\left\{\frac{\lambda_2(2m)\, x^{3/2}}{12k^{1/2}}\right\} \frac{\mathscr{V}_{0,x}\{x^{1/2} b_n(m,x)\}}{k^{n+(1/2)}}.$$

Evaluate $a_0(m,x)$, $b_0(m,x)$, $a_1(m,x)$, and $b_1(m,x)$.

***Ex. 4.6**‡ Again, let $M_{k,m}(x)$ and $W_{k,m}(x)$ denote the Whittaker functions. If $k > 0$, $m \geqslant 0$, and $x > 0$, show that for each nonnegative integer n

$$M_{-k,m}(4kx) = \frac{2\Gamma(2m+1)}{k^{m-(1/2)}} \left(\frac{x\zeta}{1+x}\right)^{1/4} \left\{ I_{2m}(4k\zeta^{1/2}) \sum_{s=0}^{n} \frac{A_s(\zeta)}{(4k)^{2s}} + \frac{\zeta^{1/2}}{4k} I_{2m+1}(4k\zeta^{1/2}) \sum_{s=0}^{n-1} \frac{B_s(\zeta)}{(4k)^{2s}} + \frac{\varepsilon_{2n+1,1}(4k,\zeta)}{\zeta^{1/2}} \right\},$$

† Similar improved bounds in the case $n = 0$ can be constructed for Theorem 3.1.

‡ An expansion of the type given in Exercise 4.5 can also be constructed for $M_{-k,m}(x)$. The advantage of the more complicated expansion of Exercise 4.6 is to provide an approximation for large k which is uniform with respect to *unbounded* x. For similar results with complex variables see Skovgaard (1966), and Erdélyi and Swanson (1957), and for some applications see Askey and Wainger (1965) and Muckenhoupt (1970). Compare also Chapter 11, Exercise 7.3.

and

$$W_{-k,m}(4kx) = \frac{2^{3/2}e^k}{\pi^{1/2}k^{k-(1/2)}\phi_n(k,m)}\left(\frac{x\zeta}{1+x}\right)^{1/4}\left\{K_{2m}(4k\zeta^{1/2})\sum_{s=0}^{n}\frac{A_s(\zeta)}{(4k)^{2s}}\right.$$

$$\left.-\frac{\zeta^{1/2}}{4k}K_{2m+1}(4k\zeta^{1/2})\sum_{s=0}^{n-1}\frac{B_s(\zeta)}{(4k)^{2s}}+\frac{\varepsilon_{2n+1,2}(4k,\zeta)}{\zeta^{1/2}}\right\},$$

provided that, in the second relation, $\phi_n(k,m) \neq 0$, where

(i) $\zeta^{1/2} = \frac{1}{2}(x^2+x)^{1/2} + \frac{1}{2}\ln\{x^{1/2}+(x+1)^{1/2}\}$;

(ii) $A_s(\zeta)$ and $B_s(\zeta)$ are continuous and bounded functions of ζ, given by (2.12) and (2.13), with $A_1(0) = A_2(0) = \cdots = 0$ and

$$\psi(\zeta) = -\frac{m^2-\frac{1}{16}}{\zeta}+\frac{m^2-\frac{1}{4}}{x(1+x)}+\frac{3+8x}{16x(1+x)^3};$$

(iii)
$$\phi_n(k,m) = \sum_{s=0}^{n}\frac{A_s(\infty)}{(4k)^{2s}} - \sum_{s=0}^{n-1}\frac{1}{(4k)^{2s+1}}\left[\lim_{\zeta\to\infty}\{\zeta^{1/2}B_s(\zeta)\}\right];$$

(iv) $\varepsilon_{2n+1,1}(4k,\zeta)$ and $\varepsilon_{2n+1,2}(4k,\zeta)$ satisfy (3.04) and (3.05), with $u = 4k$, $v = 2m$, and $\beta = \infty$.

Can results of this kind also be found for $M_{\kappa,m}(4kx)$ and $W_{\kappa,m}(4kx)$?

5 Asymptotic Properties of the Expansions

5.1 In Chapter 11, §§4.1, 4.2, and 7.4, we discussed the asymptotic nature of the Airy function approximations for Case II. Analogous observations concerning the expansions just derived for Case III are as follows.

If u is fixed and $\zeta \to 0$, then $\mathscr{V}_{0,\zeta}(\zeta^{1/2}B_n)$ is $O(\zeta^{1/2})$, and from (3.04) we have

$$\varepsilon_{2n+1,1}(u,\zeta) = I_\nu(u\zeta^{1/2})\,O(\zeta).$$

Therefore the right-hand side of (3.02), without its error term, is asymptotic to the solution $W_{2n+1,1}(u,\zeta)$ with relative error $O(\zeta^{1/2})$.† Similar conclusions hold for the solutions $W_{2n+1,2}(u,\zeta)$, $W_{2n+1,3}(u,\zeta)$, and $W_{2n+1,4}(u,\zeta)$ when $\zeta\to\beta$, 0, and α, respectively. These properties are important for the purpose of identifying the $W_{2n+1,j}(u,\zeta)$ in terms of other solutions of the differential equation, but not otherwise: simpler and more powerful approximations are available in these circumstances from the theory of earlier chapters.

5.2 Now consider the case of large u. If ν is fixed, and $\mathscr{V}_{0,\beta}(\zeta^{1/2}B_0)$ and $\mathscr{V}_{0,\beta}(\zeta^{1/2}B_n)$ are bounded functions of u, then from (3.04) we have

$$\varepsilon_{2n+1,1}(u,\zeta) = \zeta^{1/2}I_\nu(u\zeta^{1/2})\,O(u^{-2n-1}) \tag{5.01}$$

as $u\to\infty$, uniformly with respect to $\zeta\in[0,\beta)$. Accordingly, the formal series on the right-hand side of (2.09), with $\mathscr{L} = I$, is a uniform asymptotic expansion of the solution $W_{2n+1,1}(u,\zeta)$ to $2n+1$ terms. Similarly for $W_{2n+1,2}(u,\zeta)$.

† Unless, of course, $\sum_{s=0}^{n} A_s(0)\,u^{-2s} = 0$.

We further observe that if (i) ν ranges over a compact interval $\mathbf{N} \subset [0, \infty)$ which is independent of u and ζ, (ii) β and $\psi(\zeta)$ are independent of u and ν, then the coefficients $A_s(\zeta)$, $B_s(\zeta)$, and their variations are bounded functions of ν. Moreover, (5.01) holds uniformly with respect to $\nu \in \mathbf{N}$ (in addition to $\zeta \in [0, \beta)$). It is also evident that some of the conditions just stated could be relaxed without invalidating this result.

For the interval $\alpha < \zeta \leqslant 0$, if $\mathscr{V}_{\alpha,0}(|\zeta|^{1/2} B_0)$ and $\mathscr{V}_{\alpha,0}(|\zeta|^{1/2} B_n)$ are both bounded, then for large u we have uniformly

$$\varepsilon_{2n+1,3}(u,\zeta) = |\zeta|^{1/2} E_\nu^{-1}(u|\zeta|^{1/2})\, M_\nu(u|\zeta|^{1/2})\, O(u^{-2n-1}).$$

Writing $\tau = u|\zeta|^{1/2}$ and using the definitions and relations of §1, we express†

$$E_\nu^{-1}(\tau)\, M_\nu(\tau) = \mathfrak{A}_\nu(\tau)\, J_\nu(\tau) + \mathfrak{B}_\nu(\tau)\, \tau J_{\nu+1}(\tau),$$

where

$$\mathfrak{A}_\nu(\tau) = 2^{1/2}, \qquad \mathfrak{B}_\nu(\tau) = 0 \qquad (0 < \tau \leqslant X_\nu),$$

or

$$\mathfrak{A}_\nu(\tau) = -\tfrac{1}{2}\pi\tau Y_{\nu+1}(\tau)\, M_\nu(\tau), \qquad \mathfrak{B}_\nu(\tau) = \tfrac{1}{2}\pi Y_\nu(\tau)\, M_\nu(\tau) \qquad (\tau > X_\nu).$$

From (1.09) and (1.24) it is seen that $|\mathfrak{A}_\nu(\tau)|$ and $|\mathfrak{B}_\nu(\tau)|$ are bounded when $\tau \in (0, \infty)$. Hence for large u, the right-hand side of (4.01), with $\mathscr{C} = J$, provides a uniform compound expansion of $W_{2n+1,3}(u,\zeta)$ to $2n+1$ terms. Similarly for $W_{2n+1,4}(u,\zeta)$.

Again, the existence of solutions that are independent of n and have the infinite series (2.09) or (4.01) as compound asymptotic expansions may be established by the method of Chapter 10, §6.

Ex. 5.1 Assume that β and $\psi(\zeta)$ are independent of u and ν, and $\zeta^{-1/2}\psi(\zeta)$ is integrable over $(0, \beta)$. By means of Exercise 4.1 show that when $u \to \infty$

$$W_{1,1}(u,\zeta) \sim \zeta^{1/2} I_\nu(u\zeta^{1/2}), \qquad W_{1,2}(u,\zeta) \sim \zeta^{1/2} K_\nu(u\zeta^{1/2}),$$

uniformly with respect to $\zeta \in (0, \beta)$ and *unbounded* ν.

Can this result be extended to $W_{2n+1,1}(u,\zeta)$ and $W_{2n+1,2}(u,\zeta)$ for $n > 0$? Does a similar result hold for the solutions given by Theorem 4.1?

Ex. 5.2 If δ denotes an arbitrary positive constant, then by application of the theory of Chapter 10 to equation (2.05) in the interval $\delta \leqslant \zeta < \infty$ a uniform asymptotic approximation can be constructed for the solution $W_{2n+1,2}(u,\zeta)$ of §3.2 in the form

$$W_{2n+1,2}(u,\zeta) = \left(\frac{\pi\xi}{2u}\right)^{1/2} e^{-u\xi} \left\{ \sum_{s=0}^{2n} (-)^s \frac{\mathsf{A}_s(\xi)}{u^s} + O\left(\frac{1}{u^{2n+1}}\right) \right\} \qquad (u \to \infty),$$

where $\xi = \zeta^{1/2}$. Using the method indicated in Chapter 11, Exercise 7.4, obtain the identities

$$A_s(\zeta) = \sum_{j=0}^{2s} a_j(\nu+1)\, \xi^{-j} \mathsf{A}_{2s-j}(\xi), \qquad B_s(\zeta) = \sum_{j=0}^{2s+1} a_j(\nu)\, \xi^{-j-1} \mathsf{A}_{2s-j+1}(\xi),$$

in which $a_j(\nu) = (4\nu^2 - 1^2)(4\nu^2 - 3^2) \cdots \{4\nu^2 - (2j-1)^2\}/(j!\, 8^j)$.

† Compare Chapter 11, §7.4.

Ex. 5.3 By means of Theorem 3.1 show that the equation

$$\frac{d^2 w}{dx^2} + \frac{1}{x}\frac{dw}{dx} - \left(\frac{u^2}{x} + \frac{ue^{-x}}{x}\right) w = 0$$

has solutions of the form

$$\{1 + O(u^{-1})\} I_0\{2ux^{1/2} + \tfrac{1}{2}\pi^{1/2} \operatorname{erf}(x^{1/2})\}, \qquad \{1 + O(u^{-1})\} K_0\{2ux^{1/2} + \tfrac{1}{2}\pi^{1/2} \operatorname{erf}(x^{1/2})\},$$

as $u \to \infty$, uniformly with respect to $x \in (0, \infty)$.

*6 Determination of Phase Shift

6.1 Consider the radial Schrödinger equation

$$\frac{d^2 w}{dr^2} = \left\{-u^2 + \frac{l(l+1)}{r^2} + g(r)\right\} w \qquad (0 < r < \infty), \tag{6.01}$$

in which u is a positive parameter, l is a real parameter, and $g(r)$ a prescribed function of r. Physically, u represents the *momentum* of a scattering particle, and the terms in the braces not involving u are *potentials*, $l(l+1)/r^2$ being the *centrifugal potential.*

Assume that

$$\int^{\infty} |g(r)|\, dr < \infty. \tag{6.02}$$

Application of the LG approximation then shows that for large r, and fixed u and l, the general solution of (6.01) can be represented in the form

$$w = \{2/(\pi u)\}^{1/2} (1 + \alpha_l) \sin(ur - \tfrac{1}{2}\pi l + \delta_l) + o(1). \tag{6.03}$$

Here α_l and δ_l are constants which depend on the initial conditions for w, and are such that $1 + \alpha_l > 0$ and $\delta_l \in (-\pi, \pi]$. The apparently unnecessary quantities $\{2/(\pi u)\}^{1/2}$ and $\frac{1}{2}\pi l$ are introduced for later convenience.

Equation (6.01) is singular at $r = 0$. The value of δ_l corresponding to the solution which is recessive at this singularity is called the *phase shift*. In this section we investigate the phase shift when u is large.

Application of the Liouville transformation $x = r^2$, $W = x^{1/4} w$, yields

$$\frac{d^2 W}{dx^2} = \left\{-\frac{u^2}{4x} + \frac{\nu^2 - 1}{4x^2} + \frac{g(x^{1/2})}{4x}\right\} W, \tag{6.04}$$

with $\nu = l + \frac{1}{2}$. This differential equation is of the form (2.05) with $x = -\zeta$. If $g(r)$ is even in r and repeatedly differentiable throughout the interval $-\infty < r < \infty$, then $g(x^{1/2})$ and all its x derivatives are continuous when $x \in [0, \infty)$, and Theorem 4.1 is applicable. The recessive solution has the J_ν-type approximation, and from (4.05) and (4.07) asymptotic estimates for δ_l can be deduced without difficulty.

6.2 Considerable physical interest attaches to the case in which $g(r)$ is not even in r and is unbounded as $r \to 0$. To determine δ_l in these circumstances we need more delicate analysis of the error term associated with the approximation $x^{1/2} J_\nu(ux^{1/2})$.

The extension of the LG approximation given in Chapter 6, §7 supplies such analysis in the case $l = 0$. A similar result for other values of l is as follows:

Theorem 6.1 *Let u and l be parameters restricted by $u > 0$ and $l \geqslant -\frac{1}{2}$, and $g(r)$ be a real or complex function which is continuous in $(0, \infty)$ and is such that*

$$G_\nu(u, r) \equiv \tfrac{1}{2}\pi \int_0^r t M_\nu^2(ut)\,|g(t)|\,dt \tag{6.05}$$

is finite when r is finite, where $\nu = l + \frac{1}{2}$ and M_ν is defined in §1.3. Then equation (6.01) has a solution $w_l(u, r)$ which, as a function of r, is continuous in $[0, \infty)$, twice continuously differentiable in the corresponding open interval, and given by

$$w_l(u, r) = r^{1/2}\{J_\nu(ur) + \varepsilon_l(u, r)\}, \tag{6.06}$$

where

$$|\varepsilon_l(u, r)| \leqslant E_\nu^{-1}(ur)\, M_\nu(ur)[\exp\{G_\nu(u, r)\} - 1]. \tag{6.07}$$

This theorem can be established directly from (6.01) without recourse to the transformed equation (6.04). The appropriate integral equation for the error term is found to be

$$\varepsilon_l(u, r) = \tfrac{1}{2}\pi \int_0^r \mathsf{K}(r, t)\{J_\nu(ut) + \varepsilon_l(u, t)\}\, t g(t)\, dt, \tag{6.08}$$

where $\mathsf{K}(r, t) = Y_\nu(ur) J_\nu(ut) - J_\nu(ur) Y_\nu(ut)$. Bounds on $J_\nu(ut)$ and the kernel are supplied by $|J_\nu(ut)| \leqslant E_\nu^{-1}(ut)\, M_\nu(ut)$ and

$$|\mathsf{K}(r, t)| \leqslant E_\nu^{-1}(ur)\, E_\nu(ut)\, M_\nu(ur)\, M_\nu(ut) \qquad (0 < t \leqslant r);$$

compare (1.17) and (4.12). The proof is completed by applying Theorem 10.2 of Chapter 6. The essential difference from the proof of Theorem 4.1 is that we no longer use the majorants

$$\pi u t M_\nu^2(ut) \leqslant \lambda_2(\nu), \qquad \pi u t\, |J_\nu(ut)|\, E_\nu(ut)\, M_\nu(ut) \leqslant \lambda_3(\nu).$$

6.3 In applying Theorem 6.1 to determine a strict bound for the phase shift, we note first that $G_\nu(u, \infty)$ is finite; this is a consequence of Condition (6.02) and the behavior of $M_\nu(x)$ for large x, given by (1.24). Subtraction of (6.06) from (6.03) and use of the asymptotic formula (1.09) for J_ν yields, for large r,

$$(1 + \alpha_l)\sin(ur - \tfrac{1}{2}\pi l + \delta_l) - \sin(ur - \tfrac{1}{2}\pi l) = (\tfrac{1}{2}\pi u r)^{1/2}\varepsilon_l(u, r) + o(1).$$

From (6.07), (1.16), and (1.24), we see that as $r \to \infty$ the absolute value of the right-hand side of this equation is bounded by $\exp\{G_\nu(u, \infty)\} - 1 + o(1)$. Hence by analysis exactly similar to that of Chapter 6, §7.3, we derive

$$|\alpha_l|,\ 2|\delta_l|/\pi \leqslant \exp\{G_\nu(u, \infty)\} - 1, \tag{6.09}$$

provided that the right-hand side does not exceed unity.

An asymptotic approximation for the actual value of δ_l can be deduced by an extension of the analysis, and is stated in Exercise 6.3 below.

Ex. 6.1 Let Condition (6.02) be satisfied and $g(r) \sim \rho r^{-1-\sigma}$ as $r \to 0$, where ρ and σ are constants restricted by $\rho \neq 0$ and $0 \leqslant \sigma < 1$. If $l > -\frac{1}{2}$ and u is large, deduce from (6.09) that $\delta_l = O(u^{\sigma-1})$ when $\sigma \neq 0$, or $\delta_l = O(u^{-1} \ln u)$ when $\sigma = 0$.

Ex. 6.2 Compare Theorem 6.1, in the case $l = 0$, with Theorem 7.1 of Chapter 6.

Ex. 6.3 Assume that $G_\nu(u, \infty) \to 0$ as $u \to \infty$. By separating the first term in the Liouville–Neumann expansion of the solution of (6.08), establish the *Born approximation*

$$\delta_l = -\tfrac{1}{2}\pi \int_0^\infty t J_\nu^2(ut) g(t)\, dt + O\{G_\nu^2(u, \infty)\}.$$

Ex. 6.4 Assume that (i) $g(r) \sim g_0 r^{-1}$ as $r \to 0$, where $g_0 \neq 0$; (ii) $g(r) - g_0 r^{-1}$ is continuously differentiable in $[0, \infty)$; (iii) (6.02) holds; (iv) $\mathcal{V}\{g(r)\}$ is finite as $r \to \infty$. With the aid of Exercises 6.1 and 6.3 show that for large u

$$\delta_0 = -\frac{1}{2u}(g_0 \ln u + A) + O\left\{\left(\frac{\ln u}{u}\right)^2\right\},$$

where

$$A = g_0\{\ln(2a) + \gamma\} + \int_0^a \left\{g(t) - \frac{g_0}{t}\right\} dt + \int_a^\infty g(t)\, dt,$$

a being an arbitrary positive number and γ denoting Euler's constant.

*7 Zeros

7.1 The method of Chapter 11, §6 can be adapted to determine zeros of solutions of the differential equation (2.05). For simplicity, we shall suppose that ν is fixed and both α and $\psi(\zeta)$ are independent of u, but these restrictions are not essential.

Consider the solution

$$W_{1,3}(u, \zeta) = |\zeta|^{1/2} J_\nu(u|\zeta|^{1/2}) + \varepsilon_{1,3}(u, \zeta). \tag{7.01}$$

This vanishes when

$$\cos\theta_\nu(u|\zeta|^{1/2}) + \rho_3(u, \zeta) = 0, \tag{7.02}$$

where

$$\rho_3(u, \zeta) = |\zeta|^{-1/2} E_\nu(u|\zeta|^{1/2})\, M_\nu^{-1}(u|\zeta|^{1/2})\, \varepsilon_{1,3}(u, \zeta).$$

From (4.07) we derive

$$|\rho_3(u, \zeta)| \leqslant \lambda_3(\nu)\, \sigma(u),$$

where

$$\sigma(u) = \exp\left\{\frac{\lambda_2(\nu)}{u} \mathcal{V}_{\alpha,0}(|\zeta|^{1/2} B_0)\right\} \frac{\mathcal{V}_{\alpha,0}(|\zeta|^{1/2} B_0)}{u},$$

and is a decreasing function of u.† We denote by u_ν the positive root of the equation

$$\lambda_2(\nu)\, \sigma(u) = 2^{-1/2},$$

and throughout the following suppose that $u > u_\nu$.

† In consequence of Exercise 4.4 we could take $\sigma(u)$ to be $[\exp\{\lambda_2(\nu)\mathcal{V}_{\alpha,0}(|\zeta|^{1/2}B_0)/u\} - 1]/\lambda_2(\nu)$. This would sharpen the results a little.

Equation (7.02) has the same form as (6.03) of Chapter 11 and may be analyzed in the same way. We first observe from §1.3 that if $u|\zeta|^{1/2} \leqslant y_{\nu,1}$, then we have $-\frac{1}{4}\pi \leqslant \theta_\nu(u|\zeta|^{1/2}) \leqslant 0$ and $\cos\theta_\nu(u|\zeta|^{1/2}) \geqslant 2^{-1/2}$. Since

$$|\rho_3(u,\zeta)| \leqslant \lambda_3(\nu)\sigma(u) \leqslant \lambda_2(\nu)\sigma(u) < 2^{-1/2}, \tag{7.03}$$

it follows that there can be no zeros in the ζ interval $[-y_{\nu,1}^2/u^2, 0)$.

7.2 For the ζ interval $(-\infty, -y_{\nu,1}^2/u^2)$, we find from (7.02)

$$\theta_\nu(u|\zeta|^{1/2}) - (n-\tfrac{1}{2})\pi + (-)^n \sin^{-1}\{\rho_3(u,\zeta)\} = 0,$$

where n is an arbitrary positive integer. With $u > u_\nu$ the principal value of the inverse sine is numerically less than $\frac{1}{4}\pi$. Hence from the first of (1.21) we see that there is at least one zero in the interval

$$-y_{\nu,n+1}^2/u^2 < \zeta < -y_{\nu,n}^2/u^2. \tag{7.04}$$

To delimit this zero in a shorter interval, we have

$$u|\zeta|^{1/2} - j_{\nu,n} = (-)^{n-1}\sin^{-1}\{\rho_3(u,\zeta)\}/\theta_\nu'(\tau),$$

where $\tau \in (y_{\nu,n}, y_{\nu,n+1})$; compare Chapter 11, (6.06). Since $\tau > X_\nu$, it follows that

$$\{\theta_\nu'(\tau)\}^{-1} \leqslant \tfrac{1}{2}\lambda_2(\nu);$$

compare (1.20) and (4.03). Also, in consequence of (7.03) we have

$$|\sin^{-1}\{\rho_3(u,\zeta)\}| \leqslant 2^{-3/2}\pi\lambda_3(\nu)\sigma(u).$$

Combination of these results shows that, for each positive integer n, $W_{1,3}(u,\zeta)$ vanishes for at least one value of ζ satisfying

$$j_{\nu,n} - 2^{-5/2}\pi\lambda_2(\nu)\lambda_3(\nu)\sigma(u) < u(-\zeta)^{1/2} < j_{\nu,n} + 2^{-5/2}\pi\lambda_2(\nu)\lambda_3(\nu)\sigma(u).$$

Asymptotically,

$$\zeta = -\frac{j_{\nu,n}^2}{u^2} + j_{\nu,n}O\left(\frac{1}{u^3}\right)$$

as $u \to \infty$, uniformly for those values of n for which $y_{\nu,n+1} \leqslant u|\alpha|^{1/2}$. If $\alpha = -\infty$, then all values of n are included.

To state the corresponding result for the solution $W_{1,4}(u,\zeta)$, we denote X_ν conventionally by $j_{\nu,0}$. Similar analysis shows that when $u > u_\nu$ there are no zeros of $W_{1,4}(u,\zeta)$ in the ζ interval $[-j_{\nu,0}^2/u^2, 0)$, and for each positive integer n there is at least one zero satisfying both $j_{\nu,n-1} < u(-\zeta)^{1/2} < j_{\nu,n}$ and

$$y_{\nu,n} - 2^{-5/2}\pi\lambda_2(\nu)\lambda_4(\nu)\sigma(u) < u(-\zeta)^{1/2} < y_{\nu,n} + 2^{-5/2}\pi\lambda_2(\nu)\lambda_4(\nu)\sigma(u).$$

7.3 A simple illustration is provided by Exercise 4.5. For fixed m and large k, the nth positive zero of the Whittaker function $M_{k,m}(x)$ is given by

$$\frac{j_{2m,n}^2}{4k} + j_{2m,n}O\left(\frac{1}{k^{3/2}}\right). \tag{7.05}$$

Because the approximation for $M_{k,m}(x)$ is uniformly valid only in compact x intervals, the implied constant in the O term in (7.05) grows with n. More powerful approximations for the zeros could be constructed from an expansion for $M_{k,m}(4kx)$ analogous to that of Exercise 4.6.

Ex. 7.1 Let n be restricted by $y_{\nu,n+1} \leqslant u|\alpha|^{1/2}$. Show that for all sufficiently large u, uniformly with respect to n, there is exactly one zero of $W_{1,3}(u,\zeta)$ in the interval (7.04).

Ex. 7.2 (*Second approximations to zeros*) Assume the notation of Theorem 4.1 and also that $\alpha = -\infty$ and the conditions of Exercise 4.2 are satisfied. Show that when $n \geqslant 1$ the rth negative zeros of $W_{2n+1,3}(u,\zeta)$ and $W_{2n+1,4}(u,\zeta)$ are given by

$$\zeta = -\gamma_r + \frac{2\gamma_r B_0(-\gamma_r)}{u^2} + \gamma_r^{1/2} O\left(\frac{1}{u^3}\right) \qquad (u \to \infty),$$

uniformly with respect to unbounded r, where $\gamma_r = j_{\nu,r}^2/u^2$ for $W_{2n+1,3}(u,\zeta)$, and $\gamma_r = y_{\nu,r}^2/u^2$ for $W_{2n+1,4}(u,\zeta)$.

8 Auxiliary Functions for Complex Arguments

8.1 We now consider the extension of the approximation theorems of §§3 and 4 to complex values of the variable ζ and parameter u. Although extensions to complex ν are feasible, *we shall continue to suppose that* ν *is real and nonnegative*, a condition commonly satisfied in applications. Since $I_\nu(z)$ and $K_\nu(z)$ are expressible in terms of $J_\nu(iz)$ and $Y_\nu(iz)$, and *vice versa*, either modified or unmodified Bessel functions may be used as approximants when z is complex. But because $J_\nu(z)$ and $Y_\nu(z)$ do not comprise a numerically satisfactory pair when z is large and complex, we shall work in terms of the modified functions. We begin with a restatement of relevant properties of $I_\nu(z)$ and $K_\nu(z)$ established in Chapters 2 and 7.

If ν is fixed and $z \to 0$ in a bounded range of $\mathrm{ph}\, z$, then

$$I_\nu(z) \sim (\tfrac{1}{2}z)^\nu/\Gamma(\nu+1), \tag{8.01}$$

and

$$K_0(z) \sim \ln(1/z), \qquad K_\nu(z) \sim \tfrac{1}{2}\Gamma(\nu)(\tfrac{1}{2}z)^{-\nu} \qquad (\nu > 0). \tag{8.02}$$

If δ denotes an arbitrary positive constant less than π and z is large, then $I_\nu(z)$ is given by

$$(2\pi z)^{-1/2}[e^z\{1+O(z^{-1})\} + ie^{\nu\pi i}e^{-z}\{1+O(z^{-1})\}] \qquad (-\tfrac{1}{2}\pi+\delta \leqslant \mathrm{ph}\, z \leqslant \tfrac{3}{2}\pi-\delta), \tag{8.03}$$

or

$$(2\pi z)^{-1/2}[e^z\{1+O(z^{-1})\} - ie^{-\nu\pi i}e^{-z}\{1+O(z^{-1})\}] \qquad (-\tfrac{3}{2}\pi+\delta \leqslant \mathrm{ph}\, z \leqslant \tfrac{1}{2}\pi-\delta). \tag{8.04}$$

Also,

$$K_\nu(z) \sim \{\pi/(2z)\}^{1/2}e^{-z} \qquad (|\mathrm{ph}\, z| \leqslant \tfrac{3}{2}\pi-\delta). \tag{8.05}$$

Each of (8.03), (8.04), and (8.05) is uniform with respect to bounded ν.

Formulas connecting solutions in the sectors $(m-\frac{1}{2})\pi \leqslant \operatorname{ph} z \leqslant (m+\frac{1}{2})\pi$ and $|\operatorname{ph} z| \leqslant \frac{1}{2}\pi$, m being any integer, are given by

$$I_\nu(ze^{m\pi i}) = e^{m\nu\pi i} I_\nu(z), \tag{8.06}$$

and

$$K_\nu(ze^{m\pi i}) = e^{-m\nu\pi i} K_\nu(z) - \pi i \sin(m\nu\pi)\csc(\nu\pi) I_\nu(z). \tag{8.07}$$

Zeros of $I_\nu(z)$ are infinite in number and lie on the imaginary axis. $K_\nu(z)$ has no zeros in the sector $|\operatorname{ph} z| \leqslant \frac{1}{2}\pi$.

8.2 When $|\operatorname{ph} z| \leqslant \frac{1}{2}\pi$, a suitable weight function is given by

$$\mathfrak{E}_\nu(z) = \max_{t \in \mathbf{E}} |I_\nu(t)/K_\nu(t)|^{1/2}, \tag{8.08}$$

where $\mathbf{E}$ is the intersection of the strip $0 \leqslant \operatorname{Re} t \leqslant \operatorname{Re} z$ with the disk $|t| \leqslant |z|$; see Fig. 8.1. Outside $|\operatorname{ph} z| \leqslant \frac{1}{2}\pi$ we do not need to define $\mathfrak{E}_\nu(z)$. Since $K_\nu(t)$ does not vanish in $\mathbf{E}$, the ratio $|I_\nu(t)/K_\nu(t)|$ is finite and, as a consequence of the maximum-modulus theorem, attains its maximum on the boundary of $\mathbf{E}$. Clearly $\mathfrak{E}_\nu(z)$ is continuous in ν and z.

If $\nu > 0$, then

$$\frac{I_\nu(t)}{K_\nu(t)} \sim \frac{2(\frac{1}{2}t)^{2\nu}}{\nu\Gamma^2(\nu)} \qquad (t \to 0).$$

Hence

$$\mathfrak{E}_\nu(z) \sim \left(\frac{2}{\nu}\right)^{1/2} \frac{|\frac{1}{2}z|^\nu}{\Gamma(\nu)} \qquad (z \to 0). \tag{8.09}$$

Alternatively, if $\nu = 0$, then

$$\left|\frac{I_0(t)}{K_0(t)}\right|^2 \sim \left|\frac{1}{\ln(1/t)}\right|^2 = \frac{1}{(\ln|t|)^2 + (\operatorname{ph} t)^2} \qquad (t \to 0). \tag{8.10}$$

Assume that $|z| < 1$ and $0 \leqslant \operatorname{ph} z \leqslant \frac{1}{2}\pi$. Then it is easily seen that the right-hand side of (8.10) attains its maximum on the boundary of $\mathbf{E}$ somewhere on the straight

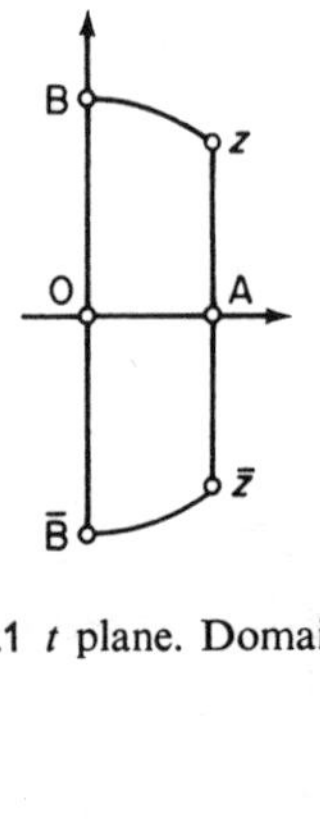

Fig. 8.1 t plane. Domain $\mathbf{E}$.

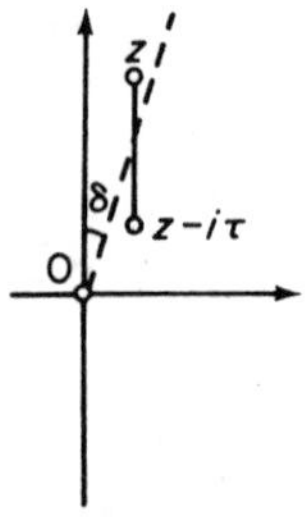

Fig. 8.2 t plane.

line $\bar{z}Az$ of Fig. 8.1. Setting $t = x + i\tau$, where $x = \operatorname{Re} z$, we find that

$$\frac{d}{d\tau}\left|\ln\left(\frac{1}{t}\right)\right|^2 = -\frac{\tau}{x^2+\tau^2}\ln\left(\frac{1}{x^2+\tau^2}\right) + \frac{2x}{x^2+\tau^2}\tan^{-1}\left(\frac{\tau}{x}\right).$$

Since $\tan^{-1}(\tau/x) \leqslant \tau/x$, this derivative is negative when $x^2+\tau^2 \leqslant 1/e^2$, that is, when $|t| \leqslant 1/e$. Thus when $|z|$ is sufficiently small, $|\ln(1/t)|$ attains its minimum at $t = z$. By symmetry, this also holds when $-\frac{1}{2}\pi \leqslant \operatorname{ph} z \leqslant 0$. Accordingly,

$$\mathfrak{E}_0(z) \sim |\ln(1/z)|^{-1/2} \qquad (z \to 0). \tag{8.11}$$

8.3 As $t \to \infty$ we have from (8.03), (8.04), and (8.05)

$$\pi I_\nu(t)/K_\nu(t) = e^{2t}\{1+O(t^{-1})\} + ie^{\nu\pi i} \qquad (0 \leqslant \operatorname{ph} t \leqslant \tfrac{1}{2}\pi), \tag{8.12}$$

or

$$\pi I_\nu(t)/K_\nu(t) = e^{2t}\{1+O(t^{-1})\} - ie^{-\nu\pi i} \qquad (-\tfrac{1}{2}\pi \leqslant \operatorname{ph} t \leqslant 0), \tag{8.13}$$

uniformly with respect to bounded ν. Also if $t = i\tau$, where τ is real, then

$$\left|\frac{I_\nu(t)}{K_\nu(t)}\right| = \frac{2}{\pi}\left|\frac{J_\nu(\tau)}{H_\nu^{(1)}(\tau)}\right| = \frac{2}{\pi}\frac{|J_\nu(\tau)|}{\{J_\nu^2(\tau)+Y_\nu^2(\tau)\}^{1/2}}. \tag{8.14}$$

From these results we now derive an upper bound for $\mathfrak{E}_\nu(z)$, and also the asymptotic form of $\mathfrak{E}_\nu(z)$ when $|z|$ is large.

First, from (8.14) it follows that

$$\left|\frac{I_\nu(i\tau)}{K_\nu(i\tau)}\right| \leqslant \frac{2}{\pi} \qquad (-\infty < \tau < \infty). \tag{8.15}$$

Consider $\Phi_\nu(t) \equiv \pi e^{-2t} I_\nu(t)/K_\nu(t)$ in the sector $|\operatorname{ph} t| \leqslant \frac{1}{2}\pi$. For large $|t|$, equations (8.12) and (8.13) yield

$$|\Phi_\nu(t)| = |1 + O(t^{-1}) \pm ie^{\pm\nu\pi i}e^{-2t}| \leqslant 2 + O(t^{-1}).$$

In consequence of (8.15) we have $|\Phi_\nu(t)| \leqslant 2$ on the imaginary axis. Application of the maximum-modulus theorem immediately shows that $|\Phi_\nu(t)| \leqslant 2$ throughout the sector. Therefore

$$\mathfrak{E}_\nu(z) \leqslant (2/\pi)^{1/2}|e^z| \qquad (|\operatorname{ph} z| \leqslant \tfrac{1}{2}\pi). \tag{8.16}$$

Equation (8.14) shows, incidentally, that this relation is an equality when $\operatorname{ph} z = \pm\frac{1}{2}\pi$ and $|z| \geqslant y_{\nu,1}$.

Next, assume that $|z|$ is large and $0 \leqslant \operatorname{ph} z \leqslant \frac{1}{2}\pi - \delta < \frac{1}{2}\pi$. On the portion $\bar{B}\bar{z}AzB$ of the boundary depicted in Fig. 8.1, we see that $|t|$ tends to infinity with $|z|$. Hence from (8.12) and (8.13) we derive

$$\left|\frac{I_\nu(t)}{K_\nu(t)}\right| \leqslant \frac{|e^{2z}|}{\pi}\left\{1 + O\left(\frac{1}{z}\right)\right\},$$

this relation being an equality on $\bar{z}Az$. From (8.15) it is seen that this bound also holds on $\bar{B}OB$, provided that $|z| \geqslant \frac{1}{2}\ln 2 \csc\delta$. Hence

$$\mathfrak{E}_\nu(z) \sim \pi^{-1/2}|e^z| \qquad (|z| \to \infty, \quad |\mathrm{ph}\, z| \leqslant \tfrac{1}{2}\pi - \delta), \tag{8.17}$$

uniformly with respect to bounded ν.

Lastly, we seek a lower bound for $\mathfrak{E}_\nu(z)$ when $|z|$ is large. Assume temporarily that $\frac{1}{2}\pi - \delta \leqslant \mathrm{ph}\, z \leqslant \frac{1}{2}\pi$, and set $t = z - i\tau$, where $\tau \in [0, \pi)$; see Fig. 8.2. If $|z| > \pi \sec\delta$, then $t \in \mathbf{E}$ and $\mathrm{Im}\, t > 0$. From (8.12)

$$\left|\frac{I_\nu(t)}{K_\nu(t)}\right| = \frac{|e^{2z}|}{\pi}\left|1 + O\left(\frac{1}{z}\right) + ie^{\nu\pi i - 2z + 2i\tau}\right| \geqslant \frac{|e^{2z}|}{\pi}\left\{1 + O\left(\frac{1}{z}\right)\right\},$$

for that value of τ for which $(\nu + \frac{1}{2})\pi - 2\,\mathrm{Im}\, z + 2\tau$ is an integer multiple of 2π. Combination of this result and the corresponding result for the conjugate sector with (8.17) produces

$$\mathfrak{E}_\nu(z) \geqslant \pi^{-1/2}|e^z|\{1 + o(1)\} \qquad (|\mathrm{ph}\, z| \leqslant \tfrac{1}{2}\pi, \quad z \to \infty), \tag{8.18}$$

uniformly with respect to bounded ν. This result should be compared with (8.16).

8.4 Having sketched the properties of the weight function $\mathfrak{E}_\nu(z)$, we now set

$$|I_\nu(z)| = \mathfrak{E}_\nu(z)\mathfrak{M}_\nu(z)\cos\vartheta_\nu(z), \quad |K_\nu(z)| = \mathfrak{E}_\nu^{-1}(z)\mathfrak{M}_\nu(z)\sin\vartheta_\nu(z) \quad (|\mathrm{ph}\, z| \leqslant \tfrac{1}{2}\pi), \tag{8.19}$$

the modulus function $\mathfrak{M}_\nu(z)$ being positive, and the phase function $\vartheta_\nu(z)$ lying in the interval $(0, \frac{1}{2}\pi]$. Thus

$$\mathfrak{M}_\nu(z) = \{\mathfrak{E}_\nu^{-2}(z)|I_\nu(z)|^2 + \mathfrak{E}_\nu^2(z)|K_\nu(z)|^2\}^{1/2}, \tag{8.20}$$

and

$$\vartheta_\nu(z) = \tan^{-1}\{\mathfrak{E}_\nu^2(z)|K_\nu(z)|/|I_\nu(z)|\}. \tag{8.21}$$

When ν is fixed and $z \to 0$, relations (8.01), (8.02), (8.09), and (8.11) yield

$$\mathfrak{M}_0(z) \sim |2\ln(1/z)|^{1/2}, \qquad \mathfrak{M}_\nu(z) \to \nu^{-1/2} \qquad (\nu > 0), \tag{8.22}$$

and

$$\vartheta_\nu(z) \to \tfrac{1}{4}\pi \qquad (\nu \geqslant 0). \tag{8.23}$$

As $z \to \infty$, we have from (8.03), (8.04), (8.05), and (8.17)

$$\mathfrak{M}_\nu(z) \sim |z|^{-1/2}, \qquad \vartheta_\nu(z) \to \tfrac{1}{4}\pi \qquad (|\mathrm{ph}\, z| \leqslant \tfrac{1}{2}\pi - \delta), \tag{8.24}$$

uniformly with respect to bounded ν. For the full range of $\mathrm{ph}\, z$ we obtain, with the aid of (8.16) and (8.18),

$$\mathfrak{M}_\nu(z) = O(z^{-1/2}), \qquad 1/\mathfrak{M}_\nu(z) = O(z^{1/2}) \qquad (|\mathrm{ph}\, z| \leqslant \tfrac{1}{2}\pi). \tag{8.25}$$

Also, $\vartheta_\nu(z)$ is bounded away from zero in these circumstances.

Corresponding to the quantities $\lambda_j(\nu)$ introduced in §§3.1 and 4.1, we define

$$\mu_1(\nu) = \sup\{2|z|\,\mathfrak{M}_\nu^2(z)\}, \qquad \mu_2(\nu) = \sup\{2|zI_\nu(z)|\,\mathfrak{E}_\nu^{-1}(z)\mathfrak{M}_\nu(z)\}, \tag{8.26}$$

$$\mu_3(\nu) = \sup\{2|zK_\nu(z)|\,\mathfrak{E}_\nu(z)\mathfrak{M}_\nu(z)\}, \qquad \mu_4(\nu) = \sup\{\mathfrak{E}_{\nu+1}^2(z)\mathfrak{E}_\nu^{-2}(z)\}, \tag{8.27}$$

all suprema being taken over the sector $|\text{ph}\, z| \leqslant \frac{1}{2}\pi$. As a consequence of relations derived in the present section, each quantity $\mu_j(v)$ is finite, and a continuous function of v.

Ex. 8.1 Show that if $z \to \infty$ along the imaginary axis, then $\vartheta_v(z)$ takes each value in the interval $[\frac{1}{4}\pi, \frac{1}{2}\pi]$ infinitely often.

Ex. 8.2 By considering imaginary values of z show that $\mu_1(v) \geqslant 4$ and $\mu_4(v) \geqslant 1$.

9 Error Bounds: Complex u and ζ

9.1 We consider now the differential equation (2.05), that is,

$$\frac{d^2W}{d\zeta^2} = \left\{\frac{u^2}{4\zeta} + \frac{v^2-1}{4\zeta^2} + \frac{\psi(\zeta)}{\zeta}\right\} W, \tag{9.01}$$

in the case when $v \geqslant 0$, u is real or complex, and $\psi(\zeta)$ is holomorphic in a complex domain $\mathbf{\Delta}$ which contains $\zeta = 0$ and may be unbounded. The coefficients $A_s(\zeta)$ and $B_s(\zeta)$ defined recursively by (2.12) and (2.13), beginning with $A_0(\zeta) = 1$, are holomorphic in $\mathbf{\Delta}$, *including* the point $\zeta = 0$. Again, if $\psi(\zeta)$ is a continuous function of v, then it can be arranged that the same is true of $A_s(\zeta)$ and $B_s(\zeta)$.

In order to give error bounds for the formal series solutions (2.09), we introduce definitions similar to those of Chapter 10, §3.1, and Chapter 11, §9.1. We denote by $\mathbf{Z}$ the intersection of $\mathbf{\Delta}$ and the sector $|\text{ph}(u\zeta^{1/2})| \leqslant \frac{1}{2}\pi$. Either branch of $\zeta^{1/2}$ may be chosen as long as it is continuous, and used consistently throughout.

For the I_v-type solution we define $\mathbf{Z}_1(u)$ to be the set of points ζ which can be linked to the origin by a path $\mathscr{Q}_1$ lying in $\mathbf{Z}$ and having the properties:

(i) *$\mathscr{Q}_1$ consists of a finite chain of R_2 arcs.*
(ii) *As v passes along $\mathscr{Q}_1$ from* 0 *to ζ, both* $\text{Re}(uv^{1/2})$ *and $|v|$ are nondecreasing.*

For the K_v-type solution we suppose α to be an arbitrary finite point of $\mathbf{Z}$, or the point at infinity on a path $\mathscr{L}$ lying in $\mathbf{Z}$. Then $\mathbf{Z}_2(u, \alpha)$ is defined to be the set of points ζ which can be linked to α by a path $\mathscr{Q}_2$ lying in $\mathbf{Z}$ and having the properties:

(i) *$\mathscr{Q}_2$ consists of a finite chain of R_2 arcs, and if α is at infinity then $\mathscr{Q}_2$ coincides with $\mathscr{L}$ in the neighborhood of α.*
(ii) *As v passes along $\mathscr{Q}_2$ from α to ζ, both* $\text{Re}(uv^{1/2})$ *and $|v|$ are nonincreasing.*

9.2 Implications of these conditions are illustrated in Figs. 9.1–9.7. The family of parabolas in Fig. 9.1 represents the curves in the ζ plane of constant $\text{Re}(u\zeta^{1/2})$. Each parabola has the origin as focus and the ray $\text{ph}\,\zeta = \pi - 2\omega$ as axis, ω denoting $\text{ph}\, u$.

A typical domain $\mathbf{\Delta}$ is shown in Fig. 9.2, C and D being given branch points of $\psi(\zeta)$. Corresponding cuts have been taken along parabolic arcs belonging to Fig. 9.1, the advantage of this arrangement being that in the $e^{i\omega}\zeta^{1/2}$ plane, depicted in

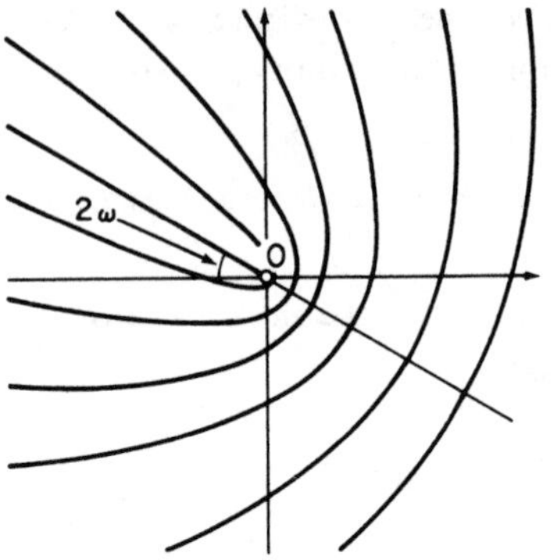

Fig. 9.1 Level curves of $\exp(e^{i\omega}\zeta^{1/2})$.

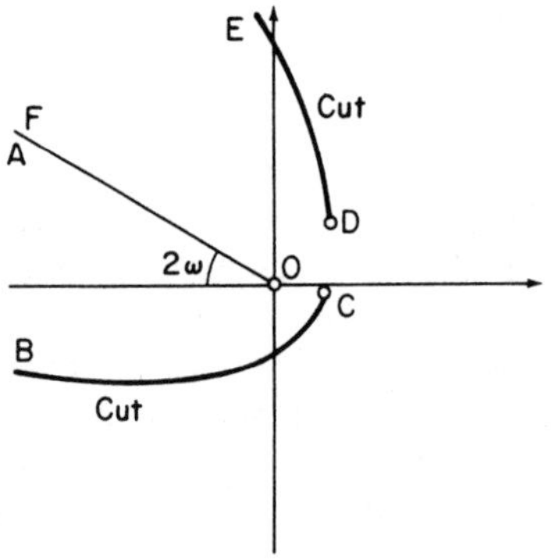

Fig. 9.2 Domain **Δ**.

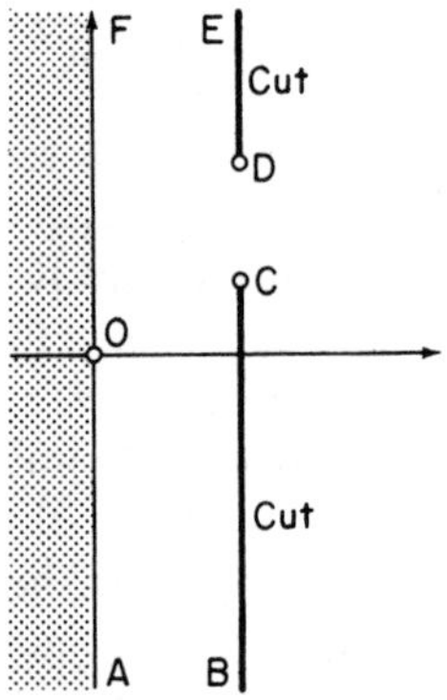

Fig. 9.3 $e^{i\omega}\zeta^{1/2}$ plane. Map of **Δ**.

Fig. 9.3, the cuts map into straight lines parallel to the imaginary axis. The region **Z** consists of **Δ** with an additional cut along the ray $\operatorname{ph}\zeta = \pi - 2\omega$.

In constructing the corresponding domains $\mathbf{Z}_1(u)$ and $\mathbf{Z}_2(u, \infty e^{-2i\omega})$, say, we note that the maps of the paths $\mathscr{Q}_j$ in the $e^{i\omega}\zeta^{1/2}$ plane must progress away from the origin and to the right when $j = 1$, or toward the origin and to the left when $j = 2$. The domains are illustrated in Figs. 9.4–9.7. In these diagrams *HC*, *DI*, and *JK* are arcs of circles centered at the origin; the other curves belong to the family of confocal parabolas. All shaded regions are excluded.

9.3 Theorem 9.1 *With the definitions of* §8 *and the present section, the differential equation* (9.01) *has, for each value of* u *and each nonnegative integer* n, *solutions* $W_{2n+1,j}(u,\zeta)$, $j = 1, 2$, *which are holomorphic in* $\Delta\backslash\{0\}$, *and are given by*

$$W_{2n+1,1}(u,\zeta) = \zeta^{1/2} I_\nu(u\zeta^{1/2}) \sum_{s=0}^{n} \frac{A_s(\zeta)}{u^{2s}} + \frac{\zeta}{u} I_{\nu+1}(u\zeta^{1/2}) \sum_{s=0}^{n-1} \frac{B_s(\zeta)}{u^{2s}} + \varepsilon_{2n+1,1}(u,\zeta) \tag{9.02}$$

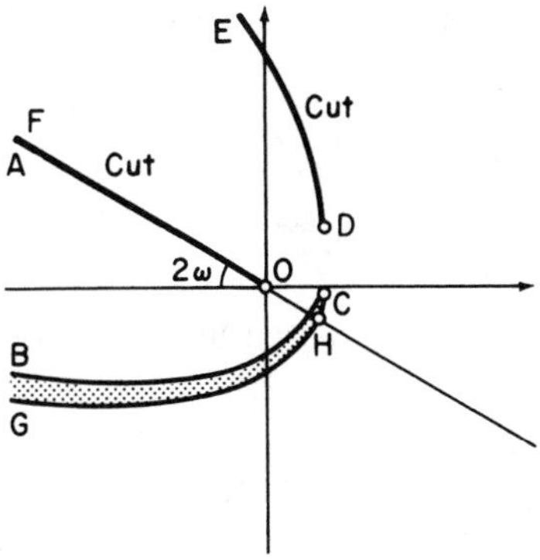

Fig. 9.4 Domain $\mathbf{Z}_1(u)$.

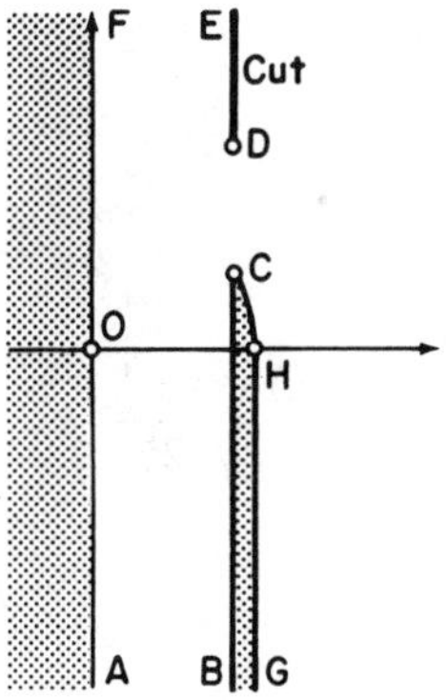

Fig. 9.5 $e^{i\omega}\zeta^{1/2}$ plane. Map of $\mathbf{Z}_1(u)$.

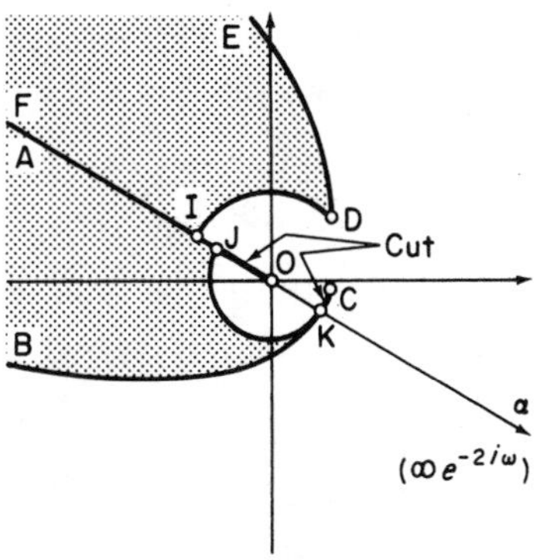

Fig. 9.6 Domain $\mathbf{Z}_2(u, \infty e^{-2i\omega})$.

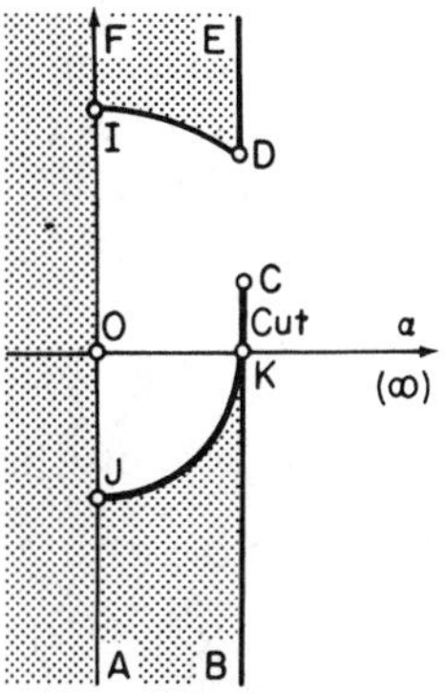

Fig. 9.7 $e^{i\omega}\zeta^{1/2}$ plane. Map of $\mathbf{Z}_2(u, \infty e^{-2i\omega})$.

and

$$W_{2n+1,2}(u,\zeta) = \zeta^{1/2} K_\nu(u\zeta^{1/2}) \sum_{s=0}^{n} \frac{A_s(\zeta)}{u^{2s}} - \frac{\zeta}{u} K_{\nu+1}(u\zeta^{1/2}) \sum_{s=0}^{n-1} \frac{B_s(\zeta)}{u^{2s}} + \varepsilon_{2n+1,2}(u,\zeta), \tag{9.03}$$

where

$$|\varepsilon_{2n+1,1}(u,\zeta)| \leqslant \mu_2(\nu)\, |\zeta|^{1/2}\, \mathfrak{E}_\nu(u\zeta^{1/2})\, \mathfrak{M}_\nu(u\zeta^{1/2}) \exp\left\{\frac{\mu_1(\nu)}{|u|} \mathcal{V}_{\mathcal{Q}_1}(\zeta^{1/2}B_0)\right\} \frac{\mathcal{V}_{\mathcal{Q}_1}(\zeta^{1/2}B_n)}{|u|^{2n+1}}, \tag{9.04}$$

$$\left| \frac{\partial \varepsilon_{2n+1,1}(u,\zeta)}{\partial \zeta} - \frac{\nu+1}{2\zeta} \varepsilon_{2n+1,1}(u,\zeta) \right| \leqslant \frac{1}{2} \mu_2(\nu)\, \mu_4(\nu) \frac{\mathfrak{E}_\nu^2(u\zeta^{1/2})}{\mathfrak{E}_{\nu+1}(u\zeta^{1/2})} \mathfrak{M}_{\nu+1}(u\zeta^{1/2})$$
$$\times \exp\left\{\frac{\mu_1(\nu)}{|u|} \mathcal{V}_{\mathcal{Q}_1}(\zeta^{1/2}B_0)\right\} \frac{\mathcal{V}_{\mathcal{Q}_1}(\zeta^{1/2}B_n)}{|u|^{2n}}, \tag{9.05}$$

when $\zeta \in \mathbf{Z}_1(u)$, *and*

$$|\varepsilon_{2n+1,2}(u,\zeta)| \leqslant \mu_3(\nu)|\zeta|^{1/2}\mathfrak{E}_\nu^{-1}(u\zeta^{1/2})\mathfrak{M}_\nu(u\zeta^{1/2}) \times \exp\left\{\frac{\mu_1(\nu)}{|u|}\mathscr{V}_{\mathscr{Q}_2}(\zeta^{1/2}B_0)\right\}\frac{\mathscr{V}_{\mathscr{Q}_2}(\zeta^{1/2}B_n)}{|u|^{2n+1}}, \tag{9.06}$$

$$\left|\frac{\partial\varepsilon_{2n+1,2}(u,\zeta)}{\partial\zeta} - \frac{\nu+1}{2\zeta}\varepsilon_{2n+1,2}(u,\zeta)\right| \leqslant \frac{1}{2}\mu_3(\nu)\mu_4(\nu)\mathfrak{E}_{\nu+1}^{-1}(u\zeta^{1/2})\mathfrak{M}_{\nu+1}(u\zeta^{1/2}) \times \exp\left\{\frac{\mu_1(\nu)}{|u|}\mathscr{V}_{\mathscr{Q}_2}(\zeta^{1/2}B_0)\right\}\frac{\mathscr{V}_{\mathscr{Q}_2}(\zeta^{1/2}B_n)}{|u|^{2n}}, \tag{9.07}$$

when $\zeta \in \mathbf{Z}_2(u,\alpha)$. *Here* α *is an arbitrary reference point in* $\mathbf{\Delta}$, *and* $W_{2n+1,2}(u,\zeta)$ *depends on* α.

This result is proved in a similar manner to Theorem 3.1. The error term $\varepsilon_{2n+1,1}(u,\zeta)$ again satisfies (3.08). Using (8.19) and (3.10), we see that the kernel $\mathsf{K}(\zeta,v)$ in the integral equation (3.09) may be expressed as

$$\mathsf{K}(\zeta,v) = 2\zeta^{-\nu/2}v^{(\nu+1)/2}\mathfrak{E}_\nu(u\zeta^{1/2})\mathfrak{E}_\nu^{-1}(uv^{1/2})\mathfrak{M}_\nu(u\zeta^{1/2})\mathfrak{M}_\nu(uv^{1/2}) \times \{\cos\vartheta_\nu(u\zeta^{1/2})\sin\vartheta_\nu(uv^{1/2}) - \mathfrak{E}_\nu^{-2}(u\zeta^{1/2})\mathfrak{E}_\nu^2(uv^{1/2}) \times \sin\vartheta_\nu(u\zeta^{1/2})\cos\vartheta_\nu(uv^{1/2})\}.$$

The conditions on the path $\mathscr{Q}_1$ imply that

$$\operatorname{Re}(u\zeta^{1/2}) \geqslant \operatorname{Re}(uv^{1/2}), \qquad |u\zeta^{1/2}| \geqslant |uv^{1/2}|.$$

Hence from (8.08) we have $\mathfrak{E}_\nu(u\zeta^{1/2}) \geqslant \mathfrak{E}_\nu(uv^{1/2})$, and therefore

$$|\mathsf{K}(\zeta,v)| \leqslant 2|\zeta^{-\nu/2}v^{(\nu+1)/2}|\mathfrak{E}_\nu(u\zeta^{1/2})\mathfrak{E}_\nu^{-1}(uv^{1/2})\mathfrak{M}_\nu(u\zeta^{1/2})\mathfrak{M}_\nu(uv^{1/2}). \tag{9.08}$$

Similarly, from (3.11), (8.27), and Exercise 8.2, we derive

$$|\partial\mathsf{K}(\zeta,v)/\partial\zeta| \leqslant \mu_4(\nu)|u(v/\zeta)^{(\nu+1)/2}|\mathfrak{E}_{\nu+1}^{-1}(u\zeta^{1/2})\mathfrak{E}_\nu^2(u\zeta^{1/2})\mathfrak{M}_{\nu+1}(u\zeta^{1/2})\mathfrak{E}_\nu^{-1}(uv^{1/2})\mathfrak{M}_\nu(uv^{1/2});$$

compare (4.13) and (4.14).

Application of Theorem 10.1 of Chapter 6, and use of (8.26) and the identity

$$\int |v^{-1/2}\psi(v)\,dv| = \mathscr{V}\{v^{1/2}B_0(v)\}$$

leads to (9.04) and (9.05). Similarly for (9.06) and (9.07).

*10 Asymptotic Properties for Complex Variables

10.1 Asymptotic properties of the expansion (9.02) when $\zeta \to 0$, or (9.03) when $\zeta \to \alpha$, or both expansions when $|u| \to \infty$, are derivable by analysis similar to §5. In the more important case of large $|u|$, suppose that ν, $\mathscr{V}_{\mathscr{Q}_1}(\zeta^{1/2}B_0)$, and $\mathscr{V}_{\mathscr{Q}_1}(\zeta^{1/2}B_n)$

are bounded. From (9.04) we have

$$\varepsilon_{2n+1,1}(u,\zeta) = \zeta^{1/2}\mathfrak{E}_\nu(u\zeta^{1/2})\,\mathfrak{M}_\nu(u\zeta^{1/2})\,O(u^{-2n-1}) \qquad (u\to\infty),$$

uniformly with respect to ζ and ν. As a consequence of (8.23) and the continuity of $\vartheta_\nu(z)$, we can assign a positive number κ such that $0 \leqslant \vartheta_\nu(z) \leqslant \frac{3}{8}\pi$ when $|z| \leqslant \kappa$. Write $\tau \equiv u\zeta^{1/2}$ and

$$\mathfrak{E}_\nu(\tau)\,\mathfrak{M}_\nu(\tau) \equiv \mathfrak{A}_\nu(\tau)\,I_\nu(\tau) + \mathfrak{B}_\nu(\tau)\,\tau I_{\nu+1}(\tau),$$

where

$$\mathfrak{A}_\nu(\tau) = |I_\nu(\tau)|/\{I_\nu(\tau)\cos\vartheta_\nu(\tau)\}, \qquad \mathfrak{B}_\nu(\tau) = 0 \qquad (|\tau| \leqslant \kappa),$$

or

$$\mathfrak{A}_\nu(\tau) = \tau\mathfrak{E}_\nu(\tau)\,\mathfrak{M}_\nu(\tau)\,K_{\nu+1}(\tau), \qquad \mathfrak{B}_\nu(\tau) = \mathfrak{E}_\nu(\tau)\,\mathfrak{M}_\nu(\tau)\,K_\nu(\tau) \qquad (|\tau| > \kappa).$$

Then from §§8.3 and 8.4 it follows that $|\mathfrak{A}_\nu(\tau)|$ and $|\mathfrak{B}_\nu(\tau)|$ are bounded when τ lies in the sector $|\mathrm{ph}\,\tau| \leqslant \frac{1}{2}\pi$ and ν is bounded. Hence for large $|u|$, the right-hand side of (9.02) is a uniform compound asymptotic expansion of $W_{2n+1,1}(u,\zeta)$ to $2n+1$ terms. Similarly for $W_{2n+1,2}(u,\zeta)$.

10.2 These asymptotic properties for large $|u|$ hold in larger ζ regions than indicated by Theorem 9.1. In the first place the conditions on the paths $\mathscr{Q}_1$ and $\mathscr{Q}_2$ can be relaxed, as follows.

In the case of $\mathscr{Q}_1$, Condition (ii) of §9.1 can be replaced by: *if v_1 and v_2 are any two points of $\mathscr{Q}_1$ in the order* $0, v_1, v_2, \zeta$, *then*

$$\mathrm{Re}(uv_1^{1/2}) \leqslant \mathrm{Re}(uv_2^{1/2}) \qquad \text{and} \qquad |v_1| \leqslant r|v_2|, \tag{10.01}$$

where r ($\geqslant 1$) is independent of u and ζ. The original version corresponds to $r = 1$. The extension to $r > 1$ depends on the following result, which is a straightforward deduction from §8. *Let t lie in the intersection of the strip* $0 \leqslant \mathrm{Re}\,t \leqslant \mathrm{Re}\,z$ *with the disk* $|t| \leqslant r^{1/2}|z|$. *Then*

$$\mathfrak{E}_\nu(t) \leqslant R(r)\,\mathfrak{E}_\nu(z),$$

where $R(r)$ is assignable independently of t and z. With the new condition we derive

$$\mathfrak{E}_\nu(uv^{1/2}) \leqslant R(r)\,\mathfrak{E}_\nu(u\zeta^{1/2}),$$

and hence

$$|\mathsf{K}(\zeta,v)| \leqslant 2R^2(r)\,|\zeta^{-\nu/2}v^{(\nu+1)/2}|\,\mathfrak{E}_\nu(u\zeta^{1/2})\,\mathfrak{E}_\nu^{-1}(uv^{1/2})\,\mathfrak{M}_\nu(u\zeta^{1/2})\,\mathfrak{M}_\nu(uv^{1/2});$$

compare (9.08). This leads to the new bound

$$|\varepsilon_{2n+1,1}(u,\zeta)| \leqslant \mu_2(\nu)\,R^2(r)\,|\zeta|^{1/2}\mathfrak{E}_\nu(u\zeta^{1/2})\,\mathfrak{M}_\nu(u\zeta^{1/2})$$
$$\times \exp\left\{\frac{\mu_1(\nu)}{|u|}R^2(r)\,\mathscr{V}_{\mathscr{Q}_1}(\zeta^{1/2}B_0)\right\}\frac{\mathscr{V}_{\mathscr{Q}_1}(\zeta^{1/2}B_n)}{|u|^{2n+1}},$$

from which the uniform asymptotic property for large $|u|$ follows as in §10.1.

Similarly, in place of Condition (ii) on $\mathscr{Q}_2$ it suffices that the inequalities (10.01) hold for any two points v_1 and v_2 of $\mathscr{Q}_2$ in the order ζ, v_1, v_2, α.

The effects of this extension on the example depicted in Figs. 9.2–9.7 are as follows. First, the shadow zone $GHCB$ associated with $\mathbf{Z}_1(u)$ can be reduced to the cut BC by taking r large enough. Secondly, the arcs DI and JK bounding the shadow zones $FIDE$ and $BKJA$ of Fig. 9.6 are replaced by concentric arcs with radius increased by the factor r. In consequence, by taking r large enough we can extend $\mathbf{Z}_2(u, \infty e^{-2i\omega})$ to include any point of the original shadow zones. It should be noticed, however, that $R(r) \to \infty$ as $r \to \infty$, implying that the error term $\varepsilon_{2n+1,2}(u, \zeta)$ may become prohibitively large.

10.3 A second extension of the regions of validity is to include sheets of the ζ plane beyond the sector $|\mathrm{ph}(u\zeta^{1/2})| \leqslant \frac{1}{2}\pi$. This is easily accomplished in the case of $W_{2n+1,1}(u, \zeta)$, as follows. If ζ lies in the sector

$$-(\tfrac{1}{2}+m)\pi \leqslant \mathrm{ph}(u\zeta^{1/2}) \leqslant (\tfrac{1}{2}-m)\pi,$$

m denoting any integer, then we set $\hat{\zeta} = \zeta e^{2m\pi i}$, so that $|\mathrm{ph}(u\hat{\zeta}^{1/2})| \leqslant \frac{1}{2}\pi$. The integral equation (3.08) for the error term may be set up with $\hat{\zeta}$ in place of ζ and solved in the same way. On restoring the original variables, we see that the expansion (9.02) holds when $\zeta \in \mathbf{Z}_1(u)$ for *any* value of $\mathrm{ph}\,\zeta$, provided that the branch of $\zeta^{1/2}$ in (9.04) and (9.05) is chosen to satisfy $|\mathrm{ph}(u\zeta^{1/2})| \leqslant \frac{1}{2}\pi$.

A simple extension of this type is unavailable for $W_{2n+1,2}(u, \zeta)$ because a change of phase in the reference point α leads to a new solution of the differential equation. By rearrangement of the kernel in the integral equation for $\varepsilon_{2n+1,2}(u, \zeta)$, it is possible to extend the asymptotic property of the expansion (9.03) to closed domains within $|\mathrm{ph}(u\zeta^{1/2})| < \frac{3}{2}\pi$; for details of this analysis see Olver (1956, 1958). The need for this extension can be avoided when solutions of (9.01) with argument $\zeta e^{2m\pi i}$ can be connected with solutions of argument ζ, as in §13.2 below. The determination of connection formulas in the general case is considered in Chapter 13, §4.

Ex. 10.1 Assume that $\mathbf{Z}_2(u, \alpha)$ contains a neighborhood of $\zeta = 0$ cut along the ray $\mathrm{ph}(u\zeta^{1/2}) = \pm\frac{1}{2}\pi$, and that $\mathscr{V}_{0,\alpha}(\zeta^{1/2}B_0)$ and $\mathscr{V}_{0,\alpha}(\zeta^{1/2}B_n)$ are both finite. Using the method of Chapter 6, §3.2, prove that $\varepsilon_{2n+1,2}(u, \zeta)/\{\zeta^{1/2}K_\nu(u\zeta^{1/2})\}$ tends to a finite limit as $\zeta \to 0$ within $\mathbf{Z}_2(u, \alpha)$.

Show also that when $\mathbf{Z}_2(u, \alpha)$ is simply connected this limit is independent of the direction in which the origin is approached.

*11 Behavior of the Coefficients at Infinity

11.1 Assume that for all values of u under consideration:

(i) $|\psi(\zeta)| \leqslant k/(1+|\zeta|^{(1/2)+\rho})$ *when* $\zeta \in \mathbf{\Delta}$, *where* k *and* ρ *are positive numbers independent of* ζ *and* u.

(ii) $|A_s(0)| \leqslant \hat{k}_s$, $s = 0, 1, \ldots$, *where* $\hat{k}_s$ *is independent of* u.

(iii) *A subdomain* $\mathbf{\Gamma}$ *of* $\mathbf{\Delta}$ *exists such that the distance between each boundary point* ζ_{B} *of* $\mathbf{\Gamma}$ *and any boundary point of* $\mathbf{\Delta}$ *is not less than* $d|\zeta_{\mathrm{B}}|^{1/2}$, *where* $d\,(>0)$ *is assignable independently of* ζ_{B} *and* u.†

† This condition implies that in the $\zeta^{1/2}$ plane there is a positive lower bound on the distance between the boundaries of the maps of $\mathbf{\Gamma}$ and $\mathbf{\Delta}$.

(iv) $\mathbf{\Gamma}$ *contains the disk* $|\zeta| \leqslant d$.

(v) *For each point* ζ *of* $\mathbf{\Gamma}$, *a path can be found in* $\mathbf{\Gamma}$ *along which*

$$\int_0^\zeta \frac{|dv|}{1+|v|^{1+\rho_1}} \leqslant l,$$

where $\rho_1 = \min(\rho, \frac{1}{2})$, *and* l *is independent of* ζ *and* u.

Theorem 11.1 *With the conditions of this subsection, there exist finite numbers* k_s, *independent of* ζ *and* u, *such that when* $\zeta \in \mathbf{\Gamma}$

$$|A_s(\zeta)| \leqslant k_s, \qquad |A_s'(\zeta)| \leqslant \frac{k_s}{1+|\zeta|^{1+\rho_1}}, \qquad |A_s''(\zeta)| \leqslant \frac{k_s}{1+|\zeta|^{(3/2)+\rho_1}},$$

$$|B_s(\zeta)| \leqslant \frac{k_s}{1+|\zeta|^{1/2}}, \qquad |B_s'(\zeta)| \leqslant \frac{k_s}{1+|\zeta|^{3/2}}, \qquad |B_s''(\zeta)| \leqslant \frac{k_s}{1+|\zeta|^{2+\rho_1}}.$$

This theorem corresponds to Theorem 5.1 of Chapter 10 and Theorem 9.2 of Chapter 11. The proof is similar but, as in Chapter 11, omitted because the result is inessential in most applications.

12 Legendre Functions of Large Degree: Real Arguments

12.1 The associated Legendre equation is given by

$$(1-x^2)\frac{d^2L}{dx^2} - 2x\frac{dL}{dx} + \left\{n(n+1) - \frac{m^2}{1-x^2}\right\}L = 0; \tag{12.01}$$

see Chapter 5, (12.02). We seek asymptotic solutions of this differential equation for *large positive values of the degree n, and fixed real values of the order m*, neither n nor m necessarily being an integer. *Throughout the present section the argument* x *is assumed to be real.*

When the term in the first derivative is removed from equation (12.01) by change of dependent variable, we obtain

$$\frac{d^2w}{dx^2} = \left\{\frac{n(n+1)}{x^2-1} + \frac{m^2-1}{(x^2-1)^2}\right\}w, \tag{12.02}$$

with solutions $w = (x^2-1)^{1/2}L$. This equation is to be identified with (2.01). In the asymptotic analysis we shall concentrate on positive x, hence x_0 is taken to be $+1$ and not -1. The most obvious choice for the parameter u is $\{n(n+1)\}^{1/2}$, but for reasons which become clear shortly, we define

$$u = n + \tfrac{1}{2}. \tag{12.03}$$

Then

$$f(x) = \frac{1}{x^2-1}, \qquad g(x) = -\frac{1}{4(x^2-1)} + \frac{m^2-1}{(x^2-1)^2}.$$

Relations (2.03) yield

$$\zeta^{1/2} = \int_1^x \frac{dt}{(t^2-1)^{1/2}} = \cosh^{-1} x \qquad (x \geqslant 1), \tag{12.04}$$

or

$$(-\zeta)^{1/2} = \int_x^1 \frac{dt}{(1-t^2)^{1/2}} = \cos^{-1} x \qquad (-1 \leqslant x \leqslant 1). \tag{12.05}$$

Since $(x-1)^2 g(x) \to \frac{1}{4}(m^2-1)$ as $x \to 1$, it follows that $\nu = m$. Accordingly, from (2.05) and (2.06) we find that

$$W \equiv \left(\frac{\zeta}{x^2-1}\right)^{1/4} w = \{\zeta(x^2-1)\}^{1/4} L$$

satisfies the equation

$$\frac{d^2W}{d\zeta^2} = \left\{\frac{u^2}{4\zeta} + \frac{m^2-1}{4\zeta^2} + \frac{\psi(\zeta)}{\zeta}\right\} W, \tag{12.06}$$

in which

$$\psi(\zeta) = \frac{4m^2-1}{16}\left(\frac{1}{x^2-1} - \frac{1}{\zeta}\right). \tag{12.07}$$

The function $\psi(\zeta)$ is analytic at the point $\zeta = 0$ corresponding to the original pole $x = 1$: this is readily confirmed by constructing the leading terms in its power-series expansion.

When $x \to \infty$, we have $\zeta \to \infty$, $\frac{1}{2}e^{\sqrt{\zeta}} \sim x$, and $\psi^{(s)}(\zeta) = O(\zeta^{-s-1})$. In consequence, Theorem 3.1 may be applied meaningfully to the infinite interval $0 < \zeta < \infty$; compare Exercise 4.2. This is the reason underlying the choice (12.03); had we taken $u^2 = n(n+1)$, then the constant term $\frac{1}{16}$ would be added to the right-hand side of (12.07), causing the variations of the quantities $\zeta^{1/2} B_s$ to diverge at infinity.

12.2 Theorem 3.1 is now applied to equation (12.06) with $\nu = m$ and $\beta = \infty$, and to avoid a clash in notation we replace the symbol n appearing in the statement of the theorem by p.

The solution $W_{2p+1,2}(u,\zeta)$ is recessive at $\zeta = \infty$; accordingly, its ratio to $\{\zeta(x^2-1)\}^{1/4} Q_n^m(x)$ is independent of x. The value of the ratio is conveniently found by letting $x \to 1+$. Assume, temporarily, that $m > 0$. Then from Chapter 5, (12.21) and (13.14), we have

$$Q_n^m(x) \sim e^{m\pi i} 2^{(m/2)-1} \Gamma(m) (x-1)^{-m/2},$$

and since $\zeta \sim 2(x-1)$, this yields

$$\{\zeta(x^2-1)\}^{1/4} Q_n^m(x) \sim e^{m\pi i} 2^{m-1} \Gamma(m) \zeta^{(1-m)/2} \qquad (m > 0). \tag{12.08}$$

On the other hand, from Theorem 3.1 and the second of (1.02) we have

$$W_{2p+1,2}(u,\zeta) = 2^{m-1}\Gamma(m) \frac{\zeta^{(1-m)/2}}{u^m} \left\{ \sum_{s=0}^{p} \frac{A_s(0)}{u^{2s}} - \frac{2m}{u^2} \sum_{s=0}^{p-1} \frac{B_s(0)}{u^{2s}} + \delta_{2p+1} + o(1) \right\}$$

$$(m > 0), \tag{12.09}$$

where δ_{2p+1} is a constant bounded by†

$$|\delta_{2p+1}| \leqslant \lambda_1(m) \exp\left\{\frac{\lambda_1(m)}{u} \mathscr{V}_{0,\infty}(\zeta^{1/2} B_0)\right\} \frac{\mathscr{V}_{0,\infty}(\zeta^{1/2} B_p)}{u^{2p+1}}.$$

We shall suppose u to be large enough to ensure that $|\delta_{2p+1}| < 1$.

Let the arbitrary constants in the relation (2.13) be determined by the condition

$$A_s(0) = 2mB_{s-1}(0) \qquad (s = 1, 2, \ldots). \tag{12.10}$$

The content of the braces in (12.09) then reduces to $1 + \delta_{2p+1} + o(1)$; accordingly

$$W_{2p+1,2}(u,\zeta) = e^{-m\pi i} u^{-m} (1 + \delta_{2p+1}) \{\zeta(x^2-1)\}^{1/4} Q_n^m(x).$$

Although we have assumed that $m > 0$ in establishing this equation, the fact that $W_{2p+1,2}(u,\zeta)$ and $Q_n^m(x)$ are continuous functions of m assures its validity for $m = 0$.

In collecting together the results of this subsection we find it convenient to denote $A_s(\zeta)$ and $B_s(\zeta)$—when standardized by (12.10)—by $A_s^m(\zeta)$ and $B_s^m(\zeta)$, respectively. Thus $A_0^m(\zeta) = 1$, and

$$B_s^m(\zeta) = -A_s^{m\prime}(\zeta) + \frac{1}{\zeta^{1/2}} \int_0^\zeta \left[\frac{4m^2-1}{16}\left\{\operatorname{csch}^2(v^{1/2}) - \frac{1}{v}\right\} A_s^m(v) - \left(m + \frac{1}{2}\right) A_s^{m\prime}(v)\right] \frac{dv}{v^{1/2}}, \tag{12.11}$$

$$A_{s+1}^m(\zeta) = m\{B_s^m(\zeta) + B_s^m(0)\} - \zeta B_s^{m\prime}(\zeta) + \frac{4m^2-1}{16} \int_0^\zeta \left\{\operatorname{csch}^2(v^{1/2}) - \frac{1}{v}\right\} B_s^m(v)\, dv, \tag{12.12}$$

when $s \geqslant 0$. Then writing $\xi \equiv \zeta^{1/2}$ and replacing $\varepsilon_{2p+1,2}(u,\xi^2)$ by $\xi \eta_{2p+1,2}(u,\xi^2)$, we arrive at

$$Q_n^m(\cosh\xi) = \frac{e^{m\pi i} u^m}{1 + \delta_{2p+1}} \left(\frac{\xi}{\sinh\xi}\right)^{1/2} \left\{K_m(u\xi) \sum_{s=0}^{p} \frac{A_s^m(\xi^2)}{u^{2s}} - \frac{\xi}{u} K_{m+1}(u\xi) \sum_{s=0}^{p-1} \frac{B_s^m(\xi^2)}{u^{2s}} + \eta_{2p+1,2}(u,\xi^2)\right\}, \tag{12.13}$$

where $u = n + \frac{1}{2}$,

$$|\delta_{2p+1}| \leqslant \lambda_1(m) \exp\left[\frac{\lambda_1(m)}{u} \mathscr{V}_{0,\infty}\{\xi B_0^m(\xi^2)\}\right] \frac{\mathscr{V}_{0,\infty}\{\xi B_p^m(\xi^2)\}}{u^{2p+1}}, \tag{12.14}$$

and

$$|\eta_{2p+1,2}(u,\xi^2)| \leqslant \lambda_1(m) K_m(u\xi) \exp\left[\frac{\lambda_1(m)}{u} \mathscr{V}_{\xi,\infty}\{\xi B_0^m(\xi^2)\}\right] \frac{\mathscr{V}_{\xi,\infty}\{\xi B_p^m(\xi^2)\}}{u^{2p+1}}. \tag{12.15}$$

† That δ_{2p+1} *is* a constant can be proved by the method of Chapter 6, §3.2. Or we may simply use the fact that the ratio of the left-hand sides of (12.08) and (12.09) is known to be independent of ζ.

These are the first of the required results. They are valid when $\xi > 0$, $m \geqslant 0$, and $n+\frac{1}{2}$ is positive and large enough to ensure that $|\delta_{2p+1}| < 1$.

12.3 Continuing the identification of the solutions of (12.06), we observe that when $m \geqslant 0$ both $\{\zeta(x^2-1)\}^{1/4} P_n^{-m}(x)$ and $W_{2p+1,1}(u,\zeta)$ are recessive as $x \to 1+$; compare Chapter 5, (12.08). To evaluate their ratio we use the rearranged form

$$W_{2p+1,1}(u,\zeta) = \zeta^{1/2} I_m(u\zeta^{1/2}) \sum_{s=0}^{p} \frac{\hat{A}_s(\zeta)}{u^{2s}} + \frac{\zeta}{u} I_{m-1}(u\zeta^{1/2}) \sum_{s=0}^{p-1} \frac{\hat{B}_s(\zeta)}{u^{2s}} + \varepsilon_{2p+1,1}(u,\zeta),$$

in which

$$\hat{A}_s(\zeta) = A_s(\zeta) - 2mB_{s-1}(\zeta), \qquad \hat{B}_s(\zeta) = B_s(\zeta);$$

compare Exercise 3.1. Letting $\zeta \to 0+$ and referring to (1.01) and (3.04), we obtain

$$W_{2p+1,1}(u,\zeta) = \frac{u^m \zeta^{(1+m)/2}}{2^m \Gamma(1+m)} \left\{ \sum_{s=0}^{p} \frac{\hat{A}_s(0)}{u^{2s}} + \frac{2m}{u^2} \sum_{s=0}^{p-1} \frac{\hat{B}_s(0)}{u^{2s}} + o(1) \right\}.$$

The content of the braces reduces to $1+o(1)$ if we prescribe

$$\hat{A}_s(0) + 2m\hat{B}_{s-1}(0) = 0 \qquad (s = 1, 2, \ldots).$$

Comparing this equation with (12.10) and remembering that $\hat{A}_s(\zeta)$ and $\hat{B}_s(\zeta)$ satisfy the same recurrence relations as $A_s(\zeta)$ and $B_s(\zeta)$—except for a change of sign in m—we see that $\hat{A}_s(\zeta) = A_s^{-m}(\zeta)$ and $\hat{B}_s(\zeta) = B_s^{-m}(\zeta)$, where $A_s^{-m}(\zeta)$ and $B_s^{-m}(\zeta)$ are defined by (12.11) and (12.12) with m replaced by $-m$.

Using (12.08) of Chapter 5, we arrive at the second of the required expansions, given by

$$P_n^{-m}(\cosh\xi) = \frac{1}{u^m}\left(\frac{\xi}{\sinh\xi}\right)^{1/2} \left\{ I_m(u\xi) \sum_{s=0}^{p} \frac{A_s^{-m}(\xi^2)}{u^{2s}} \right.$$
$$\left. + \frac{\xi}{u} I_{m-1}(u\xi) \sum_{s=0}^{p-1} \frac{B_s^{-m}(\xi^2)}{u^{2s}} + \eta_{2p+1,1}(u,\xi^2) \right\}, \qquad (12.16)$$

where $\eta_{2p+1,1}(u,\xi^2) = \xi^{-1}\varepsilon_{2p+1,1}(u,\xi^2)$, and is therefore bounded by

$$|\eta_{2p+1,1}(u,\xi^2)| \leqslant \lambda_1(m) I_m(u\xi) \exp\left[\frac{\lambda_1(m)}{u} \mathcal{V}_{0,\xi}\{\xi B_0^{-m}(\xi^2)\}\right] \frac{\mathcal{V}_{0,\xi}\{\xi B_p^{-m}(\xi^2)\}}{u^{2p+1}}, \qquad (12.17)$$

when $\xi > 0$, $m \geqslant 0$, and $n > -\frac{1}{2}$.

From (12.13) and (12.16) it is possible to construct analogous expansions for any other solution of the associated Legendre equation by use of connection formulas. In particular, this includes $Q_n^{-m}(\cosh\xi)$ and $P_n^m(\cosh\xi)$ with $m > 0$; see Exercise 12.4 below.†

† Compare Chapter 5, Theorem 12.1. It should be noted that the error bounds (12.14), (12.15), and (12.17) do not apply when m is negative.

12.4 The next problem is identification of solutions in the interval $-1 < x < 1$, that is, when $-\pi^2 < \zeta < 0$. Real solutions of equation (12.06) for negative ζ are furnished by Theorem 4.1. The solution $W_{2p+1,3}(u,\zeta)$ is recessive as $\zeta \to 0-$; it is therefore directly related to the Ferrers function $\mathrm{P}_n^{-m}(x)$ (Chapter 5, §15).

Denoting $(-\zeta)^{1/2}$ by θ and carrying out analysis similar to §12.3, we derive

$$\mathrm{P}_n^{-m}(\cos\theta) = \frac{1}{u^m}\left(\frac{\theta}{\sin\theta}\right)^{1/2}\left\{J_m(u\theta)\sum_{s=0}^{p}\frac{A_s^{-m}(-\theta^2)}{u^{2s}} + \frac{\theta}{u}J_{m-1}(u\theta)\sum_{s=0}^{p-1}\frac{B_s^{-m}(-\theta^2)}{u^{2s}} + \eta_{2p+1,3}(u,-\theta^2)\right\}, \tag{12.18}$$

where

$$|\eta_{2p+1,3}(u,-\theta^2)| \leqslant \lambda_3(m)\frac{M_m(u\theta)}{E_m(u\theta)}\exp\left[\frac{\lambda_2(m)}{u}\mathscr{V}_{0,\theta}\{\theta B_0^{-m}(-\theta^2)\}\right]\frac{\mathscr{V}_{0,\theta}\{\theta B_p^{-m}(-\theta^2)\}}{u^{2p+1}}, \tag{12.19}$$

valid when $0 < \theta < \pi$, $m \geqslant 0$, and $n > -\frac{1}{2}$. In this result E_m and M_m are the auxiliary functions of §1.3, and $A_s^{-m}(\zeta)$, $B_s^{-m}(\zeta)$ are the analytic continuations (through $\zeta = 0$) of the coefficients defined in §12.2; thus

$$B_s^{-m}(\zeta) = -A_s^{-m\prime}(\zeta) - \frac{1}{|\zeta|^{1/2}}\int_\zeta^0\left[\frac{4m^2-1}{16}\left\{\csc^2(|v|^{1/2}) + \frac{1}{v}\right\}A_s^{-m}(v) + \left(\frac{1}{2}-m\right)A_s^{-m\prime}(v)\right]\frac{dv}{|v|^{1/2}},$$

$$A_{s+1}^{-m}(\zeta) = -m\{B_s^{-m}(\zeta)+B_s^{-m}(0)\} - \zeta B_s^{-m\prime}(\zeta) + \frac{4m^2-1}{16}\int_\zeta^0\left\{\csc^2(|v|^{1/2}) + \frac{1}{v}\right\}B_s^{-m}(v)\,dv.$$

When $m = 0$ and n is a positive integer, (12.18) furnishes an expansion of the Legendre polynomial $P_n(\cos\theta)$. The advantage of this result compared with the simpler approximations of Chapter 4, §8.1 and Chapter 8, §10.2, is uniformity in a closed interval containing $\theta = 0$.

12.5 The identification of the solution $W_{2p+1,4}(u,\zeta)$ is more difficult because it is not recessive anywhere in the ζ interval under consideration. Since the associated Legendre equation is even in x, we take the reference point α of Theorem 4.1 to correspond to $x = 0$; this gives $\alpha = -\pi^2/4$. For simplicity, consider the case $p = 0$. From Theorem 4.1 we have

$$W_{1,4}(u,-\theta^2) = \theta\{Y_m(u\theta)+\eta_{1,4}(u,-\theta^2)\}, \tag{12.20}$$

where

$$|\eta_{1,4}(u, -\theta^2)|$$

$$\leqslant \lambda_4(m)\, E_m(u\theta)\, M_m(u\theta) \exp\left[\frac{\lambda_2(m)}{u} \mathscr{V}_{\theta, \pi/2}\{\theta B_0^m(-\theta^2)\}\right] \frac{\mathscr{V}_{\theta,\pi/2}\{\theta B_0^m(-\theta^2)\}}{u}. \quad (12.21)$$

When $m-n \neq 1, 2, 3, \ldots$, the last solution is expressible as a linear combination of the Ferrers functions, in the form

$$(\theta \sin\theta)^{-1/2} W_{1,4}(u, -\theta^2) = A\mathrm{P}_n^{-m}(\cos\theta) + B\mathrm{Q}_n^{-m}(\cos\theta). \quad (12.22)$$

If we set $\theta = \frac{1}{2}\pi$, then $\eta_{1,4}(u, -\theta^2)$ vanishes and from the results of Chapter 5, §15.2, we obtain

$$2^{-1/2}\pi G(u)\, Y_m(\tfrac{1}{2}\pi u) = A\cos\chi - \tfrac{1}{2}\pi B\sin\chi, \quad (12.23)$$

where

$$G(u) = 2^m\Gamma(\tfrac{1}{2}u+\tfrac{1}{2}m+\tfrac{3}{4})/\Gamma(\tfrac{1}{2}u-\tfrac{1}{2}m+\tfrac{1}{4}), \qquad \chi = (\tfrac{1}{2}u-\tfrac{1}{2}m-\tfrac{1}{4})\pi.$$

Again, differentiating (12.22) with respect to θ, setting $\theta = \frac{1}{2}\pi$ in the result, and using (1.11), we find that

$$2^{-1/2}H(u)\{(m-\tfrac{1}{2})\, Y_m(\tfrac{1}{2}\pi u) - \tfrac{1}{2}\pi u Y_{m-1}(\tfrac{1}{2}\pi u)\} = A\sin\chi + \tfrac{1}{2}\pi B\cos\chi, \quad (12.24)$$

where

$$H(u) = 2^m\Gamma(\tfrac{1}{2}u+\tfrac{1}{2}m+\tfrac{1}{4})/\Gamma(\tfrac{1}{2}u-\tfrac{1}{2}m+\tfrac{3}{4}).$$

Solving the linear equations (12.23) and (12.24) for A and B, we obtain the required identification:

$$A = 2^{-1/2}\pi G(u)\, Y_m(\tfrac{1}{2}\pi u)\cos\chi$$
$$+\, 2^{-1/2}H(u)\{(m-\tfrac{1}{2})\, Y_m(\tfrac{1}{2}\pi u) - \tfrac{1}{2}\pi u Y_{m-1}(\tfrac{1}{2}\pi u)\}\sin\chi,$$

$$B = -2^{1/2}G(u)\, Y_m(\tfrac{1}{2}\pi u)\sin\chi$$
$$+\, 2^{1/2}\pi^{-1}H(u)\{(m-\tfrac{1}{2})\, Y_m(\tfrac{1}{2}\pi u) - \tfrac{1}{2}\pi u Y_{m-1}(\tfrac{1}{2}\pi u)\}\cos\chi.$$

By means of (1.09) and Chapter 4, §5.1, we find that as $u \to \infty$,

$$A = O(u^{m-1}), \qquad B = -(2/\pi)\, u^m\{1+O(u^{-1})\},$$

and hence, with the aid of (12.18), that

$$\mathrm{Q}_n^{-m}(\cos\theta) = -\frac{\pi}{2u^m}\left(\frac{\theta}{\sin\theta}\right)^{1/2}\{Y_m(u\theta) + E_m(u\theta)\, M_m(u\theta)\, O(u^{-1})\} \qquad (u \to \infty), \quad (12.25)$$

uniformly with respect to $\theta \in (0, \frac{1}{2}\pi]$ and bounded nonnegative m.

The approximation (12.25) could be extended to higher terms by use of Theorem 4.1 with $p \geqslant 1$. But it would be difficult to determine a satisfactory bound for the error term by this method. Such a bound will be constructed via complex-variable theory in the next section.

Ex. 12.1 Verify that

$$B_0^m(\xi^2) = \frac{1-4m^2}{8\xi}\left(\coth\xi - \frac{1}{\xi}\right), \qquad \mathscr{V}_{0,\infty}\{\xi B_0^m(\xi^2)\} = \tfrac{1}{2}|m^2 - \tfrac{1}{4}|,$$

$$A_1^m(\xi^2) = \tfrac{1}{2}\xi^2\{B_0^m(\xi^2)\}^2 + (m+\tfrac{1}{2})B_0^m(\xi^2) - \psi(\xi^2) + \tfrac{1}{6}m(\tfrac{1}{4}-m^2).$$

Ex. 12.2 If m has any fixed real value and u is large and positive, then by letting $\xi \to \infty$ in (12.13) and (12.16) show that

$$\frac{\Gamma(u+m+\frac{1}{2})}{u^{m+(1/2)}\Gamma(u)} \sim \sum_{s=0}^{\infty}\frac{A_s^m(\infty)}{u^{2s}} - \sum_{s=0}^{\infty}\frac{\{\zeta^{1/2}B_s^m(\zeta)\}_{\zeta=\infty}}{u^{2s+1}},$$

and

$$\frac{\Gamma(u)}{u^{m-(1/2)}\Gamma(u-m+\frac{1}{2})} \sim \sum_{s=0}^{\infty}\frac{A_s^m(\infty)}{u^{2s}} + \sum_{s=0}^{\infty}\frac{\{\zeta^{1/2}B_s^m(\zeta)\}_{\zeta=\infty}}{u^{2s+1}}.$$

Ex. 12.3 Prove that

$$\frac{\Gamma(u+m+\frac{1}{2})}{u^{2m}\Gamma(u-m+\frac{1}{2})} \sim \sum_{s=0}^{\infty}\frac{A_s^m(0)}{u^{2s}} \qquad (u\to\infty).$$

Ex. 12.4 Using Exercise 12.2 and the connection formulas of Chapter 5, §12.3, show that for any positive integer p and any given real value of m,

$$P_n^m(\cosh\xi) = u^m\left(\frac{\xi}{\sinh\xi}\right)^{1/2}\left[I_{-m}(u\xi)\left\{\sum_{s=0}^{p}\frac{A_s^m(\xi^2)}{u^{2s}} + O\left(\frac{1}{u^{2p+2}}\right)\right\}\right.$$
$$\left. + \frac{\xi}{u}I_{-m-1}(u\xi)\left\{\sum_{s=0}^{p-1}\frac{B_s^m(\xi^2)}{u^{2s}} + \frac{1}{1+\xi}O\left(\frac{1}{u^{2p}}\right)\right\}\right],$$

$$Q_n^m(\cosh\xi) = e^{m\pi i}u^m\left(\frac{\xi}{\sinh\xi}\right)^{1/2}\left[K_m(u\xi)\left\{\sum_{s=0}^{p}\frac{A_s^m(\xi^2)}{u^{2s}} + O\left(\frac{1}{u^{2p+2}}\right)\right\}\right.$$
$$\left. - \frac{\xi}{u}K_{m+1}(u\xi)\left\{\sum_{s=0}^{p-1}\frac{B_s^m(\xi^2)}{u^{2s}} + \frac{1}{1+\xi}O\left(\frac{1}{u^{2p}}\right)\right\}\right],$$

as $n\to\infty$, uniformly with respect to $\xi \in (0,\infty)$.

Ex. 12.5 Show that for fixed nonnegative m, the rth zero of $\mathrm{P}_n^{-m}(x)$ in the interval $-1 < x < 1$, counted from the right, is given by

$$x = \cos\left\{t + \frac{(4m^2-1)(1-t\cot t)}{8t(n+\frac{1}{2})^2} + tO\left(\frac{1}{n^4}\right)\right\} \qquad (n\to\infty),$$

where $(n+\frac{1}{2})t = j_{m,r}$, the rth positive zero of $J_m(x)$.

Show also that the O term is uniform for $r = 1, 2, \ldots, [\alpha n]$, where α is any constant in the interval $(0, 1)$.

13 Legendre Functions of Large Degree: Complex Arguments

13.1 If ζ is a complex variable, $m \geqslant 0$, and u is a large complex parameter, then solutions of equation (12.06) are furnished by Theorem 9.1. For simplicity, we confine the text to positive real values of $u \equiv n+\frac{1}{2}$.

Writing $\xi = \zeta^{1/2}$ (as before) and replacing x by the complex variable z, we have

$$z = \cosh \xi. \tag{13.01}$$

Let **D** denote the sector $|\mathrm{ph}\, z| \leqslant \frac{1}{2}\pi$. To pass to the ξ plane we temporarily introduce a cut along the interval $0 \leqslant z \leqslant 1$. Then (13.01) maps **D** on the half strip **S**: $\mathrm{Re}\,\xi \geqslant 0$, $|\mathrm{Im}\,\xi| \leqslant \frac{1}{2}\pi$; see Figs. 13.1 and 13.2. The points C_1 and C_2 correspond to $z = 0 \pm i0$. In the ζ plane the corresponding domain $\boldsymbol{\Delta}$ is bounded by a parabola; see Fig. 13.3. The $z \leftrightarrow \zeta$ transformation is free from singularity at $z = 1$, and the temporary cut is no longer needed. Evidently,

$$\psi(\zeta) \equiv \frac{4m^2-1}{16}\left\{\operatorname{csch}^2(\zeta^{1/2}) - \frac{1}{\zeta}\right\} \tag{13.02}$$

is holomorphic within $\boldsymbol{\Delta}$.

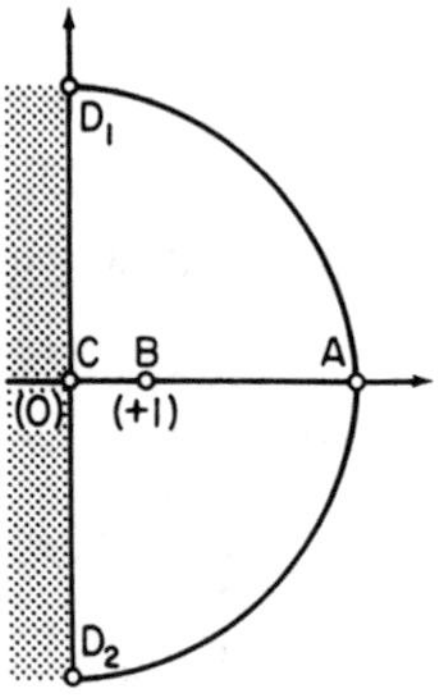

Fig. 13.1 z plane. Domain **D**.

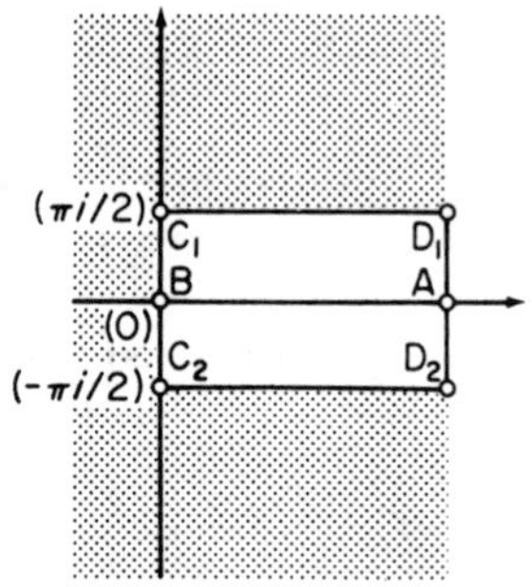

Fig. 13.2 ξ plane. Domain **S**.

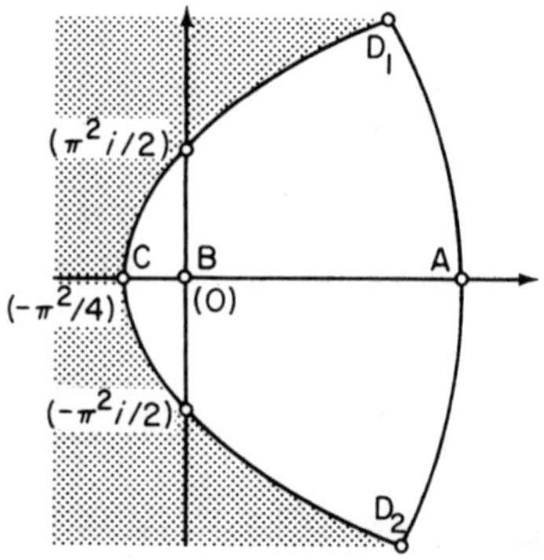

Fig. 13.3 ζ plane. Domain $\boldsymbol{\Delta}$.

Applying Theorem 9.1 and identifying the solutions in the manner of §§12.2 and 12.3, we arrive at

$$P_n^{-m}(\cosh \xi) = \frac{1}{u^m}\left(\frac{\xi}{\sinh \xi}\right)^{1/2}\left\{I_m(u\xi)\sum_{s=0}^{p}\frac{A_s^{-m}(\xi^2)}{u^{2s}} + \frac{\xi}{u}I_{m-1}(u\xi)\sum_{s=0}^{p-1}\frac{B_s^{-m}(\xi^2)}{u^{2s}} + \eta_{2p+1,1}(u,\xi^2)\right\}, \tag{13.03}$$

$$Q_n^m(\cosh \xi) = \frac{e^{m\pi i}u^m}{1+\delta_{2p+1}}\left(\frac{\xi}{\sinh \xi}\right)^{1/2}\left\{K_m(u\xi)\sum_{s=0}^{p}\frac{A_s^m(\xi^2)}{u^{2s}} - \frac{\xi}{u}K_{m+1}(u\xi)\sum_{s=0}^{p-1}\frac{B_s^m(\xi^2)}{u^{2s}} + \eta_{2p+1,2}(u,\xi^2)\right\}, \tag{13.04}$$

where

$$|\eta_{2p+1,1}(u,\xi^2)| \leqslant \mu_2(m)\mathfrak{E}_m(u\xi)\mathfrak{M}_m(u\xi) \times \exp\left[\frac{\mu_1(m)}{u}\mathcal{V}_{0,\xi}\{\xi B_0^{-m}(\xi^2)\}\right]\frac{\mathcal{V}_{0,\xi}\{\xi B_p^{-m}(\xi^2)\}}{u^{2p+1}} \tag{13.05}$$

$$|\eta_{2p+1,2}(u,\xi^2)| \leqslant \mu_3(m)\mathfrak{E}_m^{-1}(u\xi)\mathfrak{M}_m(u\xi) \times \exp\left[\frac{\mu_1(m)}{u}\mathcal{V}_{\xi,\infty}\{\xi B_0^m(\xi^2)\}\right]\frac{\mathcal{V}_{\xi,\infty}\{\xi B_p^m(\xi^2)\}}{u^{2p+1}}. \tag{13.06}$$

In these results the coefficients $A_s^m(\zeta)$ and $B_s^m(\zeta)$ are defined by (12.11) and (12.12), $\zeta^{1/2}$ having its principal value; δ_{2p+1} is the quantity which appeared in (12.13) and is therefore bounded by (12.14); $\mathfrak{E}_m$, $\mathfrak{M}_m$, $\mu_1(m)$, $\mu_2(m)$, and $\mu_3(m)$ are the auxiliary functions introduced in §8; and, lastly, the variational paths lie within **S** and are subject to the condition that both $\operatorname{Re}\xi$ and $|\xi|$ change monotonically along them.

13.2 The restrictions adopted in establishing the foregoing results are $\xi \in \mathbf{S}$, $m \geqslant 0$, and (in the case of (13.04)) n is sufficiently large to ensure that $|\delta_{2p+1}| < 1$. The first of these conditions is easily relaxed. From (13.02) it is evident that the only singularities of $\psi(\zeta)$ are poles at the points corresponding to $\xi = \pm\pi i, \pm 2\pi i, \pm 3\pi i, \ldots$. Therefore by Theorem 9.1 the inequalities (13.05) and (13.06) are valid throughout the sector $|\operatorname{ph}\xi| \leqslant \frac{1}{2}\pi$, with the sole exclusion of the singular points and, in the case of (13.05), the segments of the rays $\operatorname{ph}\xi = \pm\frac{1}{2}\pi$ outside the circle $|\xi| = \pi$.

The effect of increasing ξ by πi is to increase the phase of z by π. Accordingly, the extended results apply to the branches of $P_n^{-m}(z)$ and $Q_n^m(z)$ obtained by making any number of circuits of the cut joining $z = \pm 1$, without crossing this cut.

This extension is not entirely satisfactory, however, because the resulting error bounds are large in the neighborhoods of the points $z = \pm 1$. Alternatively, and preferably, we may confine the use of (13.03) and (13.04) to $\xi \in \mathbf{S}$, and elsewhere rely on the continuation formulas stated in Exercise 13.1 of Chapter 5. This exercise also supplies formulas for continuation across the cut $-1 \leqslant z \leqslant 1$.

Lastly, extensions of the foregoing results to negative values of m, or $n+\frac{1}{2}$, or both, can be achieved by means of connection formulas given in Chapter 5, §12.3.

13.3 We are now in a position to construct an expansion for the Ferrers function $\mathsf{Q}_n^{-m}(\cos\theta)$ analogous to (12.18). From Chapter 5, equations (13.15) and (15.02), we have

$$\mathsf{Q}_n^{-m}(x) = \frac{\Gamma(n-m+1)}{2\Gamma(n+m+1)}\{e^{-m\pi i/2}Q_n^m(x+i0) + e^{-3m\pi i/2}Q_n^m(x-i0)\}. \tag{13.07}$$

If we substitute $\xi = i\theta$ in (13.04), where $0<\theta<\pi$, and use the identities†

$$K_m(iu\theta) = -\tfrac{1}{2}\pi i e^{-m\pi i/2}\{J_m(u\theta)-iY_m(u\theta)\},$$

$$K_{m+1}(iu\theta) = -\tfrac{1}{2}\pi e^{-m\pi i/2}\{J_{m+1}(u\theta)-iY_{m+1}(u\theta)\},$$

then the following expression is obtained for $e^{-m\pi i}Q_n^m(\cos\theta+i0)$:

$$-\frac{1}{2}\pi i\frac{e^{-m\pi i/2}u^m}{1+\delta_{2p+1}}\left(\frac{\theta}{\sin\theta}\right)^{1/2}\left[\{J_m(u\theta)-iY_m(u\theta)\}\sum_{s=0}^{p}\frac{A_s^m(-\theta^2)}{u^{2s}}\right.$$

$$\left.-\frac{\theta}{u}\{J_{m+1}(u\theta)-iY_{m+1}(u\theta)\}\sum_{s=0}^{p-1}\frac{B_s^m(-\theta^2)}{u^{2s}}+\frac{2}{\pi}ie^{m\pi i/2}\eta_{2p+1,2}(u,-\theta^2+i0)\right].$$

The corresponding expression for $e^{-m\pi i}Q_n^m(\cos\theta-i0)$ is found by replacing i by $-i$. Substitution in (13.07) yields

$$\mathsf{Q}_n^{-m}(\cos\theta) = -\frac{\frac{1}{2}\pi u^m}{1+\delta_{2p+1}}\frac{\Gamma(u-m+\frac{1}{2})}{\Gamma(u+m+\frac{1}{2})}\left(\frac{\theta}{\sin\theta}\right)^{1/2}$$

$$\times\left\{Y_m(u\theta)\sum_{s=0}^{p}\frac{A_s^m(-\theta^2)}{u^{2s}}-\frac{\theta}{u}Y_{m+1}(u\theta)\sum_{s=0}^{p-1}\frac{B_s^m(-\theta^2)}{u^{2s}}+\hat{\eta}_{2p+1}(u,-\theta^2)\right\}, \tag{13.08}$$

where

$$\hat{\eta}_{2p+1}(u,-\theta^2) = -(2/\pi)\,\mathrm{Re}\{e^{m\pi i/2}\eta_{2p+1,2}(u,-\theta^2+i0)\},$$

and is therefore subject to the bound

$$|\hat{\eta}_{2p+1}(u,-\theta^2)| \leqslant \frac{2}{\pi}\mu_3(m)\frac{\mathfrak{M}_m(iu\theta)}{\mathfrak{E}_m(iu\theta)}\exp\left[\frac{\mu_1(m)}{u}\mathcal{V}_{i\theta,\infty}\{\xi B_0^m(\xi^2)\}\right]\frac{\mathcal{V}_{i\theta,\infty}\{\xi B_p^m(\xi^2)\}}{u^{2p+1}}. \tag{13.09}$$

This result is valid when $0<\theta<\pi$, $m \geqslant 0$, and n is large enough to ensure that $u > m-\frac{1}{2}$ and $|\delta_{2p+1}|<1$.

† Chapter 7, (4.13), (5.03), and (8.02).

It should be noticed that although the maps in the z and ζ planes of the endpoints of the variational paths in (13.09) lie on the real axis, the mapped path cannot be taken along the axis because this would violate the monotonicity condition on $|\xi|$.

Ex. 13.1 Using the recurrence formulas of Chapter 5, §13, verify that

$$(t-z)Q_0(t)P_0(z) = 1 + Q_1(t)P_0(z) - Q_0(t)P_1(z),$$

and, when $s \geqslant 1$,

$$(2s+1)(t-z)Q_s(t)P_s(z) = (s+1)\{Q_{s+1}(t)P_s(z) - Q_s(t)P_{s+1}(z)\} - s\{Q_s(t)P_{s-1}(z) - Q_{s-1}(t)P_s(z)\}.$$

Deduce that

$$\sum_{s=0}^{n}(2s+1)Q_s(t)P_s(z) = \frac{1}{t-z} + \frac{n+1}{t-z}\{Q_{n+1}(t)P_n(z) - Q_n(t)P_{n+1}(z)\},$$

and thence establish *Heine's formula*

$$\sum_{s=0}^{\infty}(2s+1)Q_s(t)P_s(z) = \frac{1}{t-z} \qquad (t \in \mathscr{E}_1, \quad z \in \mathscr{E}_2),$$

where $\mathscr{E}_1$ and $\mathscr{E}_2$ are ellipses with foci at ± 1, $\mathscr{E}_2$ being properly interior to $\mathscr{E}_1$. Show also that the series converges uniformly for t outside or on $\mathscr{E}_1$, and z within or on $\mathscr{E}_2$.

Ex. 13.2 Using the preceding exercise and Cauchy's integral formula, prove *Neumann's expansion theorem*: if $f(z)$ is holomorphic within the domain bounded by an ellipse with foci at ± 1 and continuous in the closure of this domain, then within and on any smaller confocal ellipse $f(z)$ can be expanded as a uniformly convergent series of Legendre polynomials, in the form

$$f(z) = \sum_{s=0}^{\infty} a_s P_s(z).$$

Show also that $a_s = (s+\tfrac{1}{2})\int_{-1}^{1} f(t)P_s(t)\,dt$.

***Ex. 13.3** Using Exercises 12.3 and 12.4, prove that when n is large and positive and m has *any* fixed real value

$$\mathrm{P}_n^{-m}(\cos\theta) \sim \frac{1}{u^m}\left(\frac{\theta}{\sin\theta}\right)^{1/2}\left\{J_m(u\theta)\sum_{s=0}^{\infty}\frac{A_s^{-m}(-\theta^2)}{u^{2s}} + \frac{\theta}{u}J_{m-1}(u\theta)\sum_{s=0}^{\infty}\frac{B_s^{-m}(-\theta^2)}{u^{2s}}\right\},$$

$$\mathrm{Q}_n^{-m}(\cos\theta) \sim -\frac{\pi}{2u^m}\left(\frac{\theta}{\sin\theta}\right)^{1/2}\left\{Y_m(u\theta)\sum_{s=0}^{\infty}\frac{A_s^{-m}(-\theta^2)}{u^{2s}} + \frac{\theta}{u}Y_{m-1}(u\theta)\sum_{s=0}^{\infty}\frac{B_s^{-m}(-\theta^2)}{u^{2s}}\right\},$$

uniformly for $\theta \in (0,\delta]$, δ being any fixed number in $(0,\pi)$.

Ex. 13.4 When $-1 < x < 1$ and τ is a real parameter, the functions $\mathrm{P}^m_{-\frac{1}{2}+i\tau}(x)$ and $\mathrm{Q}^m_{-\frac{1}{2}+i\tau}(x)$ are known as the *conical functions*. Show that if $0 < \theta < \pi$, $m \geqslant 0$, and $\tau > 0$, then

$$\mathrm{P}^{-m}_{-\frac{1}{2}+i\tau}(\cos\theta) = \frac{1}{\tau^m}\left(\frac{\theta}{\sin\theta}\right)^{1/2}\left\{I_m(\tau\theta)\sum_{s=0}^{p}(-)^s\frac{A_s^{-m}(-\theta^2)}{\tau^{2s}}\right.$$
$$\left. -\frac{\theta}{\tau}I_{m-1}(\tau\theta)\sum_{s=0}^{p-1}(-)^s\frac{B_s^{-m}(-\theta^2)}{\tau^{2s}} + \kappa_{2p+1,1}(\tau,-\theta^2)\right\},$$

where

$$|\kappa_{2p+1,1}(\tau,-\theta^2)| \leqslant \lambda_1(m)I_m(\tau\theta)\exp\left[\frac{\lambda_1(m)}{\tau}\mathscr{V}_{0,\theta}\{\theta B_0^{-m}(-\theta^2)\}\right]\frac{\mathscr{V}_{0,\theta}\{\theta B_p^{-m}(-\theta^2)\}}{\tau^{2p+1}}.$$

Ex. 13.5 With the notation and conditions of the preceding exercise, prove that $Q^{m}_{-\frac{1}{2}+i\tau}(\cos - i0$ is given by

$$e^{3m\pi i/2}\frac{\tau^m}{1+\sigma_{2p+1}}\left(\frac{\theta}{\sin\theta}\right)^{1/2}$$

$$\times\left\{K_m(\tau\theta)\sum_{s=0}^{p}(-)^s\frac{A_s^m(-\theta^2)}{\tau^{2s}}+\frac{\theta}{\tau}K_{m+1}(\tau\theta)\sum_{s=0}^{p-1}(-)^s\frac{B_s^m(-\theta^2)}{\tau^{2s}}+\kappa_{2p+1,2}(\tau,-\theta^2)\right\},$$

where

$$|\sigma_{2p+1}|\leqslant 2^{1/2}\mu_3(m)\exp\left[\frac{\mu_1(m)}{\tau}\mathscr{V}_{0,i\infty}\{\theta B_0^m(-\theta^2)\}\right]\frac{\mathscr{V}_{0,i\infty}\{\theta B_p^m(-\theta^2)\}}{\tau^{2p+1}}\qquad(<1),$$

and

$$|\kappa_{2p+1,2}(\tau,-\theta^2)|\leqslant\mu_3(m)\frac{\mathfrak{M}_m(\tau\theta)}{\mathfrak{E}_m(\tau\theta)}\exp\left[\frac{\mu_1(m)}{\tau}\mathscr{V}_{\theta,\theta+i\infty}\{\theta B_0^m(-\theta^2)\}\right]\frac{\mathscr{V}_{\theta,\theta+i\infty}\{\theta B_p^m(-\theta^2)\}}{\tau^{2p+1}},$$

the variational paths being parallel to the imaginary axis in the θ plane.

*14 Other Types of Transition Points

14.1 In Chapters 10 and 11 and in the present chapter, we have been concerned primarily with differential equations of the form

$$d^2w/dz^2 = \{u^2f(z)+g(z)\}\,w,\tag{14.01}$$

in which u is a large real or complex parameter, the functions $f(z)$ and $g(z)$ are independent of u, and the independent variable z ranges over a real or complex region **D** containing at most one transition point z_0 (say) of the differential equation. Uniform approximations to the solutions have been sought successfully in the following circumstances:

Case I. **D** is free from transition points (Chapter 10).

Case II. z_0 is a simple zero of $f(z)$, and $g(z)$ is analytic at $z = z_0$ (Chapter 11).

Case III. z_0 is a simple pole of $f(z)$, and $(z-z_0)^2g(z)$ is analytic at $z = z_0$ (present chapter).

It is natural to enquire whether the theory can be extended to include other types of transition points. The essence of the method used is to compare (14.01), or a transformation of this equation, with a similar equation

$$d^2w/dz^2 = \{u^2f(z)+h(z)\}\,w,\tag{14.02}$$

say, and to satisfy (14.01) formally by a series

$$w = P(u,z)\left\{1+\sum_{s=1}^{\infty}\frac{A_s(z)}{u^{2s}}\right\}+\frac{1}{u^2}\left\{\frac{\partial}{\partial z}P(u,z)\right\}\sum_{s=0}^{\infty}\frac{B_s(z)}{u^{2s}},\tag{14.03}$$

in which $P(u, z)$ is a solution of (14.02), and the coefficients $A_s(z)$ and $B_s(z)$ are independent of u. We may refer to (14.02) as the *basic equation* and to its solutions as the *basic functions*.

Success of the method hinges on two conditions. First, the coefficients $A_s(z)$ and $B_s(z)$ must be finite at the transition point. Secondly, $P(u, z)$ must be less recondite than the solutions of the original equation (14.01), otherwise little is achieved.

It is not easy to attach a precise meaning to the second condition. What we have in mind is that the variables u and z are separated in some sense. In Cases I, II, and III the basic functions were all of the form $P\{z\varpi(u)\}$, where P and ϖ are functions of a single variable. In considering other cases it is convenient to adopt this form as a criterion.†

14.2 Theorem 14.1 *Let $f(z)$ and $h(z)$ be holomorphic in a neighborhood of $z = 0$, except possibly for poles at $z = 0$. If $f(z) \not\equiv 0$, then necessary and sufficient conditions for equation* (14.02) *to possess a nontrivial solution of the form*

$$w = P\{z\varpi(u)\}, \tag{14.04}$$

are given by

$$f(z) = az^m, \qquad h(z) = bz^m + cz^{-2}, \tag{14.05}$$

where m is an integer other than -2, and a, b, and c are constants.

To establish this result, write $t = z\varpi(u)$ and substitute (14.04) in (14.02). Then

$$\frac{P''(t)}{P(t)} = \frac{u^2 f(z) + h(z)}{\{\varpi(u)\}^2}. \tag{14.06}$$

Near $z = 0$ the functions $f(z)$ and $h(z)$ can be expanded in convergent series

$$f(z) = z^{-l} \sum_{s=0}^{\infty} f_s z^s, \qquad h(z) = z^{-l} \sum_{s=0}^{\infty} h_s z^s,$$

in which l is an integer. Without loss of generality we may suppose that $l \geqslant 2$. Then

$$\frac{u^2 f(z) + h(z)}{\{\varpi(u)\}^2} = \sum_{s=0}^{\infty} \frac{u^2 f_s + h_s}{\{\varpi(u)\}^{s-l+2}} t^{s-l}.$$

This is a function of t alone, in consequence of (14.06). Hence

$$u^2 f_s + h_s = k_s \{\varpi(u)\}^{s-l+2}, \tag{14.07}$$

where k_s is independent of u. Taking $s = l-2$, we see immediately that $f_{l-2} = 0$. Let f_j, $j \neq l-2$, be the first nonvanishing member of the sequence $f_0, f_1, f_2, \ldots$. From (14.07) with $s = j$ we derive $k_j \neq 0$ and $\varpi(u) = \{(u^2 f_j + h_j)/k_j\}^{1/(j-l+2)}$. Therefore

$$u^2 f_s + h_s = k_s \{(u^2 f_j + h_j)/k_j\}^{(s-l+2)/(j-l+2)}.$$

† See also p. 479.

For this relation to hold we require $k_s = 0$ when $s \neq l-2$ or j. Therefore

$$f_s = 0 \quad (s \neq j), \qquad h_s = 0 \quad (s \neq l-2 \text{ or } j);$$

in consequence

$$f(z) = f_j z^{j-l}, \qquad h(z) = h_j z^{j-l} + h_{l-2} z^{-2}.$$

This is (14.05), with $m = j-l$.

14.3 The next step is to apply the Liouville transformation to convert the given differential equation (14.01) into a form closely resembling the basic equation admitted by Theorem 14.1. We first note that by rescaling u we can arrange that $a = 1$ in the basic equation.

In (14.01) assume that

$$f(z) = (z-z_0)^m F(z), \tag{14.08}$$

where m is an integer other than -2† and $F(z)$ is analytic and nonvanishing at $z = z_0$. Following Chapter 10, §1.2 we take a new variable ξ such that

$$\dot{z}^2 f(z) = \xi^m, \tag{14.09}$$

the dot denoting differentiation with respect to ξ. Then

$$d^2W/d\xi^2 = \{u^2\xi^m + \chi(\xi)\}\, W, \tag{14.10}$$

where $W = \dot{z}^{-1/2} w$, and

$$\chi(\xi) = \dot{z}^2 g(z) + \dot{z}^{1/2} \frac{d^2}{d\xi^2}(\dot{z}^{-1/2}). \tag{14.11}$$

Integration of (14.09) produces

$$\xi = \left\{\tfrac{1}{2}(m+2)\int_{z_0}^{z} f^{1/2}(t)\, dt\right\}^{2/(m+2)}. \tag{14.12}$$

This maps **D** onto a certain interval or complex region $\mathbf{\Delta}$ in the ξ plane. Substitution of (14.08) in (14.12) gives

$$\xi = \{F(z_0)\}^{1/(m+2)}(z-z_0)\{1 + O(z-z_0)\} \qquad (z \to z_0). \tag{14.13}$$

Accordingly, $z = z_0$ corresponds to $\xi = 0$, and $\xi = \xi(z)$ is holomorphic in $\mathbf{\Delta}$. From (14.11) and (14.13) we see that if $g(z)$ is analytic at $z = z_0$ then $\chi(\xi)$ is analytic at $\xi = 0$, or if $g(z)$ has a pole at $z = z_0$ then $\chi(\xi)$ has a pole of the same order at $\xi = 0$.

Let b and c be the coefficients of ξ^m and ξ^{-2}, respectively, in the expansion of $\chi(\xi)$ in ascending powers of ξ, and write

$$\chi(\xi) = b\xi^m + c\xi^{-2} + \phi(\xi).$$

† The case $m = -2$ was treated in Chapter 6, §5.3, and Chapter 10, §4.1.

Then (14.10) becomes

$$\frac{d^2W}{d\xi^2} = \left\{u^2\xi^m + b\xi^m + \frac{c}{\xi^2} + \phi(\xi)\right\}W. \tag{14.14}$$

This equation is to be compared with the basic equation

$$\frac{d^2P}{d\xi^2} = \left(u^2\xi^m + b\xi^m + \frac{c}{\xi^2}\right)P,$$

solutions of which are of the form $P = P(v\xi)$, where $v = (u^2 + b)^{1/(m+2)}$. We seek a solution of (14.14) in the form

$$W = P(v\xi)\sum_{s=0}^{\infty}\frac{A_s(\xi)}{u^{2s}} + \frac{v}{u^2}P'(v\xi)\sum_{s=0}^{\infty}\frac{B_s(\xi)}{u^{2s}}. \tag{14.15}$$

By differentiating with respect to ξ and then equating coefficients of like powers of u (as in earlier chapters) we find that (14.14) is satisfied formally when

$$A_0(\xi) = \text{constant}, \qquad B_0(\xi) = \tfrac{1}{2}A_0(\xi)\,\xi^{-m/2}\int \xi^{-m/2}\phi(\xi)\,d\xi,$$

and, for $s \geqslant 1$,

$$A_s(\xi) = -\tfrac{1}{2}B'_{s-1}(\xi) + \tfrac{1}{2}\int \phi(\xi)\,B_{s-1}(\xi)\,d\xi,$$

$$B_s(\xi) = -bB_{s-1}(\xi) + \tfrac{1}{2}\xi^{-m/2}\int \xi^{-m/2}\left\{\phi(\xi)\,A_s(\xi) - A''_s(\xi) - \frac{2c}{\xi}\frac{d}{d\xi}\left(\frac{B_{s-1}(\xi)}{\xi}\right)\right\}d\xi.$$

14.4 The problem now is to determine the circumstances in which $A_s(\xi)$ and $B_s(\xi)$ are analytic at the transformed transition point $\xi = 0$. This is resolvable by considering their Maclaurin expansions for the various possible values of m in turn. The analysis has been supplied by Olver (1956), but because details are straightforward and somewhat lengthy they are omitted here. The final result is that in addition to Cases I, II, and III, only the following forms of equation (14.14)

$$\frac{d^2W}{d\xi^2} = \left\{\frac{u^2}{\xi^m} + \frac{b}{\xi^m} + \frac{c}{\xi^2} + \frac{\psi(\xi)}{\xi^{[(m+1)/2]}}\right\}W \quad (m = 3, 4, 5, \ldots), \tag{14.16}$$

$$\frac{d^2W}{d\xi^2} = \left\{u^2\xi^m + \frac{c}{\xi^2} + \xi^m\psi(\xi^{m+2})\right\}W \quad (m = 0, 1, 2, \ldots), \tag{14.17}$$

in which ψ is an arbitrary analytic function, satisfy the conditions:

(i) *The corresponding basic equation fulfills the conditions of Theorem* 14.1.

(ii) *The coefficients* $A_s(\xi)$ *and* $B_s(\xi)$ *in the formal solution* (14.15) *are holomorphic functions of* ξ *in the region under consideration, including the transition point.*

(iii) *Equation* (14.14) *is not atypical in the sense that it involves special relations between* b, c, *and the coefficients in the expansion of* $\phi(\xi)$ *in powers of* ξ.

Equation (14.16) can be included in Case I by replacing u^2+b by u^2 and absorbing the term $c\xi^{-2}$ in $\psi(\xi)\xi^{-[(m+1)/2]}$; compare Chapter 10, §4.1.

For equation (14.17) we simply apply the Liouville transformation

$$\xi^m\left(\frac{d\xi}{d\hat{\xi}}\right)^2=\frac{1}{\hat{\xi}},\qquad \hat{W}=\left(\frac{d\xi}{d\hat{\xi}}\right)^{-1/2}W.$$

Then $\hat{\xi}=\xi^{m+2}/(m+2)^2$, and

$$\frac{d^2\hat{W}}{d\hat{\xi}^2}=\left[\frac{u^2}{\hat{\xi}}+\left\{\frac{c+\frac{1}{4}}{(m+2)^2}-\frac{1}{4}\right\}\frac{1}{\hat{\xi}^2}+\frac{\psi\{(m+2)^2\hat{\xi}\}}{\hat{\xi}}\right]\hat{W}.$$

This equation is clearly of the form (2.05).

We may summarize the results of this section as follows:

Except for certain rather special cases, all forms of equation (14.14) *that are amenable to the general method employed in the last three chapters can be transformed into Case I, Case II, or Case III.*

Ex. 14.1 If ψ denotes an arbitrary function which is analytic at $\xi=0$, show that the following equations belonging to Case III:

$$\frac{d^2W}{d\xi^2}=\left\{\frac{u^2}{\xi}-\frac{3}{16\xi^2}+\frac{\psi(\xi)}{\xi}\right\}W,\qquad \frac{d^2W}{d\xi^2}=\left\{\frac{u^2}{\xi}-\frac{2}{9\xi^2}+\frac{\psi(\xi)}{\xi}\right\}W,$$

can be transformed into Cases I and II, respectively.

Ex. 14.2† Let f_1 and f_2 be arbitrary functions of z. Show that the solutions of the equations

$$\frac{d^2w_1}{dz^2}=\left\{f_1f_2+f_2^{1/2}\frac{d^2(f_2^{-1/2})}{dz^2}\right\}w_1,\qquad \frac{d^2w_2}{dz^2}=\left\{f_1f_2+f_1^{1/2}\frac{d^2(f_1^{-1/2})}{dz^2}\right\}w_2,$$

are related by $f_1f_2^{1/2}w_1=d(f_1^{1/2}w_2)/dz$ and $f_1^{1/2}f_2w_2=d(f_2^{1/2}w_1)/dz$.

Apply this result to transform the atypical equation

$$\frac{d^2w}{dz^2}=\left\{u^2f(z)+\frac{3}{4z^2}\right\}w,$$

in which $f(z)$ has a simple zero at $z=0$, into Case II.

Historical Notes and Additional References

The earliest investigation of second-order linear differential equations belonging to Case III appears to be that of Langer (1935). For large values of the parameter u (in our notation) he derived asymptotic approximations for solutions in terms of Bessel functions, valid in a neighborhood of the singular point. This neighborhood becomes vanishingly small as $|u|\to\infty$, but Langer also gave asymptotic approximations in terms of elementary functions in an abutting region. Asymptotic approximations in terms of Bessel functions which are uniformly valid in fixed intervals or regions of the z plane were established by Swanson (1956), Kostomarov (1958), and Olver (1956, 1958). The present chapter is an expanded and considerably improved version of the last two papers; in particular, explicit bounds for the error terms in the asymptotic expansions are supplied for the first time.

† Communicated by E. R. Pike.

As in Chapters 10 and 11, the methods of the present chapter can be extended to inhomogeneous differential equations, and also to differential equations of the type

$$\frac{d^2 w}{dz^2} = u^{2l}\left\{\sum_{s=0}^{\infty} \frac{f_s(z)}{u^s}\right\} w,$$

in which l is a positive integer. For $l = 1$, Langer's results (1935) have been extended by Kazarinoff (1957b).

§2 Equation (2.05) also admits real solutions when ν is purely imaginary and the other variables are real. A similar theory for this case could be constructed by the methods of §§3 and 4.

§6 The method used here, and also in §7 of Chapter 6, is essentially due to Levinson (1949) and Newton (1960). See also McLeod (1961a), Bertocchi, Fubini, and Furlan (1965), Sabatier (1965), Calogero (1967), and Fedoryuk (1968). The author is indebted to W. M. Frank for drawing attention to these references. The present error bounds are a refinement of earlier results.

§9 For complex ν, asymptotic analyses (without error bounds) have been given by Olver (1956, 1958).

§§12–13 Uniform asymptotic expansions of the Legendre functions for large degree and bounded order have not been given previously with the present generality. Accounts of earlier results are included in the books of Robin (1958) and Szegö (1967); see also Karmazina (1960), Chuhrukidze (1966, 1968), and Schindler (1971). All of the earlier results are essentially included in the present expansions, or can be derived from them by suitable rearrangement or reexpansion.

§14 The restriction of basic functions to the form $P\{z\varpi(u)\}$ is not entirely satisfactory. It excludes, for example, the equation

$$\frac{d^2 w}{dz^2} = \left(\frac{u^2}{z^2} - \frac{1}{4z^2}\right) w,$$

whose solutions $z^{\pm u + (1/2)}$ can hardly be called recondite. Indeed, this basic equation was used, in effect, in the case when the coefficient $f(z)$ in (14.01) has a double pole at $z = z_0$; compare Chapter 6, §5.3. To include this equation, and possibly others, we would need an extension of Theorem 14.1 which admits solutions of the form $w = P\{q(z)\varpi(u)\}$, in which P, q, and ϖ are functions of one variable.

13

CONNECTION FORMULAS FOR SOLUTIONS OF DIFFERENTIAL EQUATIONS

1 Introduction

1.1 In preceding chapters we have constructed asymptotic solutions of linear differential equations of the second order in a variety of circumstances. A constantly recurring and unavoidable feature of the asymptotic approximations (or expansions) is that the independent variable is restricted to certain real intervals or complex regions. It is often of importance to determine approximations for the same solutions in other regions. In examples we showed how this could be achieved by use of an appropriate *connection formula*, that is, an equation expressing one solution of the differential equation linearly in terms of other solutions. Most of the connection formulas that were used were obtained from parametric integral representations of the solutions.

The purpose of the present chapter is to investigate ways in which connection formulas can be derived directly from the differential equation without recourse to integral representations. We begin with formulas that relate different branches of solutions in the neighborhood of a singularity of the differential equation.

2 Connection Formulas at a Singularity

2.1 Consider the equation

$$\frac{d^2w}{dz^2}+f(z)\frac{dw}{dz}+g(z)w=0, \tag{2.01}$$

in which $f(z)$ and $g(z)$ are analytic functions of the complex variable z which are single valued and holomorphic within an annulus $\mathbf{A}$: $0<|z|<r$. Let $w_1(z)$ and $w_2(z)$ be an independent pair of solutions. In consequence of Theorem 3.1 of Chapter 5, these solutions may be continued analytically along any path which begins at z, encircles the origin once in the positive sense within $\mathbf{A}$, and returns to z.

When $z=0$ is an ordinary point of the differential equation, the continuations are given by

$$w_1(ze^{2\pi i})=w_1(z), \qquad w_2(ze^{2\pi i})=w_2(z).$$

But when $z = 0$ is a singularity, these identities are inapplicable, in general, notwithstanding the fact that $f(z)$ and $g(z)$ are single valued in **A**. Both $w_1(ze^{2\pi i})$ and $w_2(ze^{2\pi i})$ are solutions of the differential equation, however; hence there exist constants c_{ij} such that

$$w_1(ze^{2\pi i}) = c_{11}w_1(z) + c_{12}w_2(z), \qquad w_2(ze^{2\pi i}) = c_{21}w_1(z) + c_{22}w_2(z). \tag{2.02}$$

These relations are the desired connection formulas.

2.2 When the singularity at $z = 0$ is regular, a method for finding the connection formulas is as follows. In matrix notation let

$$\begin{bmatrix} w_1(z) \\ w_2(z) \end{bmatrix} = \mathbf{B} \begin{bmatrix} \hat{w}_1(z) \\ \hat{w}_2(z) \end{bmatrix},$$

where $\hat{w}_1(z)$ and $\hat{w}_2(z)$ are fundamental solutions constructed as in Chapter 5, §§4 and 5, and **B** is a constant 2×2 matrix. The elements of **B** are obtainable by comparing leading terms of the ascending series expansions for the solutions. By hypothesis, $w_1(z)$ and $w_2(z)$ are linearly independent; hence by considering their Wronskian it is seen that **B** is nonsingular. If the respective roots α_1 and α_2 of the indicial equation do not differ by an integer or zero, then from Theorem 4.1 of Chapter 5 we derive

$$\begin{bmatrix} w_1(ze^{2\pi i}) \\ w_2(ze^{2\pi i}) \end{bmatrix} = \mathbf{B} \begin{bmatrix} e^{2\pi i\alpha_1}\hat{w}_1(z) \\ e^{2\pi i\alpha_2}\hat{w}_2(z) \end{bmatrix} = \mathbf{B} \begin{bmatrix} e^{2\pi i\alpha_1} & 0 \\ 0 & e^{2\pi i\alpha_2} \end{bmatrix} \mathbf{B}^{-1} \begin{bmatrix} w_1(z) \\ w_2(z) \end{bmatrix}.$$

This is equivalent to (2.02).

The procedure can be suitably modified when $\alpha_1 - \alpha_2$ is an integer or zero.

2.3 If the singularity at $z = 0$ is irregular, then in general it is still true that fundamental solutions exist with the properties

$$\hat{w}_j(ze^{2\pi i}) = e^{2\pi i\alpha_j}\hat{w}_j(z) \qquad (j = 1, 2), \tag{2.03}$$

where α_1 and α_2 are certain constants. For let c_{ij} again denote the coefficients appearing in (2.02), and suppose that the latent roots (eigenvalues) l_1 and l_2 of the matrix

$$\mathbf{C} = \begin{bmatrix} c_{11} & c_{12} \\ c_{21} & c_{22} \end{bmatrix}$$

are distinct. By an elementary theorem in matrix algebra **C** may be factorized in the form

$$\mathbf{C} = \mathbf{B} \begin{bmatrix} l_1 & 0 \\ 0 & l_2 \end{bmatrix} \mathbf{B}^{-1},$$

in which $\mathbf{B}$ is nonsingular. Solutions with the properties (2.03) are therefore given by

$$\begin{bmatrix} \hat{w}_1(z) \\ \hat{w}_2(z) \end{bmatrix} = \mathbf{B}^{-1} \begin{bmatrix} w_1(z) \\ w_2(z) \end{bmatrix},$$

with $e^{2\pi i \alpha_j} = l_j$. A similar result can also be found when $l_1 = l_2$.

The latent roots l_j are called the *circuit roots* of the singularity. It is not difficult to show that they are independent of the choice of $w_1(z)$ and $w_2(z)$. The numbers α_j, which are undetermined to the extent of an arbitrary additive integer, are called the *exponents* (or *circuit exponents*) by analogy with regular singularities. Perhaps it should be stressed that the single-valuedness of the functions $f(z)$ and $g(z)$ is essential to the foregoing analysis. If either $f(z)$ or $g(z)$ has a branch point at $z = 0$, then as a rule there are no solutions with the property (2.03).

Unlike the case of a regular singularity, the existence result just established has no immediate use in the determination of connection formulas, because explicit expressions are unavailable for the circuit roots in the general case. In consequence, numerical methods may be needed to evaluate the c_{ij}, necessary values of the solutions being calculable, in part, from the asymptotic expansions given in Chapter 7.

Ex. 2.1 In the case when there is an irregular singularity of rank 1 at infinity, it might be thought that the numbers μ_j defined in Chapter 7, §1.2 are the same as the indices α_j of §2.3 above. Show that this supposition is false by means of Bessel's equation.

Ex. 2.2 If $z = 0$ is a regular or irregular singularity with exponents α_1 and α_2, prove that the general solution of (2.01) satisfies

$$2\cos\{(\alpha_1-\alpha_2)\pi\}\, w(z) = \exp\{(\alpha_1+\alpha_2)\pi i\}\, w(ze^{-2\pi i}) + \exp\{-(\alpha_1+\alpha_2)\pi i\}\, w(ze^{2\pi i}).$$

3 Differential Equations with a Parameter

3.1 Throughout this book special emphasis has been given to differential equations of the form

$$d^2w/dz^2 = \{u^2 f(z) + g(z)\}\, w, \tag{3.01}$$

in which u is a large parameter, and the coefficients $f(z)$ and $g(z)$ are analytic in a complex domain $\mathbf{D}$. If $\mathbf{D}$ contains a singularity z_0, say, of the differential equation, then asymptotic solutions are generally valid only in subregions of $\mathbf{D}$, and we need connection formulas to pass around the singularity.

For *regular singularities*, asymptotic approximations to the solutions for large $|u|$ are given in Chapter 10 (especially §4) and Chapter 12, the former chapter applying when $f(z)$ has a double pole and the latter when $f(z)$ has a simple pole. No continuation problem exists in the case of a double pole, because the pertinent conditions impose no restriction on $\mathrm{ph}(z-z_0)$ in the neighborhood of z_0. For a simple pole, however, a continuation problem arises. This is solved in §§4 and 5 below by obtaining asymptotic approximations for the coefficients in the connection formula when $|u|$ is large.

For many types of *irregular singularities*, asymptotic approximations for the solutions are available in terms of elementary functions; see Chapter 6, §§4 and 5.† But because explicit general formulas are unavailable for the circuit roots the determination of connection formulas is difficult, and again numerical methods may be required.

3.2 A related problem arises when the domain **D** is free from singularities but contains a *simple turning point* z_0. Asymptotic approximations for the solutions in terms of Airy functions are given in Chapter 11. Connection formulas are not needed for these approximations because their regions of validity include the full neighborhood of z_0. However, there also exist the simpler LG asymptotic approximations which represent the solutions in subregions of **D** not containing z_0. In wave-penetration problems, for example, the LG approximations have direct physical significance, and it is of greater importance to connect them than to study the actual behavior of the solutions at the turning point.

One approach to this problem is to take the approximations of Chapter 11 and replace the Airy functions by their asymptotic approximations for large arguments. Another is to compare the LG approximations in the intersections of their regions of validity. Both methods are developed in later sections.

3.3 Throughout the rest of this chapter we concentrate on first approximations to solutions of the differential equation. Extensions of the results to higher approximations are quite feasible, but in most applications first approximations suffice. The object is to obtain first approximations for the coefficients in the connection formulas. Frequently we shall derive strict error bounds for the results and in these cases the presentation is simplified by omitting explicit reference to u, and then reintroducing this parameter as needed in applications. It will be recalled that this procedure was used in Chapter 6, and emphasizes that the general theory is directly applicable to certain equations of the form (3.01) in which $f(z)$ and $g(z)$ may depend on u.

4 Connection Formula for Case III

4.1 Following Chapter 12, §9 and setting $u = 1$ (in accordance with §3.3 above), we consider the equation

$$\frac{d^2W}{d\zeta^2} = \left\{\frac{1}{4\zeta} + \frac{\nu^2-1}{4\zeta^2} + \frac{\psi(\zeta)}{\zeta}\right\}W \tag{4.01}$$

in the case when $\nu \geqslant 0$ and $\psi(\zeta)$ is holomorphic in a simply connected complex domain $\boldsymbol{\Delta}$. This domain includes $\zeta = 0$ and may be unbounded. If $\boldsymbol{\Delta}_1$ denotes $\boldsymbol{\Delta}$ cut along the negative real axis, then Theorem 9.1 of Chapter 12 shows that in $\boldsymbol{\Delta}_1$ there are single-valued analytic solutions of the form

$$W_{1,1}(\zeta) = \zeta^{1/2} I_\nu(\zeta^{1/2}) + \varepsilon_{1,1}(\zeta), \tag{4.02}$$

† These results apply directly only to real values of z, but as indicated in §12.2 of the same chapter, the extension to complex z is straightforward.

and

$$W_{1,2}(\zeta) = \zeta^{1/2}K_\nu(\zeta^{1/2}) + \varepsilon_{1,2}(\zeta). \tag{4.03}$$

Here the branch of $\zeta^{1/2}$ has its principal value, except on the lower edge of the cut where it is determined by continuity. The second solution depends on an arbitrary reference point α of $\mathbf{\Delta}_1$.

The error terms in (4.02) and (4.03) are bounded by†

$$|\varepsilon_{1,1}(\zeta)| \leqslant \{\mu_2(\nu)/\mu_1(\nu)\}\,|\zeta|^{1/2}\mathfrak{E}_\nu(\zeta^{1/2})\,\mathfrak{M}_\nu(\zeta^{1/2})[\exp\{\mu_1(\nu)\mathscr{V}_{0,\zeta}(\Psi)\}-1], \tag{4.04}$$

and

$$|\varepsilon_{1,2}(\zeta)| \leqslant \{\mu_3(\nu)/\mu_1(\nu)\}\,|\zeta|^{1/2}\mathfrak{E}_\nu^{-1}(\zeta^{1/2})\,\mathfrak{M}_\nu(\zeta^{1/2})[\exp\{\mu_1(\nu)\mathscr{V}_{\zeta,\alpha}(\Psi)\}-1]. \tag{4.05}$$

In these inequalities the auxiliary functions $\mathfrak{E}_\nu$ and $\mathfrak{M}_\nu$ and the associated quantities $\mu_j(\nu)$ are defined in Chapter 12, §8, and $\Psi(\zeta)$ is the function previously denoted by $\zeta^{1/2}B_0(\zeta)$:

$$\Psi(\zeta) = \int_0^\zeta v^{-1/2}\psi(v)\,dv. \tag{4.06}$$

4.2 The bound (4.04) is valid for those points ζ that can be joined to the origin by a path which lies in $\mathbf{\Delta}_1$ and is progressive: in the present context this means that both $\mathrm{Re}(v^{1/2})$ and $|v|$ are nondecreasing as v proceeds away from the origin. The variation of Ψ has to be evaluated along a path of this kind. If $\mathbf{\Delta}$ contains a neighborhood of the origin, then all points of the corresponding cut neighborhood are included in the region of validity.

The indices associated with the regular singularity of the differential equation (4.01) at $\zeta = 0$ are $\frac{1}{2}(1 \pm \nu)$; compare Chapter 5, (4.03). When $\zeta \to 0$ we derive from (4.04) and the properties of $\mathfrak{E}_\nu$ and $\mathfrak{M}_\nu$ given in Chapter 12, §8, $\varepsilon_{1,1}(\zeta) = o\{\zeta^{(1+\nu)/2}\}$. Hence

$$W_{1,1}(\zeta) \sim \zeta^{(1+\nu)/2}/\{2^\nu\Gamma(\nu+1)\} \qquad (\zeta \to 0), \tag{4.07}$$

implying that $W_{1,1}(\zeta)$ is recessive at the origin. Therefore the continuation problem for $W_{1,1}(\zeta)$ is solved trivially by the formula

$$W_{1,1}(\zeta e^{2m\pi i}) = e^{(1+\nu)m\pi i}W_{1,1}(\zeta),$$

in which m is an arbitrary integer. This result was mentioned in Chapter 12, §10.3.

4.3 For the solution $W_{1,2}(\zeta)$, the inequality (4.05) holds whenever ζ can be linked to α by a path in $\mathbf{\Delta}_1$ on which both $\mathrm{Re}(v^{1/2})$ and $|v|$ are nonincreasing as v proceeds away from α. This does not mean that the neighborhood of $\zeta = 0$ is necessarily included, but we confine attention to this case. Let $W_{1,2}(\zeta e^{2\pi i})$ be the solution obtained from $W_{1,2}(\zeta)$ by analytic continuation, within $\mathbf{\Delta}$, once around the origin in a positive sense. The essential problem is to determine the constants Λ_1 and Λ_2 in the linear relation

$$W_{1,2}(\zeta e^{2\pi i}) = \Lambda_1 W_{1,1}(\zeta) + \Lambda_2 W_{1,2}(\zeta). \tag{4.08}$$

† The inequalities (4.04) and (4.05) are slight improvements on those obtained from the theorem simply by setting $n = 0$; compare Chapter 12, Exercise 4.4.

For this purpose it is convenient to rewrite (4.02) and (4.03) in the forms

$$W_{1,1}(\zeta) = \zeta^{1/2} I_\nu(\zeta^{1/2})\{1+\eta_1(\zeta)\}, \qquad W_{1,2}(\zeta) = \zeta^{1/2} K_\nu(\zeta^{1/2})\{1+\eta_2(\zeta)\}, \tag{4.09}$$

where the relative errors $\eta_1(\zeta)$ and $\eta_2(\zeta)$ are bounded by

$$|\eta_1(\zeta)| \leqslant \frac{\mu_2(\nu)}{\mu_1(\nu)} \frac{\exp\{\mu_1(\nu)\mathscr{V}_{0,\zeta}(\Psi)\}-1}{\cos\vartheta_\nu(\zeta^{1/2})}, \qquad |\eta_2(\zeta)| \leqslant \frac{\mu_3(\nu)}{\mu_1(\nu)} \frac{\exp\{\mu_1(\nu)\mathscr{V}_{\zeta,\alpha}(\Psi)\}-1}{\sin\vartheta_\nu(\zeta^{1/2})}; \tag{4.10}$$

compare Chapter 12, (8.19).

4.4 The coefficient Λ_2 is found by setting $\zeta = e^{-\pi i}\tau^2$, and letting $\tau \to 0$ through positive values. We have

$$I_\nu(\zeta^{1/2}) = I_\nu(e^{-\pi i/2}\tau) = e^{-\nu\pi i/2} J_\nu(\tau), \qquad K_\nu(\zeta^{1/2}) = K_\nu(e^{-\pi i/2}\tau) = \tfrac{1}{2}\pi i e^{\nu\pi i/2}\{J_\nu(\tau)+iY_\nu(\tau)\}. \tag{4.11}$$

Substitution in (4.09) and use of (4.10) yields, in the first instance,

$$W_{1,1}(\zeta) = e^{-(\nu+1)\pi i/2}\tau J_\nu(\tau)\{1+o(1)\} \qquad (\tau \to 0). \tag{4.12}$$

Next, as $\zeta \to 0$ along a progressive path $\eta_2(\zeta)$ tends to a constant, $\eta_2(0)$, say.† From the second of (4.11) we derive

$$K_\nu(\zeta^{1/2}) \sim -\tfrac{1}{2}\pi e^{\nu\pi i/2} Y_\nu(\tau) \qquad (\tau \to 0).$$

Hence

$$W_{1,2}(\zeta) = \tfrac{1}{2}\pi e^{(\nu+1)\pi i/2}\tau Y_\nu(\tau)\{1+\eta_2(0)+o(1)\} \qquad (\tau \to 0). \tag{4.13}$$

Again, the point $\zeta e^{2\pi i} \equiv e^{\pi i}\tau^2$ lies within the region of validity of the second of (4.10) when τ is sufficiently small, and similar analysis yields

$$W_{1,2}(\zeta e^{2\pi i}) = \tfrac{1}{2}\pi e^{-(\nu+1)\pi i/2}\tau Y_\nu(\tau)\{1+\eta_2(0)+o(1)\} \qquad (\tau \to 0), \tag{4.14}$$

the value of $\eta_2(0)$ being the same as in (4.13).

Substituting (4.12), (4.13), and (4.14) in (4.08), and letting $\tau \to 0$, we obtain the *exact* expression

$$\Lambda_2 = -e^{-\nu\pi i}, \tag{4.15}$$

provided that $\eta_2(0) \neq -1$. This condition is certainly satisfied when

$$2^{1/2}\{\mu_3(\nu)/\mu_1(\nu)\}[\exp\{\mu_1(\nu)\mathscr{V}_{0,\alpha}(\Psi)\}-1] < 1; \tag{4.16}$$

compare Chapter 12, (8.23).

4.5 To evaluate Λ_1, we have from (4.08) and (4.15)

$$\Lambda_1 = \{W_{1,2}(\zeta e^{2\pi i}) + e^{-\nu\pi i} W_{1,2}(\zeta)\}/W_{1,1}(\zeta). \tag{4.17}$$

† See Exercise 10.1 of Chapter 12.

We again set $\zeta = e^{-\pi i}\tau^2$ ($\tau > 0$). This time, however, the use of vanishingly small values of τ is unfruitful because of cancellation of the terms in the numerator. To obtain a meaningful result we select τ in such a way that all three solutions appearing in (4.17) are of comparable magnitude. A convenient choice is $\tau = y$, where y is any positive zero of $Y_\nu(z)$, provided that, of course, the points $e^{\pm\pi i}y^2$ lie within the regions of validity of the inequalities (4.10).

From (4.11) we obtain

$$I_\nu(e^{-\pi i/2}y) = e^{-\nu\pi i/2}J_\nu(y), \qquad K_\nu(e^{\pm\pi i/2}y) = \mp\tfrac{1}{2}\pi i e^{\mp\nu\pi i/2}J_\nu(y). \tag{4.18}$$

Accordingly,

$$W_{1,1}(e^{-\pi i}y^2) = e^{-(\nu+1)\pi i/2}yJ_\nu(y)\{1+\eta_1(e^{-\pi i}y^2)\},$$
$$W_{1,2}(e^{\pi i}y^2) = \tfrac{1}{2}\pi e^{-\nu\pi i/2}yJ_\nu(y)\{1+\eta_2(e^{\pi i}y^2)\},$$

and

$$W_{1,2}(e^{-\pi i}y^2) = \tfrac{1}{2}\pi e^{\nu\pi i/2}yJ_\nu(y)\{1+\eta_2(e^{-\pi i}y^2)\}.$$

Substitution in (4.17) by means of these equations yields the desired expression

$$\Lambda_1 = \pi i\frac{1+\frac{1}{2}\eta_2(e^{\pi i}y^2)+\frac{1}{2}\eta_2(e^{-\pi i}y^2)}{1+\eta_1(e^{-\pi i}y^2)}. \tag{4.19}$$

From Chapter 12, (8.08) and (8.14), we have $\mathfrak{E}_\nu^2(\pm iy) = 2/\pi$. Substitution of this result and (4.18) in Chapter 12, (8.21) gives $\vartheta_\nu(\pm iy) = \frac{1}{4}\pi$. Hence bounds for the error terms in (4.19) are furnished by

$$|\eta_1(e^{-\pi i}y^2)| \leqslant 2^{1/2}\{\mu_2(\nu)/\mu_1(\nu)\}[\exp\{\mu_1(\nu)\mathcal{V}_{0,e^{-\pi i}y^2}(\Psi)\}-1], \tag{4.20}$$

and

$$|\eta_2(e^{\pm\pi i}y^2)| \leqslant 2^{1/2}\{\mu_3(\nu)/\mu_1(\nu)\}[\exp\{\mu_1(\nu)\mathcal{V}_{\alpha,e^{\pm\pi i}y^2}(\Psi)\}-1], \tag{4.21}$$

the variations being evaluated along progressive paths. Maps in the $\zeta^{1/2}$ plane of suitable paths are indicated in Fig. 4.1.

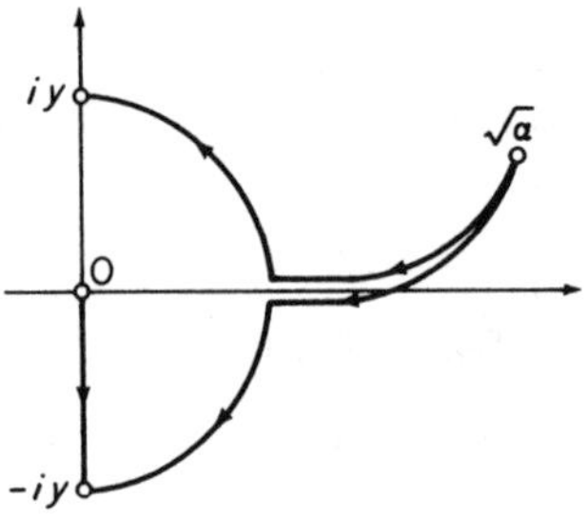

Fig. 4.1 $\zeta^{1/2}$ plane. Variational paths for error terms of Λ_1. $y =$ zero of Y_ν.

Ex. 4.1 Assume that Δ contains the real axis and the disk $|\zeta| \leqslant r$. If $\alpha = +\infty$ and $y^2 < r$, show that the variations appearing in (4.20) and (4.21) are bounded by

$$\mathcal{V}_{0,\exp(-\pi i)y^2}(\Psi) \leqslant 2y|\psi(0)| + \tfrac{2}{3}Cy^3r^{-1}, \qquad \mathcal{V}_{\infty,\exp(\pm\pi i)y^2}(\Psi) \leqslant \mathcal{V}_{\mathscr{A}}(\Psi) + \pi y|\psi(0)| + \pi Cy^3r^{-1},$$

where C is the maximum value of $|\psi(\zeta)-\psi(0)|$ on $|\zeta| = r$, and $\mathscr{A}$ is the positive real axis.

5 Application to Simple Poles

5.1 In this section the results of §4 are applied to the equation

$$d^2w/dz^2 = \{u^2 f(z)+g(z)\}\, w \qquad (z \in \mathbf{D}), \tag{5.01}$$

in which $\mathbf{D}$ is a simply connected complex domain, and $f(z)$ and $g(z)$ are analytic functions which are independent of u. For simplicity, we suppose the parameter u to be real and positive, but this restriction could be eased with little complication. We assume that at a point z_0 of $\mathbf{D}$, $f(z)$ has a simple pole and $(z-z_0)^2 g(z)$ is analytic, and we also assume that there are no other transition points in $\mathbf{D}$. Let the Laurent expansions of $f(z)$ and $g(z)$ in the neighborhood of z_0 be denoted by

$$f(z) = (z-z_0)^{-1} \sum_{s=0}^{\infty} f_s (z-z_0)^s, \qquad g(z) = (z-z_0)^{-2} \sum_{s=0}^{\infty} g_s (z-z_0)^s,$$

where $f_0 \neq 0$.

The Liouville transformation

$$\zeta = u^2 \left\{ \int_{z_0}^{z} f^{1/2}(t)\, dt \right\}^2, \qquad w = \zeta^{-1/4} f^{-1/4}(z)\, W, \tag{5.02}$$

throws (5.01) into the form (4.01) with

$$\nu = (1+4g_0)^{1/2};$$

compare Chapter 12, §2.1. We again restrict $\nu \geqslant 0$. If $z \to z_0$, then

$$\zeta \sim 4u^2 f_0 (z-z_0), \tag{5.03}$$

and ζ is an analytic function of z at this point.

The solutions of (5.01) have a branch point at $z = z_0$. We make them single valued by introducing a cut in the z plane along the curve $\mathscr{C}$ having the equation

$$\mathrm{Re}\left\{ \int_{z_0}^{z} f^{1/2}(t)\, dt \right\} = 0.$$

From (5.02) and (5.03) it is seen that $\mathscr{C}$ emerges from z_0 in the direction

$$\mathrm{ph}(z-z_0) = \pm\pi - \mathrm{ph} f_0.$$

The theory of conformal mapping shows that $\mathscr{C}$ is an R_∞ arc which can terminate only at z_0 or at the boundary of $\mathbf{D}$. We call $\mathscr{C}$ the *principal curve* associated with the transition point z_0, and denote $\mathbf{D}$ cut along $\mathscr{C}$ by $\mathbf{D}_1$. In what follows we require $\zeta^{1/2}$ and $\zeta^{1/4}$ to have their principal values within $\mathbf{D}_1$ and be determined by continuity on the boundaries of $\mathbf{D}_1$. On the other hand, any branch may be selected for $f^{1/4}(z)$, provided that it is continuous in the closure of $\mathbf{D}_1$; $f^{-1/4}(z)$ denotes the reciprocal of this branch.

Let a be a reference point situated either at infinity or on the boundary of $\mathbf{D}_1$, having the following properties: (i) *a and z_0 can be linked by a path $\mathscr{L}$ which lies in $\mathbf{D}_1$ and has a progressive ζ map (in the sense of* §4); (ii) *if $z \to a$ along $\mathscr{L}$, then*

$\zeta \to \infty$ *within the sector* $|\mathrm{ph}\,\zeta| \leqslant \pi - \delta$, *where* δ *is a positive constant*. In the ζ plane a corresponds to a point at infinity α such that $|\mathrm{ph}(\alpha^{1/2})| \leqslant \frac{1}{2}\pi - \frac{1}{2}\delta$.

5.2 Let us interpret the connection formula (4.08) in terms of the variables w, u, and z. It is convenient to use the following solutions of (5.01):

$$w_1(u,z) = \frac{\Gamma(\nu+1)}{2^{1/2}u^{\nu+(1/2)}} \frac{W_{1,1}(\zeta)}{\zeta^{1/4} f^{1/4}(z)}, \qquad w_2(u,z) = \left(\frac{2}{\pi}\right)^{1/2} \frac{W_{1,2}(\zeta)}{\zeta^{1/4} f^{1/4}(z)}.$$

From (4.07) and (5.03) we see that $w_1(u,z)$ has the recessive property

$$w_1(u,z) \sim f^{-1/4}(z)\{f_0(z-z_0)\}^{(2\nu+1)/4} \qquad (z \to z_0), \tag{5.04}$$

where the second factor has its principal value. Similarly,

$$w_2(u,z) \sim f^{-1/4}(z) \exp\left\{-u\int_{z_0}^{z} f^{1/2}(t)\,dt\right\} \qquad (z \to a \text{ along } \mathscr{L}), \tag{5.05}$$

the branch of the integral having positive real part in the neighborhood of a.

From (4.09) we derive in a similar manner

$$w_1(u,z) \sim \tfrac{1}{2}\pi^{-1/2}\Gamma(\nu+1)\{1+\eta_1(\alpha)\}\,u^{-\nu-(1/2)} f^{-1/4}(z) \exp\left\{u\int_{z_0}^{z} f^{1/2}(t)\,dt\right\}$$
$$(z \to a \text{ along } \mathscr{L}),$$

and

$$w_2(u,z) \sim -\pi^{-1/2}\{1+\eta_2(0)\}\,u^{1/2} f^{-1/4}(z)\{f_0(z-z_0)\}^{1/4} \ln\{f_0(z-z_0)\}$$
$$(z \to z_0, \quad \nu = 0),$$

or

$$w_2(u,z) \sim \pi^{-1/2}\Gamma(\nu)\{1+\eta_2(0)\}\,u^{-\nu+(1/2)} f^{-1/4}(z)\{f_0(z-z_0)\}^{(1-2\nu)/4}$$
$$(z \to z_0, \quad \nu > 0),$$

with similar specifications of branches. The numbers $\eta_1(\alpha)$ and $\eta_2(0)$ are both $O(u^{-1})$ for large u.

Lastly, let $w_3(u,z)$ be the solution obtained by continuing $w_2(u,z)$ analytically around a closed contour in **D** which encircles z_0 once in the positive sense. Then on substituting in (4.08) and using (4.15) and (4.19), we arrive at the desired connection formula

$$w_3(u,z) = \frac{2i\pi^{1/2}u^{\nu+(1/2)}}{\Gamma(\nu+1)} \frac{1+\delta_2}{1+\delta_1} w_1(u,z) - e^{-\nu\pi i} w_2(u,z), \tag{5.06}$$

with

$$\delta_1 = \eta_1(e^{-\pi i}y^2), \qquad \delta_2 = \tfrac{1}{2}\eta_2(e^{\pi i}y^2) + \tfrac{1}{2}\eta_2(e^{-\pi i}y^2).$$

5.3 It remains to reexpress the bounds (4.20) and (4.21). From (4.06) and Chapter 12, §2.1, we see that $\Psi(\zeta) = u^{-1}H(z)$, where $H(z)$ is independent of u and given by

$$H(z) = \frac{(4\nu^2-1)u}{8\zeta^{1/2}} + \frac{1}{2}\int \left\{g(z) + \frac{4f(z)f''(z) - 5f'^2(z)}{16f^2(z)}\right\} \frac{dz}{f^{1/2}(z)}, \tag{5.07}$$

the branches of $\zeta^{1/2}$ and $f^{1/2}(z)$ being continuous in $\mathbf{D}_1$ and chosen so that

$$H(z) = O\{(z-z_0)^{1/2}\} \qquad (z \to z_0). \tag{5.08}$$

From (4.20) and (4.21), we derive

$$|\delta_1| \leqslant 2^{1/2}\{\mu_2(\nu)/\mu_1(\nu)\}[\exp\{u^{-1}\mu_1(\nu)\mathscr{V}_{z_0,z_{-1}}(H)\}-1], \tag{5.09}$$

$$|\delta_2| \leqslant 2^{1/2}\{\mu_3(\nu)/\mu_1(\nu)\}[\tfrac{1}{2}\exp\{u^{-1}\mu_1(\nu)\mathscr{V}_{a,z_1}(H)\} + \tfrac{1}{2}\exp\{u^{-1}\mu_1(\nu)\mathscr{V}_{a,z_{-1}}(H)\}-1], \tag{5.10}$$

where z_1 and z_{-1} are the points on opposite sides of the cut $\mathscr{C}$ corresponding to $\zeta = e^{\pm\pi i}y^2$. Suppose that we set $y = y_{\nu,1}$, that is, the smallest positive zero of Y_ν. From (5.03) we then have

$$z_{\pm 1} = z_0 + O(u^{-2}) \qquad (u \to \infty).$$

Taking the variational path along $\mathscr{C}$ and using (5.08), we derive

$$\delta_1 = O(u^{-2}) \qquad (u \to \infty). \tag{5.11}$$

A slightly simpler bound for δ_2 can be constructed as follows. Let b_1 and b_{-1} be any two points on opposite sides of $\mathscr{C}$ chosen independently of u. Provided that u is large enough to ensure that z_1 lies between z_0 and b_1, we have

$$|\delta_2| \leqslant 2^{1/2}\{\mu_3(\nu)/\mu_1(\nu)\}[\tfrac{1}{2}\exp\{u^{-1}\mu_1(\nu)\mathscr{V}_{\mathscr{P}_1}(H)\}+\tfrac{1}{2}\exp\{u^{-1}\mu_1(\nu)\mathscr{V}_{\mathscr{P}_{-1}}(H)\}-1], \tag{5.12}$$

where $\mathscr{P}_1$ is a path which begins at a, passes through b_1, terminates at z_0, and has a map in the ζ plane which is progressive. Similarly for $\mathscr{P}_{-1}$; see Fig. 5.1. Since both paths $\mathscr{P}_1$ and $\mathscr{P}_{-1}$ are independent of u, it follows that

$$\delta_2 = O(u^{-1}) \qquad (u \to \infty). \tag{5.13}$$

We also note that the condition (4.16) is fulfilled for all sufficiently large u.

Lastly, it will be noticed that when u is large the coefficient of $w_1(u, z)$ in (5.06) cannot vanish. Hence $w_3(u, z)$ is dominant as $z \to a$, unlike its parent $w_2(u, z)$.

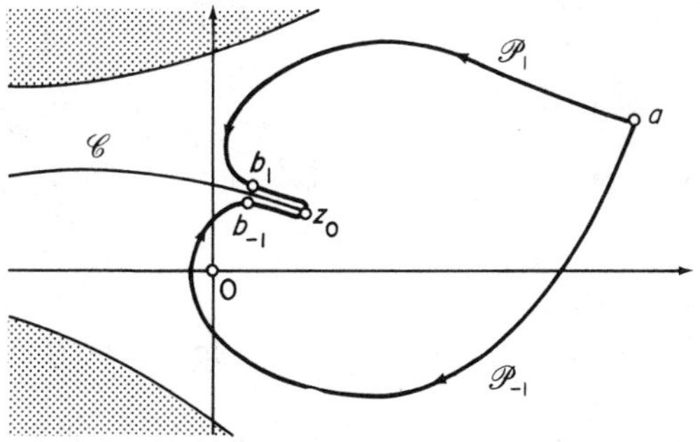

Fig. 5.1 z plane. Variational paths for δ_2. Unshaded domain is $\mathbf{D}$.

5.4 The results of this section may be summarized as follows. Assumptions and principal definitions are stated in §5.1. The fundamental connection formula is (5.06). In this formula (a) the solutions $w_1(u,z)$ and $w_2(u,z)$ are specified by the conditions (5.04) and (5.05); (b) the solution $w_3(u,z)$ is obtained from $w_2(u,z)$ by analytic continuation around z_0 once in the positive sense; (c) the error terms δ_1 and δ_2 are bounded by (5.09) and (5.12), where $H(z)$ is given by (5.07) and (5.08), and the ζ maps of the variational paths are progressive (in the sense of §4). For large u,

$$(1+\delta_2)/(1+\delta_1) = 1 + O(u^{-1}).$$

Ex. 5.1 With the conditions of §5, let $w_{2,s}(u,z)$ be the solution obtained from $w_2(u,z)$ by making, within **D**, s circuits around z_0 in the positive sense (so that $w_3(u,z) = w_{2,1}(u,z)$). Show that

$$w_{2,s}(u,z) \sim (-)^s e^{-s\nu\pi i} w_2(u,z) \qquad (z \to z_0 \text{ within } \mathbf{D}_1),$$

and

$$w_{2,s}(u,z) \sim (-)^{s-1} \frac{\sin(s\nu\pi)}{\sin(\nu\pi)} \frac{2i\pi^{1/2}u^{\nu+(1/2)}}{\Gamma(\nu+1)} \frac{1+\delta_2}{1+\delta_1} w_1(u,z) \qquad (z \to a \text{ along } \mathscr{L}).$$

6 Example: The Associated Legendre Equation

6.1 To illustrate the method of §5, consider the equation

$$\frac{d^2w}{dz^2} = \left\{\frac{(n+\frac{1}{2})^2}{z^2-1} - \frac{1}{4(z^2-1)} + \frac{m^2-1}{(z^2-1)^2}\right\} w \tag{6.01}$$

satisfied by the functions $(z^2-1)^{1/2}P_n^{\pm m}(z)$ and $(z^2-1)^{1/2}Q_n^{\pm m}(z)$; compare Chapter 12, §12.1. Take **D** to be the half-plane $\operatorname{Re} z \geqslant 0$, and $u = n+\frac{1}{2}$. Then $z_0 = 1$, $f_0 = \frac{1}{2}$, $\nu = m$ ($\geqslant 0$), and the cut $\mathscr{C}$ is the join of the points 0 and 1. Make $f^{1/4}(z)$ the principal branch of $(z^2-1)^{-1/4}$.

When $z \to 1$, we have from Chapter 5, (12.08),

$$(z^2-1)^{1/2}P_n^{-m}(z) \sim \frac{(z-1)^{(m+1)/2}}{2^{(m-1)/2}\Gamma(m+1)}.$$

Comparison with (5.04) shows that

$$w_1(u,z) = 2^{-1/2}\Gamma(m+1)(z^2-1)^{1/2}P_n^{-m}(z). \tag{6.02}$$

As $z \to +\infty$, we have

$$\int_1^z f^{1/2}(t)\,dt = \ln\{z+(z^2-1)^{1/2}\} = \ln(2z) + O(z^{-2}).$$

From this it follows that $\zeta \sim u^2(\ln z)^2$, and hence that $\mathscr{V}_{1,\infty}(H)$ converges; compare (5.07) and Chapter 6, §4.3. Accordingly, it is legitimate to take the reference point

a to be $+\infty$. From Chapter 5, (12.09) and (13.14) we derive

$$(z^2-1)^{1/2}Q_n^m(z) \sim \pi^{1/2}\frac{e^{m\pi i}\Gamma(n+m+1)}{2^{n+1}\Gamma(n+\frac{3}{2})}\frac{1}{z^n} \qquad (z\to\infty).$$

Comparison with (5.05) yields

$$w_2(u,z) = \frac{2^{1/2}\Gamma(n+\frac{3}{2})}{\pi^{1/2}e^{m\pi i}\Gamma(n+m+1)}(z^2-1)^{1/2}Q_n^m(z). \tag{6.03}$$

Lastly, let $Q_{n,1}^m(z)$ denote the branch of $Q_n^m(z)$ obtained by making a single positive circuit of $z=1$ within the right half-plane. Since $(z^2-1)^{1/2}$ changes sign on completing the circuit, it follows that

$$w_3(u,z) = -\frac{2^{1/2}\Gamma(n+\frac{3}{2})}{\pi^{1/2}e^{m\pi i}\Gamma(n+m+1)}(z^2-1)^{1/2}Q_{n,1}^m(z). \tag{6.04}$$

Substitution of (6.02), (6.03), and (6.04) in (5.06) yields the required formula

$$Q_{n,1}^m(z) = -\pi i e^{m\pi i}(n+\tfrac{1}{2})^{m+(1/2)}\frac{\Gamma(n+m+1)}{\Gamma(n+\frac{3}{2})}\frac{1+\delta_2}{1+\delta_1}P_n^{-m}(z) + e^{-m\pi i}Q_n^m(z). \tag{6.05}$$

This equation is to be compared with the known formula

$$Q_{n,1}^m(z) = e^{-m\pi i}Q_n^m(z) - \pi i e^{m\pi i}\frac{\Gamma(n+m+1)}{\Gamma(n-m+1)}P_n^{-m}(z), \tag{6.06}$$

given in Chapter 5, Exercise 13.1. Using (5.11), (5.13), and the asymptotic approximation for the ratio of Gamma functions (Chapter 4, §5), we find the coefficients of $P_n^{-m}(z)$ in (6.05) and (6.06) both have the form $-\pi i e^{m\pi i}n^{2m}\{1+O(n^{-1})\}$ as $n\to\infty$. This verifies (6.05).

7 The Gans–Jeffreys Formulas: Real-Variable Method

7.1 Consider the differential equation

$$d^2w/dx^2 = \{f(x)+g(x)\}\,w, \tag{7.01}$$

in which x ranges over a finite or infinite interval (a,b) containing just one zero x_0, say, of $f(x)$. Assume that

(i) *The function $f(x)/(x-x_0)$ is positive and twice continuously differentiable.*†
(ii) *The function $g(x)$ is real and continuous.*
(iii) *As $x\to a$ or b, $\int_{x_0}^x |f(t)|^{1/2}\,dt$ diverges and $\mathscr{V}(F)$ converges, where*

$$F(x) = \int\left\{\frac{1}{|f|^{1/4}}\frac{d^2}{dx^2}\left(\frac{1}{|f|^{1/4}}\right) - \frac{g}{|f|^{1/2}}\right\}dx. \tag{7.02}$$

† Since $f(x)/(x-x_0)$ is positive at x_0, this zero necessarily is simple.

With these conditions it follows from Chapter 6, §§2 and 3 that equation (7.01) has solutions $w_1(x)$, $w_2(x)$, and $w_3(\theta, x)$ with the properties

$$w_1(x) \sim f^{-1/4}(x) \exp\left\{-\int_{x_0}^{x} f^{1/2}(t)\,dt\right\} \qquad (x \to b), \tag{7.03}$$

$$w_2(x) \sim f^{-1/4}(x) \exp\left\{\int_{x_0}^{x} f^{1/2}(t)\,dt\right\} \qquad (x \to b), \tag{7.04}$$

and

$$w_3(\theta, x) = |f(x)|^{-1/4}\left[\sin\left\{\int_{x}^{x_0} |f(t)|^{1/2}\,dt + \theta\right\} + o(1)\right] \qquad (x \to a), \tag{7.05}$$

θ being an arbitrary parameter. The solution $w_1(x)$ is unique, as is $w_3(\theta, x)$ when θ is assigned. The solution $w_2(x)$ is not unique, however. Our object is to relate $w_1(x)$ and $w_2(x)$ to $w_3(\theta, x)$.

7.2 Theorem 3.1 of Chapter 11 shows that equation (7.01) has solutions of the form†

$$w_4(x) = (\zeta/f)^{1/4}\{\mathrm{Ai}(\zeta) + \varepsilon_4(x)\}, \qquad w_5(x) = (\zeta/f)^{1/4}\{\mathrm{Bi}(\zeta) + \varepsilon_5(x)\},$$

in which

$$\zeta = \pm\left|\tfrac{3}{2}\int_{x_0}^{x} f^{1/2}\,dt\right|^{2/3}, \tag{7.06}$$

$$|\varepsilon_4(x)| \leqslant \lambda^{-1} E^{-1}(\zeta)\, M(\zeta)\left[\exp\{\lambda \mathcal{V}_{x,b}(H)\} - 1\right] \qquad (a < x < b),$$

$$|\varepsilon_5(x)| \leqslant \lambda^{-1} E(\zeta)\, M(\zeta)\left[\exp\{\lambda \mathcal{V}_{a,x}(H)\} - 1\right] \qquad (a < x < b),$$

and

$$H(x) = \int_{x_0}^{x}\left\{\frac{1}{|f|^{1/4}}\frac{d^2}{dx^2}\left(\frac{1}{|f|^{1/4}}\right) - \frac{g}{|f|^{1/2}} - \frac{5|f|^{1/2}}{16|\zeta|^3}\right\} dx = F(x) \pm \frac{5}{24|\zeta|^{3/2}}. \tag{7.07}$$

In (7.06) and (7.07) the upper or lower signs are taken according as $x \gtrless x_0$. The auxiliary functions $E(\zeta)$, $M(\zeta)$, and the constant $\lambda = 1.04\ldots$ are defined in Chapter 11, §2.

Hypothesis (iii) implies that if $x \to b$ then $\zeta \to +\infty$, and if $x \to a$ then $\zeta \to -\infty$. On substituting for $\mathrm{Ai}(\zeta)$ and the auxiliary functions by means of their asymptotic forms for large arguments (Chapter 11, §§1,2), we obtain

$$w_4(x) = \tfrac{1}{2}\pi^{-1/2} f^{-1/4}\{1 + o(1)\} \exp\left(-\int_{x_0}^{x} f^{1/2}\,dt\right) \qquad (x \to b), \tag{7.08}$$

and

$$w_4(x) = \pi^{-1/2}|f|^{-1/4}\left\{\sin\left(\int_{x}^{x_0} |f|^{1/2}\,dt + \tfrac{1}{4}\pi\right) + \eta_4(x) + o(1)\right\} \qquad (x \to a), \tag{7.09}$$

† From here on the arguments of f and g are suppressed, for brevity.

where $\eta_4(x) \equiv \pi^{1/2}|\zeta|^{1/4}\varepsilon_4(x)$, and is bounded.

Comparison of (7.03) and (7.08) immediately reveals that

$$w_4(x) = \tfrac{1}{2}\pi^{-1/2}w_1(x).$$

Next, because the general solution of (7.01) can be represented by $Aw_3(\theta, x)$, where A and θ are disposable constants, we can express

$$w_4(x) = \pi^{-1/2}|f|^{-1/4}(1+\alpha)\left\{\sin\left(\int_x^{x_0}|f|^{1/2}\,dt + \tfrac{1}{4}\pi + \delta\right) + o(1)\right\} \qquad (x \to a), \tag{7.10}$$

where α and δ are independent of x. Comparing (7.09) with (7.10) and following the steps of Chapter 6, §7.3, we derive

$$|\alpha|,\ 2|\delta|/\pi \leqslant \lambda^{-1}[\exp\{\lambda\mathscr{V}_{a,b}(H)\}-1], \tag{7.11}$$

provided that the right-hand side of this inequality does not exceed unity.

Therefore

$$w_1(x) = 2(1+\alpha)\,w_3(\tfrac{1}{4}\pi+\delta, x). \tag{7.12}$$

This is the *first Gans–Jeffreys formula*: its error terms α and δ are bounded by (7.11). If, for example, f contains a multiplicative parameter u^2, then both α and δ are $O(u^{-1})$ as $u \to \infty$.

7.3 In a similar way, we find that

$$w_5(x) = \pi^{-1/2}f^{-1/4}\{1+\eta_5(x)+o(1)\}\exp\left(\int_{x_0}^{x} f^{1/2}\,dt\right) \qquad (x \to b), \tag{7.13}$$

and

$$w_5(x) = \pi^{-1/2}|f|^{-1/4}\left\{\cos\left(\int_x^{x_0}|f|^{1/2}\,dt + \tfrac{1}{4}\pi\right) + o(1)\right\} \qquad (x \to a), \tag{7.14}$$

where

$$\eta_5(x) = \pi^{1/2}\zeta^{1/4}\exp(-\tfrac{2}{3}\zeta^{3/2})\,\varepsilon_5(x).$$

Comparison of (7.05) and (7.14) shows that

$$w_5(x) = -\pi^{-1/2}w_3(-\tfrac{1}{4}\pi, x).$$

Also, if $x \to b$ then $\eta_5(x)$ tends to a finite limit $\eta(b)$, say; compare Chapter 11, Exercise 4.1. Therefore we have

$$w_3(-\tfrac{1}{4}\pi, x) = -\{1+\eta(b)\}\,w_2(x), \tag{7.15}$$

in the sense that there is a particular solution $w_2(x)$ with the property (7.04) which satisfies this equation. This is the *second Gans–Jeffreys formula*. Its error term is bounded by

$$|\eta(b)| \leqslant 2^{1/2}\lambda^{-1}[\exp\{\lambda\mathscr{V}_{a,b}(H)\}-1]. \tag{7.16}$$

7.4 The precise meaning of the two connection formulas needs careful understanding. Equation (7.12) really only gives information concerning the solution which is recessive as $x \to b$: it informs us approximately how this solution behaves when $x \to a$. Because the actual value of δ is unknown (7.12) yields no information concerning solutions specified by their behavior at $x = a$: the slightest perturbation in δ, in fact, would change the behavior of $w_3(\frac{1}{4}\pi+\delta, x)$ at $x = b$ from recession to dominance.

In the case of (7.15) the situation is reversed. It is correct to conclude that the solution having the property (7.05), with $\theta = -\frac{1}{4}\pi$, has the asymptotic form

$$-\{1+\eta(b)\} f^{-1/4} \exp\left(\int_{x_0}^{x} f^{1/2}\, dt\right)$$

as $x \to b$, but not *vice versa.*

Ex. 7.1 If the equation

$$d^2 w/dx^2 = u^2 x(x^2+1)^2 w \qquad (u > 0)$$

has a nontrivial solution satisfying $w = 0$ at $x = -1$ and $w \to 0$ as $x \to +\infty$, establish that $u = \frac{21}{80}(4n-1)\pi + v(n)$, where n is an arbitrary positive integer, and $v(n) = O(n^{-1})$ as $n \to \infty$.

Ex. 7.2 Let $u > 0$ and $w(u,x)$ be the solution of the equation

$$d^2 w/dx^2 = \{u(u+1)x + \tfrac{1}{4}(x-1)\}(x+1)^{-2} w \qquad (-1 < x < \infty)$$

satisfying $w(u,x) \sim x^{1/4} \exp\{-(2u+1)x^{1/2}\}$ as $x \to \infty$, u being fixed. Show that

$$w(u,x) = 2e^{-\pi/2} e^{-\pi u}(1+\alpha)(x+1)^{1/2}[\cos\{(u+\tfrac{1}{2})\ln(\tfrac{1}{4}x+\tfrac{1}{4}) + 2u + \tfrac{1}{4}\pi + 1 + \delta\} + o(1)]$$

as $x \to -1$, where both α and δ are independent of x and have the form $O(u^{-1})$ as $u \to \infty$.

Ex. 7.3 If u is a positive parameter and the equation

$$d^2 w/dx^2 = u^2 x(x^2+1)^{-1/2} w$$

has a solution with the properties $w = \cos(ux) + o(1)$ as $x \to -\infty$, and $w \to 0$ as $x \to +\infty$, show that $u = (n-\frac{1}{4})\pi k^{-1} + v(n)$, where n is an arbitrary positive integer,

$$k = \int_0^\infty \left\{1 - \frac{x^{1/2}}{(1+x^2)^{1/4}}\right\} dx,$$

and $v(n) = O(n^{-1})$ as $n \to \infty$.

Ex. 7.4 In the preceding exercise, show that if $n - \frac{3}{4} > \lambda hk/\{\pi \ln(1+\lambda)\}$, then

$$|v(n)| \leqslant \frac{\pi}{2\lambda k}\left\{\exp\left(\frac{\lambda hk}{n\pi - \frac{3}{4}\pi}\right) - 1\right\},$$

where $h = \mathscr{V}_{-\infty,\infty}(H)$, and H is defined by (7.06) and (7.07) with $f = x(x^2+1)^{-1/2}$ and $g = 0$.

8 Two Turning Points

8.1 We now consider a problem in which $f(x)$ has two real zeros x_0 and $\hat{x}_0$, say, with $\hat{x}_0 < x_0$. The differential equation is taken in the form

$$d^2 w/dx^2 = \{u^2 f(x) + g(x)\} w, \tag{8.01}$$

in which u is a large positive parameter, and $f(x)$ and $g(x)$ are independent of u. We assume that the following conditions are satisfied on the real axis:

(i) *The function* $f(x)/\{(x-\hat{x}_0)(x-x_0)\}$ *is positive and twice continuously differentiable.*

(ii) *The real or complex function* $g(x)$ *is continuous.*

(iii) *As* $x \to \pm\infty$, $\int^x f^{1/2}(t)\,dt$ *diverges and* $\mathscr{V}(F)$ *converges,* $F(x)$ *again being defined by* (7.02).

For large $|x|$, the solutions of (8.01) are of exponential character, and it is of interest to determine the eigenvalues u for which there exists a solution which is recessive at both $-\infty$ and $+\infty$. Following Jeffreys (1962, §3.61) we shall solve this problem by matching uniform Airy approximations in their common interval of validity $(\hat{x}_0, x_0)$.

8.2 Let x_1 be any interior point of $(\hat{x}_0, x_0)$, chosen independently of u. Theorem 3.1 of Chapter 11 shows that in the interval $[x_1, \infty)$ equation (8.01) has a solution of the form

$$w(u,x) = \pi^{1/2}u^{1/6}\phi^{-1/4}(x)\{\mathrm{Ai}(u^{2/3}\zeta)+\varepsilon(u,x)\}, \tag{8.02}$$

in which

$$\zeta \equiv \zeta(x) = \pm\left|\tfrac{3}{2}\int_{x_0}^{x} f^{1/2}(t)\,dt\right|^{2/3} \qquad (x \gtrless x_0),$$

and

$$\phi(x) = f(x)/\zeta(x).$$

The factor $\pi^{1/2}u^{1/6}$ is introduced in (8.02) for later convenience, and the error term is subject to the bounds

$$\frac{|\varepsilon(u,x)|}{M(u^{2/3}\zeta)},\ \frac{|\partial\varepsilon(u,x)/\partial x|}{u^{2/3}\phi^{1/2}(x)N(u^{2/3}\zeta)} \leqslant \frac{E^{-1}(u^{2/3}\zeta)}{\lambda}\left[\exp\left\{\frac{\lambda\mathscr{V}_{x,\infty}(H)}{u}\right\}-1\right], \tag{8.03}$$

in which $H(x)$ is defined by the expression (7.07), with the present $f(x)$ and $g(x)$.

Similarly, in the interval $(-\infty, x_1]$ a solution which is recessive at $-\infty$ is given by

$$\hat{w}(u,x) = \pi^{1/2}u^{1/6}\hat{\phi}^{-1/4}(x)\{\mathrm{Ai}(u^{2/3}\hat{\zeta})+\hat{\varepsilon}(u,x)\}, \tag{8.04}$$

where

$$\hat{\zeta} \equiv \hat{\zeta}(x) = \mp\left|\frac{3}{2}\int_{\hat{x}_0}^{x} f^{1/2}(t)\,dt\right|^{2/3} \quad (x \gtrless \hat{x}_0), \qquad \hat{\phi}(x) = \frac{f(x)}{\hat{\zeta}(x)},$$

$$\frac{|\hat{\varepsilon}(u,x)|}{M(u^{2/3}\hat{\zeta})},\ \frac{|\partial\hat{\varepsilon}(u,x)/\partial x|}{u^{2/3}\hat{\phi}^{1/2}(x)N(u^{2/3}\hat{\zeta})} \leqslant \frac{E^{-1}(u^{2/3}\hat{\zeta})}{\lambda}\left[\exp\left\{\frac{\lambda\mathscr{V}_{-\infty,x}(\hat{H})}{u}\right\}-1\right], \tag{8.05}$$

and

$$\hat{H}(x) = \int_{\hat{x}_0}^{x}\left\{\frac{1}{|f|^{1/4}}\frac{d^2}{dx^2}\left(\frac{1}{|f|^{1/4}}\right)-\frac{g}{|f|^{1/2}}-\frac{5|f|^{1/2}}{16|\hat{\zeta}|^3}\right\}dx. \tag{8.06}$$

Setting $x = x_1$ and substituting for the Airy function in (8.02) by means of its asymptotic approximation for large negative argument, we obtain

$$w(u, x_1) = \gamma^{-1}\left[\cos\left\{u\int_{x_1}^{x_0} |f(t)|^{1/2}\,dt - \tfrac{1}{4}\pi\right\} + \eta(u, x_1) + O(u^{-1})\right]$$

as $u \to \infty$, where $\gamma = |f(x_1)|^{1/4}$ and

$$\eta(u, x) = \pi^{1/2}u^{1/6}|\zeta|^{1/4}\varepsilon(u, x).$$

Since $\eta(u, x_1) = O(u^{-1})$ as $u \to \infty$, it follows that

$$w(u, x_1) = \gamma^{-1}\cos(uh - \tfrac{1}{4}\pi) + O(u^{-1}),$$

where we have denoted

$$\int_{x_1}^{x_0} |f(t)|^{1/2}\,dt = h. \tag{8.07}$$

In a similar way, the differentiated form of (8.02) yields

$$w'(u, x_1) = u\gamma\sin(uh - \tfrac{1}{4}\pi) + O(1).$$

The corresponding results for the solution $\hat{w}(u, x)$ are

$$\hat{w}(u, x_1) = \gamma^{-1}\cos(u\hat{h} - \tfrac{1}{4}\pi) + O(u^{-1}), \qquad \hat{w}'(u, x_1) = -u\gamma\sin(u\hat{h} - \tfrac{1}{4}\pi) + O(1),$$

where

$$\hat{h} = \int_{\hat{x}_0}^{x_1} |f(t)|^{1/2}\,dt. \tag{8.08}$$

The condition that the Wronskian of $w(u, x)$ and $\hat{w}(u, x)$ vanishes at $x = x_1$ gives

$$\sin\{u(h + \hat{h}) - \tfrac{1}{2}\pi\} = O(u^{-1}).$$

Accordingly,

$$u = (n + \tfrac{1}{2})\pi l^{-1} + O(n^{-1}), \tag{8.09}$$

where n is an arbitrary positive integer, and

$$l = h + \hat{h} = \int_{\hat{x}_0}^{x_0} |f(t)|^{1/2}\,dt. \tag{8.10}$$

Equation (8.09) is the required approximation for the eigenvalues.

8.3 It is feasible to extend this method to obtain a strict bound for the error term in (8.09) in terms of $\mathcal{V}_{x_1,\infty}(H)$ and $\mathcal{V}_{-\infty,x_1}(\hat{H})$, but because of the cumbersome nature of the analysis we shall not pursue this. The problem is also solvable by the method of §11 below and this yields explicit error bounds more easily; for details see Olver (1965b).

Ex. 8.1 In the *Mathieu equation*

$$d^2w/dx^2 = \{2q\cos(2x) - a\}\,w$$

let $a = 2qk$, where k is a fixed number such that $-1 < k < 1$, and $q > 0$. Show that the values of q for which there exists a solution with period π are given by

$$q = \tfrac{1}{8}\pi^2(2n+1)^2\left\{\int_{\cos^{-1}k}^{\pi}(k-\cos t)^{1/2}\,dt\right\}^{-2}\{1+O(n^{-2})\},$$

where n is an arbitrary positive integer.

*9 Bound States

9.1 The result of §8 can be extended to cases in which $f(x)$ and $g(x)$ depend on the parameter u. Assume that:

(i) *The conditions of* §8.1 *are satisfied for all sufficiently large* u.
(ii) *Each of the following functions is bounded when* $u \to \infty$:

$$\left|\frac{f'(x_1)}{f^{3/2}(x_1)}\right|, \quad \left|\int_{x_1}^{x_0} f^{1/2}(t)\,dt\right|^{-1}, \quad \left|\int_{\hat{x}_0}^{x_1} f^{1/2}(t)\,dt\right|^{-1}, \quad \mathscr{V}_{x_1,\infty}(H), \quad \mathscr{V}_{-\infty,x_1}(\hat{H}),$$

where $x_1 = x_1(u)$ *is an arbitrarily chosen interior point of* $(\hat{x}_0, x_0)$.

Then retracing the analysis we may verify without difficulty that the large eigenvalues satisfy

$$u\int_{\hat{x}_0}^{x_0}|f(t)|^{1/2}\,dt + O(u^{-1}) = (n+\tfrac{1}{2})\pi, \tag{9.01}$$

where n is again a positive integer.

9.2 An important problem in quantum mechanics concerns the eigenvalues of the system

$$d^2w/dx^2 = \{q(x)-u^2\}\,w, \qquad w(-\infty) = w(\infty) = 0, \tag{9.02}$$

in which $q(x)$ tends to $+\infty$ as $x \to \pm\infty$.† We shall solve this problem by showing that the result of §9.1 applies. Our assumptions are as follows:

(i) $q(x)$ *is independent of* u *and thrice continuously differentiable in* $(-\infty, \infty)$.
(ii) *As* $x \to \infty$

$$q(x) \sim kx^{\alpha}, \qquad q'(x) \sim \alpha k x^{\alpha-1}, \qquad q''(x) = O(x^{\alpha-2}), \qquad q'''(x) = O(x^{\alpha-3}), \tag{9.03}$$

where k *and* α *are positive constants.*
(iii) *As* $x \to -\infty$

$$q(x) \sim \hat{k}(-x)^{\hat{\alpha}}, \quad q'(x) \sim -\hat{\alpha}\hat{k}(-x)^{\hat{\alpha}-1}, \quad q''(x) = O(x^{\hat{\alpha}-2}), \quad q'''(x) = O(x^{\hat{\alpha}-3}), \tag{9.04}$$

where $\hat{k}$ *and* $\hat{\alpha}$ *are positive constants.*

In the notation of §8, take

$$f(x) = u^{-2}q(x) - 1, \qquad g(x) = 0.$$

† Kemble (1937, §21).

In consequence of (ii) and (iii), $f(x)$ has exactly two real zeros for all sufficiently large u, asymptotically given by

$$x_0 \sim (u^2/k)^{1/\alpha}, \qquad \hat{x}_0 \sim -(u^2/\hat{k})^{1/\hat{\alpha}}.$$

The essential difficulty is to demonstrate that $\mathscr{V}_{x_1,\infty}(H)$ and $\mathscr{V}_{-\infty,x_1}(\hat{H})$ are bounded when $u \to \infty$. Without loss of generality we may take $x_1 = 0$.

9.3† Consider first $\mathscr{V}_{0,\infty}(H)$. The assumptions of §9.2 imply that

(a) There exists a positive number X_0, such that

$$q(x) < q(X_0) \quad (0 \leqslant x < X_0), \qquad q'(x) > 0 \quad (x \geqslant X_0). \tag{9.05}$$

(b) For each $\varepsilon \in (0, \frac{1}{2})$ there exists $X(\varepsilon) \geqslant X_0$, such that when $x > X(\varepsilon)$

$$(1-\varepsilon)kx^\alpha < q(x) < (1+\varepsilon)kx^\alpha, \qquad (1-\varepsilon)\alpha kx^{\alpha-1} < q'(x) < (1+\varepsilon)\alpha kx^{\alpha-1}. \tag{9.06}$$

(c) There exists a positive number $U(\varepsilon)$ such that if $u > U(\varepsilon)$, then x_0 is the only positive zero of $q(x) - u^2$, and

$$(1-\varepsilon)u^2 < kx_0^\alpha < (1+\varepsilon)u^2. \tag{9.07}$$

In the following δ denotes an arbitrary number in the interval $(0,1)$, and A is a generic positive number which is independent of x, u, ε, and δ. It is assumed throughout that

$$u > U(\varepsilon), \qquad u^2 > (1-\varepsilon)^{-1}(1-\delta)^{-\alpha}k\{X(\varepsilon)\}^\alpha. \tag{9.08}$$

Then from (9.07) and the second of (9.08), we have

$$(1-\delta)x_0 > X(\varepsilon). \tag{9.09}$$

We treat the interval $[0, \infty)$ in four parts.

Part 1. $x_0 \leqslant x \leqslant (1+\delta)x_0$. In the notation of §8.2, we have from (7.07)

$$H(x) = u\int_{x_0}^{x}\left\{\frac{5q'^2}{16(q-u^2)^{5/2}} - \frac{q''}{4(q-u^2)^{3/2}} - \frac{5(q-u^2)^{1/2}}{36\xi^2}\right\}dx, \tag{9.10}$$

where

$$\xi \equiv \xi(x) = \tfrac{2}{3}u\zeta^{3/2} = \int_{x_0}^{x}(q-u^2)^{1/2}\,dx. \tag{9.11}$$

Two integrations by parts yield

$$\xi = \int_{x_0}^{x} q'(q-u^2)^{1/2}\frac{dx}{q'} = \frac{2(q-u^2)^{3/2}}{3q'}\left\{1 + \frac{2(q-u^2)q''}{5q'^2} - S\right\}, \tag{9.12}$$

where

$$S = \frac{2q'}{5(q-u^2)^{3/2}}\int_{x_0}^{x}(q-u^2)^{5/2}q'Q\,dx, \qquad Q = \frac{q'q''' - 3q''^2}{q'^5}.$$

† The analysis in this subsection follows that of Titchmarsh (1954).

From the mean-value theorem, it follows that

$$q(x) - u^2 = (x - x_0)q'(\tau) \qquad (x_0 \leqslant \tau \leqslant x).$$

Applying (9.06) and (9.07)—as is legitimate in consequence of (9.08) and (9.09)—we find that

$$q(x) - u^2 < Au^2\delta. \tag{9.13}$$

Next, from (9.03) and (9.06) we derive

$$q''/q'^2 < Ax^{-\alpha} < Ax_0^{-\alpha} < Au^{-2},$$
$$|2(q-u^2)q''/(5q'^2)| < Au^{-2}(q-u^2), \tag{9.14}$$

and

$$|Q| < Ax^{1-3\alpha} < Ax_0^{1-3\alpha} < Au^{(2/\alpha)-6}. \tag{9.15}$$

Substitution of (9.15) in the expression for S establishes that

$$|S| < Au^{(2/\alpha)-6}q'(q-u^2)^2 < Au^{-4}(q-u^2)^2. \tag{9.16}$$

Comparing (9.14) and (9.16) with (9.12), and then using (9.13), we see that for sufficiently small δ

$$\frac{5(q-u^2)^{1/2}}{36\xi^2} = \frac{5q'^2}{16(q-u^2)^{5/2}}\left\{1 - \frac{4(q-u^2)q''}{5q'^2} + \left(\frac{q-u^2}{u^2}\right)^2 O(1)\right\},$$

uniformly with respect to x in the given interval as $u \to \infty$. Substituting this result in (9.10), and using again (9.06), (9.07), and (9.13), we obtain

$$\mathscr{V}_{x_0,(1+\delta)x_0}(H) = O(u^{-3})\int_{x_0}^{(1+\delta)x_0} \frac{q'^2\,dx}{(q-u^2)^{1/2}}$$
$$= O(u^{-1-(2/\alpha)})\int_{x_0}^{(1+\delta)x_0} \frac{q'\,dx}{(q-u^2)^{1/2}} = O(u^{-2/\alpha}).$$

Part 2. $(1+\delta)x_0 \leqslant x < \infty$. We apply the inequality

$$\mathscr{V}_{(1+\delta)x_0,\infty}(H) \leqslant \frac{5u}{16}\int_{(1+\delta)x_0}^{\infty} \frac{q'^2\,dx}{(q-u^2)^{5/2}} + \frac{u}{4}\int_{(1+\delta)x_0}^{\infty} \frac{|q''|\,dx}{(q-u^2)^{3/2}} + \frac{5u}{36\xi\{(1+\delta)x_0\}}, \tag{9.17}$$

obtained from (9.10) and the differentiated form of (9.11). From (9.06) and (9.07) it follows that

$$q > (1-\varepsilon)k(1+\delta)^\alpha x_0^\alpha > (1-\varepsilon)^2(1+\delta)^\alpha u^2 > u^2,$$

provided that ε is small enough to ensure that

$$(1-\varepsilon)^2(1+\delta)^\alpha > 1. \tag{9.18}$$

Next,

$$\frac{q-u^2}{kx^\alpha} > \frac{(1-\varepsilon)kx^\alpha - u^2}{kx^\alpha}.$$

The right-hand side is an increasing function of x^α, and therefore has the lower bound

$$\frac{(1-\varepsilon)k(1+\delta)^\alpha x_0^\alpha - u^2}{k(1+\delta)^\alpha x_0^\alpha} > \frac{(1-\varepsilon)^2(1+\delta)^\alpha u^2 - u^2}{(1+\varepsilon)(1+\delta)^\alpha u^2}.$$

Application of (9.18) shows that

$$q - u^2 > Ax^\alpha. \tag{9.19}$$

Again, using the results for Part 1, and also (9.06), (9.07), and (9.19), we see that

$$\xi\{(1+\delta)x_0\} > \frac{A[q\{(1+\delta)x_0\} - u^2]^{3/2}}{q'\{(1+\delta)x_0\}} > Ax_0^{(\alpha/2)+1} > Au^{1+(2/\alpha)}. \tag{9.20}$$

Then substituting in (9.17) by means of (9.19) and (9.20) we arrive at

$$\mathscr{V}_{(1+\delta)x_0,\infty}(H) = O(u^{-2/\alpha}).$$

Part 3. $(1-\delta)x_0 \leqslant x \leqslant x_0$. This can be treated in a similar manner to Part 1.

Part 4. $0 \leqslant x \leqslant (1-\delta)x_0$. For this range we require

$$(1+\varepsilon)^2(1-\delta)^\alpha < 1;$$

compare (9.18). Key inequalities are

$$u^2 - q(x) \geqslant u^2 - q\{(1-\delta)x_0\} > Au^2,$$

$$(1-\delta)x_0 < (u^2/k)^{1/\alpha}, \qquad |\xi\{(1-\delta)x_0\}| > Au^{1+(2/\alpha)}.$$

Substituting these results in the bound

$$\frac{5u}{16}\int_0^{(1-\delta)x_0} \frac{q'^2\,dx}{(u^2-q)^{5/2}} + \frac{u}{4}\int_0^{(1-\delta)x_0} \frac{|q''|\,dx}{(u^2-q)^{3/2}} + \frac{5u}{36|\xi\{(1-\delta)x_0\}|}$$

for $\mathscr{V}_{0,(1-\delta)x_0}(H)$, we find that this variation is $O(u^{-2})$, $O(u^{-2}\ln u)$, or $O(u^{-2/\alpha})$, according as $\alpha < 1$, $\alpha = 1$, or $\alpha > 1$.

Combination of Parts 1–4 yields the desired result:

$$\mathscr{V}_{0,\infty}(H) = O(1) \qquad (u \to \infty).$$

Similarly for $\mathscr{V}_{-\infty,0}(\hat{H})$.

9.4 The results of the last two subsections may be summarized as follows. With Conditions (i), (ii), and (iii) of §9.2 the large eigenvalues of the differential system (9.02) satisfy

$$\int_{\hat{x}_0}^{x_0} \{u^2 - q(t)\}^{1/2}\,dt + O(u^{-1}) = (n+\tfrac{1}{2})\pi,$$

where $\hat{x}_0$ and x_0 are the zeros of the integrand, and n is a positive integer. Further progress depends on properties of the actual function $q(x)$ under consideration; compare Exercise 9.1 below.

The reader will already have noticed that a more delicate error analysis would sharpen the order of the error term. It is also evident that the conditions of §9.2 could be relaxed to some extent without invalidating the final result.

Ex. 9.1 Show that the large eigenvalues of the system

$$d^2 w/dx^2 = (x^4+x^2-u^2)\,w, \qquad w(-\infty) = w(\infty) = 0,$$

are given by $u = \pi 2^{1/3}\{\Gamma(\frac{1}{4})\}^{-4/3}(3n+\frac{3}{2})^{2/3} + O(n^{1/3})$, where n is a positive integer.

10 Wave Penetration through a Barrier. I

10.1 Consider the differential equation

$$d^2w/dx^2 = \{-u^2f(x)+g(x)\}\,w, \tag{10.01}$$

in which $f(x)$ and $g(x)$ satisfy the conditions of §8.1. The solutions have exponential character between the turning points $\hat{x}_0$ and x_0, and oscillate as $x\to\pm\infty$. A physical problem represented by this mathematical model is the penetration of a one-dimensional wave through a potential barrier.†

Of special interest are the four solutions having the properties

$$w_{\pm 1}(u,x) \sim f^{-1/4}(x)\exp\left\{\pm iu\int_{x_0}^{x} f^{1/2}(t)\,dt\right\} \qquad (x\to+\infty), \tag{10.02}$$

$$\hat{w}_{\pm 1}(u,x) \sim f^{-1/4}(x)\exp\left\{\pm iu\int_{x}^{\hat{x}_0} f^{1/2}(t)\,dt\right\} \qquad (x\to-\infty). \tag{10.03}$$

Their existence and uniqueness is established in Chapter 6, §3.5. In the wave-penetration problem the *incident*, *reflected*, and *transmitted waves* are represented by $\hat{w}_{-1}(u,x)$, $\hat{w}_1(u,x)$, and $w_1(u,x)$, respectively. There is a connecting equation

$$\hat{w}_{-1}(u,x) = A\hat{w}_1(u,x) + Bw_1(u,x), \tag{10.04}$$

in which A and B are independent of x. Our object is to determine these coefficients, and in the present section we again follow the analysis of Jeffreys (1962, §3.62). Another method is given in §15 below.

10.2 As in §8.2 let x_1 be a fixed interior point of $(\hat{x}_0, x_0)$. Theorem 3.1 of Chapter 11 shows that equation (10.01) has solutions of the form

$$w_2(u,x) = \pi^{1/2}u^{1/6}\phi^{-1/4}(x)\{\mathrm{Bi}(-u^{2/3}\zeta)+\varepsilon_2(u,x)\}, \tag{10.05}$$

and

$$w_3(u,x) = \pi^{1/2}u^{1/6}\phi^{-1/4}(x)\{\mathrm{Ai}(-u^{2/3}\zeta)+\varepsilon_3(u,x)\}, \tag{10.06}$$

when $x\in[x_1,\infty)$, and solutions of the form

$$\hat{w}_2(u,x) = \pi^{1/2}u^{1/6}\hat{\phi}^{-1/4}(x)\{\mathrm{Bi}(-u^{2/3}\hat{\zeta})+\hat{\varepsilon}_2(u,x)\}, \tag{10.07}$$

† See, for example, Morse and Feshbach (1953, p. 1099) or Jeffreys (1956).

and

$$\hat{w}_3(u,x) = \pi^{1/2}u^{1/6}\hat{\phi}^{-1/4}(x)\{\mathrm{Ai}(-u^{2/3}\hat{\zeta})+\hat{\varepsilon}_3(u,x)\}, \tag{10.08}$$

when $x \in (-\infty, x_1]$. Here ζ, $\hat{\zeta}$, ϕ, and $\hat{\phi}$ are defined in §8.2, and the error terms ε_2, $\hat{\varepsilon}_2$, ε_3, and $\hat{\varepsilon}_3$ satisfy inequalities similar to (8.03) and (8.05).

If $x \to +\infty$, then $\zeta \to +\infty$. Substitution for the Airy functions by means of their asymptotic approximations for large negative arguments produces

$$w_2(u,x) = -f^{-1/4}(x)\left[\sin\left\{u\int_{x_0}^{x} f^{1/2}(t)\,dt - \tfrac{1}{4}\pi\right\} + o(1)\right] \qquad (x\to\infty), \tag{10.09}$$

and

$$w_3(u,x) = f^{-1/4}(x)\left[\cos\left\{u\int_{x_0}^{x} f^{1/2}(t)\,dt - \tfrac{1}{4}\pi\right\} + \eta_3(u,x) + o(1)\right] \qquad (x\to\infty). \tag{10.10}$$

Here $\eta_3(u,x) = \pi^{1/2}u^{1/6}\zeta^{1/4}\varepsilon_3(u,x)$, and is $O(u^{-1})$ as $u\to\infty$, uniformly with respect to $x \in [x_0, \infty)$. Comparing (10.09) with (10.02), we immediately see that

$$2w_2(u,x) = e^{\pi i/4}w_1(u,x) + e^{-\pi i/4}w_{-1}(u,x). \tag{10.11}$$

The solution $w_3(u,x)$ can be identified by letting $x\to\infty$ through the sequence of values for which the cosine term in (10.10) is alternately unity and zero. This yields

$$2w_3(u,x) = \{e^{-\pi i/4}+O(u^{-1})\}\,w_1(u,x) + \{e^{\pi i/4}+O(u^{-1})\}\,w_{-1}(u,x). \tag{10.12}$$

For the solutions with hats, the corresponding results are found to be

$$2\hat{w}_2(u,x) = e^{\pi i/4}\hat{w}_1(u,x) + e^{-\pi i/4}\hat{w}_{-1}(u,x), \tag{10.13}$$

and

$$2\hat{w}_3(u,x) = \{e^{-\pi i/4}+O(u^{-1})\}\,\hat{w}_1(u,x) + \{e^{\pi i/4}+O(u^{-1})\}\,\hat{w}_{-1}(u,x). \tag{10.14}$$

10.3 Setting $x = x_1$ and substituting for the Airy functions in equations (10.05) to (10.08) by means of their asymptotic forms for large positive arguments, we find that

$$w_2(u,x_1) = \gamma^{-1}e^{uh}\{1+O(u^{-1})\}, \qquad w_3(u,x_1) = \tfrac{1}{2}\gamma^{-1}e^{-uh}\{1+O(u^{-1})\},$$
$$\hat{w}_2(u,x_1) = \gamma^{-1}e^{u\hat{h}}\{1+O(u^{-1})\}, \qquad \hat{w}_3(u,x_1) = \tfrac{1}{2}\gamma^{-1}e^{-u\hat{h}}\{1+O(u^{-1})\},$$

where γ, h, and $\hat{h}$ are defined in §8.2. The corresponding derivatives are given by

$$w_2'(u,x_1) = -u\gamma e^{uh}\{1+O(u^{-1})\}, \qquad w_3'(u,x_1) = \tfrac{1}{2}u\gamma e^{-uh}\{1+O(u^{-1})\},$$
$$\hat{w}_2'(u,x_1) = u\gamma e^{u\hat{h}}\{1+O(u^{-1})\}, \qquad \hat{w}_3'(u,x_1) = -\tfrac{1}{2}u\gamma e^{-u\hat{h}}\{1+O(u^{-1})\}.$$

The matching process at $x = x_1$ is straightforward, and yields the connection formulas

$$w_2(u,x) = e^{u(h-\hat{h})}O(u^{-1})\,\hat{w}_2(u,x) + 2e^{ul}\{1+O(u^{-1})\}\,\hat{w}_3(u,x), \tag{10.15}$$
$$w_3(u,x) = \tfrac{1}{2}e^{-ul}\{1+O(u^{-1})\}\,\hat{w}_2(u,x) + e^{u(\hat{h}-h)}O(u^{-1})\,\hat{w}_3(u,x), \tag{10.16}$$

in which the O terms are independent of x, and l is given by (8.10). Finally, elimination of w_2, w_3, $\hat{w}_2$, $\hat{w}_3$, and w_{-1} from the numbered equations (10.11)–(10.16) yields the following expressions for the coefficients in (10.04):

$$A = i + O(u^{-1}), \qquad B = e^{-ul}\{1+O(u^{-1})\}.$$

The *reflection coefficient* R and *transmission coefficient* T are defined to be the squares of the limiting ratios of the wave amplitudes. Accordingly, we have proved that for large u

$$R = |A|^2 = 1 + O(u^{-1}), \qquad T = |B|^2 = e^{-2ul}\{1+O(u^{-1})\}, \tag{10.17}$$

where $l = \int_{\hat{x}_0}^{x_0} |f(t)|^{1/2}\,dt$.

Ex. 10.1 Assume the conditions of the present section and also that $g(x)$ is real. By considering the Wronskian of $w_1(u,x)$ and $w_{-1}(u,x)$ establish the *energy-conservation equation* $R+T=1$.

11 Fundamental Connection Formula for a Simple Turning Point in the Complex Plane

11.1 Consider the differential equation

$$d^2w/dz^2 = \{f(z)+g(z)\}\,w, \tag{11.01}$$

in which $f(z)$ and $g(z)$ are holomorphic in a simply connected domain $\mathbf{D}$ (not necessarily one-sheeted), and $f(z)$ has just one zero in $\mathbf{D}$, a simple zero at an interior point z_0.

As in Chapter 6, write

$$\xi(z) = \int_{z_0}^{z} f^{1/2}(t)\,dt, \qquad F(z) = \int \left[\frac{1}{f^{1/4}(z)}\frac{d^2}{dz^2}\left\{\frac{1}{f^{1/4}(z)}\right\} - \frac{g(z)}{f^{1/2}(z)}\right]dz. \tag{11.02}$$

These functions are analytic in $\mathbf{D}$, their only singularity being a branch point at $z = z_0$. Theorem 11.1 of Chapter 6 informs us that if z_0 is excluded from $\mathbf{D}$, then equation (11.01) has a solution which depends on an arbitrarily chosen point a of $\mathbf{D}$, and is given by

$$w(z) = f^{-1/4}(z)e^{-\xi(z)}\{1+\varepsilon(z)\}, \tag{11.03}$$

where

$$|\varepsilon(z)| \leqslant \exp\{\mathscr{V}_{z,a}(F)\} - 1 \qquad (z \in \mathbf{H}(a)). \tag{11.04}$$

Here $\mathbf{H}(a)$ comprises the points z that can be linked to a by a ξ-progressive path lying in $\mathbf{D}$, and the variation in (11.04) is evaluated along such a path. Depending on the choice of a and the branches of $\xi(z)$ and $f^{-1/4}(z)$, we obtain differing solutions of the differential equation.

Of obvious importance in determining the regions of validity $\mathbf{H}(a)$ are the points for which

$$\operatorname{Re}\{\xi(z)\} = 0.$$

These points are independent of the choice of branch of $\xi(z)$; we call their aggregate set the *principal curves*† associated with the turning point z_0. As in §5.1 each principal curve is an R_∞ arc which can terminate only at z_0 or at the boundary of **D**; moreover, no principal curve can intersect itself or any other principal curve, except at z_0.

Denoting the Taylor-series expansion of $f(z)$ in the neighborhood of z_0 by

$$f(z) = \sum_{s=1}^{\infty} f_s(z-z_0)^s,$$

we obtain from (11.02)

$$\xi(z) = \tfrac{2}{3}f_1^{1/2}(z-z_0)^{3/2}\{1+O(z-z_0)\}. \tag{11.05}$$

Hence three principal curves emerge from z_0, initially inclined at angle $2\pi/3$ to each other. We denote them by $\mathscr{C}_1$, $\mathscr{C}_2$, and $\mathscr{C}_3$, enumerated in the positive rotational sense. These and other principal curves divide **D** into a number of subdomains. Those with z_0 on their boundary we call the *principal subdomains*, and denote them by $\mathbf{D}_1$, $\mathbf{D}_2$, and $\mathbf{D}_3$, as indicated in Fig. 11.1. Clearly each $\mathbf{D}_j$ is simply connected. For terminological convenience j is enumerated with modulo 3; thus $\mathscr{C}_0 = \mathscr{C}_3$, $\mathbf{D}_4 = \mathbf{D}_1$.

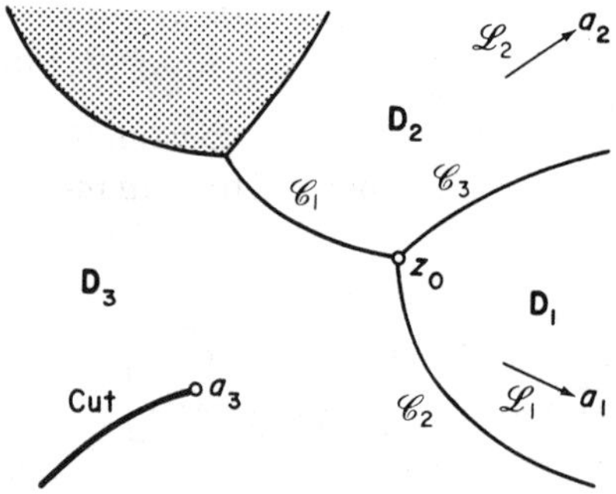

Fig. 11.1 Principal curves and subdomains.

11.2 We now specify precisely the solutions of equation (11.01) for which a connection formula is sought. Let $\xi_j(z)$, $j = 1, 2, 3$, be the branch of $\xi(z)$ which is continuous in **D** cut along the corresponding $\mathscr{C}_j$, and has positive real part in $\mathbf{D}_j$ and negative real part in the other two principal subdomains. Analytic continuation across the cut $\mathscr{C}_j$ is expressed by

$$\xi_j\{(z-z_0)e^{\pm 2\pi i}\} = -\xi_j(z-z_0);$$

compare (11.05).

Define $\phi_j(z)$, $j = 1, 2, 3$, to be a constant multiple of the function $f^{-1/4}(z)$ which is continuous in **D** cut along the corresponding $\mathscr{C}_j$, except at $z = z_0$. In order to achieve a symmetric form of final result, we stipulate that the three multiples are related by

$$\phi_j(z) = e^{\pi i/6}\phi_{j-1}(z) \qquad (z \in \mathbf{D}_{j-1} \cup \mathbf{D}_j). \tag{11.06}$$

† Or *anti-Stokes lines*; see p. 518.

Since

$$\phi_j\{(z-z_0)e^{2\pi i}\} = -i\phi_j(z-z_0)$$

equation (11.06) is self-consistent for $j = 1, 2, 3$, and there is one degree of freedom in the choice of the set $\phi_j(z)$.

Let a_j be a reference point which is either at infinity in $\mathbf{D}_j$ or is on the boundary of $\mathbf{D}_j$. In all cases we require $\operatorname{Re}\{\xi_j(z)\} \to +\infty$ as $z \to a_j$ along a progressive path $\mathscr{L}_j$ lying within $\mathbf{D}_j$. Applying the result of §11.1, we see that (11.01) has solutions $w_j(z)$, $j = 1, 2, 3$, given by

$$w_j(z) = \phi_j(z)e^{-\xi_j(z)}\{1+\varepsilon_j(z)\}, \tag{11.07}$$

where

$$|\varepsilon_j(z)| \leqslant \exp\{\mathscr{V}_{z,a_j}(F)\} - 1 \qquad (z \in \mathbf{H}(a_j)). \tag{11.08}$$

Here $F(z)$ is defined by (11.02): any branches of $f^{1/4}(z)$ and $f^{1/2}(z)$ may be used provided that they are continuous in $\mathbf{D}_j$ and the latter is the square of the former. We observe that the condition that z can be linked to a_j by a ξ-progressive path excludes $\mathscr{C}_j$ from $\mathbf{H}(a_j)$, because passage of the path through z_0 is not permitted.

11.3 Although each solution $w_j(z)$ satisfies (11.07) and (11.08) in only part of $\mathbf{D}$, the theory of Chapter 5 shows that $w_j(z)$ exists and is holomorphic throughout $\mathbf{D}$ (including z_0). Moreover, there is an identity of the form

$$\Lambda_1 w_1(z) + \Lambda_2 w_2(z) + \Lambda_3 w_3(z) = 0, \tag{11.09}$$

in which the coefficients Λ_j are independent of z. This is the desired connection formula. To approximate the Λ_j we assume that $\mathbf{D}$ contains ξ-progressive paths linking each pair of reference points a_j and a_k, and also that $\mathscr{V}(F)$ is finite along these paths.

Consider (11.09) in $\mathbf{D}_1$. Here $\xi_2(z) = \xi_3(z) = -\xi_1(z)$. Substituting by means of (11.06) and (11.07), we derive

$$\Lambda_1 e^{-\xi_1(z)}\{1+\varepsilon_1(z)\} + \Lambda_2 e^{(\pi i/6)+\xi_1(z)}\{1+\varepsilon_2(z)\} + \Lambda_3 e^{-(\pi i/6)+\xi_1(z)}\{1+\varepsilon_3(z)\} = 0.$$

Now let $z \to a_1$ along $\mathscr{L}_1$. Then $\varepsilon_1(z) \to 0$. Also, by hypothesis, $\operatorname{Re}\{\xi_1(z)\} \to +\infty$. Hence by Theorem 12.1 of Chapter 6, $\varepsilon_2(z)$ and $\varepsilon_3(z)$ tend to constant values $\varepsilon_2(a_1)$ and $\varepsilon_3(a_1)$, respectively. Thus

$$e^{\pi i/6}\{1+\varepsilon_2(a_1)\}\Lambda_2 + e^{-\pi i/6}\{1+\varepsilon_3(a_1)\}\Lambda_3 = 0. \tag{11.10}$$

Similarly,

$$e^{-\pi i/6}\{1+\varepsilon_1(a_2)\}\Lambda_1 + e^{\pi i/6}\{1+\varepsilon_3(a_2)\}\Lambda_3 = 0, \tag{11.11}$$

and

$$e^{\pi i/6}\{1+\varepsilon_1(a_3)\}\Lambda_1 + e^{-\pi i/6}\{1+\varepsilon_2(a_3)\}\Lambda_2 = 0. \tag{11.12}$$

The consistency of these three homogeneous equations is verifiable by means of the identity $\varepsilon_j(a_k) = \varepsilon_k(a_j)$ given in the cited theorem. Solving for the Λ_j and substituting in (11.09), we arrive at

$$\{1+\varepsilon_2(a_3)\}\, w_1(z) + e^{-2\pi i/3}\{1+\varepsilon_3(a_1)\}\, w_2(z) + e^{2\pi i/3}\{1+\varepsilon_1(a_2)\}\, w_3(z) = 0. \tag{11.13}$$

Again, Theorem 12.1 of Chapter 6 shows that the error terms are bounded by

$$|\varepsilon_j(a_k)| \leqslant \tfrac{1}{2}[\exp\{\mathcal{V}_{a_j,a_k}(F)\}-1], \tag{11.14}$$

that is,

$$|\varepsilon_j(a_k)| \leqslant \frac{1}{2}\exp\left\{\int_{a_j}^{a_k}\left|\frac{1}{f^{1/4}}\frac{d^2}{dz^2}\left(\frac{1}{f^{1/4}}\right)-\frac{g}{f^{1/2}}\right||dz|\right\}-\frac{1}{2}. \tag{11.15}$$

The variation (or integral) is evaluated along a ξ-progressive path.

11.4 We now summarize this section. Assumed conditions on $f(z)$ and $g(z)$ are stated in the opening paragraph of §11.1. Later in the same subsection the principal subdomains $\mathbf{D}_j$ associated with the turning point z_0 are defined. Three solutions $w_j(z)$ of the differential equation (11.01) are then constructed, given by (11.07) and (11.08) with $j = 1, 2, 3$. Here $\xi_j(z)$ is the branch of the first integral in (11.02) which has positive real part in $\mathbf{D}_j$ and negative real part in the other two principal subdomains; $\phi_j(z)$ is an arbitrary constant multiple of the function $f^{-1/4}(z)$ which is continuous within $\mathbf{D}$ cut along the corresponding $\mathscr{C}_j$ and satisfies (11.06); a_j is any reference point of $\mathbf{D}_j$ such that $\mathrm{Re}\{\xi_j(z)\}\to+\infty$ as $z\to a_j$ along a ξ-progressive path lying within $\mathbf{D}_j$. The error bound (11.08) is valid at all points z which can be linked to a_j by a ξ-progressive path lying in $\mathbf{D}$, $F(z)$ being defined by the second of (11.02). Lastly, *provided that the a_j can be linked in pairs by paths of this kind* the fundamental connection formula between the $w_j(z)$ is expressed by (11.13) and (11.15).

When the given differential equation is of the form

$$d^2w/dz^2 = \{u^2 f(z)+g(z)\}\, w,$$

in which u is a large parameter, we simply replace $f(z)$ throughout the analysis by $u^2 f(z)$. This causes ξ to be replaced by $u\xi$; in consequence, the monotonicity condition on the paths of variation requires $\mathrm{Re}(e^{i\omega}\xi)$ to be nondecreasing, where $\omega = \mathrm{ph}\, u$. The other significant change is that the error-control function $F(z)$ is replaced by $F(z)/u$; accordingly, all error terms $\varepsilon_j(a_k)$ in (11.13) are $O(u^{-1})$ as $|u|\to\infty$.

Ex. 11.1 (*Higher approximations to the connection formula*) In (11.01) let $f(z)$ be replaced by $u^2 f(z)$, where u is a large complex parameter, and $f(z)$ and $g(z)$ are independent of u. Define $h_j(z)$ to be a continuous branch of $f^{-1/4}(z)$ such that $\mathrm{Re}\{u\int_{z_0}^{z} h_j^{-2}(t)\, dt\}\to+\infty$ as $z\to a_j$. Also, define $A_{0,j}(z), A_{1,j}(z), \ldots$, recursively by $A_{0,j}(z) = 1$, and

$$A_{s+1,j}(z) = \tfrac{1}{2}h_j^2(z)A'_{s,j}(z) - \tfrac{1}{2}\int\{h_j(z)g(z)-h_j''(z)\}\, h_j(z)A_{s,j}(z)\, dz,$$

the constants of integration being chosen to make the right-hand side vanish as $z\to a_j$. Assume that each pair of points a_j, a_k can be linked by a $(u\xi)$-progressive path along which $A_{s,j}(z)$ is of bounded variation. Show that $A_{s,j}(a_k) = A_{s,k}(a_j)$, and also that as $|u|\to\infty$

$$\Lambda_1 \sim \sum_{s=0}^{\infty}\frac{A_{s,2}(a_3)}{u^s}, \qquad \Lambda_2 \sim e^{-2\pi i/3}\sum_{s=0}^{\infty}\frac{A_{s,3}(a_1)}{u^s}, \qquad \Lambda_3 \sim e^{2\pi i/3}\sum_{s=0}^{\infty}\frac{A_{s,1}(a_2)}{u^s}.$$

12 Example: Airy's Equation

12.1 The simplest differential equation to which the theory of the preceding section applies is

$$d^2w/dz^2 = zw.$$

In the notation of §11.1, set $f(z) = z$ and $g(z) = 0$. Then $z_0 = 0$ and $\xi(z) = \frac{2}{3}z^{3/2}$. $\mathbf{D}$ comprises the whole z plane, and the principal subdomains $\mathbf{D}_j$ are the sectors of angle $2\pi/3$ indicated in Fig. 12.1.

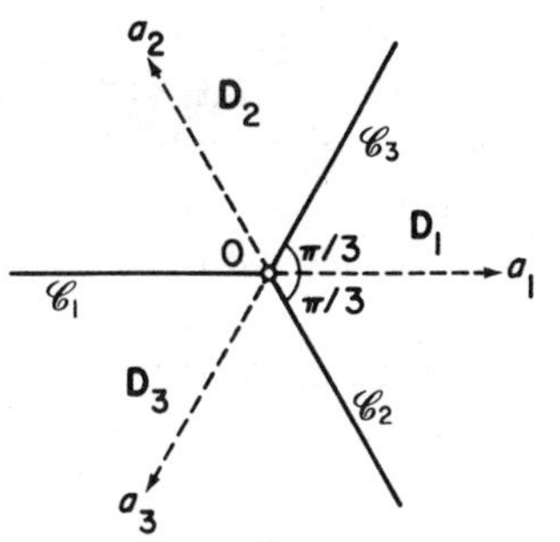

Fig. 12.1 Principal subdomains for Airy's equation.

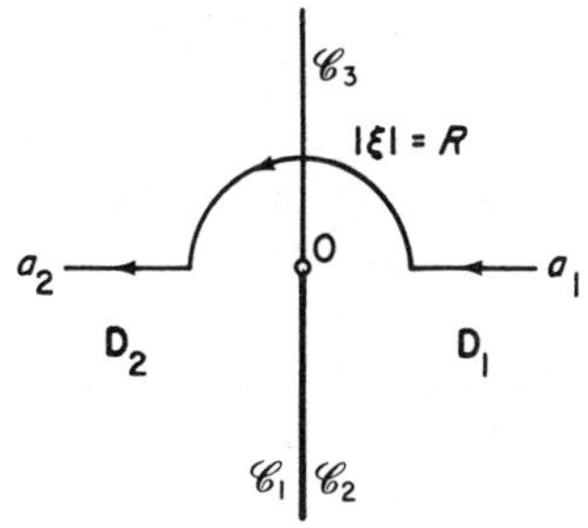

Fig. 12.2 ξ plane.

We define $\phi_1(z)$ to be the principal branch of $z^{-1/4}$. From (11.06)

$$\phi_2(z) = e^{\pi i/6}z^{-1/4} \quad (z \in \mathbf{D}_1 \cup \mathbf{D}_2), \qquad \phi_3(z) = e^{-\pi i/6}z^{-1/4} \quad (z \in \mathbf{D}_1 \cup \mathbf{D}_3).$$

We take $a_1 = +\infty$, $a_2 = \infty e^{2\pi i/3}$, and $a_3 = \infty e^{-2\pi i/3}$. By considering the ξ map it is seen that each region $\mathbf{H}(a_j)$ is obtained from the z plane by deletion of the corresponding principal curve $\mathscr{C}_j$. From (11.02)

$$F(z) = -\frac{5}{24z^{3/2}} = -\frac{5}{36\xi}.$$

The ξ map of a suitable progressive path linking a_1 with a_2 is shown in Fig. 12.2. When the radius R of the semicircular arc tends to infinity, the variation of F vanishes. Therefore, from (11.14), $\varepsilon_1(a_2) = 0$. Similarly both $\varepsilon_2(a_3)$ and $\varepsilon_3(a_1)$ are zero. Accordingly, (11.13) reduces to

$$w_1(z) + e^{-2\pi i/3}w_2(z) + e^{2\pi i/3}w_3(z) = 0. \tag{12.01}$$

By comparing (11.07) with recessive properties of the Airy functions given in Chapter 11, §§1.1 and 8.1, we identify

$$w_1(z) = 2\pi^{1/2}\,\mathrm{Ai}(z), \qquad w_2(z) = 2\pi^{1/2}\,\mathrm{Ai}(ze^{-2\pi i/3}), \qquad w_3(z) = 2\pi^{1/2}\,\mathrm{Ai}(ze^{2\pi i/3});$$

accordingly, equation (12.01) agrees with (8.06) of Chapter 2. The point of the example is that the present theory yields the connection formula without recourse to integral representations.

Ex. 12.1 Let u be a positive parameter and $w_j(u,z)$, $j = 0, \pm 1$, be the solutions of the equation

$$d^2w/dz^2 = (uz+z^2+1)(u+z)^{-1}w$$

which are holomorphic in the z plane cut along the real axis from $z = -\infty$ to $z = -u$, and have the property $w_j(u, ze^{2\pi ij/3}) \sim z^{-1/4}\exp(-\frac{2}{3}z^{3/2})$ as $z \to +\infty$, u being fixed. Show that for large u

$$e^{\pi i/3}w_1(u,z) + e^{-\pi i/3}w_{-1}(u,z) = \{1+O(u^{-1/2})\}\,w_0(u,z).$$

13 Choice of Progressive Paths

13.1 An outstanding problem in the application of the connection formula (11.13) is the construction of ξ-progressive paths linking the reference points a_j. This is an unavoidable feature of the method, and generally necessitates a study of the conformal transformation between the planes of ξ and z. In applications it may not be difficult to demonstrate the *existence* of suitable ξ-progressive paths; this is often sufficient to establish the asymptotic nature of the connection formula in the case when the differential equation contains a variable parameter. The ξ-progressive paths are not unique, however, and if minimum bounds are desired for the error terms $\varepsilon_j(a_k)$, then we need paths along which the total variation of the error-control function F is least. As in earlier chapters, this implies that the paths are kept as far away as possible from the transition points of the differential equation, including z_0, subject to fulfillment of the monotonicity condition.

In the example of §12 each error term $\varepsilon_j(a_k)$ vanished. This is essentially a consequence of a_j and a_k being at infinity without being separated by a boundary of **D**. We now formulate sufficient conditions for the vanishing of $\varepsilon_j(a_k)$ in the general case. We continue to use the notation of §11, and, without loss of generality, suppose that the turning point is at the origin.

13.2 Theorem 13.1 *For all sufficiently large* $|z|$ *in the sector* $\mathbf{S}$: $\gamma_1 \leqslant \operatorname{ph} z \leqslant \gamma_2$, *let* $f(z)$ *and* $g(z)$ *be holomorphic and*

$$f(z) \sim kz^{2\alpha-2}, \qquad g(z) = O(z^{\alpha-\beta-2}), \tag{13.01}$$

where k *is a real or complex nonzero constant, and* α *and* β *are positive real constants. Also, let* $\mathscr{L}_1$ *and* $\mathscr{L}_2$ *be rays from the origin which for large* $|z|$ *lie within* $\mathbf{S}$ *and the principal subdomains* $\mathbf{D}_1$ *and* $\mathbf{D}_2$, *respectively, and are not parallel to the limiting directions of the principal curves. Then* $\mathscr{L}_1$ *and* $\mathscr{L}_2$ *are* ξ*-progressive paths for all sufficiently large* $|z|$. *Moreover, if* a_1 *and* a_2 *are the points at infinity on* $\mathscr{L}_1$ *and* $\mathscr{L}_2$, *respectively, then* $\varepsilon_1(a_2) = 0$.

To prove this result, we first observe that

$$\phi(z) \equiv f^{-1/4}(z) \sim k^{-1/4}z^{(1-\alpha)/2} \qquad (z \to \infty \text{ in } \mathbf{S}).$$

Hence from Ritt's theorem (Chapter 1, §4.3),

$$\phi(z)\phi''(z) = O(z^{-1-\alpha})$$

as $z \to \infty$ in any sector $\mathbf{S}'$: $\gamma_1' \leqslant \operatorname{ph} z \leqslant \gamma_2'$ properly interior to $\mathbf{S}$. Substitution in

(11.02) leads to

$$F'(z) = O(z^{-1-\delta}), \tag{13.02}$$

where $\delta \equiv \min(\alpha, \beta) > 0$. Hence the integral in (11.15) converges. The essential problem is to construct a suitable ξ-progressive path linking a_1 and a_2.

Again, from (11.02)

$$\xi = k^{1/2}\alpha^{-1}z^{\alpha} + o(z^{\alpha}) \qquad (z \to \infty \text{ in } \mathbf{S}). \tag{13.03}$$

Therefore in **S** the limiting directions in which the principal curves approach infinity are given by

$$\operatorname{ph} z = \{(s+\tfrac{1}{2})\pi - \tfrac{1}{2}\kappa\}/\alpha, \tag{13.04}$$

where $\kappa = \operatorname{ph} k$, and $s = 0, \pm 1, \pm 2, \ldots$. We call these the *principal directions*. Consider the behavior of $\operatorname{Re}\xi$ along a ray with parametric equation $z = \tau e^{i\gamma}$, γ being real and fixed, and $0 < \tau < \infty$. We have

$$d\xi/d\tau = f^{1/2}(z)(dz/d\tau) = k^{1/2}\tau^{\alpha-1}e^{i\alpha\gamma}\{1+o(1)\}$$

as $\tau \to \infty$. Hence

$$d(\operatorname{Re}\xi)/d\tau = |k|^{1/2}\tau^{\alpha-1}\{\cos(\alpha\gamma+\tfrac{1}{2}\kappa) + o(1)\}. \tag{13.05}$$

The constant $\cos(\alpha\gamma+\frac{1}{2}\kappa)$ vanishes only when γ has one of the values given by (13.04). *Therefore any ray in* **S** *not in a principal direction is a ξ-progressive path for all sufficiently large* $|z|$. In particular, this includes $\mathscr{L}_1$ and $\mathscr{L}_2$.

13.3 Next, consider the path with parametric equation

$$z = \{(b+i\tau)e^{i\gamma}\}^{1/\alpha} \qquad (-\infty < \tau < \infty), \tag{13.06}$$

where γ and b are real constants, b being large and positive. As indicated in Fig. 13.1 this path begins at infinity, sweeps to its nearest approach to the origin at $z = b^{1/\alpha}e^{i\gamma/\alpha}$ and returns to infinity, the angle between its limiting directions being π/α. When b is large, we have from (13.03)

$$\xi \sim |k|^{1/2}\alpha^{-1}(b+i\tau)e^{i\gamma+i(\kappa/2)},$$

showing that the ξ map of (13.06) is approximated by the straight line indicated in Fig. 13.2. Corresponding to (13.05), we find that

$$d(\operatorname{Re}\xi)/d\tau = -|k|^{1/2}\alpha^{-1}\{\sin(\gamma+\tfrac{1}{2}\kappa)+o(1)\}$$

uniformly with respect to τ, as $b \to \infty$. *Therefore the curve* (13.06) *is ξ-progressive for all sufficiently large b, unless its limiting directions coincide with principal directions.*

Now consider a_1 and a_2. By taking γ_1' and γ_2' sufficiently close to γ_1 and γ_2, respectively, we can ensure that **S**$'$ contains $\mathscr{L}_1$ and $\mathscr{L}_2$. From (13.04) and the hypotheses on $\mathscr{L}_1$ and $\mathscr{L}_2$, it follows that

$$|\operatorname{ph} a_2 - \operatorname{ph} a_1| < 2\pi/\alpha.$$

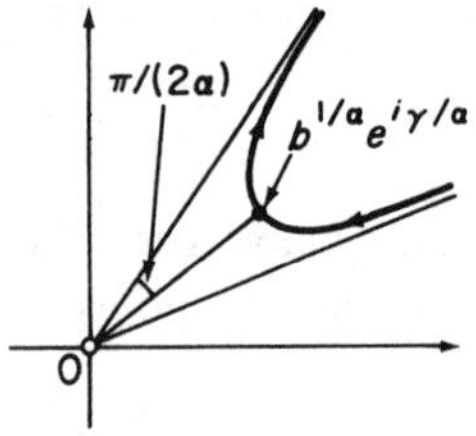

Fig. 13.1 z plane.

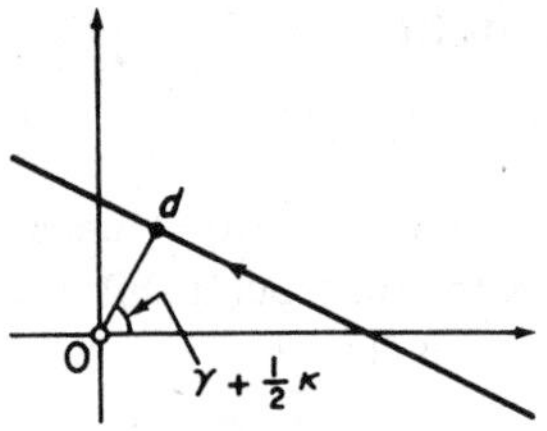

Fig. 13.2 Approximate path in ξ plane. $|d| = |k|^{1/2}b/\alpha$.

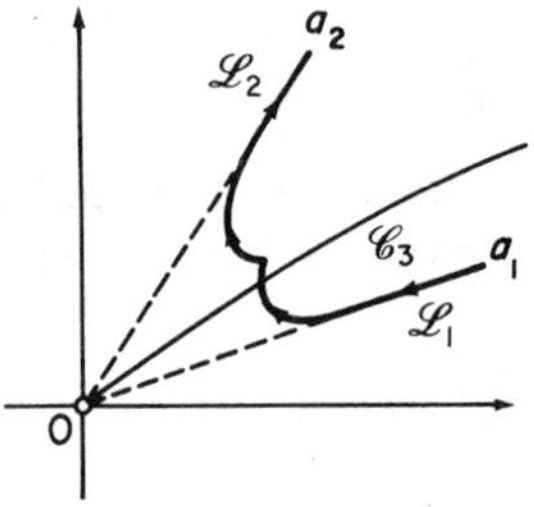

Fig. 13.3 ξ-progressive path in z plane.

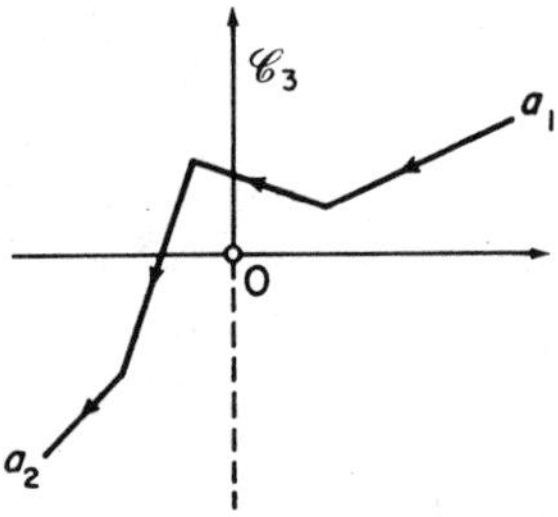

Fig. 13.4 Approximate path in ξ plane.

Figure 13.3 indicates how a_1 and a_2 can be linked by a ξ-progressive path in S′, consisting of $\mathscr{L}_1$, $\mathscr{L}_2$, and either one or two arcs of the type (13.06), depending whether or not $|\mathrm{ph}\, a_2 - \mathrm{ph}\, a_1| < \pi/\alpha$. The corresponding path in the ξ plane is indicated approximately by Fig. 13.4. On letting the b parameters of the arcs tend to infinity, it is seen from (13.02) that the variations of $F(z)$ along $\mathscr{L}_1$ and $\mathscr{L}_2$ tend to zero. For the contribution to the variation from the arcs (13.06), we observe that the corresponding values of $\int |z^{-1-\delta}\, dz|$ are bounded by

$$\frac{1}{\alpha}\int_{-\infty}^{\infty} \frac{d\tau}{|b+i\tau|^{1+(\delta/\alpha)}},$$

and therefore vanish as $b \to \infty$. The proof of Theorem 13.1 is complete.

Ex. 13.1 Assume (i) m and k are complex nonzero constants and δ is a positive constant; (ii) $f(z)$ and $g(z)$ are holomorphic in the strip **T**: $\gamma_1 \leqslant \mathrm{Im}(mz) \leqslant \gamma_2$ for all sufficiently large positive values of $\mathrm{Re}(mz)$; (iii) $f(z) \sim ke^{mz}$ and $g(z) = O(e^{mz(1-\delta)/2})$, uniformly in **T** as $\mathrm{Re}(mz) \to +\infty$; (iv) $\mathscr{L}_1$ and $\mathscr{L}_2$ are straight lines within **T** which are parallel to the boundaries of **T** and lie in the principal subdomains $\mathbf{D}_1$ and $\mathbf{D}_2$, respectively. Show that $\mathscr{L}_1$ and $\mathscr{L}_2$ are ξ-progressive for all sufficiently large $\mathrm{Re}(mz)$, unless they coincide with asymptotes of the principal curves. Show also that if a_1 and a_2 are the points at infinity on $\mathscr{L}_1$ and $\mathscr{L}_2$, respectively, then $\varepsilon_1(a_2) = 0$.

14 The Gans–Jeffreys Formulas: Complex-Variable Method

14.1 In this section the problem treated in §7 is solved by application of the fundamental formula of §11. In the equation

$$d^2w/dz^2 = \{f(z)+g(z)\}\, w \tag{14.01}$$

it is assumed that $f(z)$ and $g(z)$ are holomorphic in a simply connected domain $\mathbf{D}$ which includes the whole of the real axis. The only zero of $f(z)$ in $\mathbf{D}$ is assumed to be $z = 0$, and $f(z)$ and $g(z)$ are required to be real when z is real. Then, without loss of generality, $f(z)/z$ is taken to be positive on the real axis.

One of the principal curves emerging from the origin is the negative real axis. Call this $\mathscr{C}_1$. The other two curves emerge in directions $\pm\pi/3$ and are symmetric with respect to the real axis, in consequence of Schwarz's principle of symmetry. As in §11.1, we denote by $\mathbf{D}_1$, $\mathbf{D}_2$, and $\mathbf{D}_3$ the three subdomains bounded by the principal curves having the origin on their boundary; see Fig. 14.1.

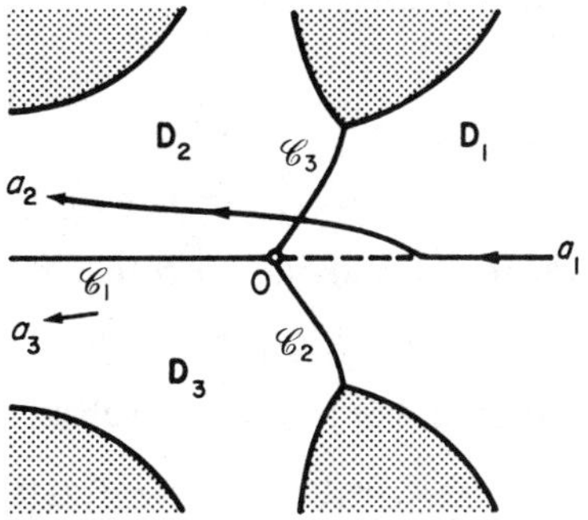

Fig. 14.1 Typical principal subdomains $\mathbf{D}_j$ for the Gans–Jeffreys formulas.

We take the reference point a_1 to be at infinity on the positive real axis, and a_2 and a_3 to be points at infinity on conjugate ξ-progressive curves $\mathscr{L}_2$ and $\mathscr{L}_3$. These curves must lie within $\mathbf{D}_2$ and $\mathbf{D}_3$, respectively, and be such that no boundary of $\mathbf{D}$ intervenes in the neighborhood of infinity. We again require that $|\mathrm{Re}\{\xi(z)\}| \to \infty$ when $z \to a_j$, $j = 1, 2, 3$, and also that the variation of the function $F(z)$, defined by (11.02), converges when $z \to a_j$. Other assumptions are:

(i) *ξ-progressive paths can be found in $\mathbf{D}$ which link a_2 or a_3 with any point of the real axis, other than the origin, and do not pass through the origin.*

(ii) $\varepsilon_2(a_3) = 0$.

(iii) *$\varepsilon_2(z)$ and $\varepsilon_3(z)$ both vanish as $z \to -\infty$ along the negative real axis.*

In applications, Assumption (i) can be tested by considering the ξ map. Conditions covering (ii) and (iii) are furnished by Theorem 13.1, and also by Exercise 13.1.

14.2 In consequence of Assumption (ii) and the fact that $\varepsilon_1(a_2) = \varepsilon_2(a_1)$ (§11.3), the connection formula (11.13) for solutions of (14.01) has the form

$$w_1(z) = e^{\pi i/3}\{1+\varepsilon_3(a_1)\}\, w_2(z) + e^{-\pi i/3}\{1+\varepsilon_2(a_1)\}\, w_3(z). \tag{14.02}$$

To fix the solutions completely, the function $\phi_1(z)$ of §11.2 is defined to be the branch of $f^{-1/4}(z)$ that is positive when $z > 0$. We now interpret the branches of the $w_j(z)$ in real form.

When z is real and positive, relations (11.07) and (11.08) yield

$$w_1(z) = \{1+\varepsilon_1(z)\}\, f^{-1/4} \exp\left(-\int_0^z f^{1/2}\, dt\right) \qquad (z > 0), \tag{14.03}$$

where

$$|\varepsilon_1(z)| \leqslant \exp\{\mathscr{V}_{z,\infty}(F)\} - 1,$$

the variation being taken along the real axis.

When z is real and negative, we have

$$\xi_2(z) = i\int_z^0 |f|^{1/2}\,dt, \qquad \phi_2(z) = e^{-\pi i/12}|f|^{-1/4}; \tag{14.04}$$

compare §11.2. Substitution of (11.07) in (14.02) accordingly gives

$$w_1(z) = \{1+\varepsilon_3(a_1)\}\,e^{\pi i/4}|f|^{-1/4}\exp\left(-i\int_z^0 |f|^{1/2}\,dt\right)\{1+\varepsilon_2(z)\} + \text{conjugate}. \tag{14.05}$$

Let

$$1+\varepsilon_2(z) = \{1+\mu(z)\}\,e^{-i\eta(z)}, \qquad 1+\varepsilon_3(z) = \{1+\mu(z)\}\,e^{i\eta(z)}, \tag{14.06}$$

where $\mu(z)$ and $\eta(z)$ are real, and $\mu(z) \geqslant -1$. Aided by Jordan's inequality, we derive

$$|\mu(z)| \leqslant |\varepsilon_2(z)|, \qquad |\eta(z)| \leqslant \sin^{-1}|\varepsilon_2(z)| \leqslant \pi|\varepsilon_2(z)|/2, \tag{14.07}$$

provided that $|\varepsilon_2(z)| \leqslant 1$. Substitution in (14.05) produces

$$w_1(z) = 2\{1+\mu(a_1)\}\{1+\mu(z)\}|f|^{-1/4} \times \sin\left\{\int_z^0 |f|^{1/2}\,dt + \tfrac{1}{4}\pi + \eta(z) - \eta(a_1)\right\} \qquad (z<0). \tag{14.08}$$

From (14.07), (11.08), and (11.14) it follows that

$$|\mu(a_1)|,\ |2\eta(a_1)/\pi| \leqslant \tfrac{1}{2}\exp\{\mathscr{V}_{a_1,a_2}(F)\} - \tfrac{1}{2}, \tag{14.09}$$

and

$$|\mu(z)|,\ |2\eta(z)/\pi| \leqslant \exp\{\mathscr{V}_{z,a_2}(F)\} - 1, \tag{14.10}$$

the variations being evaluated along ξ-progressive paths.†

The combination of (14.03) and (14.08) is the first Gans–Jeffreys connection formula; compare (7.12). The bound (14.09) for the constant error terms depends on the ξ-progressive path linking a_1 with a_2, and may be minimized by proper choice of path. Boundary conditions satisfied by $w_1(z)$ are evidently

$$\varepsilon_1(z) = o(1) \quad (z\to+\infty); \qquad \mu(z),\ \eta(z) = o(1) \quad (z\to-\infty).$$

14.3 The second formula concerns a solution which is dominant as $z\to+\infty$. Consider first $w_2(z)$. For negative z, we have from (11.07) and (14.04)

$$w_2(z) = e^{-\pi i/12}|f|^{-1/4}\exp\left(-i\int_z^0 |f|^{1/2}\,dt\right)\{1+\varepsilon_2(z)\} \qquad (z<0).$$

On the other side of the turning point

$$w_2(z) = e^{\pi i/6}f^{-1/4}\exp\left(\int_0^z f^{1/2}\,dt\right)\{1+\varepsilon_2(z)\} \qquad (z>0).$$

† In applications, the variational path in (14.10) can usually be taken along the negative real axis.

Instead of $w_2(z)$, however, we prefer the real solution $w_4(z) = \mathrm{Re}\{e^{-\pi i/6} w_2(z)\}$. Aided by (14.06) we derive

$$w_4(z) = [1+\mathrm{Re}\{\varepsilon_2(z)\}] f^{-1/4} \exp\left(\int_0^z f^{1/2}\, dt\right) \qquad (z>0), \tag{14.11}$$

and

$$w_4(z) = \{1+\mu(z)\} |f|^{-1/4} \cos\left\{\int_z^0 |f|^{1/2}\, dt + \tfrac{1}{4}\pi + \eta(z)\right\} \qquad (z<0), \tag{14.12}$$

where $|\varepsilon_2(z)|$, $|\mu(z)|$, and $|2\eta(z)/\pi|$ are all bounded by the right-hand side of (14.10). These relations constitute the second Gans–Jeffreys formula; compare (7.15). Boundary conditions satisfied by $w_4(z)$ are

$$\varepsilon_2(z) = \varepsilon_2(a_1) + o(1) \qquad (z \to +\infty),$$

where $|\varepsilon_2(a_1)|$ is bounded by the right-hand side of (14.09), and

$$\mu(z),\ \eta(z) = o(1) \qquad (z \to -\infty).$$

14.4 When the conditions of §7 and the present section are both satisfied, the two sets of connection formulas agree, except for the bounds on the error terms. In the present section the bounds are expressed in terms of the variation of a certain function along an infinite path of specified type in the complex plane. In §7 the bounds depend on the variation of a more complicated function along the real axis.

Ex. 14.1 Solve Exercise 7.1 by application of §§14.1 and 14.2.

15 Wave Penetration through a Barrier. II

15.1 The fundamental connection formula of §11 can also be used to solve the problems of §§8 to 10. For illustration we consider the wave-penetration problem.

Our assumptions are as follows. In the equation

$$d^2w/dz^2 = \{-u^2 f(z) + g(z)\}\, w \tag{15.01}$$

u is a positive parameter, and $f(z)$ and $g(z)$ are holomorphic in a simply connected complex domain **D** which includes the whole of the real axis. The only zeros of $f(z)$ in **D** are simple zeros at distinct real points $\hat{z}_0$ and z_0, and $f(z)$ is negative when $\hat{z}_0 < z < z_0$ (and therefore positive when $z < \hat{z}_0$ or $z > z_0$). The appropriate error-control function is

$$F(z) = \frac{1}{u}\int \left[\frac{1}{f^{1/4}(z)} \frac{d^2}{dz^2}\left\{\frac{1}{f^{1/4}(z)}\right\} - \frac{g(z)}{f^{1/2}(z)}\right] dz.$$

We assume that when $z \to \pm\infty$, $\mathscr{V}(F)$ converges and $\int f^{1/2}(z)\, dz$ diverges.

Consider the turning point $\hat{z}_0$. To apply the theory of §11 the other turning point must be excluded from the domain of consideration. This is effected by introducing a cut $\mathscr{A}$ along the real axis from $z = z_0$ to $z = +\infty$. We shall refer to the upper and lower sides of this cut as $\mathscr{A}_1$ and $\mathscr{A}_2$, respectively. In accordance with (11.02), we define

$$\xi(z) = u\int_{\hat{z}_0}^{z} \{-f(t)\}^{1/2}\,dt.$$

The associated principal curves $\mathscr{C}_j$ and subdomains $\mathbf{D}_j$ are indicated in Fig. 15.1. $\mathscr{C}_1$ is the real interval $(-\infty, \hat{z}_0]$. $\mathscr{C}_2$ and $\mathscr{C}_3$ are conjugate curves which emerge from $\hat{z}_0$ in directions $\mp\pi/3$, and do not intersect the real axis elsewhere.

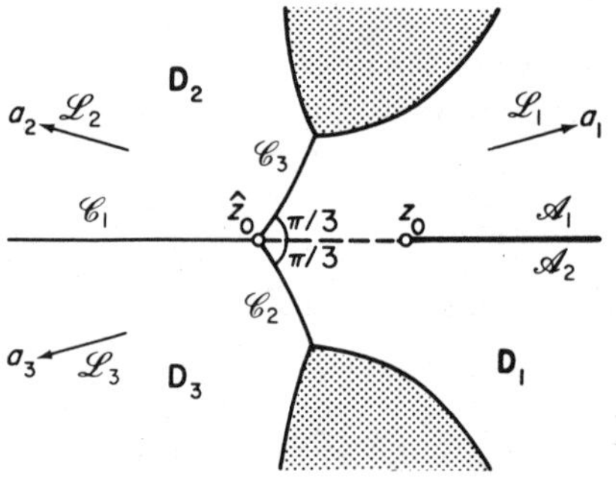

Fig. 15.1 Typical principal subdomains for the wave-penetration problem.

Let the reference points a_j be at infinity on arbitrary ξ-progressive paths $\mathscr{L}_j$ lying within $\mathbf{D}_j$, such that $|\mathrm{Re}\{\xi(a_j)\}| = \infty$. $\mathscr{L}_1$ is taken in the upper half-plane; $\mathscr{L}_2$ and $\mathscr{L}_3$ are complex conjugates, and no boundary of $\mathbf{D}$ may intervene between them in the neighborhood of infinity. Other assumptions are:

(i) *$\mathscr{V}(F)$ converges when $z \to a_j$ along $\mathscr{L}_j$, as well as when $z \to \pm\infty$ along the real axis.*

(ii) *ξ-progressive paths can be found in* $\mathbf{D}$ *linking*: (a) a_1 *with* a_2; (b) a_1 *with* a_3; (c) a_2 *or* a_3 *with any point of the interval* $(-\infty, \hat{z}_0)$; (d) a_1 *with any point of* $\mathscr{A}_1$.

(iii) *The following error terms vanish*: (a) $\varepsilon_2(a_3)$; (b) $\varepsilon_2(z)$ *and* $\varepsilon_3(z)$ *as* $z \to -\infty$; (c) $\varepsilon_1(z)$ *as* $z \to \infty$ *along* $\mathscr{A}_1$.

15.2 With these conditions, the connection formula (11.13) for the solutions

$$w_j(z) = \phi_j(z)e^{-\xi_j(z)}\{1+\varepsilon_j(z)\} \tag{15.02}$$

of equation (15.01) reduces to

$$w_1(z) = e^{\pi i/3}\{1+\varepsilon_1(a_3)\}\,w_2(z) + e^{-\pi i/3}\{1+\varepsilon_1(a_2)\}\,w_3(z). \tag{15.03}$$

We fix the $w_j(z)$ by specifying that $\phi_1(z)$ is the branch of $\{-f(z)\}^{-1/4}$ that takes its principal value on the join of $\hat{z}_0$ and z_0, and is continuous in $\mathbf{D}$ cut along $\mathscr{C}_1$ and $\mathscr{A}$. Then $\phi_2(z)$ and $\phi_3(z)$ are determined by (11.06). We now interpret the branches appearing in (15.03) in real form.

If $z \in \mathscr{A}_1$, then $\phi_1(z) = e^{\pi i/4}f^{-1/4}$. Since $\mathrm{Re}\{\xi_1(z)\}$ has to be positive in $\mathbf{D}_1$, it follows that on $\mathscr{A}_1$

$$\xi_1(z) = u\left(l - i\int_{z_0}^{z} f^{1/2}\,dt\right),$$

where, as before,

$$l = \int_{\hat{z}_0}^{z_0} |f|^{1/2}\,dt. \tag{15.04}$$

Therefore from (15.02)

$$w_1(z) = e^{-ul} f^{-1/4}\{1+\varepsilon_1(z)\} \exp\left(iu\int_{z_0}^{z} f^{1/2}\,dt + \tfrac{1}{4}\pi i\right), \tag{15.05}$$

where, from Assumption (iii), $\varepsilon_1(z) \to 0$ as $z \to \infty$ along $\mathscr{A}_1$. In the nomenclature of §10.1, (15.05) represents the transmitted wave.

When $z \in (-\infty, \hat{z}_0)$, we have

$$\phi_2(z) = e^{-\pi i/12} f^{-1/4}, \qquad \xi_2(z) = iu\int_{z}^{\hat{z}_0} f^{1/2}\,dt.$$

Hence

$$e^{\pi i/3}\{1+\varepsilon_1(a_3)\}\, w_2(z) = f^{-1/4}\{1+\varepsilon_1(a_3)\}\{1+\varepsilon_2(z)\} \exp\left(-iu\int_{z}^{\hat{z}_0} f^{1/2}\,dt + \tfrac{1}{4}\pi i\right), \tag{15.06}$$

where $\varepsilon_2(z) \to 0$ as $z \to -\infty$. This represents the incident wave. Lastly, for the reflected wave

$$e^{-\pi i/3}\{1+\varepsilon_1(a_2)\}\, w_3(z) = f^{-1/4}\{1+\varepsilon_1(a_2)\}\{1+\varepsilon_3(z)\} \exp\left(iu\int_{z}^{\hat{z}_0} f^{1/2}\,dt - \tfrac{1}{4}\pi i\right), \tag{15.07}$$

where $\varepsilon_3(z) \to 0$ as $z \to -\infty$.

Comparing (15.05), (15.06), and (15.07), and using the definitions of §10.3, we see that the reflection and transmission coefficients are given by

$$R = |1+\varepsilon_1(a_2)|^2\,|1+\varepsilon_1(a_3)|^{-2}, \qquad T = e^{-2ul}\,|1+\varepsilon_1(a_3)|^{-2}, \tag{15.08}$$

respectively. Here l is defined by (15.04), and from (11.14)

$$|\varepsilon_1(a_j)| \leqslant \tfrac{1}{2}\exp\{\mathscr{V}_{a_1, a_j}(F)\} - \tfrac{1}{2} \qquad (j = 2, 3),$$

the variation being evaluated along ξ-progressive paths. Clearly $\varepsilon_1(a_2)$ and $\varepsilon_1(a_3)$ are $O(u^{-1})$ as $u \to \infty$.

Alternative, and more effective, formulas for R are derivable by use of the equation $R+T = 1$ (Exercise 10.1). Thus

$$R = 1 - \frac{e^{-2ul}}{|1+\varepsilon_1(a_3)|^2} = \frac{|1+\varepsilon_1(a_2)|^2}{|1+\varepsilon_1(a_2)|^2 + e^{-2ul}}.$$

15.3 It will be observed that in the method of §10, connection formulas at *both* turning points $\hat{z}_0$ and z_0 were used, and the resulting approximate solutions of the differential equation matched in the interval $\hat{z}_0 < z < z_0$. In contrast, the method of the present section enables the LG solutions at $-\infty$ and $+\infty$ to be connected by use of only one formula, and no matching is required.

Ex. 15.1 For the modified Weber equation

$$d^2w/dz^2 = (a^2 - z^2)\,w,$$

in which a is a positive parameter, deduce from the results of §15.2 that $R = 1/\{1+\exp(-\pi a^2)\}$ and $T = 1/\{\exp(\pi a^2)+1\}$, exactly.

Ex. 15.2 Let a again be a positive parameter. By applying the connection formula (11.13) at $z = -a$ and $z = +a$, and matching solutions as $z \to \pm i\infty$, verify that the Weber equation

$$d^2w/dz^2 = (z^2 - a^2)\,w$$

possesses a solution which is recessive at both $z = -\infty$ and $z = +\infty$ if, and only if, a^2 is an odd positive integer.

***Ex. 15.3** In the transformed Whittaker equation

$$\frac{d^2w}{dz^2} = \left\{\frac{u^2(z-1)}{4z} + \frac{4m^2-1}{4z^2}\right\} w$$

let the parameters u and m be restricted by $u > 0$, $m \geqslant 0$, and the solution be determined by the condition

$$w(z) \sim z^{u/4} e^{-uz/2} \qquad (z \to +\infty).$$

By applying (11.13) at $z = 1$ and using the results of §5, show that when $z \to 0$ through positive values

$$w(z) = -\pi^{-1/2} u^{1/2} 2^{(1-u)/2} e^{-u/4} [\{1 + \alpha_0(u)\} \cos(\tfrac{1}{4}u\pi) + o(1)]\, z^{1/2} \ln z,$$

if $m = 0$, or

$$w(z) = \pi^{-1/2} \Gamma(2m)\, u^{(1/2)-2m} 2^{(4m+1-u)/2} e^{-u/4} [\{1 + \alpha_m(u)\} \cos(\tfrac{1}{4}u\pi - m\pi) + o(1)]\, z^{(1/2)-m},$$

if $m > 0$, where $\alpha_m(u)$ is independent of z and of order $O(u^{-1})$ for large u and fixed m.

***Ex. 15.4** (*Three turning points*) Let $u > 0$ and w_j, $j = 0, \pm 1$, be the solutions of the equation

$$d^2w/dz^2 = u^2 z(z^2 - 1)\,w$$

satisfying the condition

$$w_j \sim z^{-3/4} \exp\left\{-u \int_1^z t^{1/2}(t-1)^{1/2}(t+1)^{1/2}\,dt\right\} \qquad (u \text{ fixed}, \quad z \to \infty e^{4\pi i j/5}),$$

in which the fractional powers take their principal values. By applying (11.13) at the turning points ± 1, and denoting $\frac{2}{5}(2/\pi)^{1/2}\{\Gamma(\frac{3}{4})\}^2$ by a, show that

$$e^{-2au} w_0 + \{1 + e^{-2iau} + O(u^{-1})\}\, w_1 + \{1 + e^{2iau} + O(u^{-1})\}\, w_{-1} = 0.$$

Could this formula also be obtained by use of (11.13) at the turning points 0 and 1?

Ex. 15.5 From the result of the preceding exercise, deduce that the solution specified by

$$w \sim z^{-3/4} \exp(uz^{1/2} - \tfrac{2}{5}uz^{5/2}) \qquad (u \text{ fixed}, \quad z \to +\infty),$$

has the property

$$w = 4e^{2ua} |z|^{-3/4} \{\cos(ua) \cos(\tfrac{2}{5}u|z|^{5/2} - u|z|^{1/2} + ua - \tfrac{1}{4}\pi) + O(u^{-1})\}$$

as $u \to \infty$ uniformly with respect to $z \in (-\infty, -1-\delta]$, δ being any positive constant.

Historical Notes and Additional References

Most of this chapter is devoted to the construction and application of connection formulas for a simple turning point. The method of §7 originated in the work of Rayleigh (1912), Gans (1915), and, especially, Jeffreys (1924). In the neighborhood of a turning point x_0 of the equation

$$d^2w/dx^2 = \{u^2 f(x) + g(x)\}\, w,$$

Gans and Jeffreys approximated $f(x)$ and $g(x)$ by $(x - x_0) f'(x_0)$ and $g(x_0) + (x - x_0) g'(x_0)$, respectively. The resulting equation is solvable exactly in terms of Airy functions (or, equivalently,

Bessel functions of order one-third). Although $|x-x_0|$ has to be small, the argument of the Airy functions can be made large in absolute value by taking u sufficiently large (unless $x=x_0$). The Airy functions may then be replaced by their asymptotic approximations for positive and negative arguments, and this yields the connection formulas. As explained in the historical notes on Chapter 6, this procedure is commonly known in physics literature as the WKB or WKBJ method. Langer (1937) made the analysis much sounder by basing it on Airy-function approximations which are uniform in x, and this improvement is incorporated in §7. Extensions to complex variables were developed (in a formal manner) by Heading (1962), and following earlier writers he employs the name *phase-integral method.* Error bounds for the coefficients in the approximate formulas (also included in §7) were supplied by Olver (1965a). Further developments have been considered by Wasow (1968, 1970); he calls the process *central connection.*

The second method (§11) avoids the use of Airy-function approximations altogether, and is called by Wasow *lateral connection.* It was introduced in the thesis of Zwaan (1929) and subsequently improved by Birkhoff (1933), Kemble (1935), and Furry (1947). Firm mathematical bases for the method were established independently by Fedoryuk (1965a, b), Fröman and Fröman (1965), and Olver (1965a, b). Further developments and applications have been made by Evgrafov and Fedoryuk (1966), Fedoryuk (1969), and N. Fröman (1966, 1970, and other papers).

In the monograph of Fröman and Fröman, the general solution of equation (11.01) is expressed in the form

$$w(z) = a_1(z) f^{-1/4}(z) \exp\left\{\int f^{1/2}(z)\,dz\right\} + a_2(z) f^{-1/4}(z) \exp\left\{-\int f^{1/2}(z)\,dz\right\}.$$

Following Kemble, the authors construct a pair of simultaneous first-order differential equations for $a_1(z)$ and $a_2(z)$, and use them to determine the changes in $a_1(z)$ and $a_2(z)$ as z passes from one principal subdomain to the next. Error bounds for the changes are expressed in terms of curvilinear integrals; these are equivalent to the variations along progressive paths of our error-control function $F(z)$.

An outgrowth of the original WKBJ method of connection is the so-called *stretching-matching* method. This is not described in the text, but treatments and references may be found, for example, in the papers by Wasow (1960, 1968) and Nakano and Nishimoto (1970). This method is particularly valuable for differential equations of higher order.

Further historical information may be found in the references cited, especially Heading (1962), Fröman and Fröman (1965), and Wasow (1968). See also Langer (1934), McHugh (1971), and Dingle (1973).

§§4–5 These results appear to be new.

§7.4 Misunderstandings of the meaning of the connection formulas have occurred frequently in the literature, see discussions by Langer (1934), Jeffreys (1956), Heading (1962, §1.6), and Fröman and Fröman (1965, Chapter 8). Absences of explicit boundary conditions for the solutions and bounds for the error terms could be regarded as contributory causes.

§10 The natural formulation of the barrier penetration problem is given by

$$d^2w/dx^2 = \{V(x)-E\}w,$$

where the parameter E is the energy of the particle under consideration, and $V(x)$ represents the potential. The general shape of the graph of $V(x)$ is indicated in the accompanying diagram. As

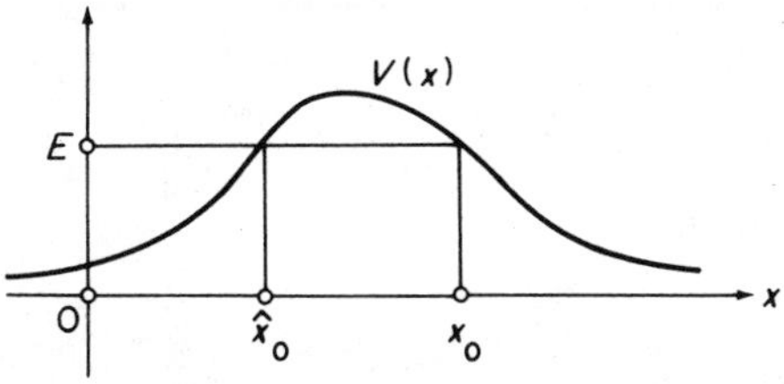

Jeffreys (1956) points out, a difficulty in treating this problem by asymptotic analysis is to specify the large parameter u. It cannot be E, because for all sufficiently large E, $V(x)-E$ is everywhere negative and there would be no barrier. Jeffreys suggests the adoption of

$$u = \int_{\hat{x}_0}^{x_0} \{V(x)-E\}^{1/2}\, dx,$$

$\hat{x}_0$ and x_0 being zeros of the integrand. The differential equation then takes the form (10.01) with $g(x) = 0$ and $f(x) = \{E-V(x)\}/u^2$. Since $f(x)$ is dependent on u, the theory of §10 is inapplicable. An appropriate extension on the lines of §9.1 could be made, however.

McLeod (1961b) has treated the case in which $V(x) \to -\infty$ as $x \to \pm\infty$, and E is large and negative. In this formulation two turning points exist however large $|E|$ may be.

§11.1 There is confusion in the literature in naming the curves $\operatorname{Re}\{\xi(z)\} = 0$. Most mathematicians call them *Stokes lines*, whereas physicists use the term *anti-Stokes lines*, reserving *Stokes lines* for the curves $\operatorname{Im}\{\xi(z)\} = 0$. As W. H. Reid has pointed out to the author, history is on the side of the physicists, for in Stokes' original treatment (1857, §16) of the Airy equation he mentions the curves $\operatorname{Im}\{\xi(z)\} = 0$—in this case $\operatorname{ph} z = 0$ and $\pm 2\pi/3$—but does not appear to label or otherwise distinguish the curves $\operatorname{Re}\{\xi(z)\} = 0$. Since "anti-Stokes" is cumbersome, the author recommends that the curves $\operatorname{Re}\{\xi(z)\} = 0$ and $\operatorname{Im}\{\xi(z)\} = 0$ be known as the principal curves and Stokes lines, respectively.

§15 This analysis has been taken from the paper by Olver (1965b); this reference also includes an analogous treatment of the eigenvalue problem of §8. Other treatments of these problems include those of Pike (1964b), Fedoryuk (1965a), and Fröman and Fröman (1965, 1970).

14

ESTIMATION OF REMAINDER TERMS

1 Numerical Use of Asymptotic Approximations

1.1 Most of the more important asymptotic approximations or expansions given in this book have been accompanied by explicit bounds for the corresponding error terms. In these cases there is generally little difficulty in assessing the accuracy of results computed from the approximation or expansion. But there are also some cases in which the complexity of the analysis renders the determination of realistic error bounds prohibitively difficult. Then the question naturally arises whether the use of the asymptotic result is permissible in numerical work.

At the outset it must be reaffirmed that an upper bound for the error term of an asymptotic expansion cannot be safely inferred simply by inspection of the rate of numerical decrease of the terms in the series at the point of truncation.† Even in the case of a convergent power-series expansion this cannot be done: the tail has to be majorized analytically (often by a geometric progression) before final accuracy can be guaranteed. For a series that is merely known to be asymptotic, the situation is much worse. First, it is impossible to majorize the tail. Secondly, the series represents not one function but an infinite class, and the error term naturally depends on which particular member of the class we have in mind.

In consequence, the numerical use of an asymptotic approximation (or expansion), without rigorous investigation of the error term, has to be regarded as being in the nature of a hypothesis. Nevertheless, it would be extravagant to reject the use of an approximation for this reason alone. The essence of progress in the physical sciences is the development and application of hypotheses that have a high probability of being correct: it would be artificial to exclude those of a purely mathematical character. Instead, what we need to do is examine the approximations by means of carefully devised tests.

1.2 In cases where the asymptotic variable x, say, is real and positive and the distinguished point is at infinity, the wanted function should be computed by an independent (preferably nonasymptotic) method at the smallest value of x for which the asymptotic approximation is intended to be used. If the results are in

† An illuminating example was provided in Chapter 3, §6.1.

agreement to S significant figures, then it is highly probable that the approximation will be accurate to at least S significant figures for all greater values of x. Again, "highly probable" is as far as we can go in our appraisal; we cannot be certain.

Consider, for example, the expansion

$$E_1(x) + 1000e^{-2x}\sin \pi x \sim e^{-x}\left(\frac{1}{x} - \frac{1!}{x^2} + \frac{2!}{x^3} - \cdots\right) \qquad (x \to +\infty),$$

in which $E_1(x)$ denotes the exponential integral.† If $[x]$ terms on the right-hand side are taken, then we would find that the suggested test works satisfactorily unless, unthinkingly, we select a trial value of x which happens to be an integer. This simple example is somewhat artificial, but the same kind of deception might occur more naturally in a complicated situation. Of course, if two (or more) trial values of x are used then the chance of deception is considerably reduced.

1.3 In the case of a complex variable z, both $|z|$ and $\operatorname{ph} z$ have to be considered when appraising accuracy. Suppose that a given asymptotic approximation is valid as $z \to \infty$ in any closed sector within $\theta_1 < \operatorname{ph} z < \theta_2$, but not within a larger open sector. In other words, the rays $\operatorname{ph} z = \theta_1$ and θ_2 are true boundaries. Examples in previous chapters indicate that accuracy deteriorates severely as these boundaries are approached. In consequence, numerical use of the approximation should be confined to a sector $\theta_1' \leqslant \operatorname{ph} z \leqslant \theta_2'$ lying well within $\theta_1 < \operatorname{ph} z < \theta_2$. In the absence of any guide the empirical choice $\theta_1' = \theta_1 + \frac{1}{6}(\theta_2 - \theta_1)$, $\theta_2' = \theta_2 - \frac{1}{6}(\theta_2 - \theta_1)$, is suggested.‡ Independent evaluations should be made at $\operatorname{ph} z = \theta_1'$ and θ_2' for one or two values of $|z|$ close to the smallest value it is intended to use.

1.4 When the regions of validity in the complex plane are not sectors appraisal is more complicated. Basically, however, the idea of keeping at a safe distance from the boundaries still applies.

For example, in Chapters 6 and 13 it was shown that if $f(z)$ and $g(z)$ are analytic functions of the complex variable z, then in a wide range of circumstances the differential equation

$$d^2w/dz^2 = \{u^2 f(z) + g(z)\}\, w$$

has asymptotic solutions of the form

$$w(u, z) \sim f^{-1/4}(z) \exp\{u\xi(z)\} \tag{1.01}$$

when $|u| \to \infty$, where

$$\xi(z) = \int f^{1/2}(z)\, dz. \tag{1.02}$$

The boundaries of the regions of validity of these approximations belong to the family of curves having an equation of the form

$$\operatorname{Re}\{u\xi(z)\} = \text{constant};$$

† Compare Chapter 1, §1.1 and Chapter 2, §3.1.

‡ Again, based upon earlier examples.

these are the so-called principal curves. A typical case is indicated in Fig. 1.1. Here z_0 is a simple turning point (zero of $f(z)$) and z_1 a simple pole of $f(z)$. Appropriate principal curves are given by (Chapter 13, §11)

$$\operatorname{Re}\{u\xi(z)-u\xi(z_0)\} = 0,$$

and are represented by the continuous curves z_0A_1, z_0A_2, and z_0A_3. The continuous curve z_1B is a suitably chosen cut, and the three principal curves and the cut subdivide the z plane into simply connected domains $\mathbf{D}_1$, $\mathbf{D}_2$, and $\mathbf{D}_3$, enumerated as indicated.

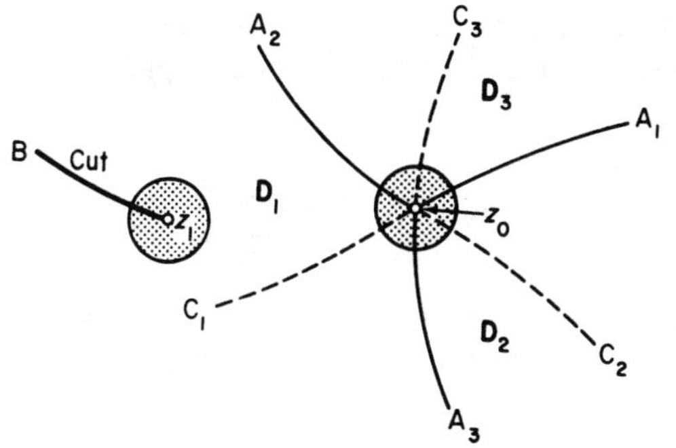

Fig. 1.1 z plane.

The theory of Chapter 6, §§11 and 12 shows that there exist solutions $w_j(u,z)$, $j=1,2,3$, having asymptotic representations of the form (1.01) for large $|u|$, the branches of $f^{-1/4}(z)$ in (1.01) and $f^{1/2}(z)$ in (1.02) in each case being chosen so that (i) they are continuous everywhere except across z_0A_j and z_1B; (ii) $u\xi(z)-u\xi(z_0)$ has negative real part in $\mathbf{D}_j$. The region of validity is obtained by deleting the points z_0 and z_1, and also the corresponding principal curve z_0A_j.

In the neighborhoods of z_0 and z_1, and also in the vicinity of z_0A_1, the approximation (1.01) for $w_1(u,z)$ is inaccurate and should be avoided. A convenient specification of this "vicinity" is the region bounded by the *Stokes lines* z_0C_2 and z_0C_3, whose equation is given by

$$\operatorname{Im}\{u\xi(z)-u\xi(z_0)\} = 0.$$

These lines are indicated by broken curves in Fig. 1.1. In this region $w_1(u, z)$ can be calculated more accurately from its linear expression in terms of $w_2(u,z)$ and $w_3(u,z)$ (Chapter 13, §11). Similarly for $w_2(u,z)$ in the region bounded by the Stokes lines z_0C_3 and z_0C_1, and for $w_3(u,z)$ in the region bounded by z_0C_1 and z_0C_2. In this way we keep as far away as possible (in the ξ plane) from the boundaries of the regions of validity of (1.01) in all three cases $w(u,z) = w_j(u,z)$, $j=1,2,3$.

Lastly, points at which the adequacy of the asymptotic representation should be tested numerically in this example are (a) intersections of the Stokes lines with the boundary of the chosen neighborhood of z_0, (b) intersection of the cut (both sides) with the boundary of the neighborhood of z_1.

2 Converging Factors

2.1 Let $f(z)$ be a function of z having the asymptotic expansion

$$f(z) \sim a_0 + \frac{a_1}{z} + \frac{a_2}{z^2} + \cdots \tag{2.01}$$

as $z \to \infty$ in a sector $\mathbf{S}$: $\theta_1 \leqslant \operatorname{ph} z \leqslant \theta_2$. As special cases, $\mathbf{S}$ can degenerate into the positive or negative real axis. Suppose that successive terms in the series (2.01) diminish in absolute value until the $(n+1)$th term is reached; thereafter they increase. Clearly $n = n(z)$, where $n(z)$ is a discontinuous function of $|z|$ which is independent of $\operatorname{ph} z$. From earlier examples for which we have explicit bounds for the remainder term

$$R_n(z) \equiv f(z) - \sum_{s=0}^{n-1} \frac{a_s}{z^s},$$

we expect the expansion (2.01) to yield its greatest accuracy when truncated at about $n(z)$ terms. We shall call $R_{n(z)}(z)$ the *optimum remainder term*, whether or not it is actually the least.

Now write

$$C(z) = R_{n(z)}(z)/\{a_{n(z)} z^{-n(z)}\},$$

so that

$$f(z) = \sum_{s=0}^{n(z)-1} \frac{a_s}{z^s} + C(z)\frac{a_{n(z)}}{z^{n(z)}}.$$

If we have a way of assessing $C(z)$ for large $|z|$, then the magnitude of the optimum remainder term in the expansion (2.01) can be estimated. In some cases it is actually possible to construct an asymptotic expansion for $C(z)$ in descending powers of z. In these fortunate circumstances $C(z)$ can be calculated to several significant figures, considerably increasing the attainable accuracy in the computed value of $f(z)$. For this reason, $C(z)$ is called a *converging factor*.

2.2 In the next three sections procedures for expanding $C(z)$ are introduced by means of examples. Instead of using z as asymptotic variable it is convenient to change to n. Suppose, for example, that $n(z) = [|z|]$. Then we write

$$f(z) = \sum_{s=0}^{n-1} \frac{a_s}{z^s} + C_n(z)\frac{a_n}{z^n},$$

and seek an asymptotic representation of $C_n\{(n+\zeta)e^{i\theta}\}$ for large n, which is uniformly valid with respect to $\theta \in [\theta_1, \theta_2]$ and $\zeta \in [0,1]$ (or some larger region). To apply the result for an assigned value of z, we take $\theta = \operatorname{ph} z$, $n = [|z|]$, and $\zeta = |z| - [|z|]$.

3 Exponential Integral

3.1 From Chapter 1, §1 and Chapter 2, §3.1, we have

$$e^z E_1(z) = \int_0^\infty \frac{e^{-zt}}{1+t}\,dt = \sum_{s=0}^{n-1} u_s(z) + R_n(z) \qquad (|\mathrm{ph}\, z| < \tfrac{1}{2}\pi), \tag{3.01}$$

where n is an arbitrary positive integer,

$$u_s(z) = (-)^s s!/z^{s+1},$$

and†

$$R_n(z) = (-)^n \int_0^\infty \frac{e^{-zt} t^n}{1+t}\,dt.$$

The term $u_s(z)$ of smallest absolute value is $u_{[|z|]}(z)$, unless $|z|$ is an integer in which event there are two equally small terms $u_{|z|-1}(z)$ and $u_{|z|}(z)$. Following §2.2, we seek the asymptotic form of

$$R_n\{(n+\zeta)e^{i\theta}\} \equiv (-)^n \int_0^\infty \frac{\exp\{-(n+\zeta)e^{i\theta}t\}\,t^n}{1+t}\,dt \tag{3.02}$$

for large n and $\zeta \in [0,1]$.

The saddle point of the integrand in (3.02) is given by

$$d\{t\exp(-e^{i\theta}t)\}/dt = 0,$$

that is, by $t = e^{-i\theta}$. In consequence, we rotate the path of integration through an angle $-\theta$. On taking a new integration variable $\tau = te^{i\theta}$, we obtain

$$R_n\{(n+\zeta)e^{i\theta}\} = (-)^n e^{-i(n+1)\theta} \int_0^\infty \frac{e^{-(n+\zeta)\tau}\tau^n}{1+\tau e^{-i\theta}}\,d\tau. \tag{3.03}$$

Although this result has been derived on the assumption that $|\theta| < \frac{1}{2}\pi$, it is easily extended to $|\theta| < \pi$ by further rotation of the path and appeal to analytic continuation; compare Chapter 4, §1.2.

To the last integral we apply Theorem 7.1 of Chapter 4, taking $z = n$, $\mathrm{ph}\, z = 0$, $p(\tau) = \tau - \ln\tau$, $q(\tau) = e^{-\zeta\tau}/(1+\tau e^{-i\theta})$, and $\tau_0 = 1$. In this way we derive

$$R_n\{(n+\zeta)e^{i\theta}\}$$
$$\sim (-)^n(1-\alpha)e^{-n-\zeta-i(n+1)\theta}\left(\frac{2\pi}{n}\right)^{1/2}\left(1 + \frac{\zeta^2 - 2\zeta + 2\alpha\zeta + \frac{1}{6} - 2\alpha + 2\alpha^2}{2n} + \cdots\right),$$

where, for conciseness, we have denoted

$$\alpha = 1/(1+e^{i\theta}).$$

† Expressed as a confluent hypergeometric function $R_n(z) = (-)^n n!\, U(n+1, n+1, z)$; compare Chapter 7, (10.04).

Moreover, from the proofs of Theorems 7.1 and 6.1 of Chapter 4 (or, more simply, the proof of Theorem 8.1 of Chapter 3) it is readily perceived that the last expansion is uniformly valid with respect to ζ in any bounded real or complex region and $\theta \in [-\pi+\delta, \pi-\delta]$, where δ is an arbitrary positive constant.

For the converging factor

$$C_n(z) = R_n(z)/u_n(z), \tag{3.04}$$

we have

$$C_n\{(n+\zeta)e^{i\theta}\} \sim \frac{(1-\alpha)(n+\zeta)^{n+1}}{n!\,e^{n+\zeta}}\left(\frac{2\pi}{n}\right)^{1/2}\left(1+\frac{\zeta^2-2\zeta+2\alpha\zeta+\frac{1}{6}-2\alpha+2\alpha^2}{2n}+\cdots\right).$$

On substituting for $(n+\zeta)^{n+1}$ by the expansion

$$n^{n+1}e^{\zeta}\left(1+\frac{\zeta-\frac{1}{2}\zeta^2}{n}+\cdots\right) \tag{3.05}$$

and for $n!$ by means of Stirling's series, we arrive at the desired expansion in the form

$$C_n\{(n+\zeta)e^{i\theta}\} \sim (1-\alpha)\left\{1+\frac{\alpha(\zeta-1+\alpha)}{n}+\cdots\right\} \tag{3.06}$$

as $n \to \infty$, uniformly with respect to $\theta \in [-\pi+\delta, \pi-\delta]$ and bounded ζ.

3.2 The first conclusion to be drawn from (3.06) is that if the expansion

$$e^z E_1(z) \sim \frac{1}{z} - \frac{1!}{z^2} + \frac{2!}{z^3} - \cdots$$

is truncated at (or near) its smallest term, then the remainder term is approximately equal to the first neglected term multiplied by $1-\alpha$, that is, $1/(1+e^{-i\theta})$. In particular, when z is real and positive this factor is $\frac{1}{2}$, and when z approaches either side of the cut along the negative real axis this factor tends to infinity.

Our next interest concerns the possibility of using (3.06) to compute accurate values of the converging factor. For this purpose additional terms are desirable. They could be found by the method already employed, but following J. C. P. Miller (1952) we use instead a less laborious method based upon a differential equation satisfied by $C_n(z)$. Yet another procedure, which yields explicit expressions for the coefficients and error term in (3.06), is indicated in Exercise 3.3 below.

From (3.01) we derive

$$\frac{d}{dz}\{e^z E_1(z)\} = -\int_0^\infty \frac{e^{-zt}t}{1+t}\,dt.$$

Hence

$$e^z E_1(z) - \frac{d}{dz}\{e^z E_1(z)\} = \int_0^\infty e^{-zt}\,dt = \frac{1}{z}.$$

Substituting in this equation by means of (3.01) and (3.04), we find, on reduction, that $C_n(z)$ satisfies the following first-order inhomogeneous differential equation:

$$\left(1+\frac{n+1}{z}\right)C_n(z) - C_n'(z) = 1.$$

Thence

$$\{(n+\zeta)e^{i\theta}+n+1\}\,C_n\{(n+\zeta)e^{i\theta}\} - (n+\zeta)e^{i\theta}C_n'\{(n+\zeta)e^{i\theta}\} = (n+\zeta)e^{i\theta}. \tag{3.07}$$

The method of §3.1 shows that there exists an expansion of the form

$$C_n\{(n+\zeta)e^{i\theta}\} \sim \sum_{s=0}^{\infty}\frac{\gamma_s}{n^s} \qquad (n\to\infty), \tag{3.08}$$

where γ_s is a polynomial in ζ, which is independent of n. Since this expansion is known to be uniformly valid in any bounded ζ domain it may be differentiated term-by-term with respect to ζ, giving

$$e^{i\theta}C_n'\{(n+\zeta)e^{i\theta}\} \sim \sum_{s=0}^{\infty}\frac{\gamma_s'}{n^s}, \tag{3.09}$$

where $\gamma_s' \equiv \partial\gamma_s/\partial\zeta$.

Substituting (3.08) and (3.09) in (3.07) and equating coefficients of like powers of n, we find that

$$(1+e^{i\theta})\gamma_0 - \gamma_0' = e^{i\theta}, \qquad (1+e^{i\theta})\gamma_1 + (1+\zeta e^{i\theta})\gamma_0 - \gamma_1' - \zeta\gamma_0' = \zeta e^{i\theta},$$

and

$$(1+e^{i\theta})\gamma_s + (1+\zeta e^{i\theta})\gamma_{s-1} - \gamma_s' - \zeta\gamma_{s-1}' = 0 \qquad (s\geqslant 2).$$

The first equation is an inhomogeneous differential equation in ζ for γ_0. The only *polynomial* solution is readily verified to be

$$\gamma_0 = e^{i\theta}/(1+e^{i\theta}) = 1-\alpha.$$

Similarly, the second equation yields

$$\gamma_1 = -\frac{e^{2i\theta}}{(1+e^{i\theta})^3} + \frac{e^{i\theta}\zeta}{(1+e^{i\theta})^2} = \alpha(1-\alpha)(\zeta-1+\alpha).$$

These results accord with §3.1.

For higher coefficients, we first express θ in terms of α:

$$\gamma_s - \alpha\gamma_s' = (\alpha\zeta-\zeta-\alpha)\gamma_{s-1} + \alpha\zeta\gamma_{s-1}' \qquad (s\geqslant 2).$$

Elementary operational methods show that the unique polynomial solution of this differential equation for γ_s is given by

$$\gamma_s = \left\{1+\alpha\frac{\partial}{\partial\zeta}+\alpha^2\frac{\partial^2}{\partial\zeta^2}+\cdots\right\}\{(\alpha\zeta-\zeta-\alpha)\gamma_{s-1}+\alpha\zeta\gamma_{s-1}'\}.$$

For $s = 2$ and 3, this formula yields

$$\gamma_2 = \alpha(1-\alpha)^2\{-\zeta^2+(1-3\alpha)\zeta+\alpha(2-3\alpha)\},$$
$$\gamma_3 = \alpha(1-\alpha)^2\{(1-\alpha)\zeta^3-(1-6\alpha+6\alpha^2)\zeta^2 - \alpha(4-17\alpha+15\alpha^2)\zeta-\alpha^2(6-20\alpha+15\alpha^2)\}.$$

In the case when z is real and positive, that is, when $\alpha = \frac{1}{2}$, the expressions for the coefficients reduce to

$$\gamma_0 = \tfrac{1}{2}, \quad \gamma_1 = \tfrac{1}{8}(2\zeta-1), \quad \gamma_2 = \tfrac{1}{32}(-4\zeta^2-2\zeta+1), \quad \gamma_3 = \tfrac{1}{128}(8\zeta^3+8\zeta^2+6\zeta+1).$$

3.3 By way of numerical illustration let us take $z = 5$. With the original expansion (3.01) the optimum procedure is to truncate at the fifth term:

$$e^z E_1(z) \doteqdot \sum_{s=0}^{4} u_s(5).$$

Exact decimal values of the required terms are

$$u_0(5) = 0.2, \qquad u_1(5) = -0.04, \qquad u_2(5) = 0.016,$$
$$u_3(5) = -0.0096, \qquad u_4(5) = 0.00768.$$

Whence

$$u_0(5)+u_1(5)+\cdots+u_4(5) = 0.17408, \tag{3.10}$$

compared with†

$$e^5 E_1(5) = 0.17042\,2176\ldots.$$

Now we compute the converging factor. We have $\theta = 0$, $\alpha = \frac{1}{2}$, and $\zeta = 0$. Hence from §3.2

$$\gamma_0 = \tfrac{1}{2}, \qquad \gamma_1 = -\tfrac{1}{8}, \qquad \gamma_2 = \tfrac{1}{32}, \qquad \gamma_3 = \tfrac{1}{128},$$

and from (3.08)

$$C_5(5) \sim 0.5 - 0.025 + 0.00125 + 0.00006\,25 + \cdots = 0.47631\,25,$$

on neglecting terms beyond the fourth. Since $u_5(5) = -0.00768$, the corresponding remainder term is given by

$$C_5(5)\,u_5(5) = -0.00365\,808.$$

Addition of this number to the right-hand side of (3.10) yields

$$0.17042\,192.$$

This is about four decimal places better than (3.10)—an impressive demonstration of the power of the method in the present example.

Ex. 3.1 For integer n and positive x, define $C_n(x)$ by

$$\pi^{1/2}\exp(x^2)\operatorname{erfc}x = \sum_{s=0}^{n-1} u_s(x) + C_n(x)\,u_n(x),$$

† N.B.S. (1964) or B.A. (1946a).

where $u_s(x) = (-)^s 1\cdot 3\cdot 5 \cdots (2s-1)/\{2^s x^{2s+1}\}$. Show that

$$C_n(x) = x^2 \int_0^\infty \frac{\exp(-x^2 t)}{(1+t)^{n+(1/2)}}\, dt,$$

and hence that as $n \to \infty$

$$C_n\{(n+\zeta)^{1/2}\} \sim \sum_{s=0}^{\infty} \frac{\gamma_s(\zeta)}{n^s},$$

uniformly for bounded ζ, where $\gamma_0(\zeta) = \frac{1}{2}$, $\gamma_1(\zeta) = \frac{1}{4}\zeta$, $\gamma_2(\zeta) = -\frac{1}{32}(4\zeta^2+2\zeta+1)$.

Ex. 3.2 Show that the coefficients in the preceding exercise are the polynomial solutions of the recurrence relation

$$\gamma_s(\zeta) - \tfrac{1}{2}\gamma_s'(\zeta) = \tfrac{1}{2}\zeta\gamma_{s-1}'(\zeta) - (\tfrac{1}{2}\zeta + \tfrac{1}{4})\gamma_{s-1}(\zeta) \qquad (s \geqslant 2).$$

Ex. 3.3 Show that in terms of the generalized exponential integral (Chapter 2, Exercises 3.5 and 3.6) the converging factor of §3.1 is given by

$$C_n(z) = ze^z E_{n+1}(z) = ze^z \int_1^{\infty e^{-i\theta}} \frac{e^{-zt}}{t^{n+1}}\, dt \qquad (|\theta| < \pi),$$

where $\theta = \operatorname{ph} z$. By repeated integrations by parts verify that

$$E_{n+1}\{(n+\zeta)e^{i\theta}\} = \exp\{-(n+\zeta)e^{i\theta}\} \sum_{s=0}^{S-1} \frac{c_s}{n^{s+1}} + R_S,$$

where ζ is an arbitrary number such that $\operatorname{Re}(n+\zeta) > 0$, S is an arbitrary positive integer,

$$c_s = \left[e^{\zeta\tau} \left(\frac{\tau}{1+\tau}\frac{d}{d\tau} \right)^s \left(\frac{e^{-\zeta\tau}}{1+\tau} \right) \right]_{\tau=\exp(i\theta)},$$

and

$$R_S = \frac{1}{n^S} \int_1^{\infty e^{-i\theta}} \frac{e^{-nat}}{t^n} \frac{d}{dt} \left\{ \left(\frac{t}{1+at}\frac{d}{dt} \right)^{S-1} \left(\frac{e^{-\zeta at}}{1+at} \right) \right\} dt,$$

with $a \equiv e^{i\theta}$. Thence obtain the following expressions for the coefficients γ_s in (3.08):

$$\gamma_0 = e^{i\theta} c_0, \qquad \gamma_s = e^{i\theta}(c_s + \zeta c_{s-1}) \qquad (s \geqslant 1).$$

[G. F. Miller, 1960.]

4 Exponential Integral (continued)

4.1 When z lies on or near the cut along the negative real axis the method of §3 fails. In the present section we show how to expand the converging factor in these circumstances. It suffices to consider one side of the cut; for the other side the sign of i is changed throughout.

Set $z = xe^{i\pi}$, where $x > 0$. By rotating the path in (3.01) through an angle $-\pi$ and taking $\tau = te^{i\pi}$ as new integration variable, we obtain

$$-e^{-x}E_1(xe^{i\pi}) = \int_0^\infty \frac{e^{-x\tau}}{1-\tau}\, d\tau,$$

where the path passes above the point 1 in the τ plane by means of an indentation. Expanding $(1-\tau)^{-1}$ in ascending powers of τ, complete with remainder term, and integrating term by term, we derive the identity

$$-e^{-x}E_1(xe^{i\pi}) = \sum_{s=0}^{n-1} u_s(x) + R_n(x),$$

in which n is an arbitrary integer, $u_s(x) = s!/x^{s+1}$, and

$$R_n(x) = \int_0^\infty \frac{e^{-x\tau}\tau^n}{1-\tau}\, d\tau. \tag{4.01}$$

In the last integral the path again passes above $\tau = 1$.

The series $\sum u_s(x)$ is the asymptotic expansion of $-e^{-x}E_1(xe^{i\pi})$ for large x. The smallest term is given by $s = [x]$, hence we are interested in the asymptotic approximation of $R_n(n+\zeta)$ for large n and bounded ζ, especially $\zeta \in [0, 1]$.

4.2 With x replaced by $n+\zeta$, the integrand in (4.01) has the special feature that the saddle point and singularity coincide. This phenomenon was previously encountered in Chapter 4, Exercise 8.4. Following the methods of that chapter we take a new integration variable v, given by

$$v = \tau - 1 - \ln \tau, \qquad \frac{dv}{d\tau} = \frac{\tau - 1}{\tau}.$$

The path transforms into a loop in the v plane:

$$R_n(n+\zeta) = e^{-n} \int_\infty^{(0-)} e^{-nv} q(v)\, dv, \tag{4.02}$$

where

$$q(v) = \frac{e^{-\zeta\tau}}{1-\tau}\frac{d\tau}{dv}.$$

Near $v = 0$ we have

$$\tau = 1 + 2^{1/2}v^{1/2} + \frac{2}{3}v + \frac{2^{1/2}}{18}v^{3/2} - \cdots,$$

where $v^{1/2}$ is positive on the upper side of the positive real axis and negative on the lower side; compare Chapter 3, §8.3. In consequence, $q(v)$ can be expanded for small $|v|$ in the form

$$q(v) = -\frac{e^{-\zeta}}{2v}\sum_{s=0}^{\infty} q_s(\zeta)\, v^{s/2}, \tag{4.03}$$

in which the coefficients $q_s(\zeta)$ are polynomials in ζ. In particular,

$$q_0(\zeta) = 1, \qquad q_1(\zeta) = 2^{1/2}(\tfrac{1}{3} - \zeta).$$

The residue theorem shows that the contribution of the first term in (4.03) to $R_n(n+\zeta)$ is given by

$$-e^{-n}\int_\infty^{(0-)} e^{-nv}\frac{e^{-\zeta}}{2v}\, dv = i\pi e^{-n-\zeta}.$$

For the difference between $q(v)$ and the first term, the path in (4.02) may be collapsed onto the positive real axis. Application of Watson's lemma then yields†

$$R_n(n+\zeta) \sim e^{-n-\zeta}\left\{i\pi - q_1(\zeta)\frac{\Gamma(\frac{1}{2})}{n^{1/2}} - q_3(\zeta)\frac{\Gamma(\frac{3}{2})}{n^{3/2}} - \cdots\right\} \qquad (n \to \infty). \tag{4.04}$$

† The expansion (4.04) can also be deduced directly from (4.02) and (4.03) by means of Watson's lemma for loop integrals (Chapter 4, §5.3).

To express the result in terms of a converging factor, we write

$$R_n(x) = i\pi e^{-x} + C_n(x)\, u_n(x),$$

so that

$$-e^{-x}E_1(xe^{i\pi}) = i\pi e^{-x} + \sum_{s=0}^{n-1} u_s(x) + C_n(x)\, u_n(x), \tag{4.05}$$

or, in terms of the Ei notation,

$$e^{-x}\,\mathrm{Ei}(x) = \sum_{s=0}^{n-1} u_s(x) + C_n(x)\, u_n(x); \tag{4.06}$$

compare Chapter 2, §3.2 and Chapter 6, Exercise 13.5. Then for large n and bounded ζ, we have

$$C_n(n+\zeta) \sim -\frac{(n+\zeta)^{n+1}}{n!} e^{-n-\zeta}\pi^{1/2}\left\{\frac{q_1(\zeta)}{n^{1/2}} + \frac{q_3(\zeta)}{2n^{3/2}} + \cdots\right\} \sim \sum_{s=0}^{\infty} \frac{\gamma_s(\zeta)}{n^s}, \tag{4.07}$$

say, where the coefficients $\gamma_s(\zeta)$ are polynomials in ζ (compare (3.05)). In particular,

$$\gamma_0(\zeta) = -q_1(\zeta)/2^{1/2} = \zeta - \tfrac{1}{3}.$$

4.3 Having established the form of the asymptotic expansion of $C_n(n+\zeta)$, we again find it easier to evaluate higher coefficients by a different method. Using analysis similar to §3.2, we may show that $C_n(x)$ satisfies the differential equation

$$\left(1 - \frac{n+1}{x}\right) C_n(x) + C_n'(x) = 1. \tag{4.08}$$

Substitution of (4.07) leads to the asymptotic identity

$$(\zeta - 1)\sum_{s=0}^{\infty} \frac{\gamma_s(\zeta)}{n^s} + (n+\zeta)\sum_{s=0}^{\infty} \frac{\gamma_s'(\zeta)}{n^s} \sim n + \zeta.$$

Therefore

$$\gamma_0'(\zeta) = 1, \qquad \gamma_1'(\zeta) = (1-\zeta)\,\gamma_0(\zeta) - \zeta\gamma_0'(\zeta) + \zeta,$$

and

$$\gamma_s'(\zeta) = (1-\zeta)\,\gamma_{s-1}(\zeta) - \zeta\gamma_{s-1}'(\zeta) \qquad (s \geqslant 2). \tag{4.09}$$

Thus we obtain

$$\gamma_0(\zeta) = \zeta + c_0, \tag{4.10}$$

where c_0 is a constant,

$$\gamma_1(\zeta) = -\tfrac{1}{3}\zeta^3 + \tfrac{1}{2}(1-c_0)\,\zeta^2 + c_0\,\zeta + c_1, \tag{4.11}$$

where c_1 is another constant, and so on. The actual values of the c_j cannot be determined in this way, however, in contrast to the analysis in §3.2.

To resolve the difficulty, we again turn to the methods of J. C. P. Miller (1952). In addition to (4.08), we employ the difference equation

$$C_n(x) = 1 + \frac{n+1}{x} C_{n+1}(x),$$

which is easily obtained from (4.06) by increasing n by a unit. Setting

$$x = n+\zeta = (n+1)+(\zeta-1)$$

and substituting by means of (4.07), we find that

$$(n+\zeta)\sum_{s=0}^{\infty} \frac{\gamma_s(\zeta)}{n^s} \sim n + \zeta + \sum_{s=0}^{\infty} \frac{\gamma_s(\zeta-1)}{(n+1)^{s-1}}.$$

The factor $1/(n+1)^{s-1}$ in the last sum can be expanded in descending powers of n. Then equating coefficients, we obtain

$$\left.\begin{aligned} \gamma_0(\zeta) &= 1 + \gamma_0(\zeta-1), & \gamma_1(\zeta) + \zeta\gamma_0(\zeta) &= \zeta + \gamma_0(\zeta-1) + \gamma_1(\zeta-1), \\ \gamma_2(\zeta) + \zeta\gamma_1(\zeta) &= \gamma_2(\zeta-1), & \gamma_3(\zeta) + \zeta\gamma_2(\zeta) &= -\gamma_2(\zeta-1) + \gamma_3(\zeta-1), \end{aligned}\right\} \tag{4.12}$$

and so on.

The first of (4.12) is satisfied by (4.10), whatever the value of c_0. But on substituting in the second equation by means of (4.10) and (4.11), we find that the terms involving c_1 and ζ disappear, leaving $c_0 = -\frac{1}{3}$. Accordingly $\gamma_0(\zeta) = \zeta - \frac{1}{3}$, in agreement with §4.2. The expression (4.11) now becomes

$$\gamma_1(\zeta) = -\tfrac{1}{3}\zeta^3 + \tfrac{2}{3}\zeta^2 - \tfrac{1}{3}\zeta + c_1.$$

Then by integration of (4.09) with $s = 2$, we have

$$\gamma_2(\zeta) = \tfrac{1}{15}\zeta^5 - \tfrac{1}{9}\zeta^3 - \tfrac{1}{2}c_1\zeta^2 + c_1\zeta + c_2.$$

Substitution of the last two results into the third of (4.12), with $\zeta = 0$, yields $c_1 = \frac{4}{135}$. Hence

$$\gamma_1(\zeta) = -\tfrac{1}{3}\zeta^3 + \tfrac{2}{3}\zeta^2 - \tfrac{1}{3}\zeta + \tfrac{4}{135}.$$

The next complete cycle produces

$$\gamma_2(\zeta) = \tfrac{1}{15}\zeta^5 - \tfrac{1}{9}\zeta^3 - \tfrac{2}{135}\zeta^2 + \tfrac{4}{135}\zeta + \tfrac{8}{2835},$$

and so on.

4.4 In applying the foregoing results to the calculation of $e^{-x}E_1(xe^{i\pi})$, it should be noticed that any possible improvement in accuracy resulting from computation of the converging factor *would be nullified if the term* $i\pi e^{-x}$ *in* (4.05) *were neglected* (on the plausible grounds that it is exponentially small compared with individual terms $u_s(x)$). This is because

$$C_n(n+\zeta)u_n(n+\zeta) = C_n(n+\zeta)\frac{n!}{(n+\zeta)^{n+1}} = (\zeta - \tfrac{1}{3})e^{-n-\zeta}\left(\frac{2\pi}{n}\right)^{1/2} + O\left(\frac{e^{-n}}{n^{3/2}}\right),$$

as $n \to \infty$. Hence for $x = n + \zeta$ and large n, $C_n(x)u_n(x)$ is smaller—by a factor $O(n^{-1/2})$—than the contribution from $i\pi e^{-x}$.

Ex. 4.1 Estimate the numerical value of $e^{-5}E_1(5e^{i\pi})$ as accurately as possible from (4.05) and (4.07).

Ex. 4.2 Let $x > 0$, $F(x)$ denote Dawson's integral (Chapter 2, §4.1), and $C_n(x)$ be defined by

$$F(x) = \sum_{s=0}^{n-1} u_s(x) + C_n(x)\,u_n(x) \qquad (n = 0, 1, 2, \ldots),$$

in which $u_s(x) = 1\cdot 3\cdot 5 \cdots (2s-1)/(2^{s+1}x^{2s+1})$. Show that

$$F(x) + \tfrac{1}{2}i\pi^{1/2}\exp(-x^2) = \frac{x}{2}\int_0^{\infty} \frac{\exp(-x^2 t)}{(1-t)^{1/2}}\,dt,$$

where the path passes above the point $t = 1$, and hence that

$$F(x) + \tfrac{1}{2}i\pi^{1/2}\exp(-x^2) = \sum_{s=0}^{n-1} u_s(x) + \{C_n(x) + iD_n(x)\}\,u_n(x),$$

where

$$C_n(x) + iD_n(x) = x^2\int_0^{\infty} \frac{\exp(-x^2 t)}{(1-t)^{n+(1/2)}}\,dt,$$

with the same path. Deduce that for large n

$$C_n\{(n+\zeta)^{1/2}\} \sim \sum_{s=0}^{\infty} \frac{\gamma_s(\zeta)}{n^s}$$

uniformly with respect to bounded ζ, where $\gamma_0(\zeta) = \zeta + \frac{1}{6}$ and higher coefficients are polynomials in ζ.

Ex. 4.3 In the notation of the preceding exercise, show that

$$xC_n'(x) + (2x^2 - 2n - 1)\,C_n(x) = 2x^2, \qquad (n+\tfrac{1}{2})\,C_{n+1}(x) = x^2 C_n(x) - x^2,$$

and hence that $\gamma_1(\zeta) = -\frac{1}{3}\zeta^3 + \frac{1}{6}\zeta^2 + \frac{1}{12}\zeta - \frac{13}{1080}$.

*5 Confluent Hypergeometric Function

5.1 In §§3 and 4 the converging factors were expressible as integrals of particularly simple type. For other asymptotic expansions, the determination of an approximation to the converging factor is usually much more difficult.

As an illustration, we consider the confluent hypergeometric function $U(a, a+b+1, z)$ for large complex values of z, and fixed real values of the parameters a and b, a being positive. From Chapter 7, §10.1, we know that in the sector $|\mathrm{ph}\, z| \leqslant \frac{3}{2}\pi - \delta$ $(< \frac{3}{2}\pi)$, the asymptotic expansion of this function is given by

$$U(a, a+b+1, z) \sim \sum_{s=0}^{\infty} u_s(z),$$

where $u_s(z) = (-)^s (a)_s (-b)_s/(s!\,z^{s+a})$. When b is a nonnegative integer, the series terminates and equals $U(a, a+b+1, z)$ exactly. We exclude this possibility in the analysis which follows.

From Chapter 7, (10.04), we have the integral representation

$$U(a, a+b+1, z) = \frac{1}{\Gamma(a)} \int_0^\infty e^{-zt} t^{a-1} (1+t)^b \, dt \qquad (|\mathrm{ph}\, z| < \tfrac{1}{2}\pi).$$

Set $z = re^{i\theta}$ and rotate the path of integration to coincide with the ray

$$t = \tau e^{-i\theta} \qquad (0 \leqslant \tau < \infty).$$

Then

$$U(a, a+b+1, z) = \frac{e^{-ia\theta}}{\Gamma(a)} \int_0^\infty e^{-r\tau} \tau^{a-1} (1+\tau e^{-i\theta})^b \, d\tau. \tag{5.01}$$

As in Chapter 4, §1.2, analytic continuation immediately extends this result to $|\theta| < \pi$.

From Taylor's theorem, any n times continuously differentiable function $\phi(\tau)$ satisfies

$$\phi(\tau) = \phi(0) + \tau\phi'(0) + \cdots + \frac{\tau^{n-1}}{(n-1)!} \phi^{(n-1)}(0) + \int_0^\tau \frac{(\tau-u)^{n-1}}{(n-1)!} \phi^{(n)}(u) \, du.$$

Therefore

$$(1+\tau e^{-i\theta})^b = \sum_{s=0}^{n-1} \binom{b}{s} \tau^s e^{-is\theta} + \binom{b}{n} \tau^n e^{-in\theta} \eta_n(\tau), \tag{5.02}$$

where

$$\eta_n(\tau) = \frac{n}{\tau^n} \int_0^\tau (\tau-u)^{n-1} (1+ue^{-i\theta})^{b-n} \, du = n \int_0^1 (1-v)^{n-1} (1+v\tau e^{-i\theta})^{b-n} \, dv. \tag{5.03}$$

Substituting (5.02) in (5.01), we obtain

$$U(a, a+b+1, z) = \sum_{s=0}^{n-1} u_s(z) + C_n(z) u_n(z),$$

where

$$C_n(z) = \frac{r^{n+a}}{\Gamma(n+a)} \int_0^\infty e^{-r\tau} \tau^{n+a-1} \eta_n(\tau) \, d\tau,$$

and is the required converging factor. Since

$$\frac{u_{s+1}(z)}{u_s(z)} = \frac{(a+s)(b-s)}{(s+1)z},$$

the term of smallest absolute value in the asymptotic expansion is again given by $s = [|z|]$, or thereabouts, when $|z|$ is large. Accordingly, we seek the asymptotic expansion, for large n and bounded ζ, of

$$C_n\{(n+\zeta)e^{i\theta}\} = \frac{(n+\zeta)^{n+a}}{\Gamma(n+a)} \int_0^\infty e^{-(n+\zeta)\tau} \tau^{n+a-1} \eta_n(\tau) \, d\tau. \tag{5.04}$$

5.2 To begin with, we restrict $|\theta| \leqslant \frac{1}{2}\pi$, deferring the more difficult case $|\theta| > \frac{1}{2}\pi$ until §5.4.

When n is large, the peak value of the factor $e^{-(n+\zeta)\tau}\tau^{n+a-1}$ is located at $\tau = 1$, approximately. Our approach is to hope that $\eta_n(\tau)$ is a sufficiently slowly varying function of τ to enable it to be replaced meaningfully in (5.04) by its truncated Taylor-series expansion at $\tau = 1$, given by

$$\eta_n(\tau) = \eta_n(1) + (\tau-1)\eta_n'(1) + \cdots + \frac{(\tau-1)^{m-1}}{(m-1)!}\eta_n^{(m-1)}(1) + \frac{(\tau-1)^m}{m!}\eta_n^{(m)}(\chi),$$

where m is an arbitrary integer and χ is some number between 1 and τ.

From (5.03) we obtain by differentiation

$$\eta_n^{(m)}(\tau) = (-)^m n(n-b)_m e^{-im\theta}\int_0^1 (1-v)^{n-1}(1+v\tau e^{-i\theta})^{b-n-m} v^m\, dv. \qquad (5.05)$$

Since $|\theta| \leqslant \frac{1}{2}\pi$ it follows that $|1+v\tau e^{-i\theta}| \geqslant 1$. Assume that m is fixed and n is sufficiently large to ensure that $b-n-m \leqslant 0$. Then

$$|\eta_n^{(m)}(\tau)| \leqslant n|(n-b)_m|\int_0^1 (1-v)^{n-1}v^m\, dv = n|(n-b)_m|\frac{\Gamma(n)\Gamma(m+1)}{\Gamma(n+m+1)} = O(1), \qquad (5.06)$$

as $n \to \infty$, uniformly with respect to $\tau \in [0, \infty)$. Therefore

$$\eta_n(\tau) = \eta_n(1) + (\tau-1)\eta_n'(1) + \cdots + \frac{(\tau-1)^{m-1}}{(m-1)!}\eta_n^{(m-1)}(1) + (\tau-1)^m O(1).$$

Changing m into $2m$ and substituting in (5.04), we obtain†

$$C_n\{(n+\zeta)e^{i\theta}\} = \sum_{j=0}^{2m-1}\frac{\eta_n^{(j)}(1)}{j!}L_j(n,\zeta) + L_{2m}(n,\zeta)\,O(1) \qquad (n\to\infty), \qquad (5.07)$$

where the O term is independent of θ and ζ, and

$$L_j(n,\zeta) = \frac{(n+\zeta)^{n+a}}{\Gamma(n+a)}\int_0^\infty e^{-(n+\zeta)\tau}\tau^{n+a-1}(\tau-1)^j\, d\tau.$$

In particular, $L_0(n,\zeta) = 1$.

For fixed j and large n, the asymptotic expansion of $L_j(n,\zeta)$ can be found by the standard procedure of subdividing the integration range at $\tau = 1$, taking $\tau - 1 - \ln\tau$ as new integration variable, and applying Watson's lemma. In this way we arrive at

$$L_{2j-1}(n,\zeta) \sim \frac{2^j\Gamma(j+\frac{1}{2})}{\pi^{1/2}n^j}\left(\frac{3a+2j-2-3\zeta}{3} + \cdots\right),$$

$$L_{2j}(n,\zeta) \sim \frac{2^j\Gamma(j+\frac{1}{2})}{\pi^{1/2}n^j}(1+\cdots),$$

† m is changed to $2m$ to facilitate the bounding of the integral of the O term: $(\tau-1)^m$ changes sign in $[0,\infty)$ when m is odd, but not when m is even.

where the series in parentheses descend in integer powers of n, the coefficients being polynomials in ζ.

5.3 We also need the asymptotic expansion of the quantity $\eta_n^{(j)}(1)$ for fixed j and large n. This is obtainable from (5.05), with $m=j$ and $\tau=1$, by taking a new integration variable w related to v by

$$w = \frac{1-v}{1+ve^{-i\theta}}, \qquad v = \frac{1-w}{1+we^{-i\theta}}, \qquad \frac{dv}{dw} = -\frac{1+e^{-i\theta}}{(1+we^{-i\theta})^2}. \tag{5.08}$$

Thus

$$\eta_n^{(j)}(1) = (-)^j n(n-b)_j e^{-ij\theta}(1+e^{-i\theta})^{b-j}\int_0^1 w^{n-1}\frac{(1-w)^j}{(1+we^{-i\theta})^{b+1}}\,dw. \tag{5.09}$$

Since the transformation from v to w is a fractional linear transformation, the new integration path is a circular arc. From the first of (5.08) it is seen that this arc lies in the right half-plane. And since the only singularity of the integrand in (5.09) is located in the left half-plane the path can be deformed into a straight line.

The required expansion of $\eta_n^{(j)}(1)$ is now derivable by Laplace's method. This is equivalent to expanding the factor $(1-w)^j/(1+we^{-i\theta})^{b+1}$ in ascending powers of $1-w$, given by

$$\frac{(1-w)^j}{(1+we^{-i\theta})^{b+1}} = \frac{(1-w)^j}{(1+e^{-i\theta})^{b+1}}\sum_{s=0}^{\infty}\binom{-b-1}{s}(-\alpha)^s(1-w)^s,$$

where, again, we have written

$$\alpha = 1/(1+e^{i\theta}), \tag{5.10}$$

and then integrating term-by-term with the aid of the Beta-function integral. The result is

$$\eta_n^{(j)}(1) \sim (-)^j(1-\alpha)\alpha^j\frac{(n-b)_j}{(n+1)_j}\sum_{s=0}^{\infty}\alpha^s\frac{(b+1)_s(s+1)_j}{(n+j+1)_s}.$$

This series is a generalized asymptotic expansion, with scale n^{-s}. It is rearrangeable in descending powers of n^{-1}, and for the cases $j=0,1,2$ these rearrangements begin

$$\eta_n(1) = (1-\alpha)\{1+\alpha(b+1)n^{-1}+O(n^{-2})\}, \qquad \eta_n'(1) = -(1-\alpha)\alpha\{1+O(n^{-1})\},$$
$$\eta_n''(1) = 2(1-\alpha)\alpha^2\{1+O(n^{-1})\}.$$

Substituting in (5.07) by means of these results and the expansions for the $L_j(n,\zeta)$ derived in §5.2, we see that the required converging factor has an asymptotic expansion of the form

$$C_n\{(n+\zeta)e^{i\theta}\} \sim \sum_{s=0}^{\infty}\frac{\gamma_s(\zeta)}{n^s} \tag{5.11}$$

as $n\to\infty$, uniformly with respect to $\theta\in[-\frac{1}{2}\pi,\frac{1}{2}\pi]$ and bounded ζ. The coefficients $\gamma_s(\zeta)$ are polynomials in ζ. Bearing in mind that $L_0(n,\zeta)=1$, we find that

$$\gamma_0(\zeta) = 1-\alpha, \qquad \gamma_1(\zeta) = \alpha(1-\alpha)(\zeta+b-a+1+\alpha), \tag{5.12}$$

where α is defined by (5.10).

As in §§3 and 4, expressions for higher coefficients can be found more easily by use of the differential equation satisfied by $U(a, a+b+1, z)$; see Exercise 5.1 below.

5.4 We show next how to extend the validity of (5.11) to the θ interval $[-\pi+\delta, \pi-\delta]$, where $\delta \in (0, \frac{1}{2}\pi)$. The main difficulty for the new range is that we cannot pass easily from (5.05) to (5.06). Instead, we use the transformation

$$w = \frac{1-v}{1+v\tau e^{-i\theta}};$$

compare (5.08) in the case $\tau = 1$. This leads to

$$\eta_n^{(j)}(\tau) = (-)^j n(n-b)_j e^{-ij\theta}(1+\tau e^{-i\theta})^{b-j} \int_0^1 w^{n-1} \frac{(1-w)^j}{(1+w\tau e^{-i\theta})^{b+1}} dw. \quad (5.13)$$

The path in the w plane is a circular arc. It may be deformed into the interval $[0, 1]$, provided that the singularity of the integrand at $w_0 \equiv -e^{i\theta}/\tau$ does not lie within the domain **S**, say, bounded by the interval and the arc. As we saw in §5.3, when $|\theta| \leqslant \frac{1}{2}\pi$ the arc and w_0 lie in opposite half-planes and no problem arises. As θ increases continuously from $\frac{1}{2}\pi$ to $\pi-\delta$, both w_0 and the boundary of **S** vary in a continuous manner. Since $\operatorname{ph} w_0 \neq 0$, the only way for w_0 to pass within **S** would be to cross the boundary arc: this is impossible, however, since in the v plane the arc corresponds to $v \in [0, 1]$, and w_0 corresponds to $v = \infty$. Similarly when θ decreases from $-\frac{1}{2}\pi$ to $-\pi+\delta$. Hence w_0 is exterior to **S** for all θ in the range of interest, and the path in (5.13) may be taken to be a straight line.

Now consider the function

$$\Psi \equiv \frac{1+\tau e^{-i\theta}}{1+w\tau e^{-i\theta}} = \frac{1}{w}\left\{1 + \frac{w-1}{1+w\tau e^{-i\theta}}\right\}.$$

If $\theta \in [-\pi+\delta, \pi-\delta]$, $\tau \in [0, \infty)$, and $w \in (0, 1]$, then $|\Psi|$ is bounded by $(1+\csc\delta)/w$. Hence

$$\left|\left(\frac{1+\tau e^{-i\theta}}{1+w\tau e^{-i\theta}}\right)^{b+1}\right| \leqslant \frac{A}{w^{b+1}} \qquad (b \geqslant -1),$$

where A denotes a generic constant. We also have

$$|1+\tau e^{-i\theta}| \geqslant \sin\delta.$$

Substituting in (5.13) by means of these bounds, we derive

$$|\eta_n^{(j)}(\tau)| \leqslant An|(n-b)_j| \int_0^1 w^{n-b-2}(1-w)^j dw = O(1) \quad (5.14)$$

as $n \to \infty$, uniformly with respect to θ and τ. When $b < -1$, we use the inequality

$$\frac{1}{|\Psi|} = \left| w + \frac{1-w}{1+\tau e^{-i\theta}} \right| \leqslant 1 + \csc\delta,$$

and the conclusion (5.14) again follows.

The analysis proceeding from (5.06) is virtually unchanged; in consequence *the final result* (5.11) *is uniformly valid with respect to* $\theta \in [-\pi+\delta, \pi-\delta]$ *and bounded* ζ.

5.5 When θ approaches $-\pi$ or π, α tends to infinity and the foregoing results break down. However, we shall not attempt an extension corresponding to §4 in the present case.

Ex. 5.1 In the notation of this section, show that

$$z^2 C_n''(z) + z(b-a-2n+1-z)\,C_n'(z) + \{(n+a)(n-b)+nz\}\,C_n(z) = nz.$$

Use this equation to confirm the values of $\gamma_0(\zeta)$ and $\gamma_1(\zeta)$ given by (5.12).

Ex. 5.2 Let ν be real and fixed, x positive, and the asymptotic expansion (8.04) of Chapter 7 for the modified Bessel function $K_\nu(x)$ truncated after n terms. By using Exercise 10.1 of Chapter 7, show that the converging factor $C_n(x)$ has the asymptotic expansion

$$C_n(\tfrac{1}{2}n+\tfrac{1}{2}\zeta) \sim \frac{1}{2} + \frac{2\zeta+1}{8n} - \frac{4\zeta^2+2\zeta+3-8\nu^2}{32n^2} + \frac{8\zeta^3-2\zeta+1-8\nu^2}{128n^3} - \cdots$$

as $n\to\infty$, uniformly with respect to bounded ζ. [Airey, 1937.]

6 Euler's Transformation

6.1 Another class of methods for increasing the accuracy obtainable from an asymptotic series consists of transformations into new series in which the initial terms decrease in a more rapid manner. Then it is often true that the optimum remainder term is smaller for the new series than for the original series. It might even happen that the new series converges and its sum is the wanted function; in this event there is no limit on the attainable accuracy.

Several transformations are available in the literature,† but we confine the text to one of the simplest and most widely used.

6.2 Consider the case in which the wanted function can be expanded in a series of inverse powers of the form

$$f(z) = \sum_{s=0}^{\infty} \frac{a_s}{z^{s+1}}, \tag{6.01}$$

and to begin with suppose that this expansion converges when $|z| > r$, say. This implies that $f(z)$ is holomorphic in this annulus and vanishes at infinity.

Let k be an arbitrary nonzero real or complex number. Since

$$|z| = |(z-k)+k| \geqslant |z-k| - |k|,$$

the region of holomorphicity includes $|z-k| > r+|k|$. Hence $f(z)$ is also capable of expansion in a series of inverse powers of $z-k$ of the form

$$f(z) = \sum_{s=0}^{\infty} \frac{b_s}{(z-k)^{s+1}}, \tag{6.02}$$

convergent in an annulus which includes $|z-k| > r+|k|$.

† See references on p. 544.

To express the coefficients b_s in terms of the a_s we employ the Binomial theorem:

$$\frac{1}{z^{s+1}} = \frac{1}{(z-k)^{s+1}} \sum_{j=0}^{\infty} \binom{-s-1}{j} \left(\frac{k}{z-k}\right)^j \qquad (|z-k| > |k|). \tag{6.03}$$

When $|z|$ is large, we have from (6.01)

$$f(z) = \sum_{s=0}^{n-1} \frac{a_s}{z^{s+1}} + O\left(\frac{1}{z^{n+1}}\right), \tag{6.04}$$

where n is an arbitrary integer. Substitution of (6.03) produces

$$f(z) = \sum_{s=0}^{n-1} \frac{a_s}{(z-k)^{s+1}} \sum_{j=0}^{n-s-1} \binom{-s-1}{j} \left(\frac{k}{z-k}\right)^j + O\left(\frac{1}{z^{n+1}}\right). \tag{6.05}$$

Comparing coefficients in this expansion with (6.02), we see that

$$b_s = a_s + \binom{-s}{1} a_{s-1} k + \binom{-s+1}{2} a_{s-2} k^2 + \cdots + \binom{-1}{s} a_0 k^s,$$

that is,

$$b_s = \sum_{j=0}^{s} (-)^j \binom{s}{j} a_{s-j} k^j \qquad (s = 0, 1, 2, \ldots). \tag{6.06}$$

Theorem 6.1 *If k is an arbitrary number and sequences $\{a_s\}$ and $\{b_s\}$ are related by (6.06), then*

$$\sum_{s=0}^{\infty} \frac{a_s}{z^{s+1}} = \sum_{s=0}^{\infty} \frac{b_s}{(z-k)^{s+1}}, \tag{6.07}$$

whenever both series converge. When one series converges and the other diverges, the sum of the convergent series is the analytic continuation of the function represented elsewhere in the z plane by the other series.

The foregoing analysis establishes this theorem, except when z lies on the circle of convergence of one (or both) of the series. In these cases the proof is completed by use of Abel's theorem on the continuity of power series.

Equation (6.07) is *Euler's transformation.*

6.3 For a given set of coefficients a_s and a given choice of k, a convenient way of evaluating the coefficients b_s is to generate the forward differences $\Delta^s(a_j k^{-j})$ of the sequence $a_0, a_1 k^{-1}, a_2 k^{-2}, \ldots$. As in Chapter 8, Exercise 5.3, the forward difference operator Δ is defined by

$$\Delta v_j = v_{j+1} - v_j, \qquad \Delta^2 v_j = \Delta v_{j+1} - \Delta v_j, \qquad \ldots.$$

From (6.06) we readily verify by induction the well-known finite-difference formula

$$b_s = k^s [\Delta^s (a_j k^{-j})]_{j=0} \qquad (s = 0, 1, \ldots). \tag{6.08}$$

The most commonly used version of the transformation in numerical work corresponds to the choice $k = 1$ and $z = -1$; thus

$$\sum_{s=0}^{\infty} (-)^s a_s = \sum_{s=0}^{\infty} (-)^s \frac{\Delta^s a_0}{2^{s+1}}, \tag{6.09}$$

where $\Delta^s a_0 \equiv [\Delta^s a_j]_{j=0}$.† Another derivation of this result is indicated in Exercise 6.4 below.

6.4 As an example, let us apply (6.09) to the expansion

$$\ln 2 = 1 - \tfrac{1}{2} + \tfrac{1}{3} - \tfrac{1}{4} + \tfrac{1}{5} - \cdots.$$

The requisite forward differences are given in the following table:

s	a_s	Δa_s	$\Delta^2 a_s$	$\Delta^3 a_s$	$\Delta^4 a_s$
0	1	$-\frac{1}{1\cdot 2}$	$\frac{2}{1\cdot 2\cdot 3}$	$-\frac{2\cdot 3}{1\cdot 2\cdot 3\cdot 4}$	$\frac{2\cdot 3\cdot 4}{1\cdot 2\cdot 3\cdot 4\cdot 5}$
1	$\frac{1}{2}$	$-\frac{1}{2\cdot 3}$	$\frac{2}{2\cdot 3\cdot 4}$	$-\frac{2\cdot 3}{2\cdot 3\cdot 4\cdot 5}$	
2	$\frac{1}{3}$	$-\frac{1}{3\cdot 4}$	$\frac{2}{3\cdot 4\cdot 5}$		
3	$\frac{1}{4}$	$-\frac{1}{4\cdot 5}$			
4	$\frac{1}{5}$				

Here (and in similar difference tables) each entry is formed by subtracting two numbers in the column immediately on its left: the one on the same line from the one just below it.

From the table it is easily seen that

$$\Delta^s a_0 = (-1)^s/(s+1) \qquad (s = 0, 1, \ldots).$$

Hence the transformed series is given by

$$\ln 2 = \frac{1}{1\cdot 2} + \frac{1}{2\cdot 2^2} + \frac{1}{3\cdot 2^3} + \frac{1}{4\cdot 2^4} + \cdots. \tag{6.10}$$

This expansion is recognizable as the power series for $-\ln(1-x)$ with $x = \frac{1}{2}$, as we expect. From the numerical standpoint the series (6.10) lends itself more readily to direct summation because of its greater rate of convergence.

For series of real terms this example typifies the circumstances in which the transformation (6.09) is highly effective, namely, the terms in the original series alternate in sign and their absolute values have smooth behavior.

6.5 In the preceding example the original series and the transformed series both converge. More striking results are sometimes obtained by applying Euler's transformation to divergent series.

† Some writers reserve the name Euler's transformation for (6.09), calling (6.07) the *generalized Euler transformation.*

Consider the expansion

$$\ln\left(\frac{z-1}{z}\right) = -\frac{1}{z} - \frac{1}{2z^2} - \frac{1}{3z^3} - \frac{1}{4z^4} - \cdots \tag{6.11}$$

in the case $z = -\frac{1}{2}$. The series becomes

$$2 - \frac{2^2}{2} + \frac{2^3}{3} - \frac{2^4}{4} + \cdots. \tag{6.12}$$

Notwithstanding divergence, we apply (6.09), this time working to an accuracy of three decimal places:

s	a_s	Δa_s	$\Delta^2 a_s$	$\Delta^3 a_s$	$\Delta^4 a_s$	$\Delta^5 a_s$	$\Delta^6 a_s$
0	2.000	0.000	0.667	−0.001	0.402	−0.003	0.289
1	2.000	0.667	0.666	0.401	0.399	0.286	
2	2.667	1.333	1.067	0.800	0.685		
3	4.000	2.400	1.867	1.485			
4	6.400	4.267	3.352				
5	10.667	7.619					
6	18.286						

From the differences we calculate the transformed series:

$$1.000 + 0.000 + 0.083 + 0.000 + 0.013 + 0.000 + 0.002 + \cdots = 1.098. \tag{6.13}$$

This sum is in close agreement with the value of the left-hand side of (6.11) at $z = -\frac{1}{2}$, that is, $\ln 3 = 1.09861\ldots$. The analysis of §6.2 confirms that this should be the case; indeed (6.13) is simply a numerical form of the expansion

$$\ln\left(\frac{z-1}{z}\right) = \frac{2}{1-2z} + \frac{2}{3(1-2z)^3} + \frac{2}{5(1-2z)^5} + \cdots$$

when $z = -\frac{1}{2}$.

To recapitulate, we arrived at a meaning and numerical value for the "sum" of the divergent series (6.12) by regarding this expansion as a power series in a complex variable. The sum is then the analytic continuation of the function represented by the power series.

Ex. 6.1 Show that successive Euler transformations with parameters k_1 and k_2 are equivalent to a single Euler transformation with parameter k_1+k_2. Deduce that in the notation of §6.2

$$a_s = \sum_{j=0}^{s} \binom{s}{j} b_{s-j} k^j \qquad (s = 0, 1, 2, \ldots).$$

Ex. 6.2 Evaluate $\frac{1}{4}\pi$ to five decimal places by summing directly the first six terms of the series $1-\frac{1}{3}+\frac{1}{5}-\frac{1}{7}+\cdots$ and then applying (6.09) to the tail.

Ex. 6.3 Prove that

$$\sum_{s=0}^{\infty} \frac{z^s}{s!} a_s = e^z \sum_{s=0}^{\infty} \frac{z^s}{s!} \Delta^s a_0$$

within the common circle of convergence of these series.

***Ex. 6.4** Let n be an arbitrary nonnegative integer and β an arbitrary real number. By the method of summation by parts (Chapter 8, §5) derive the following form of Euler's transformation

$$\sum_{s=1}^{\infty} e^{is\beta} a_s = \sum_{s=0}^{n-1} \left(\frac{e^{i\beta}}{1-e^{i\beta}}\right)^{s+1} \Delta^s a_1 + \left(\frac{e^{i\beta}}{1-e^{i\beta}}\right)^{n} \sum_{s=1}^{\infty} e^{is\beta} \Delta^n a_s,$$

with the conditions: (i) $e^{i\beta} \neq 1$; (ii) the series on the left-hand side converges; (iii) $\Delta^s a_m \to 0$ as $m \to \infty$ for each $s = 0, 1, \ldots, n-1$.

7 Application to Asymptotic Expansions

7.1 Theorem 7.1 *If*

$$f(z) \sim \sum_{s=0}^{\infty} \frac{a_s}{z^{s+1}} \tag{7.01}$$

as $z \to \infty$ in a given region **R**, *then*

$$f(z) \sim \sum_{s=0}^{\infty} \frac{b_s}{(z-k)^{s+1}} \tag{7.02}$$

as $z \to \infty$ in **R**, *where k is an arbitrary constant, and b_s is given by* (6.06) *or, equivalently,* (6.08).

The proof of this theorem is contained in the analysis of §6.2: essentially it is the step from (6.04) to (6.05).

7.2 As an illustration, consider again the asymptotic expansion of the exponential integral:

$$e^x E_1(x) \sim \sum_{s=0}^{\infty} u_s(x) \qquad (x \to +\infty),$$

where

$$u_s(x) = (-)^s s!/x^{s+1}.$$

In §3.3 we showed that numerical accuracy could be increased by means of a converging factor. Here we show how to achieve comparable improvement by truncation just before one of the smallest terms and application of Euler's transformation to subsequent terms.

We again take $x = 5$, and write $a_s = (-)^s u_{s+5}(5)$. Numerical computations yield the following values for the forward differences at $s = 0$, to eight places of decimals.[†]

$$\begin{aligned}
a_0 &= -0.00768\,000, & \Delta^5 a_0 &= -0.00433\,766, & \Delta^{10} a_0 &= -0.25517\,971,\\
\Delta a_0 &= -0.00153\,600, & \Delta^6 a_0 &= -0.00801\,915, & \Delta^{11} a_0 &= -0.74904\,262,\\
\Delta^2 a_0 &= -0.00215\,040, & \Delta^7 a_0 &= -0.01643\,201, & \Delta^{12} a_0 &= -2.35909\,768,\\
\Delta^3 a_0 &= -0.00190\,464, & \Delta^8 a_0 &= -0.03751\,802, & \Delta^{13} a_0 &= -7.93135\,618.\\
\Delta^4 a_0 &= -0.00281\,395, & \Delta^9 a_0 &= -0.09382\,364, & &
\end{aligned}$$

† The calculations in this example can be shortened by use of a recurrence relation, due to J. C. P. Miller, which is given in Exercise 7.1 below.

Thus we have

$$\sum_{s=0}^{\infty} (-)^s \frac{\Delta^s a_0}{2^{s+1}} = 10^{-8}(-384000 + 38400 - 26880 + 11904 - 8794$$
$$+ 6778 - 6265 + 6419 - 7328 + 9162 - 12460$$
$$+ 18287 - 28798 + 48409 - \cdots). \tag{7.03}$$

Truncating this expansion at the term $10^{-8}(6778)$ and summing, we obtain $10^{-8}(-362592)$. Addition to the exact sum of the first five $u_s(5)$, that is, 0.17408 (§3.3), then yields

$$0.17045\,408, \tag{7.04}$$

compared with the correct value

$$e^5 E_1(5) = 0.17042\,2176\ldots.$$

Even closer agreement is attainable by a second Euler transformation, this time applied to the neglected terms within the parentheses of (7.03). The computations of the relevant differences are given in the following table.†

s	Neglected terms	Differences						
0	−6265	−154	−755	−170	−369	−157	−407	−200
1	−6419	−909	−925	−539	−526	−564	−607	
2	−7328	−1834	−1464	−1065	−1090	−1171		
3	−9162	−3298	−2529	−2155	−2261			
4	−12460	−5827	−4684	−4416				
5	−18287	−10511	−9100					
6	−28798	−19611						
7	−48409							

Thus the extra contribution is $10^{-8}(-3132+38-94+11-12+2-3+1-\cdots)$. Truncation at the term $10^{-8}(+2)$‡ gives $10^{-8}(-3187)$. Then addition to the previous result (7.04) yields 0.17042 221, which is correct to within 4 units of the last decimal place used in the working.

7.3 Summarizing so far, we have, in effect, reexpanded the remainder term of the original asymptotic series

$$R_5(x) \sim \sum_{s=5}^{\infty} u_s(x)$$

in powers of $1/(x+5)$, and truncated the new series at the optimum stage. Then we

† Because the terms a_s are rounded to eight decimal places, errors accumulate in the successive columns of differences. This does not seriously impair the accuracy of the final answer however; see, for example, N.P.L. (1961, Chapter 7).

‡ The reason for choosing this term rather than the preceding term is clarified in the next subsection.

made a similar reexpansion and truncation of the new remainder term. The fact that the terms in the transformed series are smaller than the corresponding terms in the original series does not in itself establish that the new remainder terms are smaller. But the numerical evidence points strongly in this direction.

In this particular example the conclusions in the preceding paragraph can be supported by the following analysis.† From §3.1 we have

$$R_n(x) = (-)^n \int_0^\infty \frac{e^{-xt}t^n}{1+t}\,dt. \tag{7.05}$$

The expansion of the factor $(1+t)^{-1}$ in ascending powers of $t-1$ is given by

$$\frac{1}{1+t} = \frac{1}{2} - \frac{t-1}{2^2} + \frac{(t-1)^2}{2^3} - \cdots + (-)^{m-1}\frac{(t-1)^{m-1}}{2^m} + (-)^m \frac{(t-1)^m}{2^{m+1}\{1+\frac{1}{2}(t-1)\}},$$

m being arbitrary. Substitution in (7.05) produces

$$R_n(x) = \sum_{j=0}^{m-1} v_j(x) + S_m(x),$$

where

$$v_j(x) = \frac{(-)^{n+j}}{2^{j+1}} \int_0^\infty e^{-xt}t^n(t-1)^j\,dt, \tag{7.06}$$

and

$$S_m(x) = \frac{(-)^{n+m}}{2^{m+1}} \int_0^\infty \frac{e^{-xt}t^n(t-1)^m\,dt}{1+\frac{1}{2}(t-1)}. \tag{7.07}$$

By expansion of the factor $(t-1)^j$ in (7.06), we derive

$$v_j(x) = \frac{1}{2^{j+1}}\left\{u_{n+j}(x) + \binom{j}{1}u_{n+j-1}(x) + \binom{j}{2}u_{n+j-2}(x) + \cdots + u_n(x)\right\}.$$

Hence in the case $n = x = 5$,

$$v_j(5) = (-)^j \Delta^j a_0 / 2^{j+1};$$

that is, the $v_j(5)$ are the terms in the first transformed series. And if m is even, we derive from (7.07)

$$|S_m(x)| \leqslant \frac{1}{2^m}\int_0^\infty e^{-xt}t^n(t-1)^m\,dt = 2|v_m(x)|.$$

In other words, if the series (7.03) is truncated at a term involving a difference $\Delta^s a_0$ of odd order, then the remainder term does not exceed twice the first neglected term in magnitude (and has the same sign).

By expanding the factor $\{1+\frac{1}{2}(t-1)\}^{-1}$ in the integrand of (7.07) in powers of $\{\frac{1}{2}(t-1)-1\}$, the second Euler transformation can be justified in a similar way. We find that if both transformed series are truncated at differences of odd order,

† Rosser (1951). A factor in Rosser's error bound is corrected in the present account.

then the final error is 4ϑ times the first neglected term of the second series, where ϑ is some number in the interval $[0,1]$. This accords with the numerical results of §7.2.

Ex. 7.1 Using the notation of §7.2, let

$$a_s = (-)^s u_{s+n}(n) = (-)^n (s+n)!\, n^{-n-s-1}.$$

Show that $n\,\Delta a_s = (s+1)\,a_s$, and hence by induction that $n\,\Delta^{j+1} a_s = (s+j+1)\,\Delta^j a_s + j\,\Delta^{j-1} a_s$.

Ex. 7.2 Assume that x is real and positive and the transformation (6.09) is applied to neglected terms of the following expansion for the Goodwin–Staton integral (Chapter 4, Exercise 3.5):

$$\int_0^\infty \frac{\exp(-u^2)}{u+x}\,du \sim \sum_{s=0}^{\infty} (-)^s \frac{\Gamma(\frac{1}{2}s+\frac{1}{2})}{2x^{s+1}} \qquad (x\to\infty).$$

Show that the mth remainder term associated with the transformed series is bounded in magnitude by twice the first neglected term and has the same sign, provided that m is even.

Ex. 7.3 *Aitken's Δ^2-transformation* for a sequence $S_0, S_1, S_2, \ldots$ is defined by

$$T_n = (S_{n+2}\,S_n - S_{n+1}^2)/\Delta^2 S_n \qquad (n = 0,1,2,\ldots).$$

Show that if $f(z)$ satisfies the conditions of Theorem 7.1, and $S_n = \sum_{s=0}^{n-1} a_s z^{-s-1}$, then as $z\to\infty$ in **R**, n being kept fixed, we have

$$T_n = f(z) + O(z^{-n-3}) \quad (a_n \neq 0); \qquad T_n = f(z) + O(z^{-n-2}) \quad (a_n = 0,\ a_{n+1} \neq 0).$$

Historical Notes and Additional References

§1 The danger of using divergent expansions without error analysis was recognized both before and soon after the introduction of Poincaré's definition of an asymptotic expansion, see, for example, Stokes (1848). Generally, in systematic computations independent checks were applied with extreme care. Since then confidence in the use of asymptotic expansions has risen almost too far, for it is often believed (fallaciously) that the remainder term in an asymptotic expansion is always of the same numerical order as the first neglected term.

J. C. P. Miller and R. B. Dingle have stressed the need to distinguish regions of validity in the sense of Poincaré from regions of validity in the *complete sense of Watson*; see Watson (1911), Dingle (1962), and N.B.S. (1964, Chapter 19). Essentially the distinction is that in the former sense we neglect all terms in an asymptotic expansion which are exponentially small compared with other terms, whereas in the latter sense these terms are retained whenever they have numerical significance. For example, the expansion (4.03) of Chapter 7 is regarded as being completely valid in Watson's sense only when $-\frac{1}{2}\pi \leqslant \operatorname{ph} z \leqslant \frac{3}{2}\pi$, and would not be used for numerical purposes outside this sector.† Unfortunately a satisfactory definition of complete validity is unavailable. Another drawback to Watson's theory is the need for properties of the remainder term which are likely to be available only when a realistic bound for the remainder term is known. The theory is then largely unnecessary. In particular, this applies to the expansion of the logarithmic integral treated as an example by Watson.

§2 The idea of reexpanding the remainder term with a view to enhancing numerical accuracy was originated by Stokes (1857) and Stieltjes (1886). It has been developed (mostly in a formal manner) by several writers, especially Airey (1937), J. C. P. Miller (1952), and Dingle (1958, 1973). The present chapter is confined to cases in which rigorous analyses can be supplied.

§§3–4 These derivations are similar to those of Stieltjes (1886), Rosser (1955), and Wynn (1963). Many additional terms in the expansions of the converging factors of $\mathrm{Ei}(\pm x)$ in the case when

† Compare also §13 of Chapter 7.

x is an integer have been given by Murnaghan and Wrench (1963). Similar results for the error function and Dawson's integral have been given by Murnaghan (1965) and Wrench (1971).

§5 This analysis is based, in part, on the paper by Jeffreys (1958). Here Jeffreys indicates (in a formal manner) how the result may be used to estimate the remainder term when Watson's lemma is applied to an integral of the form

$$\int_0^\infty e^{-zt}t^{a-1}F(t)\,dt,$$

in which z is a large complex parameter, a is a positive constant, and $F(t)$ is an analytic function whose nearest singularity to the origin is a pole or branch point.

§§6–7 Many other valuable transformations of slowly convergent or asymptotic series into more powerful form have been described in the literature; see, for example, Watson (1912), Macfarlane (1949), Cherry (1950b), van Wijngaarden (1953), Shanks (1955), Franklin and Friedman (1957), and Wynn (1961).

ANSWERS TO EXERCISES

Chapter 1

2.4 False. The right-hand side should be replaced by $o(e^x)$.
2.6 4, 1, $4/e^2$.
3.3 $p^p/(e\sin\delta)^p$.
7.3 $(z+1)^{-1}+e^{-z}$.
11.2 (i) $2|n|$. (ii) 1. (iii) 2. (iv) $2/e$.
11.3 (i) 2. (ii) 2. (iii) $4e^2-2$.

Chapter 3

2.6 $\sigma = 0.11$.
8.4 (a) Yes. (b) No.

Chapter 4

1.2 $|\mathrm{ph}\, x| \leqslant \frac{3}{2}\pi-\delta < \frac{3}{2}\pi$.
2.1 $|\mathrm{ph}\, x| \leqslant \frac{3}{4}\pi-\delta < \frac{3}{4}\pi$.
8.4 Not if the analytic continuation of the original integral is used.

Chapter 5

3.1 $\frac{1}{2}\pi$.
4.1 (i) $z^i \sum a_s z^s$ and the conjugate series, where $a_0 = 1$ and

$$a_s/a_{s-1} = (2s^2+4is-3s+1-3i)/(2s^2+4is) \qquad (s \geqslant 1).$$

(ii) $\sum b_s(z-1)^s$ and $\sum c_s(z-1)^{s+(1/2)}$ where $b_0 = 1$, $b_1 = 0$, $c_0 = 1$, $c_1 = -\frac{1}{3}$, and for $s \geqslant 2$,

$$s(2s-1)\,b_s + 4(s-1)^2 b_{s-1} + (2s^2-7s+8)\,b_{s-2} = 0,$$

$$s(2s+1)\,c_s + (2s-1)^2 c_{s-1} + (2s^2-5s+5)\,c_{s-2} = 0.$$

6.1 (i) Irregular singularity of rank 1. (ii) Irregular singularity of infinite rank. (iii) Regular singularity with exponents $(3\pm 5^{1/2})/2$.

6.2 $5z^3 - 3z$ and $\dfrac{1}{z^4} + \dfrac{4\cdot 5}{2\cdot 9}\dfrac{1}{z^6} + \dfrac{4\cdot 5\cdot 6\cdot 7}{2\cdot 4\cdot 9\cdot 11}\dfrac{1}{z^8} + \cdots.$

9.5 Yes, by suitable deformation of the integration path.

Chapter 6

2.4 Approximate $w(2) = \frac{1}{2}e+\frac{1}{2}e^{-1} = 1.54\ldots$,

$$|\text{error}| \fallingdotseq |\tfrac{1}{2}e\{\varepsilon_1(2)-\tfrac{1}{2}\varepsilon_2'(1)\} + \tfrac{1}{2}e^{-1}\{-\varepsilon_2(1)+\tfrac{1}{2}\varepsilon_2'(1)\}| \leqslant 0.06.$$

Chapter 6 (cont.)

4.4 Suppose that $\int_x^\infty f^{-5/2}f'^2\,dx = \infty$, which would falsify the result. Since $\frac{3}{2}\int f^{-5/2}f'^2\,dx = -f^{-3/2}f' + \int f^{-3/2}f''\,dx$, we have $f^{-3/2}f' \to -\infty$ as $x \to \infty$. Therefore $f' < 0$ when x is sufficiently large. Hence f decreases monotonically to a constant value, which must be zero otherwise $f' \to -\infty$. From $f' = \text{constant} - \int_x^\infty (f^{-3/2}f'')f^{3/2}\,dx$ it follows that $f' = -c + o(f^{3/2})$ as $x \to \infty$, where c is a nonnegative constant. If $c > 0$, then integration would give $f \sim -cx$, which is impossible. Or if $c = 0$, we would have $f^{-3/2}f' = o(1)$, which is again a contradiction.

The second result is obtainable by integrating $f^{-3/2}f' = \text{constant} + o(1)$.

11.1

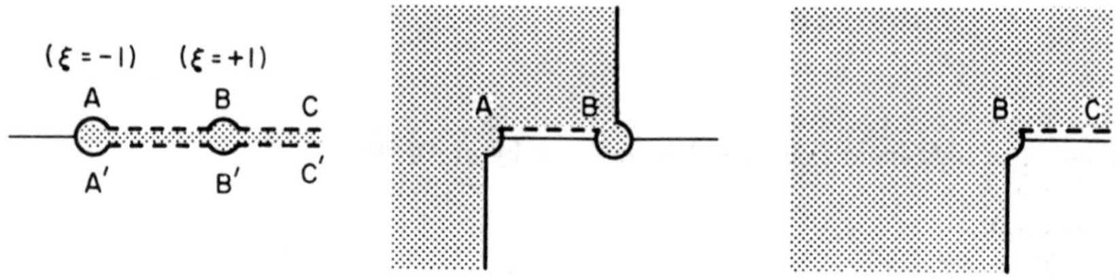

(i) (ii) Continuation across AB (iii) Continuation across BC

Continuations across $A'B'$ and $B'C'$ are the regions conjugate to (ii) and (iii).

12.1 m.

Chapter 7

1.1 Put $\zeta = z^2$ or z^4.

1.2 $w = A(z^{-3/4} - z^{-5/4})\exp(z^{1/2}) + B(z^{-3/4} + z^{-5/4})\exp(-z^{1/2})$.

1.3 $z^{1/4}\exp\{\pm i(8z)^{1/2}\}\sum_{s=0}^{\infty}(\pm i)^s \dfrac{(2L-s+\frac{3}{2})(2L-s+\frac{5}{2})\cdots(2L+s+\frac{1}{2})}{s!(32z)^{s/2}}$.

14.1 When $f_0 = g_1 = 0$, or when $f_0 \neq 0$ and $(g_1/f_0) + \alpha + 2$ is a positive integer.

Chapter 10

3.5 Yes, by continuation to the right across the imaginary axis above $z = i + i\delta$ and below $z = -i - i\delta$.

8.3 No.

Chapter 11

3.1 Yes.

10.4 By proper choice of $\zeta(z)$ and progressive path, relations (10.24) and (10.25) and the corresponding error bounds (9.03) hold without restriction on $\operatorname{ph} z$ when $z \in \mathbf{J}$. Outside the closure of $\mathbf{J}$ we must have $-\pi < \operatorname{ph} z < 2\pi$ for (10.24) and $-2\pi < \operatorname{ph} z < \pi$ for (10.25). (Compare §§8.1 and 8.2 of Chapter 10.)

10.7 $z(\zeta)$, $\quad -\dfrac{5z}{48(-\zeta)^{3/2}(z^2-1)^{1/2}} + \dfrac{3z^3+2z}{24(z^2-1)^2}, \quad \dfrac{2^{1/2}}{z}\left(\dfrac{1-z^2}{\zeta}\right)^{1/4}$.

Chapter 12

4.5 $a_0 = 1, \quad b_0 = \dfrac{x}{12}, \quad a_1 = \dfrac{(1-2m)x}{24} - \dfrac{x^3}{288}, \quad b_1 = \dfrac{4m^2-1}{24} + \dfrac{x^2}{120} - \dfrac{x^4}{10368}$.

Chapter 12 (cont.)

4.6 Yes, provided that $0 < x < 1$.
5.1 (i) Only when $\psi(\zeta)$ is identically zero, otherwise $A_1(\zeta)$ and higher coefficients are unbounded functions of ν. (ii) No, because $\lambda_2(\nu)$ is unbounded as $\nu \to \infty$; compare Exercise 4.1.
6.2 Theorem 7.1 of Chapter 6 gives the sharper error bound.

Chapter 13

15.4 No, because the coefficient of w_0 appears only as $O(u^{-1})$.

Chapter 14

4.1 Correct value $= -0.27076\,62556\ldots - (0.02116\,78848\ldots)i$. Given formulas yield almost seven decimals in real part.

REFERENCES

AIREY, J. R. (1937). The "converging factor" in asymptotic series and the calculation of Bessel, Laguerre and other functions. *Philos. Mag.* [7], **24**, 521–552.

AIRY, G. B. (1838). On the intensity of light in the neighbourhood of a caustic. *Trans. Cambridge Philos. Soc.* **6**, 379–402.

AIRY, G. B. (1849). Supplement to a paper "On the intensity of light in the neighbourhood of a caustic", *Trans. Cambridge Philos. Soc.* **8**, 595–599.

APOSTOL, T. M. (1957). *Mathematical analysis*. Addison-Wesley, Reading, Massachusetts.

ASKEY, R. A., and WAINGER, S. (1965). Mean convergence of expansions in Laguerre and Hermite series. *Amer. J. Math.* **87**, 695–708.

B. A. *See* BRITISH ASSOCIATION FOR THE ADVANCEMENT OF SCIENCE (1946a,b; 1952).

B.M.P. *See* BATEMAN MANUSCRIPT PROJECT (1953a,b).

BAKHOOM, N. G. (1933). Asymptotic expansions of the function $F_k(x) = \int_0^\infty e^{-u^k+xu}\,du$. *Proc. London Math. Soc.* [2], **35**, 83–100.

BALOGH, C. B. (1967). Asymptotic expansions of the modified Bessel function of the third kind of imaginary order. *SIAM J. Appl. Math.* **15**, 1315–1323.

BARNES, E. W. (1905). The Maclaurin sum-formula. *Proc. London Math. Soc.* [2], **3**, 253–272.

BARNES, E. W. (1906). The asymptotic expansion of integral functions defined by Taylor's series. *Philos. Trans. Roy. Soc. London Ser. A* **206**, 249–297.

BARNES, E. W. (1908). A new development of the theory of the hypergeometric functions. *Proc. London Math. Soc.* [2], **6**, 141–177.

BATEMAN MANUSCRIPT PROJECT (1953a). *Higher transcendental functions* (A. Erdélyi, ed.), Vol. I. McGraw-Hill, New York.

BATEMAN MANUSCRIPT PROJECT (1953b). *Higher transcendental functions* (A. Erdélyi, ed.), Vol. II. McGraw-Hill, New York.

BERG, L. (1958). Asymptotische Darstellungen für Integrale und Reihen mit Anwendungen. *Math. Nachr.* **17**, 101–135.

BERG, L. (1968). *Asymptotische Darstellungen und Entwicklungen.* VEB Deutscher Verlag der Wissenschaften, Berlin.

BERTOCCHI, L., FUBINI, S., and FURLAN, G. (1965). The short-wavelength approximation to the Schrödinger equation. *Nuovo Cimento* [10], **35**, 599–632.

BIRKHOFF, G. D. (1908). On the asymptotic character of the solutions of certain linear differential equations containing a parameter. *Trans. Amer. Math. Soc.* **9**, 219–231.

BIRKHOFF, G. D. (1933). Quantum mechanics and asymptotic series. *Bull. Amer. Math. Soc.* **39**, 681–700.

BLEISTEIN, N. (1966). Uniform asymptotic expansions of integrals with stationary point near algebraic singularity. *Comm. Pure Appl. Math.* **19**, 353–370.

BLEISTEIN, N. (1967). Uniform asymptotic expansions of integrals with many nearby stationary points and algebraic singularities. *J. Math. Mech.* **17**, 533–559.

BLEISTEIN, N., and HANDELSMAN, R. A. (1974). *Asymptotic expansions of integrals.* To be published.

BLEISTEIN, N., HANDELSMAN, R. A., and LEW, J. S. (1972). Functions whose Fourier transforms decay at infinity: An extension of the Riemann-Lebesgue lemma. *SIAM J. Math. Anal.* **3**, 485–495.

Boin, P. W. M. (1965). *On the method of stationary phase for double integrals.* Waltman, Delft.

Börsch-Supan, W. (1961). On the evaluation of the function $\phi(\lambda) = (1/2\pi i)\int_{\sigma-i\infty}^{\sigma+i\infty} e^{u \ln u + \lambda u}\, du$ for real values of λ. *J. Res. Nat. Bur. Standards Sect. B* **65**, 245–250.

Boyer, T. H. (1969). Concerning the zeros of some functions related to Bessel functions. *J. Mathematical Phys.* **10**, 1729–1744.

Braaksma, B. L. J. (1963). *Asymptotic expansions and analytic continuations for a class of Barnes-integrals.* Noordhoff, Groningen.

Brillouin, L. (1926). Remarques sur la méchanique ondulatoire. *J. Phys. Radium* [6], **7**, 353–368.

British Association for the Advancement of Science (1946a). *Mathematical tables*, Vol. I (2nd ed.). Cambridge Univ. Press, London and New York.

British Association for the Advancement of Science (1946b). *Mathematical tables*, Part-Vol. B, *The Airy integral*, prepared by J. C. P. Miller. Cambridge Univ. Press, London and New York.

British Association for the Advancement of Science (1952). *Mathematical tables*, Vol. X, *Bessel functions. Pt. II. Functions of positive integer order.* Cambridge Univ. Press, London and New York.

Bromwich, T. J. I'a. (1926). *An introduction to the theory of infinite series*, 2nd ed. Macmillan, London.

Buchholz, H. (1969). *The confluent hypergeometric function*, translated by H. Lichtblau and K. Wetzel from 1953 German ed. Springer-Verlag, Berlin and New York.

Burkhardt, H. (1914). Über Funktionen grosser Zahlen, insbesondere über die näherungsweise Bestimmung entfernter Glieder in den Reihenentwicklungen der Theorie der Keplerschen Bewegung. *S.-B. München Akad. Math.-Phys.* 1–11.

Calogero, F. (1967). *Variable phase approach to potential scattering.* Academic Press, New York.

Carathéodory, C. (1960). *Theory of functions of a complex variable*, Vol. II (2nd ed.), translated by F. Steinhardt. Chelsea, Bronx, New York.

Carleman, T. (1926). *Les fonctions quasi analytiques.* Gauthier-Villars, Paris.

Chako, N. (1965). Asymptotic expansions of double and multiple integrals occurring in diffraction theory. *J. Inst. Math. Appl.* **1**, 372–422.

Chakravarti, P. C. (1970). *Integrals and sums.* Oxford Univ. Press (Athlone), London and New York.

Cherry, T. M. (1950a). Uniform asymptotic formulae for functions with transition points. *Trans. Amer. Math. Soc.* **68**, 224–257.

Cherry, T. M. (1950b). Summation of slowly convergent series. *Proc. Cambridge. Philos. Soc.* **46**, 436–449.

Chester, C., Friedman, B., and Ursell, F. (1957). An extension of the method of steepest descents. *Proc. Cambridge Philos. Soc.* **53**, 599–611.

Chowla, S., Herstein, I. N., and Moore, W. K. (1951). On recursions connected with symmetric groups I. *Canad. J. Math.* **3**, 328–334.

Chuhrukidze, N. K. (1966). Asymptotic formulae for Legendre functions (Russian). *Ž. Vyčisl. Mat. i Mat. Fiz.* **6**, 61–70. [English Transl.: *U.S.S.R. Computational Math. and Math. Phys.* **6**, No. 1, 86–99 (1966).]

Chuhrukidze, N. K. (1968). Asymptotic expansions for spherical Legendre functions of imaginary argument (Russian). *Ž. Vyčisl. Mat. i Mat. Fiz.* **8**, 3–12. [English transl.: *U.S.S.R. Computational Math. and Math. Phys.* **8**, No. 1, 1–13 (1968).]

Cīrulis, T. (1968). An application of the saddle-point method in some singular cases. I (Russian). *Latvijas Valsts Univ. Zinātn. Raksti* **91**, *laid.* 3, 75–87.

Cīrulis, T. (1969). A certain generalization of the stationary phase method (Russian). *Latvian Math. Yearbook* **5**, 175–194.

Clark, R. A. (1963). Asymptotic solutions of a nonhomogeneous differential equation with a turning point. *Arch. Rational Mech. Anal.* **12**, 34–51.

Clausen, Th. (1828). Ueber die Fälle, wenn die Reihe von der Form . . . . *J. Reine Angew. Math.* **3**, 89–91.

CLEMMOW, P. C. (1950). Some extensions to the method of integration by steepest descents. *Quart. J. Mech. Appl. Math.* **3**, 241–256.

COCHRAN, J. A. (1965). The zeros of Hankel functions as functions of their order. *Numer. Math.* **7**, 238–250.

COCHRAN, J. A. (1967). The monotonicity of modified Bessel functions with respect to their order. *J. Math. and Phys.* **46**, 220–222.

COHN, J. H. E. (1967). Large eigenvalues of a Sturm-Liouville problem. *Proc. Cambridge Philos. Soc.* **63**, 473–475.

COPPEL, W. A. (1965). *Stability and asymptotic behavior of differential equations.* Heath, Boston, Massachussetts.

COPSON, E. T. (1935). *Theory of functions of a complex variable.* Oxford Univ. Press, London and New York.

COPSON, E. T. (1963). On the asymptotic expansion of Airy's integral. *Proc. Glasgow Math. Assoc.* **6**, 113–115.

COPSON, E. T. (1965). *Asymptotic expansions*, Cambridge Tracts in Math. and Math. Phys. No. 55. Cambridge Univ. Press, London and New York.

CURTIS, A. R. (1964). *Coulomb wave functions*, Roy. Soc. Math. Tables, Vol. 11. Cambridge Univ. Press, London and New York.

DARBOUX, M. G. (1878). Mémoire sur l'approximation des fonctions de très-grands nombres, et sur une classe étendue de développements en série. *J. Math. Pures Appl.* [3], **4**, 5–56, 377–416.

DAVIS, P. J. (1953). Existence and uniqueness theorems for asymptotic expansions. In *Selected topics in the theory of asymptotic expansions.* Nat. Bur. Standards Rep. No. 2392, pp. 65–102. U.S. Nat. Bur. Standards, Washington, D. C.

DAVIS, P. J. (1957). Uniqueness theory for asymptotic expansions in general regions. *Pacific J. Math.* **7**, 849–859.

DAVIS, P. J. (1959). Leonhard Euler's integral: A historical profile of the Gamma function. *Amer. Math. Monthly* **66**, 849–869.

DE BRUIJN, N. G. (1961). *Asymptotic methods in analysis*, 2nd ed. Wiley (Interscience), New York.

DE KOK, F. (1971). On the method of stationary phase for multiple integrals. *SIAM J. Math. Anal.* **2**, 76–104.

DEBYE, P. (1909). Näherungsformeln für die Zylinderfunktionen für grosse Werte des Arguments und unbeschränkt veränderliche Werte des Index. *Math. Ann.* **67**, 535–558.

DIEUDONNÉ, J. (1968). *Calcul infinitésimal.* Hermann, Paris.

DINGLE, R. B. (1958). Asymptotic expansions and converging factors I. General theory and basic converging factors. *Proc. Roy. Soc. London Ser. A* **244**, 456–475.

DINGLE, R. B. (1962). Asymptotic expansions and converging factors. *Bul. Inst. Politehn. Iaşi* **8**, 53–60.

DINGLE, R. B. (1973). *Asymptotic expansions: Their derivation and interpretation.* Academic Press, New York.

DOETSCH, G. (1950). *Handbuch der Laplace-Transformation*, Vol. I. Birkhäuser, Basel.

DOETSCH, G. (1955). *Handbuch der Laplace-Transformation*, Vol. II. Birkhäuser, Basel.

DORNING, J. J., NICOLAENKO, B., and THURBER, J. K. (1969). An integral identity due to Ramanujan which occurs in neutron transport theory. *J. Math. Mech.* **19**, 429–438.

DUNCAN, C. E. (1957). On the asymptotic behavior of trigonometric sums. *Nederl. Akad. Wetensch. Proc. Ser. A* **60**, 261–264, 369–373, 374–380.

ERDÉLYI, A. (1946). Asymptotic representation of Laplace transforms with an application to inverse factorial series. *Proc. Edinburgh Math. Soc.* [2], **8**, 20–24.

ERDÉLYI, A. (1950). Note on the paper "On a definite integral" by R. H. Ritchie. *Math. Tables Aids Comput.* **4**, 179.

ERDÉLYI, A. (1955). Asymptotic representations of Fourier integrals and the method of stationary phase. *J. Soc. Indust. Appl. Math.* **3**, 17–27.

ERDÉLYI, A. (1956a). *Asymptotic expansions.* Dover, New York.

ERDÉLYI, A. (1956b). Asymptotic expansions of Fourier integrals involving logarithmic singularities. *J. Soc. Indust. Appl. Math.* **4**, 38–47.

ERDÉLYI, A. (1960). Asymptotic forms for Laguerre polynomials. *J. Indian Math. Soc.* (*Golden Jubilee Commemoration Volume*) **24**, 235–250.

ERDÉLYI, A. (1961). General asymptotic expansions of Laplace integrals. *Arch. Rational Mech. Anal.* **7**, 1–20.

ERDÉLYI, A. (1964). The integral equations of asymptotic theory. In *Asymptotic solutions of differential equations and their applications* (C. H. Wilcox, ed.), pp. 211–229. Wiley, New York.

ERDÉLYI, A. (1970). Uniform asymptotic expansion of integrals. In *Analytic methods in mathematical physics* (R. P. Gilbert and R. G. Newton, eds.), pp. 149–168. Gordon & Breach, New York.

ERDÉLYI, A., and SWANSON, C. A. (1957). Asymptotic forms of Whittaker's confluent hypergeometric functions. *Mem. Amer. Math. Soc.* **No. 25.**

ERDÉLYI, A., and WYMAN, M. (1963). The asymptotic evaluation of certain integrals. *Arch. Rational Mech. Anal.* **14**, 217–260.

EVGRAFOV, M. A. (1961). *Asymptotic estimates and entire functions*, translated by A. L. Shields. Gordon & Breach, New York.

EVGRAFOV, M. A., and FEDORYUK, M. V. (1966). Asymptotic behavior as $\lambda \to \infty$ of solutions of the equation $w''(z) - p(z, \lambda) w(z) = 0$ in the complex z-plane (Russian). *Uspehi Mat. Nauk* **21**, 3–50. [Engl. transl.: *Russian Math. Surveys* **21**, 1–48 (1966).]

FAXÉN, H. (1921). Expansion in series of the integral $\int_y^\infty e^{-x(t \pm t^{-\mu})} t^\nu \, dt$. *Ark. Mat. Astronom. Fys.* **15**, No. 13, 1–57.

FEDORYUK, M. V. (1964). The stationary phase method. Nearby saddle-points in the higher-dimensional case (Russian). *Ž. Vyčisl. Mat. i Mat. Fiz.* **4**, 671–682. [English transl.: *U.S.S.R. Computational Math. and Math. Phys.* **4**, No. 4, 66–81 (1964).]

FEDORYUK, M. V. (1965a). One-dimensional scattering in the quasiclassical approximation (Russian). *Differencial'nye Uravnenija* **1**, 631–646, 1525–1536. [English transl.: *Differential Equations* **1**, 483–495, 1201–1210 (1965).]

FEDORYUK, M. V. (1965b). Asymptotics of the discrete spectrum of the operator $w''(x) - \lambda^2 p(x) w(x)$ (Russian). *Mat. Sb.* **68**, 81–110.

FEDORYUK, M. V. (1968). Analytic properties of the scattering amplitude in the one-dimensional case. I (Russian). *Differencial'nye Uravnenija* **4**, 1842–1853. [Engl. transl.: *Differential Equations* **4**, 948–954 (1968).]

FEDORYUK, M. V. (1969). *Asymptotic expansions of solutions of differential linear equations of the second order in a complex domain*, translated by F. Czyzewski and W. Wasow. Tech. Summary Rep. No. 993, Math. Res. Center, Univ. of Wisconsin, Madison.

FEDORYUK, M. V. (1970). The method of stationary phase in the multidimensional case. Contribution from the boundary of the domain (Russian). *Ž. Vyčisl. Mat. i Mat. Fiz.* **10**, 286–299. [English transl.: *U.S.S.R. Computational Math. and Math. Phys.* **10**, No. 2, 4–23 (1970).]

FERRERS, N. M. (1877). *Spherical harmonics.* Macmillan, London.

FIELDS, J. L. (1968). A uniform treatment of Darboux's method. *Arch. Rational Mech. Anal.* **27**, 289–305.

FIELDS, J. L., and ISMAIL, M. E. (1974). On the positivity of some ${}_1F_2$'s. To be published.

FIX, G. (1967). Asymptotic eigenvalues of Sturm-Liouville systems. *J. Math. Anal. Appl.* **19**, 519–525.

FORD, W. B. (1916). *Studies on divergent series and summability.* Macmillan, New York. Reprinted by Chelsea, Bronx, New York, 1960 (and bound with Ford, 1936).

FORD, W. B. (1936). *The asymptotic developments of functions defined by Maclaurin series.* Univ. of Michigan Press, Ann Arbor. Reprinted by Chelsea, Bronx, New York, 1960 (and bound with Ford, 1916).

FRANKLIN, J., and FRIEDMAN, B. (1957). A convergent asymptotic representation for integrals. *Proc. Cambridge Philos. Soc.* **53**, 612–619.

FROBENIUS, G. (1873). Ueber die Integration der linearen Differentialgleichungen durch Reihen. *J. Reine Angew. Math.* **76**, 214–235.

FRÖMAN, N. (1966). The energy levels of double-well potentials. *Ark. Fys.* **32**, 79–97.

FRÖMAN, N. (1970). Connection formulas for certain higher order phase-integral approximations. *Ann. Physics* **61**, 451–464.

FRÖMAN, N., and FRÖMAN, P. O. (1965). *JWKB approximation–Contributions to the theory*. North-Holland Publ., Amsterdam.

FRÖMAN, N., and FRÖMAN, P. O. (1970). Transmission through a real potential barrier treated by means of certain phase-integral approximations. *Nuclear Phys. Ser. A* **147**, 606–626.

FUCHS, L. (1866). Zur Theorie der linearen Differentialgleichungen mit veränderlichen Coefficienten. *J. Reine Angew. Math.* **66**, 121–160.

FURRY, W. H. (1947). Two notes on phase-integral methods. *Phys. Rev.* **71**, 360–371.

GANS, R. (1915). Fortpflanzung des Lichts durch ein inhomogenes Medium. *Ann. Physik* [4], **47**, 709–736.

GELLER, M., and NG, E. W. (1969). A table of integrals of the exponential integral. *J. Res. Nat. Bur. Standards Sect. B* **73**, 191–210.

GOODWIN, E. T., and STATON, J. (1948). Table of $\int_0^\infty \{e^{-u^2}/(u+x)\}\,du$. *Quart. J. Mech. Appl. Math.* **1**, 319–326.

GREEN, G. (1837). On the motion of waves in a variable canal of small depth and width. *Trans. Cambridge Philos. Soc.* **6**, 457–462.

HAAR, A. (1926). Über asymptotische Entwicklungen von Funktionen. *Math. Ann.* **96**, 69–107.

HANDELSMAN, R. A., and BLEISTEIN, N. (1973). Asymptotic expansions of integral transforms with oscillatory kernels: A generalization of the method of stationary phase. *SIAM J. Math. Anal.* **4**, 519–535.

HANDELSMAN, R. A., and LEW, J. S. (1971). Asymptotic expansion of a class of integral transforms with algebraically dominated kernels. *J. Math. Anal. Appl.* **35**, 405–433.

HANKEL, H. (1864). Die Euler'schen Integrale bei unbeschränkter Variabilität des Argumentes. *Z. Math. Physik* **9**, 1–21.

HANKEL, H. (1869). Die Cylinderfunctionen erster und zweiter Art. *Math. Ann.* **1**, 467–501.

HANSON, R. J. (1966). Reduction theorems for systems of ordinary differential equations with a turning point. *J. Math. Anal. Appl.* **16**, 280–301.

HARDY, G. H. (1949). *Divergent series*. Oxford Univ. Press (Clarendon), London and New York.

HARRIS, B., and SCHOENFELD, L. (1968). Asymptotic expansions for the coefficients of analytic functions. *Illinois J. Math.* **12**, 264–277.

HARTMAN, P. (1964). *Ordinary differential equations*. Wiley, New York.

HEADING, J. (1962). *An introduction to phase-integral methods*. Wiley, New York.

HETHCOTE, H. W. (1970a). Bounds for zeros of some special functions. *Proc. Amer. Math. Soc.* **25**, 72–74.

HETHCOTE, H. W. (1970b). Error bounds for asymptotic approximations of zeros of transcendental functions. *SIAM J. Math. Anal.* **1**, 147–152.

HOBSON, E. W. (1931). *The theory of spherical and ellipsoidal harmonics*. Cambridge Univ. Press, London and New York.

HOCHSTADT, H. (1961). *Special functions of mathematical physics*. Holt, New York.

HORN, J. (1899). Ueber eine lineare Differentialgleichung zweiter Ordnung mit einem willkürlichen Parameter. *Math. Ann.* **52**, 271–292.

HORN, J. (1903). Untersuchung der Integrale einer linearen Differentialgleichung in der Umgebung einer Unbestimmtheitsstelle vermittelst successiver Annäherungen. *Arch. Math. Physik Leipzig* [3], **4**, 213–230.

HSIEH, P.-F., and SIBUYA, Y. (1966). On the asymptotic integration of second order linear ordinary differential equations with polynomial coefficients. *J. Math. Anal. Appl.* **16**, 84–103.

HSU, L. C. (1951). On the asymptotic evaluation of a class of multiple integrals involving a parameter. *Amer. J. Math.* **73**, 625–634.

HUKUHARA, M. (1937). Sur les propriétés asymptotiques des solutions d'un système d'équations différentielles linéaires contenant un paramètre. *Mem. Fac. Engrg. Kyushu Imp. Univ.* **8**, 249–280.

HULL, T. E., and FROESE, C. (1955). Asymptotic behaviour of the inverse of a Laplace transform. *Canad. J. Math.* **7**, 116–125.

INCE, E. L. (1927). *Ordinary differential equations.* Longmans, Green, London and New York. Reprinted by Dover, New York, 1956.

JEFFREYS, H. (1924). On certain approximate solutions of linear differential equations of the second order. *Proc. London Math. Soc.* [2], **23**, 428–436.

JEFFREYS, H. (1928). The effect on Love waves of heterogeneity in the lower layer. *Monthly Notices Roy. Astronom. Soc. Geophys. Suppl.* **2**, 101–111.

JEFFREYS, H. (1942). Asymptotic solutions of linear differential equations. *Philos. Mag.* [7], **33**, 451–456.

JEFFREYS, H. (1953). On approximate solutions of linear differential equations. *Proc. Cambridge Philos. Soc.* **49**, 601–611.

JEFFREYS, H. (1956). On the use of asymptotic approximations of Green's type when the coefficient has zeros. *Proc. Cambridge Philos. Soc.* **52**, 61–66.

JEFFREYS, H. (1958). The remainder in Watson's lemma. *Proc. Roy. Soc. London Ser. A* **248**, 88–92.

JEFFREYS, H. (1962). *Asymptotic approximations.* Oxford Univ. Press, London and New York.

JEFFREYS, H., and JEFFREYS, B. S. (1956). *Methods of mathematical physics*, 3rd ed. Cambridge Univ. Press, London and New York.

JONES, A. L. (1968). An extension of an inequality involving modified Bessel functions. *J. Math. and Phys.* **47**, 220–221.

JONES, D. S. (1966). Fourier transforms and the method of stationary phase. *J. Inst. Math. Appl.* **2**, 197–222.

JONES, D. S. (1972). Asymptotic behavior of integrals. *SIAM Rev.* **14**, 286–317.

JORNA, S. (1964a). Derivation of Green-type, transitional, and uniform asymptotic expansions from differential equations I. General theory, and application to modified Bessel functions of large order. *Proc. Roy. Soc. London Ser. A* **281**, 99–110.

JORNA, S. (1964b). Derivation of Green-type, transitional, and uniform asymptotic expansions from differential equations II. Whittaker functions $W_{k,m}$ for large k, and for large $|k^2-m^2|$. *Proc. Roy. Soc. London Ser. A* **281**, 111–129.

JORNA, S. (1965). Derivation of Green-type, transitional and uniform asymptotic expansions from differential equations III. The confluent hypergeometric function $U(a,c,z)$ for large $|c|$. *Proc. Roy. Soc. London Ser. A* **284**, 531–539.

KARMAZINA, L. N. (1960). Asymptotic formulas for the functions $P_{-\frac{1}{2}+i\tau}(x)$ as $\tau\to\infty$ (Russian). *Vyčisl. Mat.* **6**, 3–16.

KAZARINOFF, N. D. (1955). Asymptotic expansions for the Whittaker functions of large complex order *m*. *Trans. Amer. Math. Soc.* **78**, 305–328.

KAZARINOFF, N. D. (1957a). Asymptotic forms for the Whittaker functions with both parameters large. *J. Math. Mech.* **6**, 341–360.

KAZARINOFF, N. D. (1957b). Asymptotic solutions with respect to a parameter of ordinary differential equations having a regular singular point. *Michigan Math. J.* **4**, 207–220.

KELVIN (LORD) (1887). On the waves produced by a single impulse in water of any depth, or in a dispersive medium. *Philos. Mag.* [5], **23**, 252–255. Reprinted in *Mathematical and physical papers*, Vol. 4, pp. 303–306. Cambridge Univ. Press, London and New York, 1910.

KEMBLE, E. C. (1935). A contribution to the theory of the BWK method. *Phys. Rev.* **48**, 549–561.

KEMBLE, E. C. (1937). *The fundamental principles of quantum mechanics.* McGraw-Hill, New York.

KIYEK, K. H. (1963). Zur Theorie der linearen Differentialgleichungssysteme mit einem grossen Parameter. Doctoral Thesis, Univ. of Würzburg.

KLEIN, F. (1894). *Über lineare Differentialgleichungen der zweiten Ordnung.* Ritter, Göttingen. Reprinted by Teubner, Leipzig, 1906.

KNOPP, K. (1951). *Theory and application of infinite series*, 2nd Engl. ed. Blackie, Glasgow and London.

KOSTOMAROV, D. P. (1958). On the asymptotic behavior of solutions of certain linear differential equations of second order containing a large parameter (Russian). *Mat. Sb.* **45**(**87**), 17–30.

KRAMERS, H. A. (1926). Wellenmechanik und halbzahlige Quantisierung. *Z. Physik* **39**, 828–840.

LANDAU, E. (1927). *Vorlesungen über Zahlentheorie*, Sec. 1. Hirzel, Stuttgart. Reprinted by Chelsea, Bronx, New York, 1947.

LANGER, R. E. (1931). On the asymptotic solutions of ordinary differential equations, with an application to the Bessel functions of large order. *Trans. Amer. Math. Soc.* **33**, 23–64.

LANGER, R. E. (1932). On the asymptotic solutions of differential equations, with an application to the Bessel functions of large complex order. *Trans. Amer. Math. Soc.* **34**, 447–480.

LANGER, R. E. (1934). The asymptotic solutions of ordinary linear differential equations of the second order, with special reference to the Stokes phenomenon. *Bull. Amer. Math. Soc.* **40**, 545–582.

LANGER, R. E. (1935). On the asymptotic solutions of ordinary differential equations, with reference to the Stokes' phenomenon about a singular point. *Trans. Amer. Math. Soc.* **37**, 397–416.

LANGER, R. E. (1937). On the connection formulas and the solutions of the wave equation. *Phys. Rev.* **51**, 669–676.

LANGER, R. E. (1949). The asymptotic solutions of ordinary linear differential equations of the second order, with special reference to a turning point. *Trans. Amer. Math. Soc.* **67**, 461–490.

LANGER, R. E. (1962). On the construction of related equations for the asymptotic theory of linear ordinary differential equations about a turning point. *Enseignement Math.* [2], **8**, 218–237.

LAPLACE (LE MARQUIS DE) (1820). *Théorie analytique des probabilités*, 3rd ed. Courcier, Paris. Reprinted in *Complete works*, Vol. 7. Gauthier-Villars, Paris, 1886.

LAUWERIER, H. A. (1966). *Asymptotic expansions*, Math. Centre Tracts No. 13. Mathematisch Centrum, Amsterdam.

LERCH, M. (1887). Note sur la fonction $\mathscr{K}(w,x,s) = \sum_{k=0}^{\infty} e^{2k\pi ix}/(w+k)^s$. *Acta Math.* **11**, 19–24.

LEVINSON, N. (1949). On the uniqueness of the potential in a Schrödinger equation for a given asymptotic phase. *Danske Vid. Selsk. Mat.-Fys. Medd.* **25**, No. 9, 1–29.

LEVINSON, N., and REDHEFFER, R. M. (1970). *Complex variables.* Holden-Day, San Francisco, California.

LINDELÖF, E. (1905). *Le calcul des résidus.* Gauthier-Villars, Paris. Reprinted by Chelsea, Bronx, New York, 1947.

LIOUVILLE, J. (1837). Sur le développement des fonctions ou parties de fonctions en séries .... *J. Math. Pures Appl.* [1], **2**, 16–35.

LOMMEL, E. C. J. VON (1868). *Studien über die Bessel'schen Functionen.* Teubner, Leipzig.

LORCH, L., and SZEGO, P. (1955). A singular integral whose kernel involves a Bessel function. *Duke Math. J.* **22**, 407–418. For corrections see *Duke Math. J.* **24**, 683.

LUDWIG, D. (1967). An extension of the validity of the stationary phase formula. *SIAM J. Appl. Math.* **15**, 915–923.

LUKE, Y. L. (1962). *Integrals of Bessel functions.* McGraw-Hill, New York.

LUKE, Y. L. (1968). An asymptotic expansion. *SIAM Rev.* (problems sect.) **10**, 229–232.

LUKE, Y. L. (1969a). *The special functions and their approximations*, Vol. I. Academic Press, New York.

LUKE, Y. L. (1969b). *The special functions and their approximations*, Vol. II. Academic Press, New York.

LYNESS, J. N. (1970). The calculation of Fourier coefficients by the Möbius inversion of the Poisson summation formula. Pt. I. Functions whose early derivatives are continuous. *Math. Comp.* **24**, 101–135.

LYNESS, J. N. (1971a). The calculation of Fourier coefficients by the Möbius inversion of the Poisson summation formula. Pt. II. Piecewise continuous functions and functions with poles near the interval [0, 1]. *Math. Comp.* **25**, 59–78.

LYNESS, J. N. (1971b). Adjusted forms of the Fourier coefficient asymptotic expansion and applications in numerical quadrature. *Math. Comp.* **25**, 87–104.

LYNESS, J. N. (1971c). The calculation of Fourier coefficients by the Möbius inversion of the Poisson summation formula. Pt. III. Functions having algebraic singularities. *Math. Comp.* **25**, 483–493.

LYNESS, J. N., and NINHAM, B. W. (1967). Numerical quadrature and asymptotic expansions. *Math. Comp.* **21**, 162–178.

MACDONALD, H. M. (1899). Zeroes of the Bessel functions. *Proc. London Math. Soc.* **30**,165–179.

MacFarlane, G. G. (1949). The application of Mellin transforms to the summation of slowly convergent series. *Philos. Mag.* [7], **40**, 188–197.

McHugh, J. A. M. (1971). An historical survey of ordinary linear differential equations with a large parameter and turning points. *Arch. History Exact Sci.* **7**, 277–324.

McKenna, J. (1967). Note on asymptotic expansions of Fourier integrals involving logarithmic singularities. *SIAM J. Appl. Math.* **15**, 810–812.

McLeod, J. B. (1961a). The determination of phase shift. *Quart. J. Math. Oxford Ser.* **12**, 17–32.

McLeod, J. B. (1961b). The determination of the transmission coefficient. *Quart. J. Math. Oxford Ser.* **12**, 153–158.

MacRobert, T. M. (1967). *Spherical harmonics*, 3rd ed. Pergamon, Oxford.

Medhurst, R. G., and Roberts, J. H. (1965). Evaluation of the integral $I_n(b) = (2/\pi)\int_0^\infty (\sin x/x)^n \cos(bx)\, dx$. *Math. Comp.* **19**, 113–117.

Miller, G. F. (1960). *Tables of generalized exponential integrals*, N.P.L. Math. Tables, Vol. 3. H.M. Stationery Office, London.

Miller, J. C. P. (1950). On the choice of standard solutions for a homogeneous linear differential equation of the second order. *Quart. J. Mech. Appl. Math.* **3**, 225–235.

Miller, J. C. P. (1952). A method for the determination of converging factors, applied to the asymptotic expansions for the parabolic cylinder functions. *Proc. Cambridge Philos. Soc.* **48**, 243–254.

Miller, J. C. P. (1955). *Tables of Weber parabolic cylinder functions.* H.M. Stationery Office, London.

Milne-Thomson, L. M. (1933). *The calculus of finite differences.* Macmillan, London.

Moriguchi, H. (1959). An improvement of the WKB method in the presence of turning points and the asymptotic solutions of a class of Hill equations. *J. Phys. Soc. Japan* **14**, 1771–1796.

Morse, P. M., and Feshbach, H. (1953). *Methods of theoretical physics.* McGraw-Hill, New York.

Mott, N. F., and Massey, H. S. W. (1965). *The theory of atomic collisions*, 3rd ed. Oxford Univ. Press, London and New York.

Muckenhoupt, B. (1970). Mean convergence of Hermite and Laguerre series: I, II. *Trans. Amer. Math. Soc.* **147**, 419–431, 433–460.

Muldoon, M. E. (1970). Singular integrals whose kernels involve certain Sturm-Liouville functions I. *J. Math. Mech.* **19**, 855–873.

Murnaghan, F. D. (1965). Evaluation of the probability integral to high precision. U.S. David Taylor Model Basin, Appl. Math. Lab. Rep. No. 1861. Washington, D. C.

Murnaghan, F. D., and Wrench, J. W., Jr. (1963). The converging factor for the exponential integral. U.S. David Taylor Model Basin, Appl. Math. Lab. Rep. No. 1535. Washington, D. C.

N. B. S. *See* National Bureau of Standards.

N. P. L. *See* National Physical Laboratory.

Nakano, M., and Nishimoto, T. (1970). On a secondary turning point problem. *Kōdai Math. Sem. Rep.* **22**, 355–384.

National Bureau of Standards (1964). *Handbook of mathematical functions.* Appl. Math. Ser. No. 55 (M. Abramowitz and I. A. Stegun, eds.). U.S. Govt. Printing Office, Washington, D. C.

National Physical Laboratory (1961). *Modern computing methods.* Notes on Appl. Sci. No. 16, 2nd ed. H.M. Stationery Office, London.

Natterer, F. (1969). Einschliessungen für die grossen Eigenwerte gewöhnlicher Differentialgleichungen zweiter und vierter Ordnung. *Numer. Math.* **13**, 78–93.

Navot, I. (1961). An extension of the Euler-Maclaurin summation formula to functions with a branch singularity. *J. Math. and Phys.* **40**, 271–276.

Nayfeh, A. H. (1973). *Perturbation methods.* Wiley (Interscience), New York.

Newton, R. G. (1960). Analytic properties of radial wave functions. *J. Mathematical Phys.* **1**, 319–347.

Ng, E. W., and Geller, M. (1969). A table of integrals of the error functions. *J. Res. Nat. Bur Standards Sect. B* **73**, 1–20.

Nörlund, N. E. (1924). *Vorlesungen über Differenzenrechnung.* Springer–Verlag, Berlin and New York. Reprinted by Chelsea, Bronx, New York, 1954.

NOSOVA, L. N., and TUMARKIN, S. A. (1965). *Tables of generalized Airy functions for the asymptotic solution of the differential equations* $\varepsilon(py')' + (q + \varepsilon r)y = f$, translated by D. E. Brown. Macmillan, New York.

OBERHETTINGER, F. (1953). Some general theorems and methods. In *Selected topics in the theory of asymptotic expansions.* Nat. Bur. Standards Rep. No. 2392, pp. 2–35. U.S. Nat. Bur. Standards, Washington, D. C.

OBERHETTINGER, F. (1959). On a modification of Watson's lemma. *J. Res. Nat. Bur. Standards Sect. B* **63**, 15–17.

OBERHETTINGER, F. (1972). *Tables of Bessel transforms*. Springer-Verlag, Berlin and New York.

OLVER, F. W. J. (1949). Transformation of certain series occurring in aerodynamic interference calculations. *Quart. J. Mech. Appl. Math.* **2**, 452–457.

OLVER, F. W. J. (1950). A new method for the evaluation of zeros of Bessel functions and of other solutions of second-order differential equations. *Proc. Cambridge Philos. Soc.* **46**, 570–580.

OLVER, F. W. J. (1952). Some new asymptotic expansions for Bessel functions of large orders. *Proc. Cambridge Philos. Soc.* **48**, 414–427.

OLVER, F. W. J. (1954a). The asymptotic solution of linear differential equations of the second order for large values of a parameter. *Philos. Trans. Roy. Soc. London Ser. A* **247**, 307–327.

OLVER, F. W. J. (1954b). The asymptotic expansion of Bessel functions of large order. *Philos. Trans. Roy. Soc. London Ser. A* **247**, 328–368.

OLVER, F. W. J. (1956). The asymptotic solution of linear differential equations of the second order in a domain containing one transition point. *Philos. Trans. Roy. Soc. London Ser. A* **249**, 65–97.

OLVER, F. W. J. (1958). Uniform asymptotic expansions of solutions of linear second-order differential equations for large values of a parameter. *Philos. Trans. Roy. Soc. London Ser. A* **250**, 479-517.

OLVER, F. W. J. (1959). Uniform asymptotic expansions for Weber parabolic cylinder functions of large orders. *J. Res. Nat. Bur. Standards Sect. B* **63**, 131–169.

OLVER, F. W. J. (1961). Error bounds for the Liouville-Green (or WKB) approximation. *Proc. Cambridge Philos. Soc.* **57**, 790–810.

OLVER, F. W. J. (1963). Error bounds for first approximations in turning-point problems. *J. Soc. Indust. Appl. Math.* **11**, 748–772.

OLVER, F. W. J. (1964a). Error bounds for asymptotic expansions, with an application to cylinder functions of large argument. In *Asymptotic solutions of differential equations and their applications* (C. H. Wilcox, ed.), pp. 163–183. Wiley, New York.

OLVER, F. W. J. (1964b). Error bounds for asymptotic expansions in turning-point problems. *J. Soc. Indust. Appl. Math.* **12**, 200–214.

OLVER, F. W. J. (1965a). Error analysis of phase-integral methods. I. General theory for simple turning points. *J. Res. Nat. Bur. Standards Sect. B* **69**, 271–290.

OLVER, F. W. J. (1965b). Error analysis of phase-integral methods. II. Application to wave-penetration problems. *J. Res. Nat. Bur. Standards Sect. B* **69**, 291–300.

OLVER, F. W. J. (1965c). Error bounds for asymptotic expansions of special functions in the complex plane. In *Error in digital computation* (L. B. Rall, ed.), Vol. 2, pp. 55–75. Wiley, New York.

OLVER, F. W. J. (1965d). On the asymptotic solutions of second-order differential equations having an irregular singularity of rank one, with an application to Whittaker functions. *SIAM J. Numer. Anal. Ser. B* **2**, 225–243.

OLVER, F. W. J. (1968). Error bounds for the Laplace approximation for definite integrals. *J. Approximation Theory* **1**, 293–313.

OLVER, F. W. J. (1970a). Why steepest descents? In *Studies in applied mathematics*, No. 6 (D. Ludwig and F. W. J. Olver, eds.), pp. 44–63. Soc. Indust. and Appl. Math., Philadelphia. Reprinted in *SIAM Rev.* **12**, 228–247.

OLVER, F. W. J. (1970b). A paradox in asymptotics. *SIAM J. Math. Anal.* **1**, 533–534.

OLVER, F. W. J. (1974). Error bounds for stationary phase approximations. *SIAM J. Math. Anal.* **5**. To be published.

OLVER, F. W. J., and STENGER, F. (1965). Error bounds for asymptotic solutions of second-order differential equations having an irregular singularity of arbitrary rank. *SIAM J. Numer. Anal. Ser. B* **2**, 244–249.

PIKE, E. R. (1964a). On the related-equation method of asymptotic approximation (W.K.B. or A-A method) I. A proposed new existence theorem. *Quart. J. Mech. Appl. Math.* **17**, 105–124.

PIKE, E. R. (1964b). On the related-equation method of asymptotic approximation II. Direct solutions of wave-penetration problems. *Quart. J. Mech. Appl. Math.* **17**, 369–379.

PITTNAUER, F. (1969). Holomorphic functions with prescribed asymptotic expansions. *SIAM J. Appl. Math.* **17**, 607–613.

POCHHAMMER, L. (1890). Zur Theorie der Euler'schen Integrale. *Math. Ann.* **35**, 495–526.

POINCARÉ, H. (1886). Sur les intégrales irrégulières des équations linéaires. *Acta Math.* **8**, 295–344.

POINCARÉ, H. (1904). Sur la diffraction des ondes électriques. *Proc. Roy. Soc. London* **72**, 42–52.

POLLAK, H. O., and SHEPP, L. (1964). An asymptotic expansion. *SIAM Rev.* (problems sect.) **6**, 60. See also *SIAM Rev.* **8**, 383–384, 1966.

PÓLYA, G., and SZEGÖ, G. (1925). *Aufgaben und Lehrsätze aus der Analysis*, Vol. 1 (3rd ed.), pp. 77–83, 242–250. Springer-Verlag, Berlin and New York.

R. S. *See* ROYAL SOCIETY.

RAYLEIGH (LORD) (1912). On the propagation of waves through a stratified medium, with special reference to the question of reflection. *Proc. Roy. Soc. London Ser. A* **86**, 207–226.

REUDINK, D. O. J. (1965). An extension of the method of Haar for determining the asymptotic behavior of integrals of the inverse Laplace transform type. Thesis, Oregon State Univ., Corvallis.

REUDINK, D. O. J. (1968). On the signs of the ν-derivatives of the modified Bessel functions $I_\nu(x)$ and $K_\nu(x)$. *J. Res. Nat. Bur. Standards Sect. B* **72**, 279–280.

RICE, S. O. (1968). Uniform asymptotic expansions for saddle point integrals—application to a probability distribution occurring in noise theory. *Bell System Tech. J.* **47**, 1971–2013.

RIEDEL, R. (1965). Asymptotische Darstellungen von Doppelintegralen mit stationärem Maximum. *Z. Angew. Math. Mech.* **45**, 323–332.

RIEKSTIŅA, V. (1968). Asymptotic expansions of some entire functions and of Fourier's integral (Russian). *Latvijas Valsts Univ. Zinātn. Raksti* **91**, *laid.* 3, 47–74.

RIEKSTIŅŠ, E. (1966). On the use of neutrices for asymptotic representation of some integrals (Russian). *Latvian Math. Yearbook*, pp. 5–21.

RIEKSTIŅŠ, E. (1968). Asymptotic expansions of sequences generated by iteration (Russian). *Latvian Math. Yearbook* **4**, 291–311.

RIEKSTIŅŠ, E. (1973). Asymptotic expansions for some types of integrals involving logarithms (Russian). *Latvian Math. Yearbook* **15**. To be published.

RIEKSTIŅŠ, E., and CĪRULIS, T. (1970). Methods that can be used for the asymptotic representation of functions defined by integrals when the parameters take large values (Russian). *Latvian Math. Yearbook* **7**, 193–253.

RIEMANN, B. (1857). Beiträge zur Theorie der durch die Gauss'sche Reihe $F(\alpha, \beta, \gamma, x)$ darstellbaren Functionen. *Abh. Kgl. Gesellsch. Wiss. Göttingen* **7**. Reprinted in *Complete works* 2nd ed., pp. 67–83. Teubner, Leipzig, 1892, or Dover, New York, 1953.

RIEMANN, B. (1859). Ueber die Anzahl der Primzahlen unter einer gegebenen Grösse. *Monatsb. Berliner Akad.* Reprinted in *Complete works* 2nd ed., pp. 145–155. Teubner, Leipzig, 1892, or Dover, New York, 1953.

RIEMANN, B. (1863). Sullo svolgimento del quoziente di due serie ipergeometriche in frazione continua infinita. *Complete works*, 2nd ed., pp. 424–430. Teubner, Leipzig, 1892, or Dover, New York, 1953.

RITCHIE, R. H. (1950). On a definite integral. *Math. Tables Aids Comput.* **4**, 75–77.

RITT, J. F. (1916). On the derivatives of a function at a point. *Ann. of Math.* **18**, 18–23.

RITT, J. F. (1918). On the differentiability of asymptotic series. *Bull. Amer. Math. Soc.* **24**, 225–227.

ROBIN, L. (1957). *Fonctions sphériques de Legendre et fonctions sphéroïdales*, Vol. I. Gauthier-Villars, Paris.

ROBIN, L. (1958). *Fonctions sphériques de Legendre et fonctions sphéroïdales*, Vol. II. Gauthier-Villars, Paris.

ROBIN, L. (1959). *Fonctions sphériques de Legendre et fonctions sphéroïdales*, Vol. III. Gauthier-Villars, Paris.

ROSSER, J. B. (1951). Transformations to speed the convergence of series. *J. Res. Nat. Bur. Standards* **46**, 56–64.

ROSSER, J. B. (1955). Explicit remainder terms for some asymptotic series. *J. Rational Mech. Anal.* **4**, 595–626.

ROYAL SOCIETY (1960). *Mathematical tables*, Vol. 7, *Bessel functions Pt. III Zeros and associated values* (F. W. J. Olver, ed.). Cambridge Univ. Press, London and New York.

RUBIN, R. J. (1967). Random walk with an excluded origin. *J. Mathematical Phys.* **8**, 576–581.

SABATIER, P. C. (1965). On the asymptotic approximation for the elastic scattering by a potential. I. Monotonic potential and uncritical range. *Nuovo Cimento* [10], **37**, 1180–1227.

SCHINDLER, S. (1971). Some transplantation theorems for the generalized Mehler transform and related asymptotic expansions. *Trans. Amer. Math. Soc.* **155**, 257–291.

SCHLESINGER, L. (1907). Über asymptotische Darstellungen der Lösungen linearer Differentialsysteme als Funktionen eines Parameters. *Math. Ann.* **63**, 277–300.

SCHMIDT, H. (1937). Beiträge zu einer Theorie der allgemeinen asymptotischen Darstellungen. *Math. Ann.* **113**, 629–656.

SCORER, R. S. (1950). Numerical evaluation of integrals of the form $I = \int_{x_1}^{x_2} f(x) e^{i\phi(x)}\, dx$ and the tabulation of the function $\mathrm{Gi}(z) = (1/\pi) \int_0^\infty \sin(uz + \frac{1}{3}u^3)\, du$. *Quart. J. Mech. Appl. Math.* **3**, 107–112.

SHANKS, D. (1955). Non-linear transformations of divergent and slowly convergent sequences. *J. Math. and Phys.* **34**, 1–42.

SKOVGAARD, H. (1966). *Uniform asymptotic expansions of confluent hypergeometric functions and Whittaker functions*. Gjellerups Publ., Copenhagen.

SLATER, L. J. (1960). *Confluent hypergeometric functions*. Cambridge Univ. Press, London and New York.

SLATER, L. J. (1966). *Generalized hypergeometric functions*. Cambridge Univ. Press, London and New York.

SNOW, C. (1952). *Hypergeometric and Legendre functions with applications to integral equations of potential theory*. Nat. Bur. Standards Appl. Math. Ser. No. 19. U.S. Govt. Printing Office, Washington, D. C.

SONI, K., and SONI, R. P. (1973). Asymptotic behavior of a class of integral transforms. *SIAM J. Math. Anal.* **4**, 466–481.

SPIRA, R. (1971). Calculation of the Gamma function by Stirling's formula. *Math. Comp.* **25**, 317–322.

STEFFENSEN, J. F. (1927). *Interpolation*. Chelsea, Bronx, New York.

STEINIG, J. (1970). The real zeros of Struve's function. *SIAM J. Math. Anal.* **1**, 365–375.

STENGER, F. (1966a). Error bounds for asymptotic solutions of differential equations I. The distinct eigenvalue case. *J. Res. Nat. Bur. Standards Sect. B* **70**, 167–186.

STENGER, F. (1966b). Error bounds for asymptotic solutions of differential equations II. The general case. *J. Res. Nat. Bur. Standards Sect. B* **70**, 187–210.

STENGER, F. (1970a). The asymptotic approximation of certain integrals. *SIAM J. Math. Anal.* **1**, 392–404.

STENGER, F. (1970b). On the asymptotic solution of two first order linear differential equations with large parameter. *Funkcial. Ekvac.* **13**, 1–18.

STENGER, F. (1972). Transform methods for obtaining asymptotic expansions of definite integrals. *SIAM J. Math. Anal.* **3**, 20–30.

STIELTJES, T. J. (1886). Recherches sur quelques séries semi-convergentes. *Ann. Sci. École Norm. Sup.* [3], **3**, 201–258. Reprinted in *Complete works*, Vol. 2, pp. 2–58. Noordhoff, Groningen, 1918.

STOKES, G. G. (1848). *Memoir and scientific correspondence*, selected and arranged by J. Larmor, Vol. II, pp. 159–160. Cambridge Univ. Press, London and New York, 1907.

STOKES, G. G. (1850). On the numerical calculation of a class of definite integrals and infinite series. *Trans. Cambridge Philos. Soc.* **9**, 166–187. Reprinted in *Mathematical and physical papers*, Vol. 2, pp. 329–357. Cambridge Univ. Press, London and New York, 1883.

STOKES, G. G. (1857). On the discontinuity of arbitrary constants which appear in divergent developments. *Trans. Cambridge Philos. Soc.* **10**, 105–128. Reprinted in *Mathematical and physical papers*, Vol. 4, pp 77–109. Cambridge Univ. Press, London and New York, 1904.

SWANSON, C. A. (1956). *Differential equations with singular points.* Tech. Rep. No. 16. Dept. of Math., California Inst. of Technol., Pasadena.

SWEENEY, D. W. (1963). On the computation of Euler's constant. *Math. Comp.* **17**, 170–178.

SZEGÖ, G. (1967). *Orthogonal polynomials*, 3rd ed. Amer. Math. Soc. Colloq. Publ., Vol. XXIII. Amer. Math. Soc., New York.

THORNE, R. C. (1960). Asymptotic formulae for solutions of linear second-order differential equations with a large parameter. *J. Austral. Math. Soc.* **1**, 439–464.

TITCHMARSH, E. C. (1939). *The theory of functions*, 2nd ed. Oxford Univ. Press, London and New York.

TITCHMARSH, E. C. (1948). *Introduction to the theory of Fourier integrals*, 2nd ed. Oxford Univ. Press, London and New York.

TITCHMARSH, E. C. (1951). *The theory of the Riemann Zeta-function.* Oxford Univ. Press (Clarendon), London and New York.

TITCHMARSH, E. C. (1954). On the asymptotic distribution of eigenvalues. *Quart J. Math. Oxford Ser.* **5**, 228–240.

TOLSTOV, G. P. (1962). *Fourier series*, translated by R. A. Silverman. Prentice-Hall, Englewood Cliffs, New Jersey.

TRICOMI, F. G. (1954). *Funzioni ipergeometriche confluenti.* Consiglio Nazionale delle Richerche, Monogr. Mate. 1, Edizioni Cremonese, Rome.

TRICOMI, F. G., and ERDÉLYI, A. (1951). The asymptotic expansion of a ratio of Gamma functions. *Pacific J. Math.* **1**, 133–142.

TURRITTIN, H. L. (1936). Asymptotic solutions of certain ordinary differential equations associated with multiple roots of the characteristic equation. *Amer. J. Math.* **58**, 364–376.

TURRITTIN, H. L. (1964). Solvable related equations pertaining to turning point problems. In *Asymptotic solutions of differential equations and their applications* (C. H. Wilcox, ed.), pp. 27–52. Wiley, New York.

URSELL, F. (1965). Integrals with a large parameter. The continuation of uniformly asymptotic expansions. *Proc. Cambridge Philos. Soc.* **61**, 113–128.

URSELL, F. (1970). Integrals with a large parameter: Paths of descent and conformal mapping. *Proc. Cambridge Philos. Soc.* **67**, 371–381.

VAN DER CORPUT, J. G. (1934). Zur Methode der stationären Phase. I. *Compositio Math.* **1**, 15–38.

VAN DER CORPUT, J. G. (1936). Zur Methode der stationären Phase. II. *Compositio Math.* **3**, 328–372.

VAN DER CORPUT, J. G. (1956). Asymptotic developments I. Fundamental theorems of asymptotics. *J. Analyse Math.* **4**, 341–418.

VAN DER CORPUT, J. G., and FRANKLIN, J. (1951). Approximation of integrals by integration by parts. *Nederl. Akad. Wetensch. Proc. Ser. A* **54**, 213–219.

VAN DER WAERDEN, B. L. (1951). On the method of saddle points. *Appl. Sci. Res. Ser. B* **2**, 33–45.

VAN WIJNGAARDEN, A. (1953). A transformation of formal series. I, II. *Nederl. Akad. Wetensch. Proc. Ser. A* **56**, 522–533, 534–543.

WASOW, W. (1960). A turning point problem for a system of two linear differential equations. *J. Math. and Phys.* **38**, 257–278.

WASOW, W. (1965). *Asymptotic expansions for ordinary differential equations.* Wiley (Interscience), New York.

WASOW, W. (1966). On turning point problems for systems with almost diagonal coefficient matrix. *Funkcial. Ekvac.* **8**, 143–171.

WASOW, W. (1968). Connection problems for asymptotic series. *Bull. Amer. Math. Soc.* **74**, 831–853.

WASOW, W. (1970). Simple turning-point problems in unbounded domains. *SIAM J. Math. Anal.* **1**, 153–170.

WATSON, G. N. (1911). A theory of asymptotic series. *Philos. Trans. Roy. Soc. London Ser. A* **211**, 279–313.

WATSON, G. N. (1912). The transformation of an asymptotic series into a convergent series of inverse factorials. *Rend. Circ. Mat. Palermo* **34**, 1–48.

WATSON, G. N. (1913). A class of integral functions defined by Taylor's series. *Trans. Cambridge Philos. Soc.* **22**, 15–37.

WATSON, G. N. (1918a). The harmonic functions associated with the parabolic cylinder. *Proc. London Math. Soc.* **17**, 116–148.

WATSON, G. N. (1918b). The limits of applicability of the principle of stationary phase. *Proc. Cambridge Philos. Soc.* **19**, 49–55.

WATSON, G. N. (1918c). Asymptotic expansions of hypergeometric functions. *Trans. Cambridge Philos. Soc.* **22**, 277–308.

WATSON, G. N. (1944). *A treatise on the theory of Bessel functions*, 2nd ed. Cambridge Univ. Press, London and New York.

WENTZEL, G. (1926). Eine Verallgemeinerung der Quantenbedingungen für die Zwecke der Wellenmechanik. *Z. Physik* **38**, 518–529.

WHITTAKER, E. T., and WATSON, G. N. (1927). *A course of modern analysis*, 4th ed. Cambridge Univ. Press, London and New York.

WIDDER, D. V. (1941). *The Laplace transform.* Princeton Univ. Press, Princeton, New Jersey.

WILKINS, J. E. (1948). Nicholson's integral for $J_n^2(z) + Y_n^2(z)$. *Bull. Amer. Math. Soc.* **54**, 232–234.

WRENCH, J. W., JR. (1968). Concerning two series for the Gamma function. *Math. Comp.* **22**, 617–626.

WRENCH, J. W., JR. (1971). Converging factors for Dawson's integral and the modified Bessel function of the first kind. U.S. Naval Ship Res. and Develop. Center, Comput. and Math. Dept. Rep. No. 3517. Washington, D. C.

WRIGHT, E. M. (1948). The asymptotic expansion of integral functions and of the coefficients in their Taylor series. *Trans. Amer. Math. Soc.* **64**, 409–438.

WYMAN, M. (1959). The asymptotic behaviour of the Laurent coefficients. *Canad. J. Math.* **11**, 534–555.

WYMAN, M. (1964). The method of Laplace. *Trans. Roy. Soc. Canada* **2**, 227–256.

WYMAN, M., and WONG, R. (1969). The asymptotic behaviour of $\mu(z, \beta, \alpha)$. *Canad. J. Math.* **21**, 1013–1023.

WYNN, P. (1961). The numerical transformation of slowly convergent series by methods of comparison. *Chiffres* **4**, 177–210.

WYNN, P. (1963). A numerical study of a result of Stieltjes. *Rev. Française Traitement Informat.* [*Chiffres*] **6**, 175–196.

ZWAAN, A. (1929). Intensitäten im *Ca*-Funkenspektrum. *Arch. Néerlandaises Sci. Exactes Natur. Ser.* 3A **12**, 1–76.

INDEX OF SYMBOLS

Numbers refer to the pages on which the symbols are defined

GENERAL INDEX

D